RHEOLOGY

ADVANCES IN INTERFACIAL ENGINEERING SERIES

Microstructures constitute the building blocks of the interfacial systems upon which many vital industries depend. These systems share a fundamental knowledge base—the molecular interactions that occur at the boundary between two materials.

Where microstructures dominate, the manufacturing process becomes the product. At the Center for Interfacial Engineering, a National Science Foundation Research Center, researchers are working together to develop the control over molecular behavior needed to manufacture reproducible and reliable interfacial products.

The books in this series represent an intellectual collaboration rooted in the disciplines of modern engineering, chemistry, and physics that incorporates the expertise of industrial managers as well as engineers and scientists. They are designed to make the most recent information available to the students and professionals in the field who will be responsible for future optimization of interfacial processing technologies.

Other Titles in the Series

Edward Cohen and Edgar Gutoff (editors)
Modern Coating and Drying Technology

D. Fennell Evans and Håkan Wennerström
The Colloidal Domain: Where Physics, Chemistry, Biology, and Technology Meet

Forthcoming

Neil A. Dotson, Rafael Galván, Robert L. Laurence, and Matthew Tirrell
Polymerization Reaction Engineering

RHEOLOGY
Principles, Measurements, and Applications

Christopher W. Macosko

NEW YORK • CHICHESTER • WEINHEIM • BRISBANE • SINGAPORE • TORONTO

This book is printed on acid-free paper.

Originally published as ISBN 1-56081-579-5.

Published simultaneously in Canada.

Library of Congress Cataloging-in-Publication Data:
Macosko, Christopher W.
Rheology : principles, measurements, and applications / by Christopher W. Macosko : with contributions by Ronald G. Larson . . . [et al.].
p. cm.--(Advances in interfacial engineering series)
Includes bibliographical references and index.
ISBN 0-471-18575-2 (alk. paper)
1. Rheology. I. Larson, Ronald G. II. Title. III. Series.
QC189.5.M33 1993
531'.11--dc20 93-31652
CIP

Printed in the United States of America.

10 9 8 7 6 5

Even the mountains flowed before the Lord.

From the song of Deborah after her victory over the Philistines, Judges 5:5, translated by M. Reiner (*Physics Today,* January 1964, p. 62).

The Soudan Iron Formation exposed in Tower-Soudan State Park near Tower, Minnesota. This rock was originally deposited as horizontal layers of iron-rich sediments at the bottom of a sea. Deposition took place more than a billion years ago, in the Precambrian era of geologic time. Subsequent metamorphism, deformation, and tilting of the rocks have produced the complex structures shown. (Photo by A.G. Frederickson, University of Minnesota.)

DEDICATION

A.M.D.G.

This book has been written in the spirit that energized far greater scientists. Some of them express that spirit in the following quotations.

"This most beautiful system of the sun, planets and comets could only proceed from the counsel and dominion of an intelligent and powerful Being."

Isaac Newton

"Think what God has determined to do to all those who submit themselves to His righteousness and are willing to receive His gift."

James C. Maxwell
June 23, 1864

"In the distance tower still higher peaks, which will yield to those who ascend them still wider prospects, and deepen the feeling whose truth is emphasized by every advance in science, that 'Great are the works of the Lord'".

J.J. Thomson,
Nature, 81, 257 (1909).

CONTENTS

PREFACE

Today a number of industrial and academic researchers would like to use rheology to help solve particular problems. They really don't want to become full-time rheologists, but they need rheological measurements to help them characterize a new material, analyze a non-Newtonian flow problem, or design a plastic part. I hope this book will meet that need. A number of sophisticated instruments are available now for making rheological measurements. My goal is to help readers select the proper type of test for their applications, to interpret the results, and even to determine whether or not rheological measurements can help to solve a particular problem.

One of the difficult barriers between much of the rheology literature and those who would at least like to make its acquaintance, if not embrace it, is the *tensor.* That monster of the double subscript has turned back many a curious seeker of rheological wisdom. To avoid tensors, several applied rheology books have been written in only one dimension. This can make the barrier seem even higher by avoiding even a glimpse of it. Furthermore, the one-dimensional approach precludes presentation of a number of useful, simplifying concepts.

I have tried to expose the tensor monster as really quite a friendly and useful little man-made invention for transforming vectors. It greatly simplifies notation and makes the three-dimensional approach to rheology practical. I have tried to make the incorporation of tensors as simple and physical as possible. Second-order tensors, Cartesian coordinates, and a minimum of tensor manipulations are adequate to explain the basic principles of rheology and to give a number of useful constitutive equations. With what is presented in the first four chapters, students will be able to read and use the current rheological literature. For curvilinear coordinates and detailed development of constitutive equations, several good texts are available and are cited where appropriate.

Who should read this book, and how should it be used? For the seasoned rheologist or mechanicist, the table of contents should serve as a helpful guide. These investigators may wish to skim over the *first section* but perhaps will find its discussion of *constitutive relations* and material functions with the inclusion of both solids and liquids helpful and concise. I have found these four chapters on constitutive relations a very useful introduction to rheology for first- and second-year engineering graduate students. I have also used portions in a senior course in polymer processing. The rubbery solid examples are particularly helpful for later development of such processes as thermoforming and blow molding. There are a number of worked examples which students report are helpful, especially if they attempt to do them before reading the solutions. There are additional exercises at the end of each chapter. Solutions to many of these are found at the end of the text.

In Part I of the book we only use the simplest deformations, primarily simple shear and uniaxial elongation, to develop the important constitutive equations. In Part II the text describes *rheometers,* which can measure the material functions described in Chapters 1 through 4. How can the assumed kinematics actually be achieved in the laboratory? This rheometry material can serve the experienced rheologist as a useful reference to the techniques presently available. Each of the major test geometries is described with the working equations, assumptions, corrections, and limitations summarized in convenient tables. Both shear and extensional rheometers are described. Design principles for measuring stress and strain in the various rheometers should prove helpful to the new user as well as to those trying to build or modify instruments. The important and growing application of optical methods in rheology is also described.

The reader who is primarily interested in using rheology to help solve a specific and immediate problem can go directly to a chapter of interest in Part III of the book on *applications of rheology.* These chapters are fairly self-contained. The reader can go back to the *constitutive equation* chapters as necessary for more background or to the appropriate rheometer section to learn more about a particular test method. These chapters are not complete discussions of the application of rheology to suspensions and polymeric liquids; indeed an entire book could be, and some cases has been, written on each one. However, useful principles and many relevant examples are given in each area.

ACKNOWLEDGMENTS

This text has grown out of a variety of teaching and consulting efforts. I have used part of the material for the past several years in a course on polymer processing at the University of Minnesota and nearly all of it in my graduate course, Principles and Applications of Rheology. Much of my appreciation for the needs of the industrial rheologist has come from teaching a number of short courses on rheological measurements at Minnesota and for the Society of Rheology and Society of Plastics Engineers. The University of Minnesota summer short course has been taught for nearly 20 years with over 800 attendees. Many of the examples, the topics, and the comparisons of rheological methods included here were motivated by questions from short course students. Video tapes of this course which follows this text closely are available. My consulting work, particularly with Rheometrics, Inc., has provided me the opportunity to evaluate many rheometer designs, test techniques, and data analysis methods, and fortunately my contacts have not been shy about sharing some of their most difficult rheological problems. I hope that the book's approach and content have benefited from this combination of academic and industrial applications of rheology.

As indicated in the Contents, two of the chapters were written by my colleagues at the University of Minnesota, Tim Lodge and Matt Tirrell. With Skip Scriven, we have taught the Rheological Measurements short course at Minnesota together for several years. Their contributions of these chapters and their encouragement and suggestions on the rest of the book have been a great help. Ron Larson, a Minnesota alumnus and distinguished member of the technical staff at ATT Bell Labs, contributed Chapter 4 on nonlinear viscoelasticity. We are fortunate to have this expert contribution, a distillation of key ideas from his recent book in this area. I collaborated with Jan Mewis of the Katholieke Universiteit Leuven in Belgium on Chapter 10 on suspensions. Jan's expertise and experience in concentrated suspensions is greatly appreciated. Robert Secor, now of 3M, prepared Appendix A to Chapter 3, concerned with fitting linear viscoelastic spectra, during his graduate studies here. Mahesh Padmanabhan was very helpful in preparation of much of the final version, particularly in writing and editing parts of Chapters 6 and 7 as well as in preparing the index.

This manuscript has evolved over a number of years, and so many people have read and contributed that it would be impossible to acknowledge them all. My present and past students have been particularly helpful in proofreading and making up examples. In addition, my colleagues Gordon Beavers and Roger Fosdick read early versions of Chapters 1 and 2 carefully and made helpful suggestions.

A major part of the research and writing of the second section on rheometry was accomplished while I was a guest of Martin

Laun in the Polymer Physics Laboratory, Central Research of BASF in Ludwigshafen, West Germany. The opportunity to discuss and present this work with Laun and his co-workers greatly benefited the writing. Extensive use of their data throughout this book is a small acknowledgment of their large contribution to the field of rheology.

A grant from the Center for Interfacial Engineering has been very helpful in preparing the manuscript. Julie Murphy supervised this challenging activity and was ably assisted by Bev Hochradel, Yoav Dori, Brynne Macosko, and Sang Le. The VCH editorial and production staff, particularly Camille Pecoul, did a fine job. I apologize in advance for any errors which we all missed and welcome corrections from careful readers.

Chris Macosko
August 1993

PART I

CONSTITUTIVE RELATIONS

The development of the chemical industry at the beginning of this century, followed by the advent of large-scale synthetic polymer production, resulted in a host of new materials with "strange" flow behavior. In 1920 the study of such materials prompted a chemistry professor at Lehigh University, Eugene Bingham, to coin a new word, *rheology*. It comes from the Greek verb $\rho\epsilon\iota\nu$, to flow. Thus rheology means the study of flow and deformation. In principle then, rheology includes everything dealing with flow behavior: aeronautics, hydraulics, fluid dynamics, and even solid mechanics. However, in practice rheology has usually been restricted to the study of the fundamental relations, called *constitutive relations*, between force and deformation in materials, primarily liquids. Bingham himself was a colloid chemist and was moved to invent his new word by the unusual flow behavior he observed in concentrated suspensions like paint. Thus, the rheologist focuses on material behavior, using very simple deformations, while the mechanicist studies the forces developed in complex deformation, applying the constitutive relations developed by the rheologist. Clearly to do a good job the rheologist must understand mechanics, and vice versa, as we shall see in the following chapters.

The simplest and probably the first relation between force and deformation, the first constitutive equation, is Hooke's law: the force is proportional to the deformation or

$$\tau = G\gamma \tag{I.1}$$

where τ is the force per unit area or *stress* and γ is the relative length change or *strain*. G is the constant of proportionality called the *elastic modulus*. It is an intrinsic property of a solid.

For liquids, the simplest constitutive equation is Newton's

law of viscosity: the stress is proportional to the rate of straining, or $\dot{\gamma} = d\gamma/dt$,

$$\tau = \eta\dot{\gamma} \tag{I.2}$$

where η, the Newtonian *viscosity*, is the constant of proportionality.

Many real materials obey these ideal laws. Hooke's law is the basic constitutive equation for solid mechanics and most metals and ceramics at small strains are "Hookean" or ideally elastic. The Newtonian fluid is the basis for classical fluid mechanics. Gases and most small molecule liquids like water and oils are Newtonian. However, if all materials obeyed these simple laws, the word rheology would not have been invented and there would be no need for this book. Many important materials, such as blood, polymers, paint, and foods, lie between the ideal elastic solid and the ideal viscous fluid.

Perhaps the simplest example of one of these "in between" materials is the common toy Silly Putty or "bouncing putty." As shown in Figure I.1, when Silly Putty is dropped, it experiences a rapid deformation and bounces back like a rubber ball. But if left to sit on a surface for several hours, the material will flow out like a liquid. At short times Silly Putty behaves like a Hookean solid, but at long times it behaves more like a Newtonian liquid. We will see in Chapter 3 how this behavior can be nicely modeled by a modulus that relaxes with time, $G(t)$.

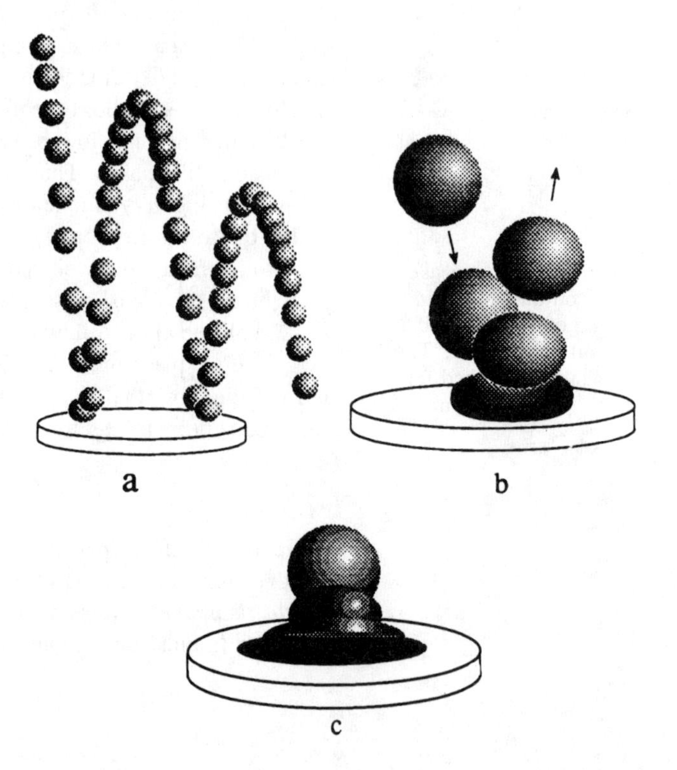

Figure I.1.
When a stress acts on Silly Putty for only a short time (a, b), the material behaves like a solid, bouncing back when dropped; but when stressed for several hours (c), it flows out like a liquid.

Figure I.2.
Mayonnaise spreads easily at the high shear stress of the moving knife, but left alone it will sit in a lump on a piece of bread.

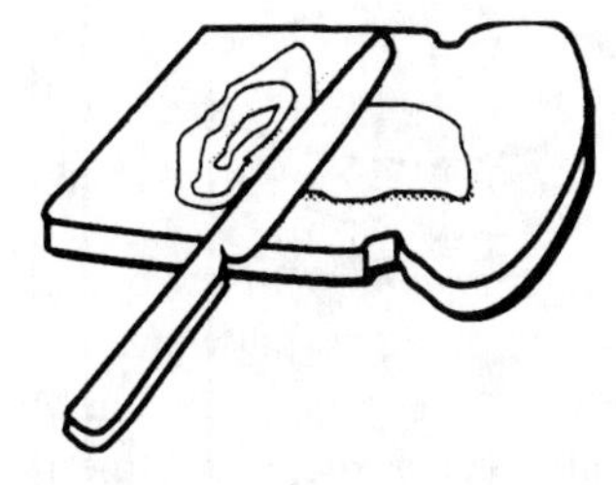

Another "in between" material is mayonnaise, Figure 1.2. If mayonnaise is left on a piece of bread subject only to stresses due to gravity, it will barely move, but when sheared by a knife it spreads easily. In contrast, honey, which is more viscous than mayonnaise at high stresses, will slowly run through holes in the bread. The viscosity of honey is constant, but the viscosity of mayonnaise depends strongly on the shear stress. Shear thinning viscosity, perhaps the most common phenomenon of non-Newtonian liquids, is the main subject of Chapter 2.

More dramatic non-Newtonian phenomena are normal stresses in shear flow. When a rod is rotated in a beaker of a Newtonian fluid like water or oil, inertial forces cause it to move away from the rod. However, as Figure I.3 shows, if a small amount of high polymer is added to the oil, it will climb up the rod. This phenomenon also happens when cake batter is mixed with an egg beater. We can think of the batter or the polymeric fluid as containing rubber bands, which when stretched pull back in the direction of the flow. This tension along the circular lines of flow generates a pressure toward the center, driving the fluid up the rod. We can best understand these *normal stresses* by looking first at solid rubber. In fact, if we twist a rubber rod or tube, it will get longer. To see how this can occur, we need to understand stress and strain in all three dimensions. We do this in Chapter 1, using large deformations of rubber.

An equally dramatic phenomenon is the "ductless siphon" shown in Figure I.4. Some non-Newtonian fluids can be stretched out for short times like a solid. You may have experienced this when trying to pull rubber cement or another adhesive from your fingers. Polymeric and other complex liquids can show much higher viscosities in extension than in shear. In Chapter 4 we develop constitutive equations to predict this extensional thickening.

Figure I.3.
A rod is rotated in a beaker of motor oil on the left and in a solution of 1% polyisobutylene in oil on the right. This experiment is described further in Chapter 5 (Figure 5.3.3).

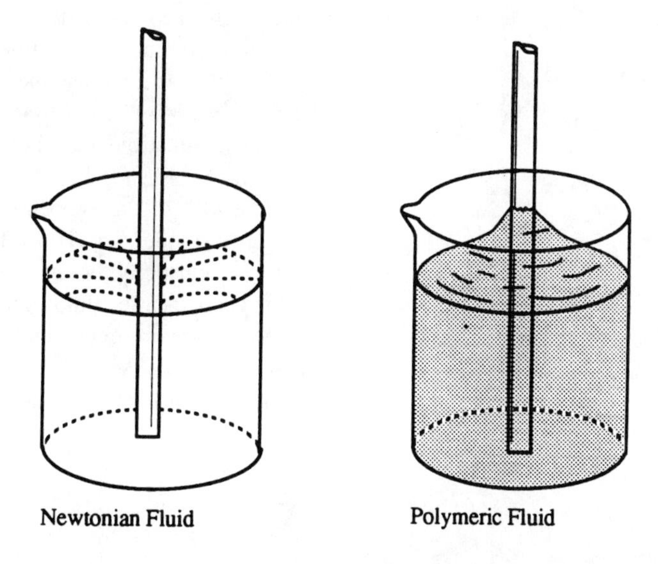

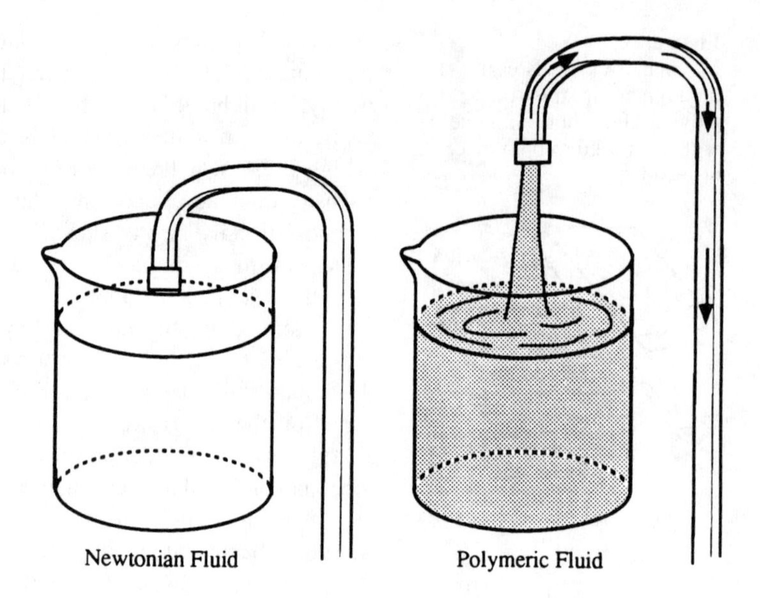

Figure I.4.
The "ductless siphon." When a siphon is slowly removed from a Newtonian fluid, it breaks very close to the surface, but some complex fluids can be pulled up many centimeters above the free surface. This experiment is described further in Section 7.5.

These experiments illustrate the four most important phenomena in rheology. They are the phenomena that the ideal constitutive equation should describe:

1. Time dependence: a relaxing modulus $G(t)$, or recoverable strain
2. Shear thinning (or thickening) viscosity, $\eta(\dot{\gamma})$
3. Normal stresses in shear, $T_{11} - T_{22} > 0$
4. Extensional thickening viscosity, $\eta_u(\dot{\epsilon})$

Although most rheologists worry about liquidlike materials, to understand much about the phenomena listed above, it is essential to understand the behavior of solids. Thus in the first chapter we study the elastic solid under large deformation. This long chapter turns out to be an excellent and useful way to introduce fundamental tools like stress in three dimensions, the finite strain tensor, invariants, and material functions, which we will need later.

In Chapter 2 we use the viscous liquid to motivate the rate of strain tensor and then look at several simple non-Newtonian constitutive equations like the power law and Bingham plastic as models for the shear-dependent viscosity, $\eta(\dot{\gamma})$. Chapter 3 introduces the idea of time-dependent liquids and develops the linear viscoelastic model for $G(t)$. The most complex behavior, nonlinear viscoelasticity, is saved for Chapter 4. Differential models like the second-order fluid, rate models like Maxwell's, and integral models are presented and tested against experimental data. A strategy for model selection is given. The goal of these constitutive equations is to explain most or all of the phenomena described above.

1

The power of any spring is in the same proportion with the tension thereof.
Robert Hooke (1678)

ELASTIC SOLID

1.1 Introduction

Figure 1.1.1 shows some of the experimental apparatuses Robert Hooke used to prove his law. When he doubled the weight attached to the springs or to the long wire, the extension doubled. Thus he proposed that the force was proportional to the change in length:

$$f \sim \Delta L \tag{1.1.1}$$

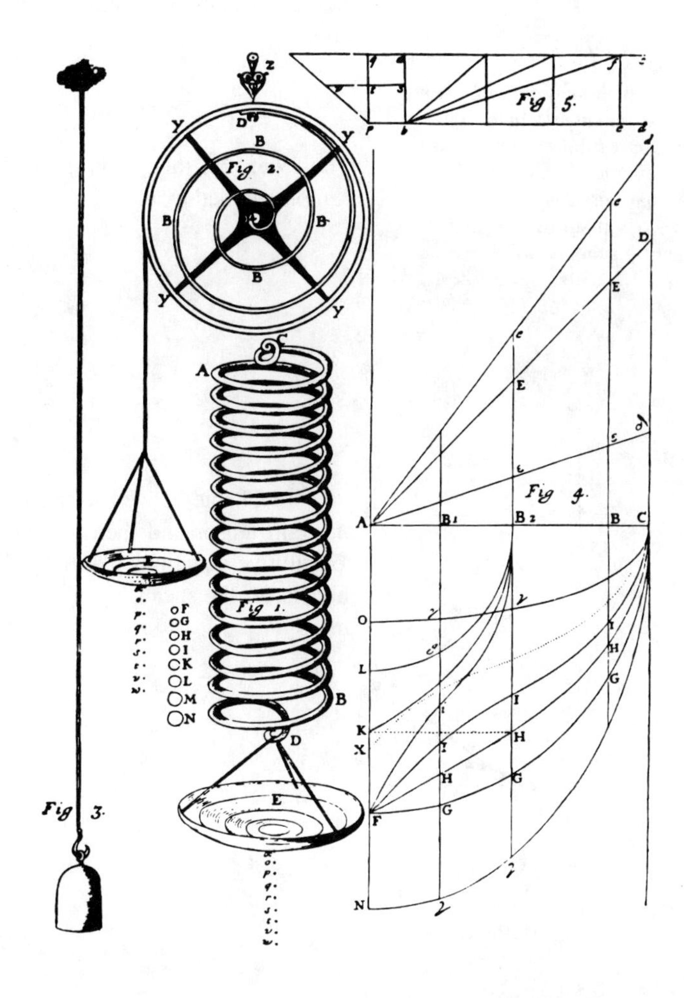

Figure 1.1.1.
Some of the experiments Hooke used to establish his law of extension. Note the marks (o, p, q, r, etc.) he used to indicate that displacement goes linearly with force. From Hooke, 1678.

He was certainly on the right track to the constitutive equation for the ideal elastic solid, but if he used a different length wire or a different diameter of the same material, he found a new constant of proportionality. Thus his constant was not uniquely a material property but also depended on the particular geometry of the sample. To find the true material constant—the elastic modulus—of his wires, Hooke needed to develop the concepts of stress, force per unit area, and strain. Stress and strain are key concepts for rheology and are the main subjects of this chapter.

If crosslinked rubber had been available in 1678, Hooke might well have also tried rubber bands in his experiments. If so he would have drawn different conclusions. Figure 1.1.2 shows results for a rubber sample tested in tension and in compression. We see that for small deformations near zero the stress is linear with deformation, but at large deformation the stress is larger than is predicted by Hooke's law. A relation that fits the data reasonably well is

$$T_{11} = G\left(\alpha^2 - \frac{1}{\alpha}\right) \tag{1.1.2}$$

where T_{11} is the tensile force divided by the area a, which it acts upon.

$$T_{11} = \frac{f}{\mathrm{a}} \tag{1.1.3}$$

The extension ratio α is defined as the length of the deformed sample divided by the length of the undeformed one:

$$\alpha = \frac{L}{L'} \tag{1.1.4}$$

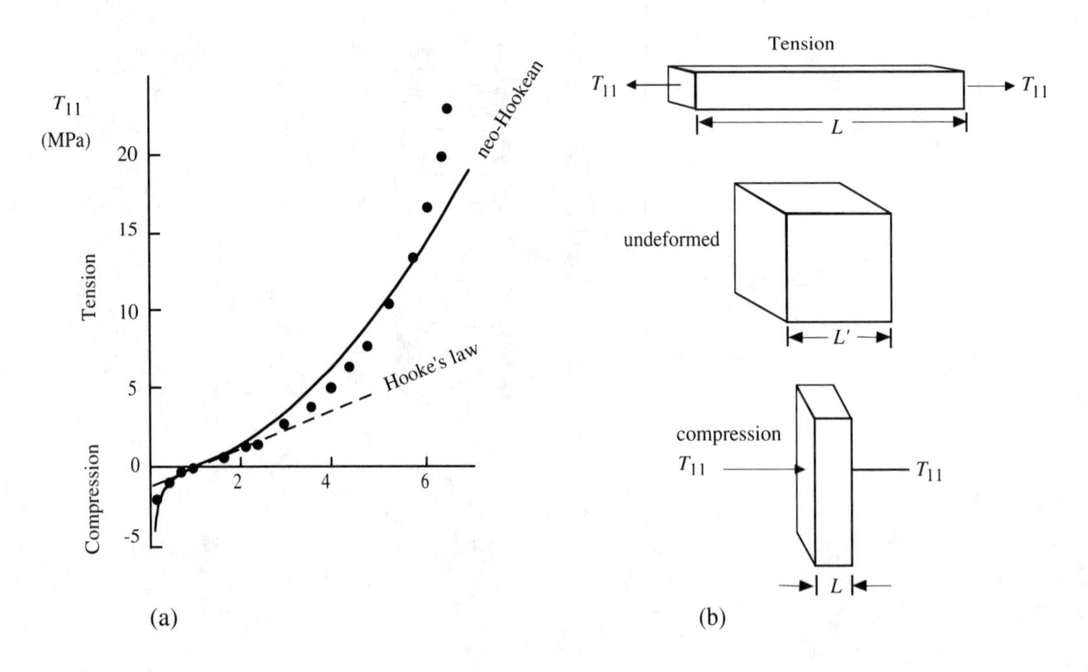

Figure 1.1.2. (a) Tensile and compressive stress versus extension ratio for a rubber sample. (b) Schematic diagram of the deformation. Data from Treloar (1975) on sulfur-vulcanized natural rubber. Solid line is eq. 1.1.2 with $G = 3.9 \times 10^5$ Pa.

Figure 1.1.3 shows the results of a different kind of experiment on a similar rubber sample. Here the sample is sheared between two parallel plates maintained at the same separation x_2. We see that the shear stress is *linear* with the strain over quite a wide range; however, additional stress components, normal stresses T_{11} and T_{22}, act on the block at large strain. In the introduction to this part of the text, we saw that elastic liquids can also generate normal stresses (Figure I.3). In rubber, the normal stress difference depends on the shear strain squared

$$T_{11} - T_{22} = G\gamma^2 \tag{1.1.5}$$

where the shear strain is defined as displacement of the top surface of the block over its thickness

$$\gamma = \frac{s}{x_2} \tag{1.1.6}$$

while the shear stress is linear in shear strain with the same coefficient

$$T_{21} = G\gamma \tag{1.1.7}$$

These apparently quite different results for different deformations of the same sample can be shown to come from Hooke's law when it is written properly in three dimensions. We will do this in the next several sections of this chapter, calling on a few ideas from vector algebra, mainly the vector summation and the dot or scalar products. For a good review of vector algebra Bird et al. (1987a, Appendix A), Malvern (1969) or Spiegel (1968) is helpful. In the following sections we develop the idea of a tensor and some basic notions of continuum mechanics. It is a very simple

Figure 1.1.3.
(a) Shear and normal stresses versus shear strain for a silicone rubber sample subject to simple shear shown schematically in (b). The open points indicate the normal stress difference $T_{11} - T_{22}$ necessary to keep the block at constant thickness x_2, while the solid points are for the shear stress. Notice that the normal stress stays positive when the shear changes sign. Data are for torsion of a cylinder (DeGroot, 1990; see also Example 1.7.1).

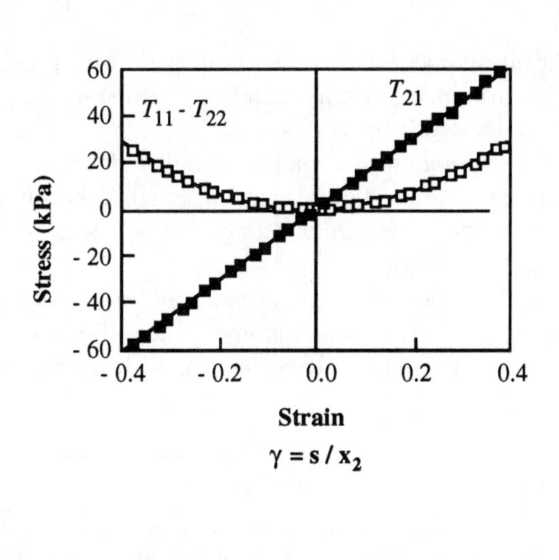

(a)

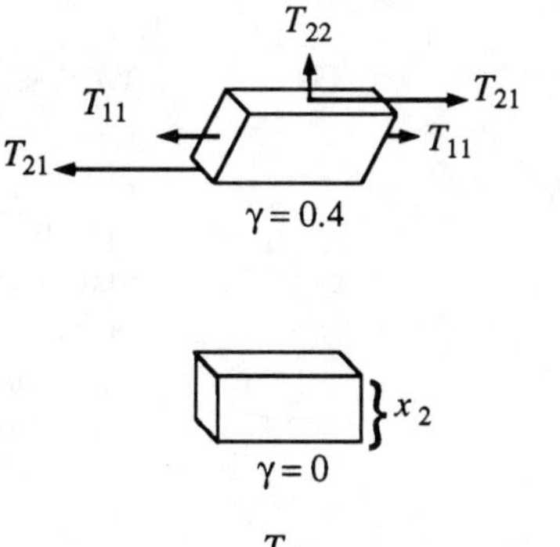

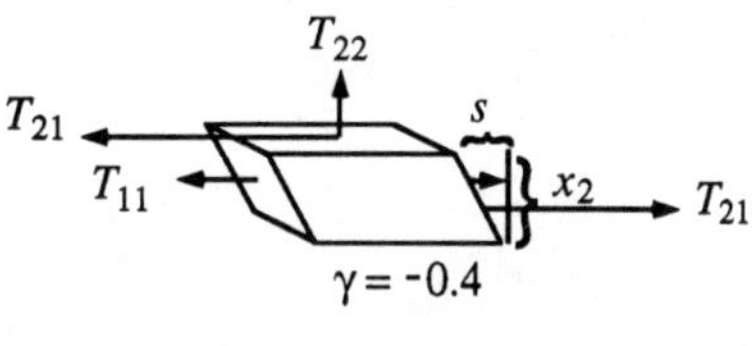

(b)

development, yet adequate for the rest of this text and for starting to read other rheological literature. More detailed studies of continuum mechanics can be found in the references above and in books by Astarita and Marrucci (1974), Billington and Tate (1981), Chadwick (1976), and Lodge (1964, 1974).

1.2 The Stress Tensor*

To help us see how both shear and normal stresses can act in a material, consider the body shown in Figure 1.2.1a. Let us cut through a point P in the body with a plane. We identify the direction of a plane by the vector acting normal to it, in this case the unit vector $\hat{\mathbf{n}}$. If there are forces acting on the body, a force component $\mathbf{f}_n$ will act on the cutting plane at point P. In general $\mathbf{f}_n$ and $\hat{\mathbf{n}}$ will have different directions. If we divide the force by a small area $d\mathbf{a}$ of the cut surface around point P, then we have the stress or traction vector $\mathbf{t}_n$ per unit area acting on the surface at point P. Figure 1.2.1b shows a cut that leads to a normal stress, while Figure 1.2.1c shows another that gives a shear stress $\mathbf{t}_m$. Note that Figure 1.2.1 shows two stress vectors of the same magnitude acting in opposite directions. This is required by Newton's law of motion to keep the body at rest. Both vectors are manifestations of the same stress component. In the discussion that follows we usually show the positive vector only.

As we have seen in Figures 1.1.2 and 1.1.3, materials may respond differently in shear and tension, so it is useful to break the stress vector $\mathbf{t}_\mathrm{n}$ into components that act normal (tensile) to the plane $\hat{\mathbf{n}}$ and those that act tangent or shear to the plane. If we pick a Cartesian coordinate system with one direction $\hat{\mathbf{n}}$, the other two directions $\hat{\mathbf{m}}$ and $\hat{\mathbf{o}}$ will lie in the plane. Thus, $\mathbf{t}_\mathrm{n}$ is the vector sum of three stress components.

$$\mathbf{t}_\mathrm{n} = \hat{\mathbf{n}}T_{nn} + \hat{\mathbf{m}}T_{nm} + \hat{\mathbf{o}}T_{no} \tag{1.2.1}$$

We designate the magnitude of these stress components with a capital T and use two subscripts to identify each one. The first subscript refers to the *plane* on which the components are acting; the second indicates the *direction* of the component on that plane. If we take another cut, say with a normal vector $\hat{\mathbf{m}}$, through the same point in the body, then the stress vector acting on $\hat{\mathbf{m}}$ will be $\mathbf{t}_m$ with components T_{mm}, T_{mo}, and T_{mn}.

So what we have now is a logical notation for describing the normal and shear stresses acting on any surface. But will it be necessary to pass an infinite number of planes through P to

**Many students with engineering or physics backgrounds are already familiar with the stress tensor. They may skip ahead to the next section. The key concepts in this section are understanding (1) that tensors can operate on vectors (eq. 1.2.10), (2) standard index notation (eq. 1.2.21), (3) symmetry of the stress tensor (eq. 1.2.37), (4) the concept of pressure (eq. 1.2.44), and (5) normal stress differences (eq. 1.2.45).*

Figure 1.2.1.
(a) A force acting on a body. (b) A cut through point P nearly perpendicular to the direction of force . The normal to the plane of this cut is $\hat{\mathbf{n}}$. The stress on this plane is $\mathbf{t}_n = f/a$, where a equals the area of the cut. (c) Another cut nearly parallel to the force direction. The equal and opposite forces acting at point P are represented by the single component $\mathbf{t}_\mathbf{m}$.

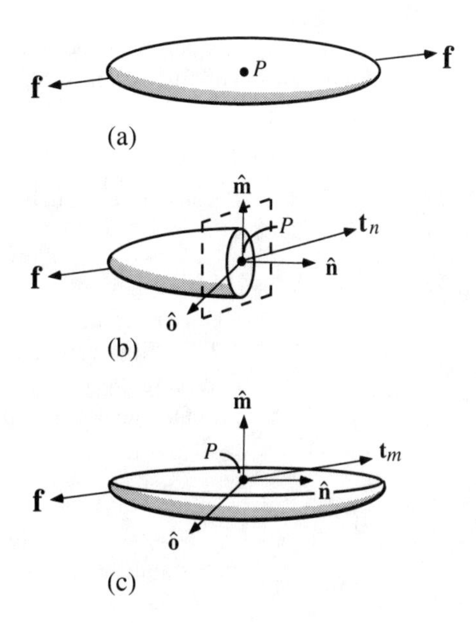

characterize the state of stress at this point? No, because in fact, the stresses acting on all the different planes are related. The stress on any plane through P can be determined from a quantity called the *stress tensor*. The stress tensor is a special mathematical operator that can be used to describe the state of stress at any point in the body.

To help visualize the stress tensor, let us set up three mutually perpendicular planes in the body near point P, as shown in Figure 1.2.2a. Let the normals to each plane be $\hat{\mathbf{x}}$, $\hat{\mathbf{y}}$, and $\hat{\mathbf{z}}$ respectively. On each plane there will be a stress vector. These planes will form the Cartesian coordinate system $\hat{\mathbf{x}}$, $\hat{\mathbf{y}}$, $\hat{\mathbf{z}}$. As shown in Figure 1.2.2b, three stress components will act on each of the three perpendicular planes. Now, if we cut a plane $\hat{\mathbf{n}}$ across these three planes, we will form a small tetrahedron around P (Figure 1.2.2c). The stress $\mathbf{t}_\mathrm{n}$

Figure 1.2.2.
(a) Three mutually perpendicular planes intersecting at the point P with their associated stress vectors. (b) Stress components acting on each of these planes. (c) A plane $\hat{\mathbf{n}}$ is cut across the three planes to form a tetrahedron. As in Figure 1.2.1, $\mathbf{t}_n$ is the stress vector acting on this plane with area a_n. For any plane $\hat{\mathbf{n}}$, $\mathbf{t}_n$ can be determined from the components on the three perpendicular planes.

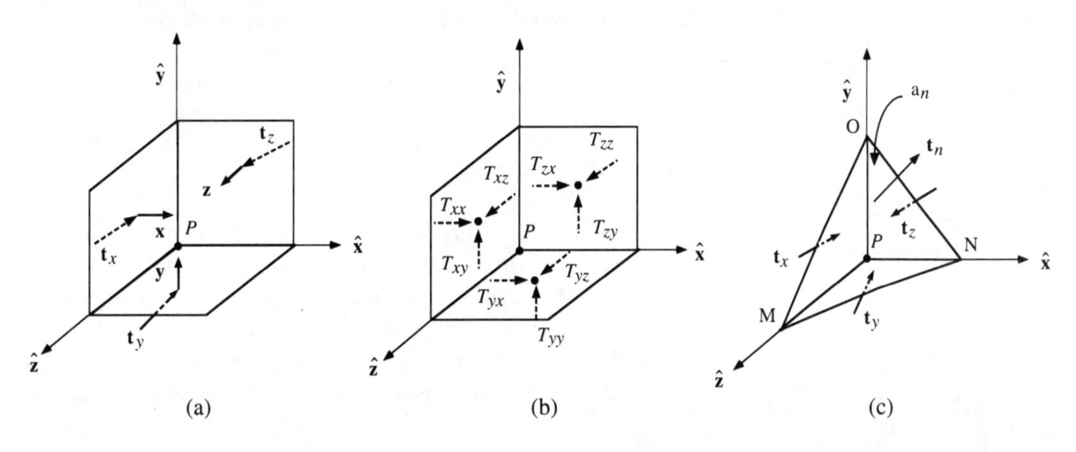

on plane $\hat{\mathbf{n}}$ can be determined by a force balance on the tetrahedron. The force on $\hat{\mathbf{n}}$ is the vector sum of the forces on the other planes

$$-\mathbf{f}_n = \mathbf{f}_x + \mathbf{f}_y + \mathbf{f}_z \tag{1.2.2}$$

Because force equals stress times area, the balance becomes

$$\mathbf{f}_n = \mathrm{a}_n\mathbf{t}_n = \mathrm{a}_x\mathbf{t}_x + \mathrm{a}_y\mathbf{t}_y + \mathrm{a}_z\mathbf{t}_z \tag{1.2.3}$$

where a_x is the area of the triangle *MOP* as indicated in Figure 1.2.2c. From geometry we know that the area a_x can be calculated by taking the projection of a_n on the $\hat{\mathbf{x}}$ plane. The projection is given by the dot or scalar product of the two unit normal vectors to each plane

$$\mathrm{a}_x = \mathrm{a}_n\hat{\mathbf{n}} \cdot \hat{\mathbf{x}} \tag{1.2.4}$$

and similarly for a_y and a_z. Thus, the force balance becomes

$$\mathrm{a}_n\mathbf{t}_n = (\mathrm{a}_n\hat{\mathbf{n}} \cdot \hat{\mathbf{x}})\mathbf{t}_x + (\mathrm{a}_n\hat{\mathbf{n}} \cdot \hat{\mathbf{y}})\mathbf{t}_y + (\mathrm{a}_n\hat{\mathbf{n}} \cdot \hat{\mathbf{z}})\mathbf{t}_z \tag{1.2.5}$$

In the limit as the area shrinks down to zero around *P*, the stresses become constant, and we can divide out a_n to give

$$\mathbf{t}_n = \hat{\mathbf{n}} \cdot [\hat{\mathbf{x}}\mathbf{t}_x + \hat{\mathbf{y}}\mathbf{t}_y + \hat{\mathbf{z}}\mathbf{t}_z] \tag{1.2.6}$$

Performing the dot operation gives

$$\mathbf{t}_n = n_x\mathbf{t}_x + n_y\mathbf{t}_y + n_z\mathbf{t}_z \tag{1.2.7}$$

where n_x is the magnitude of the projection of $\hat{\mathbf{n}}$ onto $\hat{\mathbf{x}}$. Figure 1.2.2b indicates the three components of each stress. These components with their directions can be substituted into the balance above to give

$$\begin{aligned}\mathbf{t}_n = \hat{\mathbf{n}} \cdot [&\hat{\mathbf{x}}\hat{\mathbf{x}}T_{xx} + \hat{\mathbf{x}}\hat{\mathbf{y}}T_{xy} + \hat{\mathbf{x}}\hat{\mathbf{z}}T_{xz}\\ + &\hat{\mathbf{y}}\hat{\mathbf{x}}T_{yx} + \hat{\mathbf{y}}\hat{\mathbf{y}}T_{yy} + \hat{\mathbf{y}}\hat{\mathbf{z}}T_{yz}\\ + &\hat{\mathbf{z}}\hat{\mathbf{x}}T_{zx} + \hat{\mathbf{z}}\hat{\mathbf{y}}T_{zy} + \hat{\mathbf{z}}\hat{\mathbf{z}}T_{zz}]\end{aligned} \tag{1.2.8}$$

which, when we take the dot products, reduces to

$$\begin{aligned}\mathbf{t}_n = \; &\hat{\mathbf{x}}(n_xT_{xx} + n_yT_{yx} + n_zT_{zx}) + \hat{\mathbf{y}}(n_xT_{xy} + n_yT_{yy} + n_zT_{zy})\\ &\hat{\mathbf{z}}(n_xT_{xz} + n_yT_{yz} + n_zT_{zz})\end{aligned} \tag{1.2.9}$$

It is rather cumbersome to write out all these components each time, so a shorthand was invented by Gibbs in the 1880s (see Gibbs, 1960). He defined a new quantity called a *tensor* to represent all the terms in the brackets in eq. 1.2.8. Following Gibbs and modern continuum mechanics notation, we generally use boldface capital letters to denote tensors, while boldface lowercase letters are vectors. Thus, the stress tensor becomes **T**, and when we use it eq. 1.2.8 certainly looks less forbidding

$$\mathbf{t}_n = \hat{\mathbf{n}} \cdot \mathbf{T} \tag{1.2.10}$$

Here the dot means the vector product of a tensor with a vector to generate another vector.

Perhaps the simplest way to think of a tensor dot product with a vector is as a machine (see Figure 1.2.3) that linearly transforms a vector to another vector. Push the unit vector $\hat{\mathbf{n}}$ into one side of the stress tensor machine and out comes the stress vector $\mathbf{t}_n$ acting on the surface with the normal vector $\hat{\mathbf{n}}$. **T** is a mathematical operator that acts on vectors. It is the quantity that completely characterizes the state of stress at a point. We can not draw it on the blackboard like a vector, but we can see what it can do by letting it act on any plane through eq. 1.2.10.

1.2.1. Notation

By comparing eq. 1.2.10 with eq. 1.2.6, we see that the stress tensor can be viewed as the sum of three "double vectors"

$$\mathbf{T} = \hat{\mathbf{x}}\mathbf{t}_x + \hat{\mathbf{y}}\mathbf{t}_y + \hat{\mathbf{z}}\mathbf{t}_z \tag{1.2.11}$$

These double vectors are called *dyads*. The dyad carries two directions, the first being that of the plane on which the stress vector is acting and the other the direction of the vector itself. Thus another way to visualize the stress tensor is as the dyad product, the special combination of the forces (or stress) vectors with the surface that they act on.

In eq. 1.2.8 we see the tensor represented as the sum of nine scalar components, each associated with two directions. This is the usual way to write out a tensor because the dyads are now expressed in terms of the unit vectors

$$\begin{aligned} \mathbf{T} = {} & \hat{\mathbf{x}}\hat{\mathbf{x}}T_{xx} + \hat{\mathbf{x}}\hat{\mathbf{y}}T_{xy} + \hat{\mathbf{x}}\hat{\mathbf{z}}T_{xz} \\ & + \hat{\mathbf{y}}\hat{\mathbf{x}}T_{yx} + \hat{\mathbf{y}}\hat{\mathbf{y}}T_{yy} + \hat{\mathbf{y}}\hat{\mathbf{z}}T_{yz} \\ & + \hat{\mathbf{z}}\hat{\mathbf{x}}T_{zx} + \hat{\mathbf{z}}\hat{\mathbf{y}}T_{zy} + \hat{\mathbf{z}}\hat{\mathbf{z}}T_{zz} \end{aligned} \tag{1.2.12}$$

Figure 1.2.3.
The tensor as a machine for transforming vectors.

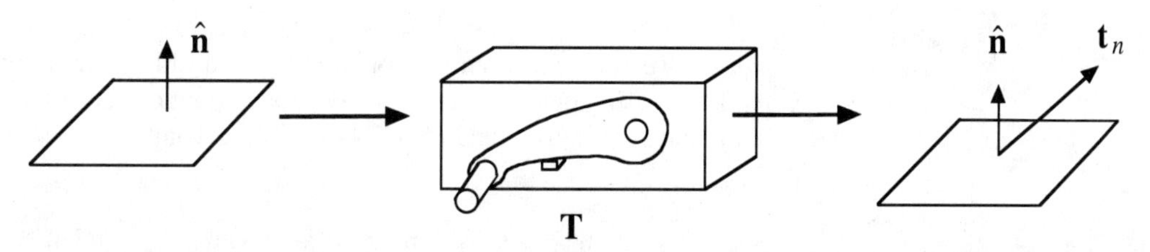

Often matrix notation is used to display the scalar components.

$$\mathbf{T} = \begin{bmatrix} T_{xx} & T_{xy} & T_{xz} \\ T_{yx} & T_{yy} & T_{yz} \\ T_{zx} & T_{zy} & T_{zz} \end{bmatrix} \tag{1.2.13}$$

Here we have left out the unit dyads that belong with each scalar component, so the = sign does not really signify "equals" but rather should be interpreted as "scalar components are." Usually the unit dyads are understood. Matrix notation is convenient because the "dot" operations correspond to standard matrix multiplication. In matrix notation eq. 1.2.10 becomes a row matrix times a 3×3 matrix.

$$\mathbf{t}_n = \hat{\mathbf{n}} \cdot \mathbf{T} = (n_x \, n_y \, n_z) \cdot \begin{bmatrix} T_{xx} & T_{xy} & T_{xz} \\ T_{yx} & T_{yy} & T_{yz} \\ T_{zx} & T_{zy} & T_{zz} \end{bmatrix} = \begin{bmatrix} n_x T_{xx} & + & n_y T_{yx} & + & n_z T_{zx} \\ n_x T_{xy} & + & n_y T_{yy} & + & n_z T_{zy} \\ n_x T_{xz} & + & n_y T_{yz} & + & n_z T_{zz} \end{bmatrix} \tag{1.2.14}$$

Remember again that we have left out the unit dyads ($\hat{\mathbf{x}}\hat{\mathbf{x}}$, etc). In matrix notation the vector scalar product of eq. 1.2.4 becomes the multiplication of a row with a column matrix.

$$\mathrm{a}_x = \mathrm{a}_n \hat{\mathbf{n}} \cdot \hat{\mathbf{x}} = \mathrm{a}_n (n_x, n_y, n_z) \begin{bmatrix} 1 \\ 0 \\ 0 \end{bmatrix} = \mathrm{a}_n n_x \tag{1.2.15}$$

Usually in rheology we use numbered coordinate directions. Under this notation scheme the unit vectors $\hat{\mathbf{x}}$, $\hat{\mathbf{y}}$, $\hat{\mathbf{z}}$ become $\hat{\mathbf{x}}_1$, $\hat{\mathbf{x}}_2$, $\hat{\mathbf{x}}_3$, and the components of the stress tensor are written

$$T_{ij} = \begin{bmatrix} T_{11} & T_{12} & T_{13} \\ T_{21} & T_{22} & T_{23} \\ T_{31} & T_{32} & T_{33} \end{bmatrix} \tag{1.2.16}$$

This numbering of components leads to a convenient index notation. As indicated the nine scalar components of the stress tensor can be represented by T_{ij}, where i and j can take the values from 1 to 3 and the unit vectors $\hat{\mathbf{x}}_1$, $\hat{\mathbf{x}}_2$, $\hat{\mathbf{x}}_3$ become $\hat{\mathbf{x}}_i$. Thus, we can write the stress tensor with its unit dyads as

$$\mathbf{T} = \sum_{i=1}^{3} \sum_{j=1}^{3} \hat{\mathbf{x}}_i \, \hat{\mathbf{x}}_j \, T_{ij} \tag{1.2.17}$$

If we evaluate the summation, we will obtain all nine terms in eq. 1.2.8. Using index notation, the "dot" operations can be written as simple summations, and eq. 1.2.4 or 1.2.15 becomes

$$\mathrm{a}_x = \mathrm{a}_1 = \mathrm{a}_n \hat{\mathbf{n}} \cdot \hat{\mathbf{x}} = \mathrm{a}_n \sum_{i=1}^{3} n_i \hat{x}_i = \mathrm{a}_n n_1 \tag{1.2.18}$$

and eq. 1.2.9 or eq. 1.2.14 becomes

$$\begin{aligned}\mathbf{t}_n = \hat{\mathbf{n}} \cdot \mathbf{T} &= \sum_{i=1}^{3} \hat{\mathbf{x}}_i \sum_{j=1}^{3} \hat{n}_j T_{ji} \\ &= \hat{\mathbf{x}}_1(\hat{n}_1 T_{11} + \hat{n}_2 T_{21} + \hat{n}_3 T_{31}) \\ &\quad + \hat{\mathbf{x}}_2(\hat{n}_1 T_{12} + \hat{n}_2 T_{22} + \hat{n}_3 T_{32}) \\ &\quad + \hat{\mathbf{x}}_3(\hat{n}_1 T_{13} + \hat{n}_2 T_{23} + \hat{n}_3 T_{33})\end{aligned} \tag{1.2.19}$$

When index notation is used, usually the summation signs are dropped and the unit vectors and unit dyads are understood. Here is how it works:

$$\mathbf{t} = \sum_{i=1}^{3} \hat{\mathbf{x}}_i t_i = t_i \tag{1.2.20}$$

$$\mathbf{T} = \sum_{i=1}^{3} \sum_{i=1}^{3} \hat{\mathbf{x}}_i \hat{\mathbf{x}}_j \mathrm{T}_{ij} = T_{ij} \tag{1.2.21}$$

$$\hat{\mathbf{n}} \cdot \mathbf{t} = \sum_{i=1}^{3} n_i t_i = n_i t_i \tag{1.2.22}$$

$$\hat{\mathbf{n}} \cdot \mathbf{T} = \sum_{i=1}^{3} \hat{x}_i \sum_{j=1}^{3} n_j T_{ji} = n_j T_{ji} = n_i T_{ij} \tag{1.2.23}$$

If an index is not repeated, multiplication of each component by a unit vector is implied (e.g., $\mathbf{t}_i$ or $\mathbf{n}_i \mathbf{T}_{ij}$). If two indices are not repeated, we will have two unit vectors or a unit dyad (e.g., $\mathbf{T}_{ij}$). If an index is repeated, summation before multiplication by a unit vector, if any, is implied. Since the indices all go from 1 to 3, the choice of which index letters is arbitrary, as indicated in eq. 1.2.23.

To summarize, three types of notation are used in vector and tensor manipulations. The simplest to write is the Gibbs form (e.g., $\mathbf{n} \cdot \mathbf{T}$), which is convenient for writing equations and seeing the physics of things quickly. The index notation in its expanded form (e.g., $\sum_i \hat{x}_i \sum_j n_j T_{ji}$), or as abbreviated (e.g., $n_j T_{ji}$), indicates all the components explicitly, but it is harder to write down and to read all the indices. Matrix notation (e.g., eq. 1.2.14) is convenient for actually carrying out "dot" operations but is even more tedious to write out and tends to obscure the physics.

All this notation associated with tensors has been known to cause a severe headache upon first reading; however, there have been no reported fatalities. In fact, when students realize that it *is* mostly notation, they usually attack tensor analysis with new confidence.

Perhaps the following examples will serve as a helpful aspirin tablet. Several more examples appear at the end of the chapter.

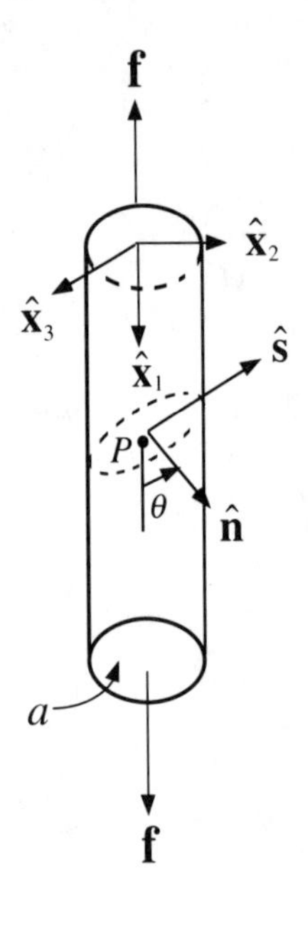

Figure 1.2.4.
Stress on a plane cutting through a cylindrical rod.

Example 1.2.1 Stress on a Shear Plane in a Rod

It is helpful to consider the simple example of the force acting on a cylindrical rod of cross-sectional area a as illustrated in Figure 1.2.4.

(a) What is the state of stress at point P?

(b) What are the normal and shear stresses acting on a plane that cuts across the rod? The normal $\hat{\mathbf{n}}$ to the cutting plane lies in the $\hat{\mathbf{x}}_1\hat{\mathbf{x}}_2$ plane and makes an angle θ with $\hat{\mathbf{x}}_1$; $\hat{\mathbf{n}} = \cos\theta\hat{\mathbf{x}}_1 + \sin\theta\hat{\mathbf{x}}_2$. The tangent $\hat{\mathbf{s}}$ is the intersection of the $\hat{\mathbf{x}}_1\hat{\mathbf{x}}_2$ plane and the cutting plane; $\hat{\mathbf{s}} = \sin\theta\hat{\mathbf{x}}_1 - \cos\theta\hat{\mathbf{x}}_2$.

Solution

(a) Using Cartesian coordinates the stress tensor everywhere in the rod is just (from eq. 1.2.11)

$$\mathbf{T} = \hat{\mathbf{x}}_1\mathbf{t}_1 + \hat{\mathbf{x}}_2\mathbf{t}_2 + \hat{\mathbf{x}}_3\mathbf{t}_3 \tag{1.2.24}$$

Since $\mathbf{t}_1 = \hat{\mathbf{x}}_1(f/a)$ and $\mathbf{t}_2 = \mathbf{t}_3 = 0$, then

$$\mathbf{T} = \left(\frac{f}{a}\right)\hat{\mathbf{x}}_1\hat{\mathbf{x}}_1 \tag{1.2.25}$$

or

$$T_{ij} = \begin{bmatrix} f/a & 0 & 0 \\ 0 & 0 & 0 \\ 0 & 0 & 0 \end{bmatrix}$$

Note that there are mostly zero components in this matrix. This is typical in rheological measurements. The rheologist needs simple stress fields to characterize complex materials.

(b) From eq. 1.2.10 the stress vector $\mathbf{t}_n$ acting on the plane whose normal is $\hat{\mathbf{n}}$ is just $\mathbf{t}_n = \hat{\mathbf{n}} \cdot \mathbf{T}$, where $\hat{\mathbf{n}} = \cos\theta\hat{\mathbf{x}}_1 + \sin\theta\hat{\mathbf{x}}_2$. Using matrix multiplication gives

$$\mathbf{t}_n = \hat{\mathbf{n}} \cdot \mathbf{T} = (\cos\theta, \sin\theta, 0)\begin{bmatrix} f/a & 0 & 0 \\ 0 & 0 & 0 \\ 0 & 0 & 0 \end{bmatrix} = \begin{bmatrix} f/a\cos\theta \\ 0 \\ 0 \end{bmatrix}$$

$$= \left(\frac{f}{a}\right)\cos\theta\hat{\mathbf{x}}_1 \tag{1.2.26}$$

The normal stress on the $\hat{\mathbf{n}}$ plane is just the projection of $\mathbf{t}_n$ on $\hat{\mathbf{n}}$

$$T_{nn} = \mathbf{t}_n \cdot \hat{\mathbf{n}} = \left(\frac{f}{a}\right)\cos\theta, 0, 0)\begin{bmatrix} \cos\theta \\ \sin\theta \\ 0 \end{bmatrix} = \left(\frac{f}{a}\right)\cos^2\theta \tag{1.2.27}$$

The shear stress comes from vector subtraction

$$\mathbf{t}_n - T_{nn}\hat{\mathbf{n}} = T_{ns}\hat{\mathbf{s}} \tag{1.2.28}$$

where $\hat{\mathbf{s}}$ is the vector in the $\hat{\mathbf{x}}_1\hat{\mathbf{x}}_2$ plane tangent to the plane $\hat{\mathbf{n}}$. Substituting on the left-hand side gives

$$\left(\frac{f}{a}\cos\theta, 0, 0\right) - \left(\frac{f}{a}\cos^3\theta, \frac{f}{a}\cos^2\theta\sin\theta, 0\right) = \left(\frac{f}{a}\cos\theta\sin^2\theta, -\frac{f}{a}\cos^2\theta\sin\theta, 0\right) \tag{1.2.29}$$

Since $\hat{\mathbf{s}} = \sin\theta\hat{\mathbf{x}}_1 - \cos\theta\hat{\mathbf{x}}_2$, then

$$T_{ns} = \left(\frac{f}{a}\right)\cos\theta\sin\theta \tag{1.2.30}$$

This shear stress can be important in failure. For example, if a certain crystal plane in a material has a lower slip or yield stress, this stress may be exceeded although the tensile strength between the planes may not.

Example 1.2.2 Stress on a Surface

Measurements of force per unit area were made on three mutually perpendicular test surfaces at point P, (Figure 1.2.2a), with the following results:

Direction of Vector Normal to Test Surface	$\mathbf{t}_i$ *Measured Force/Area* $(Pa = kN/m^2)$
$\hat{\mathbf{x}}_1$	$\hat{\mathbf{x}}_1$
$\hat{\mathbf{x}}_2$	$3\hat{\mathbf{x}}_2 - \hat{\mathbf{x}}_3$
$\hat{\mathbf{x}}_3$	$-\hat{\mathbf{x}}_2 + 3\hat{\mathbf{x}}_3$

(a) What is the state of stress at P?

(b) Find the magnitude of the stress vector acting on a surface whose normal is

$$\hat{\mathbf{n}} = \frac{1}{\sqrt{2}}(\hat{\mathbf{x}}_1 + \hat{\mathbf{x}}_2) \tag{1.2.31}$$

(c) What is the normal stress acting on this interface?

Solution

(a) The state of stress at a point is determined by the stress tensor (eq. 1.2.24)

$$\mathbf{T} = \hat{\mathbf{x}}_1\mathbf{t}_1 + \hat{\mathbf{x}}_2\mathbf{t}_2 + \hat{\mathbf{x}}_3\mathbf{t}_3$$

where the measured tractions $\mathbf{t}_i$ are

$$\mathbf{t}_1 = \hat{\mathbf{x}}_1, \ \mathbf{t}_2 = 3\hat{\mathbf{x}}_2 - \hat{\mathbf{x}}_3, \ \text{and } \mathbf{t}_3 = -\mathbf{x}_2 - 3\mathbf{x}_3$$

Thus,

$$\mathbf{T} = \hat{\mathbf{x}}_1\hat{\mathbf{x}}_1 + 0 + 0 + 0 + 3\hat{\mathbf{x}}_2\hat{\mathbf{x}}_2 - \hat{\mathbf{x}}_2\hat{\mathbf{x}}_3 + 0 - \hat{\mathbf{x}}_3\hat{\mathbf{x}}_2 + 3\hat{\mathbf{x}}_3\hat{\mathbf{x}}_3 \quad (1.2.32)$$

or

$$T_{ij} = \begin{bmatrix} 1 & 0 & 0 \\ 0 & 3 & -1 \\ 0 & -1 & 3 \end{bmatrix}$$

(b) From eq. 1.2.10 we know that the stress vector $\mathbf{t}_n$ acting on the plane whose normal is $\hat{\mathbf{n}}$ is just

$$\mathbf{t}_n = \hat{\mathbf{n}} \cdot \mathbf{T} \quad (1.2.10)$$

Using matrix multiplication (eq. 1.2.14) we find

$$\hat{\mathbf{n}} \cdot \mathbf{T} = \frac{1}{\sqrt{2}}(1, \ 1, \ 0) \begin{bmatrix} 1 & 0 & 0 \\ 0 & 3 & -1 \\ 0 & -1 & 3 \end{bmatrix} = \frac{1}{\sqrt{2}} \begin{bmatrix} 1 \\ 3 \\ -1 \end{bmatrix} \quad (1.2.33)$$

Remembering the omitted unit vectors, we write

$$\mathbf{t}_n = \frac{1}{\sqrt{2}}(\hat{\mathbf{x}}_1 + 3\hat{\mathbf{x}}_2 - \hat{\mathbf{x}}_3)$$

The magnitude of this stress vector is

$$|\mathbf{t}_n| = (\mathbf{t}_n \cdot \mathbf{t}_n)^{1/2} = \left[\frac{1}{2}(1, \ 3, \ -1) \begin{bmatrix} 1 \\ 3 \\ -1 \end{bmatrix}\right]^{1/2} = \sqrt{\frac{11}{2}} \ Pa \quad (1.2.34)$$

(c) The normal stress T_{nn} is just the projection of $\mathbf{t}_n$ onto the unit vector

$$T_{nn} = \hat{\mathbf{n}} \cdot \mathbf{t}_n = \frac{1}{\sqrt{2}}(1, \ 1, \ 0) \cdot \frac{1}{\sqrt{2}} \begin{bmatrix} 1 \\ 3 \\ -1 \end{bmatrix} = 2Pa \quad (1.2.35)$$

1.2.2 Symmetry

Notice that the stress tensor in each of the examples above is symmetric; that is, the rows and columns of the matrix for the components of **T** can be interchanged without changing **T**. The components of the traction vectors t_i were picked that way intentionally. The symmetry of the stress tensor can be shown by considering the shear stresses acting on the small tetrahedron sketched in Figure 1.2.5. The component T_{23} gives rise to a moment about the x_1 axis. To conserve angular momentum, this moment must be balanced by the one caused by T_{32}. Thus, $T_{32} = T_{23}$, and by similar arguments the other pairs of shear components $T_{12} = T_{21}$ and $T_{13} = T_{31}$ are equal. Thus, the stress tensor is symmetric with only six independent components, which when written in matrix form (again leaving out the dyads), become

$$T_{ij} = \begin{bmatrix} T_{11} & T_{12} & T_{13} \\ T_{12} & T_{22} & T_{23} \\ T_{13} & T_{23} & T_{33} \end{bmatrix} \qquad (1.2.36)$$

In Gibbs notation, we show that a tensor is symmetric by writing

$$\mathbf{T} = \mathbf{T}^T \qquad (1.2.37)$$

where $\mathbf{T}^T$ is called the *transpose* of $\mathbf{T}$. In the tensor $\mathbf{T}^T$ the scalar components of the rows and columns of $\mathbf{T}$ have been interchanged. This interchange may be clearer when $\mathbf{T}^T$ is written in index notation

$$\mathbf{T}^T = \sum_i \sum_j (\hat{\mathbf{x}}_i \hat{\mathbf{x}}_j T_{ij})^T = \sum_i \sum_j \hat{\mathbf{x}}_i \hat{\mathbf{x}}_j T_{ji} \qquad (1.2.38)$$

The transpose has wider utility in tensor analysis. For example, we can use it to reverse the order of operations in the vector product of eq. 1.2.10

$$\mathbf{t}_n = \mathbf{T}^T \cdot \hat{\mathbf{n}} \qquad (1.2.39)$$

This result can be verified by using matrix multiplication. Try it yourself. Follow eq. 1.2.14, switch rows with columns in $\mathbf{T}$, and make $\hat{\mathbf{n}}$ a column vector on the other side. Of course, in the end this manipulation does not matter for $\mathbf{T}$ because it is symmetric, but we will find the operation useful later.

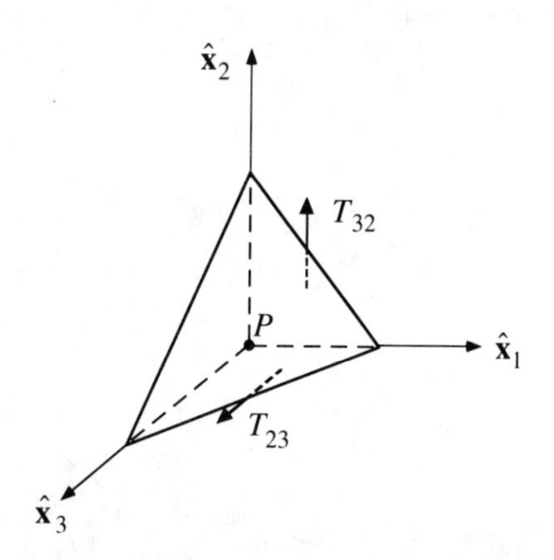

Figure 1.2.5.
To balance angular momentum about the $\hat{\mathbf{x}}_1$ axis, the two shear stress components T_{32} and T_{23} acting on the tetrahedron must be equal.

The possibility of a nonsymmetrical stress tensor is discussed by Dahler and Scriven (1961, 1963). Asymmetry has not been observed experimentally for amorphous liquids. Body torques do exist on suspension particles, but these can be treated by calculating the stress distribution over the particle surface for each orientation (see Chapter 10).

1.2.3 Pressure

One particularly simple stress tensor is that of uniform pressure. A fluid is a material that cannot support a shear stress without flowing. When a fluid is at rest, it can support only a uniform normal stress, $T_{11} = T_{22} = T_{33}$, as indicated in Figure 1.2.6. This normal stress is called the hydrostatic pressure p. Thus, for a fluid at rest, the stress tensor is

$$T_{ij} = \begin{bmatrix} -p & 0 & 0 \\ 0 & -p & 0 \\ 0 & 0 & -p \end{bmatrix} = -p \begin{bmatrix} 1 & 0 & 0 \\ 0 & 1 & 0 \\ 0 & 0 & 1 \end{bmatrix} \tag{1.2.40}$$

where the minus sign is used because compression is usually considered to be negative.

The matrix or tensor with all ones on the diagonal given in eq. 1.2.40 has a special name. It is called the *identity* or unit tensor. When multiplied by another tensor, it always generates the same tensor back again:

$$\mathbf{T} \cdot \mathbf{I} = \mathbf{T} \tag{1.2.41}$$

The Gibbs notation for the identity tensor is $\mathbf{I}$. Its components are

$$\mathbf{I} = \sum_i \hat{\mathbf{x}}_i \hat{\mathbf{x}}_i = \hat{\mathbf{x}}_1 \hat{\mathbf{x}}_1 + \hat{\mathbf{x}}_2 \hat{\mathbf{x}}_2 + \hat{\mathbf{x}}_3 \hat{\mathbf{x}}_3 \tag{1.2.42}$$

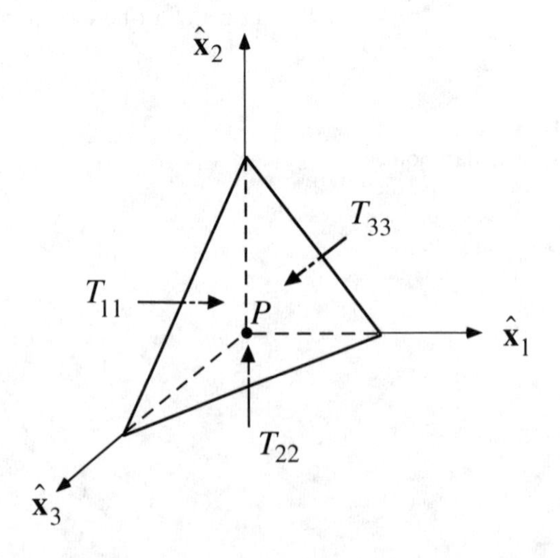

Figure 1.2.6. A hydrostatic state of stress on a tetrahedron of fluid.

Thus, the stress tensor for a fluid at rest is

$$\mathbf{T} = -p\mathbf{I} \tag{1.2.43}$$

When dealing with fluids in motion, it is convenient to retain p. Thus, we write the total stress tensor as the sum of two parts

$$\mathbf{T} = -p\mathbf{I} + \boldsymbol{\tau} \tag{1.2.44}$$

where $\boldsymbol{\tau}$ is known as the *extra* or viscous stress tensor.* Often $\mathbf{T}$ is referred to as the *total stress tensor* and $\boldsymbol{\tau}$ as just the stress tensor.

In rheology we generally assume that a material is incompressible. The deviations from simple Hookean or Newtonian behavior due to nonlinear dependence on deformation or deformation history are usually much greater than the influence of compressibility. We discuss the influence of pressure briefly in Chapters 2 and 6. For incompressible materials the overall pressure cannot influence material behavior. In other words, increasing the barometric pressure in the room should not change the reading from a rheometer. For incompressible materials the isotropic pressure is determined solely by the boundary conditions and the equations of motion (see Sections 1.7 and 1.8).

Thus, it makes sense for incompressible materials to subtract p. The remaining stress tensor $\boldsymbol{\tau}$ contains all the effects of deformation on a material. Constitutive equations are usually written in terms of $\boldsymbol{\tau}$. However, experimentally we can measure only forces which, when divided by the area, give components of the total stress $\mathbf{T}$. Since $\mathbf{T}$ includes p and $\boldsymbol{\tau}$, we would like to remove the pressure term from $\boldsymbol{\tau}$. This presents no problem for the shear stress components (because $T_{12} = \tau_{12}$, etc.), but the normal stress components will differ by p; $T_{ii} = -p + \tau_{ii}$. As we said, determination of p requires boundary conditions for the particular problem. Thus, normal stress *differences* are used to eliminate p since

$$\begin{aligned} T_{11} - T_{22} &= \tau_{11} - \tau_{22} \\ T_{22} - T_{33} &= \tau_{22} - \tau_{33} \end{aligned} \tag{1.2.45}$$

As an example of how we use the normal stress difference, consider the simple uniaxial extension shown in Figure 1.1.2b. The figure shows that tension acting on the $\pm\hat{\mathbf{x}}_1$ faces of the rubber cube will extend it. However, as shown in Figure 1.2.7, a compression, $-T_{22} = -T_{33}$, on the $\pm\hat{\mathbf{x}}_2$ and $\pm\hat{\mathbf{x}}_3$ faces could generate the same deformation. The rubber cube does not know the difference. The deformation is caused by $T_{11} - T_{22}$, the net difference between tension in the $\hat{\mathbf{x}}_1$ direction and the compression in the $\hat{\mathbf{x}}_2$ direction.

Figure 1.2.7.
Uniaxial extension generated by a uniform compression.

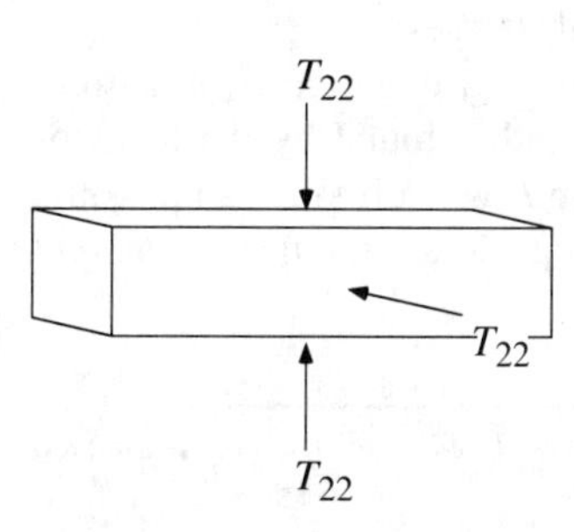

*$\boldsymbol{\tau}$ *is an exception to the general rule for using boldface Latin capital letters for tensors. However, it is in such common use in rheology that we retain it here.*

For the special case of *simple shear* we use

$$T_{11} - T_{22} = N_1 \quad \text{and} \quad T_{22} - T_{33} = N_2 \tag{1.2.46}$$

We call N_1 the *first normal stress difference* and N_2 the *second normal stress difference*. Some authors use the difference $T_{11} - T_{33}$. However, there are only two independent quantities because $T_{11} - T_{33}$ is just the sum of the other two. The reader should also be aware that other notations for stress are common: $\mathbf{P}$ or $\boldsymbol{\pi}$ for $\mathbf{T}$ and $\boldsymbol{\sigma}$ or $\mathbf{T}'$ for $\boldsymbol{\tau}$. Also, several authors use the opposite sign for $\mathbf{T}$ and $\boldsymbol{\tau}$ (See Bird et al., 1987a, p. 7, who consider compression, eq. 1.2.40, to be positive).

It is perhaps consoling to the student struggling with the stress tensor to learn that although Hooke wrote his force extension law before 1700, it took many small and painful steps until Cauchy in the 1820s was able to write the full three-dimensional state of stress at a point in a material.

1.3 Principal Stresses and Invariants*

Later in this and subsequent chapters we will want to make constitutive equations independent of the coordinate system. In particular we will need to make scalar rheological parameters like the modulus or viscosity a function of a tensor.

How can a scalar depend on a tensor? Let us start by considering a simpler but similar problem: How does a scalar depend on a vector? In particular, consider how scalar kinetic energy depends on the vector velocity. Recall the equation for kinetic energy $E_K = 1/2mv^2$, where $v^2 = \mathbf{v} \cdot \mathbf{v}$. Kinetic energy is a function of the dot or scalar product of the velocity vector, the magnitude of the vector squared. Thus, $\mathbf{v} \cdot \mathbf{v}$ is independent of the coordinate system; it is the *invariant* of the vector $\mathbf{v}$.

There is only one commonly used invariant of a vector: its magnitude. However there are *three* possible invariant scalar functions of a tensor. For the stress tensor we can give these three invariants physical meaning through the principal stresses.

It is always possible to take a special cut through a body such that only a normal stress acts on the plane through the point P. This is called a *principal plane*, and the stress acting on it is a *principal stress* σ. As demonstrated below, there are three of these planes through any point and three principal stresses.

We can visualize the principal stresses in terms of a stress ellipsoid. The surface of this ellipsoid is found by the locus of the end of the traction vector $\mathbf{t}_n$ from P when $\hat{\mathbf{n}}$ takes all possible directions. The three axes of the ellipsoid are the three *principal*

**The reader may skip to Section 1.4 on a first reading. The concept of invariants is used in Section 1.6.*

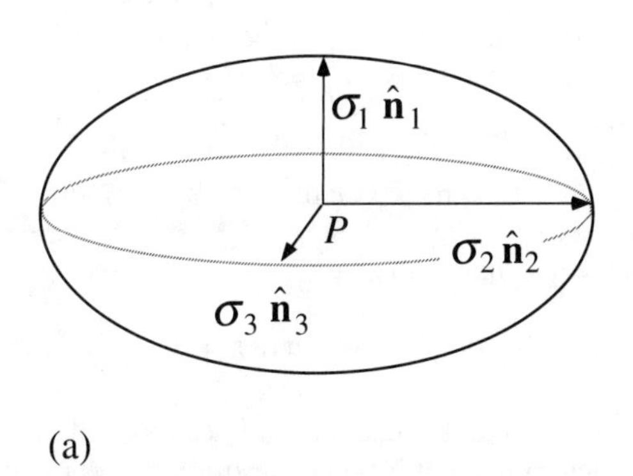

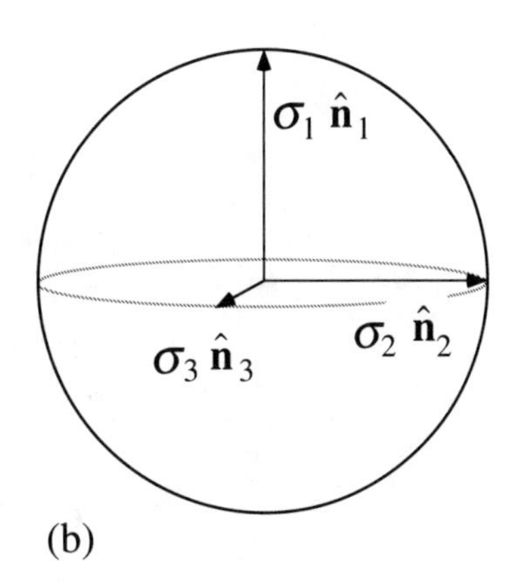

Figure 1.3.1.
(a) A section of the stress ellipsoid at P through two principal axes $\sigma_1\hat{\mathbf{n}}_1$ and $\sigma_2\hat{\mathbf{n}}_2$.
(b) The stress ellipsoid for a hydrostatic state of stress.

stresses and their directions the *principal directions*. A section of such an ellipsoid through two of the axes is shown in Figure 1.3.1.

Note that in the simplest case all the principal stresses are equal: $\sigma_1 = \sigma_2 = \sigma_3 = \sigma$. This equivalence represents the hydrostatic pressure $p = -\sigma$. As we saw at the end of the Section 1.2.3, a hydrostatic state is the only kind of stress that can exist in a fluid at rest.

If we line up our coordinate system with the three principal stresses, all the shear components in the stress tensor will vanish. This is nice because it reduces the stress tensor to just three diagonal components:

$$\text{components of the principal stress tensor} = T_{ij}^{p}$$

$$= \begin{bmatrix} \sigma_1 & 0 & 0 \\ 0 & \sigma_2 & 0 \\ 0 & 0 & \sigma_3 \end{bmatrix} \tag{1.3.1}$$

However, in practice it is often difficult to figure out the rotations of the coordinate system at every point in the material, so as to line it up with the principal directions. Furthermore, it is usually more convenient to leave the coordinates in the laboratory frame. Thus, we normally do not measure the principal stresses (except for purely extensional deformations) but rather calculate them from the measured stress tensor.* We show this next.

Because a principal plane is defined as one on which there is only a normal stress, the traction vector and the unit normal to that plane must be in the same direction:

$$\mathbf{t}_n = \sigma\hat{\mathbf{n}} \tag{1.3.2}$$

An exception is flow birefringence where differences in the principal stresses and their angle of rotation are measured directly; see Section 9.4.

Thus, σ is the magnitude of the principal stress and $\hat{\mathbf{n}}$ its direction. As we saw earlier (eq. 1.2.10), the stress tensor is the machine that gives us the traction vector on any plane through the dot operation. Thus,

$$\mathbf{t}_n = \hat{\mathbf{n}} \cdot \mathbf{T} = \sigma \hat{\mathbf{n}} \tag{1.3.3}$$

This equation can be rearranged to give

$$\hat{\mathbf{n}} \cdot (\mathbf{T} - \sigma \mathbf{I}) = 0 \quad \text{or} \quad n_i(T_{ij} - \sigma I_{ij}) = 0 \tag{1.3.4}$$

Since $\hat{\mathbf{n}}$ is not zero, to solve this equation we need to find values of σ such that the determinant of $\mathbf{T} - \sigma \mathbf{I}$ vanishes. This is usually called an eigenvalue problem.

$$\det(\mathbf{T} - \sigma \mathbf{I}) = \det \begin{bmatrix} T_{11} - \sigma & T_{12} & T_{13} \\ T_{21} & T_{22} - \sigma & T_{23} \\ T_{31} & T_{32} & T_{33} - \sigma \end{bmatrix} = 0$$

Expanding this determinant yields the characteristic equation of the matrix

$$\sigma^3 - I_T \sigma^2 + II_T \sigma - III_T = 0 \tag{1.3.5}$$

where the coefficients are

$$I_T = \text{tr}\,\mathbf{T} = T_{11} + T_{22} + T_{33} = \sigma_1 + \sigma_2 + \sigma_3 \tag{1.3.6}$$

$$\begin{aligned} II_T &= \frac{1}{2}[I_T^2 - \text{tr}\mathbf{T}^2] = T_{11}T_{22} + T_{11}T_{33} + T_{22}T_{33} \qquad (1.3.7) \\ &\quad - T_{12}T_{21} - T_{13}T_{31} - T_{23}T_{32} = \sigma_1\sigma_2 + \sigma_1\sigma_3 + \sigma_2\sigma_3 \\ III_T &= \det\mathbf{T} \qquad (1.3.8) \\ &= T_{11}T_{22}T_{33} + T_{12}T_{23}T_{31} + T_{13}T_{32}T_{21} \\ &\quad - T_{11}T_{23}T_{32} - T_{21}T_{12}T_{33} - T_{31}T_{22}T_{13} = \sigma_1\sigma_2\sigma_3 \end{aligned}$$

I_T is called the first invariant of the tensor $\mathbf{T}$, II_T the second invariant, and III_T the third invariant. They are called invariants because no matter what coordinate systems we choose to express $\mathbf{T}$, they will retain the same value. We will see that this property is particularly helpful in writing constitutive equations. Note that other combinations of T_{ij} can be used to define invariants (cf. Bird et al., 1987a, p. 568).

Equation 1.3.5 is a cubic and will have three roots, the eigenvalues σ_1, σ_2, and σ_3. If the tensor is symmetric all these roots will be real. The roots are then the principal values of T_{ij} and n_i, the principal directions. With them $\mathbf{T}$ can be transformed to a new tensor such that it will have only three diagonal components, the principal stress tensor, eq. 1.3.1.

To help illustrate the use of eq. 1.3.5 to determine the principal stresses, consider Example 1.3.1.

Example 1.3.1 Principal Stresses and Invariants

Determine the invariants and the magnitudes and directions of the principal stresses for the stress tensor given in Example 1.2.2. Check the values for the invariants using the principal stress magnitudes.

Solution

For eq. 1.2.32 we obtain

$$T_{ij} = \begin{bmatrix} 1 & 0 & 0 \\ 0 & 3 & -1 \\ 0 & -1 & 3 \end{bmatrix} \quad \text{and} \quad T_{ij}^2 = \begin{bmatrix} 1 & 0 & 0 \\ 0 & 10 & -6 \\ 0 & -6 & 10 \end{bmatrix} \tag{1.3.9}$$

Using eqs. 1.3.6–1.3.8 we can calculate the invariants

$$\begin{aligned} I_T &= \mathrm{tr}\mathbf{T} = 7 \\ II_T &= \frac{1}{2}(I_T^2 - \mathrm{tr}\mathbf{T}^2) = 14 \\ III_T &= \det\mathbf{T} = 8 \end{aligned} \tag{1.3.10}$$

From eq. 1.3.5 we can find the principal stress magnitudes: $\sigma^3 - 7\sigma^2 + 14\sigma - 8 = 0$, which factors into

$$(\sigma - 1)(\sigma - 2)(\sigma - 4) = 0$$

Thus

$$\sigma_1 = 1 \qquad \sigma_2 = 2 \qquad \sigma_3 = 4 \tag{1.3.11}$$

Clearly most cases will not factor so easily, but the cubic can be solved by simple numerical methods. We can check the values for the invariants using these σ_i:

$$\begin{aligned} I_T &= \sigma_1 + \sigma_2 + \sigma_3 = 7 \\ II_T &= \sigma_1\sigma_2 + \sigma_1\sigma_3 + \sigma_2\sigma_3 = 14 \\ III_T &= \sigma_1\sigma_2\sigma_3 = 8 \end{aligned} \tag{1.3.12}$$

To obtain the principal directions, we seek $t_n(i)$, which are solutions to

$$\mathbf{n}^{(i)} \cdot \mathbf{T} = \sigma_i \mathbf{n}^{(i)} \qquad \text{for} \qquad i = 1, 2, 3 \tag{1.3.13}$$

For each principal magnitude eq. 1.3.13 results in three equations for the three components of each principal direction.

$$\begin{array}{ccc} i = 1 & i = 2 & i = 3 \\ n_1^{(1)} = n_1^{(1)} & n_1^{(2)} = 2n_1^{(2)} & n_1^{(3)} = 4n_1^{(3)} \\ 3n_2^{(1)} - n_3^{(1)} = n_2^{(1)} & 3n_2^{(2)} - n_3^{(2)} = 2n_2^{(2)} & 3n_2^{(3)} - n_3^{(3)} = 4n_2^{(3)} \\ -n_2^{(1)} + 3n_3^{(1)} = n_3^{(1)} & -n_2^{(2)} + 3n_3^{(2)} = 2n_3^{(2)} & -n_2^{(3)} + 3n_3^{(3)} = 4n_3^{(3)} \end{array}$$

These three sets of equations are solved for directions of unit length as follows:

$$n_1^{(1)} = 1 \qquad n_1^{(2)} = 0 \qquad n_1^{(3)} = 0$$
$$n_2^{(1)} = 0 \qquad n_2^{(2)} = \tfrac{1}{\sqrt{2}} \qquad n_2^{(3)} = \tfrac{1}{\sqrt{2}}$$
$$n_3^{(1)} = 0 \qquad n_3^{(2)} = \tfrac{1}{\sqrt{2}} \qquad n_3^{(3)} = -\tfrac{1}{\sqrt{2}}$$

Thus the principal directions are

$$\mathbf{n}^{(1)} = \hat{\mathbf{x}}_1$$
$$\mathbf{n}^{(2)} = \frac{1}{\sqrt{2}}\hat{\mathbf{x}}_2 + \frac{1}{\sqrt{2}}\hat{\mathbf{x}}_3$$
$$\mathbf{n}^{(3)} = \frac{1}{\sqrt{2}}\hat{\mathbf{x}}_2 - \frac{1}{\sqrt{2}}\hat{\mathbf{x}}_3$$

where $\mathbf{n}^{(2)}$ is rotated + 45° from the $\hat{\mathbf{x}}_2$ axis.*

1.4 Finite Deformation Tensors

Now that we have a way to determine the state of stress at any point in a material by using the stress tensor, we need a similar measure of deformation to complete our three-dimensional constitutive equation for elastic solids.

Consider the small lump of material shown in Figure 1.4.1. We have drawn a cube, but any lump will do. P is a point embedded in the body and Q is a neighboring point separated by a small distance $d\mathbf{x}'$. Note that $d\mathbf{x}'$ is a vector. The area vector $d\mathbf{a}'$ represents a small patch of area around Q. We use the $'$ to denote the rest or reference state of the material; or, if the material is continually deforming, the $'$ denotes the state of the material at some past time, t'. From here on we concentrate on deformations from a rest state. In the following chapters we treat continual deformation with time.

Now let the body be deformed to a new state as shown in Figure 1.4.1. Because the points P and Q move with the material, the small displacement between them will stretch and rotate as indicated by the direction and magnitude of the new vector $d\mathbf{x}$. Somehow we need to relate $d\mathbf{x}$ back to $d\mathbf{x}'$. Another tensor to the rescue! The change in $d\mathbf{x}$ with respect to $d\mathbf{x}'$ is called the

The rotation angle χ' of the principal stress axes is used in analyzing flow birefringence data as discussed in Section 9.4.1. In this example $\chi' = 45°C$ and $\hat{\mathbf{x}}_2 - \hat{\mathbf{x}}_3$ is the plane of shearing. Then the results above satisfy eqs. 9.4.2 and 9.4.3.

$$2T_{23} = \Delta\sigma \sin 2\chi' \tag{9.4.2}$$
$$2(-1) = (2-4)\cdot 1$$
$$T_{22} - T_{33} = \Delta\sigma \cos 2\chi' \tag{9.4.3}$$
$$3 - 3 = (-2)\cdot 0$$

deformation gradient. It is sometimes written like a dyad (recall eq. 1.2.11) $\nabla'\mathbf{x}$, but usually simply as the tensor $\mathbf{F}$. In either case it represents the derivative or change in present position $\mathbf{x}$ with respect to the past position $\mathbf{x}'$.

$$\mathbf{F} = \nabla'\mathbf{x} = \frac{\partial \mathbf{x}}{\partial \mathbf{x}'} \quad \text{or} \quad F_{ij} = \frac{\partial x_i}{\partial x'_j} \tag{1.4.1}$$

By this definition we are assuming that $\mathbf{x}$ can be expressed as a differentiable function of $\mathbf{x}'$ and time

$$\mathbf{x} = \mathbf{x}(\mathbf{x}', t) \tag{1.4.2}$$

This would not be the case, for example, if a crack developed between P and Q during the deformation.

Like the stress tensor, the deformation gradient has up to nine components, each with a scalar magnitude $\partial x_i/\partial x'_j$ and two directions for each of them. One direction comes from the unit vectors of the coordinate system used to describe $\mathbf{x}$ and the other from the $\mathbf{x}'$ unit vectors. And like the stress tensor, $\mathbf{F}$ is a machine, a mathematical operator. It transforms little material displacement vectors from their past to present state, faithfully following the material deformation.

$$d\mathbf{x} = \mathbf{F} \cdot d\mathbf{x}' \tag{1.4.3}$$

Just as the stress tensor characterizes that state of stress at any point through its ability to describe the force acting on any plane, the deformation gradient describes the state of deformation and rotation at any point through the relation above. However, unlike the stress tensor, which depends only on the current state, the deformation gradient depends on both the current and a past state of deformation.

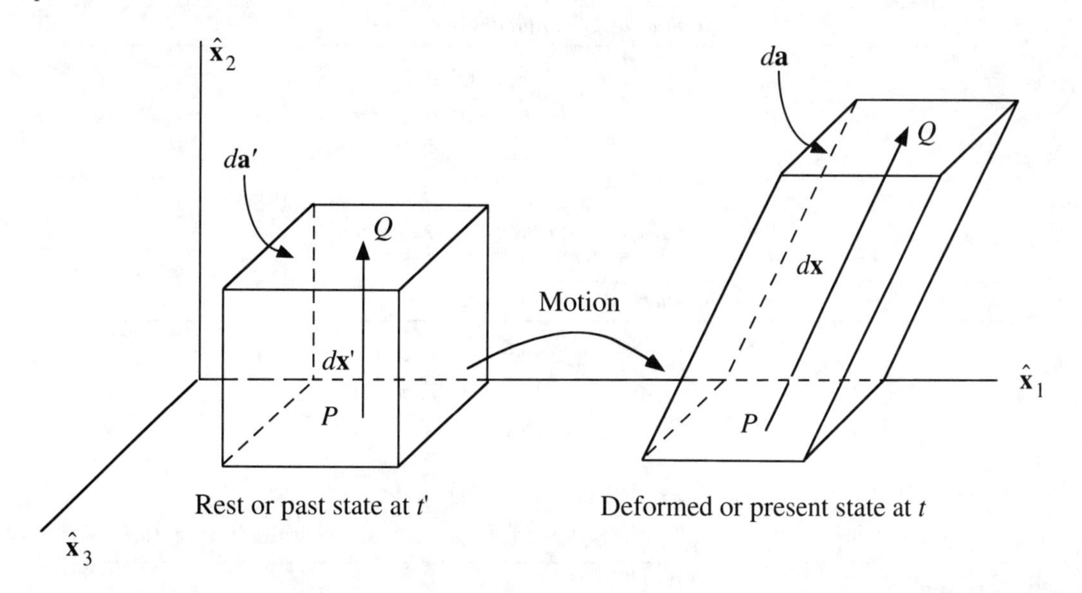

Figure 1.4.1. Deformation of a lump of material showing the motion between two neighboring points.

The following examples illustrate how the deformation gradient tensor works. Table 1.4.1 gives the components of **F** in rectangular, cylindrical, and spherical coordinates.

Example 1.4.1 Evaluation of F

Consider the block of material with dimensions $\Delta x_{1'}$, $\Delta x_{2'}$, $\Delta x_{3'}$ shown in Figure 1.4.2. Within the block is a material point P with coordinates x_1', x_2', x_3' in the reference state. Assume that the block deforms affinely (i.e., that each point within the cube moves in proportion to the exterior dimension). The block is subject to three different motions as shown: (a) uniaxial extension, (b) simple shear, and (c) solid body rotation. In each, the new coordinates of P become x_1, x_2, x_3. For each deformation, write out functions to describe the displacement of P like those given in eq. 1.4.2. From these determine the components of **F**, the deformation gradient tensor.

TABLE 1.4.1. / Components of F

Rectangular Coordinates (x, y, z)

Displacement functions:

$$x = x(x', y', z', t, t')$$
$$y = y(x', y', z', t, t')$$
$$z = z(x', y', z', t, t')$$

$F_{xx} = \partial x/\partial x'$	$F_{xy} = \partial x/\partial y'$	$F_{xz} = \partial x/\partial z'$
$F_{yx} = \partial y/\partial x'$	$F_{yy} = \partial y/\partial y'$	$F_{yz} = \partial y/\partial z'$
$F_{zx} = \partial z/\partial x'$	$F_{zy} = \partial z/\partial y'$	$F_{zz} = \partial z/\partial z'$

Cylindrical Coordinates (r, θ, z)

$$r = r(r', \theta', z', t, t')$$
$$\theta = \theta(r', \theta', z', t, t')$$
$$z = z(r', \theta', z', t, t')$$

$F_{rr} = \partial r/\partial r'$	$F_{r\theta} = \partial r/r'\partial\theta'$	$F_{rz} = \partial r/\partial z'$
$F_{\theta r} = r\partial\theta/\partial r'$	$F_{\theta\theta} = r\partial\theta/r'\partial\theta'$	$F_{\theta z} = r\partial\theta/\partial z'$
$F_{zr} = \partial z/\partial r'$	$F_{z\theta} = \partial z/r'\partial\theta'$	$F_{zz} = \partial z/\partial z'$

Spherical Coordinates (r, θ, ϕ)

$$r = r(r', \theta', \phi', t, t')$$
$$\theta = \theta(r', \theta', \phi', t, t')$$
$$\phi = \phi(r', \theta', \phi', t, t')$$

$F_{rr} = \partial r/\partial r'$	$F_{r\theta} = \partial r/r'\partial\theta'$	$F_{r\phi} = \partial r/r'\sin\theta'\partial\phi'$
$F_{\theta r} = r\partial\theta/\partial r'$	$F_{\theta\theta} = r\partial\theta/r'\partial\theta'$	$F_{\theta\phi} = r\partial\theta/r'\sin\theta'\partial\phi'$
$F_{\phi r} = r\sin\theta\partial\phi/\partial r'$	$F_{\phi\theta} = r\sin\theta\partial\phi/r'\partial\theta'$	$F_{\phi\phi} = r\sin\theta\partial\phi/r'\sin\theta'\partial\phi'$

Solutions

(a) *Uniaxial Extension.* Since the deformation is affine, the change in x_i is just proportional to the changes in the exterior of the block. Thus, the displacement functions become

$$
\begin{aligned}
x_1 &= \frac{\Delta x_1}{\Delta x_1'} x_1' = \alpha_1 x_1' \\
x_2 &= \frac{\Delta x_2}{\Delta x_2'} x_2' = \alpha_2 x_2' \\
x_3 &= \frac{\Delta x_3}{\Delta x_3'} x_3' = \alpha_3 x_3'
\end{aligned} \tag{1.4.4}
$$

where the α_i are the extension ratios, eq. 1.1.4.

Using $F_{ij} = \partial x_i / \partial x_j'$, we obtain

$$
F_{ij} = \begin{bmatrix} \alpha_1 & 0 & 0 \\ 0 & \alpha_2 & 0 \\ 0 & 0 & \alpha_3 \end{bmatrix} \tag{1.4.5}
$$

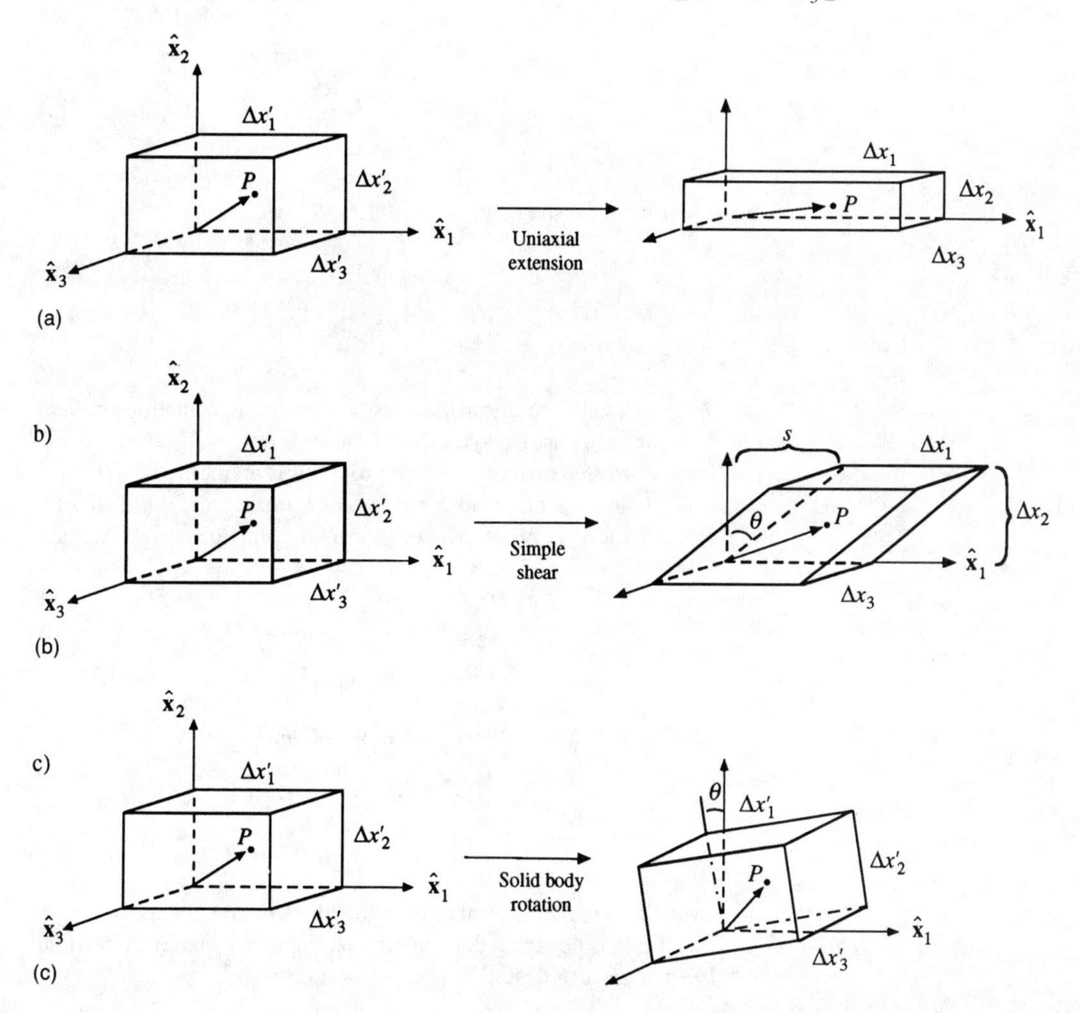

Figure 1.4.2.
A block of material subject to three different motions: (a) uniaxial extension in the $\hat{\mathbf{x}}_1$ direction, (b) simple shear in the $\hat{\mathbf{x}}_1$ direction, and (c) solid body rotation about the $\hat{\mathbf{x}}_3$ axis with no change in $\Delta x_i'$.

Because uniaxial extension is symmetric about the $\hat{\mathbf{x}}_1$ axis, $\alpha_2 = \alpha_3$. If the material is incompressible, then the volume must be constant (see Section 1.7.1 for mass balance) and

$$V' = V$$

$$\Delta x_1' \Delta x_2' \Delta x_3' = \Delta x_1 \Delta x_2 \Delta x_3 \quad \text{or} \quad \alpha_1 \alpha_2 \alpha_3 = 1 \tag{1.4.6}$$

$$\alpha_1 \alpha_2^2 = 1 \qquad \alpha_2 = \frac{1}{\alpha_1^{1/2}}$$

$$F_{ij} = \begin{bmatrix} \alpha_1 & 0 & 0 \\ 0 & \alpha_1^{-1/2} & 0 \\ 0 & 0 & \alpha_1^{-1/2} \end{bmatrix} \tag{1.4.7}$$

(b) *Simple shear.* In simple shear, material planes slide over each other in the $\hat{\mathbf{x}}_1$ direction. Thus, the x_2' and x_3' coordinates of P remain unchanged, while $\hat{\mathbf{x}}_1$ coordinate are displaced by an amount proportional to $s/\Delta x_2' = \theta$. The displacement functions become

$$\begin{aligned} x_1 &= x_1' + \frac{s}{\Delta x_2'} x_2' = x_1' + \gamma x_2' \\ x_2 &= x_2' \\ x_3 &= x_3' \end{aligned} \tag{1.4.8}$$

where γ is the shear strain, eq. 1.1.6. F_{ij} becomes

$$F_{ij} = \begin{bmatrix} 1 & \gamma & 0 \\ 0 & 1 & 0 \\ 0 & 0 & 1 \end{bmatrix} \tag{1.4.9}$$

We note that in contrast to the stress tensor, the deformation gradient $\mathbf{F}$ is not necessarily a symmetric tensor.

(c) *Solid Body Rotation.* Since rotation is about the $\hat{\mathbf{x}}_3$ axis, this coordinate does not change, and point P rotates along the arc of a circle in the $x_1 x_2$ plane. The displacement functions can be written

$$\begin{aligned} x_1 &= x_1' \cos\theta - x_2' \sin\theta \\ x_2 &= x_1' \sin\theta + x_2' \cos\theta \\ x_3 &= x_3' \end{aligned} \tag{1.4.10}$$

Solving for the components of $\mathbf{F}$, we obtain

$$F_{ij} = \begin{bmatrix} \cos\theta & -\sin\theta & 0 \\ \sin\theta & \cos\theta & 0 \\ 0 & 0 & 1 \end{bmatrix} \tag{1.4.11}$$

Even though the material lines in the block do not change in length (i.e., there is no actual deformation or change of shape) we see that $\mathbf{F}$ is not zero. $\mathbf{F}$ describes both deformation and rotation.

In Example 1.4.1 we see that the deformation gradient tensor describes rotation as well as shape change. Somehow we must eliminate this rotation. Material response is determined only by stretching or rate of stretching, not by a solid body rotation. Imagine if we did the tensile test illustrated in Figure 1.1.2 while standing on a turntable. We would not expect the rotation to change our results.

This principle is called *frame indifference*. It helps us select the proper tensor forms in constitutive equations.

1.4.1 Finger Tensor

To express the idea of both stretch and rotation in the deformation gradient, we write it as the tensor product of **V** for stretching and **R** for rotation

$$\mathbf{F} = \mathbf{V} \cdot \mathbf{R} \tag{1.4.12}$$

To remove the rotation, we can multiply **F** by its transpose (interchanging rows and columns).* We know from matrix algebra that if we multiply a matrix times its transpose, we always get a symmetric matrix. Recall that transpose simply means interchanging rows and columns of the matrix.

$$\mathbf{F} \cdot \mathbf{F}^T \equiv \mathbf{B} \quad \text{or} \quad F_{ik}F_{jk} = \frac{\partial x_i}{\partial x'_k}\frac{\partial x_j}{\partial x'_k} = B_{ij} \tag{1.4.13}$$

The dot product of two tensors is called a tensor product because it generates a new tensor just as matrix multiplicaton of one 3 × 3 matrix by another generates a new 3 × 3 matrix. In this case the new tensor is called the Finger deformation tensor after J. Finger (1894), who was the first to use it.

Physically this tensor gives us relative local change in area within the sample. The relative local area change squared is just

$$\mu^2 = \frac{d\mathbf{a}' \cdot d\mathbf{a}'}{d\mathbf{a} \cdot d\mathbf{a}} \tag{1.4.14}$$

Note that $d\mathbf{a}' \cdot d\mathbf{a}' = |da'|^2$, the square of the magnitude of the original or undeformed area. To relate this to **F**, we need to determine the volume associated with $d\mathbf{a}'$ and $d\mathbf{x}'$. If we look back at Figure 1.4.1, the volume of the material element is the scalar product of the area vector $d\mathbf{a}'$ with the length $d\mathbf{x}'$.

$$dV' = d\mathbf{a}' \cdot d\mathbf{x}' \tag{1.4.15}$$

**This can be shown by noting that* $\mathbf{F}^T = \mathbf{R}^T \cdot \mathbf{V}^T$ *and* $\mathbf{R} \cdot \mathbf{R}^T = \mathbf{I}$*; a rotation followed by a reverse rotation leads to no change.*

Because mass is always conserved in a deformation, the density ρ times volume must be constant

$$\rho' dV' = \rho\, dV$$

or in terms of area and length from eq. 1.4.15,

$$\rho' d\mathbf{a}' \cdot d\mathbf{x}' = \rho\, d\mathbf{a} \cdot d\mathbf{x} \tag{1.4.16}$$

Using the deformation gradient tensor to express $d\mathbf{x}$ in terms of $d\mathbf{x}'$, eq. 1.4.3 gives

$$d\mathbf{a}' = \frac{\rho}{\rho'} d\mathbf{a} \cdot \mathbf{F} \tag{1.4.17}$$

For incompressible materials $\rho/\rho' = 1$.*

Combining eq. 1.4.17 with 1.4.14 gives

$$\begin{aligned} \mu^2 &= \frac{(d\mathbf{a} \cdot \mathbf{F}) \cdot (d\mathbf{a} \cdot \mathbf{F})}{d\mathbf{a} \cdot d\mathbf{a}} \\ &= \frac{d\mathbf{a} \cdot (\mathbf{F} \cdot \mathbf{F}^T) \cdot d\mathbf{a}}{|d\mathbf{a}|^2} = \frac{d\mathbf{a} \cdot \mathbf{B} \cdot d\mathbf{a}}{d\mathbf{a} \cdot d\mathbf{a}} \end{aligned} \tag{1.4.18}$$

The unit normal to the surface around Q is just

$$\frac{d\mathbf{a}}{|d\mathbf{a}|} = \hat{\mathbf{n}} \tag{1.4.19}$$

So we can write eq. 1.4.18

$$\mu^2 = \hat{\mathbf{n}} \cdot \mathbf{B} \cdot \hat{\mathbf{n}} \tag{1.4.20}$$

Thus, physically the Finger tensor describes the area change around a point on a plane whose normal is $\hat{\mathbf{n}}$. $\mathbf{B}$ can give the deformation at any point in terms of area change by operating on the normal to the area defined in the present or deformed state. Because area is a scalar, we need to operate on the vector twice.

We can also express deformation in terms of length change. This comes from the Green or Cauchy–Green tensor

$$\mathbf{C} = \mathbf{F}^T \cdot \mathbf{F} \quad \text{or} \quad C_{ij} = F_{ki} F_{kj} = \frac{\partial x_k}{\partial x'_i} \frac{\partial x_k}{\partial x'_j} \tag{1.4.21}$$

Note that we have merely switched the order of the tensor product from that given in the Finger tensor, but as we will see, in general, this switch gives us different results. By a similar derivation, (see Exercise 1.10.4) as given for μ, the relative area change, we can

---

**Note that ρ/ρ' can be calculated from the determinant of $\mathbf{F}$. Since ρ/ρ' is a scalar multiplier, it can be readily carried along if desired (Malvern, 1969).*

use $\mathbf{C}$ to calculate the extension ratio α at any point in the material by the equation

$$\alpha^2 = \frac{d\mathbf{x} \cdot d\mathbf{x}}{d\mathbf{x}' \cdot d\mathbf{x}'} = \hat{\mathbf{n}}' \cdot \mathbf{C} \cdot \hat{\mathbf{n}}' \tag{1.4.22}$$

Length, area, and volume change can also be expressed in terms of the invariants of $\mathbf{B}$ or $\mathbf{C}$ (see eqs. 1.4.45–1.4.47). Note that the Cauchy tensor operates on unit vectors that are defined in the past state. In the next section we will see that the Cauchy tensor is not as useful as the Finger tensor for describing the stress response at large strain for an elastic solid. But first we illustrate each tensor in Example 1.4.2. This example is particularly important because we will use the results directly in the next section with our neo-Hookean constitutive equation.

Example 1.4.2 Evaluation of B and C

For the deformations illustrated in Figure 1.4.2 and Example 1.4.1, evaluate the components of the Finger and the Cauchy deformation tensors.

Solutions

This is straightforward using the definitions of $\mathbf{B}$ (eq. 1.4.13) and $\mathbf{C}$ (eq. 1.4.21) and the results we obtained for F_{ij} in Example 1.4.1.
(a) *Uniaxial Extension*

$$B_{ij} = F_{ik}F_{jk} = \begin{bmatrix} \alpha_1 & 0 & 0 \\ 0 & \alpha_1^{-1/2} & 0 \\ 0 & 0 & \alpha_1^{-1/2} \end{bmatrix} \cdot \begin{bmatrix} \alpha_1 & 0 & 0 \\ 0 & \alpha_1^{-1/2} & 0 \\ 0 & 0 & \alpha_1^{-1/2} \end{bmatrix} = \begin{bmatrix} \alpha_1^2 & 0 & 0 \\ 0 & \alpha_1^{-1} & 0 \\ 0 & 0 & \alpha_1^{-1} \end{bmatrix} \tag{1.4.23}$$

Since $\mathbf{F} = \mathbf{F}^T$, then $\mathbf{B} = \mathbf{C} = \mathbf{F}\cdot\mathbf{F} = \mathbf{F}^2$ and it does not matter which deformation measure is used.
(b) *Simple Shear*

$$B_{ij} = \begin{bmatrix} 1 & \gamma & 0 \\ 0 & 1 & 0 \\ 0 & 0 & 1 \end{bmatrix} \cdot \begin{bmatrix} 1 & 0 & 0 \\ \gamma & 1 & 0 \\ 0 & 0 & 1 \end{bmatrix} = \begin{bmatrix} 1+\gamma^2 & \gamma & 0 \\ \gamma & 1 & 0 \\ 0 & 0 & 1 \end{bmatrix} \tag{1.4.24}$$

$$C_{ij} = \begin{bmatrix} 1 & 0 & 0 \\ \gamma & 1 & 0 \\ 0 & 0 & 1 \end{bmatrix} \cdot \begin{bmatrix} 1 & \gamma & 0 \\ 0 & 1 & 0 \\ 0 & 0 & 1 \end{bmatrix} = \begin{bmatrix} 1 & \gamma & 0 \\ \gamma & 1+\gamma^2 & 0 \\ 0 & 0 & 1 \end{bmatrix} \tag{1.4.25}$$

Here we see that $\mathbf{B}$ and $\mathbf{C}$ do have different components for a shear deformation. Note that both tensors are symmetric, as they must be.

(c) *Solid Body Rotation*

$$B_{ij} = \begin{bmatrix} \cos\theta & -\sin\theta & 0 \\ \sin\theta & \cos\theta & 0 \\ 0 & 0 & 1 \end{bmatrix} \cdot \begin{bmatrix} \cos\theta & \sin\theta & 0 \\ -\sin\theta & \cos\theta & 0 \\ 0 & 0 & 1 \end{bmatrix} = \begin{bmatrix} 1 & 0 & 0 \\ 0 & 1 & 0 \\ 0 & 0 & 1 \end{bmatrix} \qquad (1.4.26)$$

C_{ij} also gives the same result, just the identity tensor **I**.

The last result in Example 1.4.2 says that there is no area or length change in the sample for the solid body rotation (i.e., there is no deformation). This is what we expect; deformation tensors should not respond to rotation. They are useful candidates for constitutive equations, for predicting stress from deformation.

1.4.2 Strain Tensor

So far we have defined deformation in terms of extension, the ratio of deformed to undeformed length, $\alpha = L/L'$. Thus when deformation does not occur, the extensions are unity and $\mathbf{B} = \mathbf{I}$. Frequently deformation is described in terms of strain, the ratio of change in length to undeformed length

$$\epsilon = \frac{L - L'}{L'} = \alpha - 1 \qquad (1.4.27)$$

When there is no deformation, the strains are zero. A finite strain tensor can be defined by subtracting the identity tensor from **B**

$$\mathbf{E} = \mathbf{B} - \mathbf{I} \qquad (1.4.28)$$

Thus from Example 1.4.2 the strain tensor in uniaxial extension becomes

$$E_{ij} = \begin{bmatrix} \alpha_1^2 - 1 & 0 & 0 \\ 0 & \alpha_1^{-1} - 1 & 0 \\ 0 & 0 & \alpha_1^{-1} - 1 \end{bmatrix}$$

$$\text{and in simple shear } E_{ij} = \begin{bmatrix} \gamma^2 & \gamma & 0 \\ \gamma & 0 & 0 \\ 0 & 0 & 0 \end{bmatrix}$$

Since it is often simpler to write the Finger deformation tensor, and since it only differs from the strain tensor by unity, we usually write constitutive equations in terms of **B**.

1.4.3 Inverse Deformation Tensors*

As we noted earlier, the stress tensor depends only on the current state, while the deformation gradient term depends on two states.

**The reader may skip to Section 1.5 on a first reading.*

the present $\mathbf{x}$ and the past $\mathbf{x}'$. It is a relative tensor. We have defined the deformation gradient as the change of the present configuration with respect to the past, $d\mathbf{x} = \mathbf{F} \cdot d\mathbf{x}'$. However, we can reverse the process and describe the past state in terms of the present. This tensor is called the inverse of the deformation gradient; it is like reversing the tensor machine.

$$d\mathbf{x}' = \mathbf{F}^{-1} \cdot d\mathbf{x} \tag{1.4.29}$$

where $F_{ij}^{-1} = \partial x_i'/\partial x_j$

Using the inverse of the deformation gradient, we can also define inverses of the deformation tensors $\mathbf{B}$ and $\mathbf{C}$

$$\mathbf{B}^{-1} = (\mathbf{F}^{-1})^T \cdot \mathbf{F}^{-1} \quad \text{or} \quad B_{ij}^{-1} = F_{ki}^{-1} F_{kj}^{-1} = \frac{\partial x_k'}{\partial x_i} \frac{\partial x_k'}{\partial x_j} \tag{1.4.30}$$

and

$$\mathbf{C}^{-1} = \mathbf{F}^{-1} \cdot (\mathbf{F}^{-1})^T \quad \text{or} \quad C_{ij}^{-1} = F_{ik}^{-1} F_{jk}^{-1} = \frac{\partial x_i'}{\partial x_k} \frac{\partial x_j'}{\partial x_k} \tag{1.4.31}$$

Physically $\mathbf{B}^{-1}$ is like $\mathbf{B}$; it operates on unit vectors in the deformed or present state $\mathbf{n}$, but instead of area change it gives the inverse of length change at any point in a material

$$\frac{1}{\alpha^2} = \hat{\mathbf{n}} \cdot \mathbf{B}^{-1} \cdot \hat{\mathbf{n}} \tag{1.4.32}$$

Similarly $\mathbf{C}^{-1}$ operates on unit vectors in the undeformed or past state of the material to give the inverse of the area change (as defined in eq. 1.4.14).

$$\frac{1}{\mu^2} = \hat{\mathbf{n}}' \cdot \mathbf{C}^{-1} \cdot \hat{\mathbf{n}}' \tag{1.4.33}$$

These results are proven in Exercise 1.10.5.

Thus we have four deformation tensor operators that can describe local length or area change, eliminating any rotation involved in the deformation. It is also possible to derive these four tensors directly from $\mathbf{F}$ by breaking it down into a pure deformation and a pure rotation (Astarita and Marrucci, 1974; Malvern, 1969). Other deformation tensors can also be defined, but clearly all are derived from the same information so they are not independent. We can convert from one to another, although this operation may be difficult. Which deformation tensor we use in a particular constitutive equation depends on convenience and on predictions that compare favorably with real materials.

Example 1.4.3 Evaluation of $\mathbf{B}^{-1}$ and $\mathbf{C}^{-1}$

Evaluate the inverse deformation tensors for the deformations given in Example 1.4.1.

Solutions

(a) *Uniaxial Extension.* The inverse of a diagonal matrix is simply the inverse of the components. Thus,

$$B_{ij}^{-1} = C_{ij}^{-1} = \begin{bmatrix} 1/\alpha_1^2 & 0 & 0 \\ 0 & \alpha_1 & 0 \\ 0 & 0 & \alpha_1 \end{bmatrix} \tag{1.4.34}$$

Verify this by inverting the displacement functions ($x_1' = a_1^{-1}x_1$, etc.), and determining the components (B_{11}^{-1}, etc.) directly from the definition of B_{ij}^{-1} in eq. 1.4.30.

(b) *Simple Shear.* The inverted displacement functions are

$$x_1' = x_1 - \gamma x_2$$
$$x_2' = x_2$$
$$x_3' = x_3$$

$$F_{ij}^{-1} = \frac{\partial x_i'}{\partial x_j} = \begin{bmatrix} 1 & -\gamma & 0 \\ 0 & 1 & 0 \\ 0 & 0 & 1 \end{bmatrix} \tag{1.4.35}$$

$$B_{ij}^{-1} = (F_{ij}^{-1})^T F_{jk}^{-1} = F_{ji}^{-1} F_{jk}^{-1} = \begin{bmatrix} 1 & -\gamma & 0 \\ -\gamma & 1+\gamma^2 & 0 \\ 0 & 0 & 1 \end{bmatrix} \tag{1.4.36}$$

$$C_{ij}^{-1} = F_{ij}^{-1} (F_{jk}^{-1})^T = F_{ij}^{-1} F_{kj}^{-1} = \begin{bmatrix} 1+\gamma^2 & -\gamma & 0 \\ -\gamma & 1 & 0 \\ 0 & 0 & 1 \end{bmatrix} \tag{1.4.37}$$

(c) *Solid Body Rotation.* No change: $B_{ij}^{-1} = C_{ij}^{-1} = I_{ij}$.

1.4.4 Principal Strains

For any state of deformation at a point, we can find three planes on which there are only normal deformations (tensile or compressive). As with the stress tensor, the directions of these three planes are called *principal directions* and the deformations are called principal deformations α_i, or *principal extensions*. Determining of the principal extensions is an eigenvalue problem comparable to determining the principal stresses in the preceding section. All the same equations hold. Thus from eq. 1.3.5 principal extensions are the three roots or eigenvalues of

$$\alpha^3 - I_F \alpha^2 + II_F \alpha - III_F = 0 \tag{1.4.38}$$

where the invariants of the deformation gradient tensor **F** are the same as those defined for **T**, eqs. 1.3.6–1.3.8. To help illustrate the principal extensions and invariants, consider the following example.

Example 1.4.4 Principal Extensions and Invariants of **B** and **C**

Determine the principal extensions and the invariants for each of the tensors **B** and **C** given in Example 1.4.2.

Solutions

(a) *Uniaxial Extension.* Taking eq. 1.4.23 and setting $\alpha_1 = a$, we obtain

$$B_{ij} = C_{ij} = \begin{bmatrix} a^2 & 0 & 0 \\ 0 & a^{-1} & 0 \\ 0 & 0 & a^{-1} \end{bmatrix} \tag{1.4.39}$$

Here the components of B and C form a diagonal matrix; the principal directions are already lined up with the laboratory coordinates. There is no rotation in this deformation. We can find the invariants by using eqs. 1.3.6–1.3.8

$$\begin{aligned} I_B &= \text{tr}\,\mathbf{B} = a^2 + \frac{2}{a} \\ II_B &= \frac{1}{2}(I_B^2 - \text{tr}\,\mathbf{B}^2) = \frac{1}{2}\left(a^4 + 4a + \frac{4}{a^2} - a^4 - \frac{2}{a^2}\right) \\ &= 2a + \frac{1}{a^2} \\ III_B &= a^2 a^{-1} a^{-1} = 1 \end{aligned} \tag{1.4.40}$$

or for a general extensional deformation

$$\begin{aligned} I_B &= \alpha_1^2 + \alpha_2^2 + \alpha_3^2 \\ II_B &= \alpha_1^2\alpha_2^2 + \alpha_2^2\alpha_3^2 + \alpha_3^2\alpha_1^2 \\ III_B &= \det\mathbf{B} = 1 = \alpha_1^2\alpha_2^2\alpha_3^2 \end{aligned} \tag{1.4.41}$$

(b) *Simple Shear*

$$B_{ij} = \begin{bmatrix} 1+\gamma^2 & \gamma & 0 \\ \gamma & 1 & 0 \\ 0 & 0 & 1 \end{bmatrix} \tag{1.4.24}$$

$$C_{ij} = \begin{bmatrix} 1 & \gamma & 0 \\ \gamma & 1+\gamma^2 & 0 \\ 0 & 0 & 1 \end{bmatrix} \tag{1.4.25}$$

Calculate invariants

$$I_B = \text{tr}\mathbf{B} = 3 + \gamma^2 = I_C \tag{1.4.42}$$
$$II_B = \frac{1}{2}(I_B^2 - \text{tr}\mathbf{B}^2)$$
$$= \frac{1}{2}(9 + 6\gamma^2 + \gamma^4 - 3 - 4\gamma^2 - \gamma^4)$$
$$II_B = 3 + \gamma^2 = II_C$$
$$III_B = \det\mathbf{B} = (1 + \gamma^2) - \gamma^2 = 1 = III_C$$

We see that both **B** and **C** have the same invariants. Furthermore, $I_B = II_B$ for simple shear; the average length change is the same as the average area change.

We can solve eq. 1.3.5, the characteristic equation, for the principal extensions

$$\alpha^3 - (3 + \gamma^2)\alpha^2 + (3 + \gamma^2)\alpha - 1 = 0 \tag{1.4.43}$$

Factor out $\alpha - 1$ and solve the resulting quadratic equation to give

$$\alpha_1 = \frac{1}{2}(2 + \gamma^2 + \gamma\sqrt{4 + \gamma^2})$$
$$\alpha_2 = \frac{1}{2}(2 + \gamma^2 - \gamma\sqrt{4 + \gamma^2}) \tag{1.4.44}$$
$$a_3 = 1$$

Thus in simple shear there is no deformation in one direction. We can see that this is $\hat{\mathbf{x}}_3$ in Figure 1.4.2. The extension occurs entirely within the plane perpendicular to $\hat{\mathbf{x}}_3$, but the principal extensions rotate as γ changes.

(c) *Solid Body Rotation.* There is no deformation, and thus $\alpha_i = 1$

$$I_B = I_C = II_B = II_C = 3 \quad \text{and} \quad III_B = III_C = 1$$

Note that for the strain tensor $\mathbf{E} = \mathbf{B} - \mathbf{I}$, eq. 1.4.28, all the invariants will be zero.

In Example 1.4.4 the invariants of both **B** and **C** are the same. This will always be true. It is useful to plot the invariants for the different types of deformation shown in Figure 1.4.3 as II_B versus I_B. All possible deformations are bounded by simple tension and compression. Since $I_B = II_B$ for simple shear, the deformation lies along the diagonal. So does planar extension, stretching of a sheet in which the sides are held constant (see Example 1.8.2).

The three invariants of **B** can be given physical meanings. The root of the first invariant over three is the average change in length of a line element at point P in the material averaged over all possible orientations (recall Figure 1.4.1).

$$\sqrt{\frac{I_B}{3}} = \langle\frac{dL'}{dL}\rangle \tag{1.4.45}$$

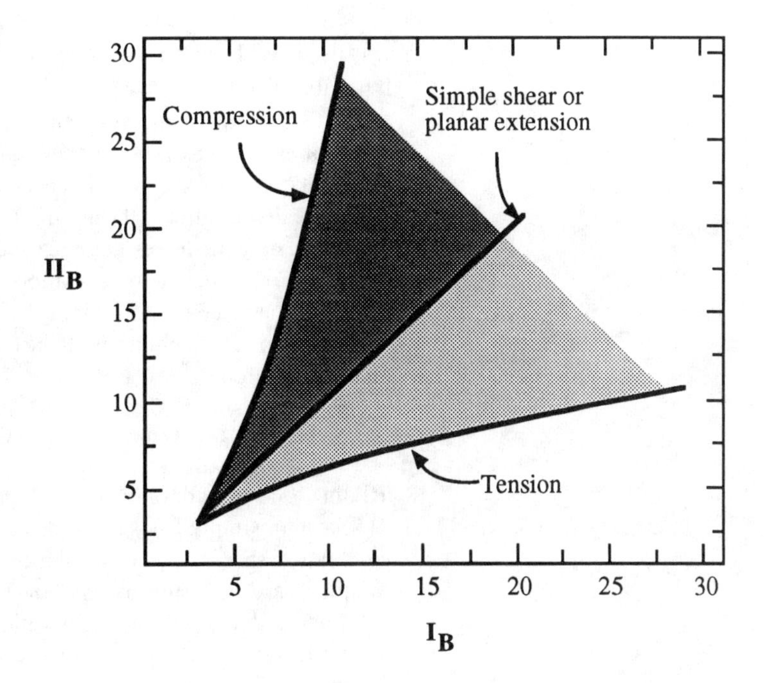

Figure 1.4.3.
Map of invariants of the Finger tensor for deformation of an incompressible material, $III_B = 1$. All types of deformation must occur in the shaded regions and thus may be considered to be a combination of the three simple ones indicated by the lines.

The root of the second invariant is the average area change on all planes around P.

$$\sqrt{\frac{II_B}{3}} = \langle\frac{da'}{da}\rangle \tag{1.4.46}$$

The third invariant, as you might have guessed by now from the dimensions of each invariant, is the volume change for an element of material around P.

$$\sqrt{III_B} = \det \mathbf{F} = \alpha_1\alpha_2\alpha_3 = \frac{dV'}{dV} = \frac{\rho}{\rho'} \tag{1.4.47}$$

From eq. 1.4.6 we know that the product of the principal extensions is unity for an incompressible material. Note also the footnote to eq. 1.4.17.

1.5 Neo-Hookean Solid

In the introduction to this chapter we noted that in 1678 Hooke proposed that the force in a "springy body" was proportional to its extension. It took about 150 years to develop the proper way to determine the three-dimensional state of force or stress and of deformation at any point in a body. In the 1820s Cauchy completed the three-dimensional formulation of Hooke's law. However, because metals and ceramics, which fracture or yield at small deformations, were the main interest at that time only a tensor for small strains was used. With such a tensor, terms like γ^2 in shear deformations (eq. 1.4.24) do not appear.

On the other hand, rubber can deform elastically up to extensions of as much as 7. As rubber came into use as an engineering material during World War II, a need arose to express Hooke's law for large deformations. Using the Finger deformation tensor, we can come up with the result quite easily. If the stress at any point is linearly proportional to deformation and if the material is isotropic (i.e., has the same proportionality in all directions), then the extra stress due to deformation should be determined by a constant times the deformation.

$$\boldsymbol{\tau} = G\mathbf{B} \qquad \text{or} \qquad \mathbf{T} = -p\mathbf{I} + G\mathbf{B} \tag{1.5.1}$$

Rivlin (1948) first developed this equation, and it is called the neo-Hookean or simply Hookean constitutive equation. G is the elastic modulus in shear. It can be a function of the deformation, but in the simplest case we assume it to be constant.

Note that when no deformation is present, $\mathbf{B} = \mathbf{I}$ (recall Example 1.4.2) and eq. 1.5.1 give $\boldsymbol{\tau} = G\mathbf{I}$. Because the pressure is arbitrary for an incompressible material, we can set $p = G$ and thus make the total stress $\mathbf{T} = \mathbf{0}$ in the rest state. An alternative way to write the neo-Hookean model is in terms of the strain tensor $\mathbf{E}$ (cf. Bird et al., 1987b, p. 365)

$$\mathbf{T} = -p\mathbf{I} + G\mathbf{E} \tag{1.5.2}$$

where $\mathbf{E} = \mathbf{B} - \mathbf{I}$, eq. 1.4.28. Then $p = 0$ in the rest state; but clearly the value of p in a constitutive equation for an incompressible material is arbitrary. For any particular problem, the boundary conditions determine p. Only normal stress differences cause deformation.

1.5.1 Uniaxial Extension

We can test out our large strain Hookean or neo-Hookean model with the deformations shown in Figure 1.4.2. For uniaxial extension we can calculate the stress components by substituting the components of $\mathbf{B}$ given in eq. 1.4.23 into eq. 1.5.2.

$$T_{11} = -p + G\alpha_1^2 \tag{1.5.3}$$

$$T_{22} = T_{33} = -p + \frac{G}{\alpha_1} \tag{1.5.4}$$

This deformation can be achieved by applying a force $\mathbf{f}$ on the $\hat{\mathbf{x}}_1$ planes, at the ends of the sample. The stress acting on the ends is divided by the area of the deformed sample $\Delta x_2 \Delta x_3 = a_1$, so that

$$\frac{f_1}{a_1} = T_{11} = -p + G\alpha_1^2 \tag{1.5.5}$$

If no tractions act on the sides of the block, then $T_{22} = T_{33} = 0$ and $p = G/\alpha_1$. Substituting this expression in eq. 1.5.5, we obtain the result given for rubber in eq. 1.1.2 and Figure 1.1.2

$$\frac{f_1}{a_1} = T_{11} = G\left(\alpha_1^2 - \frac{1}{\alpha_1}\right) \tag{1.5.6}$$

Often the force divided by original area a_1' is reported. For an incompressible material

$$\text{volume} = a_1 \Delta x_1 = a_1' \Delta x_1' \qquad \text{or} \qquad a_1 = \frac{a_1'}{\alpha_1}$$

Thus,

$$\frac{f_1}{a_1'} = G\left(\alpha_1 - \frac{1}{\alpha_1^2}\right) \tag{1.5.7}$$

We can also express eq. 1.5.6 in terms of the strain ϵ. Recalling that $\alpha = 1 + \epsilon$ and substituting, we have

$$T_{11} = G\left(\frac{3\epsilon + 3\epsilon^2 + 3\epsilon^3}{1 + \epsilon}\right) \tag{1.5.8}$$

We see that in the limit of small strain

$$T_{11} = 3G\epsilon \quad \text{for} \quad \epsilon << 1 \tag{1.5.9}$$

which is the linear region in Figure 1.1.2. We can define the tensile or Young's modulus as

$$E = \lim_{\epsilon \to 0} \frac{T_{11} - T_{22}}{\epsilon} \tag{1.5.10}$$

which gives

$$E = 3G \tag{1.5.11a}$$

Thus, the tensile modulus is three times that measured in shear, a well-known result for incompressible, isotropic materials.

For compressible, isotropic materials, a parameter μ, Poisson's ratio, is required to relate the tensile to shear modulus

$$E = 2G(\mu + 1) \tag{1.5.11b}$$

where μ ranges from 0.5 for the incompressible case to 0 (e.g., Malvern, 1969; Ward, 1972).

1.5.2 Simple Shear

For simple shear, the results from eq. 1.4.24 give $B_{12} = B_{21} = \gamma$, $B_{11} = 1 + \gamma^2$, and $B_{22} = B_{33} = 1$. Applying the neo-Hookean model gives quite different results from extension

$$T_{11} = -p + G(1 + \gamma^2) \tag{1.5.12}$$

$$T_{12} = T_{21} = G\gamma \tag{1.5.13}$$

$$T_{22} = T_{33} = -p + G \tag{1.5.14}$$

We see that the shear stress T_{12} is linear in strain even to large deformations. This finding agrees with the experimental results for rubber given in Figure 1.1.3. For an incompressible material, p is arbitrary, and only measurements of normal stress differences are meaningful (recall eq. 1.2.45). So we combine eqs. 1.5.12 and 1.5.14 to give

$$T_{11} - T_{22} = G\gamma^2 \tag{1.5.15}$$

$$T_{22} - T_{33} = 0 \tag{1.5.16}$$

Thus, in simple shear the neo-Hookean model predicts a first normal stress difference that increases quadratically with strain. This also agrees with experimental results for rubber (note Figure 1.1.3). We will see in Chapter 4 that the same kinds of normal stress appear in shear of elastic liquids (recall the rod climbing in Figure I.3). Note that there is only one normal stress difference, N_1, for the neo-Hookean solid in shear.

If we had used the Green tensor **C** instead of **B** in writing the constitutive equation, we would have obtained different results for the normal stresses in shear. (note eq. 1.4.25). Using **B** gives results that agree with observations for rubber.

The neo-Hookean model has been applied to many other large strain deformation problems. Several are given in the examples in Section 1.8 and in the exercises in Section 1.10.

1.6 General Elastic Solid

The neo-Hookean model gives a good but not perfect fit to tensile data on real rubber samples. As shown in Figures 1.1.2 and 1.6.1, tensile stress deviates from the model at high extensions. Is there some logical way to generalize the idea of an elastic solid to better describe experimental data?

In the neo-Hookean model, stress is linearly proportional to deformation. We can generalize our elastic model by letting stress be a *function* of the deformation. Thus

$$\mathbf{T} = f(\mathbf{B}) \tag{1.6.1a}$$

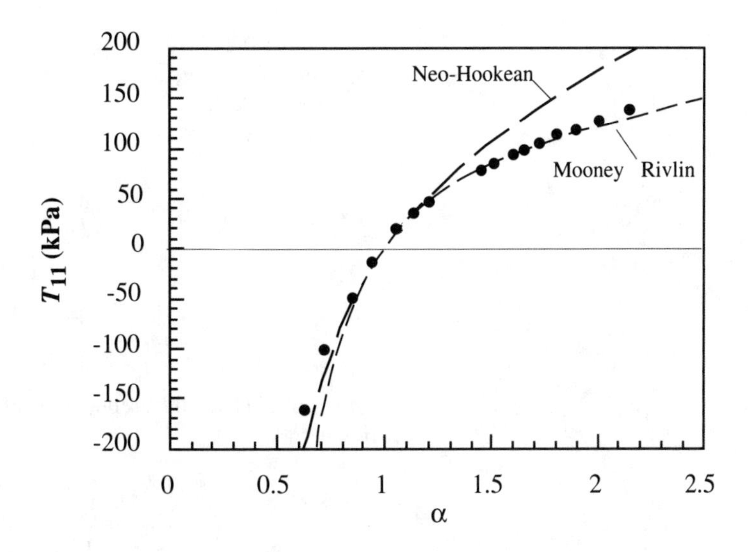

Figure 1.6.1.
Plot of tensile and compressive data for a silicone rubber sample (replotted from Gottlieb, Macosko, and Lepsch, 1981) compared with predictions of the neo-Hookean (G = 101 kPa) and Mooney–Rivlin models (C_1 = 17.5 kPa, $C_2 = 33$ kPa).

Expanding in a power series gives

$$\mathbf{T} = f_o\mathbf{B}^0 + f_1\mathbf{B} + f_2\mathbf{B}^2 + f_3\mathbf{B}^3 + \cdots \tag{1.6.1b}$$

where $\mathbf{B}^0 = \mathbf{I}$ and f_i are scalar constants. The Cayley Hamilton theorem (Chadwick, 1976) states that any tensor satisfies its own characteristic equation (eq. 1.3.5). This theorem allows us to write

$$\mathbf{B}^3 - I_B\mathbf{B}^2 + II_B\mathbf{B} - III_B\mathbf{I} = 0 \tag{1.6.2a}$$

Similarly $\mathbf{B}^4$ and higher powers can be expressed in terms of the lower powers and invariants of $\mathbf{B}$. An alternative to eq. 1.6.2a is to multiply it by $\mathbf{B}^{-1}$ and use the result to eliminate $\mathbf{B}^2$, giving

$$\mathbf{B}^3 = (I_B^2 - II_B)\mathbf{B} + (III_B - I_B II_B)\mathbf{I} + I_B III_B\mathbf{B}^{-1} \tag{1.6.2b}$$

This later form was used by Rivlin (1948, 1956), and following him we can reduce eq. 1.6.1b to

$$\mathbf{T} = g_0\mathbf{I} + g_1\mathbf{B} + g_2\mathbf{B}^{-1} \tag{1.6.3}$$

where g_i are scalar *functions* of the invariants of $\mathbf{B}$. If the material is incompressible, then $g_0 = -p$. Thus, a general incompressible, isotropic elastic solid can be described by two functions of the invariants of $\mathbf{B}$. Since $III_B = 1$ for an incompressible material, g_i can depend only on I_B and II_B. These are called *material functions*.

$$\mathbf{T} = -p\mathbf{I} + g_1(I_B, II_B)\mathbf{B} + g_2(I_B, II_B)\mathbf{B}^{-1} \tag{1.6.4}$$

The neo-Hookean solid is a special case in which $g_2 = 0$ and $g_1 = G$, a single material constant.

Let us apply the general elastic solid to the case of uniaxial extension. We have already worked out the components of **B** and $\mathbf{B}^{-1}$ for this deformation in Example 1.4.2, eq. 1.4.23, and Example 1.4.3, eq. 1.4.34. Substituting these into eq. 1.6.4 gives

$$T_{11} = g_0 + g_1\alpha_1^2 + \frac{g_2}{\alpha_1^2} \tag{1.6.5}$$

$$T_{22} = T_{33} = g_0 + \frac{g_1}{\alpha_1} + g_2\alpha_1 \tag{1.6.6}$$

The normal stress difference, the net stress on the ends of the sample, is then

$$\frac{f_1}{a_1} = T_{11} - T_{22} = \left(g_1 - \frac{g_2}{\alpha_1}\right)\left(\alpha_1^2 - \frac{1}{\alpha_1}\right) \tag{1.6.7}$$

When g_1 and g_2 are replaced by constants, this equation is known as the Mooney or Mooney–Rivlin equation

$$T_{11} - T_{22} = \left(2C_1 + \frac{2C_2}{\alpha_1}\right)\left(\alpha_1^2 - \frac{1}{\alpha_1}\right) \tag{1.6.8}$$

The Mooney–Rivlin equation has been found to fit rubber tensile data better than the neo-Hookean model and is frequently used for engineering calculations. Figure 1.6.1 illustrates how it fits data for silicon rubber. We see that it does quite well in tension, but not as well in compression. Functions rather than constants appear to be necessary to fit experimental data for a general deformation. In experiments on crosslinked natural rubber, Rivlin and Saunders (1951) found that g_1 was constant, but g_2 was quadratic in II_B. Work continues on finding better empirical material functions (Kawabata and Kawai, 1977; Tschoegl, 1979; Ogden, 1984) and on relating them to molecular structure (e.g., Eichinger, 1983).

1.6.1 Strain–Energy Function

Another way to derive the constitutive relation for a general elastic solid (eq. 1.6.4) is to start from an energy balance. We discuss the energy equation in the next chapter, but the basic idea is that for a perfectly elastic solid at equilibrium, the stress can only be a function of the change in the internal energy U of the sample away from its reference state due to a deformation

$$\mathbf{T} = \rho\frac{\partial U}{\partial \mathbf{B}} \tag{1.6.9}$$

Usually a strain energy function is defined as $W = \rho_0 U$. For an incompressible, isotropic material, the relation becomes

$$\mathbf{T} = -p\mathbf{I} + 2\frac{\partial W}{\partial I_B}\mathbf{B} - 2\frac{\partial W}{\partial II_B}\mathbf{B}^{-1} \tag{1.6.10}$$

Here the material functions are derivatives of the strain–energy function with respect to the invariants of **B**. This equation is identical in form to eq. 1.6.4, but because the strain–energy function can be derived from molecular arguments, it provides a connection to molecular theory.

The strain–energy function that gives the neo-Hookean model is

$$W = \frac{1}{2} G I_B \tag{1.6.11}$$

and for the Mooney–Rivlin model

$$W = C_1 I_B + C_2 II_B \tag{1.6.12}$$

Valanis and Landel (1967) proposed that for many materials, the strain–energy function is separable into the sum of the same function of each of the principal extensions α_i

$$W(\alpha_i) = w(\alpha_1) + w(\alpha_2) + w(\alpha_3) = \sum w(\alpha_i) \tag{1.6.13}$$

They found that an exponential function fit data

$$w(\alpha_i) = k_i \alpha_i^{m_i} \tag{1.6.14}$$

Example 1.6.1 shows that both the neo-Hookean and Mooney–Rivlin models are of the Valanis–Landel form.

Example 1.6.1 Separable Strain–Energy Function

Show that the neo-Hookean and Mooney–Rivlin models satisfy the Valanis–Landel form, eq. 1.6.14

Solution

(a) *Neo-Hookean*

$$W = \frac{1}{2} G I_B$$

Recall that in terms of the principal strains (see eq. 1.4.40) the first invariant of **B** is

$$I_B = \alpha_1^2 + \alpha_2^2 + \alpha_3^2$$

Thus

$$W = \sum \frac{1}{2} G\alpha_i^2 \tag{1.6.15}$$

which satisfies eq. 1.6.14 if $k = G/2$ and $m = 2$.
(b) *Mooney–Rivlin*

$$W = C_1 I_B + C_2 II_B$$

Recall eq. 1.4.40 once more

$$II_B = \alpha_1^2\alpha_2^2 + \alpha_2^2\alpha_3^2 + \alpha_3^2\alpha_1^2$$

Using the fact that for an incompressible material

$$III_B = \alpha_1^2\alpha_2^2\alpha_3^2 = 1$$

we can write

$$\alpha_1^2 = \alpha_1^{-2}\alpha_3^{-2}, \quad \alpha_2^2 = \alpha_1^{-2}\alpha_3^{-2}, \quad \alpha_3^2 = \alpha_1^{-2}\alpha_2^{-2}$$

Thus

$$W = C_1(\alpha_1^2 + \alpha_2^2 + \alpha_3^2) + C_2(\alpha_1^{-2} + \alpha_2^{-2} + \alpha_3^{-2})$$

or

$$W = \sum \left[C_1\alpha_i^2 + \frac{C_2}{\alpha_i^2} \right] \tag{1.6.16}$$

which satisfies eq. 1.6.14 if $k_1 = C_1, m_1 = 2$, $k_2 = C_2$, and $m_2 = -2$.

1.6.2 Anisotropy

Most crystalline solids are anisotropic. Since our main concern is polymeric liquids and rubbery solids, we generally do not need to worry about anisotropy. The general approach to constitutive equations for anisotropic materials is to use a different elastic constant for each direction. In general, to relate stress to deformation requires a fourth rank tensor with 3^4 components

$$T_{ij} = C_{ijkl} B_{kl} \tag{1.6.17}$$

However, symmetry and energy considerations usually reduce these to 21 or less (Malvern, 1969, Section 6.2). Ward (1983) discusses some anisotropic models for crystalline polymers.

1.6.3 Rubber–like Liquids

In a number of polymer processing operations, such as blow molding, film blowing, and thermoforming, deformations are rapid and the polymer melt behaves more like a crosslinked rubber than a viscous liquid. Figure I.1 showed typical deformation and recovery of a polymeric liquid. As the time scale of the experiment is shortened, the viscoelastic liquid looks more and more like the Hookean solid. In Chapters 3 and 4 we develop models for the full viscoelastic response, but in many cases of rapid deformations, the simplest and often most realistic model for the stress response of these polymeric liquids is in fact the elastic solid.

1.7 Equations of Motion

The deformation of a material is governed not only by a constitutive relation between deformation and stress, like the neo-Hookean equation discussed above, it also must obey the principles of conservation of mass and conservation of momentum. We have already used the mass conservation principle (conservation of volume for an incompressible material) in solving the uniaxial extension example, eq. 1.4.1. We have not yet needed the momentum balance because the balance was satisfied automatically for the simple deformations we chose: that is, they involved no gravity, no flow, nor any inhomogeneous stress fields. However, these balances are needed to solve more complex deformations. They are presented for a flowing system because we will use these results in the following chapters. Here we see how they simplify for a solid. Detailed derivations of these equations are available in nearly every text on fluid or solid mechanics.

1.7.1 Mass Balance

The mass of a body can be expressed as the integral of its density over its volume V.

$$m = \int_V \rho \, dV \tag{1.7.1}$$

In a flowing system V becomes a control volume and velocity can carry mass into and out of this volume, as illustrated in Figure 1.7.1. The rate of change of mass in this volume must equal the net flux of mass across the surface S

$$\frac{dm}{dt} = \frac{d}{dt} \int_V \rho \, dV = \text{mass flux across } S \tag{1.7.2}$$

The volume flux through a small surface element dS will be just the velocity component normal to the surface times the area

Figure 1.7.1.
Mass balance on control volume V in a flowing system: $\hat{\mathbf{n}} \cdot \rho\mathbf{v}$ is the mass flux through the area dS.

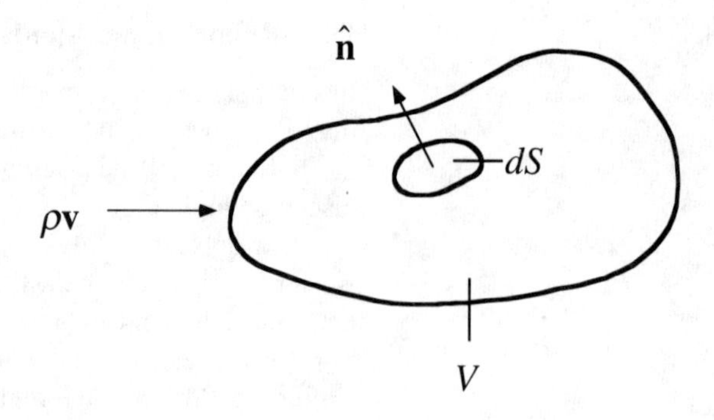

$-\hat{\mathbf{n}} \cdot \mathbf{v} dS$. We use $-\hat{\mathbf{n}}$, the negative of the surface normal, because we want to get the flux *into* the volume. The mass flux is the density times the volume flux, $-\hat{\mathbf{n}} \cdot \rho\mathbf{v}\, dS$. If we integrate over the entire surface, we obtain the rate of change of mass inside the volume

$$-\int_S \hat{\mathbf{n}} \cdot \rho\mathbf{v}\, dS = \frac{d}{dt}\int_V \rho\, dV \tag{1.7.3}$$

Using the divergence theorem, the surface integral can be transformed into a volume integral

$$-\int_V (\nabla \cdot \rho\mathbf{v}) dV = \frac{d}{dt}\int_V \rho\, dV \tag{1.7.4}$$

The divergence operator $\nabla\cdot$ on a vector $(\nabla \cdot \mathbf{v})$ is given for several coordinate systems in Table 1.7.1. Since the control volume is fixed, we can bring the time derivative inside the integral to give

$$\int_V \left(\frac{\partial \rho}{\partial t} + \nabla \cdot \rho\mathbf{v}\right) dV = 0 \tag{1.7.5}$$

The control volume is arbitrary, so we can shrink it to zero, leaving the differential equation

$$\frac{\partial \rho}{\partial t} = -\nabla \cdot \rho\mathbf{v} \tag{1.7.6}$$

which is known as the *continuity equation*. The continuity equation is usually written in terms of the material derivative

$$\frac{D\rho}{Dt} = -\rho\nabla \cdot \mathbf{v} \tag{1.7.7}$$

where the material derivative is the time derivative following the motion

$$\frac{D\rho}{Dt} = \frac{\partial \rho}{\partial t} + \mathbf{v} \cdot \nabla\rho \tag{1.7.8}$$

We will discuss other types of time derivatives in Chapter 4.

For an incompressible fluid the density is constant, and eq. 1.7.7 simplifies to

$$\nabla \cdot \mathbf{v} = 0 \tag{1.7.9}$$

Components of the continuity equation for an incompressible fluid are given in Table 1.7.1 in several coordinate systems.

For a solid, the density change from the reference ρ' to the deformed state ρ is just (Malvern, 1969)

$$\rho/\rho' = \det\mathbf{F} = III_F = III_B^{1/2} = \alpha_1\alpha_2\alpha_3 \tag{1.7.10}$$

As we showed in eq. 1.4.47, for an incompressible material $III_B = 1$. For most purposes we will assume incompressibility.

1.7.2 Momentum Balance

The momentum of any body is its mass times velocity. Using eq. 1.7.1, we have

$$m\mathbf{v} = \int_V \rho\mathbf{v}\, dV \tag{1.7.11}$$

Momentum can be transferred to the body by convection through the surface, by contact forces acting on the surface by the surrounding material, and by body forces like gravity or magnetic forces acting on the volume. Each of these is illustrated in Figure 1.7.2. Convection is determined by the volume flux normal to the surface $-\hat{\mathbf{n}}\cdot\mathbf{v}\,dS$, times the momentum per unit volume $\rho\mathbf{v}$. Contact forces on the surface are determined by the stress vector acting on the surface $+\mathbf{t}_n dS$.* (Recall Figure 1.2.1.) We can use the stress tensor, eq. 1.2.10, to express $\mathbf{t}_n\, dS = \hat{\mathbf{n}}\cdot\mathbf{T}\,dS$. The change of momentum due to gravity will be $\rho\mathbf{g}\,dV$.

By integrating each contribution, we can combine all these factors into a balance of the rate of change of momentum

$$\frac{d}{dt}\int_V \rho\mathbf{v}\,dV = -\int_S (\hat{\mathbf{n}}\cdot\mathbf{v})\rho\mathbf{v}\,dS + \int_S \hat{\mathbf{n}}\cdot\mathbf{T}\,dS + \int_V \rho\mathbf{g}\,dV \tag{1.7.12}$$

rate of momentum change within V	rate of momentum addition across S due to flow	rate of momentum addition across S due to contact forces	rate of momentum addition in V due to body forces

We can also view this equation as a force balance. From Newton's first law, the left-hand term is (mass) due to momentum flux, the contact force, and the body force all acting on the volume V.

Note that we use here the sign convention that tensile stress is positive. Some other texts, particularly Bird et al. (1987a), choose tension as negative.

TABLE 1.7.1. / Equations of Motion*

Rectangular Coordinates (x, y, z)

x-Component

$$\rho\left(\frac{\partial v_x}{\partial t}+v_x\frac{\partial v_x}{\partial x}+v_y\frac{\partial v_x}{\partial y}+v_z\frac{\partial v_x}{\partial z}\right)=-\frac{\partial p}{\partial x}+\left(\frac{\partial \tau_{xx}}{\partial x}+\frac{\partial \tau_{yx}}{\partial y}+\frac{\partial \tau_{zx}}{\partial z}\right)+\rho g_x$$

y-Component

$$\rho\left(\frac{\partial v_y}{\partial t}+v_x\frac{\partial v_y}{\partial x}+v_y\frac{\partial v_y}{\partial y}+v_z\frac{\partial v_y}{\partial z}\right)=-\frac{\partial p}{\partial y}+\left(\frac{\partial \tau_{xy}}{\partial x}+\frac{\partial \tau_{yy}}{\partial y}+\frac{\partial \tau_{zy}}{\partial z}\right)+\rho g_y$$

z-Component

$$\rho\left(\frac{\partial v_z}{\partial t}+v_x\frac{\partial v_z}{\partial x}+v_y\frac{\partial v_z}{\partial y}+v_z\frac{\partial v_z}{\partial z}\right)=-\frac{\partial p}{\partial z}+\left(\frac{\partial \tau_{xz}}{\partial x}+\frac{\partial \tau_{yz}}{\partial y}+\frac{\partial \tau_{zz}}{\partial z}\right)+\rho g_z$$

The continuity equation:

$$\nabla\cdot\mathbf{v}=\frac{\partial v_x}{\partial x}+\frac{\partial v_y}{\partial y}+\frac{\partial v_z}{\partial z}=0$$

Cylindrical Coordinates (r, θ, z)

r-Component

$$\rho\left(\frac{\partial v_r}{\partial t}+v_r\frac{\partial v_r}{\partial r}+\frac{v_\theta}{r}\frac{\partial v_r}{\partial \theta}-\frac{v_\theta^2}{r}+v_z\frac{\partial v_r}{\partial z}\right)=-\frac{\partial p}{\partial r}+\left(\frac{1}{r}\frac{\partial}{\partial r}(r\tau_{rr})+\frac{1}{r}\frac{\partial \tau_{\theta r}}{\partial \theta}-\frac{\tau_{\theta\theta}}{r}+\frac{\partial \tau_{zr}}{\partial z}\right)+\rho g_r$$

θ-Component

$$\rho\left(\frac{\partial v_\theta}{\partial t}+v_r\frac{\partial v_\theta}{\partial r}+\frac{v_\theta}{r}\frac{\partial v_\theta}{\partial \theta}+\frac{v_r v_\theta}{r}+v_z\frac{\partial v_\theta}{\partial z}\right)=-\frac{1}{r}\frac{\partial p}{\partial \theta}+\left(\frac{1}{r^2}\frac{\partial}{\partial r}(r^2\tau_{r\theta})+\frac{1}{r}\frac{\partial \tau_{\theta\theta}}{\partial \theta}+\frac{\partial \tau_{\theta z}}{\partial z}+\frac{\tau_{\theta r}-\tau_{r\theta}}{r}\right)+\rho$$

z-Component

$$\rho\left(\frac{\partial v_z}{\partial t}+v_r\frac{\partial v_z}{\partial r}+\frac{v_\theta}{r}\frac{\partial v_z}{\partial \theta}+v_z\frac{\partial v_z}{\partial z}\right)=-\frac{\partial p}{\partial z}+\left(\frac{1}{r}\frac{\partial}{\partial r}(r\tau_{rz})+\frac{1}{r}\frac{\partial \tau_{\theta z}}{\partial \theta}+\frac{\partial \tau_{zz}}{\partial z}\right)+\rho g_z$$

The continuity equation:

$$\nabla\cdot\mathbf{v}=\frac{1}{r}\frac{\partial}{\partial r}(r v_r)+\frac{1}{r}\frac{\partial v_\theta}{\partial \theta}+\frac{\partial v_z}{\partial z}=0$$

Spherical Coordinates (r, θ, ϕ)

r-Component

$$\rho\left(\frac{\partial v_r}{\partial t}+v_r\frac{\partial v_r}{\partial r}+\frac{v_\theta}{r}\frac{\partial v_r}{\partial \theta}+\frac{v_\phi}{r\sin\theta}\frac{\partial v_r}{\partial \phi}-\frac{v_\theta^2+v_\phi^2}{r}\right)=-\frac{\partial p}{\partial r}$$

$$+\left(\frac{1}{r^2}\frac{\partial}{\partial r}(r^2\tau_{rr})+\frac{1}{r\sin\theta}\frac{\partial}{\partial\theta}(\tau_{\theta r}\sin\theta)+\frac{1}{r\sin\theta}\frac{\partial\tau_{r\phi}}{\partial\phi}-\frac{\tau_{\theta\theta}+\tau_{\phi\phi}}{r}\right)+\rho g_r$$

θ-Component

$$\rho\left(\frac{\partial v_\theta}{\partial t}+v_r\frac{\partial v_\theta}{\partial r}+\frac{v_\theta}{r}\frac{\partial v_\theta}{\partial \theta}+\frac{v_\theta}{r\sin\theta}\frac{\partial v_\theta}{\partial \phi}+\frac{v_r v_\theta}{r}-\frac{v_\phi^2\cot\theta}{r}\right)=-\frac{1}{r}\frac{\partial p}{\partial \theta}$$

$$+\left(\frac{1}{r^3}\frac{\partial}{\partial r}(r^3\tau_{r\theta})+\frac{1}{r\sin\theta}\frac{\partial}{\partial\theta}(\tau_{\theta\theta}\sin\theta)+\frac{1}{r\sin\theta}\frac{\partial\tau_{\phi\theta}}{\partial\phi}+\frac{(\tau_{\theta r}-\tau_{r\theta})-\cot\theta\,\tau_{\phi\phi}}{r}\right)+\rho g_\theta$$

ϕ-Component

$$\rho\left(\frac{\partial v_\phi}{\partial t}+v_r\frac{\partial v_\phi}{\partial r}+\frac{v_\theta}{r}\frac{\partial v_\phi}{\partial \theta}+\frac{v_\phi}{r\sin\theta}\frac{\partial v_\phi}{\partial \phi}+\frac{v_\phi v_r}{r}+\frac{v_\theta v_\phi}{r}\cot\theta\right)=-\frac{1}{r\sin\theta}\frac{\partial p}{\partial \phi}$$

$$+\left(\frac{1}{r^3}\frac{\partial}{\partial r}(r^3\tau_{r\phi})+\frac{1}{r\sin\theta}\frac{\partial(\tau_{\theta\phi}\sin\theta)}{\partial\theta}+\frac{1}{r\sin\theta}\frac{\partial\tau_{\phi\phi}}{\partial\phi}+\left(\frac{\tau_{\phi r}-\tau_{r\phi}+2\cot\theta\,\tau_{\theta\phi}}{r}\right)\right)+\rho g_\phi$$

The continuity equation:

$$\nabla\cdot\mathbf{v}=\frac{1}{r^2}\frac{\partial}{\partial r}(r^2 v_r)+\frac{1}{r\sin\theta}\frac{\partial}{\partial\theta}(v_\theta\sin\theta)+\frac{1}{r\sin\theta}\frac{\partial v_\phi}{\partial_\phi}=0$$

* τ has not been assumed symmetric.

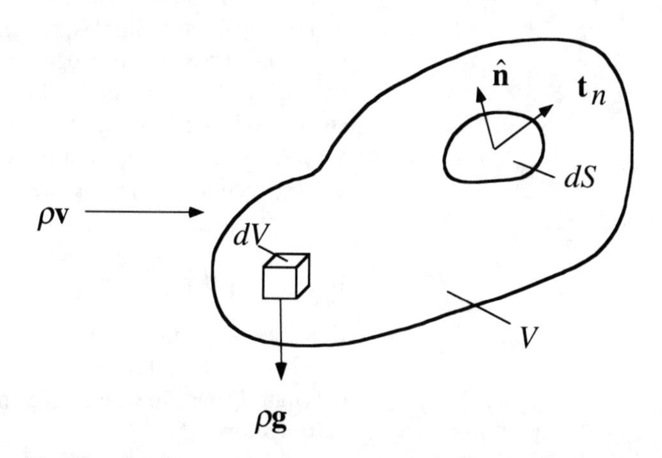

Figure 1.7.2. Momentum balance on a control volume V. Momentum is brought into the body by surface traction $\mathbf{t}_n$, by body forces $\rho\mathbf{g}$, and by momentum convected across the surface $\hat{\mathbf{n}}\cdot\mathbf{v}(\rho\mathbf{v})$.

Again we can apply the divergence theorem to change the surface integrals to volume integrals. We also can use the transport theorem to bring the time derivative inside the integral

$$\int_V \frac{\partial \rho \mathbf{v}}{\partial t} dV = -\int_V \nabla \cdot \rho \mathbf{v}\mathbf{v}\, dV + \int_V \nabla \cdot \mathbf{T}\, dV + \int_V \rho \hat{\mathbf{g}}\, dV \quad (1.7.13)$$

As with the mass balance, the control volume is arbitrary so we can shrink it to zero, leaving the differential equation

$$\frac{\partial \rho \mathbf{v}}{\partial t} = -\nabla \cdot \rho \mathbf{v}\mathbf{v} + \nabla \cdot \mathbf{T} + \rho \mathbf{g} \quad (1.7.14)$$

or in terms of the material derivative with the help of the continuity equation

$$\rho \frac{D\mathbf{v}}{Dt} = \nabla \cdot \mathbf{T} + \rho \mathbf{g} \quad (1.7.15)$$

Equations 1.7.15 and 1.7.7 together are often called *equations of motion*, and their components are given in several coordinate systems in Table 1.7.1. In deriving the table, eq. 1.2.44 was used to express $\mathbf{T}$ as the sum of pressure and the extra stress $\boldsymbol{\tau}$.

For solids there is no flow, so the balance of momentum can also be written

$$\rho \mathbf{a} = \nabla \cdot \mathbf{T} + \rho \mathbf{g} \quad (1.7.16)$$

where $\mathbf{a}$ represents the acceleration of the body. In most elasticity problems the body is in static equilibrium and the effects of gravity are negligible, so the equation reduces to just a stress balance

$$0 = \nabla \cdot \mathbf{T} \quad \text{or} \quad 0 = \frac{\partial T_{ij}}{\partial x_i} \quad (1.7.17)$$

We can write this equation in different coordinate systems using only the stress terms in Table 1.7.1. In the uniaxial extension and simple shear examples, which we worked for the neo-Hookean solid, the stress was homogeneous, so all its derivatives are zero and eq. 1.7.17 is satisfied. However, with more complex shapes, such as twisting of a cylinder shown in Figure 1.7.3, the stresses do vary across the sample and the stress balance is required to solve for the tractions on the surface.

Example 1.7.1 Twisting of a Cylinder

Consider a cylinder of a neo-Hookean rubber with diameter $2R$, which is twisted through an angle θ while its height ℓ_o is held constant. Determine the torque and normal force generated by this deformation.

Figure 1.7.3.
Torsion of cylinder. The bottom surface is fixed and the top surface is twisted through an angle θ.

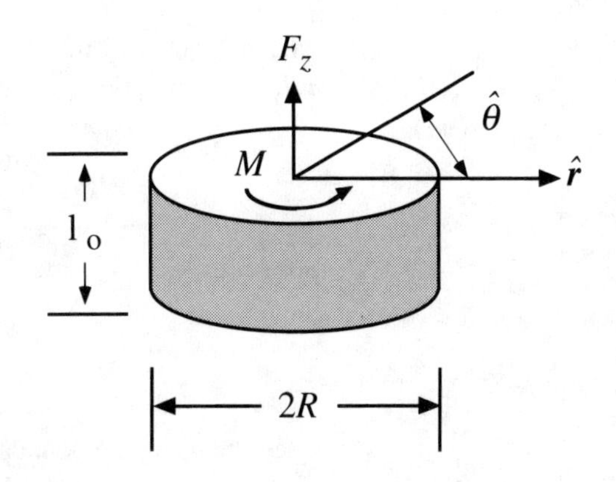

Solution

The displacement functions in cylindrical coordinates become

$$\begin{aligned} r &= r' && \text{(no change in diameter)} \\ \theta &= \theta' + \alpha z' && \text{(twist angle proportional to height above base)} \\ z &= z' && \text{(no change in height)} \end{aligned} \tag{1.7.18}$$

where $\alpha = \theta/\ell_o$. Note that α has the units of reciprocal length.

Applying these to determine **F** in cylindrical coordinates, Table 1.4.1, we obtain very similar results to simple shear, Example 1.4.2.

$$F_{ij} = \begin{bmatrix} 1 & 0 & 0 \\ 0 & 1 & \alpha r \\ 0 & 0 & 1 \end{bmatrix} \qquad B_{ij} = F_{ik}F_{jk} = \begin{bmatrix} 1 & 0 & 0 \\ 0 & 1+\alpha^2 r^2 & \alpha r \\ 0 & \alpha r & 1 \end{bmatrix} \tag{1.7.19}$$

However, because the deformation is not uniform, the stress will not be homogeneous. Applying the neo-Hookean model, eq. 1.5.2, gives

$$\begin{aligned} T_{rr} &= -p + G \\ T_{\theta\theta} &= -p + G(1 + \alpha^2 r^2) \\ T_{\theta z} &= T_{z\theta} = G\alpha r \\ T_{zz} &= -p + G \end{aligned} \tag{1.7.20}$$

We can readily calculate the torque on the cylinder by integrating the shear stress

$$M = 2\pi \int_0^R T_{\theta z} r^2 dr = \frac{1}{2}\pi G \alpha R^4 \tag{1.7.21}$$

To determine the normal force F_z we need to determine p. Because the stress field is not homogeneous, we need to use the stress balance equations in cylindrical coordinates, eq. 1.7.17 and Table 1.7.1, which in this case reduce to

$$T_{\theta\theta} = \frac{\partial}{\partial r}(rT_{rr}) = r\frac{\partial T_{rr}}{\partial r} + T_{rr} \tag{1.7.22}$$

Substituting for $T_{\theta\theta}$ and T_{rr}, we obtain

$$G\alpha^2 r = \frac{\partial T_{rr}}{\partial r}$$

Integrate using the boundary condition $T_{rr} = 0$ at $r = R$

$$\int_0^{T_{rr}} dT_{rr} = T_{rr}(r) = G\alpha^2\left(\frac{r^2}{2} - \frac{R^2}{2}\right) = T_{zz} \tag{1.7.23}$$

$$F_z = 2\pi \int_0^R rT_{zz}dr = -\frac{1}{4}\pi G\alpha^2 R^4 \tag{1.7.24}$$

Note that at large strains this deformation may not be stable. Buckling can occur as shown by Penn and Kearsley (1976). This can be demonstrated by twisting a rubber hose.

Figure 1.1.3 showed data for torsion of a silicone rubber disk cured between parallel plates of a rotational rheometer. The data are plotted as stress

$$T_{12} = T_{\theta z}\mid_R = \frac{2M}{\pi R^3} \tag{1.7.25}$$

$$T_{11} - T_{22} = T_{\theta\theta} - T_{zz} = \frac{4F_z}{\pi R^2} \tag{1.7.26}$$

as a function of maximum shear strain at the edge of the plates

$$\gamma_{\max} = \alpha R \tag{1.7.27}$$

The solid lines are calculated using eq. 1.7.20 and G = 163 kPa. The excellent agreement between data and calculations indicates that the neo-Hookean model describes this material very well in shear up to $\gamma = 0.4$.

1.8 Boundary Conditions

To actually solve problems, in addition to the equations of motion and a constitutive equation, we need the boundary conditions. In general, for problems dealing with fluids or solids, we have the following basic kinds of boundary condition:

1. *Motion of a surface*: This usually means specifying displacements of solids or velocities of fluids. Typically sample ma-

terials (a) move with solid surfaces of the rheometer and (b) do not penetrate it. If a represents the sample side and b the rheometer, then

$$\mathbf{x}_{t_a} = \mathbf{x}_{t_b} \qquad \text{and} \qquad \mathbf{x}_{n_a} = 0 \tag{1.8.1}$$

or in terms of velocities

$$\mathbf{v}_{t_a} = \mathbf{v}_{t_b} \qquad \text{and} \qquad \mathbf{v}_{n_a} = 0 \tag{1.8.2}$$

2. *Tractions at a surface*: Continuity of shear and normal forces across the interface

$$\mathbf{t}_{t_a} = \mathbf{t}_{t_b} \quad \text{or} \quad (\hat{\mathbf{n}} \cdot \mathbf{T} \cdot \hat{\mathbf{t}})_a = (\hat{\mathbf{n}} \cdot \mathbf{T} \cdot \hat{\mathbf{t}})_b$$

and (1.8.3)

$$\mathbf{t}_{n_a} = \mathbf{t}_{n_b} \quad \text{or} \quad (\hat{\mathbf{n}} \cdot \mathbf{T} \cdot \hat{\mathbf{n}})_a = (\hat{\mathbf{n}} \cdot \mathbf{T} \cdot \hat{\mathbf{n}})_b$$

3. *A mixture of motion and tractions.*

The first type of boundary condition frequently does not hold for large deformation of rubber and highly elastic or very viscous liquids. These materials can slip at solid surfaces (Schowalter, 1989).

Special conditions apply to the tractions at liquid–liquid and liquid–gas interfaces. These conditions and the concept of surface tension are discussed at the end of Chapter 2. However, the interfacial tension concept also can be useful for modeling deformations of thin rubber sheets. With interfacial tension Γ, the normal traction condition becomes

$$(\hat{\mathbf{n}} \cdot \mathbf{T} \cdot \hat{\mathbf{n}})_a = (\hat{\mathbf{n}} \cdot \mathbf{T} \cdot \hat{\mathbf{n}})_b + 2H\Gamma \tag{1.8.4}$$

where H is the arithmetic mean of the two principal curvatures that describe the interface. For example, for a sphere $2H = 2/R$ and for a cylinder $2H = 1/R$. We can treat the elastic stresses in a membrane as surface tensions (see Figure 1.8.1). The normal forces are the pressures on each side of the membrane. So then eq. 1.8.4 becomes

$$p_a = p_b + \frac{\Gamma_1}{R_1} + \frac{\Gamma_2}{R_2} \tag{1.8.5}$$

Figure 1.8.1.
Tensions in a thin membrane with two radii of curvature R_1 and R_2.

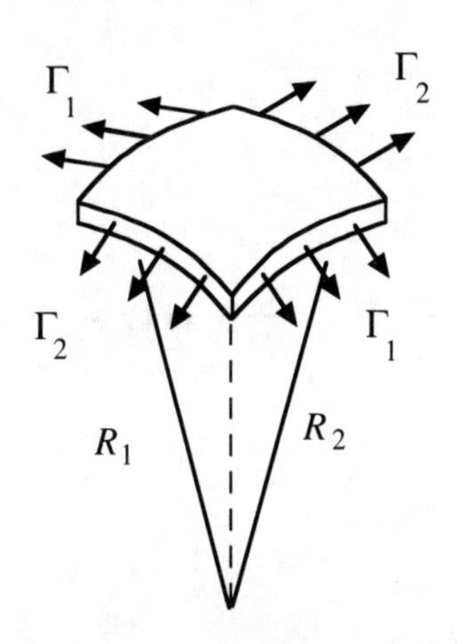

The use of these boundary conditions is illustrated for solids in Examples 1.8.1–1.8.3 and in the exercises given at the end of this chapter and for fluids in the next chapter.

Example 1.8.1 Uniaxial Extension

Illustrate the use of both types of boundary condition by reworking the uniaxial extension problem solved in eq. 1.5.5 (see also Figure

1.4.2a and Examples 1.4.1 and 1.4.2). State the boundary conditions explicitly.

Solution

In eq. 1.4.4 we were given that the displacement of the $\hat{\mathbf{x}}_1$ surface for uniaxial extension of a cube is L/L'

$$x_1 = \frac{\Delta x_1}{\Delta x_1'} x_1' = \alpha_1 x_1' \qquad (1.4.4)$$

The discussion that followed eq. 1.5.5 pointed out that the tractions on the other two surfaces are zero.

$$\mathbf{t}_2 = \mathbf{t}_3 = 0$$

These boundary conditions are illustrated in Figure 1.8.2. Since $\mathbf{t}_2 = \mathbf{t}_3$, the displacements, α_2 and α_3, must also be equal by symmetry.

Next the conservation of mass of an incompressible solid, eq. 1.4.6 (or 1.4.38), was used to solve for α_2 and α_3 in terms of α_1.

$$\alpha_1 \alpha_2 \alpha_3 = 1 \qquad (1.4.6)$$

$$\alpha_2 = \alpha_3 = \sqrt{\frac{1}{\alpha_1}}$$

The components of the stress tensor were determined from the neo-Hookean model, eqs. 1.5.3 and 1.5.4, and the arbitrary pressure p was eliminated by using the boundary tractions $\mathbf{t}_2 = \mathbf{t}_3 = 0$. Then the unknown traction $\mathbf{t}_1$ was determined from $\mathbf{t}_1 = T_{11}\hat{\mathbf{x}}_1$.

We can see that if instead of specifying the constitutive equation in this problem, we specify the boundary traction $\mathbf{t}_1$, we can

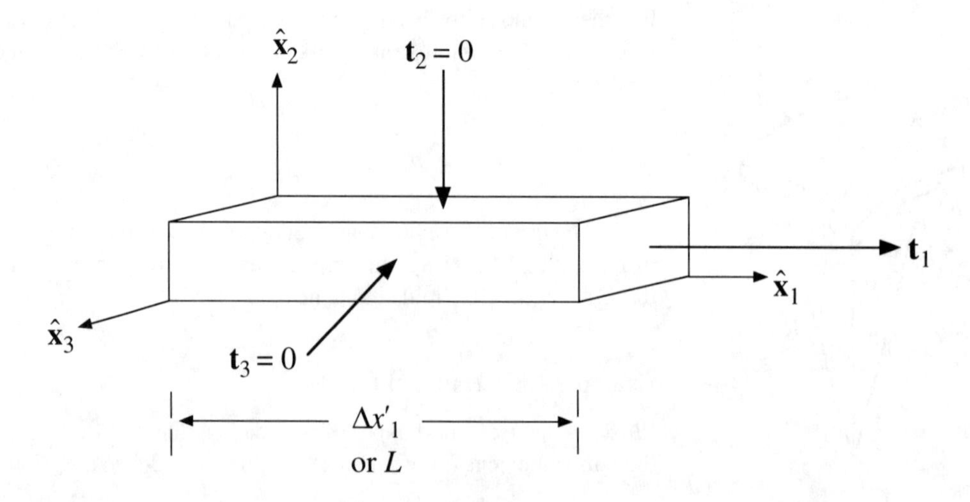

Figure 1.8.2. Boundary conditions for uniaxial extension.

Figure 1.8.3.
Planar extension of a rubber sheet.

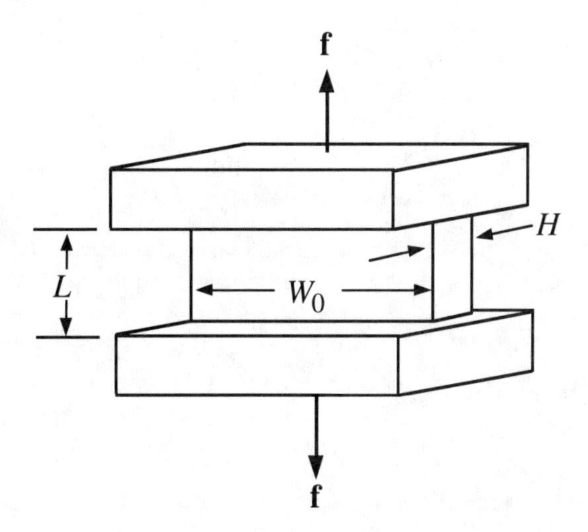

get information on the constitutive relation. If it is neo-Hookean, we have immediately G, the elastic modulus. If it is a general elastic solid, we have the dependence of the material function group $g_1(I_B, II_B) - g_2(I_B, II_B)/\alpha_1$ for $I_B = \alpha_1^2 + 2/\alpha_1$ and $II_B = 2\alpha_1 + \alpha_1^2$. This then is the rheologist's strategy: specify enough boundary conditions and satisfy the equations of motion such that both stress and deformation can be determined. Then material functions of an unknown material can be measured.

Example 1.8.2 Planar Extension

A wide, short, and thin sheet of rubber, $W_o >> L_o >> H_o$, is clamped along its wide edges and pulled from its original length L_o to L with a force $\mathbf{f}$. This deformation is called planar extension, plain strain, or sometimes pure shear. Assume neo-Hookean behavior, eq. 1.5.2.

(a) Write out the boundary conditions and justify the assumption that W_o is constant during the deformation.

(b) Calculate the components of the deformation gradient $\mathbf{F}$ and the strain tensors $\mathbf{B}$ and $\mathbf{B}^{-1}$. Determine the invariants of $\mathbf{B}$.

Figure 1.8.4.
Deformation of free edges in planar extension.

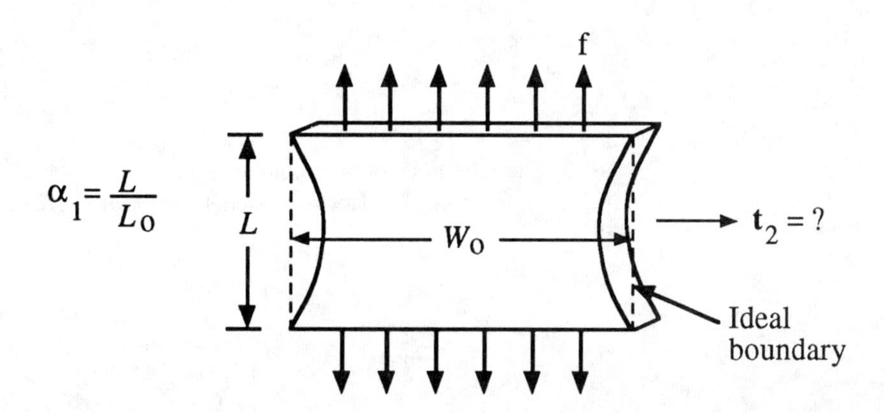

(c) Relate $\mathbf{f}$ to the initial dimensions and the shear modulus G of the sample.

Solution

(a) If the sheet is wide, short, and thin, $W_o >> L_o >> H_o$, and clamped rigidly at the top and bottom edges, W_o will change little compared to L_o and H_o. Some workers have used small clips to apply forces along the edge to maintain the sides straight (e.g., Kawabata and Kawai, 1977). Thus the boundary conditions are three displacements

$$\alpha_1 = \frac{L}{L_o}, \qquad \alpha_3 = \frac{H}{H_o}, \qquad \alpha_2 = \frac{W}{W_o} = 1 \tag{1.8.6}$$

and one force per unit area

$$\mathbf{t}_1 = \left(\frac{f}{W_o H}\right)\hat{\mathbf{x}}_1$$

(b) For an incompressible material $\alpha_1\alpha_2\alpha_3 = 1$, thus $\alpha_1 = 1/\alpha_3$. The displacement functions are (note eq. 1.4.4)

$$\begin{aligned} x_1 &= \alpha_1 x_1' \\ x_3 &= \alpha_3 x_3' = x_3'/\alpha_1 \\ x_2 &= x_2' \end{aligned} \tag{1.8.7}$$

Applying the definitions of $\mathbf{F}$ and $\mathbf{B}$ gives

$$\begin{aligned} F_{ij} &= \frac{\partial x_i}{\partial x_j'} = \begin{bmatrix} \alpha_1 & 0 & 0 \\ 0 & 1 & 0 \\ 0 & 0 & 1/\alpha_1 \end{bmatrix} \\ B_{ij} &= F_{ik}F_{jk} = \begin{bmatrix} \alpha_1^2 & 0 & 0 \\ 0 & 1 & 0 \\ 0 & 0 & 1/\alpha_1^2 \end{bmatrix} \end{aligned} \tag{1.8.8}$$

C_{ij} will be the same as B_{ij} because $F_{ij} = F_{ji}$. The invariants are

$$\begin{aligned} I_B &= \alpha_2 + \alpha_2^{-2} + 1 = II_B \\ III_B &= 1 \end{aligned} \tag{1.8.9}$$

as expected for an incompressible material.
(c) Applying the neo-Hookean model $\mathbf{T} = -p\mathbf{I} + G\mathbf{B}$, eq. 1.5.1 gives

$$\begin{aligned} T_{11} &= -p + G\alpha_1^2 \\ T_{33} &= -p + \frac{G}{\alpha_1^2} \\ T_{22} &= -p + G \end{aligned} \tag{1.8.10}$$

Since the sheet is thin and there are no forces acting on its surface, $T_{33} = 0,\ p = G/\alpha_1^2$. Thus

$$T_{11} = G\left(\frac{\alpha_1^2 - 1}{\alpha_1^2}\right) \tag{1.8.11}$$

We can also solve the problem by considering the force balance on the sheet. The force in the x_1 direction is

$$f = \int_0^H \int_0^{W_o} T_{11} dx_3 dx_2$$

Thus by substituting we have

$$f = \int_0^H \int_0^{W_o} (G\alpha_1^2 - G\alpha_1^{-2}) dx_3 dx_2 = G(\alpha_1^2 - \alpha_1^{-2}) W_o H \tag{1.8.12}$$

Note that for small strains $\alpha_1 = 1 + \epsilon$, $T_{11} = 4G\epsilon$; the "planar" tensile modulus is four times that in shear.

Example 1.8.3 Tube Inflation

A long, thin-walled rubber tube (Figure 1.8.5) is inflated with a gas pressure p_a with its length W_o held constant. Assuming that the material obeys the neo-Hookean model, determine how much the tube will inflate. Show that this is a planar extension, identical to Example 1.8.2.

Solution

Boundary conditions are

$$\alpha_2 = 1$$

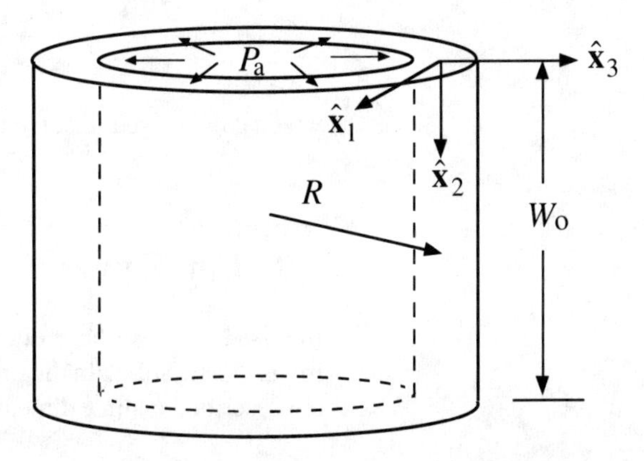

Figure 1.8.5. Inflation of a thin-walled rubber tube.

since the tube length W_o is constant and

$$T_{33} = 0$$

since we assume a thin membrane. For a cylinder, eq. 1.8.5 becomes

$$p_a - p_b = \frac{\Gamma_1}{R}$$

We can define the extensions as

$$\begin{aligned}
\alpha_1 &= \frac{2\pi R}{2\pi R_o} = \frac{R}{R_o} \\
\alpha_3 &= \frac{H}{H_o} \\
\alpha_2 &= \frac{W_o}{W_o} = 1
\end{aligned} \tag{1.8.13}$$

By conservation of volume then $\alpha_3 = 1/\alpha_1$. These are the same extension ratios found in Example 1.8.2 for planar extension with L replaced by R. Thus **F** and **B** will be identical. So will **T**, since the same constitutive equation is used. Using $T_{33} = 0$, we get the same result for T_{11}.

$$T_{11} = G(\alpha_1^2 - \alpha_1^{-2})$$

We can now relate T_{11} to the applied pressure–tension balance, eq. 1.8.5. Approximate the surface tension by $\Gamma_1 = T_{11}H$ and assume no external pressure $p_b = 0$. Thus

$$p_a = \frac{GH\left(\alpha_1^2 - 1/\alpha_1^2\right)}{R} \tag{1.8.14a}$$

or using the initial dimensions of the cylinder $R = \alpha_1 R_o$ and $H = H_o/\alpha_1$ gives

$$p_a = \frac{GH_o\left(1 - 1/\alpha_1^4\right)}{R_o} = \frac{GH_o\left(1 - R_o^4/R^4\right)}{R_o} \tag{1.8.14b}$$

which can be used to solve for the inflated radius R.

1.9 Summary

In this chapter we have developed a general constitutive equation for an elastic solid. In the process we learned how stress and strain are described in three dimensions. We saw that when large strain is properly described, extensional stresses become nonlinear in strain, and the surprising phenomenon of normal stresses in shear arises naturally. We learned that scalar material functions depend on invariants of strain (or stress).

We gave several specific constitutive equations that fit stress–deformation data for real rubber reasonably well. These are useful for design of tires and other rubber goods. In Section 1.8 we showed how to attack some elastic boundary value problems. Additional problems are given in the exercises. However, we introduced the neo-Hookean model primarily for its value in developing constitutive equations for viscolastic liquids, particularly in Chapter 4.

Furthermore, in a number of polymer processing operations, such as blow molding, film blowing, and thermoforming, deformations are rapid and the polymer melt behaves more like a crosslinked rubber than a viscous liquid. Figure I.1 showed typical deformation and recovery of a polymeric liquid. As the time scale of the experiment is shortened, the viscoelastic liquid looks more and more like the neo-Hookean solid. In Chapters 3 and 4 we develop models for the full viscoelastic response, but in fact in many cases of rapid deformations the simplest and often most realistic model for the stress response of these polymeric liquids is the elastic solid.

However, before we go on to "softening" the Hookean solid into the viscoelastic liquid, we need to look at the other extreme, the Newtonian liquid, the subject of our next chapter.

1.10 Exercises

1.10.1 Tensor Algebra

Consider a stress tensor **T** whose components are

$$T_{ij} = \begin{bmatrix} 3 & 2 & -1 \\ 2 & 2 & 1 \\ -1 & 1 & 0 \end{bmatrix}$$

Note that the tensor is symmetric. Consider also a vector v with components

$$v_i = (5, 3, 7)$$

Evaluate the following.

(a) $\mathbf{T} \cdot \mathbf{v}$	(e) $\mathbf{v}\mathbf{v}$	(dyad product)
(b) $\mathbf{v} \cdot \mathbf{T}$	(f) $\mathbf{T} \cdot \mathbf{x}_1$	
(c) $\mathbf{T} : \mathbf{T}$	(g) $\mathbf{T} \cdot \mathbf{I}$	
(d) $\mathbf{v} \cdot (\mathbf{T} \cdot \mathbf{v})$		

1.10.2 Invariants

Determine the invariants of the stress tensor **T** in Example 1.10.1.

1.10.3 Determination of the Stress Tensor

Measurements of force were made on 1 mm^2 test surfaces around a point in a fluid. The vectors normal to these test surfaces correspond to the coordinate directions $\hat{\mathbf{x}}_1$, $\hat{\mathbf{x}}_2$, and $\hat{\mathbf{x}}_3$. The measured force vectors on these surfaces are

$$
\begin{aligned}
f_1 &= 1N \ \text{ in the } \hat{\mathbf{x}}_1 \text{ direction;} \\
f_2 &= 2N \ \text{ in the } -\hat{\mathbf{x}}_3 \text{ direction} \\
f_3 &= 2N \ \text{ in the } -\hat{\mathbf{x}}_2 \text{ direction}
\end{aligned}
$$

(a) What is the state of stress at the point?
(b) What is the net force on 1 mm^2 surface whose normal is in the $\hat{\mathbf{x}}_1 + \hat{\mathbf{x}}_2$?
(c) What is the component of this force normal to the surface?
(d) Calculate the three invariants of this stress tensor.

1.10.4 **C** as Length Change

Section 1.4 mentioned that physically the Green tensor **C** is an operator that gives length changes; that is, prove eq. 1.4.22.

$$\alpha^2 = \frac{d\mathbf{x} \cdot d\mathbf{x}}{d\mathbf{x}' \cdot d\mathbf{x}'} = \hat{\mathbf{n}}' \cdot \mathbf{C} \cdot \hat{\mathbf{n}}' \tag{1.4.22}$$

where

$$\mathbf{C} = \mathbf{F}^T \cdot \mathbf{F} \tag{1.4.21}$$

1.10.5 Inverse Deformation Tensors

(a) Show that the inverse of the Finger tensor operates on unit vectors in the deformed state to give inverse square of length change α^2

$$\frac{1}{\alpha_2} = \hat{\mathbf{n}} \cdot \mathbf{B}^{-1} \cdot \hat{\mathbf{n}} \tag{1.4.32}$$

where

$$\mathbf{B}^{-1} = (\mathbf{F}^{-1})^T \cdot (\mathbf{F}^{-1}) \tag{1.4.30}$$

(b) Show that the inverse of the Cauchy tensor operates on unit vectors in the undeformed or past state to give inverse area change squared

$$\frac{1}{\mu^2} = \mathbf{n}' \cdot \mathbf{C}^{-1} \cdot \mathbf{n}' \tag{1.4.33}$$

$$\mathbf{C}^{-1} = \mathbf{F}^{-1} \cdot (\mathbf{F}^{-1})^T \tag{1.4.31}$$

Figure 1.10.1.
Eccentric rotating disks.

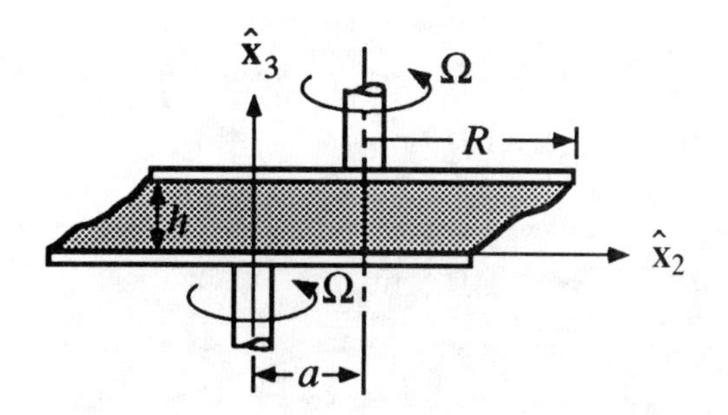

1.10.6 Planar Extension of a Mooney–Rivlin Rubber

Redo Example 1.8.2 for a general elastic model, eq. 1.6.3, with $g_1 = 2C_1$ and $g_2 = -2C_2$ (i.e., a Mooney–Rivlin material).

1.10.7 Eccentric Rotating Disks (Macosko and Davis, 1974; Walters, 1975; see also Section 5.7)

Figure 1.10.1 shows two parallel disks rotating at velocity Ω. The disks are separated by a distance h and their axes of rotation are eccentric or displaced by an amount a.

The displacement functions for a material confined between the disks are

$$x_1 = x_1' \cos \Omega t - x_2' \sin \Omega t$$
$$x_2 = x_1' \sin \Omega t + x_2' \cos \Omega t + \gamma x_3'$$
$$x_3 = x_3'$$

where $\gamma = \mathrm{a}/h$ and x_i' are the coordinates in the undeformed state.

(a) Calculate the components of **F** and **B** for this deformation.

(b) If the material between the disks is a neo-Hookean solid, determine the components of the force it exerts on the disks due to the deformation.

1.10.8 Sheet Inflation

A thin sheet of rubber is clamped over a circular hole and inflated by a pressure p into a hemispherical bubble (Figure 1.10.2).

Figure 1.10.2.
Inflation of a thin sheet.

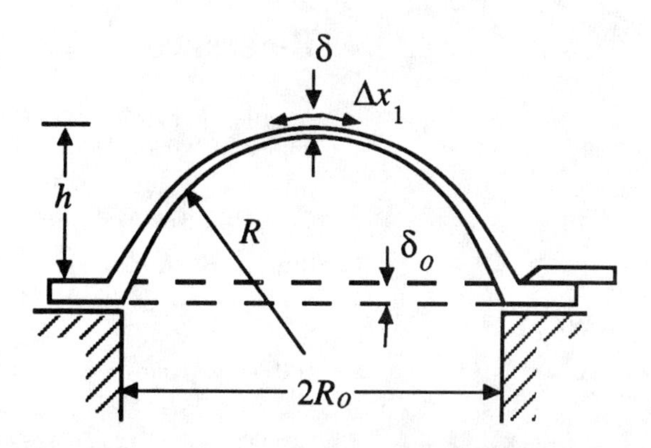

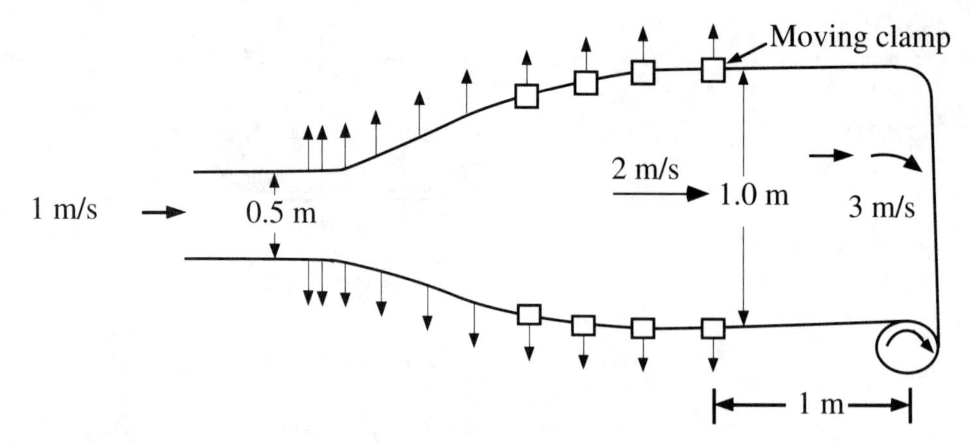

Figure 1.10.3.
Stretching of a polymer film.

(a) Relate the bubble height h, which is relatively easy to measure, to the bubble radius R and the hole radius R_o.
(b) Relate bubble thickness near the pole δ to relative length changes on the surface of the bubble.
(c) Relate bubble height to the pressure, pole thickness, and the modulus G of the rubber. Assume that the neo-Hookean model, eq. 1.5.2, is valid.
(d) Compare this result with the Mooney–Rivlin model, eq. 1.6.3.

1.10.9 Film Tenter

Polymer film can be oriented with a device called tenter, a set of moving clamps on the edge of the film that pull on the film as indicated schematically in Figure 1.10.3. The film is usually held at just above its transition temperature by infrared heaters.
(a) If the initial film thickness is 150 μm (0.006 in.), what is the final film thickness?
(b) If the film can be treated as a neo-Hookean solid with $G = 5 \times 10^5$ Pa, what is the stress it exerts on the last pair of clamps?
(c) What is the torque needed to turn the take-up roll (diameter 0.3 m)?

References

Astarita, G.; Marrucci, G., *Principles of Non-Newtonian Fluid Mechanics*; McGraw-Hill: New York, 1974.

Billington, E. W.; Tate, A., *The Physics of Deformation and Flow*; McGraw Hill: New York, 1981.

Bird, R. B.; Armstrong, R. C.; Hassager, O., *Dynamics of Polymeric Liquids,* Vol. 1*: Fluid Mechanics*, 2nd ed.; Wiley: New York, 1987a.

Bird, R. B.; Curtiss, C. F.; Hassager, O.; Armstrong, R. C., *Dynamics of Polymeric Liquids:* Vol. 2*: Kinetic Theory*, 2nd ed.; Wiley: New York, 1987b.

Chadwick, P., *Continuum Mechanics*; Wiley: New York, 1976.

Dahler, J. S.; Scriven, L. E., *Nature* 1961, *192*, 36; *Proc. R. Soc.* 1963, *A275*, 505.

DeGroot, J., Report for ChE8105, University of Minnesota, 1990.

Eichinger, B., *Annu. Rev. Phys. Chem.* 1983, *34*, 359.

Finger, J., *Akad. Wiss. Wein. Sitzber.* 1894, *103*, 1073.

Gibbs, J. W., *Vector Analysis*; Dover Reprints: New York, 1960.

Gottlieb, M.; Macosko, C. W.; Lepsch, T. C., *J. Polym. Sci., Phys. Ed.* 1981, *19*, 1603.

Hooke, R., *Lectures de Potentia Restitution*; John Martyn: London; p. 1678; reprinted by R. T. Gunter *Early Science in Oxford*, Vol. 8; Oxford University Press: London, 1931; pp. 331–388.

Kawabata, S.; Kawai, T., *Adv. Polym. Sci.* 1977, *24*, 89.

Lodge, A. S., *Elastic Liquids*; Academic Press: London, 1964.

Lodge, A. S., *Body Tensor: Fields in Continuum Mechanics*; Academic Press: London, 1974.

Macosko, C. W.; Davis, W. M., *Rheol. Acta* 1974, *13*, 814.

Malvern, L. E., *An Introduction to the Mechanics of a Continuous Medium*; Prentice-Hall: Englewood Cliffs, NJ, 1969.

Ogden, R.W., *Non-Linear Elastic Deformations*; Wiley: New York, 1984.

Penn, R.; Kearsley, E., *Trans. Soc. Rheol.* 1976, *20*, 227.

Rivlin, R. S., *Phil. Trans. R. Soc.* 1948, *A241*, 379.

Rivlin, R. S., in *Rheology*, Vol. 1; Eirich, F. R., Ed.; Academic Press: New York, 1956.

Rivlin, R. S.; Saunders, D. W., *Phil. Trans. R. Soc.* 1951, *A243*, 251; *Trans. Faraday Soc.* 1952, *48*, 200.

Schowalter, W. S., *J. Non-Newtonian Fluid Mech.* 1989, *29*, 25.

Spiegel, A. B., *Vectors and Tensors*; McGraw-Hill, Schaum Series: New York: 1968.

Treloar, L. R. G., *The Physics of Rubber Elasticity*, 3rd ed.; Oxford University Press: London, 1975.

Tschoegl, N. W., *Polymer* 1979, *20* 1365.

Valanis, K. C.; Landel, R. R., *J. Appl. Phys.* 1967, *38*, 2997.

Walters, K., *Rheometry*; Chapman & Hall: London, 1975.

Ward, I. M., *Mechanical Properties of Solid Polymers*; 2nd ed. Wiley: Chichester, 1983.

2

VISCOUS LIQUID

The resistance which arises from the lack of slipperiness originating in a fluid, other things being equal, is proportional to the velocity by which the parts of the fluid are being separated from each other.

Isaac S. Newton (1687)

2.1 Introduction

Only a few years after Hooke expressed the concept that eventually led to the constitutive equation for the ideal elastic solid, Newton (Figure 2.1.1) wrote his famous *Principia Mathematica*. Here Newton expressed, among many other things, the basic idea for a viscous fluid. His "resistance" means local stress; "velocity by which the parts of the fluid are being separated" means velocity

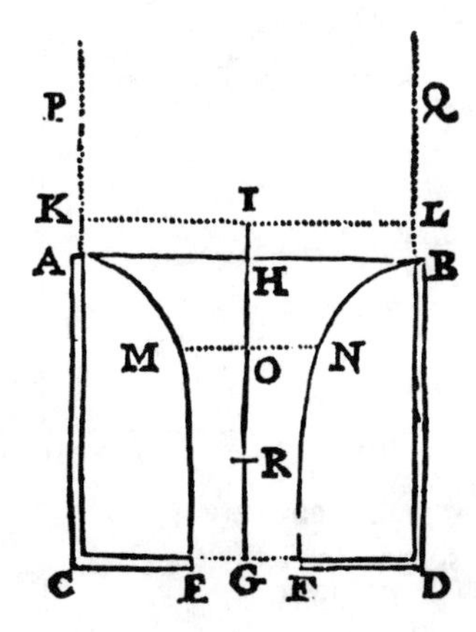

Figure 2.1.1. Portrait of Newton in 1702 and sketch from the *Principia* in which he illustrated some of his ideas about fluid flow.

gradient or the change of velocity with position in the fluid. The proportionality between them is the viscosity or "lack of slipperiness." In one dimension this can be written as

$$\tau_{yx} = \eta \frac{dv_x}{dy} \tag{2.1.1}$$

Although Newton had the right physical insight, it was not until 1845 that Stokes finally was able to write out this concept in three-dimensional mathematical form. Only in 1856 were Poiseuille's capillary flow data analyzed to prove Newton's relation experimentally. Couette tested the relation carefully, using the concentric cylinder apparatus shown in Chapter 5 (Figure 5.1.1), and found that his results agreed with the viscosities he measured in capillary flow experiments (Couette, 1890; Markovitz, 1968).

After Couette described his apparatus, several researchers used the design to study a wide variety of fluids. They soon found that many colloidal suspensions and polymer solutions did not obey this simple linear relation. Nearly all these materials give a viscosity that decreases with increasing velocity gradient in shear. Figure 2.1.2 shows that *shear thinning* occurs in a wide range of materials

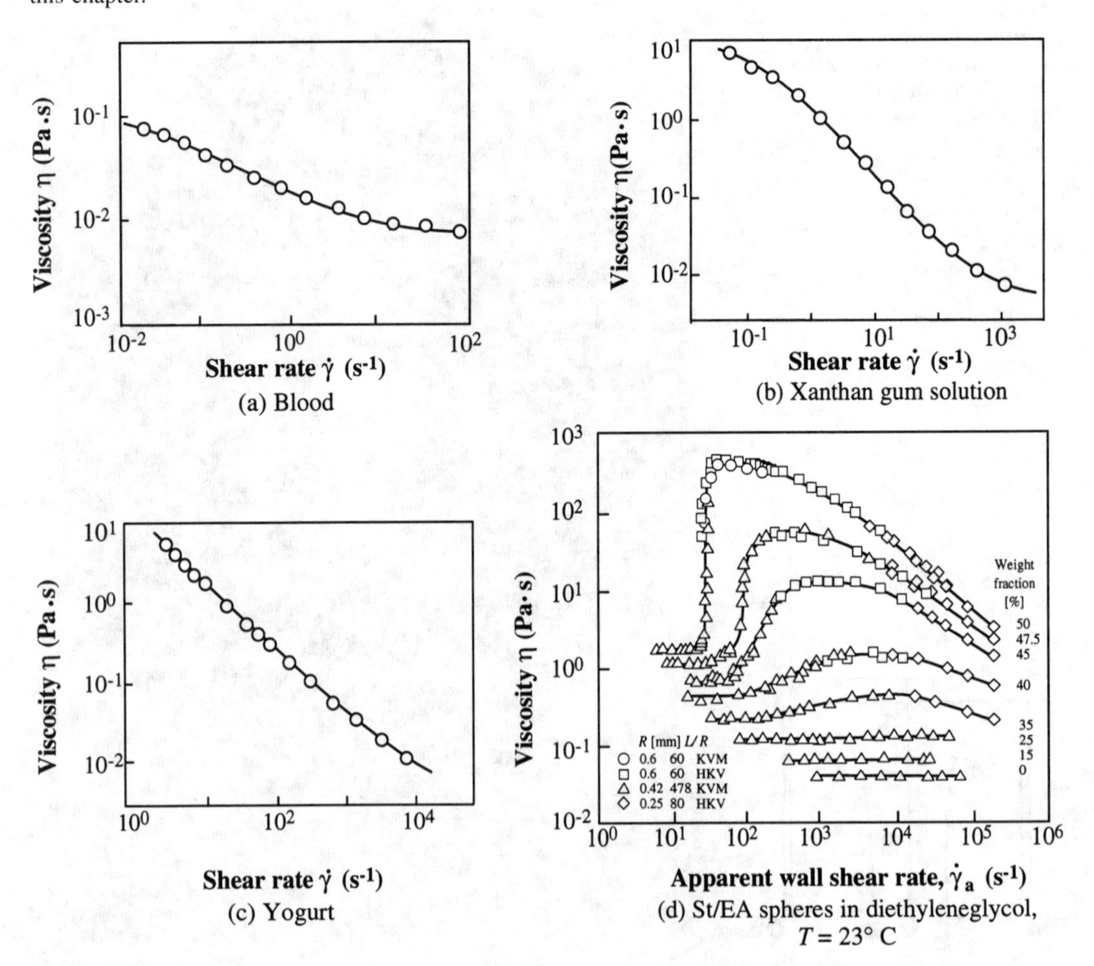

Figure 2.1.2.
Examples of shear-dependent viscosity. (a) Red blood cells (normal human, Hb 37%; Mills et al., 1980). (b) A 0.35 wt % aqueous solution of xanthan gum, a stiff biopolymer (Whitcomb and Macosko, 1978). (c) A commercial yogurt (deKee et al., 1980). (d) Polystyrene–ethylacrylate latex spheres $D_W = 0.5\ \mu$m at various concentrations in diethylene glycol, $T = 23°$ C (Laun, 1988). The lines in (a)–(c) are fits using models described in this chapter.

(see also Figures 2.4.1, 2.5.2, 2.5.3, and 2.5.4). It was this unusual flow behavior, first observed in the early 1900s, that led Bingham to coin the word *rheology*. Some concentrated suspensions also show *shear thickening* behavior as illustrated in Figure 2.1.2d.

Other experimenters in the early 1900s measured viscosity on very stiff liquids such as pitch and molten glass. Because these materials are so viscous, Trouton (1906) was able to test them like a solid in tension with little sagging due to gravity. His apparatus is shown in Chapter 7 (Figure 7.1.1). He found that the proportion between stress and velocity gradient was constant but three times larger than the value he measured in shear. This result turns out to be quite consistent with Newton's viscosity law when it is written properly in three dimensions. However, in the 1960s, workers exploring higher velocity gradients and other materials found that the viscosity in extension could *increase* with rate, although the shear viscosity *decreased*. A typical result for a polystyrene melt is shown in Figure 2.1.3. We see that the ratio of extensional to shear viscosity is 3 at low rates but becomes much greater with increasing deformation rate. We saw this qualitative difference between shear and extension with the elastic solid in Chapter 1. It is also typical of polymeric liquids.

In this chapter we show how Newton's viscosity law can be written in three dimensions with the rate of deformation tensor to give Trouton's result. Then we will generalize this ideal viscous model to explain the shear thinning and thickening behavior shown in Figure 2.1.2 in a manner similar to our derivation of the general elastic solid in Chapter 1. However, as we can see in Figures 2.1.2 and 2.1.3, viscous liquids exhibit a wider degree of departure from ideality. For example, some materials show a sudden shear thinning or plastic behavior. We will find that it is difficult to describe shear

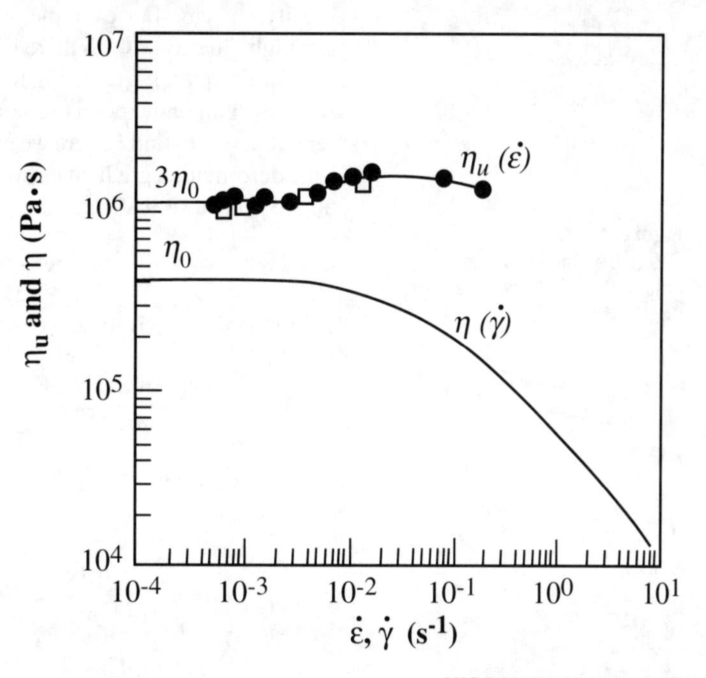

Figure 2.1.3.
Viscosity measured in uniaxial extension and in shear for a polystyrene melt at 160°C. Note that at low rates $\eta_u = 3\eta$. Adapted from Munstedt (1980).

thinning and extensional thickening in the same sample (Figure 2.1.3). In Section 2.6.4, we examine the temperature dependence of viscosity and the role of viscous dissipation in flows.

2.2 Velocity Gradient

To extend eq. 2.1.1 to three dimensions, we can use the stress tensor developed in Chapter 1, but we need a way to determine the velocity gradient in any direction at a point in the fluid. Newton said that the resistance depends on the "velocity by which parts of the fluid are being separated," so let us consider P and Q, two points embedded in a flowing fluid, separated by a small distance dx. In general, the velocity in the fluid is a function of position and time.

$$\mathbf{v} = \mathbf{v}(\mathbf{x}, t) \qquad (2.2.1)$$

The velocity at point P is $\mathbf{v}$ and at Q is $\mathbf{v} + \mathbf{dv}$ (Figure 2.2.1). We are not concerned with the absolute velocity but with the relative rate of separation of $d\mathbf{v}$ of point P and Q, which can be calculated at any point in the fluid by taking the gradient of the velocity function

$$d\mathbf{v} = \frac{\partial \mathbf{v}}{\partial \mathbf{x}} \cdot d\mathbf{x} \quad \text{or} \quad d\mathbf{v} = \mathbf{L} \cdot d\mathbf{x} \qquad (2.2.2)$$

where $\mathbf{L}$ is the velocity gradient tensor. This new tensor like the other tensors we have seen has two directions: one the direction of the velocity and the other of the gradient. Like the other tensors we have seen, it is an operator. It operates on a local displacement vector at a point to generate the magnitude and direction of the velocity change. For example, a uniform flow field may display very high $\mathbf{v}$ but $\mathbf{L} = \mathbf{0}$, so there is no rate of separation of points.

In Chapter 1 (recall Figure 1.4.1) we were also concerned with describing how points separate. There we used the displacement of points to find $\mathbf{F}$; here we need rate of displacement. Clearly $\mathbf{F}$, the deformation gradient tensor, and $\mathbf{L}$, the velocity gradient tensor, are related. Recall eq. 1.4.3

Figure 2.2.1. Relative velocity, $d\mathbf{v}$, between points P and Q moving with the fluid.

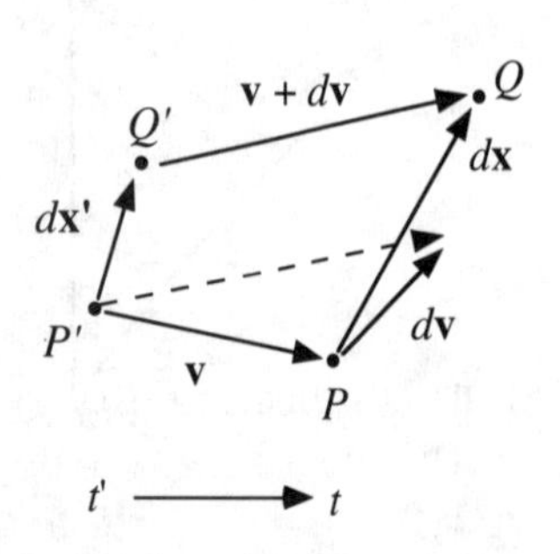

$$d\mathbf{x} = \mathbf{F} \cdot d\mathbf{x}' \qquad (1.4.3)$$

Taking the time derivative, we obtain

$$\frac{\partial(d\mathbf{x})}{\partial t} = \frac{\partial \mathbf{F}}{\partial t} \cdot d\mathbf{x}' + \mathbf{F} \cdot \frac{\partial(d\mathbf{x}')}{\partial t} \qquad (2.2.3)$$

But because $d\mathbf{x}'$ is fixed at t' then $\partial(d\mathbf{x}')/\partial t = 0$. Thus eq. 2.2.3 becomes

$$\frac{\partial(d\mathbf{x})}{\partial t} = d\mathbf{v} = \dot{\mathbf{F}} \cdot d\mathbf{x}'$$

Equating this result with eq. 2.2.2 gives

$$\dot{\mathbf{F}} \cdot d\mathbf{x}' = \mathbf{L} \cdot d\mathbf{x}$$

Using eq. 1.4.3 to substitute for $d\mathbf{x}$ gives

$$\dot{\mathbf{F}} = \mathbf{L} \cdot \mathbf{F} \tag{2.2.4}$$

Since we are only interested in the instantaneous rate of separation of points right now, t' becomes t and $d\mathbf{x}'$ becomes $d\mathbf{x}$. In the limit, as the past displacement is brought up to the present, we can write

$$\lim_{\mathrm{x}' \to \mathrm{x}} \mathbf{F} = \mathbf{I} \quad \text{and} \quad \lim_{\mathrm{x}' \to \mathrm{x}} \dot{\mathbf{F}} = \mathbf{L} \tag{2.2.5}$$

The velocity gradient $\mathbf{L}$ is often written out as the dyad product of the gradient vector (symbol ∇) and the velocity vector

$$\nabla \mathbf{v} = \mathbf{L}^T = \sum_i \sum_j \hat{\mathbf{x}}_i \hat{\mathbf{x}}_j \frac{\partial v_j}{\partial x_i} \tag{2.2.6}$$

Dropping the summation and the direction vectors as discussed in Chapter 1, we can write simply

$$\nabla \mathbf{v} = \mathbf{L}^T = \frac{\partial v_j}{\partial x_i}$$

Note that the index order is the reverse of that for $\mathbf{L}$, and thus $\mathbf{L}$ is the transpose of $\nabla \mathbf{v}$ in the usual notation.

To illustrate the velocity gradient tensor further, let us apply it to several simple flow examples. To help us, Table 2.2.1 gives the components of $\mathbf{L}$ in several coordinate systems.

TABLE 2.2.1 / Components of L or $(\nabla\ \mathbf{v})^{\mathrm{T}}$

Cartesian Coordinates (x, y, z)		
$\mathbf{L}_{xx} = \frac{\partial v_x}{\partial x}$	$\mathbf{L}_{xy} = \frac{\partial v_x}{\partial y}$	$\mathbf{L}_{xz} = \frac{\partial v_x}{\partial z}$
$\mathbf{L}_{yx} = \frac{\partial v_y}{\partial x}$	$\mathbf{L}_{yy} = \frac{\partial v_y}{\partial y}$	$\mathbf{L}_{yz} = \frac{\partial v_y}{\partial z}$
$\mathbf{L}_{zx} = \frac{\partial v_z}{\partial x}$	$\mathbf{L}_{zy} = \frac{\partial v_z}{\partial y}$	$\mathbf{L}_{zz} = \frac{\partial v_z}{\partial z}$
Cylindrical Coordinates (r, θ, z)		
$\mathbf{L}_{rr} = \frac{\partial v_r}{\partial r}$	$\mathbf{L}_{r\theta} = \frac{1}{r}\frac{\partial v_r}{\partial \theta} - \frac{v_\theta}{r}$	$\mathbf{L}_{rz} = \frac{\partial v_r}{\partial z}$
$\mathbf{L}_{\theta r} = \frac{\partial v_\theta}{\partial r}$	$\mathbf{L}_{\theta\theta} = \frac{1}{r}\frac{\partial v_\theta}{\partial \theta} + \frac{v_r}{r}$	$\mathbf{L}_{\theta z} = \frac{\partial v_\theta}{\partial z}$
$\mathbf{L}_{zr} = \frac{\partial v_z}{\partial r}$	$\mathbf{L}_{z\theta} = \frac{1}{r}\frac{\partial v_z}{\partial \theta}$	$\mathbf{L}_{zz} = \frac{\partial v_z}{\partial z}$
Spherical Coordinates (r, θ, ϕ)		
$\mathbf{L}_{rr} = \frac{\partial v_r}{\partial r}$	$\mathbf{L}_{r\theta} = \frac{1}{r}\frac{\partial v_r}{\partial \theta} - \frac{v_\theta}{r}$	$\mathbf{L}_{r\phi} = \frac{1}{r\sin\theta}\frac{\partial v_r}{\partial \phi} - \frac{v_\phi}{r}$
$\mathbf{L}_{\theta r} = \frac{\partial v_\theta}{\partial r}$	$\mathbf{L}_{\theta\theta} = \frac{1}{r}\frac{\partial v_\theta}{\partial \theta} + \frac{v_r}{r}$	$\mathbf{L}_{\theta\phi} = \frac{1}{r\sin\theta}\frac{\partial v_\theta}{\partial \phi} - \frac{v_\phi}{r}\cot\theta$
$\mathbf{L}_{\phi r} = \frac{\partial v_\phi}{\partial r}$	$\mathbf{L}_{\phi\theta} = \frac{1}{r}\frac{\partial v_\phi}{\partial \theta}$	$\mathbf{L}_{\phi\phi} = \frac{1}{r\sin\theta}\frac{\partial v_\phi}{\partial \phi} + \frac{v_r}{r} + \frac{v_\theta}{r}\cot\theta$

Example 2.2.1 Evaluation of L for Steady Extension, Shear, and Rotation

Consider the same types of deformation given in Example 1.4.1 but now let them be steady motions (rather than step deformations), whose velocities are independent of time.

Solutions

(a) Uniaxial extension is illustrated in Figure 2.2.2. Fluid enters in the x_2x_3 plane and exits along the x_1 axis. Following eq. 2.2.5 to find **L**, we take time derivatives of the displacement functions, eq. 1.4.4, and then evaluate them at $x_1' \to x_1$ to obtain

$$\frac{dx_1}{dt} = \frac{d\alpha_1}{dt}x_1 \quad \text{or} \quad v_1 = \frac{d\alpha_1}{dt}x_1 = \dot{\alpha}_1 x_1$$

Similarly

$$\begin{aligned} v_2 &= \dot{\alpha}_2 x_2 \\ v_3 &= \dot{\alpha}_3 x_3 \end{aligned} \tag{2.2.7}$$

Since $dv_1/dx_1 = \dot{\alpha}_1$, etc., the components of **L** for extension are

$$L_{ij} = \begin{bmatrix} \dot{\alpha}_1 & 0 & 0 \\ 0 & \dot{\alpha}_2 & 0 \\ 0 & 0 & \dot{\alpha}_3 \end{bmatrix} \tag{2.2.8}$$

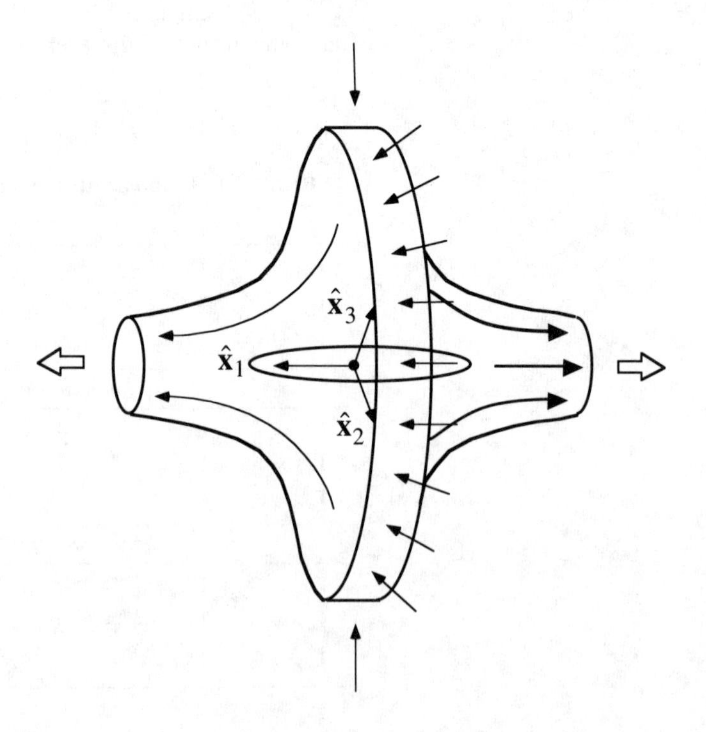

Figure 2.2.2. Streamlines for steady uniaxial extension.

Because uniaxial extension is symmetric, $v_2 = v_3$ and thus $\dot{\alpha}_2 = \dot{\alpha}_3$. For an incompressible fluid the continuity equation (eq. 1.7.9) becomes

$$\nabla \cdot \mathbf{v} = 0 \tag{1.7.9}$$

$$\text{or} \quad \dot{\alpha}_1 + \dot{\alpha}_2 + \dot{\alpha}_3 = 0$$

Combining results gives

$$\dot{\alpha}_2 = \dot{\alpha}_3 = -\frac{\dot{\alpha}_1}{2}$$

Substituting into eq. 2.2.8 gives the components of $\mathbf{L}$ for uniaxial extension

$$L_{ij} = \begin{bmatrix} \dot{\alpha}_1 & 0 & 0 \\ 0 & -\dot{\alpha}_1/2 & 0 \\ 0 & 0 & -\dot{\alpha}_1/2 \end{bmatrix} = \begin{bmatrix} \dot{\epsilon} & 0 & 0 \\ 0 & -\dot{\epsilon}/2 & 0 \\ 0 & 0 & -\dot{\epsilon}/2 \end{bmatrix} \tag{2.2.9}$$

Frequently called the extension rate, $\dot{\epsilon}$ is used for $\dot{\alpha}_1$ in steady extensional flows.

Thus eq. 2.2.9 gives the velocity gradient tensor $\mathbf{L}$ for steady uniaxial extension. When this tensor operates on displacement vectors imbedded in the liquid, it generates the velocity field for steady uniaxial extension.

(b) Steady simple shear is shown in Figure 2.2.3. Here planes of fluid slide over each other like cards in a deck. Again by taking time derivatives of the displacement functions for simple shear, eq. 1.4.8, we obtain

$$\frac{dx_1}{dt} = \frac{d\gamma}{dt} x_2 = v_1$$

or

$$v_1 = \dot{\gamma} x_2 = \frac{dv_1}{dx_2} x_2 \quad \text{and} \quad v_2 = v_3 = 0 \tag{2.2.10}$$

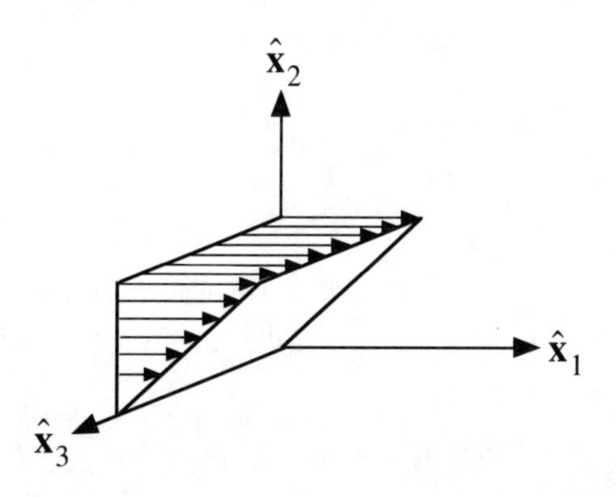

Figure 2.2.3.
Velocity profile in steady simple shear flow.

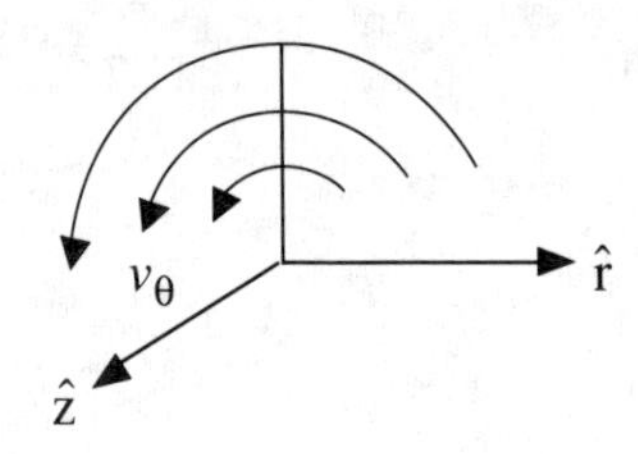

Figure 2.2.4.
Velocity in solid body rotation.

Thus **L** has only one component

$$L_{ij} = \begin{bmatrix} 0 & \dot{\gamma} & 0 \\ 0 & 0 & 0 \\ 0 & 0 & 0 \end{bmatrix} \qquad (2.2.11)$$

$\dot{\gamma}$ is called the shear rate. Note that since $\nabla \mathbf{v} = \mathbf{L}^T$, we have

$$(\nabla \mathbf{v})_{ij} = \begin{bmatrix} 0 & 0 & 0 \\ \dot{\gamma} & 0 & 0 \\ 0 & 0 & 0 \end{bmatrix} \qquad (2.2.12)$$

(c) Solid body rotation about the x_2 or z axis is illustrated in Figure 2.2.4. In cylindrical coordinates the velocity field is

$$v_\theta = \Omega r$$

$$v_r = v_z = 0$$

Evaluating **L** from Table 2.2.1 we obtain

$$\mathbf{L} = \begin{bmatrix} 0 & \Omega & 0 \\ -\Omega & 0 & 0 \\ 0 & 0 & 0 \end{bmatrix} \qquad (2.2.13)$$

If we use Cartesian coordinates (see eq. 1.4.10), $v_1 = \Omega x_2$ and $v_2 = -\Omega x_1$, and we obtain the same result.

From Example 2.2.1c we see that the velocity gradient tensor is not zero for a solid body rotation. From the statement of Newton's viscosity law, we would not expect stress to be generated as a result of the flow in solid body rotation because there is no relative separation of points. Like the deformation gradient **F**, the velocity gradient tensor **L** contains rate of rotation as well as stretching. We need a way to remove this rotation.

2.2.1 Rate of Deformation Tensor

We can remove rotation from the velocity gradient by recalling the definition of **F** in terms of the stretch and rotation tensors.

$$\mathbf{F} = \mathbf{V} \cdot \mathbf{R} \qquad (1.4.12)$$

If we differentiate with respect to time

$$\dot{\mathbf{F}} = \dot{\mathbf{V}} \cdot \mathbf{R} + \mathbf{V} \cdot \dot{\mathbf{R}}$$

and then let the past displacement x' come up to the present x, there will be no displacements, so

$$\lim_{x' \to x} \mathbf{V}(t') = \lim_{x' \to x} \mathbf{R}(t') = \mathbf{I} \qquad (2.2.14)$$

With this result and eq. 2.2.5 (i.e., $\dot{\mathbf{F}} = \mathbf{L}$), we obtain

$$\lim_{x' \to x} \dot{\mathbf{F}} = \mathbf{L} = \dot{\mathbf{V}} + \dot{\mathbf{R}} \tag{2.2.15}$$

and similarly

$$\mathbf{L}^T = (\dot{\mathbf{V}} + \dot{\mathbf{R}})^T$$

The time derivative of the stretching tensor $\dot{\mathbf{V}}$ is usually called the *rate of deformation* tensor $2\mathbf{D}$. It is symmetric.

$$2\dot{\mathbf{V}} = 2\mathbf{D} = \mathbf{L} + \mathbf{L}^T \quad \text{or} \quad 2\mathbf{D} = (\nabla \mathbf{v})^T + \nabla \mathbf{v} \tag{2.2.16}$$

$\dot{\mathbf{R}}$ is antisymmetric (or skew-symmetric) and is called $\mathbf{W}$, the *vorticity* tensor.

$$2\dot{\mathbf{R}} = 2\mathbf{W} = \mathbf{L} - \mathbf{L}^T \quad \text{or} \quad 2\mathbf{W} = (\nabla \mathbf{v})^T - \nabla \mathbf{v} \tag{2.2.17}$$

From the definitions we see that

$$\mathbf{L} = \mathbf{D} + \mathbf{W} = (\nabla \mathbf{v})^{\mathrm{T}} \tag{2.2.18}$$

A number of other symbols are frequently used in rheological literature to represent the rate of deformation and vorticity tensors, particularly $\boldsymbol{\Delta}$ or $\dot{\boldsymbol{\gamma}}$ for $2\mathbf{D}$ and $\boldsymbol{\Omega}$ for $\mathbf{W}$. Bird et al. (1987) call $\dot{\boldsymbol{\gamma}}$ the rate-of-strain tensor. Example 2.2.2 should help to illustrate $2\mathbf{D}$ and $2\mathbf{W}$.*

Example 2.2.2 Evaluation of 2D and 2W

For the flows given in Example 2.2.1, determine the components of $2\mathbf{D}$ and $2\mathbf{W}$. Also calculate their invariants.

Solutions

(a) *Steady Uniaxial Extension*. From eq. 2.2.9 we see immediately that

$$2D_{ij} = (L_{ij} + L_{ji}) = \begin{bmatrix} 2\dot{\epsilon} & 0 & 0 \\ 0 & -\dot{\epsilon} & 0 \\ 0 & 0 & -\dot{\epsilon} \end{bmatrix} \tag{2.2.19}$$

$$2W_{ij} = (L_{ij} - L_{ji}) = 0$$

This flow is called *irrotational* since $\mathbf{W} = \mathbf{0}$. Recalling eqs. 1.3.6–1.3.8, we can readily calculate the invariants of $2\mathbf{D}$.

$$I_{2D} = \mathrm{tr}2\mathbf{D} = 0$$

At this point the reader has the key concepts and may skip ahead to Section 2.3.

$$II_{2D} = \frac{1}{2}[(\text{tr}2\mathbf{D})^2 - \text{tr}(2\mathbf{D})^2] = -3\dot{\epsilon}^2 \tag{2.2.20}$$
$$III_{2D} = 2\dot{\epsilon}^3$$

(b) *Steady Simple Shear*

From eq. 2.2.11 and our definitions, we can write:

$$2D_{ij} = \begin{bmatrix} 0 & \dot{\gamma} & 0 \\ \dot{\gamma} & 0 & 0 \\ 0 & 0 & 0 \end{bmatrix} \tag{2.2.21}$$

$$2W_{ij} = \begin{bmatrix} 0 & \dot{\gamma} & 0 \\ -\dot{\gamma} & 0 & 0 \\ 0 & 0 & 0 \end{bmatrix} \tag{2.2.22}$$

$$\begin{aligned} I_{2D} &= 0 \\ II_{2D} &= \dot{\gamma} \\ III_{2D} &= 0 \end{aligned} \tag{2.2.23}$$

$$\begin{aligned} I_{2W} &= 0 \\ II_{2W} &= \dot{\gamma} \\ III_{2W} &= 0 \end{aligned} \tag{2.2.24}$$

(c) *Solid Body Rotation*

$$D_{ij} = 0 \qquad W_{ij} = \begin{bmatrix} 0 & -\Omega & 0 \\ \Omega & 0 & 0 \\ 0 & 0 & 0 \end{bmatrix} = L_{ij} \tag{2.2.25}$$

$$\begin{aligned} I_W &= 0 \\ II_W &= \Omega^2 \\ III_W &= 0 \end{aligned} \tag{2.2.26}$$

This flow is purely rotational because the rate of deformation tensor is zero.

From these examples we see that physically **W** gives the angular rotation in a material at any point. For a solid body rotation we have only **W**. **D** characterizes the rate of stretching at a point. We see that for uniaxial extension there is only stretching: $\mathbf{W} = \mathbf{0}$. Such flows are *irrotational*. From Example 2.2.2b we see that shear flow is a mixture of both stretching and rotation.

We note that in each flow the first invariant of **D** is zero.

$$I_D = 0 \tag{2.2.27}$$

This is true in general for an incompressible material. Since

$$I_D = \mathrm{tr}\mathbf{D} = \nabla_i v_i = \boldsymbol{\nabla} \cdot \mathbf{v}$$

and by eq. 1.7.9, $\boldsymbol{\nabla} \cdot \mathbf{v} = 0$ for an incompressible fluid.

The second and third invariants of $2\mathbf{D}$ are the only ones that vary during incompressible flows. As the invariants of the Finger tensor bound the possible *deformations* in a material (Figure 1.4.3), the invariants of the rate of deformation bind the possible *flows*. The domain of all possible flows is shown in Figure 2.2.5.

The fact that $\mathbf{D}$ measures stretching is demonstrated in Example 2.2.3, which shows $\mathbf{D}$ is the local rate of change in length. Example 2.2.4 shows that $\mathbf{D}$ is also the time derivative of $\mathbf{B}$ in the limit of small deformation.

Example 2.2.3 The Rate of Deformation Tensor 2D as a Rate of Length Change

Show that the rate of deformation tensor measures the rate of squared length change in a material $d|d\mathbf{x}|^2/dt$. That is, show

$$\frac{d|d\mathbf{x}|^2}{dt} = d\mathbf{x} \cdot 2\mathbf{D} \cdot d\mathbf{x} \tag{2.2.28}$$

Solution

Note that

$$\frac{d|d\mathbf{x}|^2}{dt} = \frac{d(d\mathbf{x} \cdot d\mathbf{x})}{dt} = 2d\mathbf{x} \cdot \frac{d(d\mathbf{x})}{dt} \tag{2.2.29}$$

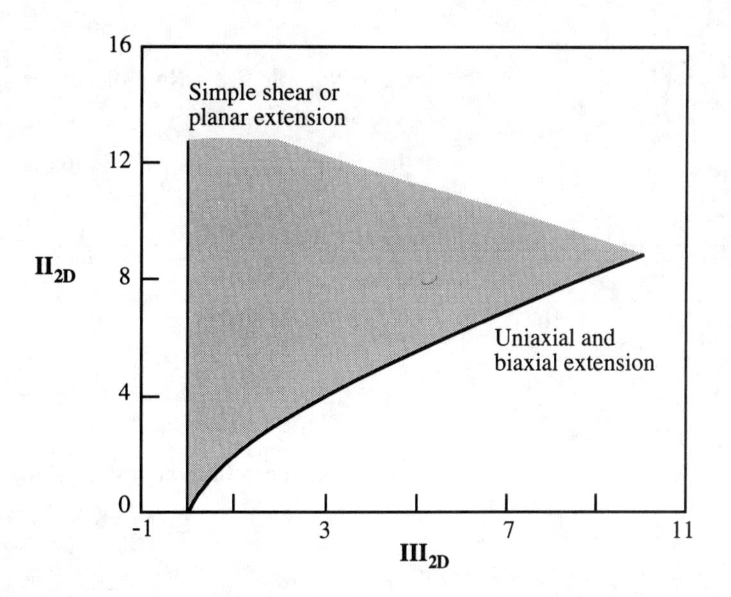

Figure 2.2.5. The map of invariants of the rate of deformation tensor lie in the shaded region bounded by simple shear and uniaxial extension for all flows of an incompressible material $I_{2D} = 0$.

From eq. 2.2.2

$$d\mathbf{v} = \mathbf{L} \cdot d\mathbf{x} \tag{2.2.2}$$

Thus

$$\frac{d(d\mathbf{x})}{dt} = \mathbf{L} \cdot d\mathbf{x} \tag{2.2.30}$$

Substitution into eq. 2.2.29 gives

$$\frac{d|d\mathbf{x}|^2}{dt} = d\mathbf{x} \cdot 2\mathbf{L} \cdot d\mathbf{x} \tag{2.2.31}$$

Breaking **L** down into **D** and **W** yields

$$\frac{d|d\mathbf{x}|^2}{dt} = d\mathbf{x} \cdot 2\mathbf{D} \cdot d\mathbf{x} + d\mathbf{x} \cdot 2\mathbf{W} \cdot d\mathbf{x} \tag{2.2.32}$$

In each of the right-hand terms the tensor is operating on two $d\mathbf{x}$ vectors. Because these vectors are identical, we can reverse the operations, operating on the other one first. However, changing the operation order requires the tensor to be symmetric. **D** is, but **W** is not (see eqn 2.2.17). Thus the last term is **0** and we have

$$\frac{d|d\mathbf{x}|^2}{dt} = d\mathbf{x} \cdot 2\mathbf{D} \cdot d\mathbf{x} \tag{2.2.33}$$

Thus we see that 2**D** is a tensor that operates on small displacement vectors around a point in a material to give the time rate of change of the squared length of those vectors.

Example 2.2.4 Rate of Deformation as a Time Derivative of **B** and **C**

Show that

$$\lim_{t' \to t} \frac{d\mathbf{B}}{dt} = 2\mathbf{D}$$

and similarly for **C**.

Solution

From the definition of **B**, eq. 1.4.13, we can write

$$\frac{d\mathbf{B}}{dt} = \dot{\mathbf{B}} = \dot{\overline{\mathbf{F} \cdot \mathbf{F}^T}} = \dot{\mathbf{F}} \cdot \mathbf{F}^T + \mathbf{F} \cdot \dot{\overline{\mathbf{F}^T}}$$

As in deriving eq. 2.2.15, we note that

$$\lim_{t' \to t} \mathbf{F} = \lim_{t' \to t} \mathbf{F}^T = \mathbf{I}$$

Thus

$$\lim_{t' \to t} \dot{\mathbf{B}} = \dot{\mathbf{F}} + \dot{\overline{\mathbf{F}^T}} \qquad (2.2.34)$$

which from eq. 2.2.5 becomes

$$\lim_{t' \to t} \dot{\mathbf{B}} = \mathbf{L} + \mathbf{L}^T = 2\mathbf{D} \qquad (2.2.35)$$

In the same way starting from the definition of $\mathbf{C} = \mathbf{F}^T \cdot \mathbf{F}$, we obtain

$$\lim_{t \to t'} \dot{\mathbf{C}} = \mathbf{L}^T + \mathbf{L} = 2\mathbf{D} \qquad (2.2.36)$$

2.3 Newtonian Fluid

We now have the tools to extend Newton's law to three dimensions. As we have seen, particularly in Example 2.2.3, $\mathbf{D}$ is the proper three-dimensional measure of the rate "by which the parts of the fluid are being separated." Thus using $\boldsymbol{\tau}$ and $\mathbf{D}$ we can turn eq. 2.1.1 into a tensor relation

$$\boldsymbol{\tau} = \eta 2\mathbf{D} \qquad (2.3.1)$$

or in terms of the total stress

$$\mathbf{T} = -p\mathbf{I} + \eta 2\mathbf{D} \qquad (2.3.2)$$

The factor of 2 arises naturally because $\dot{\mathbf{B}} = 2\mathbf{D}$, as we saw in Example 2.2.4, and also because viscosity is normally defined by the shear relation given in eq. 2.1.1. We can see this by applying the Newtonian constitutive relation to the steady simple shear flow of Example 2.2.2. Substituting eq. 2.2.21 for $2\mathbf{D}$ into 2.3.2 gives

$$T_{ij} = -p \begin{bmatrix} 1 & 0 & 0 \\ 0 & 1 & 0 \\ 0 & 0 & 1 \end{bmatrix} + \eta \begin{bmatrix} 0 & \dot{\gamma} & 0 \\ \dot{\gamma} & 0 & 0 \\ 0 & 0 & 0 \end{bmatrix} \qquad (2.3.3)$$

We see that the shear stress reduces to eq. 2.1.1

$$T_{12} = \tau_{12} = \eta \dot{\gamma} \qquad (2.3.4)$$

Recall that $\dot{\gamma}$ is the shear rate, dv_1/dx_2, eq. 2.2.10. The only normal stress in steady shear of an incompressible Newtonian fluid

TABLE 2.3.1 / The Viscosity of Some Familiar Materials at Room Temperature

Liquid	*Approximate Viscosity (Pa·s)*
Glass	10^{40}
Molten glass (500°C)	10^{12}
Asphalt	10^{8}
Molten polymers	10^{3}
Heavy syrup	10^{2}
Honey	10^{1}
Glycerin	10^{0}
Olive oil	10^{-1}
Light oil	10^{-2}
Water	10^{-3}
Air	10^{-5}

Adapted from Barnes et al. (1989).

is the arbitrary hydrostatic pressure. The two normal stress differences N_1 and N_2, eq. 1.2.46, are zero.

The SI unit of viscosity is the pascal-second (Pa·s). The cgs unit, poise=0.1 Pa·s, is also often used. One centipoise, cp=10^{-3} Pa·s or 1mPa·s is approximately the viscosity of water. Table 2.3.1 shows the tremendous viscosity range for common materials. Very different instruments are required to measure over this range.

Shear rates of common processes can also cover a very wide range, as indicated in Table 2.3.2. Typically, however, very high shear rate processes are applied to low viscosity fluids. Thus shear stresses do not range as widely.

TABLE 2.3.2 / Shear Rates Typical of Some Familiar Materials and Processes

Process	*Typical Range of Shear Rates* (s^{-1})	*Application*
Sedimentation of fine powders in a suspending liquid	$10^{-6} - 10^{-4}$	Medicines, paints
Leveling due to surface tension	$10^{-2} - 10^{-1}$	Paints, printing inks
Draining under gravity	$10^{-1} - 10^{1}$	Painting and coating; emptying tanks
Screw extruders	$10^{0} - 10^{2}$	Polymer melts, dough
Chewing and swallowing	$10^{1} - 10^{2}$	Foods
Dip coating	$10^{1} - 10^{2}$	Paints, confectionery
Mixing and stirring	$10^{1} - 10^{3}$	Manufacturing liquids
Pipe flow	$10^{0} - 10^{3}$	Pumping, blood flow
Spraying and brushing	$10^{3} - 10^{4}$	Spray-drying, painting, fuel atomization
Rubbing	$10^{4} - 10^{5}$	Application of creams and lotions to the skin
Injection mold gate	$10^{4} - 10^{5}$	Polymer melts
Milling pigments in fluid bases	$10^{3} - 10^{5}$	Paints, printing inks
Blade coating	$10^{5} - 10^{6}$	Paper
Lubrication	$10^{3} - 10^{7}$	Gasoline engines

Adapted from Barnes et al. (1989).

2.3.1 Uniaxial Extension

For steady uniaxial extension, eq. 2.2.19 gives the components of 2**D**. Using these components with eq. 2.3.2 gives the stresses in a Newtonian fluid

$$T_{11} = -p + 2\eta\dot{\epsilon} \tag{2.3.5}$$

$$T_{22} = T_{33} = -p - \eta\dot{\epsilon} \tag{2.3.6}$$

If we can neglect surface tension effects, the boundary conditions on the free surface are

$$T_{22} = T_{33} = 0 \tag{2.3.7}$$

Substituting gives

$$T_{11} = 3\eta\dot{\epsilon} \tag{2.3.8}$$

If we define an extensional viscosity as

$$\eta_u = \frac{T_{11} - T_{22}}{\dot{\epsilon}} \tag{2.3.9}$$

we obtain

$$\eta_u = 3\eta \tag{2.3.10}$$

This important result demonstrates the value of the tensor form of Newton's viscosity law. It is directly analogous to the result in Chapter 1, that the tensile modulus is three times the shear modulus, eq. 1.5.11. The three times rule for viscosity in steady uniaxial extension is often called the Trouton ratio. We see it holds true at low rates for the polymer melt in Figure 2.1.3. The following examples give applications of the Newtonian model to more complex deformations. Further examples appear at the end of the chapter. Bird, et al. (1987, Chapter 1) or any other good fluid mechanics book contains many worked Newtonian examples.

Example 2.3.1 Flow Between Eccentric Rotating Disks

Consider again the flow between eccentric rotating disks pictured in Exercise 1.10.7.

(a) Calculate the components of 2**D**, the rate of deformation tensor.

(b) If the material between the disks is a Newtonian liquid, determine the force components on the disks.

Solution

(a) To calculate the deformation gradient tensor **2D** we must determine the velocities from the displacement functions. Recall

$$x_1 = x_1' \cos \Omega t - x_2' \sin \Omega t \tag{2.3.11}$$

$$x_2 = x_1' \sin \Omega t + x_2' \cos \Omega t + \gamma x_3' \tag{2.3.12}$$

$$x_3 = x_3' \tag{2.3.13}$$

where $\gamma = a/h$ (see Exercise 1.10.7). Applying $v_i = dx_i/dt$ gives

$$v_1 = -x_1'\Omega \sin \Omega t - x_2'\Omega \cos \Omega t$$

$$v_2 = +x_1'\Omega \cos \Omega t - x_2'\Omega \sin \Omega t$$

$$v_3 = 0$$

We need to get v_i in terms of the present coordinates x_i (rather the past time positions x_i'). This can be done formally by expressing x_i' in terms of x_i, that is, by finding the inverse deformation gradient $d\mathbf{x}' = \mathbf{F}^{-1} \cdot d\mathbf{x}$ (eq. 1.4.29). However, in this case we can eliminate x_i' by substituting these results back into displacement functions. Let

$$x_2' \cos \Omega t = \frac{-v_1}{\Omega} - x_1' \sin \Omega t$$

Then with eq. 2.3.12

$$x_2 = \frac{-v_1}{\Omega} - x_1' \sin \Omega t + x_1' \sin \Omega t + \gamma x_3'$$

or

$$v_1 = \Omega \gamma x_3 - \Omega x_2 \tag{2.3.14}$$

Then let

$$x_1' \cos \Omega t = \frac{v_2}{\Omega} + x_2' \sin \Omega t$$

which substituted into eq. 2.3.11 gives

$$x_1 = \frac{v_2}{\Omega} + x_2' \sin \Omega t - x_2' \sin \Omega t$$

Thus,

$$v_2 = \Omega x_1 \tag{2.3.15}$$

Therefore,

$$2D_{ij} = \frac{\partial v_i}{\partial x_j} + \frac{\partial v_j}{\partial x_i} = \begin{bmatrix} 0 & 0 & \Omega\gamma \\ 0 & 0 & 0 \\ \Omega\gamma & 0 & 0 \end{bmatrix} \tag{2.3.16}$$

(b) Applying the Newtonian constitutive relation, eq. 2.3.2, for an incompressible material, we obtain the stress components

$$
\begin{aligned}
T_{13} &= T_{31} = \eta\Omega\gamma \\
T_{11} &= T_{22} = T_{33} = -p \\
T_{12} &= T_{23} = 0
\end{aligned}
\tag{2.3.17}
$$

The only stress component acting on the disk will be

$$\hat{\mathbf{x}}_3 \cdot \mathbf{T} = \mathbf{t}_1 = T_{31}\hat{\mathbf{x}}_3.$$

The force will equal this stress times the disk area

$$f_1 = f_x = \pi R^2 \eta\Omega\gamma \tag{2.3.18}$$

Example 2.3.2 Cone and Plate Rheometer

The cone and plate geometry shown in Figure 2.3.1 is a common one for measuring viscosity.

(a) Derive the relation between the geometry, the angular velocity, and the shear rate:

$$\dot{\gamma} = 2D_{\phi\theta} = 2D_{12}$$

(b) Derive a relationship between the torque M and the shear stress τ_{12}. Does this relation require a constitutive equation? Assume laminar, steady, isothermal flow with negligible gravity and edge effects and $\beta < 0.10$ rad.

Solution

(a) The proper coordinate system for this problem is spherical. By symmetry and as a result of the small angles with no inertial effects (slow flow), we expect $v_\theta = v_r = 0$. Thus the only ve-

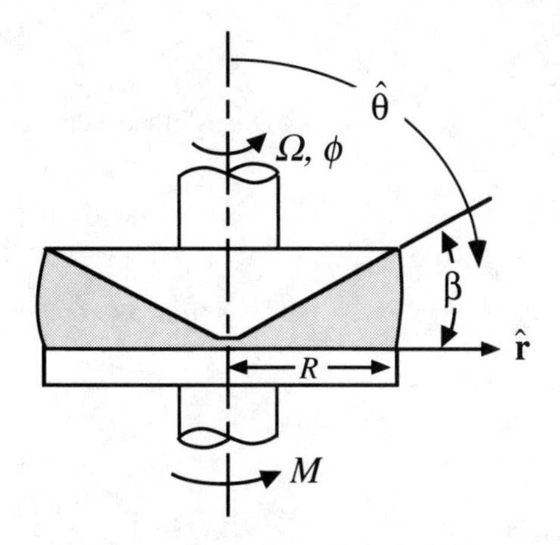

Figure 2.3.1.
Cone and plate rheometer. The small-angle cone rotates at angular velocity Ω while the plate is stationary.

locity component is $v_\phi(\theta)$, and the components for $\mathbf{L}$ in spherical coordinates reduce to (see Table 2.2.1).

$$L_{ij} = \begin{bmatrix} 0 & 0 & -\frac{v_\phi}{r} \\ 0 & 0 & -\frac{v_\phi}{r}\cot\theta \\ 0 & \frac{1}{r}\frac{\partial v_\phi}{\partial\theta} & 0 \end{bmatrix} \tag{2.3.19}$$

Thus, since $\cot(\pi/2-0.1) < 0.1$,

$$\begin{aligned} 2D_{\phi\theta} &= \frac{1}{r}\frac{\partial v_\phi}{\partial\theta} - \frac{v_\phi}{r}\cot\theta \\ &\simeq \frac{1}{r}\frac{\Delta v_\phi}{\Delta\theta} \\ &\simeq \frac{1}{r}\frac{v_\phi(\pi/2-\beta) - v_\phi(\pi/2)}{\beta} \end{aligned} \tag{2.3.20}$$

From the no-slip boundary conditions

$$v_\phi\left(\frac{\pi}{2}\right) = 0 \qquad v_\phi(\pi/2-\beta) = \Omega r\sin(\pi/2-\beta) \simeq \Omega r$$

Substituting gives

$$\dot{\gamma} = 2D_{\phi\theta} \simeq \frac{1}{r}\frac{\Omega r - 0}{\beta} \simeq \frac{\Omega}{\beta} \tag{2.3.21}$$

The error in this result due to the approximation made in eq. 2.3.20 is less that 1% for $\beta = 0.1$ rad (5.7°).

(b) The equations of motion in spherical coordinates for the ϕ direction, Table 1.7.1, reduce to

$$0 = \frac{1}{r}\frac{\partial\tau_{\phi\theta}}{\partial\theta} + \frac{2}{r}\cot\theta\,\tau_{\phi\theta}$$

Integrating gives

$$\tau_{\phi\theta} = \frac{C_1}{\sin^2\theta} \simeq \text{constant}$$

This is independent of θ and ϕ because

$$1.00 \geq \sin^2\left(\frac{\pi}{2}-\beta\right) \geq .99 \quad \text{for} \quad \beta < 0.1 \text{ rad}$$

Thus from a torque balance

$$M = \int_0^{2\pi}\int_0^R r^2\tau_{\phi\theta}|_{\pi/2}dr\,d\phi$$

$$\tau_{\phi\theta} = \frac{3M}{2\pi R^3} \tag{2.3.22}$$

Equation 2.3.22 is independent of the constitutive equation because of the small angle. The result is a homogeneous shear field, like simple shear between sliding parallel plates or between closely fitting cylinders. Because stress and deformation rates can be determined independent of a constitutive equation, these flows are very useful as rheometers and are discussed further in Chapter 5.

2.4 General Viscous Fluid

The Newtonian constitutive equation is the simplest equation we can use for viscous liquids. It (and the inviscid fluid, which has negligible viscosity) is the basis of all of fluid mechanics. When faced with a new liquid flow problem, we should try the Newtonian model first. Any other will be more difficult. In general, the Newtonian constitutive equation accurately describes the rheological behavior of low molecular weight liquids and even high polymers at very slow rates of deformation. However, as we saw in the introduction to this chapter (Figures 2.1.2 and 2.1.3) viscosity can be a strong function of the rate of deformation for polymeric liquids, emulsions, and concentrated suspensions.

A large number of models that depend on rate of deformation have been developed, but they all arise logically from the *general viscous fluid.* The general viscous model can be derived by a process very similar to the derivation of the general elastic solid in Section 1.6. Here we propose that stress depends only on the *rate* of deformation

$$\mathbf{T} = f(2\mathbf{D}) \tag{2.4.1}$$

Expanding the function in a power series gives

$$\mathbf{T} = f_o\mathbf{D}^o + f_1\mathbf{D}^1 + f_2\mathbf{D}^2 + f_3\mathbf{D}^3 + \cdots \tag{2.4.2}$$

Note that $\mathbf{D}^o = \mathbf{I}$ and for an incompressible fluid $f_o = -p$. Again we can evoke the Cayley Hamilton theorem, eq. 1.6.2. Thus

$$\mathbf{T} = -p\mathbf{I} + \eta_1 2\mathbf{D} + \eta_2 (2\mathbf{D})^2 \tag{2.4.3}$$

where η_1 and η_2 are scalar functions of the invariants of $2\mathbf{D}$.

$$\mathbf{T} = -p\mathbf{I} + \eta_1(II_{2D}, III_{2D})2\mathbf{D} + \eta_2(II_{2D}, III_{2D})(2\mathbf{D})^2 \tag{2.4.5}$$

This constitutive equation is also known as a Reiner–Rivlin fluid. The Newtonian fluid is simply a special case with $\eta_1(II_{2D}, III_{2D}) = \eta$, a constant, and $\eta_2 = 0$.

The η_2 term gives rise to normal stresses in steady shear flow, but unfortunately they are not even in qualitative agreement with experimental observations. This can be readily seen by noting the components of **2D** for steady simple shear from eq. 2.2.21 and calculating $(2\mathbf{D})^2$

$$4D_{ij}^2 = \begin{bmatrix} \dot{\gamma}^2 & 0 & 0 \\ 0 & \dot{\gamma}^2 & 0 \\ 0 & 0 & 0 \end{bmatrix} \tag{2.4.6}$$

Substituting these into eq. 2.4.5 gives for the stresses

$$T_{12} = \eta_1 \dot{\gamma} \tag{2.4.7}$$

$$T_{11} = T_{22} = -p + \eta_2 \dot{\gamma}^2 \tag{2.4.8}$$

$$T_{33} = -p \tag{2.4.9}$$

In steady simple shear flow nearly all fluids that exhibit normal stresses show a positive first normal stress difference $T_{11} - T_{22}$ and a much smaller, typically negative second normal stress difference $T_{22} - T_{33}$. From eq. 2.4.8 we see that the first normal stress will be zero but the second will not.

This little exercise is significant. It tells us that one of the four key rheological phenomena laid out in the introduction to these chapters on constitutive relations–normal stresses in steady shear flows–cannot be explained by any function of the rate of the deformation tensor. On the other hand, almost any function of **B**, the Finger tensor, does generate proper shear normal stresses. We will wait until Chapter 4 to pursue this reasoning further.

Since the η_2 term gives qualitatively the wrong result, it is usually discarded. Therefore, the general viscous fluid reduces to

$$\mathbf{T} = -p\mathbf{I} + \eta(II_{2D}, III_{2D})2\mathbf{D} \tag{2.4.10}$$

Because so much rheological work has been done with simple shear flows where $III_{2D} = 0$ (note eq. 2.2.23), most functional forms for η have assumed $\eta(II_{2D})$ only.

$$\mathbf{T} = -p\mathbf{I} + \eta(II_{2D})2\mathbf{D} \quad \text{or} \quad \boldsymbol{\tau} = 2\eta(II_{2D})\mathbf{D} \tag{2.4.11}$$

Several common expressions for $\eta(II_{2D})$ are given in Sections 2.4.1–2.4.5.

2.4.1 Power Law

The most widely used form of the general viscous constitutive relation is the power law model

$$\tau_{ij} = m|II_{2D}|^{(n-1)/2}(2D_{ij}) \tag{2.4.12}$$

This equation is most often applied to steady simple shear flows in which the absolute value of the second invariant becomes (eq. 2.2.23)

$$|II_{2D}| = \dot{\gamma}^2$$

Thus for steady shear the power law becomes

$$\tau_{12} = \tau_{21} = m\dot{\gamma}^n \qquad \text{or} \qquad \eta = m\dot{\gamma}^{n-1} \tag{2.4.13}$$

with no other stress components. Equation 2.4.13 is frequently the way the power law is written, but it is important to remember that this version is valid only for simple shear. For other flows, such as radial flow between plates (e.g., Good et al., 1974), the full three-dimensional version given in eq. 2.4.12 must be used.

In the processing range of many polymeric liquids and dispersions the power law is a good approximation to the data from viscosity versus shear rate. Figure 2.4.1 shows viscosity versus shear rate data for an acrylonitrile–butadiene–styrene (ABS) polymer melt (Cox and Macosko, 1974). At high shear rate, $\dot{\gamma} > 1$, the power law fits the data well, with m representing a function of temperature. The power law has been used extensively in polymer process models (e.g., Middleman, 1977; Tadmor and Gogos, 1979; Tanner, 1985; Bird et al., 1987). Nearly all non-Newtonian materials show shear thinning, $n < 1$, but some, particularly concentrated suspensions, show regions of shear thickening, as illustrated in Figure 2.1.2d. These can be fit locally with $n > 1$, but one should be cautious in modeling flows of shear thickening fluids. Thickening often signals such other complications as instability, phase separation, and lack of reversibility.

One of the obvious disadvantages of the power law is that it fails to describe the low shear rate region. Since n is usually less than one, at low shear rate η goes to infinity rather than to a constant η_0, as is usually observed experimentally (Figures 2.4.1 and 2.1.2a, b). It is also observed that viscosity becomes Newtonian at high shear rates for many suspensions and dilute polymer solutions (see Figure 2.1.2a).

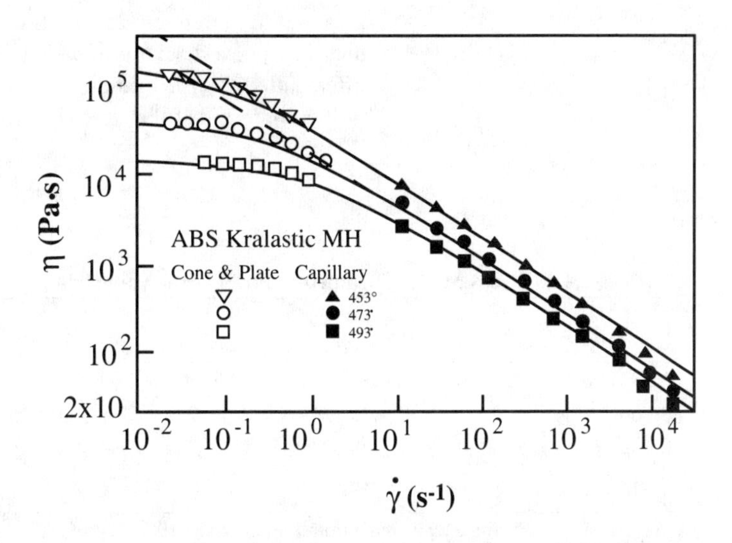

Figure 2.4.1.
Plot of viscosity versus shear rate for an ABS polymer melt at three temperatures (Cox and Macosko, 1974): dashed lines, power law fit; solid line represents a Cross or Ellis model. See Table 2.3.3.

2.4.2 Cross Model

To give Newtonian regions at both low and high shear rates, Cross (1965) proposed

$$\frac{\eta - \eta_\infty}{\eta_0 - \eta_\infty} = \frac{1}{1 + \left(K^2|II_{2D}|\right)^{(1-n)/2}} \tag{2.4.14}$$

Typically $\eta_0 \gg \eta_\infty$, so when $(II_{2D})^{1/2} = \dot{\gamma}$ is very small, η goes to η_0. At intermediate $\dot{\gamma}$ the Cross model has a power law region

$$(\eta - \eta_\infty) \simeq (\eta_0 - \eta_\infty)m\dot{\gamma}^{n-1} \quad \text{where } m = K^{1-n} \tag{2.4.15}$$

or for $\eta \gg \eta_\infty$

$$\eta \simeq \eta_0 m \dot{\gamma}^{n-1}$$

At very high shear rates the right-hand side of eq. 2.4.15 becomes very small, and η goes to the high shear rate Newtonian limit, η_∞.

The Cross model has been used to fit the data sets shown in Figures 2.4.1 and 2.1.2. The parameters used are shown in Table 2.4.1.

2.4.3 Other Viscous Models

To fit data even better Yasuda et al. (1981) have proposed

$$\frac{\eta - \eta_\infty}{\eta_0 - \eta_\infty} = \frac{1}{\left[1 + \lambda^a|II_{2D}|\right]^{(1-n)/a}} \tag{2.4.16}$$

This is equivalent to the Cross model with a fifth fitting parameter, a. With $a = 2$, eq. 2.4.16 is known as the Carreau model (Bird et al., 1987, p. 171).

Frequently the high shear rate region is not observed, and η_∞ is set to zero in eq. 2.4.14. Such a model fits the polymer melt data in Figure 2.4.1 quite well. This three-parameter version is often called the *Ellis* model. The Ellis model, however, is usually written in terms of the stress invariant

$$\frac{\eta_0}{\eta} = 1 + \left(\frac{II_\tau}{k}\right)^{a-1} \tag{2.4.17}$$

TABLE 2.4.1 / Cross Model Parameters for Several Materials[a]

Material	*Figure Number*	η_0 $(Pa \cdot s)$	η_∞ $(Pa \cdot s)$	K (s)	n
ABS (200°C)	2.4.1	45,000	–	2.5	0.40
Blood	2.1.2a	0.125	0.005	52.5	0.285
Xanthan	2.1.2b	15	0.005	10	0.20
Yogurt	2.1.2c	10	0.004	0.26	0.1

[a] Fits for data in Figure 2.1.2 adapted from Barnes et al. (1989).

The three-parameter Ellis model is still simple enough to allow analytical solution of some complex flow problems (Bird, 1976; Tadmor and Gogos, 1979). Applying the more complex viscous models requires numerical solutions.

Other models have been used (Bird et al., 1987, p. 228), but most studies have concentrated on the power law or the Cross (Carreau) models. Once one has chosen a numerical method, any of the general viscous models that depend on II_{2D} can be used.

The following example illustrates the use of the power law model.

Example 2.4.1 Flow of a Power Law Fluid Through a Tube

Determine the relationship between pressure drop $p_0 - p_L$ and flow rate for a power law fluid flowing through a circular tube (Figure 2.4.2). Assume that (a) the flow is steady, laminar incompressible, and fully developed; (b) gravity is negligible; and (c) isothermal conditions.

Solution

From the assumptions above and symmetry, we expect that

$$v_x = v_x(r), \quad v_\theta = 0, \quad \text{and} \quad v_r = 0$$

Thus the continuity equation is satisfied identically and the momentum balance reduces to (see Table 1.7.1)

$$0 = -\frac{\partial p}{\partial x} + \frac{1}{r}\frac{\partial}{\partial r}(r\tau_{rx})$$

Since $\partial p/\partial x$ is a function only of x and the second term is a function only of r, we can change the partials to ordinary derivatives

$$\frac{dp}{dx} = \frac{1}{r}\frac{d}{dr}(r\tau_{rx})$$

Figure 2.4.2.
Flow through a circular tube.

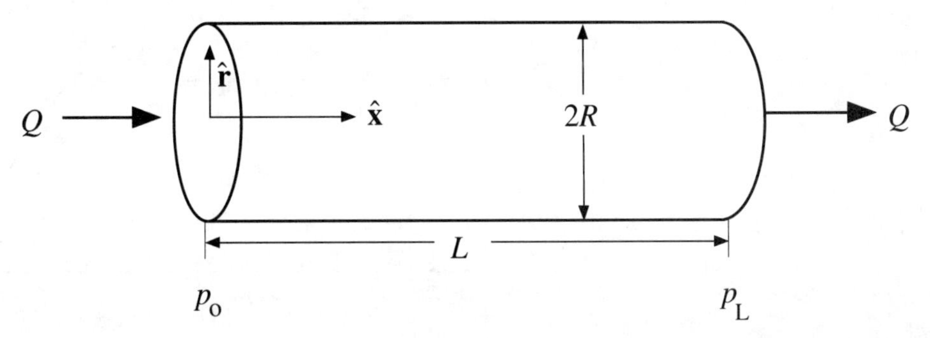

and integrate dp/dx to obtain $p = c_1 x + c_0$. From the pressure at $x = 0$ and L we get

$$p = -\left(\frac{p_0 - p_L}{L}\right)x + p_0$$

Thus

$$-\left(\frac{p_0 - p_L}{L}\right) = \frac{1}{r}\frac{d}{dr}(r\tau_{rx})$$

Integrate

$$\tau_{rx} = -\left(\frac{p_0 - p_L}{2L}\right)r + \frac{c_2}{r} \tag{2.4.18a}$$

$c_2 = 0$ because the shear stress cannot be infinite at the tube center. We will use the shear stress at the wall

$$\tau_{rx}|_R = \tau_w = \frac{(p_0 - p_L)R}{2L} \tag{2.4.18b}$$

Substituting the power law equation, which in one dimension becomes

$$\tau_{rx} = m\left|\frac{-dv_x}{dr}\right|^n \tag{2.4.19}$$

and integrating, we obtain

$$v_x = \left(\frac{\tau_w}{m}\right)^{1/n} \frac{R}{1/n + 1}\left[1 - \left(\frac{r}{R}\right)^{1/n+1}\right] \tag{2.4.20}$$

As we can see in Figure 2.4.3, this velocity profile is more pluglike than the Newtonian profile for $n < 1$.

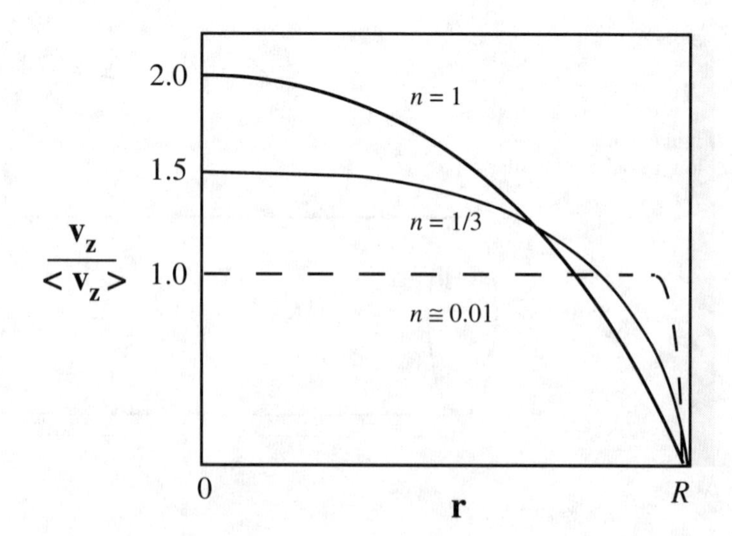

Figure 2.4.3.
Reduced velocity profiles for tube flow of a power law fluid.

We can determine the flow rate by integrating the velocity profile

$$Q = \int_0^{2\pi}\int_0^{R} v_x r\, dr\, d\theta$$

$$= 2\pi R^2 \int_0^1 v_x \frac{r}{R} d\left(\frac{r}{R}\right) \qquad (2.4.21)$$

$$= \frac{\pi R^2}{1/n + 3}\left[\frac{(p_0 - p_n)R}{2mL}\right]^{1/n}$$

which is the desired relation. We can also write an expression for the shear rate at the wall in terms of Q.

$$\left.\frac{dv_x}{dr}\right|_R = \dot{\gamma}_w = \frac{4Q}{\pi R^3}\left[\frac{3}{4} + \frac{1}{4n}\right] \qquad (2.4.22)$$

We will use these results for analyzing capillary rheometer data in Chapter 6.

2.4.4 The Importance of II_{2D}

The problem of combined flows of a shear thinning liquid in a die offers an excellent example of why the scalar material functions must depend on the *invariants* of the deformation (or stress) tensor.

Consider an axial annular die in which the liquid is pumped out under pressure *and* the central core is rotated at an angular velocity Ω (see Figure 2.4.4). Ignoring curvature the stress, balance becomes

$$0 = \frac{d\tau_{yx}}{dy}$$

$$0 = -\frac{p_1 - p_2}{L} + \frac{d\tau_{yz}}{dy}$$

with boundary conditions

$$v_x = 0 \quad v_z = 0 \quad \text{at } y = B$$

$$v_x = U \quad v_z = 0 \quad \text{at } y = 0$$

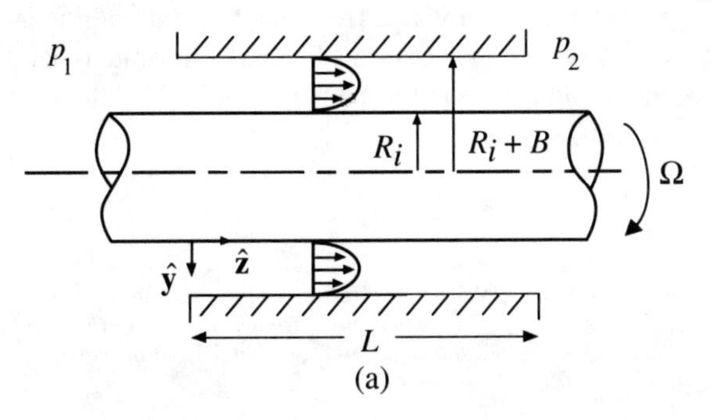

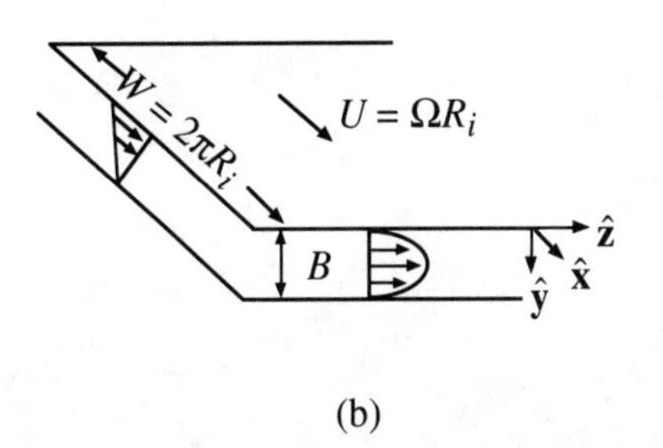

Figure 2.4.4.
(a) Pressure-driven flow through an annulus with rotation (helical annular flow). (b) The same flow "unwrapped," showing the projection of the velocity in both planes.

Figure 2.4.5. Dimensionless flow rate versus dimensionless pressure drop for helical annular flow of a power law fluid; $n = 1$ is the Newtonian solution representing the result in the absence of shear thinning. The solid lines are the results obtained numerically, while the dashed line (for $n = \frac{1}{3}$) is the analytical solution obtained using eq. 2.4.23. Note how much Q is enhanced at high pressures.

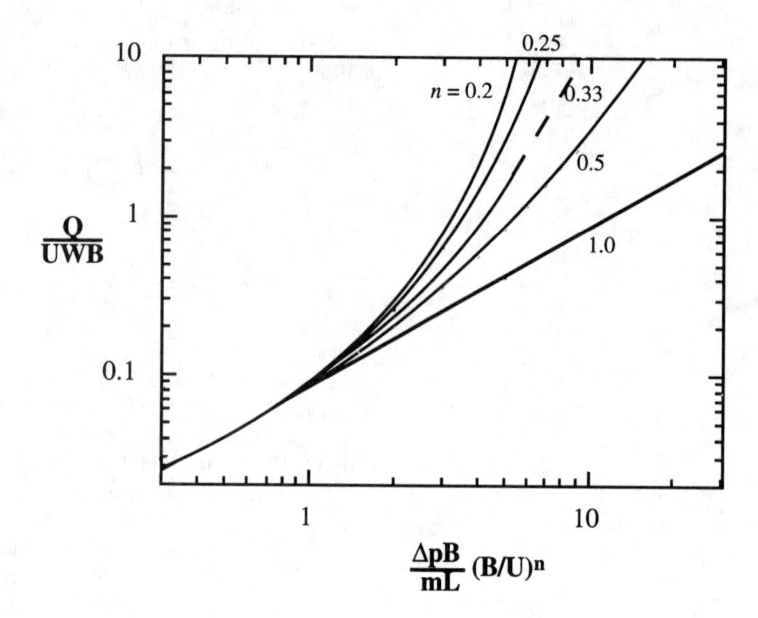

The solution of this problem for any value of n can be obtained by numerical integration (Middleman, 1977). Of special interest is the relationship between the volumetric flow rate Q and the pressure drop $\Delta p = p_1 - p_2$. The relationship between dimensionless flow rate and dimensionless pressure drop is shown in Figure 2.4.5 for different values of n.*

For $n = \frac{1}{3}$, an analytical solution can be obtained (Bird et al., 1987, p. 184). The expression is

$$Q = \frac{WB^2}{80}\left(\frac{B\Delta p}{mL}\right)^3\left[1 + 960\left(\frac{U}{B}\right)^2\left(\frac{mL}{B\Delta p}\right)^6 + \cdots\right] \quad (2.4.23)$$

The series expansion employed to obtain eq. 2.4.23 is valid for large pressure drop values. Using only the terms shown in eq. 2.4.23, we obtain the results shown by the dashed line in Figure 2.4.5, which are in good agreement with the numerical results shown by the continuous lines for large values of dimensionless pressure drop.

The cross flow due to rotation of the die shear thins the viscosity, further increasing the flow rate from the die. But nothing is free. It takes considerable power to rotate the die! Flow rate is also augmented when Δp and $\dot{\gamma}$ are in the same direction (Middleman, 1977; Bird et al., 1987).

**Obtaining the results in Figure 2.4.5 will be an interesting exercise in numerical analysis for the curious reader—a combination of numerical integration and interative solution is required; see Middleman (1977, Section 5.5) for the equation.*

2.4.5 Extensional Thickening Models

As we saw in Figure 2.1.3, viscosity measured in extension can be qualitatively different from that measured in shear. Primarily this difference constitutes a problem in nonlinear viscoelasticity, and we will wait until Chapter 4 to fully address it. However, several models that only depend on **2D** or **2W** have been proposed. These have no time dependence but are much easier to apply than viscoelastic models and are worth mentioning here.

In uniaxial extension the third invariant of **2D** is $2\dot{\epsilon}^3$, but in simple shear $III_{2D} = 0$ (recall eqs. 2.2.20 and 2.2.23). Thus a model that depends on III_{2D} may give the possibility of different behavior in extension. Debbaut et al. (1988) have proposed several models that depend upon III_{2D}. One that gives shear thinning but constant extensional viscosity as in Figure 2.1.3, is

$$\eta = \eta_0 \frac{(1 + m|II_{2D}|)^{(n-1)/2}}{(1 + 3m(3III_{2D}/II_{2D})^2)^{(n-1)/2}} \tag{2.4.24}$$

A problem with this model is that it gives effects in uniaxial extension only. In planar extension we also expect thickening, but here $III_{2D} = 0$, as shown in Exercise 2.8.1.

To solve this problem, Schunk and Scriven (1990) proposed a model that uses W_{rel}, a magnitude of the vorticity, to give extensional thickening. They define this magnitude such that it is insensitive to simple solid body rotation yet retains the rotation inherent in simple shear. They argue that this rotation is a key to understanding the difference between shear thinning and extensional thickening

$$\eta = \eta_S(II_{2D})W + \eta_u(II_{2D})(1 - W) \tag{2.4.25}$$

where

$$W = \begin{cases} \frac{2W_{\text{rel}}/S}{1+W_{\text{rel}}/S} & \frac{W_{\text{rel}}}{S} < 1 \\ 1 & \frac{W_{\text{rel}}}{S} \geq 1 \end{cases}$$

$$\mathbf{W}_{\text{rel}} = \mathbf{n}^{(i)} \times \left(\frac{\partial \mathbf{n}^{(i)}}{\partial t} + \mathbf{v} \cdot \nabla \mathbf{n}^{(i)} \right) - \frac{1}{2} \nabla \times \mathbf{v} \tag{2.4.26}$$

where $\mathbf{n}^{(i)}$ are the eigenvectors of **2D** (recall eqs. 1.3.3 and 1.3.13) and

$$S = |II_{2D}|^{1/2}$$

Typically $\eta_S(II_{2D})$ is a shear thinning function of the Cross or Carreau form and $\eta_u(II_{2D})$ is an extensionally thickening function of the same form (but $n > 1$). The ratio $W_{\text{rel}}/S = 1$ in simple shear and 0 in pure extension.

Souza Mendes et. al. (1994) have suggested that rather than the arithmetic mean (eq. 2.4.25) a geometric mean may give a more reasonable viscosity function

$$\eta = \eta_S(II_{2D})^W \eta_u(II_{2D})^{1-W} \tag{2.4.27}$$

The approaches above can describe only steady state, time-independent viscosities. In Chapter 4 we will show that for time-dependent viscoelastic models, like Maxwell's, extensional thickening arises naturally.

2.5 Plastic Behavior

A plastic material is one that shows little or no deformation up to a certain level of stress. Above this *yield stress* the material flows readily. Plasticity is common to widely different materials. Many metals yield at strains less than 1%. Concentrated suspensions of solid particles in Newtonian liquids often show a yield stress followed by nearly Newtonian flow. These materials are called viscoplastic or Bingham plastics after E. C. Bingham, who first described paint in this way in 1916. House paint and food substances like margarine, mayonnaise, and ketchup are good examples of viscoplastic materials.

A simple model for plastic material is Hookean behavior at stresses below yield and Newtonian behavior above. For one-dimensional deformations

$$\tau = G\gamma \qquad \text{for} \quad \tau < \tau_y$$

and (2.5.1)

$$\tau = \eta\dot{\gamma} + \tau_y \quad \text{for} \quad \tau \geq \tau_y$$

The model also can be written as allowing no motion below the yield stress

$$\dot{\gamma} = 0 \qquad \text{for} \quad \tau < \tau_y$$

and (2.5.2)

$$\tau = \eta\dot{\gamma} + \tau_y \quad \text{for} \quad \tau \geq \tau_y$$

This latter form is the one Bingham used in his original paper.

Figure 2.5.1 illustrates eqs. 2.5.1 and 2.5.2 and compares the Bingham model to Newtonian and power law fluids. We can see why strong shear thinning behavior is frequently called *pseudo*-plastic. In fact the data shown in Figure 2.1.2c for yogurt, which is very shear thinning, $n = 0.1$, can be fit nearly as well with a Bingham model using $\tau_y = 60$ Pa and $\eta = 4$ mPa·s.

An important feature of plastic behavior is that if the stress is not constant over a body, parts of it may flow while the rest acts like a solid. Consider flow in a tube: the shear stress goes linearly from zero at the center of the tube to a maximum at the wall (eq. 2.4.18a). Thus the central portion of the material flows like a solid plug (Frederickson, 1964, p. 178). The shape is shown in Figure 2.4.3, the curve for $n = 0.01$. Neck formation during uniaxial extension of a solid at constant strain rate is another example of local flow. At the smallest sample cross section or at an inhomogeneity, the stress during the test will just exceed the yield stress and large deformation can occur.

To handle deformations occurring in more than one direction, eq. 2.5.1 should be put into three-dimensional form. The only significant change required is to replace the one-dimensional yield criterion with some scalar function of the invariants of $\boldsymbol{\tau}$. There are a number of yield criteria in the literature (Malvern, 1969, Sections 6.5, 6.6). The von Mises criterion, which uses the second invariant of $\boldsymbol{\tau}$, is the most common.

$$\boldsymbol{\tau} = G\mathbf{B} \quad \text{for} \quad II_\tau < \tau_y^2$$

and

$$\boldsymbol{\tau} = \left[\eta_0 + \frac{\tau_y}{|II_{2D}|^{1/2}}\right] 2\mathbf{D} \quad \text{for} \quad II_\tau \geq \tau_y^2 \tag{2.5.3}$$

The Bingham model has been applied to a wide variety of flow problems (Bird et al., 1982). One case is given in Example 2.5.1.

Example 2.5.1 The Ketchup Bottle

We have all been frustrated by that malevolent Bingham plastic, ketchup. To exceed yield stress of this substance in the neck of the

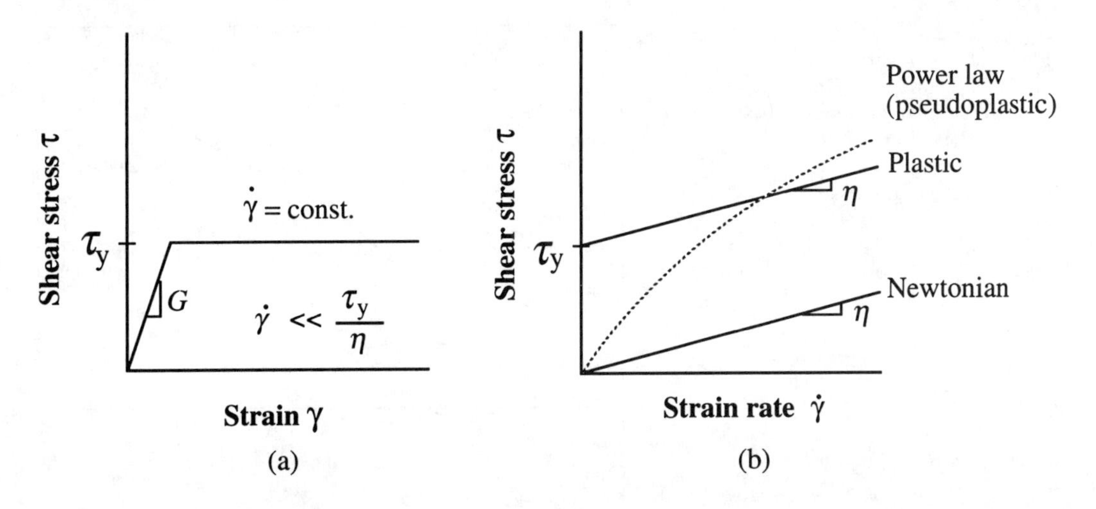

Figure 2.5.1. Bingham plastic behavior. (a) Shear stress versus strain at constant strain rate according to eq. 2.5.1. (b) Shear stress versus strain rate following eq. 2.5.2 compared to power law and Newtonian models.

bottle, one must frequently tap the bottle, and then, when the shear stress at the wall exceeds τ_y, flow is rapid. Figure 2.5.2 shows shear stress versus shear rate data for ketchup and several other food products. For this ketchup sample, $\tau_y = 200$ dynes/cm^2 (note that in contrast to Figure 2.5.1, this is a log–log plot). Will ketchup empty under gravity from a typical bottle?

Solution

In the neck of the bottle, the wall shear stress τ_w will be balanced by the pressure head of ketchup in the bottle. If we can approximate the neck of the bottle as a tube, length L and diameter D, and the body of the bottle as a cylindrical reservoir of height H, then from eq. 2.4.18

$$\tau_w(\pi DL) = p\left(\frac{\pi D^2}{4}\right) \quad \text{or} \quad \tau_w = \frac{\rho g H D}{4L} \quad \text{using } \rho g H = p \qquad (2.5.4)$$

Let density $\rho \simeq 1$ g/cm^3 and gravitational acceleration $g =$ 980 cm/s^2. Assume a bottle with a "standard" neck: $D \simeq 1.5$ cm, $L \simeq 6$ cm. If the bottle is partially full, $H \simeq 4$ cm, then $\tau_w \simeq 200$ dynes/cm^2 and the ketchup should not flow without some thumping. Note that the situation is probably worse because we have assumed atmospheric pressure above the ketchup in the bottle. The pressure will typically be less because a partial vacuum is created as the bottle is inverted. With a "wide mouth" bottle, $D \simeq 3$ cm. Then $\tau_w \simeq 400$ dynes/cm^2, which may make meal times flow more smoothly.

Figures 2.5.3 shows Bingham plots of data for an iron oxide suspension. Data over a wide shear rate range are shown in Figure 2.5.3a, while the lower shear rate range of data is shown in Figure 2.5.3b, both on a linear scale. The constants of the Bingham model

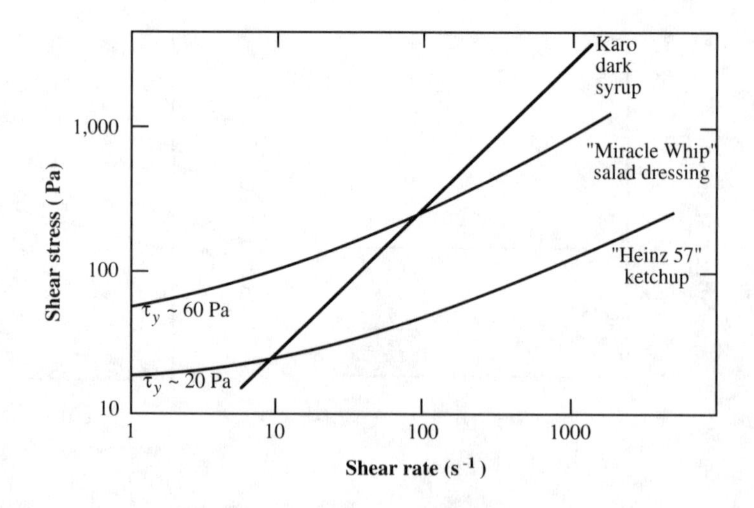

Figure 2.5.2.
Flow data for several food products (courtesy of Graco Co.). The Newtonian syrup has a higher viscosity at high shear rate, while the viscoplastic salad dressing and the ketchup show a yield stress and much higher viscosity at low rates.

fitted to these two ranges of data are considerably different. The difference is more obvious when the data are examined on a log–log plot (Figure 2.5.3c), especially in the lower shear rate range. Large errors in τ_y can result by picking the wrong shear rate range to fit the Bingham model.

2.5.1 Other Viscoplastic Models

Casson (1959) proposed an alternate model to describe the flow of viscoplastic fluids. The three-dimensional form is left as an exercise to the reader (hint: see eq. 2.5.9), but the one-dimensional form of the Casson model is given by.

$$\begin{aligned} \dot{\gamma} &= 0 \quad \text{for} \quad \tau < \tau_y \\ \tau^{1/2} &= \tau_y^{1/2} + (\eta\dot{\gamma})^{1/2} \quad \text{for} \quad \tau \geq \tau_y \end{aligned} \tag{2.5.5}$$

This model has a more gradual transition from the Newtonian to the yield region. For many materials, such as blood and food products, it provides a better fit.

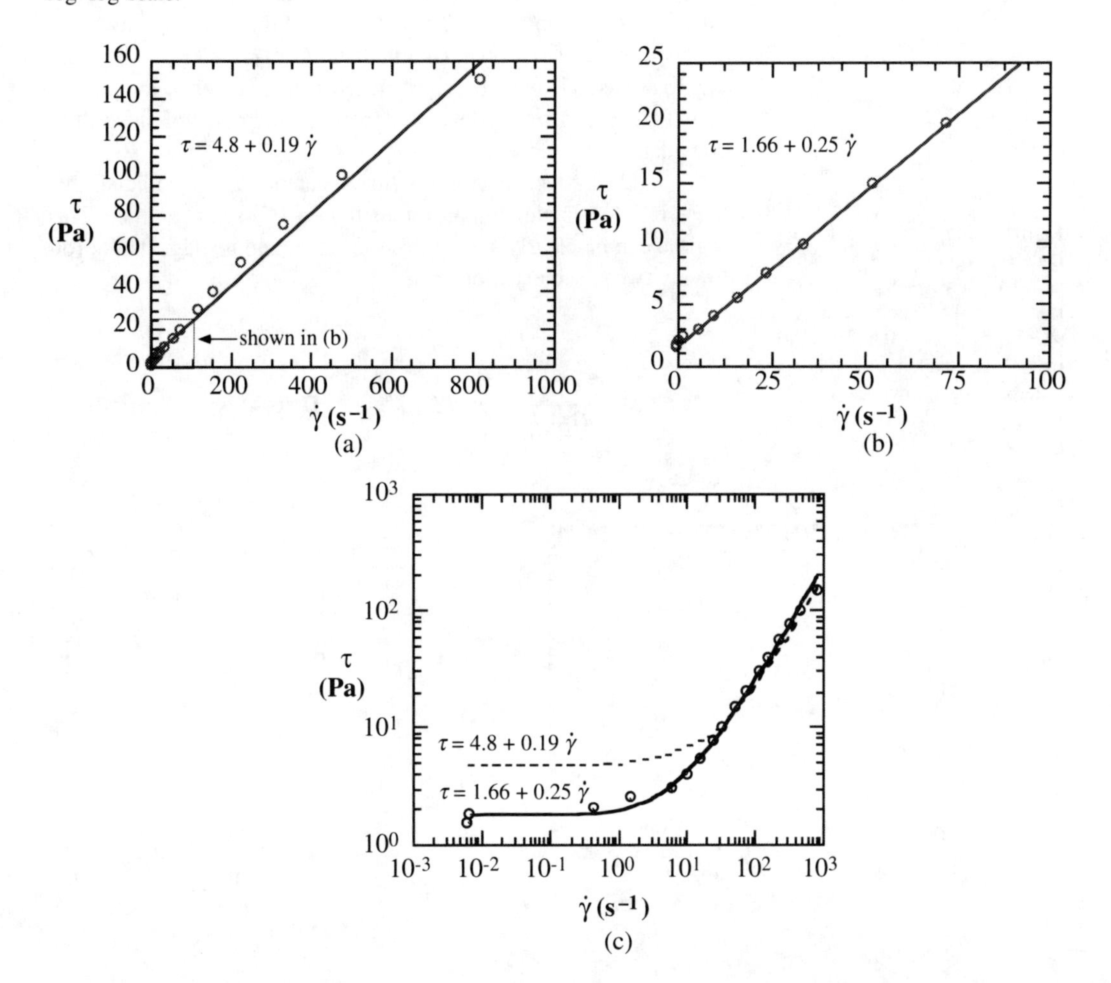

Figure 2.5.3. Bingham fits for experimental data of 6.0 vol % iron oxide suspension in mineral oil (replotted from Navarrete, 1991) in (a) the higher shear rate range, (b) the lower shear rate range, and (c) all data on a log–log scale.

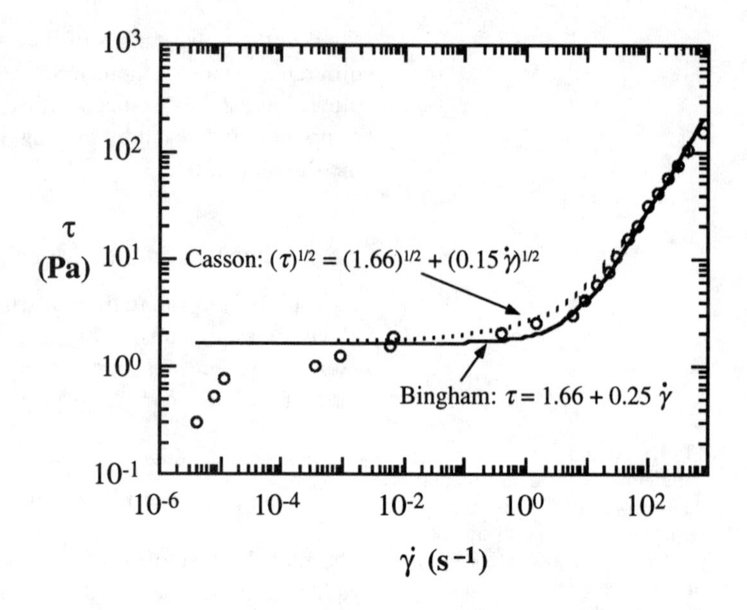

Figure 2.5.4.
Comparison of Bingham and Casson fits to the iron oxide suspension data over the entire range of experimental data obtained; parameters for the Bingham model are $\eta = 0.25$ Pa·s and $\tau_y = 1.66$ Pa, while for the Casson model they are 0.15 Pa·s and 1.66 Pa, respectively.

The iron oxide suspension data, including even lower shear rates, are shown in Figure 2.5.4. Curves of the Bingham and Casson models are also shown; Bingham model parameters from Figure 2.5.3b are used instead of a best fit. Note that values of the parameters for the Casson model also depend on the range of shear rates considered.

Studies covering a very wide shear rate range indicate that there is a lower Newtonian rather than a Hookean regime. This regime can be clearly seen in Figure 2.5.5, and it suggests the following two-viscosity model:

$$\boldsymbol{\tau} = 2\eta \mathbf{D} \quad \text{for} \quad II_{2D}^{1/2} \le \dot{\gamma}_c$$

$$\boldsymbol{\tau} = 2\left[\frac{\tau_y}{|II_{2D}|^{1/2}} + m|II_{2D}|^{(n-1)/2}\right]\mathbf{D} \text{ for } II_{2D}^{1/2} > \dot{\gamma}_c \quad (2.5.6)$$

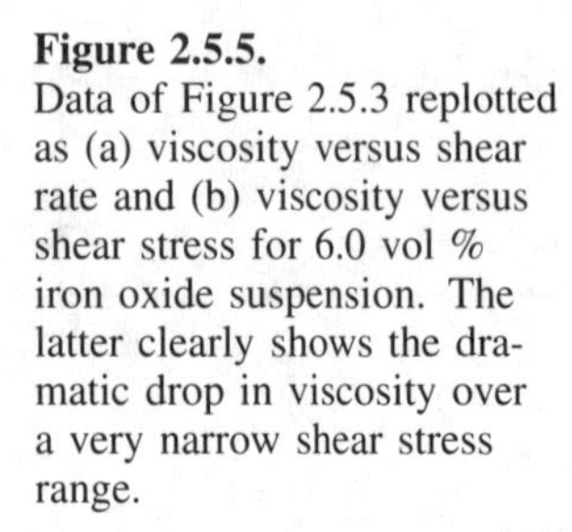

Figure 2.5.5.
Data of Figure 2.5.3 replotted as (a) viscosity versus shear rate and (b) viscosity versus shear stress for 6.0 vol % iron oxide suspension. The latter clearly shows the dramatic drop in viscosity over a very narrow shear stress range.

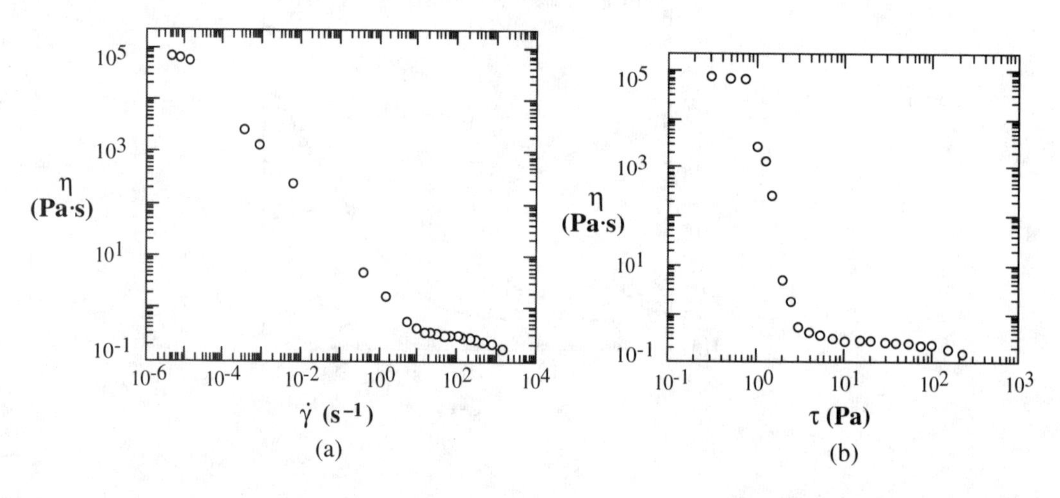

Using a critical shear rate rather than shear stress as a yield criteria makes application to numerical calculations much easier (Beverly and Tanner, 1989). Equation 2.5.6 with stress yield criteria (eq. 2.5.3) is known as Herschel–Bulkley model (Herschel and Bulkley, 1926; Bird et al., 1982). From Figure 2.5.5 we see that the two-viscosity models will better describe the iron oxide suspension data illustrated here.

Papanastasiou (1987) proposed a modification to the viscoplastic fluid models that avoids the discontinuity in the flow curve due to the incorporation of the yield criterion. Papanastasiou's modification involves the incorporation of an exponential term, thereby permitting the use of one equation for the entire flow curve, before and after yield. The one-dimensional form of Papanastasiou's modification is

$$\tau = \left\{ \eta + \frac{\tau_y[1 - \exp(-a\dot{\gamma})]}{\dot{\gamma}} \right\} \dot{\gamma} \tag{2.5.7}$$

The three-dimensional forms of the Herschel–Bulkley (eq. 2.5.6) and Casson (eq. 2.5.5) equations, with Papanastasiou's modification, are as follows.

Modified Herschel–Bulkley:

$$\boldsymbol{\tau} = \left\{ m|II_{2\mathrm{D}}|^{(n-1)/2} + \frac{\tau_y(1 - \exp(-a|II_{2\mathrm{D}}|^{1/2})]}{|II_{2\mathrm{D}}|^{1/2}} \right\} 2\mathbf{D} \tag{2.5.8}$$

Modified Casson:

$$\sqrt{|\boldsymbol{\tau}|} = \left\{ \sqrt{\eta} + \left(\frac{|\tau_y|[1 - \exp(-a|II_{2\mathrm{D}}|^{1/2})]}{|II_{2\mathrm{D}}|^{1/2}} \right)^{1/2} \right\} \sqrt{2\mathbf{D}} \tag{2.5.9a}$$

and

$$\boldsymbol{\tau} = |\tau|\mathrm{sgn}(\mathbf{D}) \tag{2.5.9b}$$

where the function "sgn" gives the corresponding sign of the **D** component. By choosing larger values for a, we can approach closer to the ideal yield stress behavior. Typically, $a \geq 100$ (Ellwood et al., 1990) is large enough to approximate the ideal viscoplastic behavior (Figure 2.5.6). Papanastasiou and co-workers (Papanastasiou, 1987; Ellwood et al., 1990) also have demonstrated that the exponential term used in eqs. 2.5.7–2.5.9 results in a flow curve that fits experimental data well.

The two-viscosity models (eq. 2.5.6) and Papanastasiou's modification (eqs. 2.5.7–2.5.9) are empirical improvements designed primarily to afford a convenient viscoplastic constitutive equation for numerical simulations (Abdali et al., 1992). Figure 2.5.6 compares them with the ideal Bingham model.

Barnes and Walters (1985) proposed that τ_y is a consequence of instrumental limitation, and given instruments capable of meas-

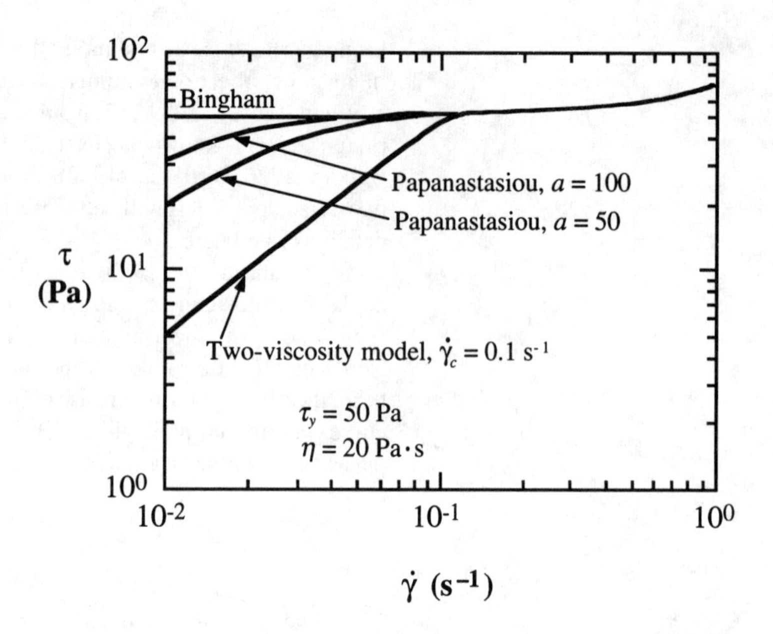

Figure 2.5.6.
Comparison of the two-viscosity models and Papanastasiou's modification for a Bingham fluid. Larger values of parameter a in Papanastasiou's modification (eq. 2.5.7) permit a better approximation to the Bingham model.

uring at extremely low shear rates, the yield stress does not exist. The iron oxide suspension data shown in Figure 2.5.5 demonstrate that measurements at $\dot{\gamma} < 10^{-5}s^{-1}$ are required to observe the existence of a high viscosity limit rather than a solidlike limit. Hartnett and Hu (1989) disputed this hypothesis by showing that a nylon ball suspended in a Carbopol solution did not fall by a measurable distance over 3 months. A compromise to this controversy may be obtained from Astarita's (1990) argument that yield stress is a convenient engineering reality, dependent on the nature of the problem under consideration.

Perhaps the best picture of a viscoplastic fluid is that of a very viscous, even solidlike, material at low stresses. Over a narrow stress range, which can often be modeled as a single yield stress, its viscosity drops dramatically. This is shown clearly in Figure 2.5.5b, where viscosity drops over five decades as shear stress increases from 1 to 3 Pa. (The drop is even more dramatic in Figure 10.7.2.) Above this yield stress the fluid flows like a relatively low viscosity, even Newtonian, liquid. Because of the different behaviors exhibited by these fluids, the model (Bingham, Casson, etc.) and the range of shear rates used to calculate the parameters must be chosen carefully. In Section 10.7 we will discuss microstructural bases for τ_y. It is also important to note that experimental problems like wall slip are particularly prevelant with viscoplastic materials. Aspects of slip are discussed in Section 5.3.

2.6 Balance Equations

To apply these viscous models to more complex problems, we need to use the equations of motion and boundary conditions in a manner similar to the way we approached the elastic solids problems in Section 1.7. The equations of motion are the same as those given there. The common boundary conditions relevant for fluids problems are also summarized. Finally, we discuss the energy equation, which can be important for viscous fluids that generate heat during flow. To solve the energy equation, relations for the temperature dependence of viscosity are needed. Two are given in this section.

2.6.1 Equations of Motion

As discussed in Section 1.7, for incompressible fluids the continuity equation becomes

$$\nabla \cdot \mathbf{v} = 0 \tag{1.7.9}$$

and the momentum equation is

$$\rho \frac{D\mathbf{v}}{Dt} = \nabla \cdot \mathbf{T} + \rho \mathbf{g} \tag{1.7.15}$$

These equations are given in component form for several coordinate systems in Table 1.7.1.

2.6.2 Boundary Conditions

At solid boundaries, the no-slip, no-penetration conditions generally hold

$$\begin{aligned} \mathbf{v}_t &= \mathbf{v}_{\text{surface}} \\ \mathbf{v}_n &= 0 \end{aligned} \tag{2.6.1}$$

At liquid–liquid interfaces, the velocities and stresses of both fluids (a and b) tangent to the interface must match

$$\begin{aligned} \mathbf{v}_{t_a} &= \mathbf{v}_{t_b} \\ (\hat{\mathbf{n}} \cdot \mathbf{T} \cdot \hat{\mathbf{t}})_a &= (\hat{\mathbf{n}} \cdot \mathbf{T} \cdot \hat{\mathbf{t}})_b \end{aligned} \tag{2.6.2}$$

The velocities normal to the liquid–liquid interfaces are again zero, and the normal stress balance must include any interfacial tension Γ and the surface curvature H

$$\begin{aligned} \mathbf{v}_{n_a} &= \mathbf{v}_{n_b} \\ (\hat{\mathbf{n}} \cdot \mathbf{T} \cdot \hat{\mathbf{n}})_a &= (\hat{\mathbf{n}} \cdot \mathbf{T} \cdot \hat{\mathbf{n}})_b + 2H\Gamma \end{aligned} \tag{2.6.3}$$

At a gas–liquid interface the same conditions hold, but usually we can assume the shear stress is zero at the interface

$$\hat{\mathbf{n}} \cdot \mathbf{T} \cdot \hat{\mathbf{t}} = 0 \tag{2.6.4}$$

2.6.3 Energy Equation

Because flowing viscous liquids can generate heat, we also need to consider the energy balance. In a manner similar to that used with the mass and momentum conservation relations, we can write a balance for the rate of change of internal energy over a control volume. This integral balance can be converted to a differential balance (Bird et al., 1987, p. 9) giving

$$\underset{\substack{\text{rate of change}\\\text{of internal}\\\text{energy/unit vol}}}{\frac{\partial \rho U}{\partial t}} = -\underset{\substack{\text{change due to}\\\text{convection}}}{\nabla \cdot \rho U \mathbf{v}} - \underset{\substack{\text{change due to}\\\text{conduction}}}{\nabla \cdot q} + \underset{\substack{\text{stress work}\\\text{(viscous}\\\text{dissipation)}}}{\mathbf{T} : \mathbf{D}} \tag{2.6.5}$$

Here U is the internal energy per unit mass and q is the conductive energy flux.

It is more convenient to express internal energy in terms of temperature because it can readily be measured for an incompressible material with constant conductivity and no chemical reaction

$$\rho \hat{c}_p \frac{DT}{Dt} = k_T \nabla^2 T + \mathbf{T} : \mathbf{D} \tag{2.6.6}$$

where $\hat{c}_p$ is the heat capacity per unit mass at constant pressure, T is the temperature (not to be confused with the stress tensor $\mathbf{T} = T_{ij}$), and k_T is the thermal conductivity (assumed to be constant). Bird, Stewart, and Lightfoot (1960) give the energy equation in a number of other forms and tabulate it in component form for rectangular, cylindrical, and spherical coordinates. In Chapters 5 and 6 we will see how the energy equation can be used to estimate rises in rheometer temperature.

Many non-Newtonian materials also have very high viscosity, with the result that the viscous dissipation term $\mathbf{T} : \mathbf{D}$ can become significant. This is illustrated in Example 2.6.1. Solution of such problems is complicated by the fact that viscosity also depends on temperature, and thus shear heating can change the velocity profile. Then the energy and momentum equations are coupled through the temperature-dependent viscosity.

2.6.4 Temperature and Pressure Dependence of Viscosity

The temperature dependence of viscosity can often be as important as its shear rate dependence for nonisothermal processing problems (e.g., Tanner, 1985). For all liquids, viscosity decreases with increasing temperature and decreasing pressure. A useful empirical model for both effects on the limiting low shear rate viscosity is

$$\eta_0 = K_1 e^{bT} e^{ap} \tag{2.6.7}$$

This equation is relatively easy to use in solving flow problems and is valid over a temperature range of about 50 K and pressure change of 1 kbar for polymers. Typical values of b range from $-0.03\ \mathrm{K}^{-1}$

for polyolefins to $-0.1\ \text{K}^{-1}$ for polystyrene and $a = 1$–$4\ \text{kbar}^{-1}$ for the same materials (van Krevelen, 1976). Pressure dependence of viscosity is discussed further in section 6.3.

A relation that is valid over a wider temperature range is the Andrade–Eyring equation (Bird et al., 1960)

$$\eta_0 = K_2 e^{E_\eta/RT} \qquad (2.6.8)$$

This equation was derived from the hypothesis that small molecules move by jumping into unoccupied sites or holes. For many small molecule liquids $E_\eta = 10-30$ kJ/mol. Polymer melts also obey this equation at temperatures well above their glass transition. With polymers, viscosity may be governed by successive jumps of segments of the chain. E_η ranges from about 25 kJ/mol for polyethylene to 60 for polystyrene and 85 for polycarbonate and polyvinyl chloride.

Near the glass transition temperature in polymers, E_η decreases. This decrease has been explained by the extra free volume created by thermal expansion, which leads to the Williams–Landel–Ferry (WLF) equation (Ferry, 1980). This equation describes the viscosity at temperature T in terms of viscosity at some reference temperature T_r.

$$\log \frac{\eta T_r \rho_r}{\eta_r T \rho} = \frac{-C_1(T - T_r)}{C_2 + T - T_r} = \log a_T \qquad (2.6.9)$$

The change in temperature times density $T_r\rho_r/T\rho$ is small and often ignored. Typically T_r is chosen as the glass transition and then $C_1 = 17.44$ and $C_2 = 51.6$ K for many polymers. Van Krevelen (1976) reports a better fit with $T_r = T_g + 43$ K and $C_1 = 8.86$ and $C_2 = 101.6$ K. The WLF equation is used extensively to make master curves of viscoelastic data; several examples are shown in Chapters 3 and 4. The equation is discussed further at the end of Chapter 11. Figure 11.6.1 is particularly useful to illustrate the transition from eqs. 2.6.8 and 2.6.9 in polymer melts.

Figure 2.6.1 shows that the data of Figure 2.4.1 can be shifted quite well with eq. 2.6.9. Van Krevelen shows that the WLF equation is most useful for amorphous polymers close to their glass

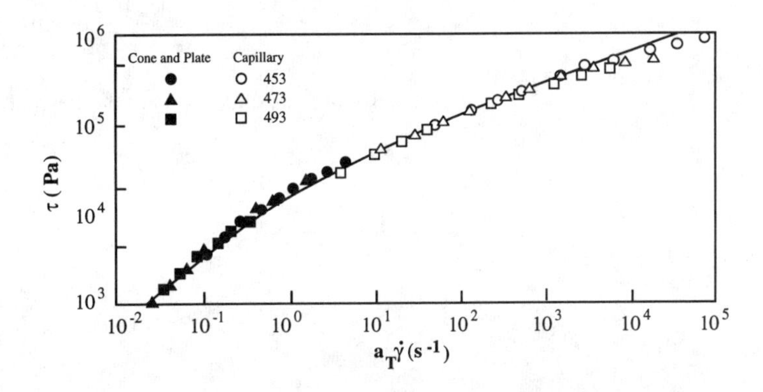

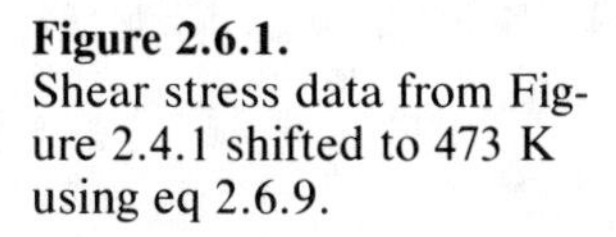
Figure 2.6.1.
Shear stress data from Figure 2.4.1 shifted to 473 K using eq 2.6.9.

transition temperature, while the exponential form appears to be more satisfactory for $T > T_g + 100^\circ$ K.

Temperature dependence of concentrated suspensions has been much less studied. Laun (1988) fit his data with a Casson model and found that over a 50 K range near room temperature both η and τ_y followed eq. 2.6.8. The viscosity decreased with increasing temperature with the same activation energy as the matrix (polyisobutylene), but the yield stress *increased*. Similar trends were found for shear thickening lattices like those in Figure 2.1.2d.

Example 2.6.1 Viscous Dissipation in Shear Flow

Consider steady simple shear flow between infinite parallel plates as in Example 2.2.1b, but do not neglect viscous dissipation.
(a) Determine temperature as a function of position for a power law fluid, assuming viscosity is independent of temperature.
(b) If the viscosity is a decreasing function of temperature, sketch the velocity profile and indicate how the force necessary to move the plate at v_0 will change.

Solution

(a) *Temperature-Independent Viscosity.* For simple shear flow between parallel plates, the velocity profile is

$$v_x = \frac{v_0}{B} y$$

where v_0 is the velocity of the upper plate, $y = B$. The lower plate, $y = 0$, is at rest. For a steady state temperature profile, the energy equation becomes

$$0 = k_T \frac{d^2 T}{dy^2} + \eta \left(\frac{dv_x}{dy} \right)^2 \tag{2.6.10}$$

For a power law fluid

$$0 = k_T \frac{d^2 T}{dy^2} + m \left(\frac{v_0}{B} \right)^{n-1} \left(\frac{v_0}{B} \right)^2 \tag{2.6.11}$$

Boundary conditions are isothermal walls or

$$T = T_0 \quad \text{at} \quad y = 0 \quad \text{and} \quad B$$

Integrating gives

$$T - T_0 = \frac{m v_0^{n+1}}{2 k_T B^{n-1}} \left[\frac{y}{B} - \left(\frac{y}{B} \right)^2 \right] \tag{2.6.12}$$

The maximum in temperature occurs at the midplane, $y = B/2$. Thus,

$$T_{\max} - T_0 = \frac{\mathrm{Br}T_0}{8} = \frac{mv_0^{n+1}}{8k_T B^{n-1}} \tag{2.6.13}$$

where Br is the Brinkman number.

(b) Because viscosity decreases with increasing temperature for liquids, if we allow viscosity to depend on temperature in this flow, we would find a lower viscosity near the flow center. Since the shear stress is constant across the flow, the velocity gradient will be higher than expected near the center and lower near each plate surface. The velocity field is sketched in Figure 2.6.2.

The force needed to keep the plate moving is just its area times the shear stress at the surface. Near the surface the temperature is T_0, so the viscosity is unchanged, but as Figure 2.6.2 indicates, the velocity gradient has decreased from the isothermal case. Thus for the case of viscous dissipation the force will be less. Another way to look at this phenomenon is to simply consider that the average viscosity will be lower as a result of the increased temperature and thus the force will be lower.

Bird et al. (1987, p. 223) solve this problem. For a power law fluid with an exponential temperature dependence,

$$m = m_0 e^{-b(T-T_0)/T_0} \tag{2.6.14}$$

but assuming n and k_T are constant, they obtain the velocity profile as a series expansion.

$$\frac{v_x}{v_0} = \frac{y}{B} - \frac{\mathrm{Br}b}{12n}\left[\frac{y}{B} - 3\left(\frac{y}{B}\right)^2 + 2\left(\frac{y}{B}\right)^3\right] + \mathrm{Br}^2(0) - \cdots \tag{2.6.15}$$

Note that they use the Nahme–Griffith number $Na = b\mathrm{Br}$. Na is a ratio of temperature rise due to viscous heating to the temperature change necessary to alter the viscosity. Thus since

$$F = (\text{area})\tau_{xy}|_{y=B} = (\text{area})m|_{T=T_0}\left(\frac{\partial v_x}{\partial y}\right)^n\Big|_{y=B} \tag{2.6.16}$$

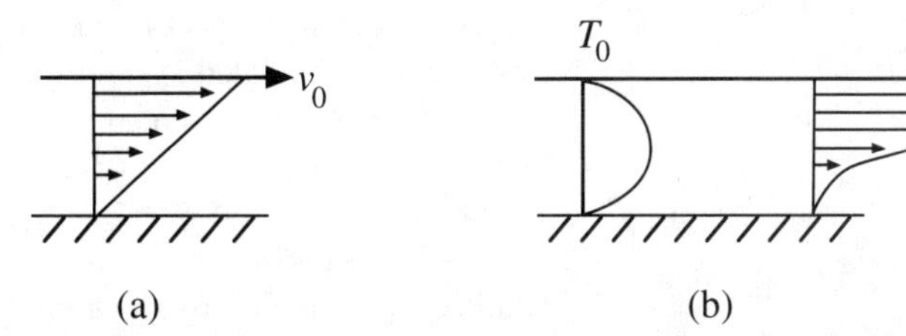

Figure 2.6.2.
Velocity field for simple shear flow (a) isothermal and (b) viscous dissipation.

the force on the plate will decrease. Considering only the first two terms in eq. 2.6.15 gives

$$\frac{F}{F_0} = \left(1 - \frac{\mathrm{Br}b}{12n}\right)^n \tag{2.6.17}$$

2.7 Summary

In this chapter we have developed the general constitutive equation for a viscous liquid. We found that by using the rate of deformation or strain rate tensor $2\mathbf{D}$, we can write Newton's viscosity law properly in three dimensions. By making the coefficient of $2\mathbf{D}$ dependent on invariants of $2\mathbf{D}$, we can derive models like the power law, Cross, and Carreau. We also showed how to introduce a three-dimensional yield stress to describe plastic materials with models like those Bingham and Casson. We saw two ways to describe the temperature dependence of viscosity and the importance of shear heating.

These models fit the shear rate dependence of viscosity very well and are very useful to engineers. They form the backbone of polymer processing flow analyses. If the problem is to predict pressure drop versus steady flow rate in channels of relatively constant cross section, or torque versus steady rotation rate, the general viscous fluid gives excellent results. We need to be sure that we pick a model that describes our particular material over the rates and stresses of concern, however. With numerical methods, the multiple parameter models are readily solved.

Although our viscous models describe shear thinning, shear thickening, and yielding well, they cannot describe the other three phenomena we said were very important in the introduction to Part I on constitutive equations, namely:

Time dependence

Normal stresses in shear

Different behaviors in extension and in shear

In fact we have lost something over Chapter 1. The Finger tensor **B** is able to give us normal stresses and extensional thickening. We will have to wait until Chapter 4 to get these factors back into our models. But in the next chapter we will see how to bring in the phenomenon of time dependence, which is so important for polymeric systems. We should note that for concentrated suspensions, especially flocculated systems, there is little elastic recovery, and time dependence is often either very short or extremely long. Thus the viscous models of this chapter are often quite adequate (recall Figures 2.5.3 and 2.5.4 and look ahead to Chapter 10).

The exercises in Section 2.8 illustrate how viscous models can be applied to different flow geometries and to some simple process problems. Many excellent applications of viscous, non-

Newtonian models can be found in Bird et al. (1987, Chapter 4) or in texts on polymer processing such as Middleman (1977, Chapter 5), Tadmor and Gogos (1979), Pearson (1984), or Tanner (1985).

2.8 Exercises

2.8.1 **B** and **D** for Steady Extension

Consider three *steady* extensional flows: (a) uniaxial, (b) equal biaxial, and (c) planar. Determine components of the tensors **B** and **D** for each and for their invariants.

2.8.2 Stresses in Steady Extension

Determine the normal stress differences in the three steady extensional flows of Exercise 2.8.1.

(a) *For a Power Law Fluid*

$$\mathbf{T} = -p\mathbf{I} + m|II_{2D}|^{(n-1)/2}2\mathbf{D} \tag{2.4.12}$$

(note: $m = \eta_0$ and $n = 1$ for a Newtonian fluid)

(b) *For a Bingham Plastic*

$$\boldsymbol{\tau} = G\mathbf{B} \quad \text{for} \quad II_{\tau} < \tau_y^2$$

and

$$\boldsymbol{\tau} = \left[\eta_0 + \frac{\tau_y}{|II_{2D}|^{1/2}}\right]2\mathbf{D} \quad \text{for} \quad II_{\tau} \geq \tau_y^2 \tag{2.5.3}$$

2.8.3 Pipe Flow of a Power Law Fluid

A power law fluid is being transported 10 m between two tanks. The process is being redesigned and you must increase the distance to 20 m yet maintain the same flow rate with the same pressure drop. Design the new pipe.

2.8.4 Yield Stress in Tension

If a uniaxial tensile stress of 10 Pa just yields a sample of a Bingham material, find τ_y (the yield stress that would be found by simple shear measurements on the same material).

2.8.5 Polymer Melt Pumped Through a Tubing Die

A polymer melt is pumped through a tubing die set at 200°C. As shown in Figure 2.8.1, the outside diameter is 10 cm and the inner mandrel is 9.5 cm. The length of the die (axial length of the annulus) is 10 cm.

The melt can be modeled with the power law

$$\eta = m(II_{2D})^{(n-1)/2}$$

where $n = \frac{1}{3}$ and $m = 10^5$ Pa·s$^{1/2}$. Density of the melt is 800 kg/m^3.

(a) Sketch the temperature profile in the land of this die near the

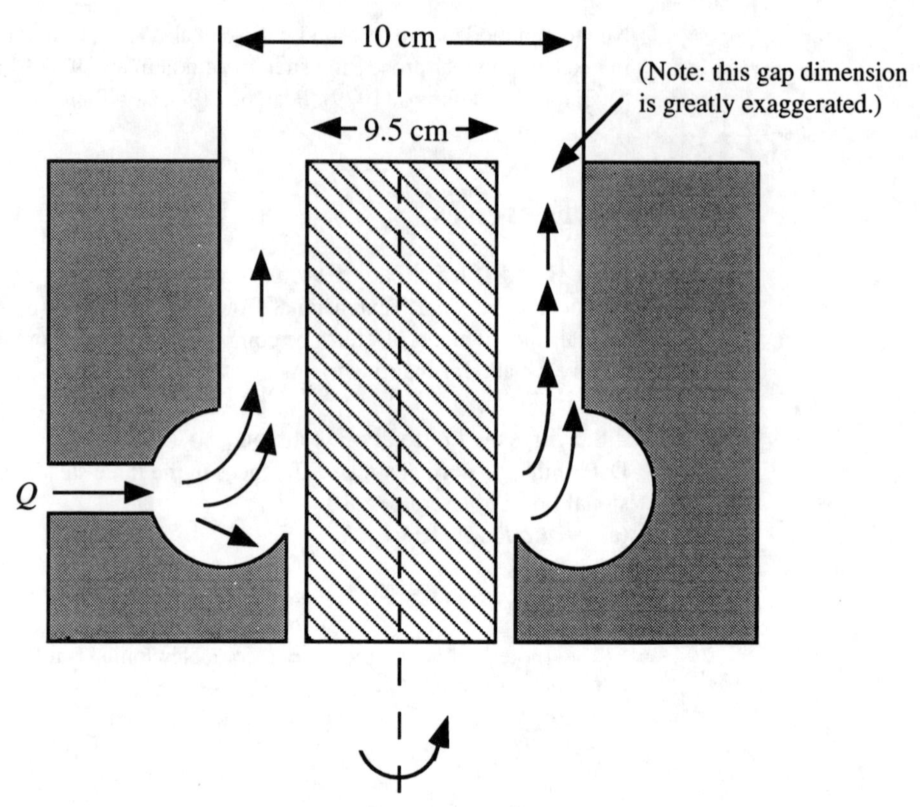

Figure 2.8.1.
Schematic of polymer melt pumped through annular die with rotating inner cylinder.

exit. Assume that little heat can be conducted out of the inner mandrel and that the outer wall is isothermal.

(b) How many kilograms per hour can be produced from this die if the extruder delivers a pressure of 2×10^6 Pa. (You can assume the gap is very small and neglect curvature.)

(c) If we double the pressure from the extruder, how much will the flow rate increase?

(d) The mandrel of the tubing die can be rotated. If it is rotated at 10 rpm, calculate II_{2D} in the land region.

(e) Estimate how the flow rate through this die will change when the mandrel is rotated in this range at 10 rpm.

2.8.6 Casson Model

Derive the three dimensional form of eq. 2.5.5.

References

Abdali, S. S.; Mitsoulis, E.; Markatos, N. C., *J. Rheol.* 1992, *36*, 389.

Astarita, G., *J. Rheol.* 1990, *34*, 275.

Barnes, H. A.; Hutton, J. F.; Walters, K., *An Introduction to Rheology*; Elsevier: Amsterdam, 1989.

Barnes, H. A.; Walters, K., *Rheol. Acta* 1985, *24*, 323.

Beverly, C. R.; Tanner, R. I., *J. Rheol.* 1989, *33*, 989.

Bingham, E. C., *U.S. Bur. Stand. Bull.* 1916, *13*, 309.

Bird, R. B., *Annu. Rev. Fluid Mech.* 1976, *8*, 13.

Bird, R. B.; Stewart, W. E.; Lightfoot, E. N., *Transport Phenomena*; Wiley: New York, 1960.

Bird, R. B.; Armstrong, R. C.; Hassager, O., *Dynamics of Polymeric Liquids,* Vol. 1, 2nd ed.; Wiley: New York, 1987.

Bird, R. B.; Dai, G. C.; Yarusso, B. J., *Rev. Chem. Eng.* 1982, *1*, 1.

Casson, N., in *Rheology of Disperse Systems*; Mill, C. C., Ed.; Pergamon: London, 1959; p. 84.

Couette, M. M., *J. Phys.* 1890, *9*, 414.

Cox, H. W.; Macosko, C. W., *AIChE J.* 1974, *20*, 785.

Cross, M. M., *J. Colloid Sci.* 1965, *20*, 417.

Debbaut, B.; Crochet, M. J.; Barnes, H.; Walters, K., in *Proceedings of the 10th International Congress on Rheology,* Vol. 1; Uhlherr, P. H. T., Ed.: Sydney, 1988, p. 291. Debbaut, B.; Crochet, M.J., *J. Non-Newtonian Fluid Mech.* 1988, *30*, 169.

deKee, D.; Turcotte, G.; Code, R. K., in *Rheology*, Vol. 3: *Applications*; Astarita, G.; Marrucci, G.; Nicolais, L.; Eds.; Plenum: New York, 1980; p. 609.

Ellwood, K. R. J.; Georgiou, G. C.; Papanastasiou, T. C.; Wilkes, J. O., *J. Rheol.* 1990, *34*, 787.

Ferry, J. D., *Viscoelastic Properties of Polymers*, 3rd ed.; Wiley: New York, 1980.

Fredrickson, A. G., *Principles and Applications of Rheology*: Prentice-Hall, Englewood Cliffs: NJ, 1964.

Good, P. A.; Schwartz, A.; Macosko, C. W., *AIChE J.* 1974, *20*, 67.

Hartnett, J. P.; Hu, R. Y. Z., *J. Rheol.* 1989, *33*, 67.

Herschel, W. H.; Bulkley, R., *Kolloid Z.* 1926, *39*, 291.

Laun, H. M., in *Proceedings of the 10th International Congress on Rheolology,* Vol. 1; Uhlherr, P. H. T.; Ed.; Sydney, 1988.

Malvern, L. E., *Introduction to the Mechanics of Continuous Media*; Prentice-Hall: Englewood Cliffs, NJ, 1969, p. 37.

Markovitz, H., *Phys. Today* 1968, *21*, 23.

Middleman, S., *Fundamentals of Polymer Processing*; McGraw-Hill: New York, 1977.

Mills, P.; Rubi, J. M.; Quemada, D., in *Rheology*, Vol. 2: *Fluids*, Astarita, G.; Marrucci, G.; Nicolais, L.; Eds.; Plenum: New York, 1980, p. 639.

Munstedt, H., *J. Rheol.* 1980, *24*, 847.

Navarrete, R. C., Ph.D. thesis, University of Minnesota, 1991.

Newton, I. S., *Principia Mathematica*, 1687. Translation of quotation preceding Section 2.1 is from M. Reiner, *Deformation, Strain and Flow*; Wiley-Interscience: New York, 1960.

Papanastasiou, T. C., *J. Rheol.* 1987, *31*, 385.

Pearson, J. R. A., *Mechanics of Polymer Processing*; Elsevier: London, 1984.

Schunk, R.; Scriven, L. E., *J. Rheol.* 1990, *34*, 1085.

Souza Mendes, P. R.; Padmanabhan, M.; Scriven, L. E.; Macosko, C. W., in preparation 1994.

Tadmor, Z.; Gogos, C. G., *Principles of Polymer Processing*; Wiley: New York, 1979.

Tanner, R. I., *Engineering Rheology*, Oxford University Press: London, 1985.

Trouton, F., *Proc. R. Soc.* 1906, *A77*, 326.

Van Krevelen, D. W., *Properties of Polymers*, 2nd ed.; Elsevier: Amsterdam, 1976.

Whitcomb, P.; Macosko, C. W., *J. Rheol.* 1978, *22*, 493.

Yasuda, K.; Armstrong, R. C.; Cohen, R. E., *Rheol. Acta* 1981, *20*, 163.

3

LINEAR VISCOELASTICITY

The state of the solid depends not only on the forces actually impressed on it, but on all the strains to which it has been subjected during its previous existences.
James C. Maxwell (1866)

3.1 Introduction

During the latter half of the nineteenth century, scientists began to note that a number of materials showed time dependence in their elastic response.

When materials like silk, gum rubber, pitch, and even glass were loaded in shear or extension, an instantaneous deformation, as expected for a Hookean solid, was followed by a continuous deformation or "creep." One of the earliest apparatuses used to measure this phenomenon is shown in Figure 3.1.1. When the load was removed, part of the deformation recovered instantly, more recovered with time, and in some materials there was a permanent set.

Figure 3.1.1.
Torsional creep apparatus used by Kolrausch (1863) to study viscoelasticity in glass fibers and rubber threads. The sample was twisted by the lever arm and then released. The time-dependent recovery was recorded by light reflected from the round mirror attached to the bottom of the sample.

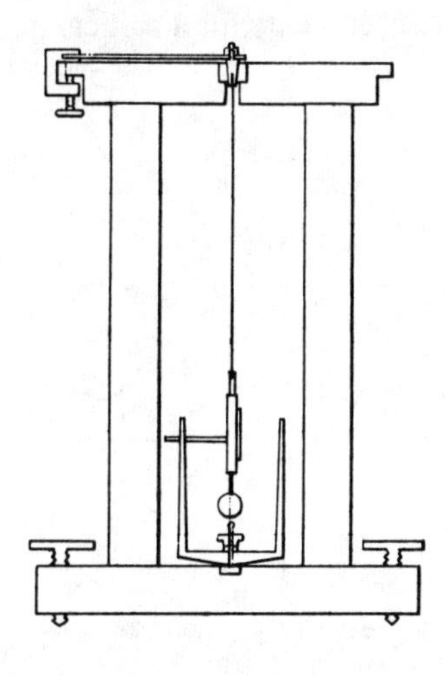

Today we call this time-dependent response *viscoelasticity*. It is typical of all polymeric materials. Another common way to measure the phenomenon is by stress relaxation. As illustrated in Figure 3.1.2, when a polymeric liquid is subject to a step increase in strain, the stress relaxes in an exponential fashion. If a purely viscous liquid is subjected to the same deformation, the stress relaxes instantly to zero as soon as the strain becomes constant. An elastic solid would show no relaxation.

If we convert stress relaxation data to a relaxation modulus

$$G(t) = \frac{\tau(t)}{\gamma} \tag{3.1.1}$$

all the data for small strains, typically $\gamma < \gamma_c \simeq 0.5$ for polymeric liquids, fall on the same curve. This is shown as a log–log plot in Figure 3.1.3b. Note that at short times the relaxation modulus approaches a constant value known as the plateau modulus G_e. This linear dependence of stress relaxation on strain, eq. 3.1.1, is called *linear viscoelasticity*. For larger strains, $\gamma > \gamma_c$, the relaxation modulus is no longer independent of strain, as illustrated in Figure 3.1.3b.

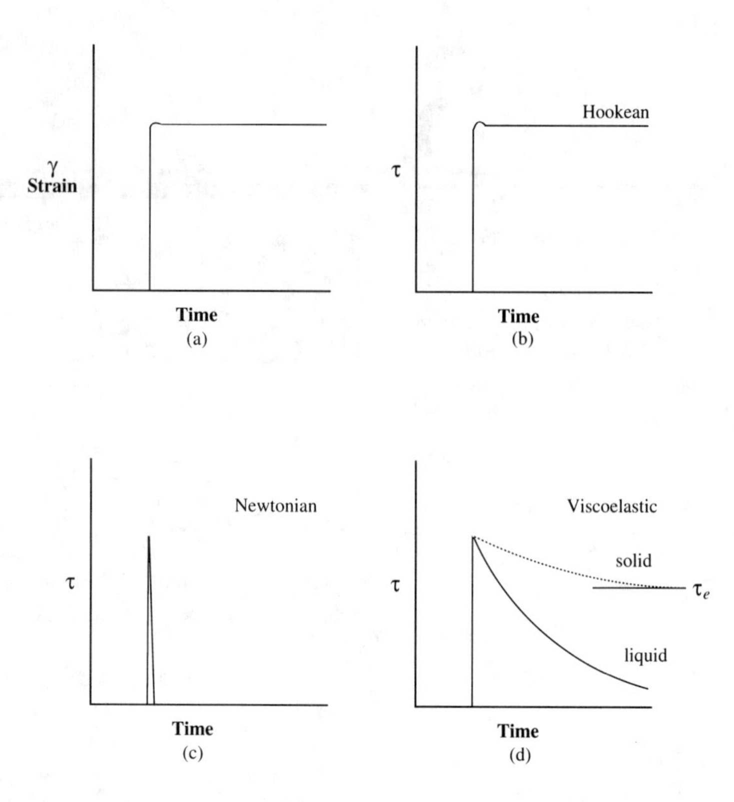

Figure 3.1.2.
Stress response τ versus time for a step input in strain γ. The Hookean solid (b) shows no stress relaxation; the Newtonian fluid (c) relaxes as soon as the strain is constant, while the viscoelastic liquid or solid shows stress relaxation over a significant time. In a viscoelastic liquid the stress relaxes to zero, while for the viscoelastic solid it asymptotically approaches an equilibrium stress τ_e. A small overshoot is shown in the strain versus time plot (a). This is typical of actual control systems, which may require 0.01 second or more to stabilize (see Chapter 8).

Figure 3.1.3a shows typical stress relaxation data for increasing strain magnitudes. We can convert this data into a relaxation modulus

$$G(t, \gamma) = \frac{\tau(t, \gamma)}{\gamma} \quad \text{for } \gamma > 1 \tag{3.1.2}$$

This is known as *nonlinear* viscoelastic behavior, and it is the subject of the next chapter. Here we consider only relatively small strains, such that the relaxation modulus is independent of strain. Note that we can also define linear viscoelasticity for a particular

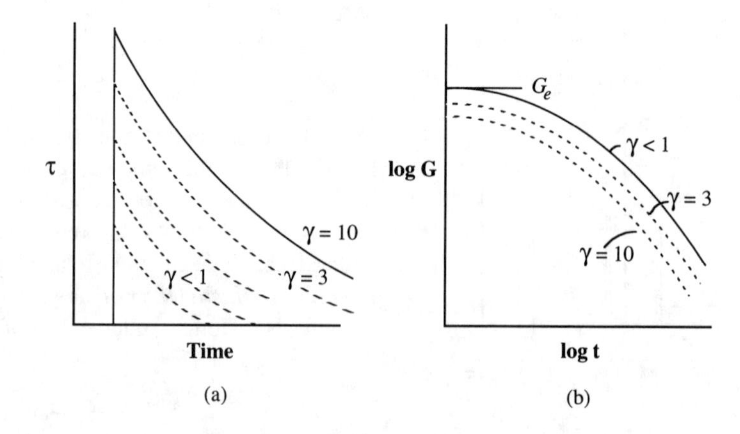

Figure 3.1.3.
Stress relaxation data for a polydimethylsiloxane sample. As in Figure 3.1.2a, shear strain is stepped nearly instantly to a constant value. (a) Shear stress jumps and then decays exponentially. (b) When a log of the time-dependent shear modulus, $G(t) = \tau(t)/\gamma$, is plotted versus log time, all the data at small strain fall on the same curve.

material as a region of stress in which strain varies linearly with stress.

Because we are dealing with relatively small strains, shear normal stresses and the type of deformation—for example, shear versus uniaxial extension—will not be important. To impose a small strain on a material in one direction only requires exerting a stress in that direction (Lockett, 1972). Thus, we could just as well define a tensile relaxation modulus from tensile stress relaxation (recall eq. 1.5.10)

$$E(t) = \frac{(T_{11} - T_{22})(t)}{\epsilon} \tag{3.1.3}$$

but for small strains this will be just three times the shear relaxation modulus

$$E(t) = 3G(t) \tag{3.1.4}$$

Thus in all our discussions for simple shear we must realize that the models can be applied to other types of deformation. In the next section we develop a general model for linear viscoelastic behavior in only one dimension. Then we extend it to three dimensions, and in Section 3.3 we examine its behavior for different deformation histories: stress relaxation, creep, and sinusoidal oscillation.

3.2 General Linear Viscoelastic Model

In an attempt to model the early experiments on viscoelastic solids, Boltzmann (1876) suggested that small changes in stress equal small changes in the modulus times the strain

$$d\tau = \gamma dG \tag{3.2.1}$$

We can define a new function, the *memory function*, as the time derivative of G

$$M(t) = -\frac{dG(t)}{dt} \tag{3.2.2}$$

Since the relaxation modulus $G(t)$ decreases with time, the derivative will be negative. Thus the minus sign is added to make $M(t)$ a positive function.

With $M(t)$ we can rewrite eq. 3.2.1 as

$$d\tau = -M\gamma\, dt \tag{3.2.3}$$

If the relaxation modulus (and thus the memory function) depend only on time, we can make up any larger deformation (but deformations still within the linear range of the material) by summing up all the small deformations. This can be expressed as the integral over all past time, as suggested by the quotation from Maxwell at the opening of this chapter

$$\int_0^\tau d\tau = \tau = -\int_{-\infty}^{t} M(t-t')\gamma(t')dt' \tag{3.2.4}$$

where t' is the past time variable running from the infinite past $-\infty$ to the present time t (also note Figure 1.4.1 for a definition of t'). The memory function M depends only on the elapsed time $t - t'$ between the remembered past and the present. Often the elapsed time is denoted by $s = t - t'$ so that eq. 3.2.4 becomes

$$\tau(t) = -\int_0^\infty M(s)\gamma(t-s)ds \tag{3.2.5}$$

Equation 3.2.5 is a one-dimensional constitutive model for linear viscoelastic behavior. We can also write the model directly in terms of the relaxation modulus. Consider a small change in stress due to a change in strain

$$d\tau = G\,d\gamma \tag{3.2.6}$$

which can also be written

$$d\tau = G\frac{d\gamma}{dt}dt = G\dot{\gamma}\,dt$$

Integrating this expression gives

$$\tau = \int_{-\infty}^{t} G(t-t')\dot{\gamma}(t')dt' \tag{3.2.7}$$

or in terms of s

$$\tau = \int_0^\infty G(s)\,\dot{\gamma}(t-s)ds \qquad s = t - t'$$

Thus the stress is an integral, over all past time, of the relaxation modulus times the rate of strain. Since the deformation might be changing with time, $\dot{\gamma}$ is a function of time. This is the form that is most frequently used because $G(t)$ can be measured directly.

What form does the relaxation modulus function have? Figure 3.1.3 indicates that we should try an exponential decay

$$G(t) = G_0\,e^{-t/\lambda} \tag{3.2.8}$$

where λ is the relaxation time. Substituting this into eq. 3.2.7 gives the single relaxation or simple Maxwell model

$$\tau = \int_{-\infty}^{t} G_0 \, e^{-(t-t')/\lambda} \dot{\gamma}(t') dt' \tag{3.2.9}$$

Figure 3.2.1a shows that this model is qualitatively reasonable, but does not fit typical data very well. A logical improvement on this model is to try several relaxation times. This can be written as a series of relaxation times λ_k multiplied by the weighting constants G_k

$$G(t) = \sum_{k=1}^{N} G_k \, e^{-t/\lambda_k} \tag{3.2.10}$$

When substituted into eq. 3.2.7, this gives the *general linear viscoelastic* model

$$\tau = \int_{-\infty}^{t} \sum_{k=1}^{N} G_k \, e^{-(t-t')/\lambda_k} \, \dot{\gamma}(t') dt' \tag{3.2.11}$$

We see that the relaxation data for polydimethylsiloxane in Figure 3.2.1 can be fit quite well with only five relaxation times. The disadvantage of this approach is that a unique set of relaxation times and weighting constants may not be available. The constants in Table 3.2.1 were determined by first selecting a set of relaxation times evenly spaced on a log scale, in this case one per decade, and then running a linear regression program to determine the G_k values that minimize the least square of the deviation of eq. 3.2.11 from the data. Appendix 3A has more information on

Figure 3.2.1.
Comparison of experimental relaxation modulus data on a polydimethylsiloxane sample to (a) the single exponential model, eq. 3.2.8, with $\lambda = 0.1$ s and $G_0 = 10^5$ Pa and (b) a five-constant model, eq. 3.2.10, with the constants given in Table 3.2.1.

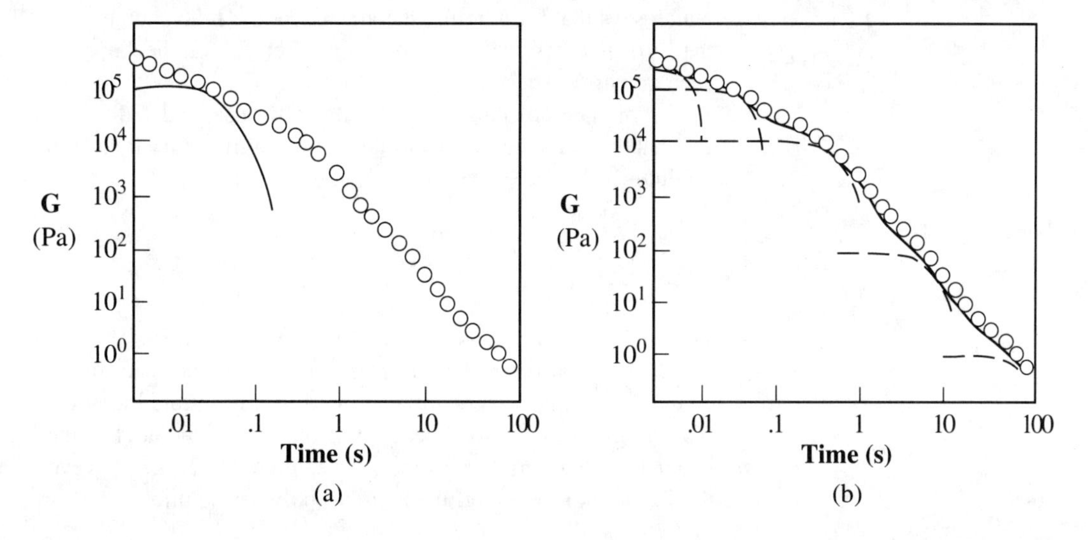

TABLE 3.2.1 / Relaxation Times for Polydimethylsiloxane at 25°C

k	λ_k (s)	G_k (Pa)
1	0.01	2×10^5
2	0.1	10^5
3	1.0	10^4
4	10	10^2
5	100	10

the calculation of these constants from $G(t)$ data, including a non-linear regression method.

The data in Figure 3.2.1 cover about five decades of time. Three to five decades are all that are practical for a typical single viscoelastic experiment. However, to fully describe $G(t)$ we often need a wider range. This can be accomplished by collecting data at different temperatures and shifting it to one reference temperature. This idea of time–temperature superposition has been described, eq. 2.6.9, with respect to viscosity and is also illustrated in Chapter 11.6. Equation 2.6.9 holds for any of the viscoelastic functions, for example, $G(t)$.

$$\log \frac{G(t, T)T_r\rho_r}{G(t, T_r)T\rho} = \log a_T = \frac{-C_1(T - T_r)}{C_2 + T - T_r} \tag{2.6.9}$$

The shift factors a_T used to construct Figures 3.3.7 and 3A.2 were determined with eq. 2.6.9. Table 11.6.1 gives C_1 and C_2 values for many polymers.

A useful test for the $G(t)$ function determined by fitting procedures is the relation

$$\eta_o = \int_o^\infty G(t)dt \tag{3.2.12}$$

which says that for a liquid, the area under the relaxation curve is the zero shear viscosity. The relation is helpful for checking λ_i at each end of the spectra.

Polymer molecular theories, like the Rouse model, discussed in Chapter 11, suggest an infinite series form for the relaxation modulus

$$G_k = G_0 \frac{\lambda_k}{\sum_k \lambda_k}; \quad \lambda_k = \frac{\lambda_0}{k^2} \tag{3.2.13}$$

Thus only two constants must be fit from the data, and the form for $G(s)$ is prescribed. Unfortunately, this model does not fit data on high molecular weight polymer melts or concentrated solutions. Its response for the polydimethylsiloxane data would not be much better than the single relaxation time, Figure 3.2.1a. However, the Rouse model is useful for dilute polymer solutions and low molecular weight systems. We examine its response and other molecular models in Chapter 11.

The stress relaxation behavior of a viscoelastic solid is shown in Figure 3.1.2. We can see that instead of relaxing to zero, the stress goes to some equilibrium value τ_e. This can readily be treated with the models already discussed by simply adding on a constant equilibrium modulus G_e. For example, eq. 3.2.11 for a viscoelastic solid becomes

$$\tau = \int_{-\infty}^{t} \left[\sum_{k=1}^{N} G_k \, e^{-(t-t')/\lambda_k} + G_e \right] \dot{\gamma}(t') dt' \qquad (3.2.11a)$$

3.2.1 Relaxation Spectrum

Another approach to linear viscoelasticity that has been widely used in the past is the relaxation spectrum $H(\lambda)$. Using it provides a continuous function of relaxation time λ rather than a discrete set. The relation between the relaxation modulus and the spectra is

$$G(s) = \int_{0}^{\infty} \frac{H(\lambda)}{\lambda} e^{-s/\lambda} d\lambda \qquad (3.2.14)$$

or for a narrow spacing of relaxation times

$$G_k \cong H(\lambda_k) \ln \frac{\lambda_k}{\lambda_{kH}}$$

There are a number of forms for $H(\lambda)$, and much of the literature is in terms of it (Ferry, 1980; note also Exercise 3.4.1 and eq. 3.4.2.).

3.2.2 Linear Viscoelasticity in Three Dimensions

We can extend eq. 3.2.7 to three dimensions by using the extra stress tensor $\boldsymbol{\tau}$ for the shear stress and the rate of deformation tensor $2\mathbf{D}$ for $\dot{\gamma}$.

$$\boldsymbol{\tau} = \int_{-\infty}^{t} G(t - t') \; 2\mathbf{D}(t') \; dt' \qquad (3.2.15)$$

As we saw in Example 2.2.2 for simple shear there are only two components of **2D**, so $2D_{ij} = 2D_{12} = 2D_{21} = \dot{\gamma}$. Thus the linear viscoelastic model does not predict normal stresses in steady shear flow. The viscosity is independent of $\dot{\gamma}$, but τ_{12} is time dependent.

Example 3.2.1 Stress Growth in Steady Shear Flow

Consider the situation in which a fluid at rest is suddenly set into simple shear flow for $t > 0$. Let the velocity gradient be $\dot{\gamma}_0$ for $t > 0$. Using the general viscoelastic model:
(a) Determine η^+, the time-dependent viscosity

(b) Show that at steady state viscosity η_0 is independent of shear rate
(c) Show that the relaxation modulus can be calculated from the start-up viscosity by $G(t) = d\eta^+(t)/dt$.

Solution

For $t \leq 0 \quad \tau_{12} = 0$

For $t > 0 \quad \tau_{12} = \dot{\gamma}_0 \int_0^t G(t - t')dt' = \dot{\gamma}_0 \int_0^t G(s)ds$

$$(a)\ \eta^+(t) \equiv \frac{\tau_{12}(t)}{\dot{\gamma}_0} = \int_0^t G(t - t')dt' = \int_0^t G(s)ds \tag{3.2.16}$$

$$= \int_0^t \sum_{k=1}^N G_k e^{-s/\lambda_k} ds = \sum_{k=1}^N G_k \lambda_k (1 - e^{-t/\lambda_k})$$

(b)

$$\eta_0 = \lim_{t\to\infty} \eta^+(t) = \int_0^\infty G(t - t')dt'$$

$$= \lim_{t\to\infty} \sum_{k=1}^N G_k \lambda_k (1 - e^{-t/\lambda_k}) = \sum_{k=1}^N G_k \lambda_k$$

Since G_k and λ_k are constants, η_0 is constant. Note that the viscosity η_0 is simply the area under the *G(s)* curve, eq. 3.2.12.

(c) From the stress equation, eq. 3.2.16,

$$G(s) = \frac{d\eta^+(s)}{ds} \tag{3.2.17}$$

3.2.3 Differential Form

We can also write the concept of linear viscoelasticity in a differential form

$$\tau + \lambda \frac{d\tau}{dt} = \eta \dot{\gamma} \tag{3.2.18}$$

This result is equivalent to eq. 3.2.9, which can be shown by determining the time derivative of eq. 3.2.9

$$\frac{d\tau}{dt} = -\int_{-\infty}^t \frac{G_0}{\lambda} e^{-(t-t')/\lambda} \dot{\gamma}(t')dt' + G_0 \dot{\gamma}(t) \tag{3.2.19}$$

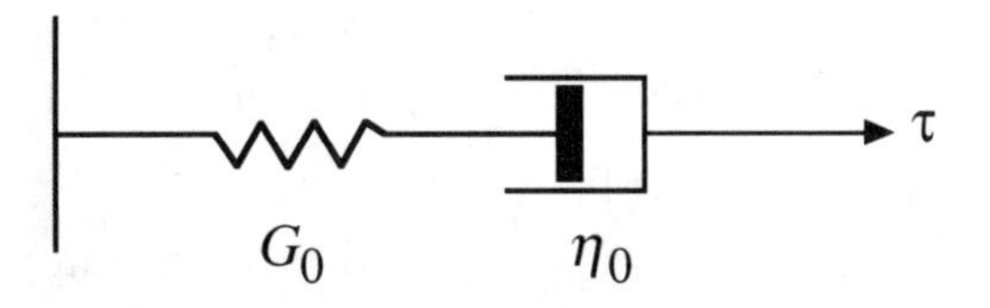

Figure 3.2.2.
Spring and dashpot representation of a Maxwell element.

Multiplying this result by λ and combining with eq. 3.2.9, the integrals cancel, giving

$$\tau + \lambda \frac{d\tau}{dt} = \lambda G_0 \dot{\gamma} = \eta \dot{\gamma} \tag{3.2.20}$$

since $\eta = G_0 \lambda$.

Maxwell (1867) first proposed this equation for the viscosity of gases! Despite his initial misapplication of a good theory, rheologists have forgiven him and embrace eq. 3.2.18 as the Maxwell model. It is often represented as a series combination of springs, elastic elements, and dashpots, viscous ones as shown in Figure 3.2.2.
We see from eq. 3.2.18 and from the spring and dashpot representation that for slow motions the dashpot or Newtonian behavior dominates. For rapidly changing stresses, the derivative term dominates, and thus at short times the model approaches elastic behavior (recall Figure 3.1.2 and the bouncing putty in Figure I.1).

In the next chapter we discuss differential models further. We will find that for large strains we need to use another type of time derivative.

3.3 Small Strain Material Functions

A number of small strain experiments are used in rheology. Some of the more common techniques are stress relaxation, creep, and sinusoidal oscillations. In the linear viscoelastic region all small strain experiments must be related to one another through $G(t)$, as indicated by the basic constitutive equation, eq. 3.2.7, or through $M(t)$, eq. 3.2.4. Different experimental methods are used because they may be more convenient or better suited for a particular material or because they provide data over a particular time range. Furthermore, it is often not easy to transform results from one type of linear viscoelastic experiment to another. For example, transformation from the creep compliance $J(t)$ to the stress relaxation modulus $G(t)$ is generally difficult. Thus both, functions are often measured.

Sections 3.3.1–3.3.3 describe each of these small strain material functions and show typical data for several rheologically different materials: lightly crosslinked rubber, polymer melts, dilute polymer solutions, and suspensions.

3.3.1 Stress Relaxation

As discussed in Section 3.2, stress relaxation after a step strain γ_0 is the fundamental way in which we define the relaxation modulus. The relaxing stress data illustrated in Figures 3.1.2 and 3.1.3 can be used to determine $G(t)$ directly

$$G(t) = \frac{\tau(t)}{\gamma_0} \tag{3.3.1}$$

Typical relaxation modulus data are illustrated in Figure 3.3.1 for several materials. We see that crosslinked rubber shows a short time relaxation followed by a constant modulus. Concentrated suspensions show the same qualitative response but only at very small strain. High molecular weight concentrated polymeric liquids show behavior similar to rubber at shorter times with a nearly constant modulus plateau G_o, eventually followed by flow at long times. Dilute solutions and suspensions show complete relaxation in short times. Molecular weight distribution (MWD) and long chain branching of the polymer also have a strong effect on the long time relaxation of $G(t)$. This is discussed further in Chapter 11.

Experimentally it is impossible to jump the strain γ instantaneously to γ_0. This is illustrated in Figure 3.1.2. A short rise and stabilization time, typically 0.01–0.1 s, is required for current instruments, as discussed in Chapter 8; therefore, it is difficult to get data for $t < 0.1$ s by stress relaxation methods. Also, it is very difficult to measure a decaying stress over more than three decades with one transducer. Thus methods are needed that can

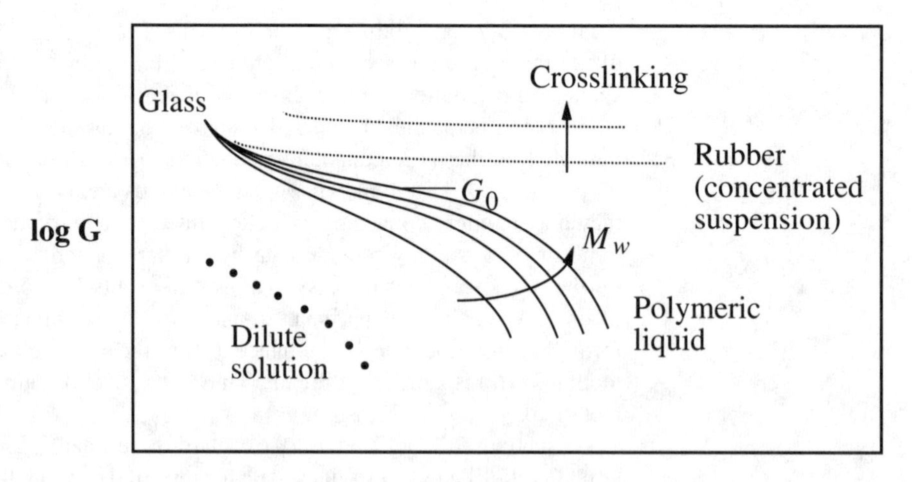

Figure 3.3.1. Typical relaxation modulus data for several different materials.

better measure the short time and long time ends of the relaxation spectra.

3.3.2 Creep

Creep experiments are particularly useful for the long time end of the relaxation spectra. As illustrated in Figures 3.1.3 and 3.3.2, in a creep experiment the stress is increased instantly from 0 to τ_0 and the strain is recorded versus time.

Typical creep data are shown in Figure 3.3.2. We note that rubber shows only short-term creep, while polymeric liquids continue to deform, eventually reaching a steady rate of straining $\dot{\gamma}_\infty$. Data are usually expressed in terms of $J(t)$, the compliance

$$J(t) = \frac{\gamma(t)}{\tau_0} \tag{3.3.2}$$

J has the units of reciprocal modulus but in general it does not equal $1/G$. However, because in the linear viscoelastic regime strain is linear with stress, strain versus time data at different τ_0 collapse into one $J(t)$ plot. This is analogous to the reduction of stress relaxation curves to one $G(t)$ plot.

As indicated in Figure 3.3.2c, a *steady state creep compliance* J_e^0 is defined by extrapolation of the limiting slope to $t = 0$. The slope is the inverse of the viscosity at low shear rate, η_o. Thus, in the steady creeping regime, we have

$$J(t) = \frac{\gamma_0}{\tau_0} + \frac{t\dot{\gamma}_\infty}{\tau_0} \tag{3.3.3}$$

or

$$J(t) = J_e^0 + \frac{t}{\eta_0} \tag{3.3.4}$$

Figure 3.3.2.
Creep experiment. (a) Stress is increased from 0 to τ_0 at $t = 0$. (b) Strain is recorded versus time. (c) Data are usually plotted as creep compliance.

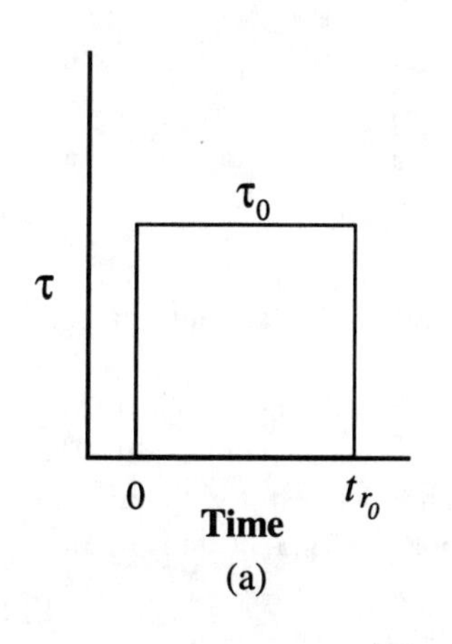

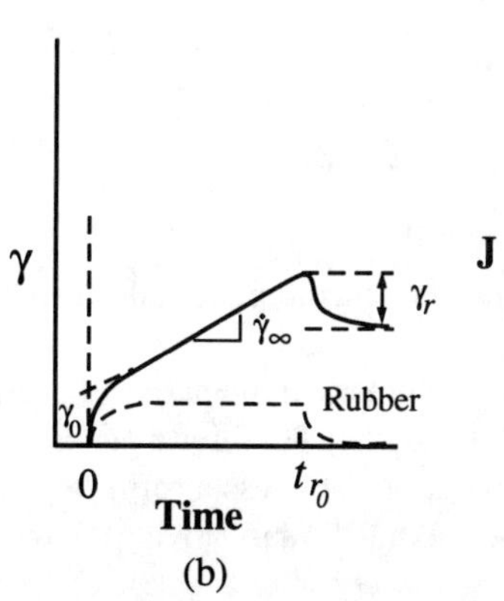

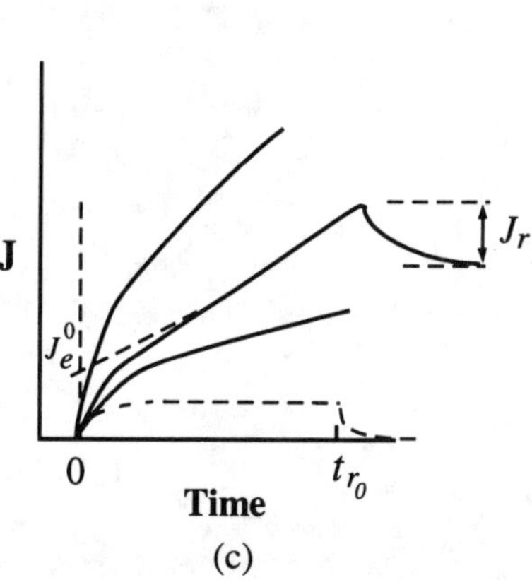

From Figure 3.3.25 we note that as molecular weight (or concentration) increases, there is less creep, lower J, and higher η_0. For rubber, which does not flow, $\eta_0 \to \infty$.

If we apply to the creep experiment the single relaxation time model, the simple Maxwell model, eq. 3.2.9, or eq. 3.2.16, we obtain

$$J(t) = \frac{1}{G_0} + \frac{t}{\eta} = \frac{1}{G_0} + \frac{t}{\lambda G_0} \tag{3.3.5}$$

Nevertheless, for the more general model it is not simple to calculate the compliance. Since τ is the independent variable, we must invert the integral in eq. 3.2.7.

$$\tau_0 = \int_{-\infty}^{t} G(t - t')\,\dot{\gamma}(t')dt' \tag{3.3.6}$$

However, we can look at some limiting relations. At long time the shear rate becomes steady $\dot{\gamma}(t) = \dot{\gamma}_\infty$, and we can write

$$\tau_0 = \int_{-\infty}^{t} G(t - t')\,\dot{\gamma}_\infty dt' \tag{3.3.7}$$

Equating these two expressions after some manipulation, we can obtain (Bird et al., 1987, p. 287)

$$\frac{\gamma_0}{\dot{\gamma}_\infty} = \lambda_0, \text{ longest relaxation time} = \frac{\int_0^\infty sG(s)ds}{\int_0^\infty G(s)ds} \tag{3.3.8}$$

or since $\gamma_0 = J_e^0 \tau_0$, then (3.3.9)

$$J_e^0 = \frac{\int_0^\infty sG(s)ds}{\left[\int_0^\infty G(s)ds\right]^2} \tag{3.3.10}$$

Thus the relaxation modulus can be used to calculate limiting portions of the creep wave.

Numerical methods are needed to actually obtain $G(t)$ from $\gamma(t)$ using eq. 3.3.6 for a general linear viscoelastic material. The program must guess a form for $G(t)$ and then test it with the $\gamma(t)$ data, taking the derivative $\dot{\gamma}(t)$, then applying eq. 3.3.6 and iterating to reduce the error. A helpful relation in this context is (Ferry, 1980, p. 68)

$$\int_0^t G(s)J(t-s)ds = t \tag{3.3.11}$$

When $J(t)$ is written as a discrete spectra

$$J(t) = \frac{t}{\eta_0} + \sum_k J_k(1 - e^{-t/\lambda_k}) \tag{3.3.12}$$

explicit relations between the coefficients of $G(t)$ can be derived (Ferry, 1980; Baumgaertel and Winder, 1989)

Another type of experiment often done in conjunction with creep is *creep recovery*, the recoil of strain after the stress is removed, as illustrated in Figure 3.3.2. After the stress has been removed from a viscoelastic material, the deformation reverses itself. We can define a recoverable creep function

$$J_r(t) = \frac{\gamma_r(t)}{\tau_0} \tag{3.3.13}$$

where τ_0 is the stress before recovery starts. If recovery is performed after steady state creep, $\dot{\gamma}(t) = \dot{\gamma}_\infty$ and the equilibrium creep recovery directly measures J_e^0

$$\lim_{t\to\infty} J_r(t) = J_e^0 \; for \; \dot{\gamma}(t) = \dot{\gamma}_\infty \tag{3.3.14}$$

If steady state creep has truly been achieved, then eqs. 3.3.14 and 3.3.4 should give the same result. In fact this equivalence is used to test for achievement of steady state and accurate determination of J_e^0. For high molecular weight and broad distribution polymer melts, the relaxation times are very long, on the order of tens of minutes, and accurately measuring J_e^0 can be difficult.

3.3.3 Sinusoidal Oscillations

In another important small strain experiment, the sample is deformed sinusoidally. Within a few cycles of start-up and often much less, the stress will also oscillate sinusoidally at the same frequency but in general will be shifted by a phase angle δ with respect to the strain wave. This is illustrated in Figure 3.3.3 and expressed mathematically as follows:

$$\gamma = \gamma_0 \sin \omega t \tag{3.3.15}$$

$$\tau = \tau_0 \sin(\omega t + \delta) \tag{3.3.16}$$

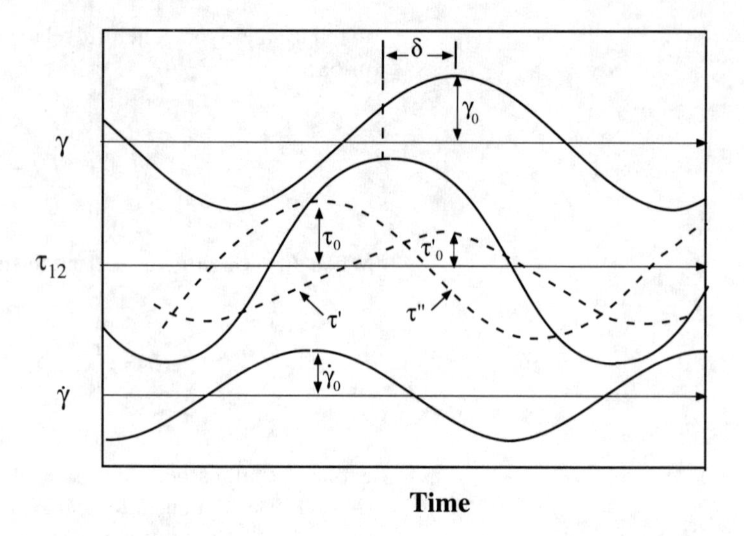

Figure 3.3.3. Sinusoidally oscillating shear strain produces a sinusoidal stress phase shifted by an amount δ. For analysis the stress wave is broken down into two waves, τ' in phase with γ and τ'' 90° out of phase. Note that τ'' is in phase with the rate of strain wave $\dot{\gamma} = d\gamma/dt$.

Such data are analyzed by decomposing the stress wave into two waves of the same frequency, one in phase with the strain wave (sin ωt) and one 90° out of phase with this wave (cos ωt). Thus,

$$\tau = \tau' + \tau'' = \tau_0' \sin \omega t + \tau_0'' \cos \omega t \tag{3.3.17}$$

As you can try for yourself, trigonometry shows that

$$\tan \delta = \frac{\tau_0''}{\tau_0'} \tag{3.3.18}$$

This decomposition suggests two dynamic moduli

$$G' = \frac{\tau_0'}{\gamma_0}, \text{ the in–phase or elastic modulus} \tag{3.3.19}$$

and

$$G'' = \frac{\tau_0''}{\gamma_0}, \text{ the out–of–phase, viscous, or loss modulus} \tag{3.3.20}$$

From eq. 3.3.18 we can also write

$$\tan \delta = \frac{G''}{G'} \tag{3.3.21}$$

The prime and double prime notation has its origin in complex numbers. Recall that

$$e^{i\theta} = \cos\theta + i \sin\theta \quad \text{where } i = \sqrt{-1} \tag{3.3.22}$$

Then we can represent γ as the imaginary part of the complex number $\gamma_0 e^{i\omega t}$ and likewise $\tau' = \mathrm{Im}\{\tau_0 e^{i\omega t}\}$ and $\tau'' = \mathrm{Re}\{\tau_0 e^{i\omega t}\}$. Then we can define G^* such that

$$\tau_0 = |G^*|\gamma_0 \tag{3.3.23}$$

where G^* is a complex number with G' as its real and G'' as its imaginary parts, respectively

$$G^* = G' + iG'' \tag{3.3.24}$$

or

$$\tau = G'\gamma_0 \sin \omega t + G''\gamma_0 \cos \omega t$$

However, there is nothing physically "imaginary about" G'' and, in fact, it is a measure of the energy dissipated per cycle of deformation per unit volume (see Exercise 3.4.4).

$$\begin{matrix}\text{energy} \\ \text{dissipated}\end{matrix} = \int_0^t \boldsymbol{\tau} : \mathbf{D}\, dt = \int_0^{2\pi/\omega} \tau\dot{\gamma}\, dt = \pi G''\gamma_0^2 \tag{3.3.25}$$

Sometimes the magnitude of the complex modulus is reported

$$|G^*| = (G'^2 + G''^2)^{1/2} = \frac{\tau_0}{\gamma_0} \tag{3.3.26}$$

Another way to view the same experiments is in terms of a sinusoidal strain rate. Then a *dynamic viscosity* material function is defined (Bird, et al., 1987). This may be more comfortable for those dealing with liquids, but we can readily convert from one to the other. Noting that the derivative of the small strain is the strain rate, we have

$$\dot{\gamma} = \frac{d\gamma}{dt} = \gamma_0\omega \cos \omega t = \dot{\gamma}_0 \cos \omega t \tag{3.3.27}$$

If we decompose the stress again, the τ'' wave will be in phase with the strain rate as shown in Figure 3.3.3. If we had started from the strain rate viewpoint, we would have called this the τ' wave, following the complex number ideas discussed above. But let us not switch notation in midstream! From the magnitudes of the viscous stress to the strain rate, we can define a dynamic viscosity

$$\eta' = \frac{\tau_0''}{\dot{\gamma}_0} = \frac{G''}{\omega} \tag{3.3.28}$$

We can also define an elastic part of the complex viscosity

$$\eta'' = \frac{\tau_0'}{\dot{\gamma}_0} = \frac{G'}{\omega} \tag{3.3.29}$$

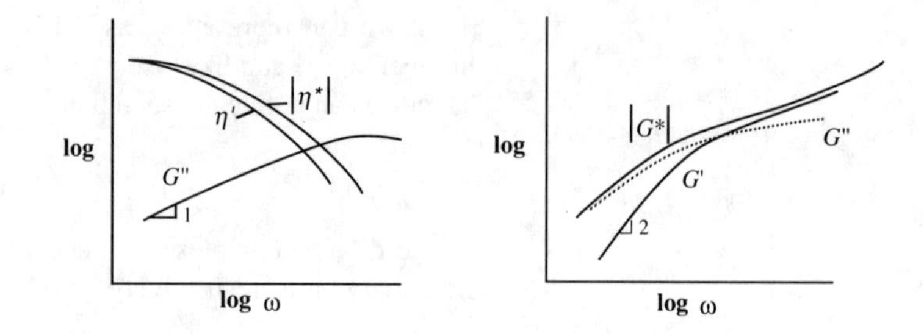

Figure 3.3.4.
Typical results from sinusoidal experiments on a polymer melt plotted as moduli G^*, G', G'' and viscosities η^*, η', η''.

where the magnitude of the complex viscosity is

$$|\eta^*| = (\eta'^2 + \eta''^2)^{1/2} = \left[\left(\frac{G''}{\omega}\right)^2 + \left(\frac{G'}{\omega}\right)^2\right]^{1/2} = \frac{1}{\omega}|G^*| \quad (3.3.30)$$

Plots of η', η^*, and G'' versus ω and of G'', G^*, and G' versus ω are shown in Figure 3.3.4 for a typical polymer melt.

Let us look at typical behavior of these material functions. In Figure 3.3.5 we see that G' versus ω looks similar to G versus $1/t$ from Figure 3.3.1. For rubber it becomes constant at low frequency (long times), and for concentrated polymeric liquids it shows the plateau modulus G_e and decreases with ω^{-2} in the limit of low frequency. The loss modulus is much lower than G' for a crosslinked rubber and sometimes can show a local maximum. This maximum is more pronounced in polymeric liquids, especially for narrow molecular weight distribution. The same features are present in dilute suspensions of rodlike particles, but not for dilute random coil polymer solutions, as Figure 3.3.3b shows. These applications of the dynamic moduli to structural characterization are discussed in Chapters 10 and 11.

Just as creep behavior must be a function of $G(t)$ in the linear region, so must sinusoidal response. If we apply the general linear viscoelastic model of eq. 3.2.11 (see Exercise 3.4.3 for a derivation), we obtain

$$G'(\omega) = \sum_k G_k \frac{\omega^2\lambda_k^2}{1 + \omega^2\lambda_k^2} \quad (3.3.31)$$

$$G''(\omega) = \sum_k G_k \frac{\omega\lambda_k}{1 + \omega^2\lambda_k^2} \quad (3.3.32)$$

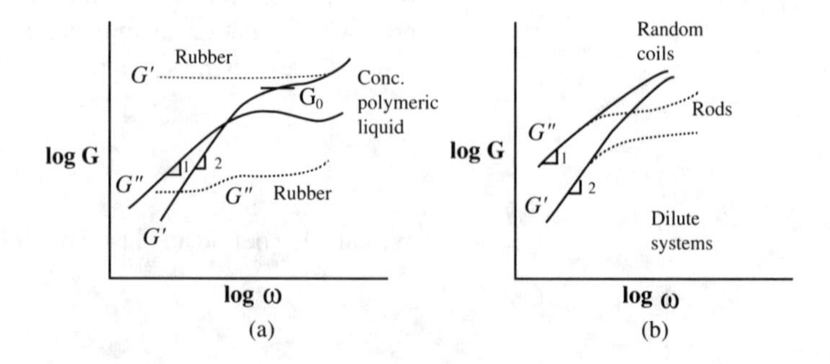

Figure 3.3.5.
Typical behavior of the dynamic moduli for (a) rubber and concentrated polymeric liquids and (b) dilute solution of random coils and rodlike particles.

Figure 3.3.6.
Shear relaxation modulus versus time for a low density polyethylene at 150°C. The heavy line is the sum of the eight exponential relaxation times given in Table 3.3.1. Replotted from Laun, (1978).

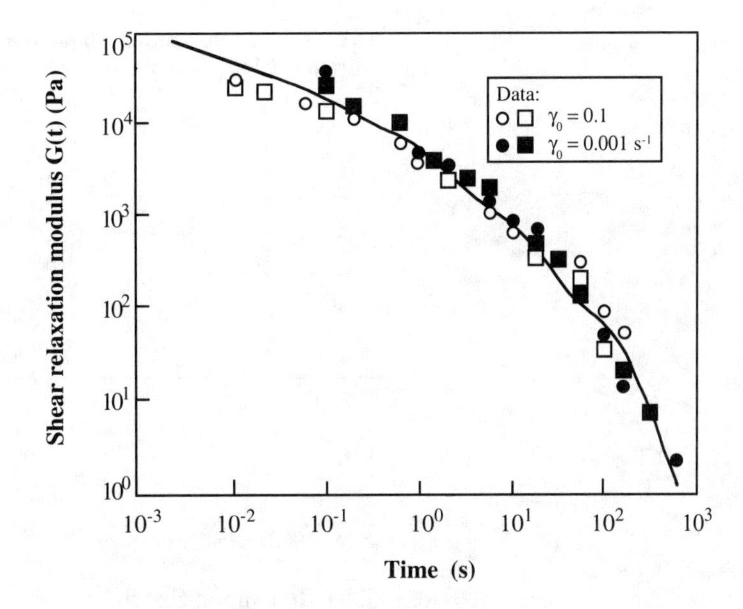

Examples of the fit of the same $G(t)$ function to both stress relaxation and sinusoidal results appear in Figures 3.3.6 and 3.3.7. Table 3.3.1 gives the relaxation times and moduli used to fit all these data. Figure 3.3.7 illustrates that short time (high frequency) data are more readily obtained with sinusoidal methods.

It is possible to define other small strain material functions, such as stress growth under constant rate of straining (Example 3.2.1) or recoverable strain after constant strain rate. However, these deformation histories are better suited for large strain studies and are discussed in Section 3.2. The small strain material functions will be seen as limits of the large strain ones. Table 3.3.2 lists some of the interrelations between the various experiments for linear viscoelastic behavior. Note that the limiting low shear rate viscosity η_0 can be calculated from

$$\eta_0 = \frac{\lambda_0}{J_e^o} \tag{3.3.33}$$

Figure 3.3.7.
Dynamic shear moduli for the same low density polyethylene as in Figure 3.3.6. Data were collected at different temperatures and shifted according to eq. 2.6.9. Lines calculated from $G(t)$ using eqs. 3.3.31 and 3.3.32, replotted from Laun (1978).

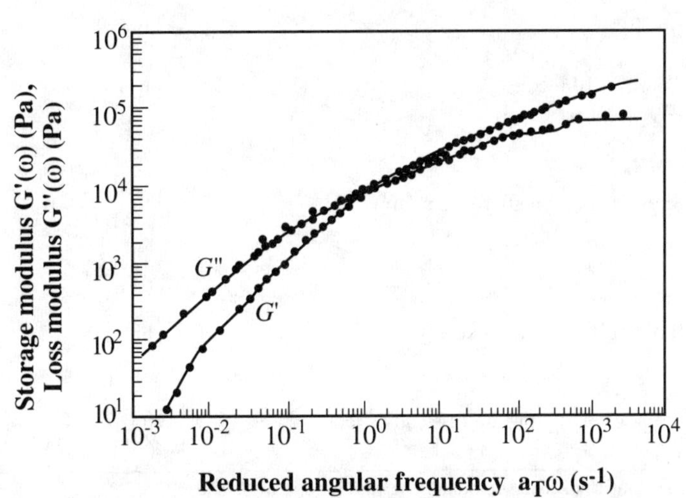

TABLE 3.3.1 / Relaxation Times and Moduli for Low Density Polyethylene at 150°C

k	$\lambda_k(s)$	$G_k(Pa)$
1	10^3	1.00
2	10^2	1.80×10^2
3	10^1	1.89×10^3
4	10^0	9.80×10^3
5	10^{-1}	2.67×10^4
6	10^{-2}	5.86×10^4
7	10^{-3}	9.48×10^4
8	10^{-4}	1.29×10^5

From Laun, (1978).

3.4 Exercises

3.4.1 Relaxation Spectrum

Using the definition of the relaxation spectrum, eq. 3.2.14, show that

$$\eta'(\omega) = \frac{G''(\omega)}{\omega} = \int_0^\infty G(s) \cos \omega s \, ds = \int_0^\infty \frac{H(\lambda)d\lambda}{1 + (\lambda\omega)^2}$$

TABLE 3.3.2 / Limiting Relations for Linear Viscoelasticity

Property	*Equilibrium Creep Compliance* $J_e^0 = \lambda_0/\eta_0$	*Longest Relaxation Time* λ_0
Steady shear[a]	$\lim_{\dot\gamma \to 0} \frac{N_1}{2\tau_{12}^2}$	$\lim_{\dot\gamma \to 0} \frac{N_1}{2\tau_{12}\dot\gamma}$
Sinusoidal oscillations	$\lim_{\omega \to 0} \frac{G'}{G''^2}$	$\lim_{\omega \to 0} \frac{G'}{G''\omega}$
Creep	$\lim_{\tau_0 \to 0} \frac{\gamma_0}{\tau_0}$	$\lim_{\tau_0 \to 0} \frac{\gamma_0}{\dot\gamma_\infty}$
Constrained recoil	$\lim_{\tau_0 \to 0} \frac{\gamma_r}{\tau_0}$	$\lim_{\tau_0 \to 0} \frac{\gamma_r}{\dot\gamma_\infty}$
Stress relaxation after step strain	$\lim_{\gamma_0 \to 0} \frac{\gamma_0 \int_0^\infty t\tau(t)dt}{\left[\int_0^\infty \tau(t)dt\right]^2}$	$\lim_{\gamma_0 \to 0} \frac{\int_0^\infty t\tau(t)dt}{\int_0^\infty \tau(t)dt}$
Stress relaxation after steady shearing	$\lim_{\dot\gamma \to 0} \int_0^\infty \frac{\dot\gamma_0 \tau(\tau)dt}{\tau_0^2}$	$\lim_{\dot\gamma \to 0} \int_0^\infty \frac{\tau(t)}{\tau_0} dt$
Stress growth (start-up of steady shear)	$\lim_{\dot\gamma \to 0} \int_0^\infty \frac{\dot\gamma_0(\tau_0 - \tau(t))}{\tau_0^2} dt$	$\lim_{\dot\gamma \to 0} \int_0^\infty \frac{\tau_0 - \tau(t)}{\tau_0} dt$

[a] The linear viscoelastic model does not predict N_1 in shear.

3.4.2 Two-Constant Maxwell Model

Extend the derivation of the Maxwell model, eq. 3.2.20, from eq. 3.2.9 to show that a two-constant integral linear viscoelastic model can also be expressed as follows:

$$\tau+(\lambda_1+\lambda_2)\frac{\partial \tau}{\partial t}+\lambda_1\lambda_2\frac{\partial^2 \tau}{\partial t^2}=(\eta_1+\eta_2)\left[\dot{\gamma}+\left(\frac{\eta_1\lambda_2+\eta_2\lambda_1}{\eta_1+\eta_2}\right)\frac{\partial \dot{\gamma}}{\partial t}\right]$$

3.4.3 Derivation of G' and G''

Derive eqs. 3.3.31 and 3.3.32, the relation between the relaxation modulus, and the dynamic moduli.

3.4.4 Energy Dissipation

Verify eq. 3.3.25 for the energy dissipation per cycle per unit volume during small-amplitude sinusoidal oscillations.

3.4.5 Zero Shear Viscosity and Compliance from G', G''

Show that $\eta_0 = \lim_{\omega\to 0} \frac{G''(\omega)}{\omega}$ and $J_e^o = \lim_{\omega\to 0} \frac{G'(\omega)}{G''^2(\omega)}$

APPENDIX 3A

Robert B. Secor

Curve Fitting of the Relaxation Modulus

For purposes of comparison and modeling, it is often necessary to fit experimental relaxation modulus data to an analytic function. The objective of the fitting process is to determine parameter values that in some sense represent the "best" fit of the approximating function to the experimental data. Consideration should be given to (1) the form of the approximating function, (2) the measure of the error between the experimental data and the approximating function, and (3) the procedure for finding the parameters that minimize the error.

Fitting of the relaxation modulus falls into a more general class of problems known as constrained optimization. The goal of constrained optimization is to optimize (minimize or maximize) some entity (the objective function) while satisfying certain constraints. In curve fitting, the objective function is the error between the approximating function and the experimental data. Constraints can be of three types: inequality constraints, equality constraints, and variable bounds. For instance, only positive relaxation times have physical significance. . . . For further discussion of fitting of relaxation spectra see Tschoegl (1989), Baumgaertel and Winter (1989, 1992), and Honerkamp and Weese (1993).

Approximating Form

The approximating form of the relaxation modulus should exhibit the following properties:

1. Decrease monotonically with time (fading memory)

2. Have a finite memory, $G(\infty) = 0$ (for liquids)
3. Have a bounded modulus, $G(0) < \infty$
4. Exhibit a bounded steady state viscosity

$$\eta_0 = \int_0^\infty G(s)ds < \infty$$

5. Exhibit a bounded steady state compliance

$$J_e^0 = \frac{\int_0^\infty sG(s)ds}{\eta_0^2} < \infty$$

The traditional form that satisfies these conditions and is similar to forms predicted by molecular theory of polymeric liquids is a sum of exponentially fading functions in time:

$$G(s) = \sum_{k=1}^{N} G_k \exp\left(\frac{-s}{\lambda_k}\right) \tag{3A.1}$$

Other approximating forms that come to mind fall short in some category. For instance, a form analogous to the Carreau viscosity function (eq. 2.4.16) that does not allow analytic conversion between relaxation modulus and the elastic modulus or loss modulus is

$$G(s) = \frac{A}{\left[1 + (s)^2\right]^n} \tag{3A.2}$$

Error Measure

The next consideration is the selection of a measure of the error between the approximating function and the experimental data. One convenient measure is the Euclidean norm of the deviations at the M data points

$$E = \sum_{i=1}^{M} [G(t_i) - G^{\text{fit}}(t_i)]^2 \tag{3A.3}$$

where $G(t_i)$ is the experimental data point and $G^{fit}(t_i)$ is the approximating function. However, since the relaxation modulus generally decays over several orders of magnitude, this error measure artificially weights the error at short times higher than the error at long times. An error measure, which avoids this defect, measures the deviations between the natural logarithms of the modulus

$$E = \sum_{i=1}^{M} \left[\ln[G(t_i)] - \ln[G^{\text{fit}}(t_i)]\right]^2 \tag{3A.4}$$

An alternative is to introduce a weighting function of $(1/G(t_i)^2)$ to eq. 3A.3 so that the error measure becomes the following:

$$E = \sum_{i=1}^{M} \left[\frac{1 - G^{\text{fit}}(t_i)^2}{G(t_i)} \right]^2 \tag{3A.5}$$

By considering the Taylor expansion of $\ln(G^{\text{fit}}/G)$ around $G^{\text{fit}}/G = 1$, it can be shown that the last two error measures differ by an amount that is third order in the quantity $(1 - G^{\text{fit}}/G)$

$$\left[\ln\left(\frac{G^{fit}}{G} \right) \right]^2 - \left[\frac{1 - G^{\text{fit}}}{G} \right]^2 = \left(\frac{1 - G^{\text{fit}}}{G} \right)^3 + HOT \tag{3A.6}$$

where HOT are Higher Order Terms

Consequently, as long as the approximating function is reasonably close to the experimental data, the two error measures are equivalent. However, the last one does not introduce nonlinearities into the problem.

Search Procedures

In constrained optimization, the task of finding the optimum point is divided into two parts. The constrained optimization problem is converted to an equivalent unconstrained problem followed by a constrained search. The methods of converting constrained problems to unconstrained ones are beyond the scope of this writing, but a description of them can be found in Pierre and Lowe (1975). The objective of an unconstrained search is to find the parameter values that minimize (or maximize) the objective function. In the case of the relaxation modulus, the objective is to find the relaxation times and relaxation strengths that minimize the error between the approximating function and the experimental data.

The efficiency of any search routine can be sensitive to the scaling of the parameters. Ordinarily, it is desirable to have all parameters of the same order of magnitude. Unfortunately, relaxation times and relaxation strengths generally vary over several orders of magnitude. A change of variables that helps scale the parameters is

$$a_k = \ln G_k \tag{3A.7}$$

$$t_k = \ln \lambda_k \tag{3A.8}$$

In these new variables the approximating form of the relaxation modulus is

$$G(s) = \sum_{k=1}^{N} \exp[a_k - se^{-t_k}] \tag{3A.9}$$

Common unconstrained search methods include steepest descent methods and Newton's method. Newton's method has the advantage of rapid rate of convergence when it converges, while steepest descent methods often converge but at a painfully slow rate. In the curve fitting of the relaxation modulus, it has been found to be difficult to obtain convergence with Newton's method. Thus, some quasi-Newtonian methods employing a combination of Newton's and steepest descent methods have been used. Both the Levenburg–Marquardt algorithm (Marquardt, 1963) and the Davidson–Fletcher–Powell (DFP) algorithm have been used with success. The DFP algorithm was designed for constrained optimization problems, and a program listing can be found in Pierre and Lowe (1975).

The results of applying the DFP program to the fitting of some oscillatory shear experiments follow. Consistent with eq. 3A.9, the appropriate approximating forms for $G'(\omega)$, $G''(\omega)$, and $|G^*(\omega)|$ are

$$G'(\omega) = \sum_{k=1}^{N} \frac{\omega^2 \exp\,[a_k + 2t_k]}{[1 + \omega^2 \exp\,(2t_k)]} \tag{3A.10}$$

$$G''(\omega) = \sum_{k=1}^{N} \frac{\omega \exp\,[a_k + t_k]}{[1 + \omega^2 \exp\,(2t_k)]} \tag{3A.11}$$

$$|G^*(\omega)| = [G'(\omega)^2 + G''(\omega)^2]^{1/2} \tag{3A.12}$$

The data shown in Table 3A.1 for a narrow distribution polybutadiene sample were fit to four relaxation times, and the broad distribution polyisobutylene data in Table 3A.2 (Ferry 1980, p. 606) were fit to ten relaxation times. The results are shown in Table 3A.3 and Figures 3A.1 and 3A.2.

TABLE 3A.1 / Polybutadiene Data

ω *(rad/s)*	G' *(Pa)*	G'' *(Pa)*	$\|G^*\|$ *(Pa)*
0.1	53.7	3460.0	3460.0
0.1778	176.0	6280.0	6285.0
0.3162	414.0	1.12×10^4	1.13×10^4
0.5623	1110.0	2.00×10^4	2.01×10^4
1.00	2970.0	3.55×10^4	3.56×10^4
1.778	8500.0	6.23×10^4	6.29×10^4
3.162	2.40×10^4	1.07×10^5	1.10×10^5
10.0	1.52×10^5	2.46×10^5	2.89×10^5
17.78	2.84×10^5	2.88×10^5	4.05×10^5
31.62	4.21×10^5	2.83×10^5	5.07×10^5
56.23	5.30×10^5	2.53×10^5	5.87×10^5
100.00	6.19×10^5	2.23×10^5	6.58×10^5
177.8	6.92×10^5	2.06×10^5	7.22×10^5
316.2	7.69×10^5	1.87×10^5	7.92×10^5

TABLE 3A.2 / Polyisobutylene Data[a]

ω *(rad/s)*	G' *(Pa)*	G'' *(Pa)*	$\lvert G^*\rvert$ *(Pa)*
0.000316	1.05×10^5	4.68×10^4	1.15×10^5
0.001	1.32×10^5	4.57×10^4	1.40×10^5
0.00316	1.62×10^5	4.37×10^4	1.68×10^5
0.01	1.78×10^5	3.98×10^4	1.82×10^5
0.0316	2.00×10^5	3.02×10^4	2.02×10^5
0.1	2.24×10^5	2.82×10^4	2.26×10^5
0.316	2.40×10^5	2.75×10^4	2.42×10^5
1.0	2.63×10^5	2.51×10^4	2.64×10^5
3.16	2.75×10^5	2.63×10^4	2.77×10^5
10.0	2.95×10^5	3.09×10^4	2.97×10^5
31.6	3.02×10^5	5.62×10^4	3.07×10^5
100.0	3.24×10^5	1.12×10^5	3.43×10^5
316.0	4.07×10^5	2.40×10^5	4.73×10^5
10^3	5.75×10^5	5.13×10^5	7.71×10^5
3.16×10^3	9.12×10^5	1.07×10^6	1.41×10^6
10^4	1.59×10^6	2.34×10^6	2.83×10^6
3.16×10^4	3.09×10^6	4.90×10^6	5.79×10^6
10^5	6.76×10^6	1.05×10^7	1.25×10^7
3.16×10^5	1.45×10^7	2.19×10^7	2.62×10^7
10^6	3.09×10^7	4.37×10^7	5.35×10^7
3.16×10^6	6.46×10^7	8.32×10^7	1.05×10^8
10^7	1.29×10^8	1.45×10^8	1.94×10^8
3.16×10^7	2.51×10^8	2.29×10^8	3.40×10^8
10^8	4.37×10^8	3.16×10^8	5.39×10^8
3.16×10^8	6.61×10^8	3.24×10^8	7.36×10^8
10^9	8.51×10^8	2.46×10^8	8.86×10^8

[a] Ferry, 1980

TABLE 3A.3 / Parameter Values

Polybutadiene		*Polyisobutylene*	
λ_k *(s)*	G_k *(Pa)*	λ_k *(s)*	G_k *(Pa)*
8.04×10^{-3}	3.00×10^5	8.33×10^{-9}	6.13×10^8
5.93×10^{-2}	4.83×10^5	1.31×10^{-7}	1.79×10^8
1.46×10^{-1}	2.98×10^4	2.23×10^{-6}	3.06×10^7
7.61×10^{-1}	1.04×10^2	4.78×10^{-5}	3.62×10^6
		1.03×10^{-3}	5.40×10^5
		4.26×10^{-2}	3.99×10^4
		9.84×10^{-1}	4.12×10^4
		9.31×10^0	4.41×10^4
		1.62×10^2	5.33×10^4
		3.83×10^3	1.46×10^5

Figure 3A.1.
Narrow distribution polybutadiene.

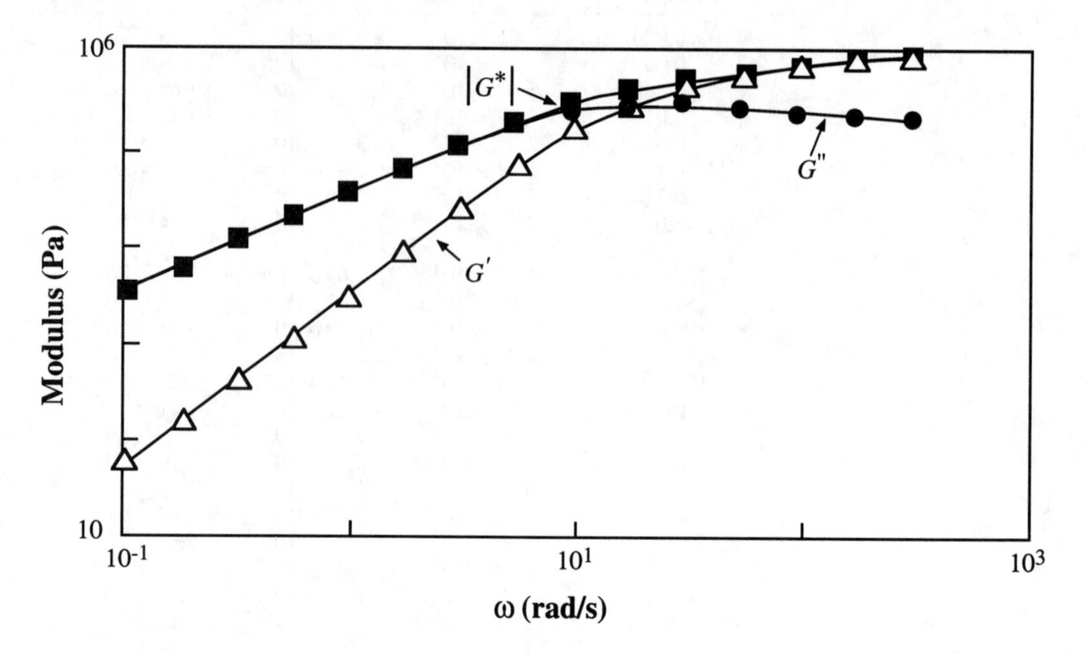

Figure 3.A.2.
Broad distribution polyisobutylene.

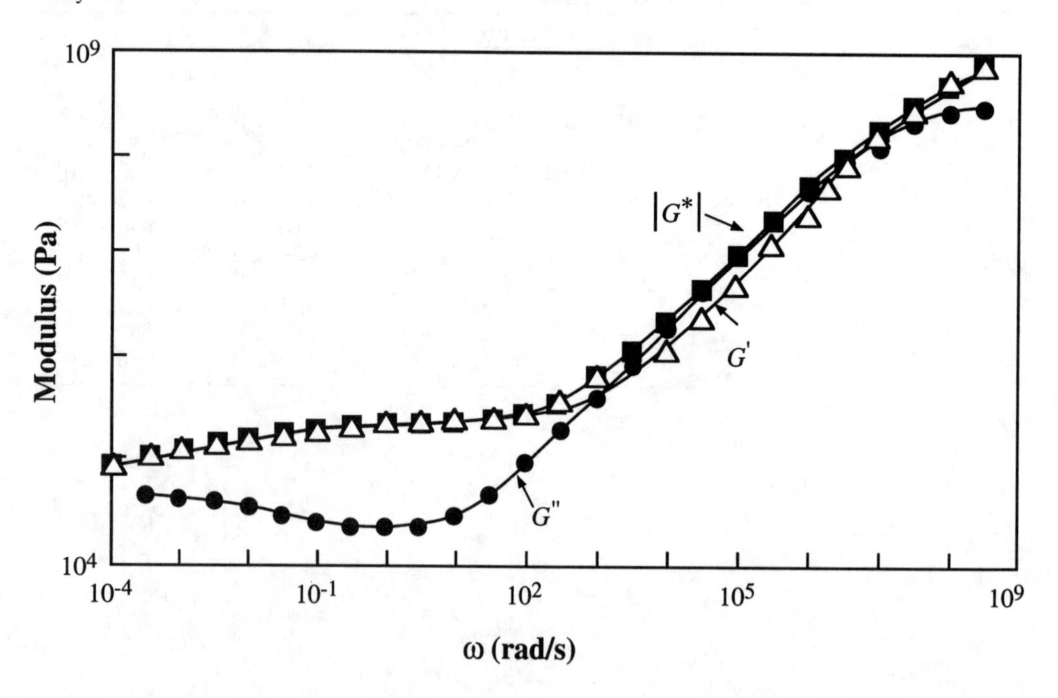

References

Baumgaertel, M.; Winter, H. H., *Rheol. Acta.* 1989, *28,* 511.

Baumgaertel, M.; Winter, H. H., *J. Non-Newt. Fluid Mech.* 1992, *44,* 15.

Bird, R. B.; Armstrong, R. C.; Hassager O., *Dynamics of Polymeric Liquids*, 2nd ed.; Wiley: New York, 1987; Chapter 5.

Boltzmann, L., *Ann. Phys. Chem.* 1876, *7*, 624.

Ferry, J. D., *Viscoelastic Properties of Polymers*, 3rd ed.; Wiley: New York, 1980.

Honerkamp, J.; Weese, J., *Rheol. Acta* 1993, *32,* 65.

Kolrausch, F., *Ann. Phys. Chem.* 1863, *119*, 337.

Laun, H. M., *Rheol. Acta* 1978, *11*, 1.

Lockett, F. J., *Nonlinear Viscoelastic Solids*; Academic Press: London, 1972.

Marquardt, D. W., *J. SIAM* 1963, *11*, 2.

Maxwell, J. C., *Phil. Trans.* 1866, *156*, 249.

Maxwell, J. C., *Phil. Trans.* 1867, *157*, 49.

Pierre, D. A., *Optimization Theory with Applications*; Wiley: New York, 1969.

Pierre; D. A., Lowe, M. J,. *Mathematical Programming via Augmented Lagrangians*; Addison-Wesley: Reading, MA, 1975.

Tschoegl, N. W., *The Phenomenological Theory of Linear Viscoelastic Behavior:* Springer: Berlin, 1989.

4

The experimental results show that . . . in addition to the shear stress [there is] a pull along the lines of flow.
Karl Weissenberg (1947)

NONLINEAR VISCOELASTICITY

4.1 Introduction

The experiments to which Weissenberg refers were done during World War II in England on materials for flame throwers. One goal of this research was to improve predictions of the pressure drop through the spray nozzles (Russell, 1946). Gum rubber in gasoline, polymethyl methacrylate in benzene, and similar materials were studied. Figure 4.1.1 shows some of the experiments that were used to demonstrate normal stress effects.

In Chapter 1 we saw that nonlinear normal stresses can arise in simple shear (eq 1.5.15) and in torsion (Example 1.7.1) of an elastic solid. In this chapter our immediate goal is to develop constitutive equations that can predict normal stresses and other nonlinear phenomena in flowing viscoelastic liquids. The eventual goal of this effort is to use these equations to predict and control viscoelastic

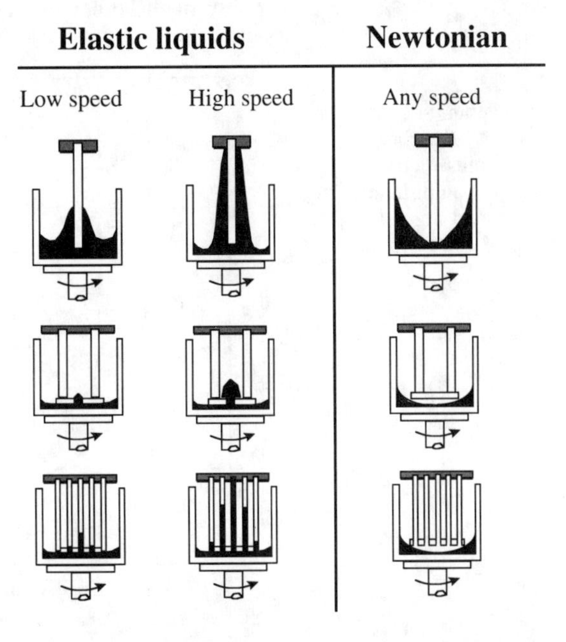

Figure 4.1.1.
Various experiments used by Weissenberg (1947) and his co-workers to demonstrate normal stress effects in liquids.

fluid flows in practical applications, such as polymer processing or even in flame throwers.

Nonlinear rheology vastly extends all the phenomena (elastic, viscous, and linear time dependent) discussed in Chapters 1–3. Elastic, viscous, and linear viscoelastic behaviors are but coastal zones on a continent of nonlinear rheology; see Figure 4.1.2. The abscissa on Figure 4.1.2 is the *Deborah number*, which is generally defined as the ratio of the material's characteristic relaxation time λ to the characteristic flow time t.

$$\mathrm{De} = \frac{\lambda}{t} \tag{4.1.1}$$

The origin of Deborah's number is indicated in the frontispiece to this text. In Figure 4.1.2 we take the characteristic flow time to be the inverse of the typical deformation rate $\dot{\gamma}^{-1}$, while in oscillatory flows we use the amplitude of the oscillatory strain times its frequency $(\gamma_o\omega)^{-1}$. The elastic, Newtonian, and linear viscoelastic limits illustrated in Figure 4.1.2 have already been discussed in Chapters 1, 2, and 3, respectively. Second-order fluids, to be covered shortly, reside in a fringe of the regime of nonlinear viscoelasticity that lies just across the border from the Newtonian domain.

The breadth of the scope of nonlinear phenomena can be grasped in part by considering the various time-dependent probes of linear viscoelasticity cited in Table 3.3.2: sinusoidal oscillation, creep, constrained recoil, stress relaxation after step strain, stress relaxation after steady shearing, and stress growth after start-up of steady shearing. In the linear regime—that is, at small strains or small strain rates—the experimental results of any *one* of these probes (in simple shear, for example) can be used to predict results for any of the other probes, not only for simple shearing defor-

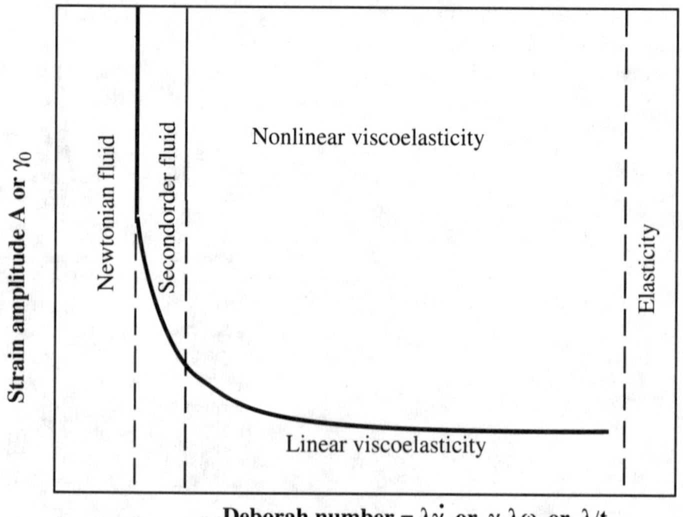

Figure 4.1.2.
Schematic diagram showing the behavior of viscoelastic fluids in the limits of low strain rates, low amplitude deformations, and high strain rates. Adapted from Pipkin (1972).

mations, but for *any* volume-preserving deformation. For strains and strain rates that are not small, each of these probes gives a nonlinear response. In principle, in the nonlinear regime *none* of these probes or combinations of probes gives a response that can be used to predict the nonlinear response to any of the other probes. Nor can any amount of nonlinear data (in shear, for example) be used to generate from first principles predictions in any other type of deformation, such as uniaxial extension. Each nonlinear test gives data that in principle are directly relevant to that test only. Therefore, at the minimum the task of the nonlinear rheologist is to develop empirical correlations among nonlinear measurements; or more ambitiously, it is to find constitutive equations that allow one to predict a useful range of nonlinear behavior, using as input rheological measurements over a small subset of that range.

Thus the general nonlinear rheological behavior of a material is characterized by finding a constitutive equation appropriate for that material. The appropriateness of a constitutive equation is a balance of many factors, including the equation's accuracy in fitting and predicting various data, the soundness of its theoretical underpinnings, its simplicity, its mathematical and computational tractability, and the range of phenomena one wishes to address using the equation. The weighting of the various factors is subjective to a significant degree; hence a rich diversity of constitutive equations has been developed over the past 50 years, many of which continue to be used.

In the interest of clarity, we shall confine our attention in this chapter to three major topics. First, we mention the most important *nonlinear* phenomena that a constitutive equation should describe, namely *normal stresses differences*, *shear thinning*, and *extensional thickening*. Restricted examples of these phenomena have been discussed in Chapters 1 and 2; normal stresses in shearing of purely elastic materials were mentioned in Section 1.5; shear thinning of inelastic fluids was described in Chapter 2; and the fact that the stress response in extension can be quite different from that in shear was touched on in Figures 1.1.3 and 2.1.3. A good nonlinear constitutive equation should describe not only general manifestations of all three of these phenomena, but also the *time dependence* of rheological material functions, such as the linear viscoelastic time dependence discussed in Chapter 3.

As our second major topic, we present the simplest equations from each of the three important classes of constitutive equations, namely the differential equations from the retarded-motion expansion, the Maxwell-type differential equations, and the integral equations. Third and finally, we summarize the more accurate constitutive equations that we feel are the most promising for simply and realistically describing viscoelastic fluids and for modeling viscoelastic flows. More complete treatments of nonlinear constitutive equations are available elsewhere (Tanner, 1985; Bird et al., 1987; Larson, 1988; Joseph, 1990). Throughout this chapter, our examples are drawn from the literature on polymeric fluids, especially polymer melts. Although suspensions, emulsions, microemulsions, and other nonpolymeric fluids also have fascinating nonlinear rheological properties and are important classes of materials for industrial applications, a discussion of the nonlinear properties unique to these materials is best deferred to Chapter 10.

4.2 Nonlinear Phenomena

4.2.1 Normal Stress Differences in Shear

When a viscoelastic material is sheared between two parallel surfaces at an appreciable rate of shear, in addition to the viscous shear stress T_{12}, there are normal stress differences $N_1 \equiv T_{11} - T_{22}$ and $N_2 \equiv T_{22} - T_{33}$. Here "1" is the flow direction, "2" is perpendicular to the surfaces between which the fluid is sheared, as defined by eq 1.4.8, and "3" is the neutral direction. The largest of the two normal stress differences is N_1, and it is responsible for the rod climbing phenomenon mentioned at the beginning of this book. For isotropic materials, N_1 has always been found to be positive in sign (unless it is zero). In a cone and plate rheometer this means that the cone and plate surfaces tend to be pushed apart. N_2 is usually found to be negative and smaller in magnitude than N_1; typically the ratio $-N_2/N_1$ lies between 0.05 and 0.3 (Keentok et al., 1980; Ramachandran et al., 1985). Figure 4.2.1 shows the

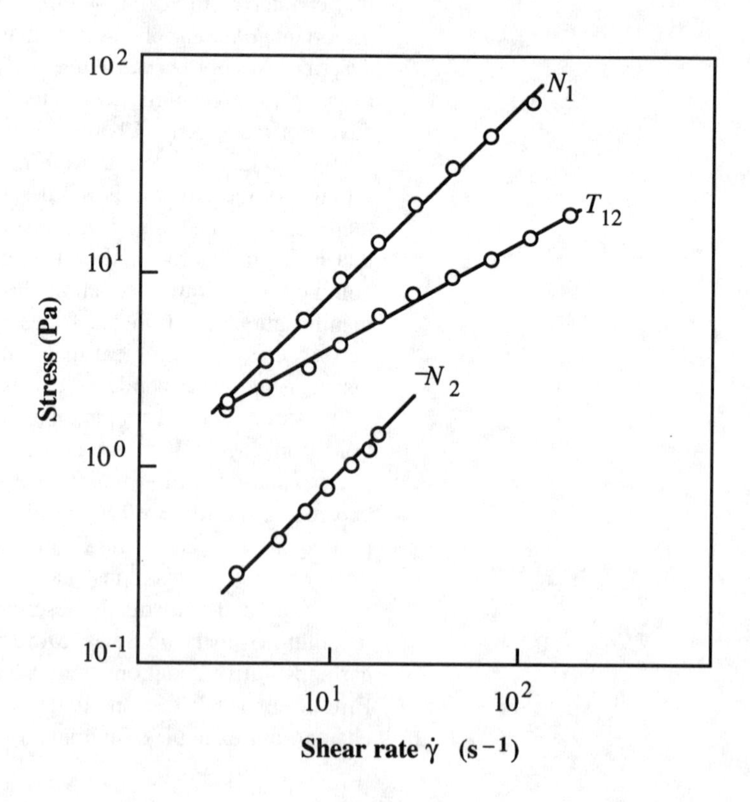

Figure 4.2.1. Shear stress τ_{12}, first normal stress difference N_1, and negative of the second normal stress difference N_2 as functions of shear rate $\dot{\gamma}$ for a 1.18% solution of polyisobutylene in decalin. Replotted from Keentok et al. (1980).

three quantities T_{12}, N_1, and $-N_2$ measured by Keentok et al. for a 1.18% solution of polyisobutylene in decalin. The ratio $-N_2/N_1$ for this fluid is small (0.11) and nearly independent of shear rate. At sufficiently low shear rates, lower than those for which data are plotted in Figure 4.2.1, T_{12} usually becomes linear in the shear rate $\dot{\gamma}$; that is, the shear viscosity $\eta \equiv T_{12}/\dot{\gamma}$ becomes independent of $\dot{\gamma}$. Similarly N_1 and N_2 approach the limits $N_1 \propto \dot{\gamma}^2$, $N_2 \propto \dot{\gamma}^2$ at small $\dot{\gamma}$, and thus the *normal stress coefficients*

$$\psi_1 \equiv \frac{T_{11} - T_{22}}{\dot{\gamma}^2} \tag{4.2.1}$$

$$\psi_2 \equiv \frac{T_{22} - T_{33}}{\dot{\gamma}^2} \tag{4.2.2}$$

approach constant values at small $\dot{\gamma}$.

4.2.2 Shear Thinning

Figure 4.2.2 shows the dependence of η and ψ_1 on $\dot{\gamma}$ at steady state for a polyethylene melt. Although η and ψ_1 approach constants defined as η_0 and $\psi_{1,0}$ at low $\dot{\gamma}$, at higher $\dot{\gamma}$, both η and ψ_1 decrease dramatically with increasing $\dot{\gamma}$. The decrease in η, and the related decreases in ψ_1 and ψ_2, are referred to as *shear thinning*. Shear thinning is a nonlinear phenomenon that is especially pronounced in polymer melts and in concentrated polymer solutions. A molecular explanation for shear thinning in polymeric fluids is given when entanglement concepts are introduced in Chapter 11.

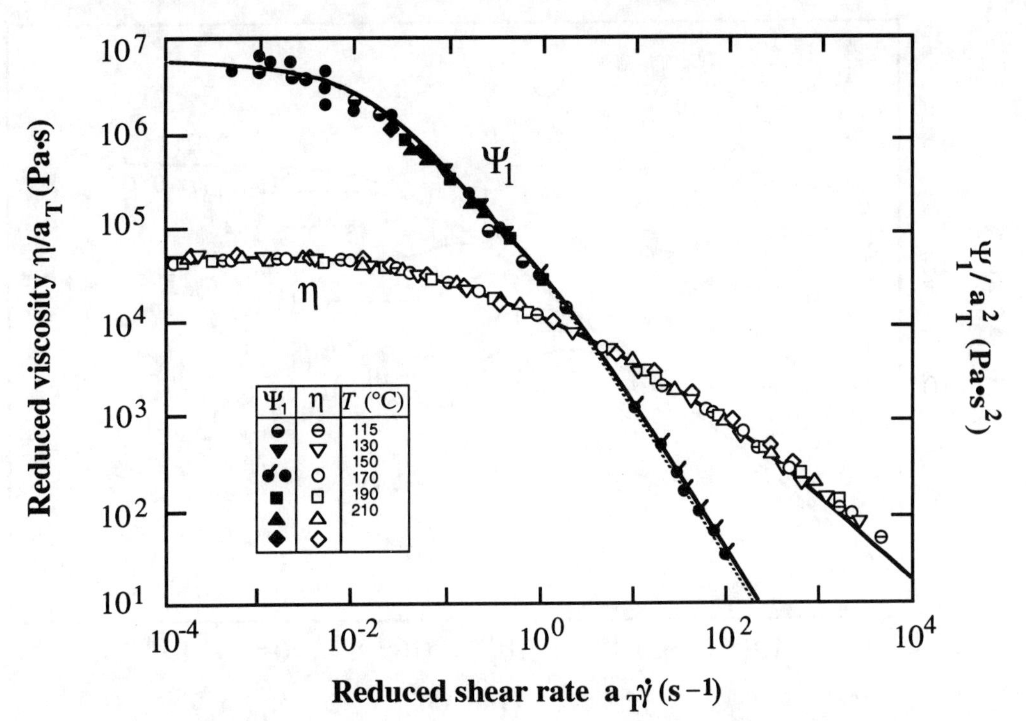

Figure 4.2.2. Steady state shear viscosity and primary normal stress coefficient for low density polyethylene "melt I" and from the Kaye–Bernstein, Kearsley, Zapas (K-BKZ) equation with the double exponential damping function, eq 4.4.13 (solid lines) and with the single exponential, eq 4.4.12 (dotted line). Data at different temperatures have been shifted to one master curve by $a_T(T)$. Replotted from Laun (1978).

Shear thinning phenomena are also evident in time-dependent measurements. Figure 4.2.3 shows the *time-dependent* viscosity after start-up of steady shearing for a polybutadiene solution (Menezes and Graessley, 1980). The time-dependent viscosity is defined analogously to the steady viscosity as

$$\eta^+(t, \dot{\gamma}) \equiv \frac{T_{12}(t, \dot{\gamma})}{\dot{\gamma}} \tag{4.2.3}$$

The lowest shear rate in Figure 4.2.3, 0.0214 s^{-1}, is nearly in the linear viscoelastic regime where $\eta^+(t, \dot{\gamma})$ becomes independent of $\dot{\gamma}$. As the shear rate increases with t fixed, $\eta^+(t, \dot{\gamma})$ decreases. Also, an *overshoot* appears; that is, η^+ at fixed $\dot{\gamma}$ passes through a maximum before the steady state value $\eta^+(t, \dot{\gamma}) \to \eta(\dot{\gamma})$ is reached at large t. Note also that even when $\dot{\gamma}$ is large, $\eta^+(t, \dot{\gamma})$ at small t does not depart from the linear viscoelastic response. Deviations from linear viscoelasticity occur only when *both* the strain rate $\dot{\gamma}$ and the strain $\gamma = \dot{\gamma} t$ are not small.

4.2.3 Interrelations Between Shear Functions

The introduction emphasized that in general one cannot use linear viscoelastic data to predict a nonlinear viscoelastic material function, nor as a rule can one use one nonlinear material function to predict another. Nevertheless, in shearing flows a few useful interrelations between material functions have often been observed to hold, at least approximately, for polymer melts and solutions.

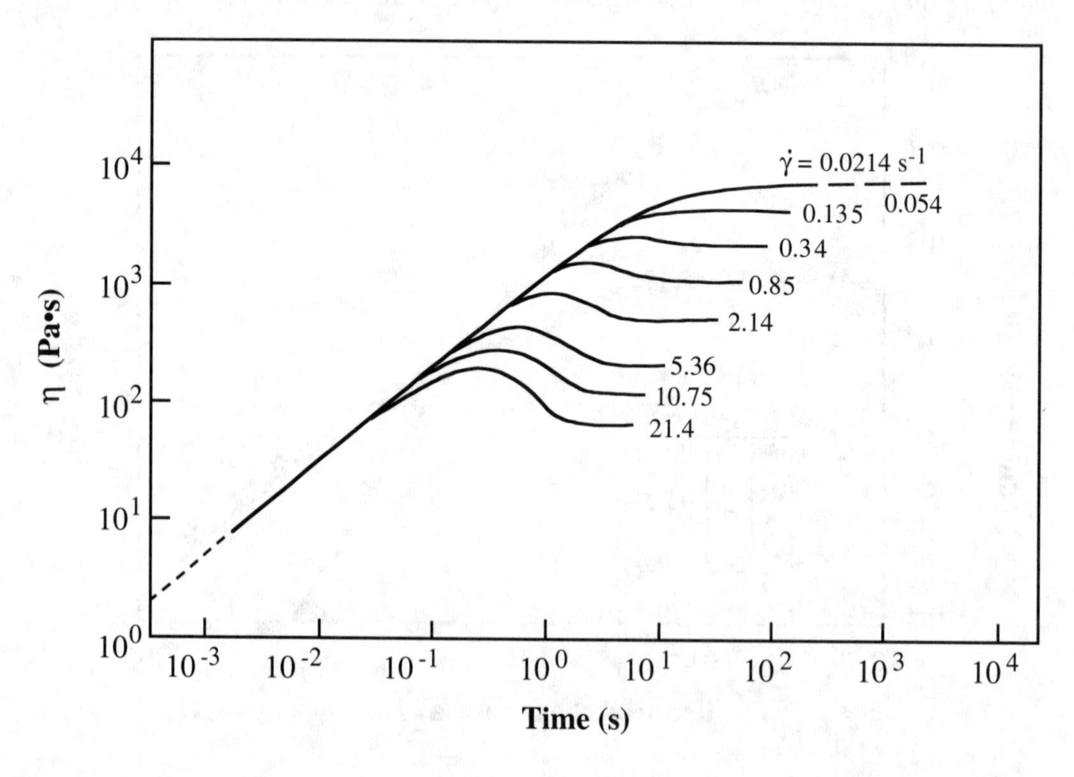

Figure 4.2.3. Growth of shear viscosity as a function of time after start-up of steady shearing at various shear rates $\dot{\gamma}$ for a 7.55% vol% solution of polybutadiene ($M_w = 350,000$) in a hydrocarbon oil. The dashed line is the prediction of linear viscoelasticity. Replotted from Menezes and Graessley (1980).

A few of these interrelationships can be derived from postulates that are general enough to be valid for most polymeric fluids. For example, the steady state shear viscosity and first normal stress coefficient at low shear rate can be derived from low frequency linear viscoelastic measurements:

$$\eta(\dot{\gamma}) = \eta'(\omega)\,; \quad \text{as} \quad \omega = \dot{\gamma} \to 0 \tag{4.2.4}$$

$$\psi_1(\dot{\gamma}) = \frac{2G'(\omega)}{\omega^2}\,; \quad \text{as} \quad \omega = \dot{\gamma} \to 0 \tag{4.2.5}$$

The first of these relationships, eq. 4.2.4, follows from little more than the definitions of η and η', while eq. 4.2.5 is less obvious (Coleman and Markovitz, 1964). Both relationships are of limited usefulness because they are relevant only for low shear rate properties. However, an empirical relationship, called the "Cox–Merz rule," often holds fairly well at high shear rates. This rule states that the shear rate dependence of the steady state viscosity η is equal to the frequency dependence of the linear viscoelastic viscosity η^*; that is,

$$\eta(\dot{\gamma}) = |\eta^*(\omega)|; \quad \text{with } \dot{\gamma} = \omega \tag{4.2.6}$$

The quantities η' and η^* were defined in eqs. 3.3.28–3.3.30. An analogous relationship, the "mirror rule" (Gleissle, 1980) between $\eta(\dot{\gamma})$ and $\eta^+(t)$, has been proposed:

$$\eta(\dot{\gamma}) = \eta^+(t); \quad \text{with } \dot{\gamma} = \frac{1}{t} \tag{4.2.7}$$

Here we remind the reader that $\eta^+(t)$ is a low shear rate, linear viscoelastic function of time only. Figure 4.2.4 shows that eqs. 4.2.6 and 4.2.7 work well for a sample of linear low density polyethylene

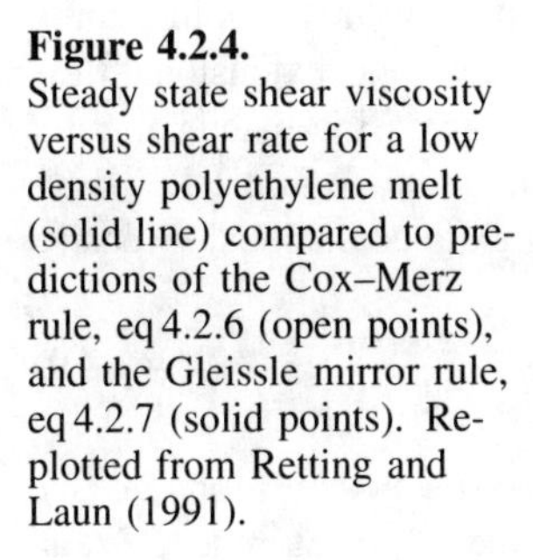
Figure 4.2.4. Steady state shear viscosity versus shear rate for a low density polyethylene melt (solid line) compared to predictions of the Cox–Merz rule, eq 4.2.6 (open points), and the Gleissle mirror rule, eq 4.2.7 (solid points). Replotted from Retting and Laun (1991).

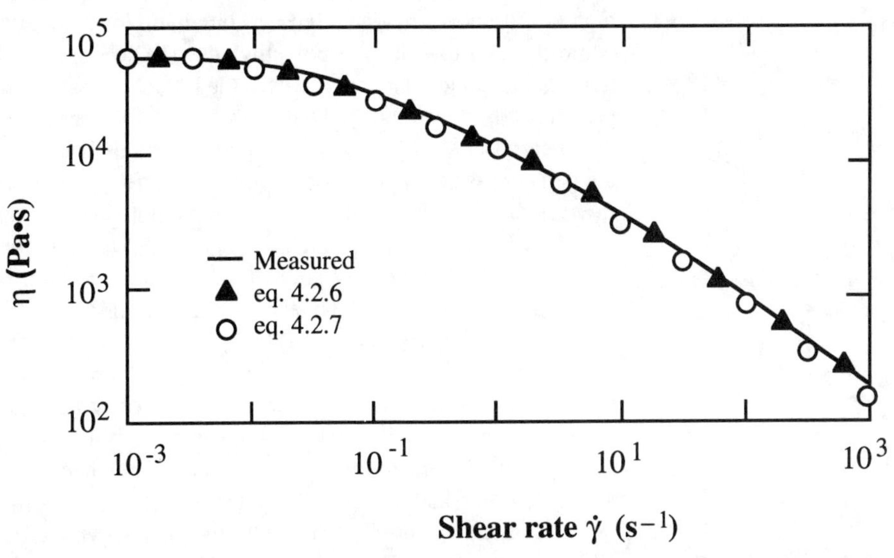

(Laun, 1990). Empirical relationships have also been proposed that predict the shear rate dependence of the first normal stress coefficient $\psi_1(\dot{\gamma})$, using either the linear viscoelastic quantities $G'(\omega)$ and G'', or $\psi_1^+(t)$ (Gleissle, 1980; Wissbrun, 1986; Al-Hadithi et al., 1988), and even using $\eta(\dot{\gamma})$ (see eq. 6.5.1). Although these "rules" cannot be derived rigorously, plausible arguments can be made that they should generally hold for polydisperse polymeric fluids (Booij et al., 1983; Larson, 1985; Wissbrun, 1986).

A "rule" that can be derived theoretically from general premises likely to be valid for polymer melts and solutions is the *Lodge–Meissner* relationship (Lodge and Meissner, 1972) between the shear stress and the first normal stress difference after a step shear strain:

$$N_1 = T_{12}\gamma \tag{4.2.8}$$

where γ is the step shear strain. Equation 4.2.8 has repeatedly been observed to hold (an example will be given in Figure 4.4.1) and is thought to be quite general (Laun, 1978; Vrentas and Graessley, 1981; Larson et al., 1988).

4.2.4 Extensional Thickening

Although in a shear flow the viscosity of a polymeric fluid usually decreases with increasing deformation rate, in an extensional flow the viscosity frequently *increases* with increasing extension rate; that is, the fluid is extensional *thickening* (recall Figure 2.1.3). Figure 4.2.5 shows the time-dependent uniaxial extensional viscosity

$$\eta_u^+(t,\dot{\epsilon}) \equiv \frac{T_{11}(t,\dot{\epsilon}) - T_{22}(t,\dot{\epsilon})}{\dot{\epsilon}} \tag{4.2.9}$$

for several polymer melts, where $\dot{\epsilon}$ is the steady extension rate (see eq. 2.3.9). The molecules in two of the melts, polystyrene (PS) and high density polyethylene (HDPE), contain no long side branches, while the two low density polyethylenes (LDPEs) are branched, treelike molecules. Note that η_u^+ for the branched materials shows pronounced extensional thickening; that is, η_u^+ for large $\dot{\epsilon}$ lies *above* the linear viscoelastic response obtained for small $\dot{\epsilon}$. This thickening behavior is the opposite of the nonlinear effect seen in shearing flows depicted in Figure 4.2.3. The same point is made more dramatically in Figure 4.2.6, which shows both $\eta^+(t,\dot{\gamma})$ and $\eta_u^+(t,\dot{\epsilon})$ for a branched low density polyethylene. For low strain rates or small strains—that is, in the linear viscoelastic regime—η^+ and η_u^+ differ only by the Trouton ratio $\eta_u^+(t,\dot{\epsilon})/\eta^+(t,\dot{\epsilon}) = 3$. As the strain rate or strain increases, the shear viscosity sinks below the linear viscoelastic result, while the extensional viscosity rises above it. Melts without long side branches, such as the polystyrene and high density polyethylene in Figure 4.2.5, often are much less extension thickening than melts with long branches.

For melts the nonlinearity in the extensional viscosity is usually more sensitive to the molecular architecture (e.g., the presence

Figure 4.2.5.
Uniaxial extensional viscosity η_u^+ versus time for different extension rates $\dot{\epsilon}$ for four different melts. The solid lines are the fits to the data of a special form of the integral equation, eq 4.4.15. The curves at the lowest extension rates correspond to linear viscoelastic response. Note that curves have been shifted vertically by the multiplier indicated along the ordinate. Replotted from Laun (1984).

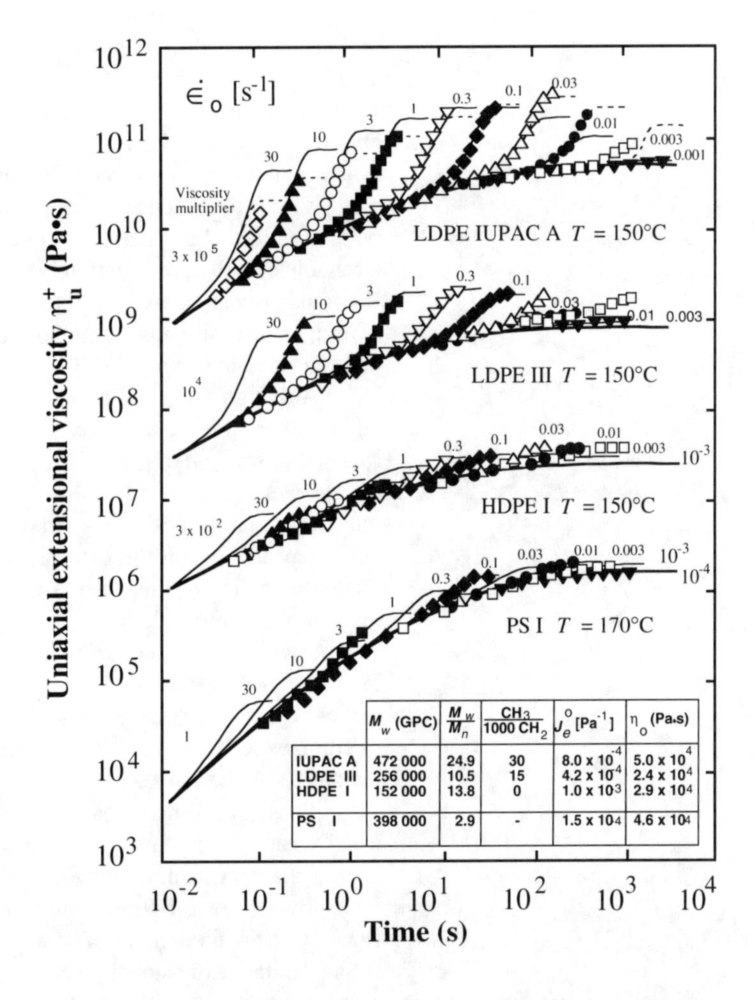

	M_w (GPC)	$\frac{M_w}{M_n}$	$\frac{CH_3}{1000\ CH_2}$	J_e^o [Pa^{-1}]	η_o (Pa.s)
IUPAC A	472 000	24.9	30	8.0×10^{-4}	5.0×10^4
LDPE III	256 000	10.5	15	4.2×10^{-4}	2.4×10^4
HDPE I	152 000	13.8	0	1.0×10^{-3}	2.9×10^4
PS I	398 000	2.9	-	1.5×10^{-4}	4.6×10^4

Figure 4.2.6.
Uniaxial extensional viscosity η_u^+ and shear viscosity η^+ as functions of time after inception of steady straining for IUPAC A low density polyethylene. The open symbols are elongational viscosities; the solid and half-open symbols are shear viscosities. Adapted from Meissner (1972).

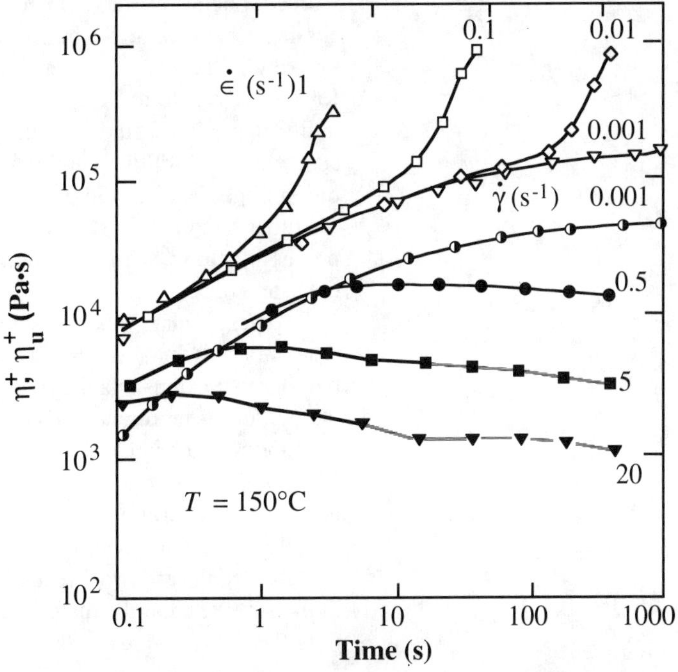

of long branches) than is the nonlinearity in the shear viscosity. The level of shear thinning generally does not strongly correlate with the magnitude of the extensional thickening behavior (Laun and Schuch, 1989). There do not seem to be any reliable interrelationships between nonlinear shear viscosities and nonlinear extensional viscosities. Hence it is important that extensional viscosities be measured if the material is to be adequately characterized and a reliable constitutive equation is to be chosen. Unfortunately, extensional viscosities are notoriously difficult to measure accurately; several of the best available methods of measuring or inferring them are described in Chapter 7.

Although the transient extensional viscosity in most of the curves of Figure 4.2.5 seems to approach a steady state at long times after start-up of extension, data of Meissner and co-workers (Wagner et al., 1979) indicate that in some cases the plateau in the extensional viscosity is a maximum from which the viscosity begins to drop when the extensional flow is continued to larger total strains. To see this drop in viscosity, Meissner had to stretch his sample to more than a hundred times its initial length. No steady state was obtained even after the sample's length had been increased a thousandfold over its initial length. Of course, the sample cannot be stretched indefinitely because if it is stretched too thin, experimental artifacts significantly degrade the quality of the data (see Section 7.2).

Thus, in the nonlinear regime of an extensional flow, one usually cannot be confident that a steady state has been reached, even if the data seem to possess a plateau at long times. By contrast, arbitrarily large strains can be imposed in rotational shearing flows, and the attainment of steady state can usually be assured. A steady state extensional viscosity, even if it could be measured with confidence, would not necessarily be directly relevant to processing flows because a material element in a processing flow is unlikely to see a steady extensional flow that persists long enough for a steady state to be approached. The extraordinary effort required to produce in the laboratory even an approximation to steady state extensional flow highlights the unlikelihood that such a condition might be achieved unintentionally in a processing flow. On the other hand nearly steady state shearing does occur in various processing flows, such as molding, extrusion, and transport through ducts and dies.

These remarks about reaching a steady state apply not only to uniaxial extensional flows, data for which appear in Figure 4.2.5, but for other extensional flows as well. Besides uniaxial extension, the two most important extensional flows are equal biaxial extension and planar extension. Kinematic tensors for these extensional flows were to have been found in Exercise 2.8.1. In uniaxial extension the material is stretched in one direction and compressed equally in the other two; in equal biaxial extension the material is stretched equally in two directions and compressed in the third; and in planar extension the material is stretched in one direction, held to the same dimension in a second, and compressed in the third. Following Meissner et al. (1982), it is useful to define the *stressing viscosities* μ_1 and μ_2 for a general extensional flow as follows:

$$\mu_1^+ = \frac{T_{11} - T_{33}}{2(2+m)\dot{\epsilon}}; \quad \mu_2^+ = \frac{T_{22} - T_{33}}{2(1+2m)\dot{\epsilon}} \tag{4.2.10}$$

where $m = -1/2$ for uniaxial extension, $m = 1$ for equal biaxial extension, and $m = 0$ for planar extension. These viscosities are defined in such a way that in the linear viscoelastic regime both of them equal the linear viscoelastic shear viscosity η^+. Thus for uniaxial extension, the Trouton ratio of 3 is divided out of the stressing viscosity μ_1^+. For uniaxial extension only μ_1^+ is defined, while $\mu_1^+ = \mu_2^+$ for biaxial extension; but for planar extension μ_1^+ and μ_2^+ are unequal. Note from eqs. 4.2.9 and 4.2.10 that for uniaxial extension $\mu_1^+ = \eta_u^+/3$.

Measurements of these viscosities in equal biaxial and planar extension are extremely rare. Laun and Schuch (1989) measured μ_1^+ and μ_2^+ in steady planar extension and μ_1^+ in steady uniaxial extension for a polyethylene with long branches, IUPAC X. For these sets of data, shown in Chapter 7, Figure 7.4.6, the planar extensional viscosity μ_1^+ is almost as extension thickening as is the corresponding viscosity in uniaxial extension. The second viscosity μ_2^+ for planar extension seems to be small and does not show extension-thickening characteristics. For a melt of polyisobutylene that presumably has few or no long side branches, Meissner et al. (1982) have found that the stressing viscosities in equal biaxial extension and in planar extension show little or no extension thickening, while significant extension thickening occurs in uniaxial extension (see Figure 4.2.7). Because of the lack of corroborating data for other melts or solutions, it would be dangerous or even foolhardy to try to draw general inferences from these limited data.

In addition to the nonlinear phenomena we have discussed in this section, namely normal stresses, shear thinning, and ex-

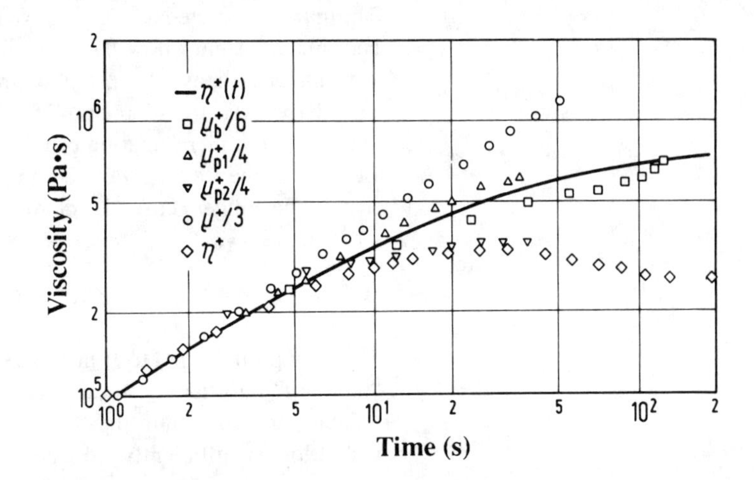

Figure 4.2.7. Stressing viscosity μ_1^+ for uniaxial, biaxial, and planar extension, stressing viscosity μ_2^+ for planar extension, and shear viscosity η^+, as functions of time after inception of steady straining for polyisobutylene. The solid line is the low shear rate limit of η^+. Extension and shear rates are 0.08s^{-1} except the biaxial which is 0.02s^{-1}. From Retting and Lawn, 1991.

tensional thickening, nonlinearity modifies the linear phenomena discussed in Chapter 3. We have seen an example in Figure 4.2.3, which illustrated how nonlinear effects can change the linear viscoelastic stress growth after start-up of steady shearing. Other ways of probing time-dependent viscoelasticity (such as stress relaxation after a step strain: Figure 4.4.1) are also strongly affected by nonlinear phenomena if the strains are large. We would like to be able to describe or predict these phenomena using nonlinear constitutive equations. As a first step, we first describe the simplest constitutive equations.

4.3 Simple Nonlinear Constitutive Equations

4.3.1 Second-Order Fluid

The simplest constitutive equation capable of predicting a first normal stress difference is the equation of the *second-order fluid* (Bird et al., 1987; Larson, 1988):

$$\mathbf{T} = -p\mathbf{I} + 2\eta_0\mathbf{D} - \psi_{1,0}\overset{\nabla}{\mathbf{D}} + 4\psi_{2,0}\mathbf{D}\cdot\mathbf{D} \tag{4.3.1}$$

Hereafter in constitutive equations such as eq 4.3.1 we will use the stress tensor τ, which does not contain the isotropic pressure term $p\mathbf{I}$. In eq 4.3.1 we have introduced the upper-convected derivative, denoted by "∇," which when acting on an arbitrary tensor $\mathbf{A}$ gives by definition

$$\overset{\nabla}{\mathbf{A}} \equiv \dot{\mathbf{A}} - (\nabla\mathbf{v})^T\cdot\mathbf{A} - \mathbf{A}\cdot\nabla\mathbf{v} \tag{4.3.2}$$

As usual, the dot ($\cdot$) over a tensor denotes the substantial or material time derivative of that tensor; that is,

$$\dot{\mathbf{A}} \equiv \frac{\partial}{\partial t}\mathbf{A} + \mathbf{v}\cdot\nabla\mathbf{A}$$

The upper-convected time derivative is a time derivative in a special coordinate system whose base coordinate vectors stretch and rotate with material lines. With this definition of the upper-convected time derivative, stresses are produced only when material elements are deformed; mere rotation produces no stress (see Section 1.4). Because of the way it is defined, the upper-convected time derivative of the Finger tensor is identically zero (see eqs. 2.2.35 and 1.4.13):

$$\overset{\nabla}{\mathbf{B}} = \mathbf{0} \tag{4.3.3}$$

The term proportional to $\overset{\nabla}{\mathbf{D}}$ in eq 4.3.1 incorporates a weak elastic "memory" into the constitutive equation. It can be shown under quite general conditions that a viscoelastic fluid will obey eq 4.3.1 if the flow is sufficiently slow and slowly varying, to ensure that departures from Newtonian behavior are small (Coleman and Noll, 1960). For example, if the flow is a shear flow, eq 4.3.1 will hold only if the shear rate is low enough to prevent the viscosity and first and second normal stress coefficients from departing from their low shear rate values η_0, $\psi_{1,0}$, and $\psi_{2,0}$; that is, if there is no shear thinning.

Example 4.3.1 The Second-Order Fluid in Simple Shear

Show for a steady simple shearing flow that the shear viscosity and first and second normal stress coefficients predicted by eq 4.3.1 are indeed η_0, $\psi_{1,0}$, and $\psi_{2,0}$.

Solution

First we evaluate $\overset{\nabla}{\mathbf{D}}$ for steady simple shearing by replacing $\mathbf{A}$ in eq 4.3.2 by expression 2.2.16 for $\mathbf{D}$. For steady flow $\partial\mathbf{D}/\partial t = 0$; for homogeneous flow $\nabla\mathbf{D} = 0$. Thus $\dot{\mathbf{D}} = 0$, and by substituting expression 2.2.12 for $\nabla\mathbf{v}$, we obtain

$$\overset{\nabla}{\mathbf{D}} = -\frac{1}{2}\begin{pmatrix} \dot{\gamma}^2 & 0 & 0 \\ 0 & 0 & 0 \\ 0 & 0 & 0 \end{pmatrix} - \frac{1}{2}\begin{pmatrix} \dot{\gamma}^2 & 0 & 0 \\ 0 & 0 & 0 \\ 0 & 0 & 0 \end{pmatrix} = -\begin{pmatrix} \dot{\gamma}^2 & 0 & 0 \\ 0 & 0 & 0 \\ 0 & 0 & 0 \end{pmatrix} \tag{4.3.4}$$

Substituting this result and eq 2.2.21 for $\mathbf{D}$ into eq 4.3.1 gives

$$\boldsymbol{\tau} = \eta_0\begin{pmatrix} 0 & \dot{\gamma} & 0 \\ \dot{\gamma} & 0 & 0 \\ 0 & 0 & 0 \end{pmatrix} + \psi_{1,0}\begin{pmatrix} \dot{\gamma}^2 & 0 & 0 \\ 0 & 0 & 0 \\ 0 & 0 & 0 \end{pmatrix} + \psi_{2,0}\begin{pmatrix} \dot{\gamma}^2 & 0 & 0 \\ 0 & \dot{\gamma}^2 & 0 \\ 0 & 0 & 0 \end{pmatrix} \tag{4.3.5}$$

By the definitions of shear viscosity, $\eta \equiv \tau_{12}/\dot{\gamma}$, and normal stress coefficients $\psi_1 \equiv (\tau_{11} - \tau_{22})/\dot{\gamma}^2$ and $\psi_2 \equiv (\tau_{22} - \tau_{33})/\dot{\gamma}^2$, it is clear that η_0 in eq. 4.3.5 is the zero shear viscosity and $\psi_{1,0}$ and $\psi_{2,0}$ are the limiting first and second normal stress coefficients.

Example 4.3.1 proves only that we have assigned the proper coefficients to the terms in the second-order fluid equation. Let us consider a more interesting example.

Example 4.3.2 Uniaxial Extensional Viscosity for a Second-Order Fluid

Suppose that for viscoelastic fluid X we measure the ratio ψ_2/ψ_1 as -0.2 at low shear rates. In a slow steady uniaxial extensional flow, should we expect fluid X to show extensional thickening or extensional thinning?

Solution

Taking $\mathbf{D}$ for uniaxial extension from eq 2.2.19 and setting $\dot{\mathbf{D}} = 0$, we find from eqs. 4.3.1 and 4.3.2:

$$\boldsymbol{\tau} = \eta_0 \begin{pmatrix} 2\dot{\epsilon} & 0 & 0 \\ 0 & -\dot{\epsilon} & 0 \\ 0 & 0 & -\dot{\epsilon} \end{pmatrix} + \frac{1}{2}\psi_{1,0} \begin{pmatrix} 4\dot{\epsilon}^2 & 0 & 0 \\ 0 & \dot{\epsilon}^2 & 0 \\ 0 & 0 & \dot{\epsilon}^2 \end{pmatrix} + \psi_{2,0} \begin{pmatrix} 4\dot{\epsilon}^2 & 0 & 0 \\ 0 & \dot{\epsilon}^2 & 0 \\ 0 & 0 & \dot{\epsilon}^2 \end{pmatrix} \tag{4.3.6}$$

From the definition of the uniaxial extensional viscosity, eq. 4.2.9, we find from eq. 4.3.6 that $\eta_u = 3\eta_0 + \frac{3}{2}(\psi_{1,0} + 2\psi_{2,0})\dot{\epsilon}$. Therefore we can conclude that as long as $-\psi_{2,0}/\psi_{1,0} < 0.5$, the first departures of η_u from the Newtonian value will be *positive*. Hence fluid X will be extension thickening at low extension rates. A similar calculation for equal biaxial extension (Larson, 1988) shows that $\eta_b = 6\eta - 6(\psi_1 + 2\psi_2)\dot{\epsilon}$.

This example makes evident the usefulness of the equation of the second-order fluid. Once the three coefficients of the equation have been specified—and measurements in simple shear alone are enough to specify them—predictions can be made for the first deviations from Newtonian behavior for any other flow. This property has proved useful in analyzing slow but complex nonuniform flows, such as those observed in rod climbing (see Chapter 5) and flow over a pressure hole (see Chapter 6).

The equation of the second-order fluid is so named because it contains all terms up to second order in the velocity gradient in a perturbation expansion about the rest state. The Newtonian term $2\eta_0\mathbf{D}$ is the first-order term, and $p\mathbf{I}$ is the zeroth-order term. There are also equations of the third-order fluid, and so on (Bird et al., 1987). These form a series of equations in the "retarded motion expansion"; so called because it assumes that the flow is a small perturbation from the state of rest. The equations that are higher in order than second are of less practical importance because of their complexity and the restricted conditions under which they are accurate.

Most flows of polymeric fluids are not slow enough for the second-order-fluid equation or any of the equations of the retarded-motion expansion to apply to them. Of course eq 4.3.1 can be made accurate for any steady shearing flow merely by replacing the constants η_0, $\psi_{1,0}$, and $\psi_{2,0}$ by the shear rate dependent coefficients $\eta(\dot{\gamma})$, $\psi_1(\dot{\gamma})$, and $\psi_2(\dot{\gamma})$. Although the resulting equation, called the *Criminale–Ericksen–Filbey equation* (1958), is valid for any steady shearing flow, it cannot be expected to predict steady extensional viscosities or the stresses in any flow besides steady simple shear. More seriously, neither the second-order-fluid equation, nor any retarded-motion equation, nor the Criminale–Ericksen–Filbey equation can predict time-dependent viscoelastic phenomena such as stress growth or stress relaxation. If the flow is suddenly stopped, the stress tensor in eq 4.3.1 or any other of the retarded-motion equations goes immediately to zero rather than relaxing gradually. This is what we mean when we say that eq 4.3.1 contains elastic effects in a weak sense only. Although eq 4.3.1 is able to predict some nonlinear effects, it is not able to describe time-dependent phenomena even in the linear viscoelastic regime.

4.3.2 Upper-Convected Maxwell Differential Equation

Perhaps the simplest way to combine time-dependent phenomena and rheological nonlinearity is to incorporate nonlinearity into the simple Maxwell equation, eq. 3.2.18. This can be done by replacing the substantial time derivative in a tensor version of eq 3.2.18 with the upper-convected time derivative of τ, using eq. 4.3.2 (Oldroyd, 1950)*:

$$\boldsymbol{\tau} + \lambda \overset{\nabla}{\boldsymbol{\tau}} = 2\eta_0 \mathbf{D} \tag{4.3.7}$$

This equation, which is called the *upper-convected Maxwell (UCM) equation*, is nonlinear because $\overset{\nabla}{\boldsymbol{\tau}}$ contains products of the velocity gradient $\nabla \mathbf{v}$ and the stress tensor $\boldsymbol{\tau}$. For small strain amplitudes, the nonlinear terms disappear and the upper-convected time derivative reduces to the substantial time derivative; eq 4.3.7 is then equivalent to the linear Maxwell model. On the other hand, if the flow is steady and the strain *rate* is small, $\overset{\nabla}{\boldsymbol{\tau}}$ is negligible and Newtonian behavior is recovered. Thus, to first order in the velocity gradient, we obtain

$$\boldsymbol{\tau} \approx 2\eta_0 \mathbf{D} \quad + \text{ second} - \text{order terms} \tag{4.3.8}$$

Suppose that we now increase the strain rate until we start to see a weak departure from Newtonian behavior. We can calculate this departure by using eq. 4.3.8 to evaluate the small term $\overset{\nabla}{\boldsymbol{\tau}}$:

$$\overset{\nabla}{\boldsymbol{\tau}} \approx 2\eta_0 \overset{\nabla}{\mathbf{D}} + \text{ third} - \text{order terms} \tag{4.3.9}$$

Then eq. 4.3.7 gives

$$\boldsymbol{\tau} \approx 2\eta_0 \mathbf{D} - 2\eta_0 \lambda \overset{\nabla}{\mathbf{D}} + \text{ third} - \text{order terms} \tag{4.3.10}$$

Comparing eq. 4.3.10 with eq. 4.3.1, we see that to second order in the velocity gradient the upper-convected Maxwell equation for small strain rates reduces to a special case of the equation of the second-order fluid with $\psi_{1,0} = 2\lambda\eta_0$ and $\psi_{2,0} = 0$. All properly formulated constitutive equations for which the stress is a smooth functional of the strain history reduce at second order in the velocity gradient to the equation of the second-order fluid. Example 4.3.3, however, illustrates that the equation of the second-order fluid cannot be trusted except for slow nearly steady flows.

**Note that it is possible to define other convected derivatives (Bird et al., 1987; Larson, 1988). The upper-convected derivative arises most naturally from molecular theory. We will see this in Chapter 11 with the elastic dumbbell model. Note also that* $\overset{\nabla}{\boldsymbol{\tau}}$ *arises naturally from using the time derivative of the neo-Hookean model, eq 1.5.2, in eq 3.2.18. See Exercise 4.6.3.:*

Example 4.3.3 Shear and Extensional Flow Predictions of the UCM Equation

Calculate the predictions of the upper-convected Maxwell equation in (a) start-up of steady shear and (b) steady state uniaxial extension for arbitrary shear rate $\dot{\gamma}$ and extension rate $\dot{\epsilon}$, and compare these predictions with those for the Newtonian and second-order fluids.

Solution

(a) Start-up of Steady Shear

Using the definition of eq. 4.3.2 for the upper-convected derivative, the upperconvected Maxwell equation (eq. 4.3.7) can be written in expanded form as follows:

$$\boldsymbol{\tau} + \lambda\frac{\partial}{\partial t}\boldsymbol{\tau} + \lambda\mathbf{v}\cdot\nabla\boldsymbol{\tau} - \lambda(\nabla\mathbf{v})^T\cdot\boldsymbol{\tau} - \lambda\boldsymbol{\tau}\cdot\nabla\mathbf{v} = 2\eta_0\mathbf{D} \qquad (4.3.11)$$

The term $\nabla\boldsymbol{\tau}$ is zero because we are considering a homogeneous flow. The symmetry of the shearing flow leads us to expect that the stress tensor will contain only the components τ_{12}, τ_{21}, τ_{11}, τ_{22}, and τ_{33}. Assuming this form for the stress tensor and using eq. 2.2.10 for the velocity gradient, we obtain

$$(\nabla\mathbf{v})^T\cdot\boldsymbol{\tau} = \begin{pmatrix}\dot{\gamma}\tau_{12} & \dot{\gamma}\tau_{22} & 0\\ 0 & 0 & 0\\ 0 & 0 & 0\end{pmatrix}; \quad \boldsymbol{\tau}\cdot\nabla\mathbf{v} = \begin{pmatrix}\dot{\gamma}\tau_{12} & 0 & 0\\ \dot{\gamma}\tau_{22} & 0 & 0\\ 0 & 0 & 0\end{pmatrix}$$

Thus eq. 4.3.11 becomes

$$\begin{pmatrix}\tau_{11} & \tau_{12} & 0\\ \tau_{12} & \tau_{22} & 0\\ 0 & 0 & \tau_{33}\end{pmatrix} + \lambda\frac{\partial}{\partial t}\begin{pmatrix}\tau_{11} & \tau_{12} & 0\\ \tau_{12} & \tau_{22} & 0\\ 0 & 0 & \tau_{33}\end{pmatrix} - \lambda\begin{pmatrix}2\dot{\gamma}\tau_{12} & \dot{\gamma}\tau_{22} & 0\\ \dot{\gamma}\tau_{22} & 0 & 0\\ 0 & 0 & 0\end{pmatrix} = \eta_0\begin{pmatrix}0 & \dot{\gamma} & 0\\ \dot{\gamma} & 0 & 0\\ 0 & 0 & 0\end{pmatrix} \qquad (4.3.12)$$

To obtain the steady state results, we set the time derivative to zero. We find immediately that $\tau_{33} = \tau_{22} = 0$. With this result for τ_{22}, we find that $\tau_{12} = \eta_0\dot{\gamma}$, from which we can obtain $\tau_{11} = 2\eta_0\lambda\dot{\gamma}^2$. This result implies that the shear viscosity is a constant η_0, the first normal stress coefficient is also a constant equal to $2\lambda\eta_0$, and the second normal stress coefficient is zero. This is the same result that we obtained in the second-order fluid limit of the UCM equation!

To obtain the stress growth predictions, we note from eq. 4.3.12 that the only nonzero stress components are τ_{11} and τ_{12} and that these satisfy

$$\tau_{11} + \lambda\frac{\partial\tau_{11}}{\partial t} - 2\lambda\tau_{12}\dot{\gamma} = 0; \qquad \tau_{12} + \lambda\frac{\partial\tau_{12}}{\partial t} = \eta_o\dot{\gamma} \qquad (4.3.13)$$

If the shearing starts at $t = 0$, we can take $\tau_{11} = \tau_{12} = 0$ at $t = 0$. Solving first for τ_{12} gives

$$\eta^+ \equiv \frac{\tau_{12}}{\dot{\gamma}} = \eta_0(1 - e^{-t/\lambda}) \qquad (4.3.14)$$

This is the same result that we obtained in Example 3.2.1 for the linear viscoelastic (LVE) model. However, unlike the LVE model, the UCM predicts shear normal stresses. Substituting this solution for τ_{12} into the equation for τ_{11} and solving the resulting equation gives

$$\psi_1^+ \equiv \frac{\tau_{11} - \tau_{22}}{\dot{\gamma}^2} = \frac{\tau_{11}}{\dot{\gamma}^2} = 2\eta_0\lambda\left[1 - e^{-t/\lambda}\left(1 + \frac{t}{\lambda}\right)\right] \quad (4.3.15)$$

Figure 4.3.1 compares $\eta^+(t)$ for the UCM equation to the predictions for the Newtonian and the second-order fluids. Although all three of these constitutive equations predict a shear rate independent viscosity at steady state, only the UCM equation predicts a gradual rather than instantaneous growth of stress after inception of shearing. Therefore the UCM equation is the most realistic of the three.

(b) *Steady State Uniaxial Extension*

Substituting the velocity gradient for uniaxial extension, eq. 2.2.9, into eq. 4.3.7, we readily find that at steady state

$$\tau_{11} = \frac{2\eta_0\dot{\epsilon}}{1 - 2\lambda\dot{\epsilon}}; \qquad \tau_{22} = \tau_{33} = \frac{-\eta_0\dot{\epsilon}}{1 + \lambda\dot{\epsilon}} \quad (4.3.16)$$

Thus we obtain

$$\bar{\eta}_u = \frac{\tau_{11} - \tau_{22}}{\dot{\epsilon}} = \frac{2\eta_0}{1 - 2\lambda\dot{\epsilon}} + \frac{\eta_0}{1 + \lambda\dot{\epsilon}} \quad (4.3.17)$$

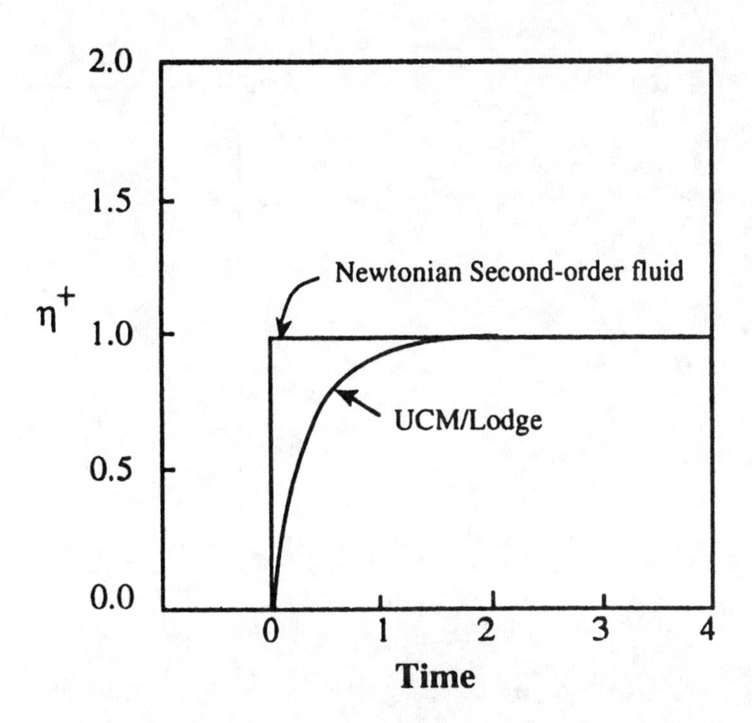

Figure 4.3.1. Growth of the shear viscosity η^+ after onset of steady shearing for the UCM/Lodge equation, compared to a Newtonian and a second-order fluid.

This viscosity is extremely extension thickening, since it rises to infinity as $\dot{\epsilon}$ approaches $1/2\lambda$. Recall that the second-order fluid equation predicts only a linear rise of η_u with $\dot{\epsilon}$, while the Newtonian fluid predicts no dependence of η_u on $\dot{\epsilon}$.

Thus for steady state uniaxial extensional flow, in contrast to steady state shearing flow, the second-order fluid result agrees with the UCM prediction only at small strain rates. The UCM, second-order fluid, and Newtonian fluid equations all differ in their predictions of the strain rate dependences of the extensional viscosity, though the strain rate dependences of the shear viscosity are the same for all three equations. This result typifies the usual finding that constitutive equations differ among themselves more strongly in their predictions of extensional viscosities than in their predictions of shear viscosities.

Following the steps we used for start-up of steady shearing, we can also use the UCM equation to calculate η_u^+, the time-dependent uniaxial extensional viscosity after start-up of steady extension. We leave it to the reader to show that the result is identical to that obtained in Section 4.4.4 for the Lodge integral equation. This time-dependent viscosity η_u^+ for the UCM equation is compared to that for the Newtonian and second-order fluids in Figure 4.3.2. Note that as was the case in shear, only the UCM equation predicts a gradual growth of the extensional viscosity after start-up of flow. Figure 4.3.2 also depicts the unbounded growth of the extensional viscosity shown by the UCM equation when $\dot{\epsilon}$ exceeds $1/2\lambda$.

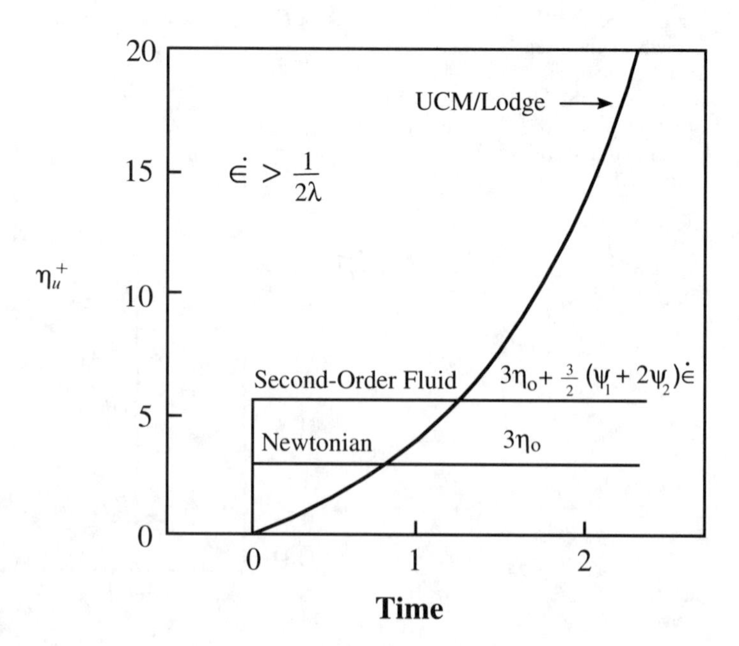

Figure 4.3.2.
Growth of the uniaxial extensional viscosity η_u^+ after onset of steady extension for the UCM/Lodge equation compared to a Newtonian and a second-order fluid.

4.3.3 Lodge Integral Equation

Recall from Chapter 3 that the linear Maxwell equation can be written in an integral form. The nonlinear upper-convected Maxwell equation can also be written in an integral form, namely

$$\boldsymbol{\tau} = \int_{-\infty}^{t} \frac{\eta_0}{\lambda^2} e^{-(t-t')/\lambda} (\mathbf{B}(t, t') - \mathbf{I}) dt' \tag{4.3.18}$$

By making use of eq. 4.3.3, it can be shown that eq. 4.3.7 can be derived from eq. 4.3.18 (Astarita and Marrucci, 1974; Larson, 1988). For an incompressible liquid, the stress tensor $\boldsymbol{\tau}$ is determined only to within an isotropic constant; thus the unit tensor $\mathbf{I}$ in eq. 4.3.18 can safely be dropped. Although equivalent to the differential UCM equation, it is worthwhile to introduce eq. 4.3.18 because some problems are easier to address with an integral equation and because eq. 4.3.18 is a simple prototype of more realistic integral equations to be discussed in Section 4.4. To illustrate how calculations are carried out with integral constitutive equations, let us calculate the shear stress in a couple of different simple shearing deformation histories. From eq. 1.4.24 we find that the shear or off-diagonal component of the tensor $\mathbf{B}$ in a simple shearing deformation is simply the shear strain γ. Thus $B_{12}(t, t') = \gamma(t, t')$, and from eq. 4.3.18 we obtain

$$\tau_{12} = \int_{-\infty}^{t} \frac{\eta_0}{\lambda^2} e^{-(t-t')/\lambda} \gamma(t, t') dt' \tag{4.3.19}$$

Here $\gamma(t, t')$ is the shear strain that accumulated between the times t' and t. Let us illustrate the meaning of $\gamma(t, t')$ with two examples.

Example 4.3.4 Step Shear Strain

Calculate $\gamma(t, t')$ for a step shear of magnitude γ_0 applied at time zero as illustrated in Figure 4.3.3a.

Solution

If $t' > 0$, then since the step occurred at time zero, no strain was added to the material between t' and t and $\gamma(t', t) = 0$. However, if $t' < 0$, then between t' and t the strain γ_0 was added to the material and $\gamma(t, t') = \gamma_0$. Thus, for a step shear strain at time zero

$$\gamma(t, t') = \begin{cases} 0; & t' > 0 \\ \gamma_0; & t' \le 0 \end{cases} \tag{4.3.20}$$

Figure 4.3.3.
(a) Strain history for a step strain γ_0 applied at time zero. (b) Strain history for start-up of steady shearing at a rate $\dot{\gamma}$, starting at time zero.

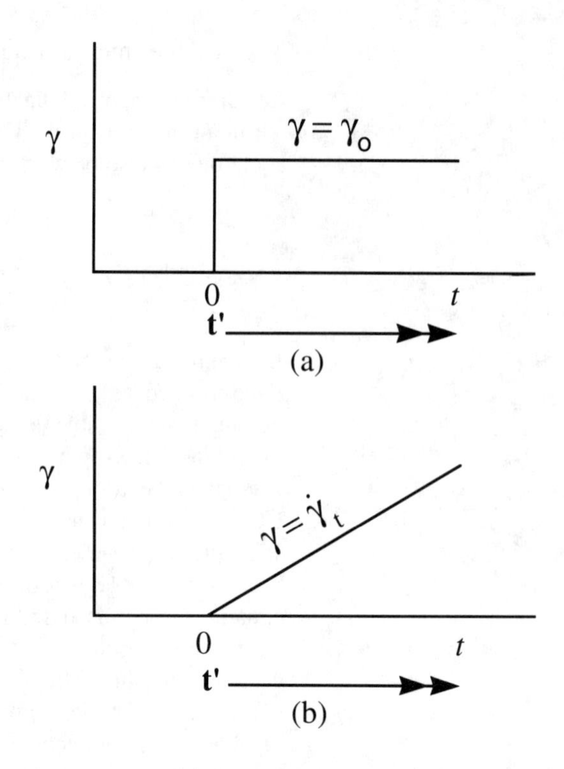

Example 4.3.5 Start-up of Steady Shearing

Calculate $\gamma(t, t')$ for steady shearing at a shear rate $\dot{\gamma}$ that began at time zero as illustrated in Figure 4.3.3b.

Solution

If $t' > 0$, then the shearing has been continuous at a rate $\dot{\gamma}$ from time t' until the present time t, so the total strain accumulated between t' and the present time t is $\dot{\gamma}(t - t')$. However, if $t' < 0$, then since no shearing occurred from time t' up to time zero, the total accumulated strain between t' and t is that which accumulated since time zero, namely $\dot{\gamma}t$. Thus

$$\gamma(t, t') = \begin{cases} \dot{\gamma}(t - t'); & t' > 0 \\ \dot{\gamma}t; & t' \leq 0 \end{cases} \tag{4.3.21}$$

We leave it as an exercise to the reader to calculate the shear stress at time t for the two strain histories given by eq.s 4.3.20 and 4.3.21.

As one more example, let us consider an extensional flow.

Example 4.3.6 Stress Growth in Start-up of Uniaxial Extension

Calculate, using eq. 4.3.18, the growth of the extensional viscosity after start-up of a steady uniaxial extension. Compare the steady state extensional viscosity from this equation with the prediction of the differential version of the UCM equation, eq. 4.3.7.

Solution

The tensor **B** for uniaxial extension is given in Exercise 2.8.1a. If the uniaxial extensional flow starts at time $t = 0$, by analogy with Example 4.3.5 we have

$$B_{11} = \begin{cases} e^{2\dot{\epsilon}(t-t')}; & t' > 0 \\ e^{2\dot{\epsilon}t}; & t' \leq 0 \end{cases}; \quad B_{22} = \begin{cases} e^{-\dot{\epsilon}(t-t')}; & t' > 0 \\ e^{-\dot{\epsilon}t}; & t' \leq 0 \end{cases} \tag{4.3.22}$$

From this and eq. 4.3.18 we obtain

$$\tau_{11} = \int_{-\infty}^{0} \frac{\eta_0}{\lambda^2} e^{-(t-t')/\lambda} e^{2\dot{\epsilon}t} dt' + \int_{0}^{t} \frac{\eta_0}{\lambda^2} e^{-(t-t')/\lambda} e^{2\dot{\epsilon}(t-t')} dt' - \int_{-\infty}^{t} \frac{\eta_0}{\lambda^2} e^{-(t-t')/\lambda} dt' \tag{4.3.23}$$

$$\tau_{22} = \int_{-\infty}^{0} \frac{\eta_0}{\lambda^2} e^{-(t-t')/\lambda} e^{-\dot{\epsilon}t} dt' + \int_{0}^{t} \frac{\eta_0}{\lambda^2} e^{-(t-t')/\lambda} e^{-\dot{\epsilon}(t-t')} dt' - \int_{-\infty}^{t} \frac{\eta_0}{\lambda^2} e^{-(t-t')/\lambda} dt' \tag{4.3.24}$$

Carrying out these integrals yields

$$\tau_{11} = \frac{\eta_0/\lambda}{1 - 2\lambda\dot{\varepsilon}} + \frac{\eta_0}{\lambda}\left(1 - \frac{1}{1 - 2\lambda\dot{\varepsilon}}\right) e^{-t(1 - 2\dot{\varepsilon}\lambda)/\lambda} - \frac{\eta_0}{\lambda} \tag{4.3.25}$$

$$\tau_{22} \frac{\eta_0/\lambda}{1 + \lambda\dot{\varepsilon}} + \frac{\eta_0}{\lambda}\left(1 - \frac{1}{1 + \lambda\dot{\varepsilon}}\right) e^{-t(1 + \dot{\varepsilon}\lambda)/\lambda} - \frac{\eta_0}{\lambda} \tag{4.3.26}$$

We find at steady state, when $t \to \infty$, that

$$\bar{\eta}_u = \frac{\tau_{11} - \tau_{22}}{\dot{\epsilon}} = \frac{\eta_0}{\lambda\dot{\epsilon}}\left(\frac{1}{1 - 2\lambda\dot{\epsilon}} - \frac{1}{1 + \lambda\dot{\epsilon}}\right) = \frac{3\eta_0}{(1 - 2\lambda\dot{\epsilon})(1 + \lambda\dot{\epsilon})} = \frac{2\eta_0}{1 - 2\lambda\dot{\epsilon}} + \frac{\eta_0}{1 + \lambda\dot{\epsilon}} \tag{4.3.27}$$

which is the same result that we obtained earlier from the Maxwell model, eq. 4.3.7 in Example 4.3.3b.

Equation 4.3.25 also shows us that when $\dot{\epsilon}$ exceeds $1/2\lambda$, the extensional stress grows without bound and no steady state is reached for this constitutive equation.

As discussed in Chapter 3, real polymeric fluids possess a distribution of relaxation times. Equation 4.3.18 can easily be generalized to include this spectrum:

$$\boldsymbol{\tau} = \int_{-\infty}^{t} M(t - t') \, \mathbf{B}(t, t') dt' \tag{4.3.28}$$

where $M(t)$ is the memory function defined in eq. 3.2.2. Equation 4.3.28 is called the equation of the *Lodge rubber like liquid* (Lodge, 1956, 1964, 1968). A differential form for this equation can be obtained if $M(t - t')$ is represented by a discrete spectrum (recall eq. 3.2.10):

$$M(t - t') = \sum_{i=1}^{N} \frac{G_i}{\lambda_i} e^{-(t-t')/\lambda_i} \tag{4.3.29}$$

With this spectrum, we can decompose the stress tensor into a sum of contributions from individual modes:

$$\boldsymbol{\tau} = \sum_{i=1}^{N} \boldsymbol{\tau}_i \tag{4.3.30}$$

with

$$\boldsymbol{\tau}_i = \int_{-\infty}^{t} \frac{G_i}{\lambda_i} e^{(t-t')/\lambda_i} \mathbf{B}(t, t')dt' \tag{4.3.31}$$

To within an added isotropic constant, the stress $\boldsymbol{\tau}_i$, given by eq. 4.3.31 for each mode, satisfies an upper-convected Maxwell equation:

$$\boldsymbol{\tau}_i + \lambda_i \overset{\nabla}{\boldsymbol{\tau}_i} = 2G_i \lambda_i \mathbf{D} \tag{4.3.32}$$

With the inclusion of the spectrum of relaxation times in Lodge's equation, or equivalently, in the upper-convected Maxwell equation, we recapture all the power of the theory of linear viscoelasticity described in Chapter 3. Of course we obtain the Newtonian behavior discussed in Chapter 2 when the strain rate is low, and we retain the nonlinear features of the UCM equation, in particular the existence of a first normal stress difference in simple shearing and extensional thickening in uniaxial extensional flows. However, as with the UCM equation, we still predict a zero second normal stress difference and strain rate independent values of the shear viscosity and first normal stress coefficient—that is, we predict no shear thinning. Also, the extensional thickening we predict is too severe, since an infinite steady state extensional viscosity is obtained above a critical value of $\dot{\epsilon}$. Thus, although the Lodge/UCM equation has some of the qualitative behavior of real polymeric fluids, it is still far from being a quantitative constitutive equation for most polymeric fluids.

For dilute solutions of polymers, however, the UCM equation, or a simple variation of it, seems to be satisfactory. The solutions we have in mind are very dilute, at most a few tenths of a percent polymer. For these solutions, polymer molecules do not entangle much with each other, and the viscoelastic properties of the polymer solution are particularly simple. In most dilute polymer solutions, the longest relaxation time is so short that the fluid shows little viscoelasticity unless strain rates are hundreds of reciprocal seconds or higher. These high strain rates occur in some processing flows such as coating flows.

On the other hand, if the solvent viscosity is very high, say a hundred poise or greater, the longest relaxation time can be a second or more, and strong viscoelastic effects are seen even at shear rates of only a few reciprocal seconds. Such viscous dilute solutions, which are called *Boger fluids* (Boger, 1977; Binnington and Boger, 1985), are often concocted as test fluids. The solvent in a Boger fluid is often itself a low molecular weight polymeric fluid. The use of Boger fluids is advantageous in that their rheology has the simplicity of dilute solutions, yet the relaxation times are long enough to permit the manifestation of strong nonlinear viscoelastic effects at shear rates low enough to be easily accessed by most rheometers.

Although the the UCM equation gives the polymer contribution to the stress in a dilute solution such as a Boger fluid, the solvent contribution to the stress cannot be neglected, and so the total stress tensor τ in these solutions is the sum of the polymeric and solvent contributions

$$\boldsymbol{\tau} = \boldsymbol{\tau}^p + \boldsymbol{\tau}^s \tag{4.3.33}$$

Here $\boldsymbol{\tau}^p$ is given by the UCM equation 4.3.7 (or equivalently by the Lodge equation, eq. 4.3.18), and $\boldsymbol{\tau}^s$ is usually just a Newtonian term $2\eta_s\mathbf{D}$, where η_s is the solvent viscosity. The combination of these two terms is the *Oldroyd-B constitutive equation* (Oldroyd, 1950; see Exercise 4.6.4). Figure 4.3.4 compares the storage mod-

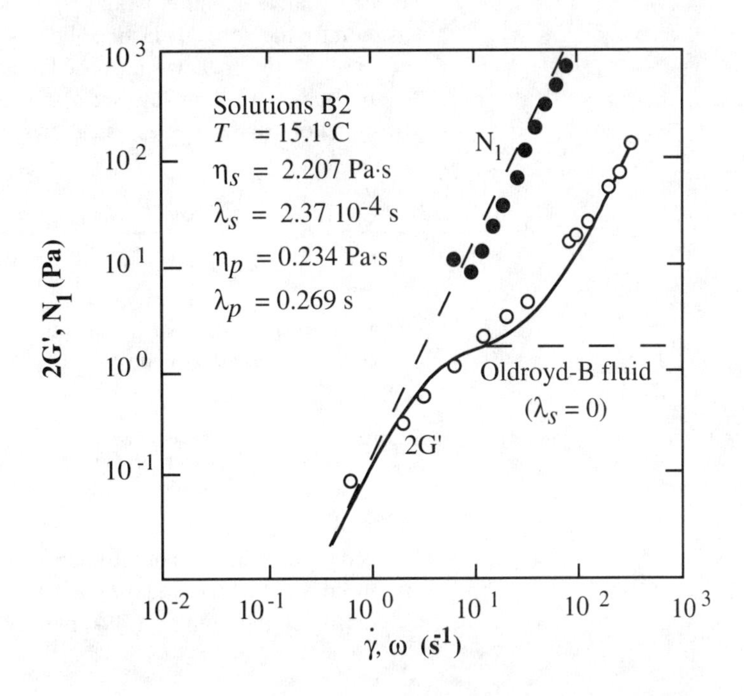

Figure 4.3.4
$2G'$ versus ω and N_1 versus $\dot{\gamma}$ for a Boger fluid, in this case a high molecular weight polyisobutylene in a solvent consisting of poly(1-butene) of low molecular weight and kerosene. η_p and λ_p are the viscosity and relaxation time of the high molecular weight polymer; η_s and λ_s are the viscosity and relaxation time of the solvent. The solid line gives the best fit to G' data of the upper-convected Maxwell equation with two relaxation times; the dashed lines give the best fit of the Oldroyd-B equation, eq. 4.3.33. Both models fit the N_1 data well. Replotted from Mackay and Boger (1987).

ulus G' and the first normal stress difference N_1 for a Boger fluid to the predictions of the Oldroyd-B and UCM equations. In this case the solvent had a small, but non negligible relaxation time, so the solvent contribution could itself be represented by the UCM equation. Thus $\boldsymbol{\tau}^p$ and $\boldsymbol{\tau}^s$ were both given by eq. 4.3.7, but the relaxation time λ_s for the solvent was some three orders of magnitude smaller than λ_p, the relaxation time for the polymer. Therefore the constitutive equation for the solution as a whole was eq. 4.3.32 with two relaxation times (Mackey and Boger, 1987). The agreement of this two-relaxation-time UCM equation with the experimental data for this Boger fluid is satisfactory.

4.4 More Accurate Constitutive Equations

4.4.1 Integral Constitutive Equations

Most polymeric fluids are not described very well by the Lodge/UCM equation. The Lodge/UCM fluid is an inadequate description for most materials because its elasticity is that of a simple Hookean material. This can be seen by supposing that the deformation occurs suddenly at time zero. Immediately after the deformation, the stress tensor of eq. 4.3.28 becomes

$$\boldsymbol{\tau} = \mathbf{B} \int_{-\infty}^{0} M(t - t')dt' = \mathbf{B}\, G(0) \tag{4.4.1}$$

which is identical to eq. 1.5.1 for a Hookean rubber. A more general constitutive equation would be a viscoelastic version of the equation for the general elastic solid, eq. 1.6.4. A viscoelastic equation based on this idea was proposed independently by Kaye (1962) and by Bernstein, Kearsley, and Zapas (1963). The so-called K-BKZ equation can be written as follows:

$$\boldsymbol{\tau} = \int_{-\infty}^{t} \left[2\frac{\partial u(I_B, II_B, t - t')}{\partial I_B} \mathbf{B}(\mathbf{t}, \mathbf{t}') - 2\frac{\partial u(I_B, II_B, t - t')}{\partial II_B} \mathbf{B}^{-1}(t, t') \right] dt' \tag{4.4.2}$$

Here $u(I_B, II_B, t - t')$ is a time-dependent elastic energy kernel function. The strain energy

$$W = \int_{-\infty}^{t} u[I_B(t, t'), II_B(t, t'), t - t']dt'$$

is the generalization of the energy function of the elastic solid given in eq. 1.6.13. I_B and II_B are the invariants of the tensor $\mathbf{B}$ as discussed in Example 1.4.4. In eq. 4.4.2, $\mathbf{B}$ and the invariants of $\mathbf{B}$

depend on t and t'. The Lodge equation is a special case of eq. 4.4.2 obtained when

$$u(I_B, II_B, t - t') = \frac{1}{2} M(t - t')\, I_B \tag{4.4.3}$$

With the introduction of the K-BKZ equation, we are at last in a position to describe shear thinning phenomena; but to do so we need to obtain the energy function $u(I_B, II_B, t - t')$. Because u depends on two strain invariants as well as time, obtaining the complete energy function requires a lot of experimental data or else some guidance from molecular theory (which will come in Chapter 11). However, things become much simpler if we restrict ourselves to shearing flows. For shearing flows, using eq 1.4.24 for **B** and 1.4.36 for its inverse, eq. 4.4.2 yields

$$\tau_{12} = \int_{-\infty}^{t} 2\left(\frac{\partial u}{\partial I_B} - \frac{\partial u}{\partial II_B}\right) \gamma(t, t') dt' \tag{4.4.4}$$

$$N_1 = \int_{-\infty}^{t} 2\left(\frac{\partial u}{\partial I_B} - \frac{\partial u}{\partial II_B}\right) \gamma^2(t, t') dt' \tag{4.4.5}$$

$$N_2 = -\int_{-\infty}^{t} 2\frac{\partial u}{\partial II_B} \gamma^2(t, t') dt' \tag{4.4.6}$$

If one wishes to obtain τ_{12} and N_1, one need only obtain the quantity

$$\phi(\gamma, t - t') \equiv 2\left(\frac{\partial u}{\partial I_B} - \frac{\partial u}{\partial II_B}\right) \tag{4.4.7}$$

This function can most readily be obtained by performing step strain deformations; if a shear strain γ is imposed on the material at time zero, eq. 4.4.4 gives

$$\tau_{12} = \gamma \int_{-\infty}^{0} \phi(\gamma, t - t')\, dt' = \gamma \int_{t}^{\infty} \phi(\gamma, s) ds \tag{4.4.8}$$

where s is defined in Chapter 3 (3.2.5). Thus the function $\phi(\gamma, s)$ can be obtained by taking a time derivative of the relaxing shear stress after imposition of a step shear strain. Once ϕ has been obtained, τ_{12} and N_1 in *any* shearing deformation history can be predicted for the K-BKZ equation from eqs. 4.4.4 and 4.4.5.

Figure 4.4.1 shows the shear stress modulus $G(\gamma, t) = \tau_{12}/\gamma$ as a function of time after imposition of a step shear strain for a commercial low density polyethylene. This particular batch of polyethylene is referred to as melt I. It is almost identical to the melts IUPAC A and IUPAC X encountered in Figures 4.2.5 and 4.2.7, respectively. The data of Figure 4.4.1 can be used to obtain the K-BKZ function $\phi(\gamma, s)$. On a log–log plot the relaxation curves for Melt I are nearly parallel; this means that the stress relaxation modulus is nearly factorable into time- and strain-dependent terms:

$$G(\gamma, t) = G(t)\, h(\gamma) \tag{4.4.9}$$

This, in turn, implies that the kernel function $\phi(\gamma, t)$ is also factorable:

$$\phi(\gamma, t) = M(t)\, h(\gamma) \tag{4.4.10}$$

At small strains $\gamma \;\; \rightarrow 0$, so $G(\gamma, t)$ must reduce to the linear viscoelastic modulus $G(t)$, and $h(\gamma)$ must approach unity for small γ. Note also in Figure 4.4.1 that the normal stress modulus N_1/γ^2 equals the shear stress modulus τ_{12}/γ; this implies that melt I obeys the Lodge–Meissner relationship, eq 4.2.8. Note that this relation follows directly from eqs. 4.4.4 and 4.4.5.

For deformations other than shear—such as step biaxial extension (Soskey and Winter, 1985) and step planar extension (Khan and Larson, 1991)—time-strain factorability has also been found to hold, at least approximately, for commercial polymer melts. If

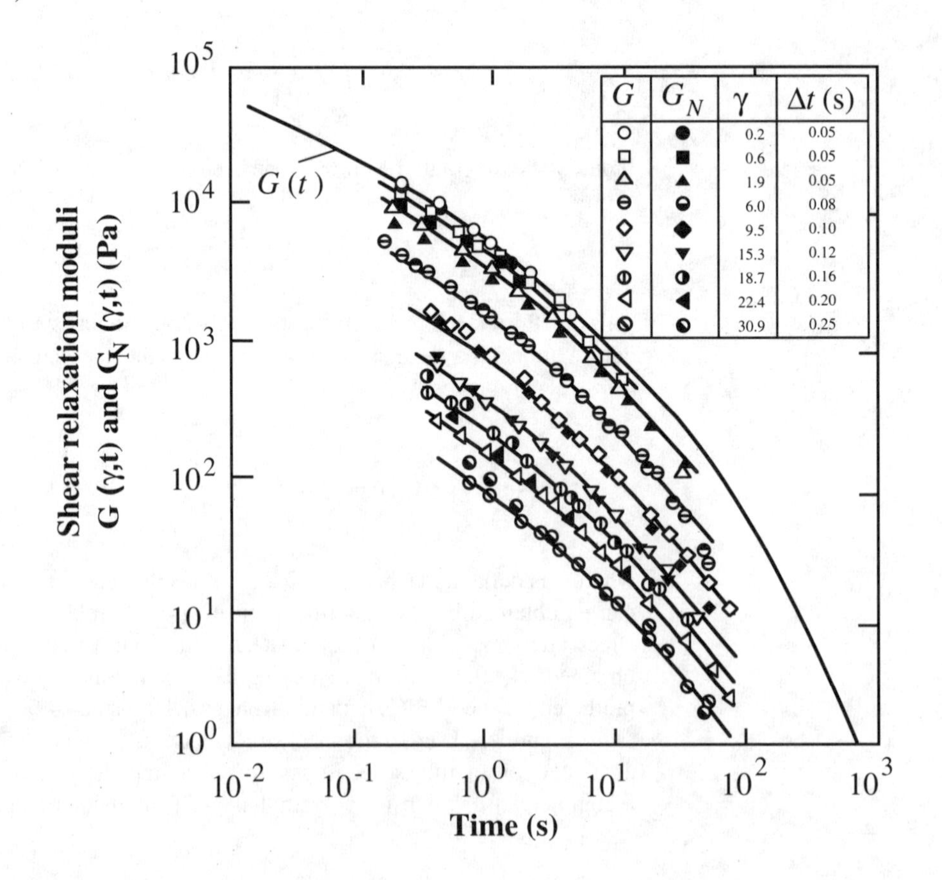

Figure 4.4.1
Modulis of shear stress $G \equiv \tau_{12}/\gamma$ (open symbols) and of the primary normal stress difference $G_N \equiv N_1/\gamma^2$ (solid symbols) of melt I as functions of time after a step shear. Δt is the time over which the "step" in strain occurred. Replotted from Laun (1978).

it holds for all types of deformation for a given material, then $u(I_B, II_B, t-t') = M(t-t')\,U(I_B, II_B)$, and the K-BKZ equation reduces to

$$\boldsymbol{\tau} = \int_{-\infty}^{t} 2M(t-t') \left[\frac{\partial U(I_B, II_B)}{\partial I_B} \mathbf{B}(t,t') - \frac{\partial U(I_B, II_B)}{\partial II_B} \mathbf{B}^{-1}(t,t') \right] dt' \quad (4.4.11)$$

Time–strain factorability, when it can be used, simplifies the K-BKZ equation considerably; since $M(t-t')$ can be obtained from simple linear viscoelastic measurements, only nonlinear experiments are needed to obtain the strain-dependent function $U(I_B, II_B)$. To predict τ_{12} or N_1 in any shearing flow, $h(\gamma)$ is the only nonlinear material function that needs to be measured. This so-called *damping function*, $h(\gamma)$, can be obtained at each γ simply by finding the amount of vertical shift on a log–log plot such as Figure 4.4.1 required to superimpose a curve of $G(\gamma, t)$ onto the linear viscoelastic curve $G(t) = G(\gamma \to 0, t)$. When this is done for melt I, one obtains the damping function plotted in Figure 4.4.2. Note that $h(\gamma)$ decreases monotonically as the strain γ increases. The negative departures of the modulus $G(\gamma, t)$ from the linear viscoelastic limit become ever greater as the strain increases, a phenomenon known as *strain softening*. The damping function of Figure 4.4.2 can be fit out to a strain γ of 10 or so by a simple exponential (Wagner, 1976)

$$h(\gamma) = \exp(-n\gamma) \quad (4.4.12)$$

Figure 4.4.2.
The damping function for melt I in step shearing fitted by the single exponential of eq 4.4.12 (dashed line) and the double exponential of eq 4.4.13 (solid line). Replotted from Laun (1978).

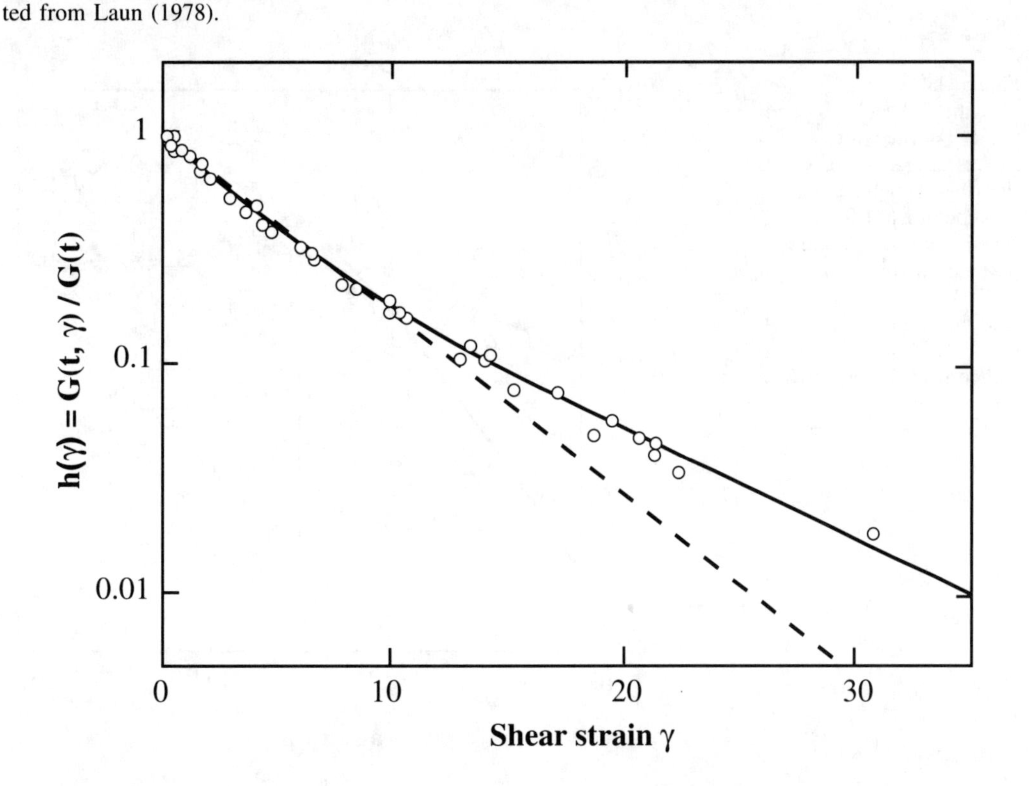

with $n = 0.18$. For higher strains a sum of two exponentials (Laun, 1978)

$$h(\gamma) = f_1 \exp(-n_1\gamma) + f_2 \exp(-n_2\gamma) \tag{4.4.13}$$

with $f_1 = 0.57$, $f_2 = 0.43$, $n_1 = 0.31$, and $n_2 = 0.106$, gives a near-perfect fit. These data can also be fit fairly well out to $\gamma = 10$ by the simple expression (Khan and Larson, 1987)

$$h(\gamma) = \frac{1}{1 + \frac{1}{3}\alpha\gamma^2} \tag{4.4.14}$$

with $\alpha = 0.21$.

With expressions 4.4.12 or 4.4.13 for $h(\gamma)$ and with $G(t)$ obtained from a sum of exponentials with parameters λ_i and G_i (given in Table 3.3.1 for melt I), predictions can now be made for various shearing flows in the nonlinear regime. Figure 4.2.2 shows that the predicted steady state shear viscosity η and first normal stress coefficient ψ_1 agree well with the experimental data.

The same can be said for the growth of τ_{12} and N_1 after start-up of steady shearing in Figure 4.4.3 and the relaxation of τ_{12} and N_1 after cessation of steady state shearing in Figure 4.4.4. The agreement of the predicted η and ψ_1 with the measured values shows that at least in simple shearing flows, there is a strong connection between strain softening and shear thinning; if the strain softening in step shear is properly incorporated into the constitutive equation, accurate predictions of the shear thinning follow directly. The good

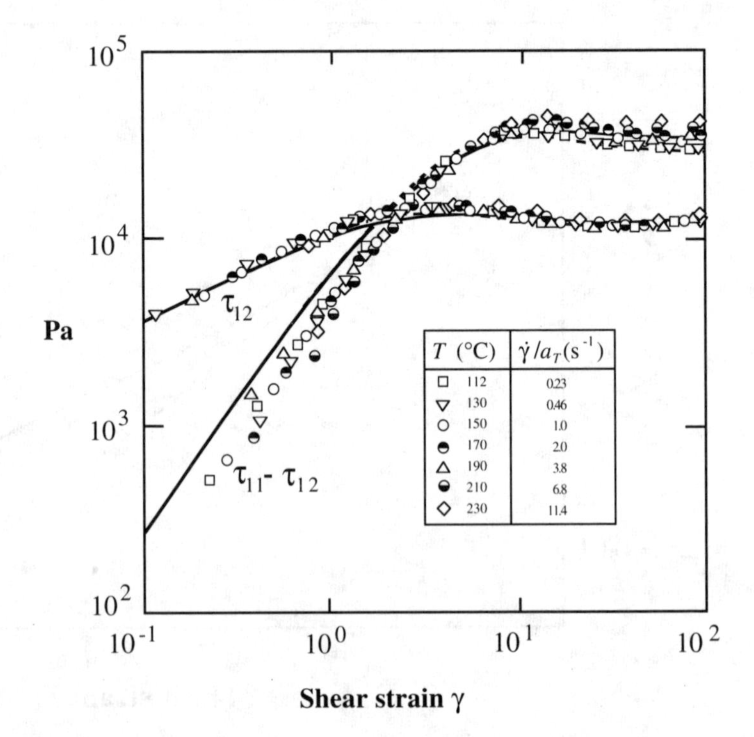

Figure 4.4.3. Growth of shear stress and primary normal stress difference after start-up of steady shearing of melt I compared to predictions of the K-BKZ equation with the double exponential damping function, eq 4.4.13 (solid lines), and with the single exponential, eq 4.4.12 (dashed line). Replotted from Laun (1978).

Figure 4.4.4. Relaxation of shear stress and primary normal stress difference after cessation of steady state shearing of melt I compared to predictions of the BKZ equation with the double exponential damping function, eq 4.4.13 (solid lines), and with the single exponential, eq 4.4.12 (dashed lines). Redrawn from Laun (1978).

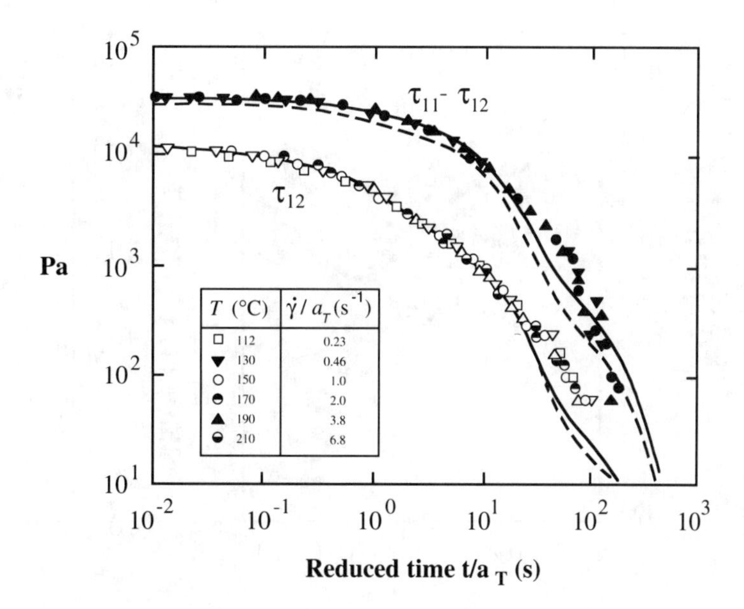

agreement obtained in Figures 4.2.2, 4.4.3, and 4.4.4 also shows that a wealth of nonlinear rheological data for melt I can be economically represented by a simple damping function that expresses the degree of strain softening this material shows in simple shearing deformations. Indeed, by fitting expressions such as eqs. 4.4.12–4.4.14 to data for other polymer melts, the nonlinear behavior of a series of different melts can be summed up or characterized, for example, by the values of α that give the best fits to the damping functions for those melts. For melts in step shear, the best-fit values of α in eq. 4.4.14 typically lie in the range 0.15–0.6 (Khan et al., 1987).

One can also extract the damping function from the shear rate dependence of the steady state shear viscosity by fitting the integral constitutive equation to the experimental data. Although the damping function one obtains this way differs significantly from that obtained directly from the step strain data (Laun, 1986), the predictions of the constitutive equation in steady shearing are not terribly sensitive to the damping function chosen. This is evident in Figures 4.2.2–4.2.4, in which accurate stress predictions are obtained from both the single and double exponential damping functions, eqs. 4.4.12 and 4.4.13.

Factorization of the function $\phi(I_B, II_B, t - t')$ in eq 4.4.8 is not always a good approximation. Figure 4.4.5 shows that for a nearly monodisperse concentrated polystyrene solution, time–strain factorability is not valid at short times after the imposition of the step shear. An accurate K-BKZ constitutive equation for shearing flows of this material will be much more complex than that for melt I. Furthermore, in strain histories in which a strain reversal takes place, such as constrained recoil (Wagner and Laun, 1978) or double-step strains with the second strain of sign opposite the first (Doi, 1980; Larson and Valesano, 1986), good agreement

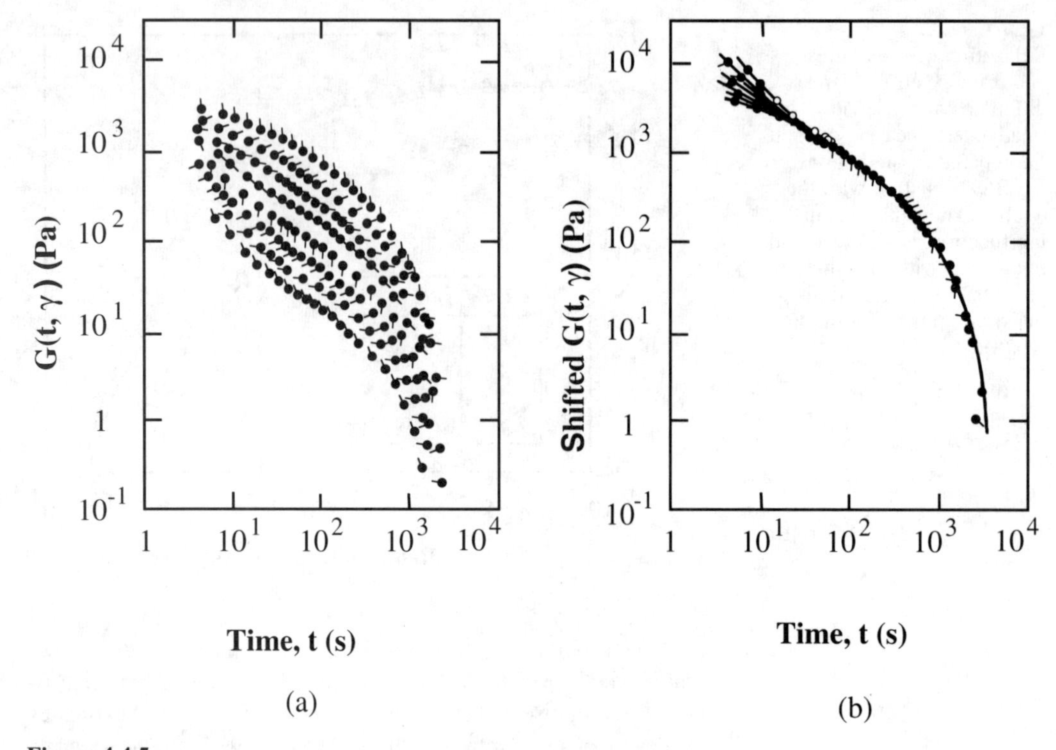

Figure 4.4.5.
(a) $G(t, \gamma) = \tau_{12}/\gamma$ as a function of time for various strain magnitudes for concentrated polystyrene solutions. Each curve represents a different strain; the lower curves correspond to higher strains. (b) The curves can be superimposed at times longer than about 20 seconds by vertical shifting by an amount, $h(\gamma)$, that depends on strain. Redrawn from Einaga et al. (1971).

with the measured stresses is sometimes not attained unless the K-BKZ integral is modified.

Even when strain reversal is not a concern and when time–strain factorability can be assumed, we still have much more work to do, because a really useful constitutive equation must be capable of predicting stresses for flows other than simple shearing. Unfortunately, there is a paucity of data for other types of flow, such as the various extensional flows. Especially rare are *sets* of data for a single fluid in a variety of different types of flow. Such sets of data are necessary to obtain the strain energy function $U(I_B, II_B)$ for a factorable K-BKZ fluid, or $u(I_B, II_B, t - t')$ for a nonfactorable fluid. Reasonably complete sets of data do exist for long chain branched melt I (Laun, 1978; Wagner et al., 1979; Wagner and Meissner, 1980) and for a polyisobutylene melt studied by Meissner and co-workers (1982). There are also somewhat less complete sets of data for a handful of other melts (Papanastasiou et al., 1983; Soskey and Winter, 1985; Khan et al., 1987; Laun and Schuch, 1989; Samurkas et al., 1989). With these limited data, theoretical workers have attempted to devise simple and yet accurate kernel functions for K-BKZ and other integral equations.

Several theoreticians have worked with an integral equation that is somewhat more general than the factored K-BKZ equation (Bird et al. 1987):

$$\boldsymbol{\tau} = \int_{-\infty}^{t} M(t - t') \left[\phi_1(I_B, II_B)\mathbf{B}(t, t') + \phi_2(I_B, II_B)\,\mathbf{B}^{-1}(t, t')\right] dt' \tag{4.4.15}$$

Here the functions ϕ_1 and ϕ_2 are not necessarily derivatives with respect to I_B and II_B of a strain–energy function. Equation 4.4.15 is easier to fit to experimental data than is the factorized K-BKZ equation, eq 4.4.11. Since eq 4.4.15 lacks a strain–energy function, a fluid obeying this constitutive equation could theoretically violate the second law of thermodynamics in flows that are much faster than any of the relaxation processes included in the linear relaxation spectrum (Larson, 1983). However, for realistic relaxation spectra and realistic flows, no such violation is likely. Perhaps it is more significant to ask whether real fluids respect the constraint of the K-BKZ theory that ϕ_1 and ϕ_2 be derivatives of a strain–energy function. This question remains unanswered.

In Table 4.4.1 we present some forms for $\phi_1(I_B, II_B)$ and $\phi_2(I_B, II_B)$ that have been found to fit data for IUPAC A, Meissner's polyisobutylene melt, and other melts. However, data for IUPAC X, a material almost identical to IUPAC A, indicates that eq 4.4.15 may be incapable of giving even an approximate fit to a full range of extensional and shear flows for this and other melts that have long chain branching (Samurkas et al., 1989). For many melts without long chain branching, good fits are obtained not only with the integral BKZ and BKZ-like equations we have just discussed, but also with simpler differential equations, discussed below.

TABLE 4.4.1 / Kernel Functions ϕ_1 and ϕ_2 for Superposition Integral Equations

Authors	$\boldsymbol{\phi_1}$	$\boldsymbol{\phi_2}$	*Fits to Data for Polymer Melts*
Wagner et al. (1979)	$f_1 e^{n_1\sqrt{I-3}} + f_2 e^{-n_2\sqrt{I-3}}$; $I = \alpha I_B + (1-\alpha) II_B$	0	For IUPAC A polyethylene, $f_1 = 0.57$; $f_2 = 0.43$; $n_1 = 0.31$; $n_2 = 0.106$; $\alpha = 0.032$
Papanastasiou, et al. (1983)	$\frac{1}{1+a(I_B-3)+b(II_B-3)}$	0	For IUPAC A polyethylene, $a = 0.0013$; $b = 0.068$ For a polystyrene, $a = 0.0021$; $b = 0.093$ For a polydimethylsiloxane, $a = 0.0040$; $b = 0.195$
Wagner and Demarmels (1990)	$\frac{1+\beta}{1+a\sqrt{(I_B-3)(II_B-3)}}$	$\frac{\beta}{1+a\sqrt{(I_B-3)(II_B-3)}}$	For Meissner's polyisobutylene, $a = 0.11$; $\beta = -0.27$
Doi and Edwards (1978); also see	$\frac{5}{I-1}$;	$\frac{-5}{(I-1)\sqrt{II_B+13/4}}$	K-BKZ kernel function[a] $U = \frac{5}{2}\ell n[(I - 1/7)]$
Currie (1980)	$I = I_B + 2\sqrt{II_B + 13/4}$		

[a]This potential is based on a molecular theory; there are no adjustable parameters. Good fits are obtained only with polymers without long-chain branching for which the polydispersity is not too high.

Before leaving the integral constitutive equations, we remark that a class of equations has been proposed in which the functions ϕ_1 and ϕ_2 of eq 4.4.15 depend on the invariants of the strain *rate* tensor **D**, rather than the strain tensor **B** (Bird et al., 1968). Many of the simpler examples of these equations do not reduce to the equation of linear viscoelasticity at small strains (Gross and Maxwell, 1972; Astarita and Marrucci, 1974); and this class of equations has not been much favored lately.

4.4.2 Maxwell-Type Differential Constitutive Equations

The K-BKZ and other integral constitutive equations discussed above can be regarded as generalizations of the Lodge integral, eq 4.3.18. The upper-convected Maxwell (UCM) equation, which is the differential equivalent of the Lodge equation, can also be generalized to make possible more realistic predictions of nonlinear phenomena.

Many differential constitutive equations of the Maxwell type have been proposed; most of them are of the form

$$\overset{\triangledown}{\boldsymbol{\tau}} + \mathbf{f}_c(\tau, \mathbf{D}) + \frac{1}{\lambda}\,\boldsymbol{\tau} + \mathbf{f}_d(\tau) = 2G\mathbf{D} \tag{4.4.16}$$

Here $\mathbf{f}_c(\tau, \mathbf{D})$, which depends on both the stress tensor τ and the strain rate tensor **D**, modifies the rate at which stress tends to *build up*, and $\mathbf{f}_d(\tau)$, which depends only on the stress tensor, modifies the rate at which the stress tends to *decay*. For the UCM fluid, $\mathbf{f}_d = \mathbf{f}_c = 0$. Shear thinning or strain softening can now be introduced into eq 4.4.16 either through $\mathbf{f}_c$, which can reduce the rate at which stress builds up in a flow field, or through $\mathbf{f}_d$, which can accelerate the rate at which stress decays. In principle, nonlinear effects introduced through $\mathbf{f}_d$ affect the time dependence of the stresses in transient flows somewhat differently from nonlinearities introduced through $\mathbf{f}_c$. But when eq 4.4.16 is generalized to allow a distribution of relaxation times typical of commercial materials, it has been found that similar nonlinear effects can be introduced through either $\mathbf{f}_d$ or $\mathbf{f}_c$ (Larson, 1988 and Chapter 7).

Table 4.4.2 lists some of the more popular constitutive equations that can be expressed in the form of eq 4.4.16. Besides the linear viscoelastic parameters G and λ, the parameters in this table are ξ, a, α, and β. Where these appear, they must be fit by nonlinear rheological experiments. Because none of the equations listed contains more than two such nonlinear parameters, it is not hard to find enough data to fit all the parameters. Each of the equations listed has its own strengths and weaknesses. As examples of weaknesses, the Johnson–Segalman and White–Metzner equations show infinite extensional viscosities at finite extension rates; the White–Metzner and Larson equations have a zero second normal stress difference in simple shearing flows; and the Johnson–Segalman and Phan Thien–Tanner equations can show spurious oscillations in start-up

TABLE 4.4.2 / Stress Buildup and Decay Functions in Maxwell-Type Differential Equation 4.4.16

Authors	Constitutive Models f_c	f_d	Fits to Data for Polymer Melts
Johnson and Segalman (1977)	$\xi(\mathbf{D}\cdot\boldsymbol{\tau}+\boldsymbol{\tau}\cdot\mathbf{D})$	0	Predicts negative shear stress in step shear. Spurious oscillations in start-up of steady shearing. Singularities in steady extensional flows.
White and Metzner (1963, 1977)	$a(2\mathbf{D}{:}\mathbf{D})^{1/2}$	0	Poor fits in step shears. $N_2 = 0$. Singularities in steady extensional flows.
Larson (1984)	$\frac{2\alpha}{3G}\mathbf{D}{:}\boldsymbol{\tau}(\boldsymbol{\tau}+G\mathbf{I})$	0	Fits data reasonably well for a variety of different types of deformation, except it predicts $N_2 = 0$
Giesekus (1966, 1982)	0	$\frac{\alpha}{\lambda G}\boldsymbol{\tau}\cdot\boldsymbol{\tau}$	Excellent fits in shearing flows; not the best for extensional flows.
Leonov (1976)	0	$\boldsymbol{\tau}\cdot\boldsymbol{\tau}/2G\lambda - \frac{(\boldsymbol{\tau}+\mathbf{I})}{6G\lambda}\,\mathrm{tr}(\boldsymbol{\tau}+\mathbf{I}) + \frac{G(\boldsymbol{\tau}+\mathbf{I})}{6\lambda}\,\mathrm{tr}(\boldsymbol{\tau}+\mathbf{I})^{-1}$	Excellent fits in shearing flows; not the best for extensional flows. No fitting parameters other than those of linear viscoelasticity.
Phan Thien and Tanner (1977, 1978)	$\xi(\mathbf{D}\cdot\boldsymbol{\tau}+\boldsymbol{\tau}\cdot\mathbf{D})$	$\frac{1}{\lambda}\exp(\frac{\beta}{G}\mathrm{tr}\,\boldsymbol{\tau})(\boldsymbol{\tau}-\mathbf{I})$	Fits data reasonably well for a variety of different types of deformation. But there are spurious oscillations in start-up of steady shearing when $\xi \neq 0$; and when $\xi = 0$, then $N_2 = 0$.

of simple shearing. None of these equations fits time-dependent experimental data well unless a spectrum of relaxation modes is introduced in a way analogous to that described earlier for the UCM equation. That is, G, λ, and τ in eq 4.4.16 are subscripted with a mode index i, and the total stress is given by a sum of the stresses from all modes, as in eq 4.3.30. When this is done, the equations of Phan Thien and Tanner and of Larson seem to agree better than the other differential equations with large sets of data for melts without long side branches; see Figure 4.4.6. However, with suitable adjustment other equations, such as that of Giesekus, can also be made to work reasonably well if the parameters in the equation that control the nonlinear phenomena are made to depend on mode number i (Bird et al., 1987; p. 413) or if refinements are made to the simplest forms of the equations given in Table 4.4.2. The reader should be aware that versions of these equations other than those given in Table 4.4.2 are to be found in the literature (White and Metzner, 1963; Phan Thien and Tanner, 1977).

In numerical simulations of complex flows, some of the equations seem to be less tractable for certain flows (Apelian et al., 1988), but no consensus has emerged that would distinctly favor one equation over the others. As a class, the differential equations can be incorporated more simply than the integral equations into numerical techniques that solve flow problems, but the integral equations admit a broad spectrum of relaxation times more

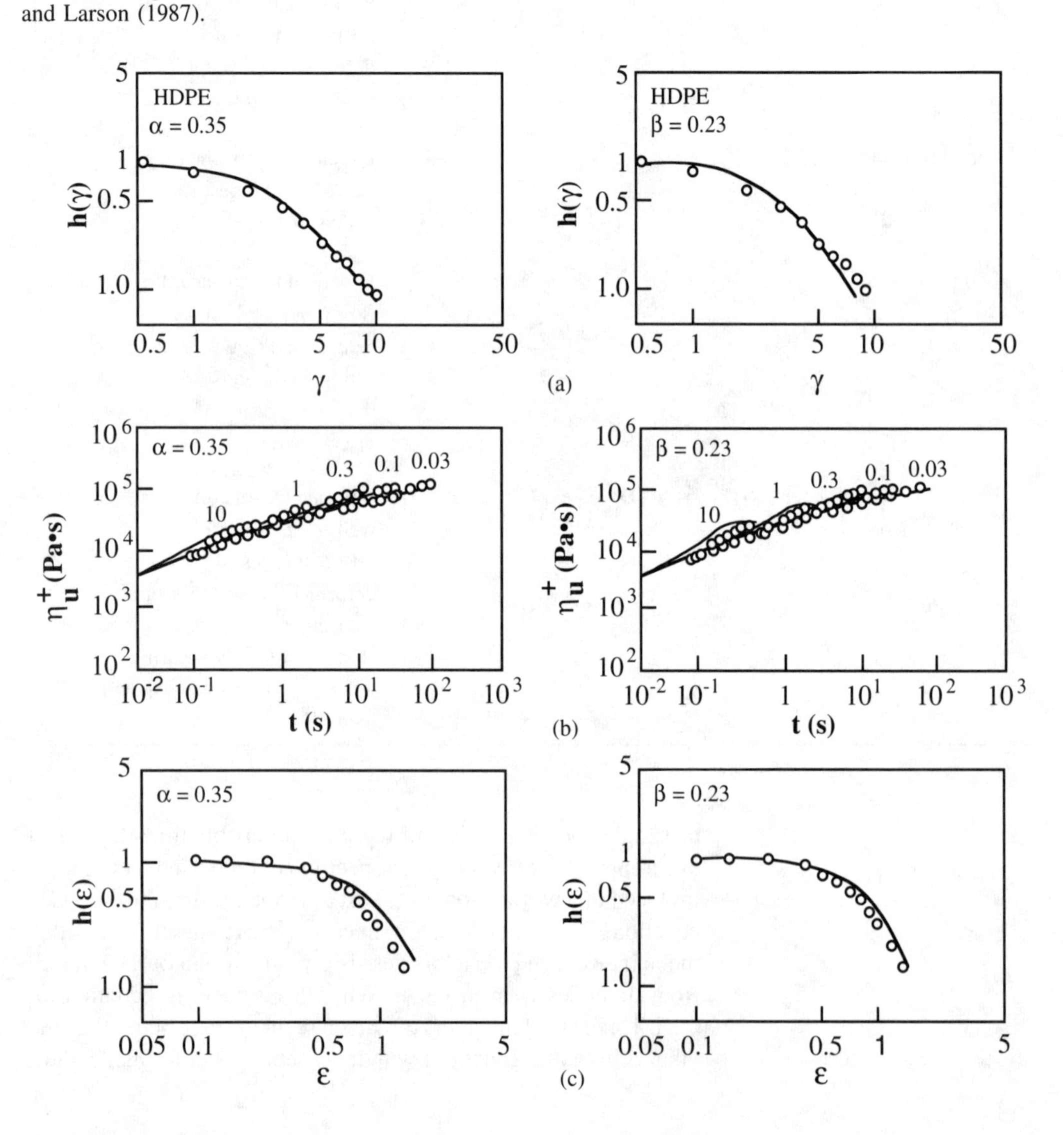

Figure 4.4.6
Best overall fit of the Larson model (left), and the Phan Thien–Tanner model (right) to data for a high density polyethylene (circles) in step shear (a), in steady elongation (b), and in step biaxial extension (c). Adapted from Khan and Larson (1987).

naturally and in a way that requires fewer primitive unknowns and thus consumes less computer storage than do the Maxwell-type differential equations. The differential equations in Table 4.4.2, some additional differential equations not expressible in the form of eq. 4.4.16, and the integral equations discussed earlier all have their advocates. And occasionally a new entree is placed on the smorgasbord of constitutive equations. No integral or differential equation has yet proven so convincingly superior that the champions of the other equations are cowed into submission. Thus, at this stage, one who is interested in modeling a polymer flow must weigh the advantages and disadvantages of the various equations for his or her application. Tables 4.4.1 and 4.4.2 may help one to choose well; much more complete comparisons of the various equations are offered elsewhere (Tanner, 1985; Bird et al., 1987; Larson, 1988).

Example 4.4.1 Viscosity and Normal Stress Coefficients for the Johnson–Segalman Equation

Calculate the steady state values of η, ψ_1, and ψ_2 for the Johnson–Segalman equation, defined by eq. 4.4.16, with $\mathbf{f}_c$ and $\mathbf{f}_d$ given in Table 4.4.2.

Solution

Using eq 2.2.10 for the velocity gradient in simple shear, the Johnson–Segalman equation reduces at steady state to

$$\begin{aligned} &-\begin{pmatrix} 2\dot{\gamma}\tau_{12} & \dot{\gamma}\tau_{22} & 0 \\ \dot{\gamma}\tau_{22} & 0 & 0 \\ 0 & 0 & 0 \end{pmatrix} \\ &+\xi\begin{pmatrix} \dot{\gamma}\tau_{12} & \frac{1}{2}\,\dot{\gamma}(\tau_{22}+\tau_{11}) & 0 \\ \frac{1}{2}\,\dot{\gamma}(\tau_{22}+\tau_{11}) & \dot{\gamma}\tau_{12} & 0 \\ 0 & 0 & 0 \end{pmatrix} \\ &+\frac{1}{\lambda}\begin{pmatrix} \tau_{11} & \tau_{12} & 0 \\ \tau_{12} & \tau_{22} & 0 \\ 0 & 0 & \tau_{33} \end{pmatrix} = G\begin{pmatrix} 0 & \dot{\gamma} & 0 \\ \dot{\gamma} & 0 & 0 \\ 0 & 0 & 0 \end{pmatrix} \end{aligned} \tag{4.4.17}$$

Some algebra suffices to show that

$$\eta = \frac{\eta_0}{1+(1-a^2)\lambda^2\dot{\gamma}^2} \tag{4.4.18}$$

$$\psi_1 = \frac{2\lambda\eta_0}{1+(1-a^2)\lambda^2\dot{\gamma}^2} \tag{4.4.19}$$

$$\psi_2 = \frac{-1(1-a)\lambda\eta_0}{1+(1-a^2)\lambda^2\dot{\gamma}^2} \tag{4.4.20}$$

where $\eta_0 = G\lambda$ and $a \equiv 1-\xi$. Thus for the Johnson–Segalman equation, there is pronounced shear thinning in all three functions, and $N_2/N_1 = -\xi/2$.

4.5 Summary

The major categories of nonlinear rheological phenomena have been described. These include the existence of nonzero first and second normal stress differences in shearing, shear thinning, and extensional thickening. Nonlinear phenomena become important when two conditions are met. These are shown schematically in Figure 4.5.1. The first condition is that the strain experienced by a material particle be appreciable; the theory of linear viscoelasticity then no longer applies. The second condition for nonlinear phenomena to be important is that the Deborah number not be small. The Deborah number is the ratio of the material relaxation time to the flow time, or equivalently, the product of the relaxation time and the characteristic strain rate.

When the Deborah number is very high, as in rapid deformations of a polymer melt, the elastic solid is appropriate even for a liquid. Polymer melts have relaxation times of 1–100 seconds. Thus, as noted in Section 1.9, in polymer processing operations such as blow molding or thermoforming and even some rapid compression molding, the neo-Hookean model is often the best choice for predicting stress response. Example 1.8.3 and Exercises 1.10.8 and 1.10.9 illustrate the application of the neo-Hookean model to flows of these type. Polymer gels, highly filled polymers, and associated colloids have even longer relaxation times but typically show

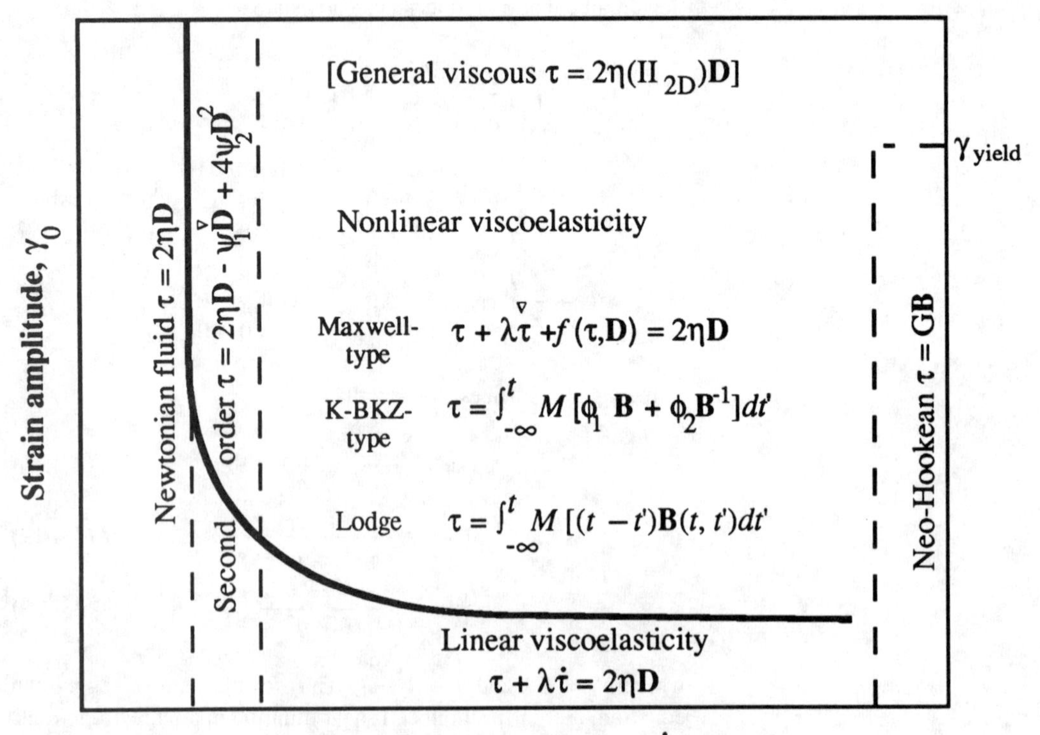

Figure 4.5.1.
The schematic diagram from Figure 4.1.2, summarizing the major viscoelastic models discussed in this chapter.

neo-Hookean behavior only up to a smaller strain limit; γ_{yield} may even be less than 1% for colloids. This leads to the Bingham plastic model described in Section 2.5, which is given a microstructural basis in Section 10.7.

When the Deborah number is vanishingly small, the fluid can be described as Newtonian. The Newtonian model should always be used first to study a new and complex flow problem and of course is appropriate for small molecule liquids. When the Deborah number is small but not negligible and the flow is steady or near steady, nonlinear effects are weak and can be described by the equation of the second-order fluid. The second-order fluid equation predicts the existence of normal stress differences in shearing and weak extension thickening but does not predict shear thinning or time-dependent rheological phenomena. It is a simple model to solve and gives qualitative trends like rod climbing (Figure 5.3.3), extrudate swell, or pressure hole errors (Figure 6.3.6).

Besides the equation of the second-order fluid, which cannot describe high Deborah number flows, perhaps the simplest nonlinear constitutive equation is the upper-convected Maxwell equation or its equivalent, the Lodge integral equation. These two equations predict the existence of normal stress differences in simple shearing flows, extreme extension thickening in extensional flows, and time-dependent phenomena, but cannot predict shear thinning and are generally inaccurate in predicting complex rheological phenomena.

If shear thinning is the main phenomenon to be described, the simplest model is the general viscous fluid, Section 2.4. It has no time dependence, nor can it predict any normal stresses or extensional thickening (however, recall eq. 2.4.24). Nevertheless, it should generally be the next step after a Newtonian solution to a complex process flow. The power law, Cross, or Carreau-type models are available on all large-scale fluid mechanics computation codes. As discussed in Section 2.7, they accurately predict pressure drops in flow through channels, forces on rollers and blades, and torques on mixing blades.

Finally, more accurate differential and integral constitutive equations were presented, and their successes and failures in describing experimental data were discussed. No single nonlinear constitutive equation is best for all purposes, and thus one's choice of an appropriate constitutive equation must be guided by the problem at hand, the accuracy with which one wishes to solve the problem, and the effort one is willing to expend to solve it. Generally differential models of the Maxwell type are easier to implement numerically, and some are available in fluid mechanics codes. Also, some constitutive equations are better founded in molecular theory, as discussed in Chapter 11.

4.6 Exercises

4.6.1 Relaxation After a Step Strain for the Lodge Equation

Calculate the relaxation of the shear stress and the first normal stress difference after imposition of a step shear of magnitude γ_0 at time zero for the Lodge equation with one relaxation time (i.e., with eq 4.3.18). Show that the Lodge–Meissner relationship, eq 4.2.8, holds for this constitutive equation.

4.6.2 Stress Growth After Start-up of Steady Shearing for the Lodge Equation

Calculate the growth of the shear stress after a steady shear with shear rate $\dot{\gamma}$ that began at time zero.

4.6.3 Derive the Maxwell Model from Neo-Hookean and Newtonian Models

Show that the Maxwell model, eq. 4.3.7, can be derived from a series combination of the neo-Hookean, $\boldsymbol{\tau} = G\mathbf{E}$, eq. 1.5.2, and Newtonian models $\boldsymbol{\tau} = 2\eta\mathbf{D}$.

4.6.4 Derive the Oldroyd-B Model

Show that the Oldroyd-B model as Oldroyd first presented it, $\tau + \lambda_1 \overset{\nabla}{\tau} = 2\eta \{\mathbf{D} + \lambda_2 \overset{\nabla}{\mathbf{D}}\}$ can be derived from 4.3.33.

References

Al-Hadithi, A. S. R.; Walters, K., Barnes, H. A., in *Proceedings of the Tenth International Congress on Rheology, Vol.1*; Uhlherr, P. H. T., Ed.; Sydney, 1988.

Apelian, M. R.; Armstrong, R. C.; Brown, R. A., *J. Non-Newtonian Fluid Mech.* 1988, *27*, 299.

Astarita, G.; Marrucci, G., *Principles of Non-Newtonian Fluid Mechanics*: McGraw-Hill: London, 1974.

Bernstein, B.; Kearsley, E. A.; Zapas, L. J., *Trans. Soc. Rheol.* 1963, *7*, 391.

Binnington, R. J.; Boger, D. V., *J. Rheol.* 1985, *29*, 887.

Bird, R. B.; Carreau, P. J., *Chem. Eng. Sci.* 1968, *23*, 427; Carreau, P. J.; MacDonald, I. F.; Bird, R. B., *ibid.* 1968, *23*, 901.

Bird, R. B.; Armstrong, R. C.; Hassager, O., *Dynamics of Polymeric Liquids*, 2nd ed., Vol. 1; Wiley: New York, 1987.

Boger, D. V., *J. Non-Newtonian Fluid Mech.* 1977, *3*, 87.

Booij, H. C.; Leblans, P.; Palmen, J.; Tiemersma-Thoone, G., *J. Polym. Sci., Polym. Phys. Ed.* 1983, *21*, 1703.

Coleman, B. D.; Noll, W., *Arch. Ration. Mech. Anal.* 1960, *6*, 355.

Coleman, B. D.; Markovitz, H., *J. Appl. Phys.* 1964, *35*, 1.

Criminale, W. O., Jr.; Ericksen, J. L.; Filbey, G. L., Jr., *Arch. Ration. Mech. Anal.* 1958, *1*, 410.

Currie, P. K., in *Rheology*, Astarita, G.; Marrucci,G.; Nicolais, L.; Eds.; Plenum: New York, 1980.

Doi, M., *J. Polym. Sci., Polym. Phys. Ed.* 1980, *18*, 1891.

Doi, M.; Edwards, S. F., *J. Chem. Soc., Faraday Trans. II*, 1978, *74*, 1789; 1802; 1818; 1979, *75*, 32.

Einaga, Y.; Osaki, K.; Kurata, M.; Kimura, S.; Tamura, M., *Polym. J.* 1971, *2*, 550.

Giesekus, H., *Rheol. Acta* 1966, *5*, 29.

Giesekus, H., *J. Non-Newtonian Fluid Mech.* 1982, *11*, 69.

Gleissle, W., in *Rheology*, Vol. II; Astarita, G.; Marrucci, G.; Nicolais, L.; Eds.; Plenum: New York, 1980.

Gross, L. H.; Maxwell, B., *Trans. Soc. Rheol.* 1972, *16*, 577.

Johnson, M. W., Jr.; Segalman, D., *J. Non-Newtonian Fluid Mech.* 1977, *2*, 225.

Joseph, D. D., *Fluid Dynamics of Viscoelastic Liquids*; Springer-Verlag: New York, 1990.

Kaye, A., College of Aeronautics, Cranford, U.K.; Note No. 134, 1962.

Keentok, M.; Georgescu, A. G.; Sherwood, A. A.; Tanner, R. I., *J. Non-Newtonian Fluid Mech.* 1980, *6*, 303.

Khan, S. A.; Larson, R. G., *J. Rheol.* 1987, *31*, 207.

Khan, S. A.; Larson, R. G., *Rheol. Acta* 1991, *30*, 1.

Khan, S. A.; Prud'homme, R. K.; Larson, R. G., *Rheol. Acta* 1987, *26*, 144.

Larson, R. G., *J. Non-Newtonian Fluid Mech.* 1983, *13*, 279.

Larson, R. G., *J. Rheol.* 1984, *28*, 545.

Larson, R. G., *Rheol. Acta* 1985, *24*, 327.

Larson, R. G., *Constitutive Equations for Polymer Melts and Solutions*; Butterworths: Boston, 1988.

Larson, R. G.; Valesano, V. A., *J. Rheol.* 1986, *30*, 1093.

Larson, R. G.; Khan, S. A.; Raju, V. R., *J. Rheol.* 1988, *32*, 145.

Laun, H. M., *Rheol. Acta* 1978, *17*, 1.

Laun, H. M., in *Proceedings of the Ninth International Congress on Rheology*; Acapulco, Mexico, 1984.

Laun, H. M., *J. Rheol.* 1986, *30*, 459.

Laun, H. M.; Schuch, H., *J. Rheol.* 1989, *33*, 119.

Leonov, A. I., *Rheol. Acta* 1976, *15*, 85.

Lodge, A. S., *Elastic Liquids*; Academic Press: New York, 1964.

Lodge, A. S., *Trans. Faraday Soc.* 1956, *52*, 120.

Lodge, A. S., *Rheol. Acta* 1968, *7*, 379.

Lodge, A. S.; Meissner, J., *J. Rheol. Acta* 1972, *11*, 351.

Mackay, M. E.; Boger, D. V., *J. Non-Newtonian Fluid Mech.* 1987, *22*, 235.

Meissner, J., *J. Appl. Polym. Sci.* 1972, *16*, 2877.

Meissner, J.; Stephenson, S. E.; Demarmels, A.; Portmann, P., *J. Non-Newtonian Fluid Mech.* 1982, *11*, 221.

Menezes, E. V.; Graessley, W. W., *Rheol. Acta* 1980, *19*, 38.

Oldroyd, J. G., *Proc. R. Soc.* 1950, *A200*, 523.

Papanastasiou, A. C.; Scriven, L. E.; Macosko, C. W., *J. Rheol.* 1983, *27*, 387.

Phan Thien, N.; Tanner, R. I., *J. Non-Newtonian Fluid Mech.* 1977, *2*, 353; the function $\mathbf{f}_d$ is given in Phan Thien, N., *J. Rheol.* 1978, *22*, 259.

Pipkin, A. C., *Lectures on Viscoelasticity Theory*; Springer-Verlag, New York, 1972.

Ramachandran, S.; Gao, H. W.; Christiansen, E. B., *Macromolecules* 1985, *18*, 695.

Retting, W.; Laun, H. M., *Kunstoff-Physik*; Hanser-Verlag, Munich, 1991, p. 138.

Russell, R. J., Ph.D. thesis, University of London, 1946.

Samurkas, T.; Larson, R. G.; Dealy, J. M., *J. Rheol.* 1989, *33*, 559.

Soskey, P. R.; Winter, H. H., *J. Rheol.* 1985, *29*, 493.

Tanner, R. I., *Engineering Rheology*; Oxford University Press: New York, 1985.

Vrentas, C. M.; Graessley, W. W., *J. Non-Newtonian Fluid Mech.* 1981, *9*, 339.

Wagner, M. H., *Rheol. Acta* 1976, *15*, 136.

Wagner, M. H.; Demarmels, A., *J. Rheol.* 1990, *34*, 943.

Wagner, M. H.; Laun, H. M., *Rheol. Acta* 1978, *17*, 138.

Wagner, M. H.; Meissner, J., *Makromol. Chem.* 1980, *181*, 1533.

Wagner, M. H.; Raible, T.; Meissner, J., *Rheol. Acta* 1979, *18*, 427.

Weissenberg, K., *Nature* 1947, *159*, 310.

White, J. L.; Metzner, A. B., *J. Appl. Polym. Sci.* 1963, *8*, 1367; the function $\mathbf{f}_c$ is given in Ide, Y., White, J. L., *J. Non-Newtonian Fluid Mech.* 1977, *2*, 281.

Wissbrun, K. F., *J. Rheol.* 1986, *30*, 1143.

PART II

MEASUREMENTS: RHEOMETRY

In the first four chapters of this book we developed constitutive equations and compared them to rheological data. Those nice, smooth-looking curves are in fact often very difficult to generate. Clever utilization of several different instruments and careful measurements are required of rheology researchers. New rheometers often generate new ideas about constitutive relations. The relatively recent extensional methods described in Chapter 7 are a case in point.

Thus the second major division of this book is rheometry, the science of making rheological measurements. What is a rheometer?* It is an instrument that measures *both stress* and *deformation history* on a material for which the constitutive relation is not known. A rheometer can be used to determine material functions. The word itself may be generic, referring to a particular deformation geometry such as the cone and plate rheometer or an extensional rheometer, or it may refer to a specific instrument for achieving such a deformation.

It is helpful to categorize rheometers by what material functions they can measure. The first major division is by kinematics: shear versus extension. Chapters 5 and 6 describe shear rheometers, while Chapter 7 describes extensional ones.

The other major division is by type of straining. There are three major regimes: *small strain*, *large strain*, and *steady straining*. Figure II.1, which illustrates these three regimes with the transient shear and extensional viscosity functions, is based on data like

**Often the word "viscometer" is used. Strictly speaking, a viscometer is a special case of a rheometer. It can measure only the steady shear viscosity function,* $\eta(\dot{\gamma})$.

those given in Figures 4.2.3 and 4.2.5–4.2.7. At small strain rates or small strains, $\gamma = \dot{\gamma} t$, both shear and extensional measurements give the same material function

$$\lim_{\dot{\gamma}\to 0} \eta^+(t, \dot{\gamma}) = \lim_{\dot{\epsilon}\to 0} \mu_1^+(t, \dot{\epsilon}) = \eta(t) \qquad (II.1)$$

$$\text{or } \gamma < \gamma_c \qquad \text{or } \epsilon < \epsilon_c$$

where $\eta(t)$ is a function of $G(t)$ and thus is the linear viscoelastic limit of the time-dependent viscosity. The other limit is very long times or very large strains. In this limit $\eta^+(t, \dot{\gamma})$ becomes independent of time and only a function of shear rate, $\eta(\dot{\gamma})$. As discussed in Section 4.2, it is not clear whether all viscoelastic materials have a comparable steady extensional viscosity. It is clear, however, that in the regime between these two limits extensional behavior can be very different from shear behavior and it depends on both time and strain or strain rate.

Figure II.1 can be compared to the Deborah number diagrams in Chapter 4, Figures 4.1.2 and 4.5.1. At short times or high Deborah number the neo-Hookean model will be valid, but at low De the Newtonian or some general viscous model will describe the shear viscosity. At small strain the linear viscoelastic model of Chapter 3 will fit the $\eta(t)$ curve well. The region between the dotted lines requires the nonlinear viscoelastic models of Section 4.4 and represents the real challenge today in rheology. We have categorized rheometers into shear and extensional classes and according to the regimes of straining they can measure. If the rheometer deformation is *homogeneous* (i.e., stress and strain are independent of position throughout the sample), it can measure all three regions: small, large, and steady straining. An example of a homogeneous rheometer is simple shear generated by one plate sliding over an-

Figure II.1.
Transient shear and extensional viscosity curves for a typical viscoelastic liquid. Response can be divided into three regimes: small strain or linear viscoelastic (LVE: the heavy central line η^+), large strain (between the dashed lines), and steady straining (outside the dashed lines).

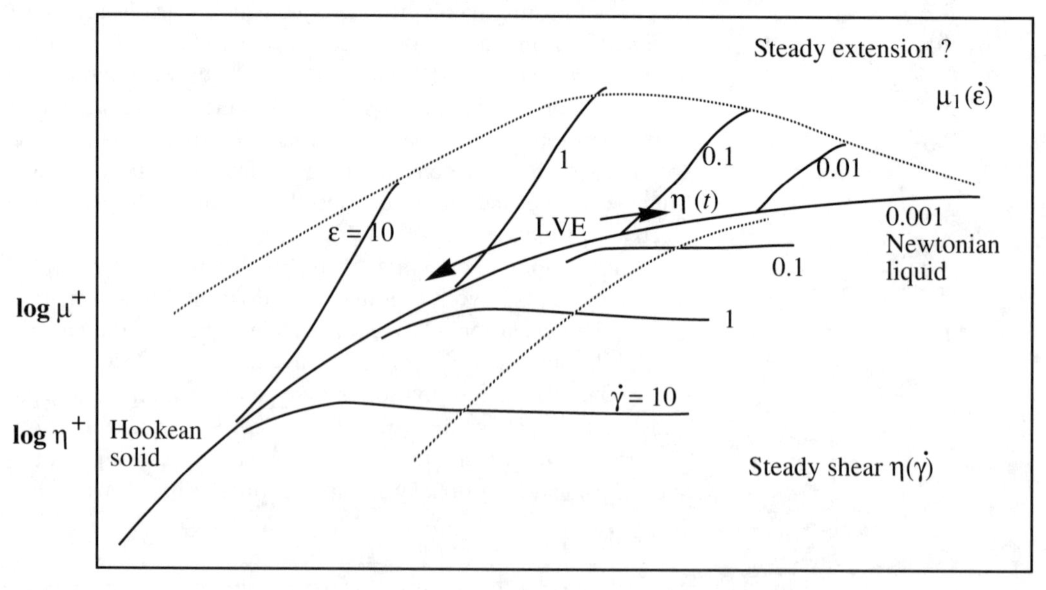

other, Figure II.2a. In Chapter 5 we will see that flow between close-fitting concentric cylinders or between a small-angle cone and plate is also homogeneous simple shear.

In a *nonhomogeneous rheometer* the stress is not uniform but represents some simple function of position. At small strain and steady straining, differences in strain history with position can be correctly analyzed to give true material function data. However, a large strain material function, such as $\eta^+(t, \dot{\gamma})$, cannot be measured. An example of a nonhomogeneous rheometer is (Figure II.2b) flow through a capillary tube. The shear rate is maximum at the wall, but zero in the center. In Chapter 6 we will see how $\eta(\dot{\gamma})$ can be determined from measures of pressure and flow rate. Another useful nonhomogeneous rheometer is torsional flow between parallel disks.

There is a third category of flow geometry consisting of *indexers*, which are not really rheometers but resemble them. In *indexers* the stress field is so complex that stress and deformation can be related only through a constitutive equation. The deformation may be a mixture of shear and extensional flow. Indexers are sometimes called "uncontrollable flows" because the deformation depends on the particular material being tested. Strain in the sample cannot be determined from simple boundary measurements. Figure II.2c shows an example: the melt indexer, flow through a contraction and into a capillary tube. Flow in the tube is simple shear, but the contraction is primarily extensional. Moreover, the capillary is short, so that flow is still developing over a significant part of its length. The flow rate under constant pressure is some complex function of both $\eta^+(t, \dot{\gamma})$ and $\mu_1(t, \dot{\epsilon})$.

In principle then indexers are not useful for fundamental rheological work because they give an index number rather than a material function. However, in practice indexers can be very useful. They are generally less expensive and simpler to operate than rheometers. Indexers are often sensitive to important material or process parameters and can be used for relative ranking. The quality of thousands of tons of polymers, bread dough, paints, and cheese is controlled by rheological indexers every year. Some index tests, such as entrance flows or the pressure hole, can be used to give material function data through approximate analyses or empirical correlations. In fact, currently the only way to measure extensional behavior for low viscosity fluids is to use an indexer.

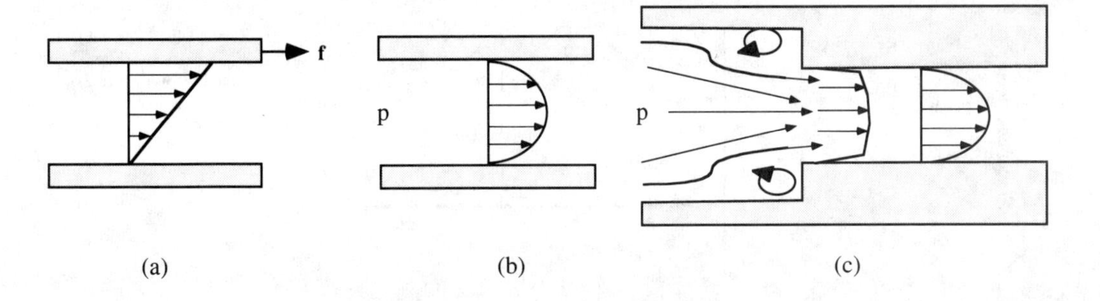

Figure II.2.
Examples of three types of geometry used in rheology: (a) homogeneous deformation, (b) nonhomogeneous, and (c) complex or indexer.

Figure II.3 summarizes the classification of rheometers into shear versus extensional, homogeneous/nonhomogeneous, and indexer. Each rheometer that is discussed in Chapters 5–7 is listed.

We divide shear rheometers into those that use drag (Chapter 5) and those that employ pressure (Chapter 6) to generate the shear deformation. The instruments and experimental problems appropriate for use with each type are quite different. Chapter 7 describes the rheometers and indexers used for extensional measurements. In these three chapters we derive the working equations required to convert boundary forces and pressures into stresses. Errors due to such *sample problems* as edge failure, shear heating, and secondary flows are discussed. Good summaries of shear and extensional rheometry and the errors one encounters in each are given at the ends of Chapters 6 and 7, respectively.

The last two chapters in this rheometry section describe how instruments control and measure these boundary motions and forces. Chapter 8, for example, discusses stress control versus rate control of rotary rheometers and tells how to measure phase angle by cross-correlation. Chapter 9 explains how birefringence can be used to measure stress. A useful summary strategy for making rheological measurements—rheometer selection, estimation of errors, sample problems—is given at the end of Chapter 8.

With the background contained in Chapters 1–4, one can read any of these rheometry chapters independently. At the end of Chapters 6 and 7 are summary tables and comparisons of methods. For further background books that focus particularly on rheometry in-

Figure II.3.
Map of material functions and classification of rheometers into homogeneous, nonhomogeneous, and index-type deformations. Only homogeneous deformations can be used to determine material functions at large strain. Numbers in parentheses indicate sections in which the respective subjects are treated.

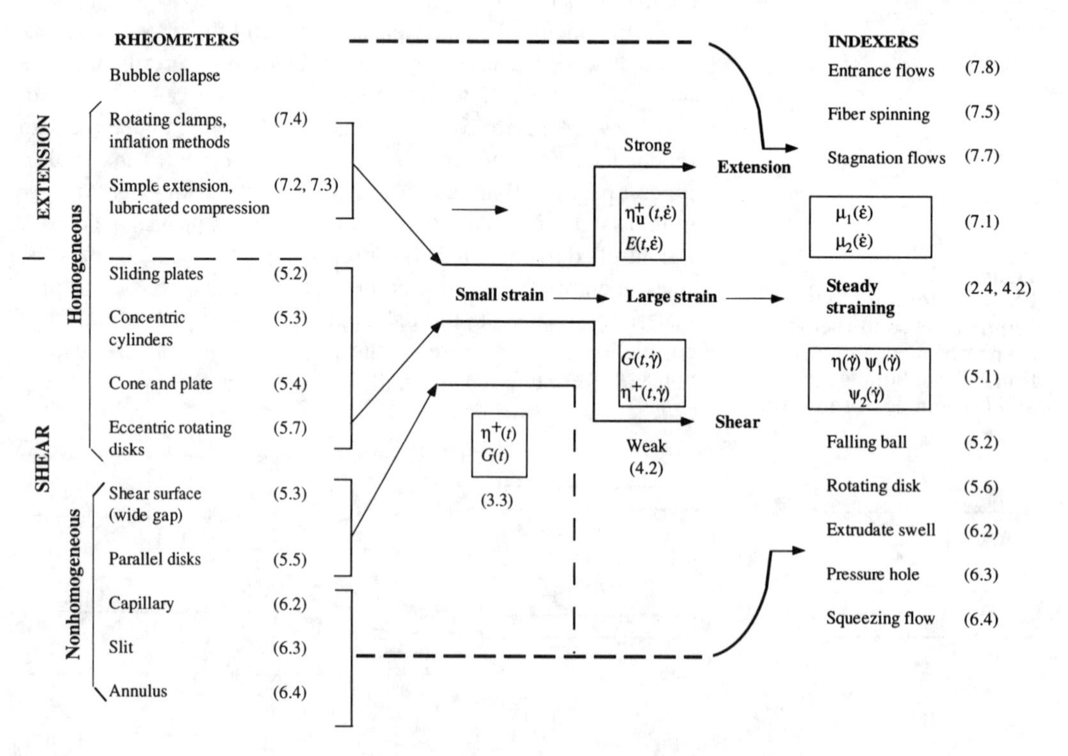

clude Walters (1975, 1980), Whorlow (1979, 1992), Dealy (1982), Collyer and Clegg (1988), and Dealy and Wissbrun (1990).

References

Collyer, A. A.; Clegg, D. W., Eds., *Rheological Measurement*: Elsevier: London, 1988.

Dealy, J. M., *Rheometers for Molten Plastics*: Van Nstrand Reinhold: New York, 1982.

Dealy, J. M.; Wissbrun, K. F., *Melt Rheology and Its Role in Plastics Processing: Theory and Applications*; Van Nostrand Reinhold: New York, 1990.

Walters, K., *Rheometry*; Chapman & Hall: London, 1975.

Walters, K., Ed., *Rheometry: Industrial Applications*; Wiley: New York, 1980.

Whorlow, R. W., *Rheological Techniques*; Wiley: New York, 1979.

Whorlow, R. W., *Rheological Techniques*, 2nd ed.; Ellis Horwood: London, 1992.

5

SHEAR RHEOMETRY: DRAG FLOWS

The comparison of [Poiseuille's results] with . . . the results given by the cylinder apparatus makes obvious the parallelism between the two phenomena.
Maurice Couette (1890)

5.1 Introduction

As discussed in the introduction to the chapters on rheometry, it is convenient to divide shear rheometers into two groups: *drag flows* like the Couette apparatus shown in Figure 5.1.1, in which shear is generated between a moving and a fixed solid surface, and *pressure-driven flows*, in which shear is generated by a pressure difference over a closed channel. The important members of each of these groups are shown schematically with the coordinate systems used for analysis in Figure 5.1.2. Because the design and measurements for pressure-driven rheometers differ greatly from those for drag flow rheometers, pressure-driven rheometers are treated separately in Chapter 6. Results from both are compared at the end of that chapter.

All these rheometers can be used to measure one or more of the shear material functions discussed in Chapters 1–4:

Relaxation modulus

$$G(t) = \frac{\tau_{12}(t)}{\gamma_0} \tag{5.1.1}$$

and related *linear* viscoelastic functions like $G^*(\omega)$, $J(t)$ or $\eta^+(t)$,

transient shear viscosity

$$\eta^+(t, \dot{\gamma}) = \frac{\tau_{12}(t, \dot{\gamma})}{\dot{\gamma}} \tag{5.1.2}$$

which in the limit of long time becomes the steady shear viscosity,

$$\eta(\dot{\gamma}) = \frac{\tau_{12}(\dot{\gamma})}{\dot{\gamma}}$$

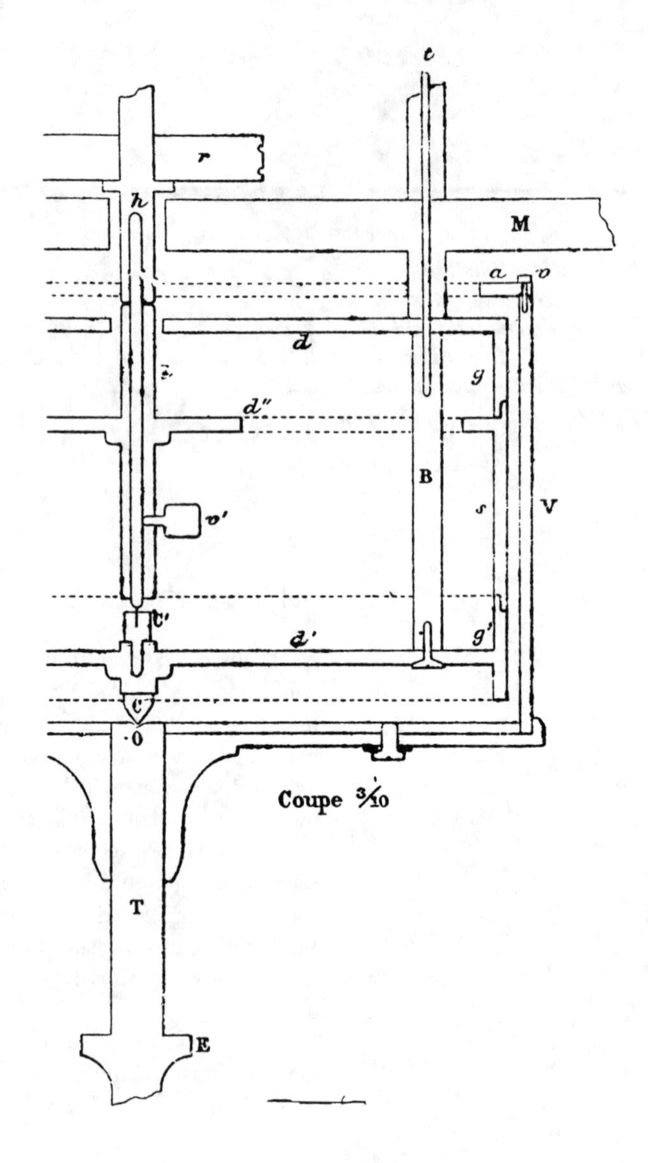

Figure 5.1.1.
The original apparatus used by Couette in 1890 to measure viscosity. Shaft T rotates the outer cylinder V. The sample is sheared in the narrow gap and exerts a torque on the cylinder s. Light reflected from the mirror v' measures the amount of twist of the torsion wire at C'. Guard rings g and g' are fixed and serve to reduce end effects.

first normal stress coefficient

$$\psi_1(\dot{\gamma}) = N_1/\dot{\gamma}^2 = \frac{\tau_{11} - \tau_{22}}{\dot{\gamma}^2} \tag{5.1.3}$$

second normal stress coefficient

$$\psi_2(\dot{\gamma}) = \frac{N_2}{\dot{\gamma}^2} = \frac{\tau_{22} - \tau_{33}}{\dot{\gamma}^2} \tag{5.1.4}$$

The sections that follow present the important working equations for each of these rheometers. The working equations give the material functions in terms of the instrument dimensions, measured motions, and forces. Derivations of these equations are presented, along with potential sources of error and comments on the circumstances in which each rheometer is most useful.

The equations for shear strain or strain distribution also are

Figure 5.1.2.
Common shear flow geometries.

Drag Flows: Chapter 5 (section number)	Coordinates x_1	x_2	x_3
Sliding plates (5.2)	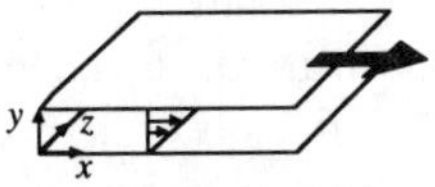 x	y	z
Concentric cylinders (5.3) (Couette flow)	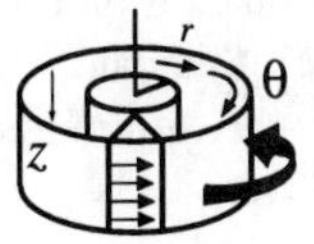 θ	r	z
Cone and plate (5.4)	ϕ	θ	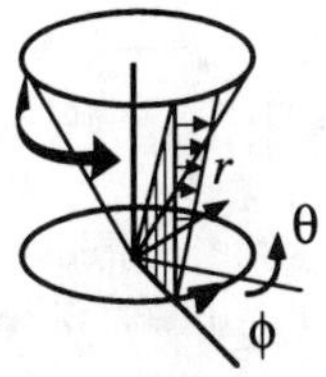 r
Parallel disks (5.5) (torsional flow)	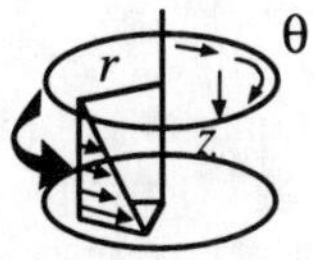 θ	z	r
Pressure Flows: Chapter 6 (section number)			
Capillary (6.2) (Poiseuille flow)	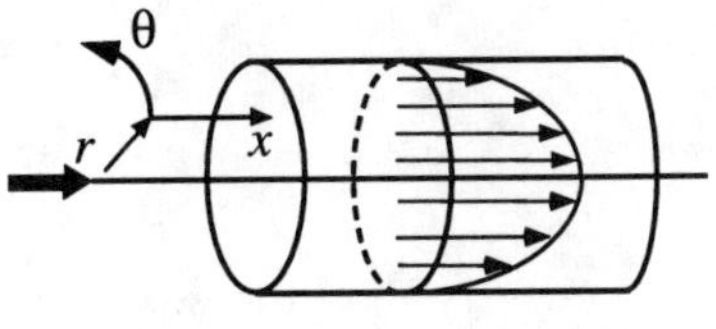 x	r	θ
Slit flow (6.3)	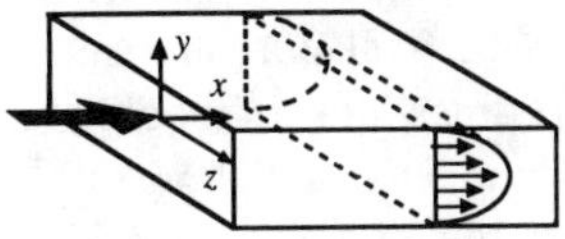 x	y	z
Axial annulus flow (6.4)	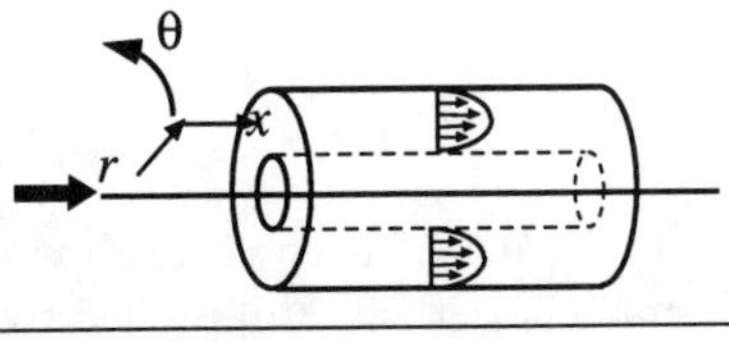 x	r	θ

given. These are needed for the small strain, linear viscoelastic, material functions like $G(t)$, $G^*(\omega)$, and $J(t)$ and for the large strain functions like $G(t, \gamma)$, $\eta^+(\dot{\gamma}, t)$, and $J(t, \gamma)$. However, the special problems associated with making these time-dependent measurements, such as transducer inertia, instrument compliance and cross correlation software, are treated in Chapter 8, where we also look at some of the instrument design features required for accuracy in drag flow rheometers.

As the chapter quotation indicates, almost the first thing Couette did after he built his famous rotational rheometer was to compare its results to those from Poiseuille's capillary instrument. If we do our measurements right and make the appropriate corrections, all the instruments shown in Figure 5.1.2 should give the same value of the viscosity. The major theme of Chapters 5–9 is determining what it takes to get absolute material function data. We will see how well this can be done by comparing results (for G^*, η, ψ_1, etc.) by the different shear methods at the end of Chapter 6.

5.2 Sliding Plates, Falling Ball

Perhaps the simplest way of generating steady simple shear is to place one material between a large fixed plate and another plate moving at constant velocity, as shown in Figure 5.2.1. If inertial and edge effects can be neglected, then the flow is homogeneous and the equations of motion are identically satisfied (Table 1.7.1). As we saw in Examples 1.4.2 and 2.2.2, the shear strain, shear rate, and shear stress, respectively, will be simply

$$B_{xy} = B_{12} = \gamma = \frac{\Delta x}{h} = \frac{v_o t}{h} \tag{5.2.a}$$

$$2D_{xy} = 2D_{12} = \dot{\gamma} = \frac{v_o}{h} \tag{5.2.1b}$$

$$T_{xy} = T_{12} = \frac{f_x}{LW} \tag{5.2.2}$$

At very short times or at high frequency oscillation inertial effects cannot be neglected; especially when testing low viscosity samples. It takes a finite time for the velocity profile to develop. At short times it looks roughly like the nonisothermal velocity profile of Figure 2.6.2b. Schrag (1977) gives a simple criteron for the time to establish homogenous, simple shear flow

$$t_c \sim \frac{1}{w_c} = \frac{10\rho h^2}{\eta} \tag{5.2.3}$$

This criterion is applicable to the other narrow gap drag flows in this chapter. It is illustrated by the birefringence measurements

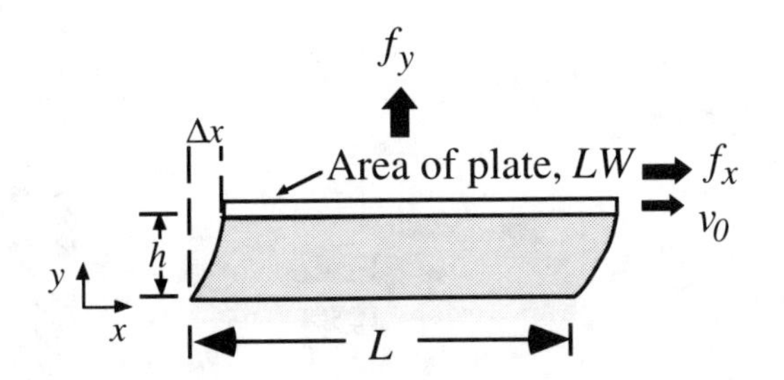

Figure 5.2.1.
Schematic of sliding plate rheometer. The displacement $\Delta x = v_0 t$ (recall Examples 1.4.1 and 2.2.1).

shown in Figure 9.5.1. It takes about 0.1s for the shear stress wave to propogate across the gap.

In principle, from the total thrust f_y or from pressure measurements with sliding plates one could also determine T_{22}, but this does not appear to have been reported in the literature. A major practical problem with this geometry is edge effects. With solids, unless L and W are much greater than h, buckling will occur (e.g., Reiner, 1960, p. 38). It is very difficult to keep the two plates parallel for large strains and also for large normal forces. With liquids, the sample must be viscous enough not to run out, although if it is very viscous, it may also show the buckling and tearing problems typical for solids. The other obvious problem associated with liquids is achieving steady shear. As strain increases, the edge effects become more severe. Thus, most shear measurements on liquids are done with rotating geometries that have closed stream lines.

However, a number of studies have used shear plates. Van Wazer et al. (1963, p. 302) describes a device for studying asphalts that uses the sliding plate shown in Figure 5.2.1. Most shear plate rheometers, however, use a double sample. This helps to prevent cocking and eliminates any normal stress effects. A shear sandwich geometry for solids that can be used in a standard tensile machine is shown in Figure 5.2.2. The sample can be machined from a large block with a thin test section on each side and a thick section of sample used for applying the force (Sternstein et al., 1968). Rubber samples can be bonded to a metal block with adhesives (e.g., Goldstein, 1974). Sliding plates and sliding cylinders rheometers have been developed to study polymer melts (Laun and Meissner, 1980; Kimura, et al., 1981; Liu et al., 1983; Sivashinsky et al., 1984; Dealy and Giacomin, 1988).

Figure 5.2.2.
Shear sandwich fixture for polymer melts, adhesives, or rubber.

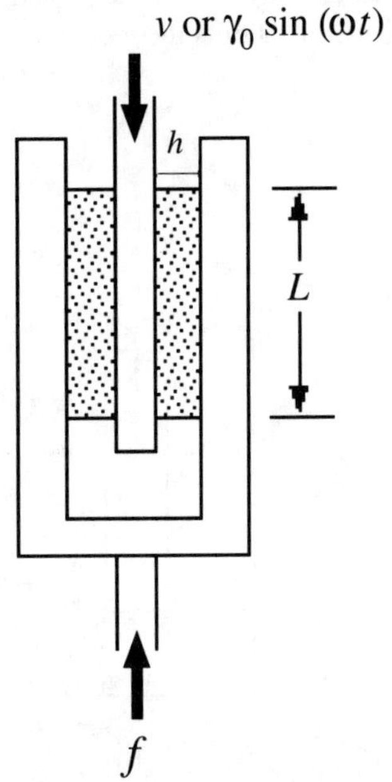

A similar concept, shown in Figure 5.2.3, has been used for some time as a high shear rheometer for printing inks (Van Wazer et al., 1963, p. 296). As indicated in Figure 5.2.3, a thin, wide film is pulled through a narrow gap filled with fluid. It is assumed that the flow will keep the film centered. The velocity is determined by timing marks on the film.

5.2.1 Falling Cylinder

To eliminate part of the edge problem in the sliding plates, we can slide a cylinder inside a tube. If both ends are open and the gap is

Figure 5.2.3.
Band viscometer used for high shear rate testing of inks and coatings.

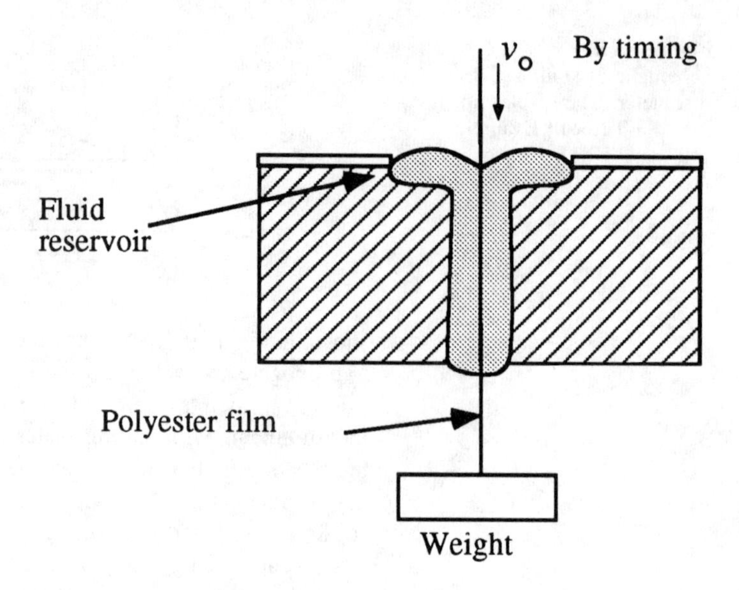

small, then we have simple shear as indicated in Figure 5.2.4a. Song and co-workers (Sivashinsky et al., 1984) have used this geometry for polymer melts with results very similar to those derived from the shear sandwich studies. As with the sandwich, high viscosity and surface tension hold the sample in the gap.

If a cylinder is dropped in a closed tube, the displaced fluid must flow back, which results in the velocity profile shown in Figure 5.2.4b. Bird et al., (1987) give relations for the narrow gap case, but the most commonly used method is the other extreme. Typically a small diameter needle is dropped in a large cylinder of the test fluid. After the needle has fallen for a distance great enough to allow the fluid to reach steady state, the terminal velocity v_∞ is determined by timing between two marks. This generally limits the technique to transparent fluids.

Figure 5.2.4.
Falling cylinder rheometers: (a) open ends for high viscosity samples and (b) closed end, free falling.

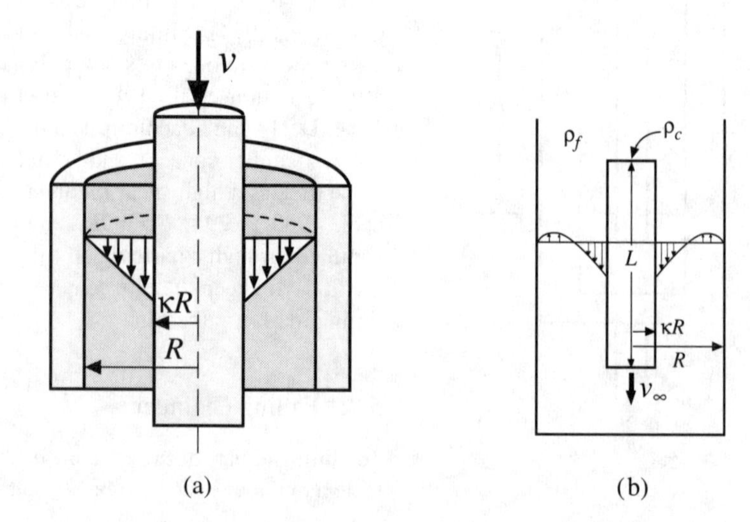

Assuming a wide gap, $\kappa \ll 1$, the relations for the shear stress and the Newtonian viscosity are

$$\tau_{\omega\kappa} = \frac{(\rho_c - \rho_f)gR}{2\kappa} \tag{5.2.4}$$

$$\eta_0 = \frac{(\rho_c - \rho_f)gR^2}{2v_\infty}(1 + \ln\kappa) \tag{5.2.5}$$

Park and Irvine (1988) give relations for a power law fluid. They also demonstrate that one can easily change the effective density of the needle and thus $\tau_{\omega\kappa}$ by using a hollow tube filled with various amounts of steel shot. In this way one can obtain an $\eta(\dot{\gamma})$ function.

5.2.2 Falling Ball

The time required for a ball to fall a given distance in a fluid is probably the simplest and certainly one of the oldest viscosity tests (Stokes, 1851). Unfortunately, creeping flow around a sphere is very complex. Thus the falling ball is really an index test and requires a constitutive equation for complete analysis. Analyses of the flow have been made for inelastic (Gottlieb, 1979; Beris et al., 1985) and viscoelastic fluids (Hassager and Bisgaard, 1983; Graham et al., 1989). Usually an apparent viscosity based on the Newtonian analysis is reported.

$$\eta_0 = \frac{2(\rho_b - \rho_f)gR^2}{9v_\infty} \tag{5.2.6}$$

Because the flow is slow, this value usually corresponds well to the zero shear viscosity even for elastic, shear thinning liquids. A test, of course, is to see whether the same η_0 is obtained for several different density or diameter (small) balls. If 10R is less than D_c wall effects can be ignored.

Like the falling needle, the falling ball viscometer can be sealed to prevent evaporation and permit measurements to be taken at high pressure.

5.2.3 Rolling Ball

Some problems with the falling ball can be reduced by tilting the tube and allowing the ball to roll down one side; see Figure 5.2.5b. When a ball is dropped, it often does not fall straight in elastic liquids or falls too fast in low viscosity liquids. The rolling ball always follows the same path, making detection easier even in translucent fluids. Tilt angle can be used to adjust the time and the effective shear stress. The price of this convenience is an even more complex flow. A result for $R \ll D_c$ is

$$\eta_0 = \frac{(\rho_b - \rho_f)gR^2 \sin\theta}{v_\infty} \tag{5.2.7}$$

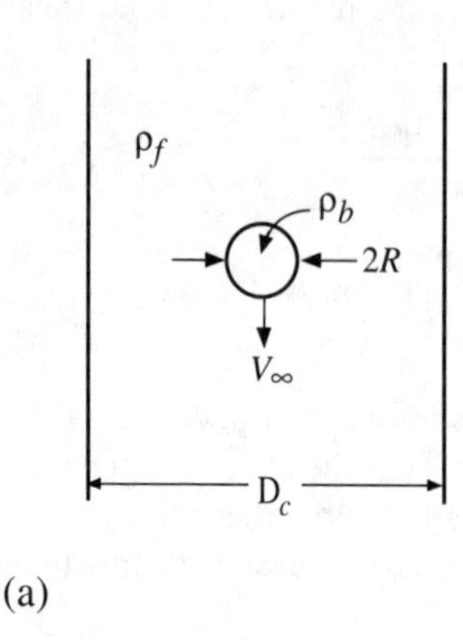

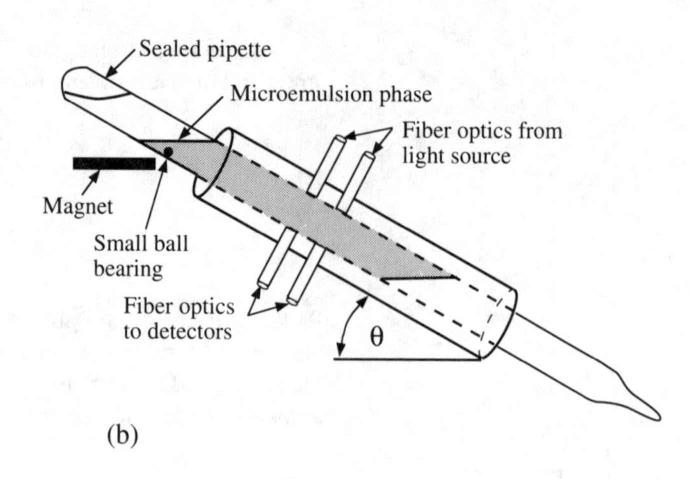

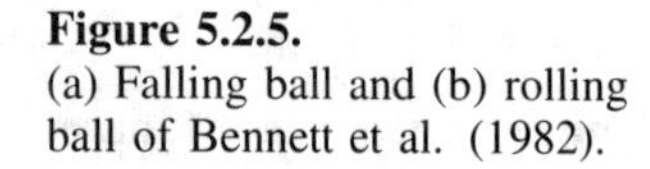

Figure 5.2.5.
(a) Falling ball and (b) rolling ball of Bennett et al. (1982).

There is disagreement in the analysis even for Newtonian fluids (Bennett, 1982).

5.3 Concentric Cylinder Rheometer

The first practical rotational rheometer was the concentric cylinder instrument of Maurice Couette (1890). As shown in Figure 5.1.1, Couette utilized a rotating outer cup and an inner cylinder suspended by a torsion wire. The angular deflection of the wire was measured by a mirror and indicated the torque on the inner cylinder. Today most commercial instruments utilize similar design concepts (see Chapter 8). Working equations relating the shear stress to torque measurements, the shear rate to angular velocity, and normal stress coefficients to radial pressure difference are given below. Their derivations and the importance of nonidealities in the flow are discussed. Table 5.3.1 summarizes the important working equations, errors, and best uses for concentric cylinder rheometers.

Consider the flow of a fluid confined between concentric cylinders with the inner cylinder rotating at Ω_i, as shown in Figure 5.3.1. If we assume:

1. Steady, laminar, isothermal flow
2. $v_\theta = r\Omega$ only and $v_r = v_z = 0$
3. Negligible gravity and end effects
4. Symmetry in θ, $\partial/\partial\theta = 0$

then the equations of motion in cylindrical coordinates (see Table 1.7.1) become

$$r : -\rho \frac{v_\theta^2}{r} = \frac{1}{r}\frac{\partial(rT_{rr})}{\partial r} - \frac{T_{\theta\theta}}{r} \tag{5.3.1}$$

$$\theta : 0 = \frac{\partial(r^2 \tau_{r\theta})}{\partial r} \tag{5.3.2}$$

TABLE 5.3.1 / Working Equations for Concentric Cylinders

Shear stress

$$T_{21} = \tau_{r\theta}(R_i) = \frac{M_i}{2\pi R_i^2 L} \tag{5.3.8a}$$

Shear strain

$$\gamma = \frac{\theta \overline{R}}{R_o - R_i} \text{ or } \frac{\Omega t \overline{R}}{R_o - R_i} \quad (\text{for } narrow \text{ gaps, } \kappa = \frac{R_i}{R_o} \geq 0.99) \tag{5.3.9}$$

where θ is the angular displacement, $\theta = \Omega t$ for steady rotation

and $\overline{R} = (R_o + R_i)/2$, the mean radius

Strain rate

$$(\text{for } \kappa > 0.99) \quad \dot{\gamma}(R_i) \cong \dot{\gamma}(R_o) = \frac{\Omega_i \overline{R}}{R_o - R_i} = \frac{2\Omega_i}{1 - \kappa^2} \tag{5.3.11}$$

$$(\text{for gaps } 0.5 < \kappa < 0.99) \quad \dot{\gamma}(R_i) = \frac{2\Omega_i}{n(1 - \kappa^{2/n})} \tag{5.3.24}$$

$$\text{and} \quad \dot{\gamma}(R_o) = \frac{-2\Omega_i}{n(1 - \kappa^{-2/n})} \quad \text{where} \quad n = \frac{d \ln M_i}{d \ln \Omega_i} \tag{5.3.25}$$

Normal stress

$$T_{11} - T_{22} = \tau_{\theta\theta} - \tau_{rr} = \frac{[\tau_{rr}(R_i) - \tau_{rr}(R_o)]\overline{R}}{R_o - R_i} \tag{5.3.29}$$

Errors
End effects, eq. 5.3.41, Figure 5.3.6
Wall slip, eq. 5.3.27
Inertia and secondary flows, eqs. 5.3.32, 5.3.42
Eccentricities, eq. 5.3.43
Viscous heating, eq. 5.3.44

Utility
Best for lower viscosity systems, $\eta_0 < 100$ Pa·s
Good for high shear rates
Gravity settling of suspensions has less effect than in cone and plate
Normal stresses hard to measure because of curvature and need to transmit signal through a rotating shaft
Rod climbing is another option, eq. 5.3.36

Figure 5.3.1.
Schematic of a concentric cylinder rheometer.

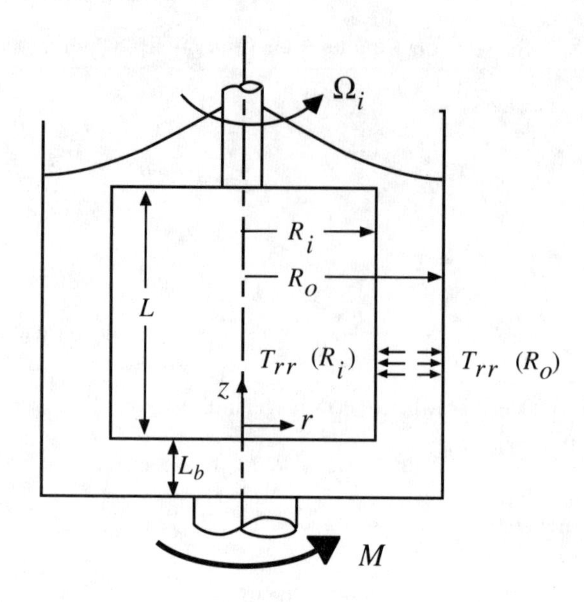

$$z : 0 = -\frac{\partial p}{\partial z} + \rho g_z \tag{5.3.3}$$

Equation 5.3.3 simply determines the hydrostatic pressure in the gap. Equation 5.3.1 governs the normal stress and eq. 5.3.2 the shear stress. The boundary conditions are

$$v_\theta = \Omega_i R_i \quad \text{at} \quad r = R_i \tag{5.3.4}$$

$$v_\theta = 0 \quad \text{at} \quad r = R_o \tag{5.3.5}$$

or in general when both cylinders can rotate

$$v_\theta = \Omega_o R_o \quad \text{at} \quad r = R_o \tag{5.3.6}$$

5.3.1 Shear Stress

The shear stress distribution across the gap between the cylinders is obtained by integrating eq. 5.3.2

$$\tau_{\theta r} = \frac{c_1}{r^2} \tag{5.3.7}$$

where the integration constant c_1 can be found from a torque balance. If torque is measured on the inner cylinder M_i, then

$$\frac{M_i}{R_i} = \tau_{\theta r}(R_i) 2\pi R_i L$$

and thus

$$\tau_{r\theta}(R_i) = \frac{M_i}{2\pi R_i^2 L} \tag{5.3.8a}$$

If torque is measured on the outer cylinder, then

$$\tau_{r\theta}(R_o) = \frac{M_o}{2\pi R_o^2 L} \tag{5.3.8b}$$

5.3.2 Shear Strain and Rate

For very narrow gaps ($\kappa = R_i/R_o > 0.99$), the curvature can be neglected and the shear strain will be the same as that between parallel plates (eq. 5.2.1)

$$B_{r\theta} = \gamma = \frac{\Delta x}{\Delta r} = \frac{\theta \overline{R}}{R_o - R_i} \tag{5.3.9}$$

where θ is the angular displacement of the cylinder and $\overline{R}$ is the midpoint between the cylinders

$$\overline{R} = \frac{R_o + R_i}{2} \tag{5.3.10}$$

Similarly, the velocity gradient is constant across the gap as in flow between parallel plates (eq. 5.2.2), and thus the shear rate is the average in the gap

$$\text{shear rate} \quad \dot{\gamma}(R_i) = \frac{\Delta v}{\Delta r} = \frac{\Omega_i \overline{R}}{R_o - R_i} \quad \text{(narrow gap)} \tag{5.3.11}$$

This result is generally used to evaluate shear in Couette viscometers. However, for many instruments R_i/R_o is less than 0.99, and thus we need to develop a more exact analysis.

From the components of the rate of deformation tensor in cylindrical coordinates, the shear rate is (see Table 2.2.1)

$$|2D_{r\theta}| = \dot{\gamma} = \frac{\partial v_\theta}{\partial r} - \frac{v_\theta}{r} = r\frac{\partial}{\partial r}\left(\frac{v_\theta}{r}\right) = r\frac{\partial \Omega}{\partial r} \tag{5.3.12}$$

The problem with a wide gap viscometer is that the shear rate changes across the gap. In principle, to evaluate the derivative in eq. 5.3.12, we need to have actual measurements of the velocity profile, but we can avoid these difficult measurements by using eq. 5.3.7 to make the shear rate a function of $\tau_{r\theta}$

$$\dot{\gamma}(r) = \left|r\frac{d\Omega}{dr}\right| = 2\tau_{r\theta}\frac{d\Omega}{d\tau_{r\theta}} \tag{5.3.13}$$

Thus (dropping the subscript $r\theta$), we have

$$\dot{\gamma}(\tau) = 2\tau \frac{d\Omega}{d\tau} \tag{5.3.14}$$

Integrating this for a rotating inner cylinder, $\Omega_i(R_i) = \Omega_i$, and stationary outer cylinder

$$\int_0^{\Omega_i} d\Omega = \Omega_i = \int_{\tau_{R_o}}^{\tau_{R_i}} \frac{\dot{\gamma}(\tau)}{2(\tau)} d\tau \tag{5.3.15}$$

Then differentiating with respect to τ_{R_i} gives

$$\frac{d\Omega_i}{d\tau_{R_i}} = \frac{1}{2}\left[\frac{\dot{\gamma}(\tau_{R_i})}{\tau_{R_i}} - \frac{\dot{\gamma}(\tau_{R_o})}{\tau_{R_o}} \frac{d\tau_{R_o}}{d\tau_{R_i}}\right] \tag{5.3.16}$$

From eqs. 5.3.8 we see that

$$\frac{d\tau_{R_o}}{d\tau_{R_i}} = \left(\frac{R_i}{R_o}\right)^2 \tag{5.3.17}$$

Substituting, we obtain

$$2\tau_{R_i} \frac{d\Omega_i}{d\tau_{R_i}} = \dot{\gamma}(\tau_{R_i}) - \dot{\gamma}(\tau_{R_o}) \tag{5.3.18}$$

For a large gap, for example, if $\kappa = R_i/R_o < 0.1$, then by eq. 5.3.7 $\tau_{R_o} < 0.01\, \tau_{R_i}$ and $\dot{\gamma}(\tau_{R_o}) \ll \dot{\gamma}\, (\tau_{R_i})$. From eq. 5.3.13 for a power law fluid (eq. 2.4.13)

$$\frac{\dot{\gamma}(\tau_{R_o})}{\dot{\gamma}(\tau_{R_i})} = (\kappa)^{2/n}$$

which for $\kappa \leq 0.1$ gives

$$\frac{\dot{\gamma}(\tau_{R_o})}{\dot{\gamma}(\tau_{R_i})} = 10^{-6} \quad \text{for} \quad n = 1/3 \quad \text{and} \quad \kappa = 0.1$$

Thus eq. 5.3.18 becomes

$$2\tau_{R_i} \frac{d\Omega_i}{d\tau_{R_i}} \cong \dot{\gamma}(\tau_{R_i}) \tag{5.3.20}$$

or

$$\dot{\gamma}_{R_i} \cong 2\Omega_i \frac{d \ln \Omega_i}{d \ln M_i} \tag{5.3.21}$$

Newton tried to derive this equation in his *Principia*, but he made an error that was not discovered until 1845, when Stokes derived the equation for calculating the true shear rate for instruments in which a rotating cylinder is immersed into a large beaker of test fluid (Reiner, 1960).

This derivation is typical of all those for nonhomogeneous rheometers. In each case, the derivative of velocity with position, which is difficult to measure, is transformed to a derivative of stress with respect to rotation or flow rate. We will see it again with the parallel disks and the pressure flow rheometers.

Many concentric cylinder rheometers, including Couette's original, use a fairly narrow gap, $0.5 < \kappa < 1.0$. To find the shear rate in this case, we expand eq. 5.3.18 in a Maclaurin series (Krieger and Elrod, 1953; Yang and Krieger, 1978)

$$\dot{\gamma}(\tau_{R_i}) = \frac{\Omega_i}{-\ln \kappa}\left[1 - \frac{1}{n}\ln \kappa + \left(\frac{1}{n}\ln \kappa\right)^{2/3} - \left(\frac{1}{n}\ln \kappa\right)^{4/45} + \ldots\right] \quad (5.3.22)$$

where n is the power law index, or in terms of torque and rotation rate,

$$n = \frac{d \ln M_i}{d \ln \Omega_i} \quad (5.3.23)$$

For - $(\ln \kappa)/n < 0.2$, the first two terms give an error of less than 1%. For $0.2 < -(\ln \kappa/n) < 1.0$, the third term should be used. Coleman and Noll (1959) give another series for $\dot{\gamma}$, but it appears to converge more slowly.

For $\kappa > 0.5$ if n is constant over the region τ_{R_i} to τ_{R_o} (usually a good approximation), then we have the power law case and eq. 5.3.18 becomes (McKelvey, 1962, p. 107)

$$\dot{\gamma}_{R_i} = \frac{2\Omega_i}{n(1 - \kappa^{2/n})} \quad (5.3.24)$$

$$\dot{\gamma}_{R_o} = \frac{-2\Omega_i}{n(1 - \kappa^{-2/n})} \quad (5.3.25)$$

A number of concentric cylinder instruments rotate the outer cylinder Ω_0 and hold the inner cylinder fixed. For this case we can just replace Ω_i with Ω_0 in eq. 5.3.24 and 5.3.25.

Many commercial devices employ a narrow gap ($\kappa > 0.9$), and their operation instructions suggest that the shear rate equation for a Newtonian fluid, $n = 1$ in the equations above, be used to reduce data. The error of this approximation for a power law fluid will be

$$\left.\frac{\dot{\gamma}_{\text{power}}}{\dot{\gamma}_{\text{Newt}}}\right|_{R_i} = \frac{1 - \kappa^2}{n(1 - \kappa^{2/n})} \quad (5.3.26)$$

For typical shear thinning fluids, $1 < n < 1/4$. For such fluids, if the gap between the cylinders is very narrow, such that $\kappa >$

0.99, then the error in using the simplified eq. 5.3.11 will be 3% or less. Otherwise, the data should be plotted as $\ln M$ versus $\ln \Omega$, n determined graphically or numerically, and then eq. 5.3.24 or 5.3.25 used to calculate $\dot{\gamma}$. Giesekus and Langer (1977) give a single point correction method similar to that for parallel disks (eq. 5.6.10).

Figure 5.3.2 illustrates the magnitude of the error involved in using eq. 5.3.11 with $\kappa = 0.794$ and a typical pseudoplastic polymer solution.

In the foregoing analysis we assumed ideal Couette flow: $\mathbf{v} = v(0, r\Omega, 0)$. For concentrated suspensions, some gels, and polymer solutions, a low viscosity layer can develop near the cylinder surfaces (note Figure 10.2.1a). This leads to an apparent wall slip. This slip velocity can be determined by making measurements with two different radii bobs, R_1 and R_2, with cups sized to give the same κ (Yoshimura and Prud'homme, 1988)

$$v_s(\theta) = \frac{\kappa}{\kappa + 1}\left[\frac{\Omega_1 - \Omega_2}{\frac{1}{R_1} - \frac{1}{R_2}}\right] \tag{5.3.27}$$

This slip velocity can be used to correct shear rate readings to give the true material viscosity. Kiljanski (1989) also proposed a two-bob method that does not require the same κ.

Wall slip and even sample fracture can be so severe in Couette flow of some concentrated suspensions that it is doubtful whether a true viscosity can be measured (Toy et al., 1991). One method to attempt to eliminate wall slip is to use the vaned bob shown later (Figure 5.6.2).

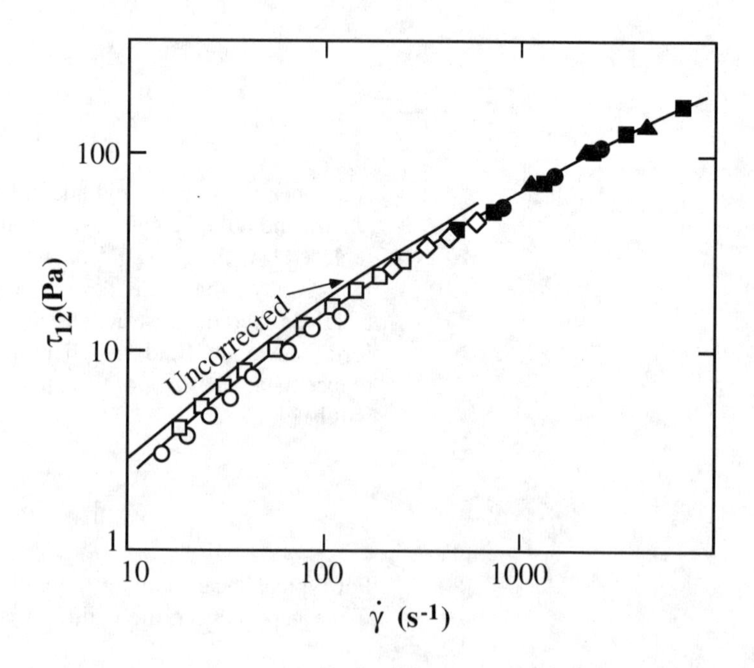

Figure 5.3.2.
Shear stress versus shear rate data for concentric cylinder viscometer (Epprecht, Model RM-15), $\kappa = 0.794$. Uncorrected shear rate (eq. 5.3.11) is compared to true shear rate (eq. 5.3.24) and with n evaluated graphically versus $\dot{\gamma}$ for a 1% carboxymethylcellulose solution From Middleman (1968, p. 25).

Besides wall slip there can be departure from ideal Couette flow due to the ends of the cylinder, to inertia, and to eccentricity. There also can be errors due to shear heating. These are discussed below, but first we examine normal stress effects.

5.3.3 Normal Stresses in Couette Flow

As mentioned in Part I, one of the remarkable phenomena of elastic liquids is that they climb a rotating cylinder. This is illustrated in Figure 5.3.3. We see that for the Newtonian oil the surface near the rod is slightly depressed. The surface acts as a sensitive manometer for the small negative pressure near the rod generated by centrifugal force. The surface of the polyisobutylene in oil solution shows a large rise. We can see how this rise can occur from the normal stress terms in the r component of the equation of motion in eq. 5.3.1.

Rearranging eq. 5.3.1, we obtain

$$-\rho\frac{v_\theta^2}{r} = \frac{\partial \tau_{rr}}{\partial r} - \frac{\tau_{\theta\theta} - \tau_{rr}}{r} \qquad (5.3.28a)$$

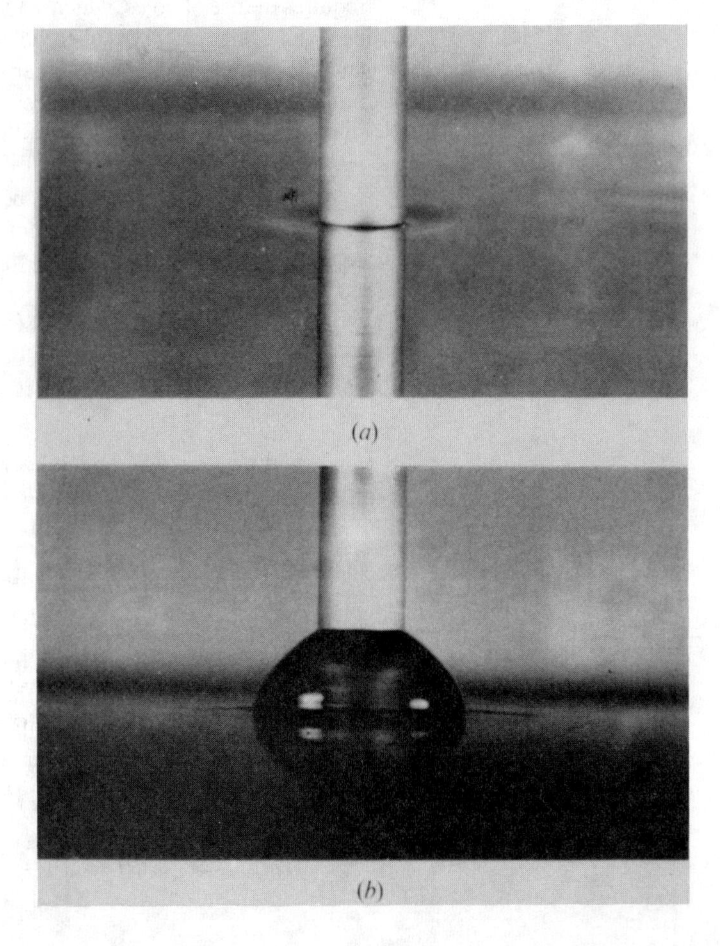

Figure 5.3.3. A 9.52 mm diameter aluminum rod rotating at 20π rad/s in a large vat of (a) Newtonian oil and (b) polyisobutylene solution. From Beavers and Joseph (1975).

We note that the numerator of the second term on the right is the first normal stress difference, $T_{11} - T_{22}$. If the gap between the cylinders is small, then

$$\frac{\partial \tau_{rr}}{\partial r} \cong \frac{\tau_{rr}(R_i) - \tau_{rr}(R_o)}{R_o - R_i} \tag{5.3.28b}$$

and if for the moment we neglect inertial effects, we can write an expression for the first normal stress difference directly

$$\tau_{11} - \tau_{22} = \tau_{\theta\theta} - \tau_{rr} = \frac{[\tau_{rr}(R_i) - \tau_{rr}(R_o)]\overline{R}}{R_o - R_i} \tag{5.3.29}$$

where $\overline{R}$ is the midpoint between the cylinders (eq. 5.3.10). $\tau_{rr}(R_i) - \tau_{rr}(R_o)$ is simply the pressure difference between the inner and outer cylinders. In Chapter 3 we saw that the first normal stress difference can be a large positive number, with the result that the pressure on the inner cylinder will be higher, causing the fluid to rise up the rod.

There are several problems with using narrow gap Couette flow for normal stress measurements. Mounting a typical flat-tipped pressure transducer on the curved surface of a cylinder requires the use of holes, which leads to pressure hole errors (Broadbent and Lodge, 1971). This problem is discussed further in Chapter 6. Transmitting the transducer signals through a rotating shaft requires slip rings, which can reduce accuracy. Finally, for narrow gaps $\tau_{rr}(R_i) \cong \tau_{rr}(R_o)$, resulting in considerable uncertainty in $\tau_{11} - \tau_{22}$.

The latter uncertainty can be reduced by operating at large gap. The relations for this situation can be obtained by integrating eq. 5.3.28a from R_o to R_i.

$$\Delta p_r = \tau_{rr}(R_i) - \tau_{rr}(R_o)$$
$$= \int_{R_o}^{R_i} (\tau_{\theta\theta} - \tau_{rr}) d \ln r - \int_{R_o}^{R_i} \frac{\rho v_\theta^2}{r} dr \tag{5.3.30}$$

If the gap is not too large, v_θ varies approximately linearly across the gap and the second integral becomes

$$\int_{R_o}^{R_i} \rho r \Omega^2(r) dr = -\frac{1}{2}\rho R^2 \Omega = \Delta p_{\text{inertia}} \tag{5.3.31}$$

Since this term is negative, for a Newtonian fluid the pressure at the inner cylinder will be less than that at the outer and the fluid will be depressed as shown in Figure 5.3.3. For larger gaps a secondary flow sets up giving an additional contribution (Bird et al., 1959).

$$\Delta p_{\text{inertia}} = \frac{1}{2}\rho\, R_o^2\, \Omega_i^2 \left[1 + \frac{1 - \kappa^2 - 4\ln\kappa}{\kappa^{-2} - 1}\right] \tag{5.3.32}$$

Figure 5.3.4 shows a test of this equation against data on a silicone oil.

Either inertial correction given above can be added to the measured pressure difference Δp_r to give an effective pressure difference, which will be a function only of $N_1 = \tau_{\theta\theta} - \tau_{rr}$

$$\Delta p_r^i = \Delta p_r + \Delta p_{\text{inertia}} \tag{5.3.33}$$

To evaluate the other integral, we can use eq. 5.3.7 to change variables from r to τ as we did in deriving the shear rate relations

$$\Delta p_r^i = \frac{1}{2}\int_{\tau_{R_o}}^{\tau_{R_i}} N_1 d\ln\tau \tag{5.3.34}$$

Differentiating this expression with respect to $\ln\Omega$, we obtain

$$\frac{d\Delta p_r^i}{d\ln\Omega} = N_1(\tau_{R_i})\frac{d\ln\tau_{R_i}}{d\ln\Omega} - N_2(\tau_{R_o})\frac{d\ln\tau_{R_o}}{d\ln\Omega} \tag{5.3.35}$$

For a power law fluid $d\ln\tau_{R_i}/d\ln\Omega = d\ln\tau_{R_o}/d\ln\Omega = n$, and N_1 can be evaluated from a plot of Δp_r^i versus $\ln\Omega$. However, it is more reliable to plot Δp_r^i versus $\ln\tau$ according to eq. 5.3.34 and do a numerical or graphical integration. Such a plot is shown in Figure 5.3.5. Note the large pressure hole correction.

An alternative to measuring small pressure differences between the curved cylinder surfaces has been developed by Giesekus and co-workers (Abdel-Wahab et al., 1990). They made the inner cylinder slightly eccentric and measured the total side thrust on the stationary outer cylinder. Their results were reasonable when compared to cone and plate measurements on two polymer solutions.

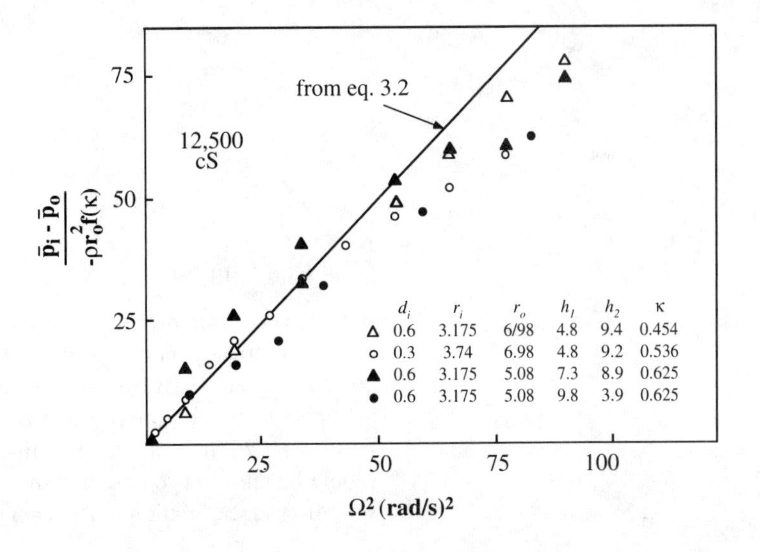

	d_i	r_i	r_o	h_1	h_2	κ
△	0.6	3.175	6/98	4.8	9.4	0.454
○	0.3	3.74	6.98	4.8	9.2	0.536
▲	0.6	3.175	5.08	7.3	8.9	0.625
●	0.6	3.175	5.08	9.8	3.9	0.625

Figure 5.3.4. Ratio of measured Δp_r to theoretical (eq. 5.3.32) $\Delta p_{\text{inertia}} / \Omega_i^2$ versus Ω_i^2 for a 12.5 $Pa{\cdot}s$ silicone fluid. From Broadbent and Lodge (1971).

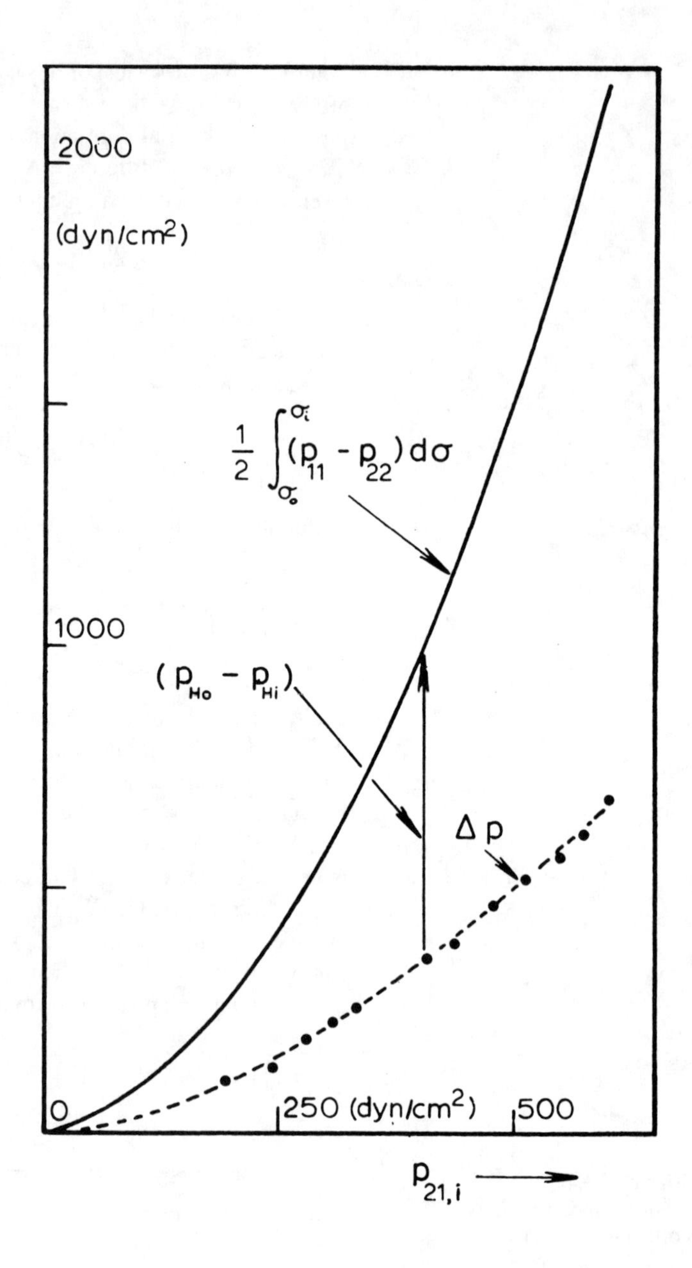

Figure 5.3.5.
Data of Broadbent and Lodge (1971) on a polyisobutene solution taken in Couette flow with recessed transducers. Solid curve from cone and plate data of Kaye et al. (1968).

5.3.4 Rod Climbing

In some ways a simpler approach to obtaining normal stresses from the concentric cylinder system is to use the rod climbing phenomenon quantitatively. Because the flow is complex, analysis requires the assumption of some constitutive relation. Joseph and Fosdick (1973) have done this using the second-order fluid, which should be an exact representation of elastic liquids in the limit of slow flows (see Section 4.3). They derive a power series for the height rise in terms of a combination of normal stresses and surface tension. For slow flow, this series should be approximated by the first two terms (Joseph et al., 1973 have solved for the next two terms)

$$h = h_s(0, R_i) + \frac{R_i \Omega_i^2}{2(\Gamma \rho g)^{1/2}} \left[\frac{4\beta}{4+\lambda} - \frac{\rho R_i^2}{2+\lambda} \right] + O(\Omega_i^4) \quad (5.3.36)$$

where h_s is the static rise due to the surface tension Γ and $\lambda^2 = R_i^2 \rho g / \Gamma$. β is a combination of the low shear rate limiting values of the normal stress coefficients

$$\beta = (1/2)\,\psi_{1,0} + 2\,\psi_{2,0} \quad (5.3.37)$$

Note that according to eq. 5.3.36 there is an optimum radius to give the maximum climbing effect.

To test this result Beavers and Joseph (1975) measured h versus Ω^2 for a polyisobutylene solution. Some of their results are reproduced in Figure 5.3.6. We see that there is an approximately constant static climb, $h_s(0)$, with different radii and a fairly large region of linearity in Ω^2. This linearity appears to justify dropping the higher order terms in eq. 5.3.36. From the slopes in Figure 5.3.6, Beavers and Joseph calculate $\beta = 0.97 \pm 0.5$ g/cm at 26°C. Davis and Macosko (1973) estimated $\psi_{1,0}$ and $\psi_{2,0}$ from the low shear rate limits of N_1 from cone and plate thrust data and $N_1 - N_2$ from parallel disk data and obtained $\beta = 0.98 \pm 0.5$ g/cm on the same material. Joseph et al. (1984) have made similar comparisons on a number of polymers.

One of the disadvantages of the rod climbing experiment is that it requires another physical property measurement: surface

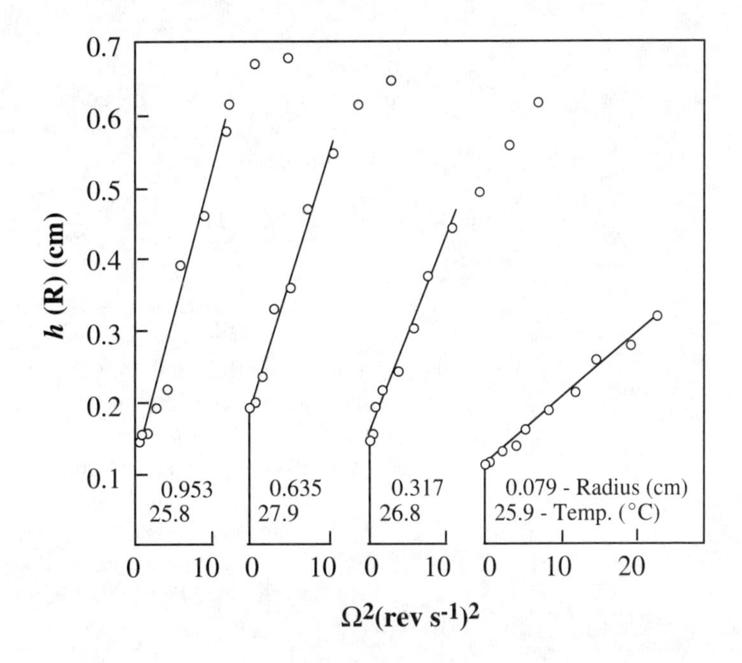

Figure 5.3.6. Experimental values of $h(R, \Omega)$ for several aluminum rods in polyisobutylene in oil (STP oil additive). Replotted from Beavers and Joseph (1975).

tension at test conditions. This is difficult to obtain on materials like polymer melts or suspensions. Fortunately, changes in Γ with respect to temperature, concentration, and molecular weight are relatively small compared to those for viscosity (van Krevelen, 1976, p. 95). Furthermore, eq. 5.3.36 is rather insensitive to Γ, since it appears as $\Gamma^{-1/2}$. A more serious disadvantage may be the ability to determine the slope accurately. For example, for many polymeric liquids β should be 10^4 or 10^5 g/cm, but constant Ψ_1 values—and thus presumably second-order fluid behavior—are achieved only for $\dot{\gamma} < 0.1\ \mathrm{s}^{-1}$. This implies very low rotation rates and long times. Dealy and Yu (1978) reported great difficulty in making climbing measurements on polymer melts. Siginer (1984) found large errors due to eccentricity when climbing was measured in a concentric cylinder apparatus.

However, the sensitivity and simplicity of the experiment are great advantages. Furthermore, the limiting normal stress values from β are useful data for molecular theories of rheological phenomena.

5.3.5 End Effects

In the assumptions for flow between concentric cylinders, we ignored any end effects. It is not too difficult to estimate their importance. At the bottom of the cylinder in Figure 5.3.1 there will also be a shear flow. This can be approximated as torsional flow between parallel disks, to be discussed in Section 5.5. From that section we can use eq. 5.5.8 and the power law, eq. 2.4.12, for τ_{12} to calculate the extra torque contributed by the end

$$M_{pp} = \frac{2\pi R_i^3 m}{3+n}\left(\frac{\Omega_i}{L_b}\right)^n \tag{5.3.38}$$

where $m = 1/n$.

If we can ignore the corners, eq. 5.3.38 can be combined with eqs. 5.3.8 and 5.3.24 to give the total torque in a flat-bottomed instrument.

$$M = M_{cc} + M_{pp} = 2\pi R_i^2 Lm\left[\frac{2\Omega}{n(1-\kappa^{2/n})}\right]^n + \frac{2\pi R_i^3 m}{3+n}\left(\frac{\Omega R_i}{L_b}\right)^n \tag{5.3.39}$$

Oka (1960) has treated the Newtonian case to include the corners.

It has been suggested that different L (and thus L_b) be used on the same fluid and M/Ω be plotted versus L to give a straight line with the intercept as the end effect (Van Wazer et al., 1963; p. 69). From eq. 5.3.39 we see that this procedure can be only generally valid for a Newtonian fluid.

Three design ideas sketched in Figure 5.3.7 are frequently used to minimize end effects. One features a conical bottom. With proper choice of cone angle, the shear rate in the bottom can be quite satisfactorily matched to that in a narrow gap on the sides. From eqs. 5.3.24 and 5.4.12 in the next section, we see that matching $\dot{\gamma}$s requires that

$$\beta_0 = \frac{1-\kappa^{2/n}}{2n\kappa^{2/n}} \cong \frac{1-\kappa^2}{2\kappa^2} \text{for} \quad 1 > \kappa > 0.99 \quad \text{and} \quad \beta < 0.1 \text{ rad} \tag{5.3.40}$$

Thus, $\tau_{cc} = \tau_{cp}$ and

$$M = M_{cc} + M_{cp} = [2\pi R^2 L + 2/3\pi R^3]\,\tau(\dot{\gamma}) \tag{5.3.41}$$

Because narrow gaps and small angles are required, the devices must be well constructed and aligned.

The other designs shown in Figure 5.3.7 are the recessed bottom and double Couette geometries used, for example, by Haake. The recessed bob traps air, which transfers essentially no torque to the fluid. Princen (1986) has used mercury, a very low viscosity liquid, instead of air in the bottom recess. The thin rotating, inverted cup shown in Figure 5.3.7c should also make $M_{\text{bottom}} \ll M_{cc}$. This design also allows better temperature control. Note, however, that unless the gap is small, the shear rates on the inside and outside of the rotating cup will not be equal.

The strategy that Couette himself used to reduce end effects was to connect only the central portion of the inner cylinder to his torque transducer. The upper and lower segments of the cylinder (g and g' in Figure 5.1.1) are called guard rings. One hundred years later Giesekus and co-workers found that the same design is still valid (Abdel-Wahab et al., 1990). The main problem with these guard rings is the difficulty of cleaning test fluids from the small gaps.

The top end of the Couette flow usually presents fewer problems, since it is open to a gas interface. Provision should be made in the design for exact maintenance of L without the need for precise volumetric filling. This can be accomplished by the spillover cup shown in Figure 5.3.7. Two problems can occur at the upper end: surface composition changes and normal stress climbing.

Figure 5.3.7.
Three common designs for eliminating end effects in concentric cylinder rheometers: (a) conicylinder, (b) recessed bottom, and (c) double Couette.

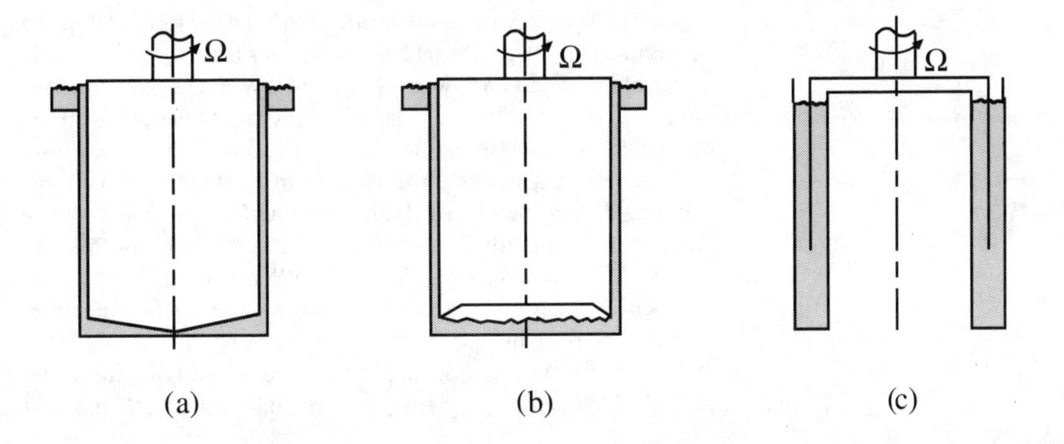

Typically, surface tension forces are too small to influence the total torque readings, but some materials tend to change at the free surface, creating a stiff, solidlike crust, which can have large effects on measurements. The denaturization of protein in blood plasma at the gas interface is a well-known example of artificially high yield stresses due to such a crust. Evaporation of solvents from polymer solutions can cause similar high torque readings. Some workers have used guard rings to reduce these surface problems (Copley and King, 1970; Cokelet, 1972). A conicylinder design with an upper cone and plate region can be used to minimize the amount of free surface (Van Wazer et al., 1963, p. 81; Leary, 1975). Other approaches include humidifying the gas above the sample with an appropriate solvent or floating a low viscosity, low density oil on the sample surface.

Normal stress climbing or the Weissenberg effect is well known. Elastic materials will climb up a rotating inner cylinder and even leave the apparatus. Using upper cover plates or filling the overflow dam can reduce this problem. Other materials-related problems in using Couette flows, such as particle migration and particle size effects, are discussed in Chapter 10 under applications of rheology to suspensions.

5.3.6 Secondary Flows

Secondary flows in the Couette geometry have been well studied following Taylor's classic work (1923). With the inner cylinder rotating at some speed, inertial forces cause a small axisymmetric cellular secondary motion known as Taylor vortices or Taylor cells. These dissipate energy and cause an increase in the measured torque. Some data from Denn and Roisum (1969) for a glycerin–water solution are shown in Figure 5.3.8. For Newtonian fluids and narrow gaps, the criterion for stability is

$$\mathrm{Ta} = \frac{\rho^2 \Omega^2 (R_o - R_i)^3 R_i}{\eta(\dot{\gamma})^2} < 3400 \qquad (5.3.42a)$$

where Ta is the Taylor number. Equation 5.3.42 is well established experimentally (Chandrasekhar, 1961). For non-Newtonian polymer solutions and narrow gaps, the stability limit seems to move to higher Taylor number. This can be seen for the polyisobutylene (PIB) solution in Figure 5.3.8. Thus, a conservative stability criterion for experimentalists appears to be the Newtonian one, eq. 5.3.42. For the outer cylinder, rotating Couette flow is stable until the onset of turbulence at a Reynolds number of about 50,000, where $N_{\mathrm{Re}} = \rho \Omega R_o (R_o - R_i)/n$ (Van Wazer 1963, p. 86). The influence of inertia and secondary flow on normal stress measurements was given in eq. 5.3.32.

Even at low Taylor number there is a purely elastic instability which causes very fine, time periodic cells. The onset of these instabilities depends on a critical Weissenberg (or Deborah) Number

$$\text{Wi}_c = \frac{\Omega\lambda}{R_o - R_i} \sim 30 \text{ for } \kappa < 0.95 \tag{5.3.42b}$$

Larson (1992) has reviewed these and other instabilities in viscoelastic liquids.

Eccentricity of the two cylinders also changes the assumed velocity profile and can lead to decreased torque. This situation is the same as that encountered in a journal bearing and has been approximately analyzed (Pinkus and Sternlicht, 1961, p. 41). For cylinders with parallel axis displaced an amount a, the reduction in torque is

$$\frac{M}{M_0} = \frac{2[1 - (a/\overline{\Delta R})^2]^{1/2}}{2 + (1/\overline{\Delta R})^2} \tag{5.3.43}$$

where $\overline{\Delta R}$ is the mean gap thickness. The effect of eccentricity on the torque with a power law fluid is expected to be less (Ehrlich and Slattery, 1968).

5.3.7 Shear Heating in Couette Flow

The shearing action in Couette flow generates heat. Many organic and polymeric fluids have rather low thermal conductivity. Thus, the viscous dissipation results in a temperature rise in the fluid. Since viscosity decreases, typically exponentially, with tempera-

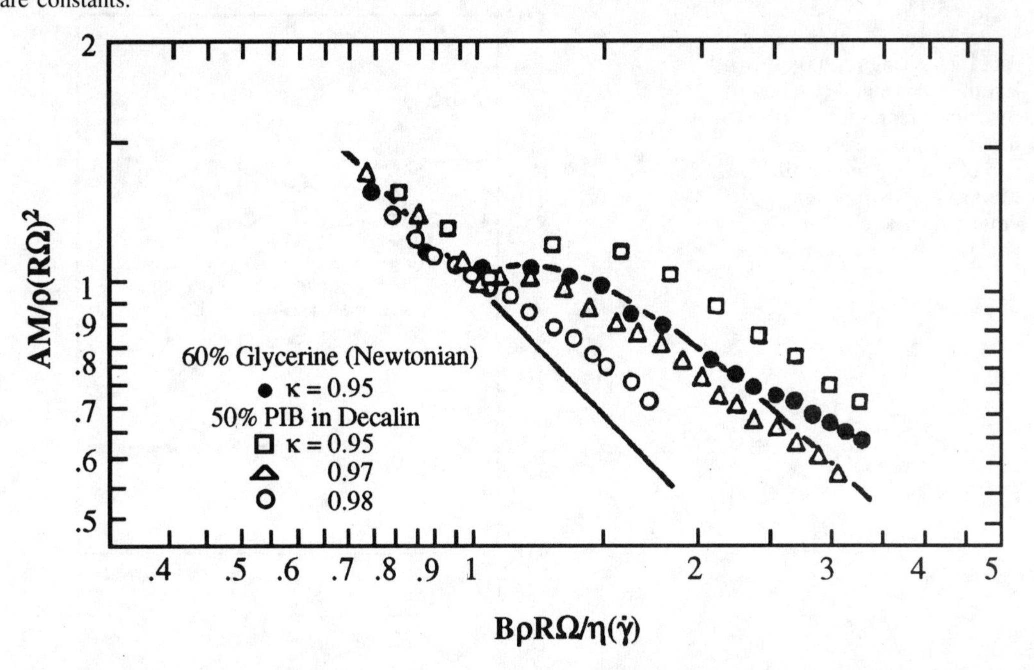

Figure 5.3.8. Effect of secondary flow (Taylor vortices) on reduced torque in a Couette viscometer. From Denn and Roisum (1969). Note that A and B are constants.

ture, this temperature rise results in a decreasing torque with time and with decreasing shear stress. Sukanek and Laurence (1974) report pronounced torque decreases for a Newtonian oil, Arochlor 1260, which has a highly temperature-sensitive viscosity. As we can see from their data in Figure 5.3.9, this Newtonian liquid appears to be pseudoplastic after reaching thermal equilibrium.

For flow in a narrow gap viscometer, the energy equation and the momentum balance are coupled together by the temperature-dependent viscosity. These equations have been solved for the equilibrium temperature profile and the effect on shear stress by Gavis and Laurence (1968) for a power-law fluid and by Turian (1969) with the Ellis model. For the power law model, the effect on torque in a narrow gap instrument can be expressed in terms of a power series in the Brinkman number (see Example 2.6.1, eq. 2.6.15). The first term of the series is helpful to the experimentalist to indicate where shear heating can affect data.

$$\frac{M}{M_o} = 1 - \left(\frac{b\text{Br}}{12n}\right)^n \tag{5.3.44}$$

where M_o is the torque under isothermal conditions and the Brinkman number is

$$\text{Br} = \frac{m_0(R_i\Omega)^{n+1}}{k_T T_0 (R_o - R_i)^{n-1}} \tag{5.3.45}$$

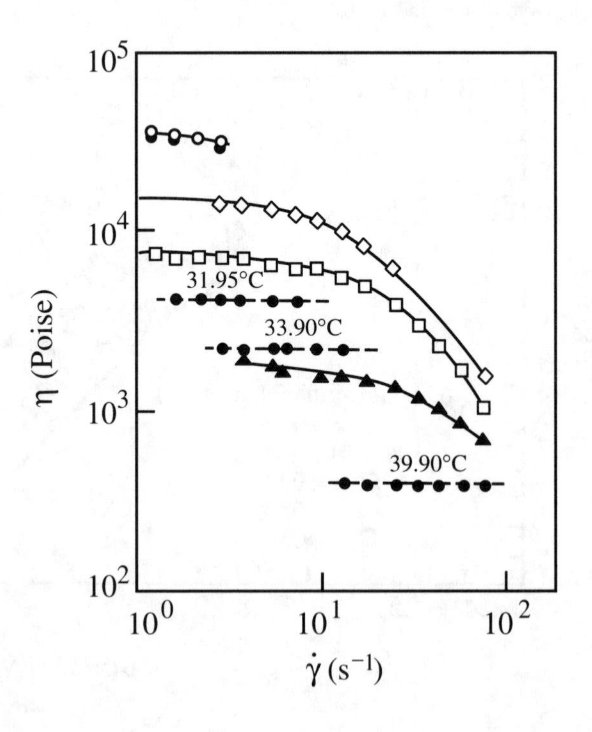

Figure 5.3.9.
Viscosity versus shear rate data for Arochlor 1260. Broken lines are initial readings; solid lines represent readings taken after equilibration of temperature profiles due to viscous dissipation. From Sukanek and Laurence (1974).

The temperature sensitivity of the viscosity is b, the exponential factor in the power law expression

$$\tau = m_0 \dot{\gamma}^n e^{-b(T-T_0)/T_0} \tag{5.3.46}$$

Some typical values of b for polymer melts are 0.05–0.1 (Tadmor and Gogos, 1979). These can be somewhat higher for solutions. Further discussion of the temperature dependence of viscosity is given in Section 2.6.

5.4 Cone and Plate Rheometer

Mooney and Ewart (1934) appear to have been the first to suggest the cone and plate geometry for viscosity measurements. Russell (1946) used the geometry for normal stress measurements. His studies eventually led to the development of the Weissenberg Rheogoniometer (Jobling and Roberts, 1959; Lammiman and Roberts, 1961) and the Ferranti Shirley instrument (McKennell, 1954). Today the cone and plate, with its constant rate of shear and direct measurement of N_1 by total thrust, is probably the most popular rotational geometry for studying non-Newtonian effects.

A sketch of the cone and plate geometry is shown in Figure 5.4.1. Spherical coordinates are the proper ones for the problem; note also Figure 5.1.2. If we assume:

1. Steady, laminar, isothermal flow
2. $v_\phi(r, \theta)$ only; $v_r = v_\theta = 0$
3. $\beta < 0.10$ rad ($\approx 6°$)
4. Negligible body forces
5. Spherical liquid boundary

then the equations of motion from Table 1.7.1 reduce to

$$r: \frac{\rho v_\phi^2}{r} = \frac{1}{r^2}\frac{\partial}{\partial r}(r^2 \tau_{rr}) - \frac{\tau_{\theta\theta} + \tau_{\phi\phi}}{r} \tag{5.4.1}$$

$$\theta: 0 = \frac{1}{r \sin\theta}\frac{\partial(\tau_{\theta\theta}\sin\theta)}{\partial\theta} - \frac{\cot\theta\,\tau_{\theta\theta}}{r} \tag{5.4.2}$$

$$\phi: 0 = \frac{1}{r}\frac{\partial \tau_{\theta\phi}}{\partial\theta} + \frac{2}{r}\cot\theta\,\tau_{\theta\phi} \tag{5.4.3}$$

Figure 5.4.1.
Schematic of cone and plate rheometer.

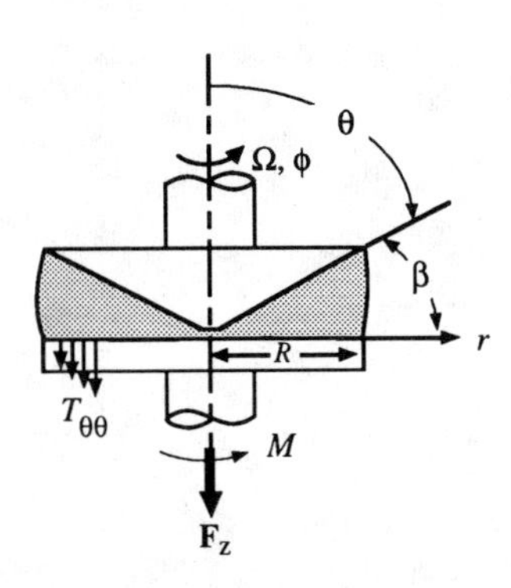

The boundary conditions are

$$v_\phi\left(\frac{\pi}{2}\right) = 0 \tag{5.4.4}$$

$$v_\phi\left(\frac{\pi}{2-\beta}\right) = \Omega r \sin\left(\frac{\pi}{2-\beta}\right) \tag{5.4.5}$$

since $\sin(\pi/2 - \beta) = \cos\beta = 1 - \frac{\beta^2}{2!} + \frac{\beta^4}{4!} \cdots \cong 1$ within 1% for $\beta \leq 0.1$, then

$$v_\phi\left(\frac{\pi}{2 - \beta}\right) = \Omega r \tag{5.4.6}$$

The solutions to these equations are summarized in Table 5.4.1 and derived in the Sections 5.4.1–5.4.6.

5.4.1 Shear Stress

As we saw in Example 2.3.2, the shear stress can be found by integrating eq. 5.4.3.

$$\tau_{\phi\theta} = \frac{C_1}{\sin^2\theta} \tag{5.4.7}$$

From a torque balance on the plate

$$M = \int_0^{2\pi}\int_0^R r^2 \tau_{\phi\theta}\bigg|_{(\pi/2)} dr\, d\phi = \frac{2\pi R^3}{3}\tau_{\phi\theta}\bigg|_{(\pi/2)} \tag{5.4.8}$$

TABLE 5.4.1 / Working Equations for Cone and Plate

Shear stress	$\tau_{12} = \tau_{\phi\theta} = \frac{3M}{2\pi R^3}$	(5.4.10)
Shear strain	$\gamma = \frac{\phi}{\beta}$ (homogeneous)	(5.4.11)
Shear rate	$\dot\gamma = \frac{\Omega}{\beta}$	(5.4.13)
Normal stress	$\tau_{11} - \tau_{22} = \tau_{\phi\phi} - \tau_{\theta\theta} = \frac{2F_z}{\pi R^2}$	(5.4.21)
Errors		
Inertia and secondary flow	$N_1 = \frac{2F_z}{\pi R^2} - 0.15\rho\,\Omega^2 R^2$	(5.4.24)
Torque correction	$\frac{M}{M_0} = 1 + 6.1\times10^{-4}\mathrm{Re}^2,\ \mathrm{Re} = \frac{\rho\Omega\beta^2R^2}{\eta_0}$	(5.4.25)
Gap opening	$\frac{6\pi R\eta_0}{\kappa\beta^3}$ < material relaxation time	(5.4.28)
Shear heating	$\frac{M}{M_0} = \frac{1 - b\mathrm{Br}}{20n}$	(5.4.32)

Utility
Most common instrument for normal stress measurements
Simple working equations: homogeneous deformation
Nonlinear viscoelasticity $G(t, \gamma)$
Useful for low and high viscosity materials
High viscosity limited by elastic edge failure
Low viscosity limited by inertia corrections, secondary flow, and loss of sample at edges

Noting from eq. 5.4.7 that $\tau_{\phi\theta}|_{\pi/2} = C_1 = \tau_{\phi\theta}(\theta)\sin^2\theta$, we substitute to obtain

$$\tau_{\phi\theta}(\theta) = \frac{3M}{2\pi R^3 \sin^2\theta} \tag{5.4.9}$$

But since $\beta < 0.1$ rad, then $1 > \sin^2(\pi/2 - \beta) \geq \sin^2(\pi/2 - 0.1) = 0.990$ and the shear stress is essentially constant throughout the fluid

$$\tau_{12} = \tau_{\phi\theta} = \frac{3M}{2\pi R^3} \tag{5.4.10}$$

5.4.2 Shear Strain Rate

Since τ_{12} is nearly constant, it follows that the shear strain and shear rate will also be nearly constant. From B in spherical coordinates (Table 1.4.1 and eq. 1.4.13)

$$B_{\phi\theta} = \gamma = \frac{\phi}{\beta} \tag{5.4.11}$$

Similarly from D in spherical coordinates (Table 2.2.1 and eq. 2.2.10)

$$|2D_{\phi\theta}| = \dot{\gamma} = \left|\frac{\sin\theta}{r}\frac{\partial}{\partial\theta}\left(\frac{v_\phi}{\sin\theta}\right)\right| = \left|\frac{1}{r}\frac{\partial v_\phi}{\partial\theta} - \frac{v_\phi}{r}\cot\theta\right|$$

Because $\theta = \pi/2 - \beta$, we can rewrite $\cot\theta$ by

$$\cot\theta = \cot\left(\frac{\pi}{2} - \beta\right) = \tan\beta$$

However, since β is small we can approximate $\tan\beta$ by

$$\tan\beta \simeq \beta + \frac{\beta^3}{3} \tag{5.4.12}$$

To a good approximation, the velocity profile is

$$v_\phi = \Omega r \frac{(\pi/2) - \theta}{\beta}$$

so the shear rate is given by

$$\dot{\gamma} = \frac{\Omega}{\beta}\left(1 - \beta^2 + \frac{\beta^4}{3}\right) \approx \frac{\Omega}{\beta} \tag{5.4.13}$$

Adams and Lodge (1964) show that the error in using eq. 5.4.13 for the shear rate at the plate is less than 0.7% at 0.1 rad and only 2% at 0.18 rad. The attractiveness of the cone and plate geometry lies in the fact that the shear rate and shear stress are independent of position and can be easily calculated. No differentiation of data is required to obtain a true $\tau_{12}(\dot{\gamma})$ relation for an unknown fluid.

5.4.3 Normal Stresses

Normal stress differences can be determined from pressure and thrust measurements on the plate. If we ignore inertial effects for the moment, eq. 5.4.1 can be written as

$$r\frac{\partial \tau_{rr}}{\partial r} = \tau_{\phi\phi} + \tau_{\theta\theta} - 2\tau_{rr} = N_1 + 2N_2 \qquad (5.4.14)$$

We recall that N_2 is a steady shear material function and only depends on shear rate. Since $\dot{\gamma}$ is independent of r

$$\frac{\partial N_2}{\partial r} = \frac{\partial \tau_{\theta\theta}}{\partial r} - \frac{\partial \tau_{rr}}{\partial r} = 0 \quad \text{or} \quad \frac{\partial \tau_{\theta\theta}}{\partial r} = \frac{\partial \tau_{rr}}{\partial r} \qquad (5.4.15)$$

Thus, eq. 5.4.14 becomes

$$\frac{\partial \tau_{\theta\theta}}{\partial \ln r} = N_1 + 2N_2 \qquad (5.4.16)$$

$\tau_{\theta\theta}$ is just the pressure measured by a transducer on the plate or the cone surface.

Thus at a fixed angular velocity Ω, plots of $\tau_{\theta\theta}$ versus $\ln r$ should be a straight line. Miller and Christiansen (1972) have mounted transducers flush to the plate surface and measured $\tau_{\theta\theta}(r)$. From their results on a polyacrylamide (PAA) solution plotted in Figure 5.4.2, we see that $\tau_{\theta\theta}(r)$ versus $\log r$ is indeed linear. We also note that there is a small positive pressure at the rim of the plate, $\tau_{\theta\theta}(R)$. If we can assume that the stress in the radial direction at the rim $\tau_{\theta\theta}(R)$ is balanced by the ambient pressure p_a (i.e., there are no surface tension or other edge effects), then $\tau_{\theta\theta}(R)$ is just the second normal stress difference.

$$N_2 = \tau_{22} - \tau_{33} = \tau_{\theta\theta}(R) - \tau_{rr}(R) = \tau_{\theta\theta}(R) - 0 \qquad (5.4.17)$$

It is more common and simpler to measure the total thrust on the plate. Making a force balance on the plate, we have

$$F_z = -\int_0^{2\pi}\int_0^R \tau_{\theta\theta}(r) dr\, d\phi - p_a \pi R^2 \qquad (5.4.18)$$

Integrating by parts and noting that $\tau_{\phi\phi} + \tau_{\theta\theta} - 2\tau_{rr} = N_1 + 2N_2$ is only a function of shear rate and thus is independent of r

$$F_z = -\pi R^2 \left[\tau_{\theta\theta}(R) - \left(\frac{\tau_{\phi\phi} + \tau_{\theta\theta} - 2\tau_{rr}}{2} \right) + p_a \right]$$

Figure 5.4.2.
Normal stress as a function of log r for a 1.49% polyacrylamide. From Miller and Christiansen (1972).

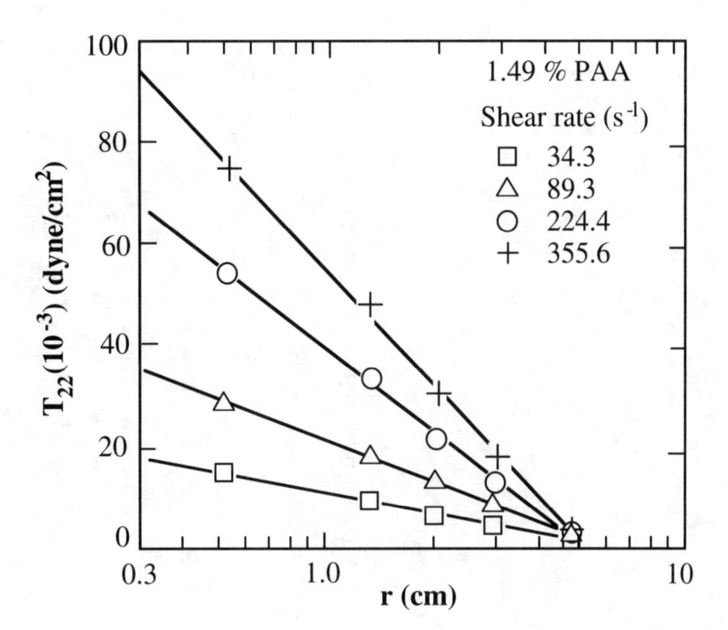

or

$$= -\pi R^2\left[\left(\frac{\tau_{\theta\theta} - \tau_{\phi\phi}}{2}\right) + \tau_{rr} + p_a\right] \tag{5.4.19}$$

As mentioned above, if the boundary has a spherical shape, and if surface tension effects are negligible, then $\tau_{rr}(R)$ must be balanced by p_a. Thus, we have the simple result

$$F_z = \frac{\pi R^2}{2}(\tau_{\phi\phi} - \tau_{\theta\theta}) = \frac{1}{2}\pi R^2 N_1 \tag{5.4.20}$$

or

$$N_1 = \frac{2F_z}{\pi R^2} \tag{5.4.21}$$

Miller and Christiansen (1972) also measured total thrust and obtained the N_1 values shown in Figure 5.4.3. By subtracting $2\tau_{\theta\theta}(R)$ from the slopes in Figure 5.4.2, N_1 can be calculated independently. Similarly subtracting N_1 (calculated from the total thrust from the slopes) gives a test of the rim pressure measurements. Results by each of these methods for two cone angles compare well, particularly for N_1.

5.4.4 Inertia and Secondary Flow

In the foregoing derivations we neglected the inertia term in eq. 5.4.1. We saw with concentric cylinders that inertia causes a depression around the inner cylinder rather than the climb due to viscoelastic normal stresses (Figure 5.3.3). In cone and plate

Figure 5.4.3.
Normal stress differences increased by two different methods. Miller and Christiansen (1972).

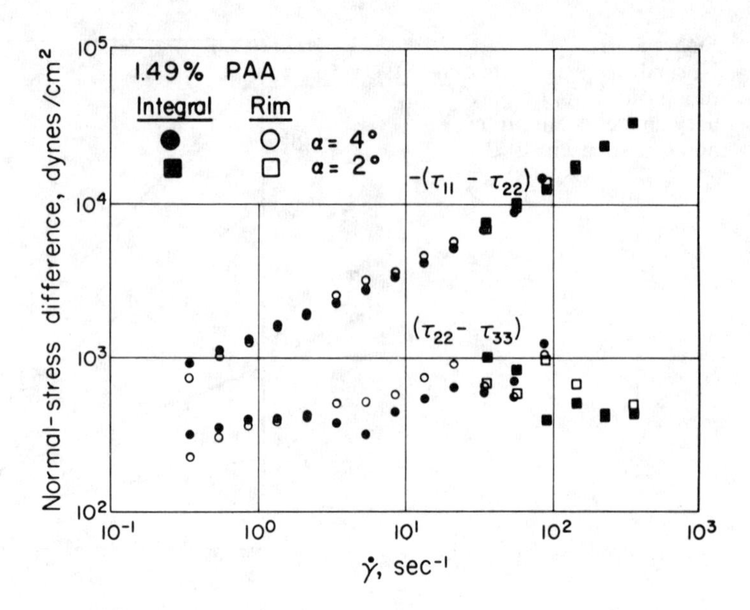

rheometers, inertia forces tend to pull the plates together rather than push them apart. This "negative" normal force can be expressed in terms of stress distribution (Walters, 1975)

$$\left(\tau_{\theta\theta}\right)_{\text{inert}} = 0.15\rho\Omega^2(r^2 - R^2) \tag{5.4.22}$$

or in terms of the total thrust

$$(F_z)_{\text{inert}} = 0.075\pi\rho\Omega^2 R^4 \tag{5.4.23}$$

When eq. 5.4.23 is combined with eq. 5.4.21, we obtain a result that is commonly used to correct measured F_z values

$$N_1 = \frac{2F_z}{\pi R^2} - 0.15\rho\Omega^2 R^2 \tag{5.4.24}$$

This result has been well confirmed experimentally for Newtonian fluids using pressure distribution data (Greensmith and Rivlin, 1953; Markovitz and Brown, 1963; Adams and Lodge, 1964; Alvarez et al., 1985) and from total thrust (Huppler et al., 1967; Miller and Christiansen, 1972; Kulicke, et al., 1977; Whitcomb and Macosko, 1979). Figure 5.4.4 compares eq. 5.4.23 to experimental data. Several of these studies indicate that the correction also seems to be valid for non-Newtonian fluids.

Inertia also generates secondary flows in the $r\theta$ plane. These have been observed for large cone angles (Giesekus, 1963; Hoppman and Miller, 1963; Walters and Waters, 1968) and modeled theoretically for the Newtonian case (Turian, 1969, 1972; Fewell and Hellums, 1977; Sdougas et al., 1984). Secondary flow will

Figure 5.4.4.
Comparison of data of Miller and Christiansen (1972) to eq. 5.4.23 (dashed line).

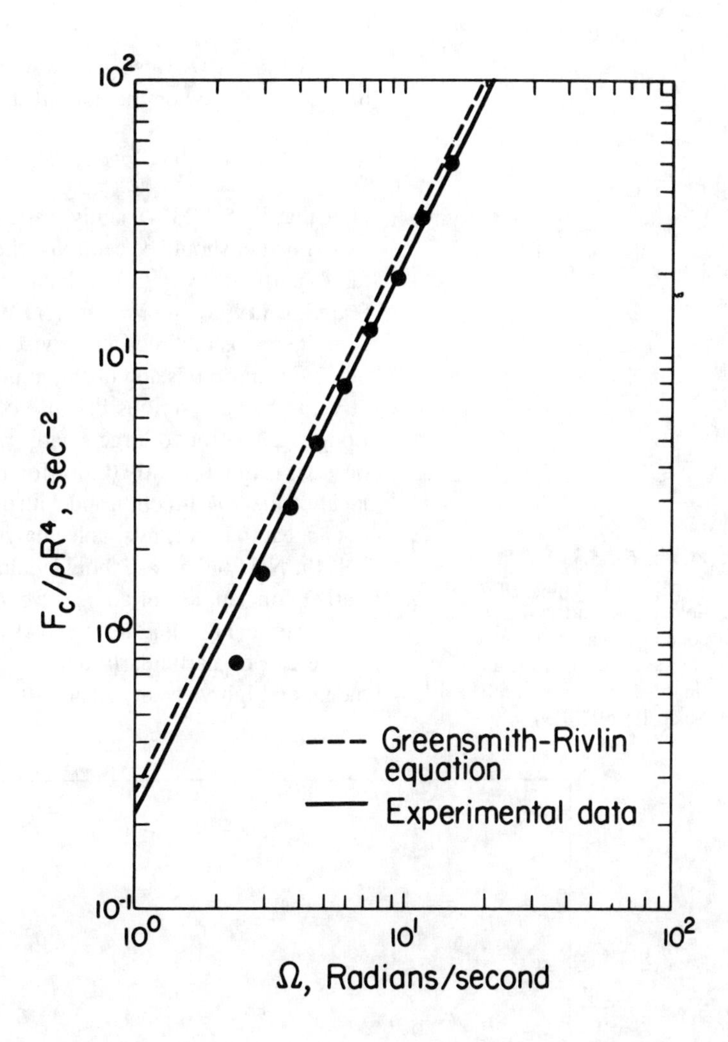

increase the torque over that given by eq. 5.4.10. Turian (1972) solves for this extra torque in terms of the Reynolds number

$$\frac{M}{M_0} = 1 + 6.1 \times 10^{-4} \mathrm{Re}^2 \tag{5.4.25a}$$

where

$$\mathrm{Re} = \frac{\rho \Omega \beta^2 R^2}{\eta_0}$$

Whitcomb and Macosko (1978) find that this result fits both their high shear rate data on Newtonian oils and the data reported by Cheng (1968) up to $M/M_0 \cong 2$. These results are shown in Figure 5.4.5. Ellenberger and Fortuin (1985) report a better fit to this data with the empirical expression

$$\frac{M}{M_0} = 1 + \frac{0.309 Re^{3/2}}{50 + Re} \tag{5.4.25b}$$

Turian (1972) also gives a similar correction for the influence of secondary flow on the normal stresses.

$$(F_z)_{\text{inert}} = 0.075\pi\rho\Omega^2 R^4[1 + 4 \times 10^{-3}\text{Re}^2] \qquad (5.4.26)$$

Note that eq. 5.4.23 is usually sufficient. In all these corrections the experimenter should be cautious whenever the adjustments become large, particularly with non-Newtonian fluids. As the pictures of Giesekus (1963) and Walters and Waters (1965) show, circulation is in the opposite direction for viscoelastic and Newtonian liquids.

Eccentricities and misalignment in the cone and plate geometry can be very serious because of the small angle. Adams and Lodge (1964) found large steady $\tau_{\theta\theta}$ values for Newtonian oils if one axis is tilted slightly (0.06° for a 3° cone). The values vary with location in the ϕ direction and with direction of rotation. There does not appear to be any available analysis of the problem, but presumably the Newtonian case should yield to lubrication approximations. Larger cone angles, of course, reduce these effects.

Vertical oscillations are always present in rotating members. These can be particularly annoying with normal stress measurements on high viscosity materials. Adams and Lodge give an es-

Figure 5.4.5.
Theoretical torque correction for secondary flow in cone and plate geometry compared to data on three fluids. From Whitcomb and Macosko (1979).

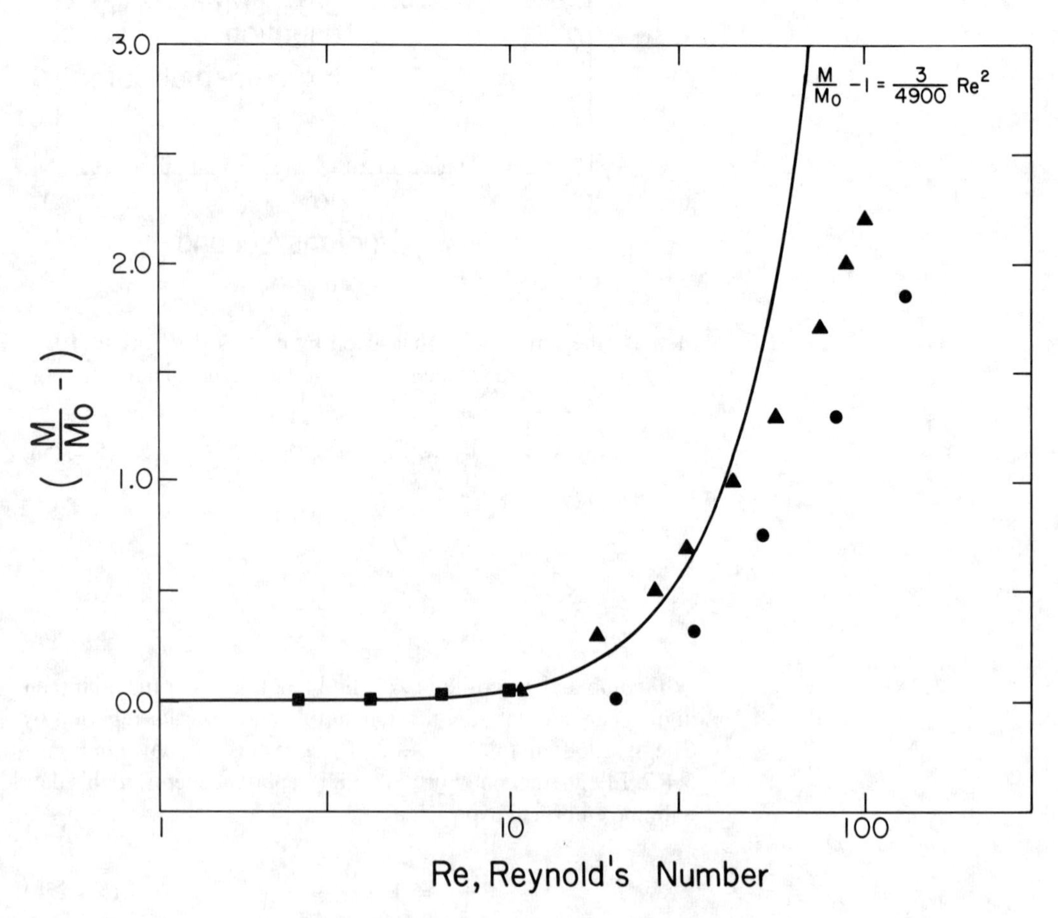

timate for the size of pressure oscillations based on the squeezing flow equations (to be discussed in Chapter 6)

$$p - p_a = -6\frac{\eta_0 v}{\beta^3}\left(\frac{1}{r} - \frac{1}{R}\right) \tag{5.4.27}$$

where v is the vertical velocity of the boundary. If the oscillations are sinusoidal $v_{\max} = \Omega\, \Delta h$, where Δh is the vertical run out of the rotating spindle. Equation 5.4.27 shows a very strong dependence on cone angle; however, it does not appear to have been tested experimentally. Tanner (1970) gives a similar result for the effect of instrument stiffness κ on normal force start-up data. He indicates that the instrument response time must be much less than the material relaxation time

$$\frac{6\pi R\eta_0}{\kappa\beta^3} \ll \lambda \tag{5.4.28}$$

This problem is discussed further in Chapter 8.

5.4.5 Edge Effects with Cone and Plate

There are two important assumptions in deriving the cone and plate equations: that the free surface is spherical and the velocity field is maintained up to the edge. Adams and Lodge (1964) discuss the consequences of this double assumption. In eq. 5.4.17, $\tau_{\theta\theta}(R)$, should be particularly sensitive to the edge shape. Miller and Christiansen (1972) report that with a 4° cone they varied the interface considerably inward and outward from spherical and found no change in $\tau_{\theta\theta}(R)$. Their good experimental agreement between various methods and cone angles in Figure 5.4.3 supports the assumption that edge effects can be neglected under proper experimental conditions.

A "drowned" edge or "sea" of liquid around the cone has been used frequently by experimenters. The flow will extend out into the sea, increasing the torque and affecting normal stresses. Vrentas et al. (1991) have analyzed the effect on torque for parallel plates. Their results (eq. 5.5.12) should be approximately valid for the cone and plate. The effect of liquid outside the gap on normal stresses can be seen if we look carefully at the data of Olabisi and Williams (1972) shown in Figure 5.4.6. The pressures near $r = R$ drop slightly below the line expected from eq. 5.4.16. Note also that some deviation occurs near the center. The authors attribute this to a slight elevation of the cone or to the finite diameter (indicated on graph) of the transducer.

Tanner (1970) has analyzed the drowned edge and finds for a second-order fluid that eq. 5.4.20 becomes

$$\frac{F_z}{\pi R^2} = \frac{1}{2}\pi R^2(N_1 + N_2) \tag{5.4.29}$$

and eq. 5.4.17

$$\tau_{\theta\theta}(R) = \frac{1}{2} N_2 \tag{5.4.30}$$

He reports that the results of Adams and Lodge (1964) and Kaye et al. (1968) seem more reasonable when corrected with eq. 5.4.30 and for pressure hole errors. Olabisi and Williams also find some experimental agreement with Tanner's equations.

A more serious edge effect occurs with high viscosity samples like polymer melts. At disappointingly low shear rates, typically $\dot{\gamma} < 10\ s^{-1}$, the sample appears to cut in at the midplane and flow out at each of the solid surfaces. Figure 5.4.7 shows this edge failure with parallel plates. The effect is very similar for parallel disks. Before it is even visible at the free surface, edge failure can be detected by a drop in the torque and normal force values (Macosko and Morse, 1976). The continuous decrease in η^+ at 5 and 20 s^{-1} in Figure 4.2.6 is due to edge failure. This defect severely limits the useful shear rate range of either geometry with polymer melts.

Turian's secondary flow analysis ignores the free surface. But it appears that in any case the failure shown in Figure 5.4.7 may be due more to elastic effects. Hutton (1969) applied an elastic energy criterion to predict the failure, while Tanner and Keentok (1983) used a fracture mechanics approach. However, the experimental results of Hutton (1969) and Broyer and Macosko (1975)

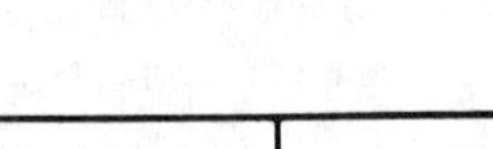

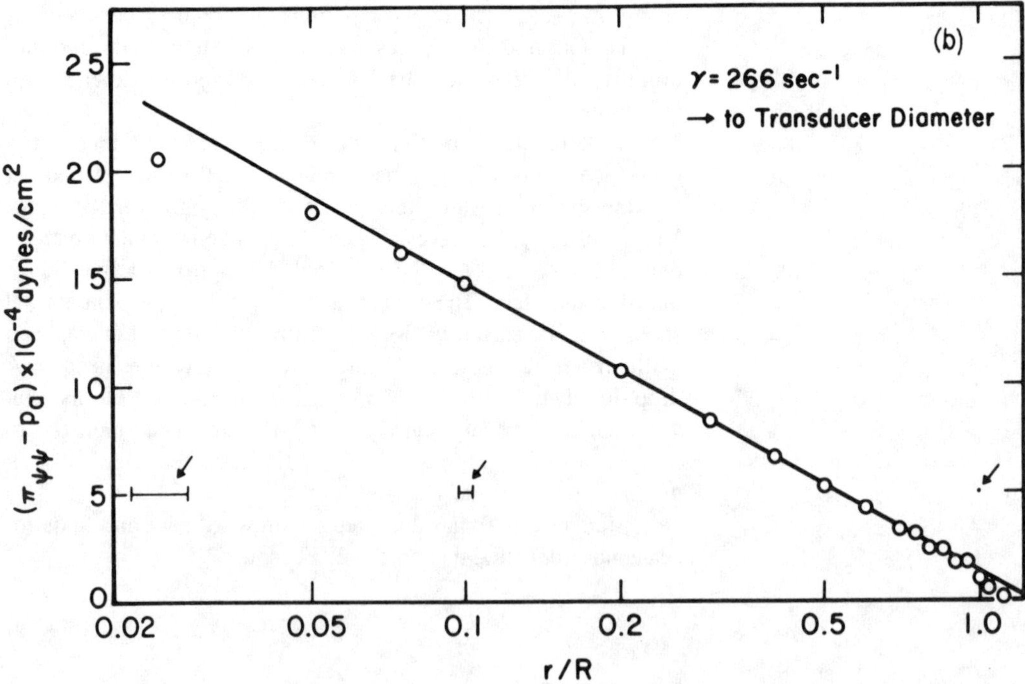

Figure 5.4.6.
Average pressure profiles (clockwise and counterclockwise) for a 3% polyethylene oxide solution. From Olabisi and Williams (1972).

Figure 5.4.7.
Edge failure with polydimethylsiloxane (η_0 = 14,000 poise at 25°C) in parallel disks. From Broyer and Macosko (1975).

suggest that there may be simply a critical edge velocity for a given material. This critical velocity seems to decrease with increasing viscosity and decreasing surface tension.

Some of Broyer and Macosko's data are shown in Figure 5.4.8. Using smaller cone angles or gaps in parallel disks (see next section) allows higher shear rate data. However, at small gaps the effects of misalignment and eccentricity on the normal forces become quite severe; note eq. 5.4.27. Quinzani and Valles (1986) have been able to increase the shear rate limit on normal force measurements by using a cup-shaped lower plate that fits closely to the edge of the cone.

Figure 5.4.8.
Critical velocity for the onset of edge failure within approximately 15 seconds of shearing. From Broyer and Macosko (1975).

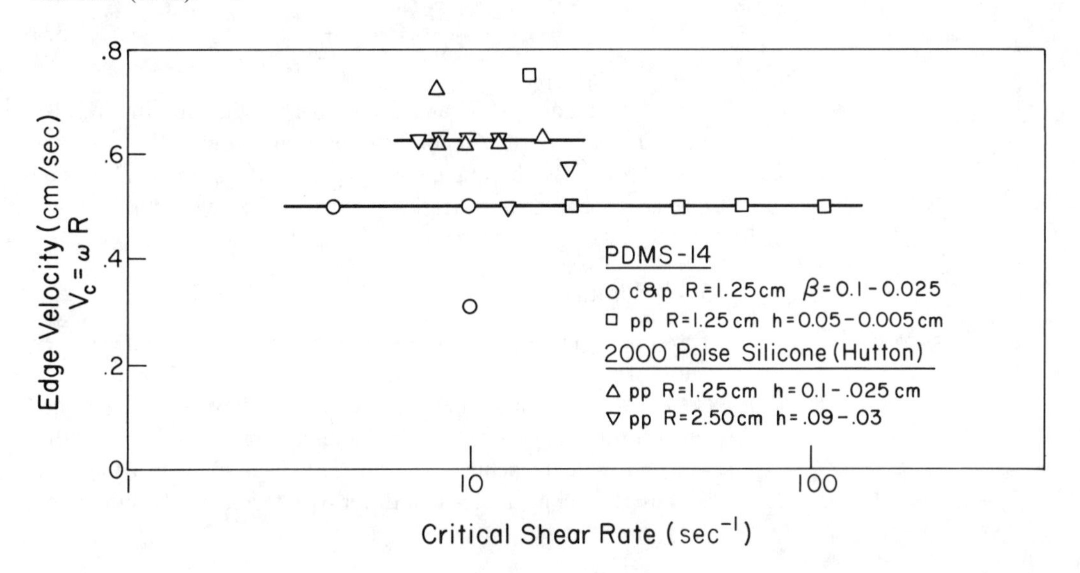

Another edge problem is the same as that discussed with the concentric cylinder end effects: change or hardening of the sample at the gas–liquid interface. To prevent evaporation, a layer of silicone or other nonvolatile oil can be spread over the free surface (Boger and Rama Murthy, 1969; Orafidiya, 1989).

5.4.6 Shear Heating

The shear heating of Newtonian (Turian and Bird, 1963) and power law fluids (Turian, 1965) has been studied in the cone and plate geometry. For the case that both cone and plate are isothermal at T_o, Turian's power law analysis predicts the maximum temperature rise (see also Bird et al., 1987, p. 226; note $b\mathrm{Br} = \mathrm{Na}$)

$$\frac{T_{\max} - T_o}{T_o} = \frac{\mathrm{Br}}{8} - \frac{b\mathrm{Br}^2}{24n}\left(\frac{n}{4} - \frac{1}{16}\right) \tag{5.4.31}$$

and the decrease in torque

$$\frac{M}{M_o} = 1 - \frac{b\mathrm{Br}}{20n} \tag{5.4.32}$$

where, as in flow between concentric cylinders, b is the temperature sensitivity of the power law viscosity in eq. 5.3.46. The Brinkman number is defined for the cone and plate as

$$\mathrm{Br} = m_o\left(\frac{\Omega}{\beta}\right)^{n-1}\frac{R^2\Omega^2}{k_T T_o}$$

From their experiments with Newtonian oils in a Ferranti Shirley viscometer, Turian and Bird find better agreement with an adiabatic cone:

$$\frac{M}{M_o} = 1 - \frac{bBr}{5n} \tag{5.4.33}$$

We see from eqs. 5.4.32 and 5.4.33 that calculating $b\mathrm{Br}$ provides a useful check on the importance of shear heating. When $b\mathrm{Br}$ is less than 1 for isothermal cone and plate and less than 0.2 for one surface adiabatic, shear heating effects will be negligible.

5.4.7 Summary

The cone and plate is a very useful and simple test geometry. Because of the small angle, it requires more precise alignment than Couette devices. A number of possible errors have been suggested, but in normal operations these problems appear to be minimal with a well constructed machine. The simplest check for all error sources is to use two cone angles that differ by a factor of 2 or more, and possibly also two different radii. The drowned edge should not be used to obtain good normal stress data. With high viscosity samples, the edge failure problem limits the high shear rate use of the device.

5.5 Parallel Disks

The parallel disk geometry was suggested by Mooney (1934). The Mooney tester, which consists of a disk rotating inside a cylindrical cavity, is used extensively in the rubber industry (ASTM D1646). Russell (1946) first measured normal forces from the total thrust between two disks. Greensmith and Rivlin (1953) measured the pressure distribution, and Kotaka et al.(1959) used total thrust to study normal stresses in polymer melts. In many ways the flow is similar to the cone and plate. Most instruments are designed to permit the use of either gometry. However, in contrast to the cone and plate, flow between parallel disks is not homogeneous.

The parallel disks geometry is sketched in Figure 5.5.1. If we assume:

1. Steady, laminar, isothermal flow
2. $v_\theta(r, z)$ only, $v_r = v_z = 0$
3. Negligible body forces
4. Cylindrical edge

then the equations of motion reduce to

$$\theta : \quad \frac{\partial \tau_{\theta z}}{\partial z} = 0 \tag{5.5.1}$$

$$z : \quad \frac{\partial \tau_{zz}}{\partial z} = 0 \tag{5.5.2}$$

$$r : \quad \frac{1}{r}\frac{\partial}{\partial r}(r\tau_{rr}) - \frac{\tau_{\theta\theta}}{r} = -\rho\frac{v_\theta^2}{r} \tag{5.5.3}$$

Figure 5.5.1.
Schematic of a parallel plate rheometer.

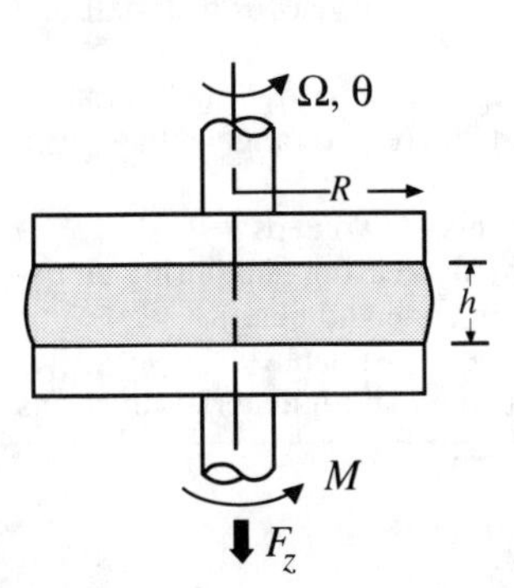

Table 5.5.1 gives the working equations for this geometry.

With one disk stationary and the other rotating at Ω, assuming no slip at these surfaces and neglecting inertial forces, the velocity must be

$$v_\theta(r, z) = \frac{r\Omega z}{h}$$

and thus

$$\dot{\gamma}(r) = \frac{r\Omega}{h} \tag{5.5.4}$$

TABLE 5.5.1 / Working Equations for Parallel Disks

Shear strain

$$\gamma = \frac{\theta r}{h} \text{ (nonhomogeneous, depends on position)} \tag{5.5.5}$$

Shear rate at $r = R$

$$\dot{\gamma}_R = \frac{R\Omega}{h} \tag{5.5.4}$$

Shear stress

$$\tau_{12} = \tau_{\theta z} = \frac{M}{2\pi R^3}\left[3 + \frac{d \ln M}{d \ln \dot{\gamma}_R}\right] \tag{5.5.8}$$

$$\tau_a = \frac{2M}{\pi R^3} \text{ apparent or Newtonian shear stress} \tag{5.5.9}$$

Representative shear stress

$$\eta(\tau) = \eta_a(\tau_a) \pm 2\% \tag{5.5.10}$$

$$\text{for } \tau = 0.76\tau_a \text{ and } \frac{d \ln M}{d \ln \dot{\gamma}_R} < 1.4$$

Normal stress

$$N_1 - N_2 = \frac{F_z}{\pi R^2}\left[2 + \frac{d \ln F_z}{d \ln \dot{\gamma}_R}\right] \tag{5.5.17}$$

$$(F_z)_{\text{inert}} = 0.075\pi\rho\Omega^2 R^4 \tag{5.4.23}$$

Errors

Inertia and secondary flow
Edge failure (same as cone and plate)
Shear heating
Nonhomogeneous strain field (correctable)

Utility

Sample preparation and loading is simpler for very viscous materials and soft solids
Can vary shear rate (and shear strain) independently by rotation rate Ω (and θ) or by changing the gap h; permits increased range with a given experimental set up
Determine wall slip by taking measurements at two gaps
Delay edge failure to higher shear rate by decreasing gap during an experiment (requires change of cone angle in cone and plate)
Measure N_2 when used with cone and plate thrust data
Preferred geometry for viscous melts for small strain material functions

Similarly the strain goes from zero at the center to maximum at the edge

$$\gamma = \frac{\theta r}{h} \tag{5.5.5}$$

As in wide gap Couette and Poiseuille flow (Chapter 6), shear rate is not constant. Thus we must use a derivative to relate shear stress to total torque. The resulting equations are given below and then derived in the remainder of this section.

From eq. 5.5.1 and recalling that the shear stress can be a function of shear rate alone, $\tau_{\theta z} = \tau_{12} = \tau_{12}(\dot{\gamma})$. From a torque balance

$$M = 2\pi \int_0^R r\tau_{12}(r)r\, dr \tag{5.5.6}$$

Changing variables, we have

$$r = \frac{h}{\Omega}\dot{\gamma} = \frac{R\dot{\gamma}}{\dot{\gamma}_R} \quad \text{where} \quad \dot{\gamma}_R = \frac{\Omega R}{h}$$

$$dr = \frac{R}{\dot{\gamma}_R} d\dot{\gamma}$$

Then

$$M = 2\pi \int_0^{\dot{\gamma}_R} \left(\frac{R}{\dot{\gamma}_R}\right)^3 \dot{\gamma}^2 \tau_{12} d\dot{\gamma} \tag{5.5.7}$$

Rearranging and differentiating using Leibnitz's rule gives

$$\tau_{12}(R) = \frac{M}{2\pi R^3}\left[3 + \frac{d \ln M}{d \ln \dot{\gamma}_R}\right] \tag{5.5.8}$$

Thus to evaluate shear stress for an unknown fluid, a sufficient amount of ln M versus ln $\dot{\gamma}_R$ data must be taken to determine the derivative accurately. In practice this is not highly difficult, since many materials have power law regions and numerical software packages are readily available for handling the data. Furthermore, the derivative is generally less than 1, and thus a 10% error results in less than 3% error in τ_{12}.

If the test liquid is Newtonian, $d \ln M / d \ln \dot{\gamma}_R = 1.0$ and the shear stress becomes

$$\tau_a(R) = \frac{2M}{\pi R^3} \tag{5.5.9}$$

This *apparent* shear stress often is used to calculate an apparent viscosity, since only a single torque measurement is required. However, a simple, approximate single point method was developed by Geisekus and Langer (1977). It is similar to the method for approximating true shear rate in a capillary rheometer given in Example 6.2.1. Here the idea is that the true and apparent shear stresses must be equal at some radial position. This occurs at nearly the same point, $r/R = 0.76$, for a wide range of liquids, i.e. those for which $d\ln M/d\ln\dot{\gamma}_R < 1.4$. Thus the true viscosity is equal to the apparent viscosity evaluated at

$$\eta(\tau) = \eta_a(\tau_a) \pm 2\% \quad \text{for} \quad \tau = 0.76\tau_a \tag{5.5.10}$$

It is easier to load and unload viscous or soft solid samples with the parallel disk geometry than with cone and plate or concentric cylinders. Thus parallel disks are usually preferred for measuring viscoelastic material functions like $G(t, \gamma)$, $G'(\omega)$, or $J(t, \tau)$ on polymer melts. To evaluate moduli or compliance, we use the strain and stress at the edge of the disk (eqs. 5.5.5 and 5.5.8), but now the stress must be corrected by $d\ln M/d\ln\gamma$ (Soskey and Winter, 1984). In the linear viscoelastic region $G(t, \gamma) = G(t)$ and $d\ln M/d\ln\gamma = 1$.

The parallel disk rheometer is also very useful for obtaining viscosity and normal stress data at high shear rates. As eq. 5.4.4 indicates, shear rate can be increased by either increasing rotation rate or decreasing gap. Errors due to secondary flows (similar to those shown in Figure 5.4.5), edge effects (Figure 5.4.7), and shear heating (Figure 5.3.9) are all reduced by operating at small gaps. Binding and Walters (1976) report reaching $10^5\ \text{s}^{-1}$ with $h = 3.2\ \mu\text{m}$. Connelly and Greener (1985) and Kramer et al. (1985) report that below about 300 μm they needed to correct for an error in the measured gap. A simple constant error term ($h_e = 10$ to 40 μm) was found adequate; that is, for a Newtonian liquid the relation

$$\frac{\dot{\gamma}_h h}{\tau_R} = \frac{h}{\eta} + \frac{h_e}{\eta} \tag{5.5.11}$$

fit their data in which $\dot{\gamma}_h$ is the apparent shear rate calculated at the measured gap.

Another use for data collected at different gaps in the parallel disk geometry is in determining wall slip. Yoshimura and Prud'homme (1988) have shown that the difference in apparent stress versus shear rate at two different gaps can be related to wall slip, analogous to eq. 5.3.27.

Vrentas et al. (1991) have analyzed the effect on the torque of using a cup instead of the lower disk. The flow extends out into the surrounding fluid. For small gaps and large cups $R_{\text{cup}}/R > 1.1$, the effect of the cup radius can be ignored if $R/R_{\text{cup}} > 1.1$. Then the extra torque depends solely on the ratio of gap to radius

$$\frac{M}{M_o} = 1 + 1.9\frac{h}{R} \tag{5.5.12}$$

5.5.1 Normal Stresses

The normal stresses arise from eq. 5.5.3. Neglecting the inertial term and noting that

$$\frac{d\tau_{22}}{dr} = \frac{d\tau_{33}}{dr} + \frac{d}{dr}(\tau_{22} - \tau_{33})$$

$$\frac{d\tau_{22}}{dr} = -\frac{\tau_{33} - \tau_{11}}{r} + \frac{d}{dr}(\tau_{22} - \tau_{33}) \tag{5.5.13}$$

Integrating from r to R

$$\int_{\tau_{22}(r)}^{\tau_{22}(R)} d\tau_{22} = \int_{N_2(r)}^{N_2(R)} d(\tau_{22} - \tau_{33}) + \int_r^R \frac{N_1 + N_2}{\zeta} d\zeta$$

gives, after substituting for N_1 and N_2,

$$\tau_{22}(r) = N_2(r) - \int_r^R \frac{N_1 + N_2}{\zeta} d\zeta \tag{5.5.14}$$

As in our cone and plate discussion, we assume that $\tau_{33}(R)$ is exactly balanced by atmospheric pressure and the free surface is cylindrical with negligible surface tension effects. This parallel disk pressure distribution is not as useful as eq. 5.4.12 was for the cone and plate. There is no simple linear relation, as can be seen in the data of Greensmith and Rivlin (1953). Note that $p_{\text{rim}} = N_2$, eq. 5.4.16, should still hold.

More frequently, total thrust is measured. Integrating eq. 5.5.10 for F_z gives

$$F_z = -2\pi \int_0^R N_2 r\, dr + 2\pi \int_0^R \frac{N_1 + N_2}{\zeta} \left(\int_r^R d\zeta \right) r\, dr \tag{5.5.15}$$

If we are careful about the limits of integration, we can change the order to give

$$F_z = -2\pi \int_0^R N_2 r\, dr + 2\pi \int_0^R \frac{N_1 + N_2}{\zeta} \int_0^\zeta r\, dr\, d\zeta$$
$$= -\pi \int_0^R (N_2 - N_1) r\, dr \tag{5.5.16}$$

Changing variables again by eq. 5.5.6 and differentiating with respect to $\dot{\gamma}_R$ we obtain

$$\frac{2\dot{\gamma}_R F_z}{\pi R^2} + \frac{\dot{\gamma}_R^2}{\pi R^2} \frac{dF_z}{d\dot{\gamma}_R} = \dot{\gamma}_R (N_1 - N_2)$$

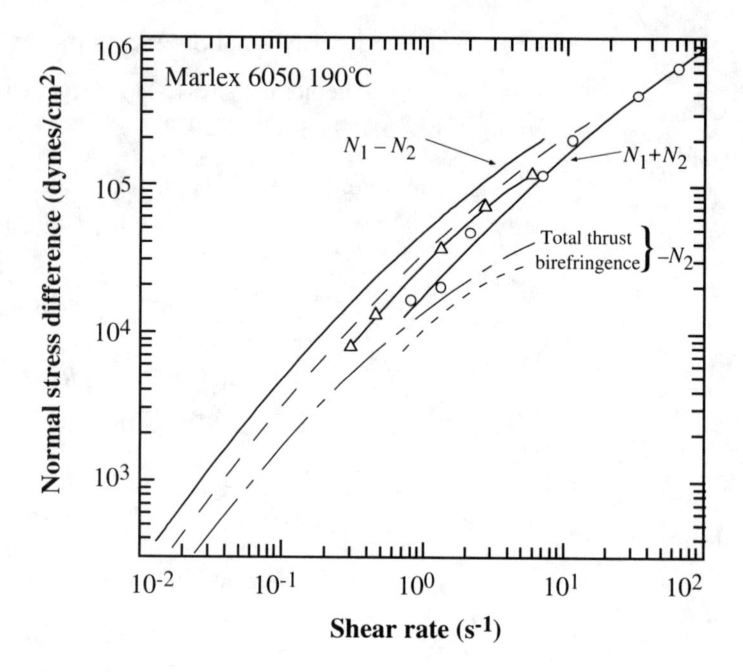

Figure 5.5.2.
Normal stress differences for a high density polyethylene (Marlex 6050) at 190°C. Total thrust between parallel disks (solid line) $N_1 - N_2$ and cone and plate (— — —) N_1 and their difference (— - - —) $-N_2$ (data taken from Macosko, 1970, 1974). Birefringence in cone and plate (triangles) N_1 and slit flow (circles) $N_1 + N_2$ and their difference (- - - -) $-N_2$ (data from Wales 1969, 1970).

or

$$(N_1 - N_2)|_{\dot{\gamma}_R} = \frac{F_z}{\pi R^2}\left[2 + \frac{d \ln F_z}{d \ln \dot{\gamma}_R}\right] \qquad (5.5.17)$$

Secondary flow effects are similar to the cone and plate. Turian (1972) reports that his eqs. 5.4.25a and 5.4.26 can be used for parallel disks with the substitution of h/R for β. The experimental results of Greensmith and Rivlin described in connection with eq. 5.4.22 are for the parallel disks and agree with the theory.

Eccentricities or misalignments are not as important except for very narrow gaps. One advantage of the parallel disk system is that the shear rate can readily be changed by both Ω and h. Broyer and Macosko (1975) have used small h values to delay edge failure effects to higher shear rates than can be achieved in typical cone and plate geometries. Some of their results were shown in Figure 5.4.7.

Accurate data can be obtained with the parallel disk geometry. Figure 5.5.2 shows some parallel disk total thrust data for $N_1 - N_2$ calculated by eq. 5.5.17. There is reasonable agreement between N_2 by the difference between parallel disks and cone and plate and by birefringence studies.

5.6 Drag Flow Indexers

There are a number of index tests in which the flow is driven by drag. Most are based on rotating a disk or other complex shape in a large quantity of the sample. These geometries are used because they are easy to load or because they help sample mixing.

Here we will focus on two major types of drag flow indexers. The first consists of rotating disk or vaned fixtures used for fluid systems, dispersions, and soft gels. The other group is made up of instrumented mixing devices, typically consisting of a rotating screw or two counterrotating blades in a heated chamber, which monitor torque as a sample is melted. This equipment is used to mix both food products such as bread dough, and polymer powder or pellets. The machines are mainly process simulators equipped with a torque monitor to get some idea of viscosity.

In the rubber industry a double parallel plate device, the Mooney Viscometer (ASTM D1646) is used to measure viscosity, Data is normally reported in Mooney units from 0 to 100, but these can be translated to torque and analyzed by eq. 5.5.8 (Nakajima and Collins, 1974). Another rubber indexer is the oscillating disk curometer, which is typically a bicone oscillating at 3 Hz in a disk-shaped cavity (ASTM D2704). Peak torque is reported on a scale of 0–100, but if the amplitude is known, this can be translated to G^*. However, at high G^* levels and the usual operating conditions wall slip can occur.

5.6.1 Rotating Disk in a Sea of Fluid

For quality control, and even for formulation development of many industrial fluids from paints to pizza sauce, the rotating disk indexer is the mainstay. As shown in Figure 5.6.1, the hand-held rheometer can be easily placed in a beaker of the sample and a viscosity number recorded. The device is often called a Brookfield after a manufacturer of commonly used equipment.

Shear rate varies in both the r and z directions from the disk surface. The flow field can be analyzed for a Newtonian fluid

$$\eta_0 = \frac{3M}{32R^3\Omega} \tag{5.6.1}$$

Williams (1979) reviews various correction factors and gives a numerical method for using the rotating disk to obtain true $\eta(\dot{\gamma})$ data from curves of M versus Ω. Geometries other than a disk are sometimes used. With a long cylinder shear rate varies only in the τ direction and true viscosity can be determined (eq 5.3.21). The vane fixture for fluids with a yield stress is described below. A T-shaped fixture is used for soft gels. It is lowered into the sample as it rotates, to ensure that unsheared sample is always being tested.

Figure 5.6.1.
Side view of a disk rotating in a large container of test fluid.

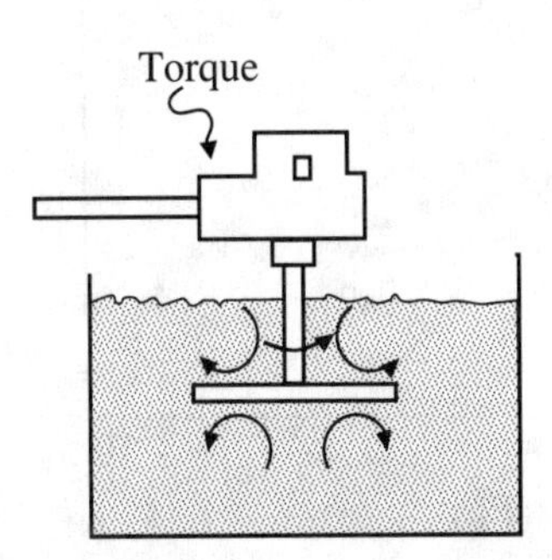

5.6.2 Rotating Vane

Figure 5.6.2 shows the vane fixture designed for samples with a yield stress. It can be rotated in a large container in the same way as the disk discussed above, but more recently it has been used with a close-fitting cylindrical cup (Barnes and Carnali, 1990, with review of previous work). A vane inserted into a fluid will create

Figure 5.6.2.
Schematic of vane fixture.

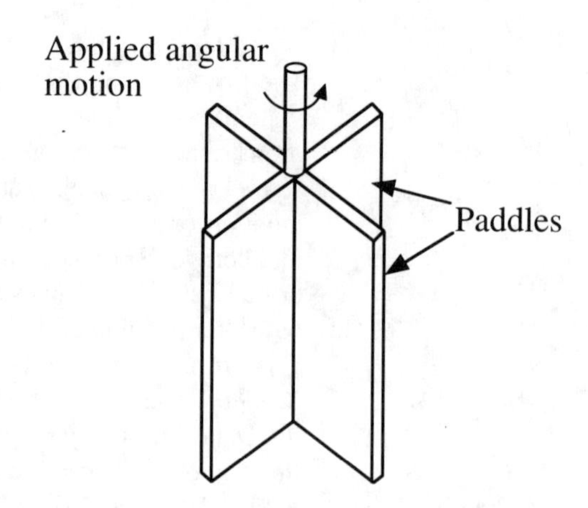

far less disturbance than a solid bob. Large particles present a less severe problem than they do for a narrow gap Couette device, but, most significantly, for strongly shear thinning fluids ($n < 0.5$), the fluid within the vanes moves as a solid plug. This helps to prevent wall slip. For less shear thinning fluids there will be secondary flow between the vanes, and the geometry will not give correct viscosity shear rate data.

Figure 5.6.3 compares vane-in-cup to bob-in-cup measurements for a very shear thinning polymer solution. The bob-in-cup data indicate that there is yield stress, but this is actually due to wall slip.

5.6.3 Helical Screw Rheometer

If the die end of a screw extruder is closed off, fluid will recirculate between the fights in a complex shear flow pattern. Kraynik and co-workers (1984) have shown that this flow can be analyzed to give viscosity versus shear rate data from pressure rise over the screw and rotation rate. An advantage of this geometry is that it can keep

Figure 5.6.3.
Viscosity versus shear stress for a 5.5% sodium carboxymethylcellulose in water solution. Solid points are for bob-in-cup geometry, open points are vane-in-cup. From Barnes and Carnali (1990).

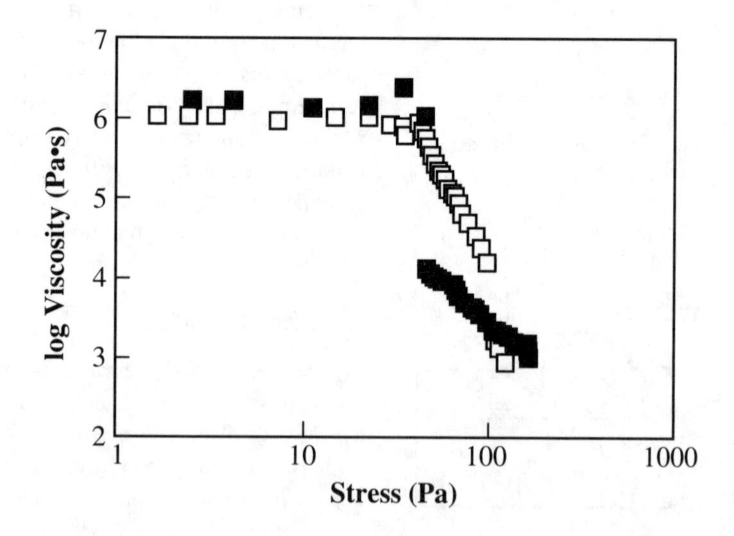

concentrated suspensions well mixed during measurement. It can also be quickly pumped out by opening the end, whereupon a new sample can be tested. This makes the geometry suitable for on-line measurements (see Chapter 8). Scott and Macosko (1993) have used a screw with a central hole to provide better mixing. They also report that pressure versus screw speed measurements could give accurate data on viscosity versus shear rate for several polymer melts.

5.6.4 Instrumented Mixers

Related to the helical screw devices are the instrumented batch mixers often called torque rheometers. They consist of a chamber or mixing bowl with two counterrotating blades. There are many different blade configurations. A common one is shown in Figure 5.6.5. These devices are primarily used as simulators of larger batch mixers and extruders and to prepare compounded samples. One of the rotating shafts has a transducer that records torque as polymers melt and mix with other additives. Blyler and Daane (1967) have shown that when the sample has reached steady state, one can obtain a good measure of n (but not m) for power law fluids from torque and screw speed.

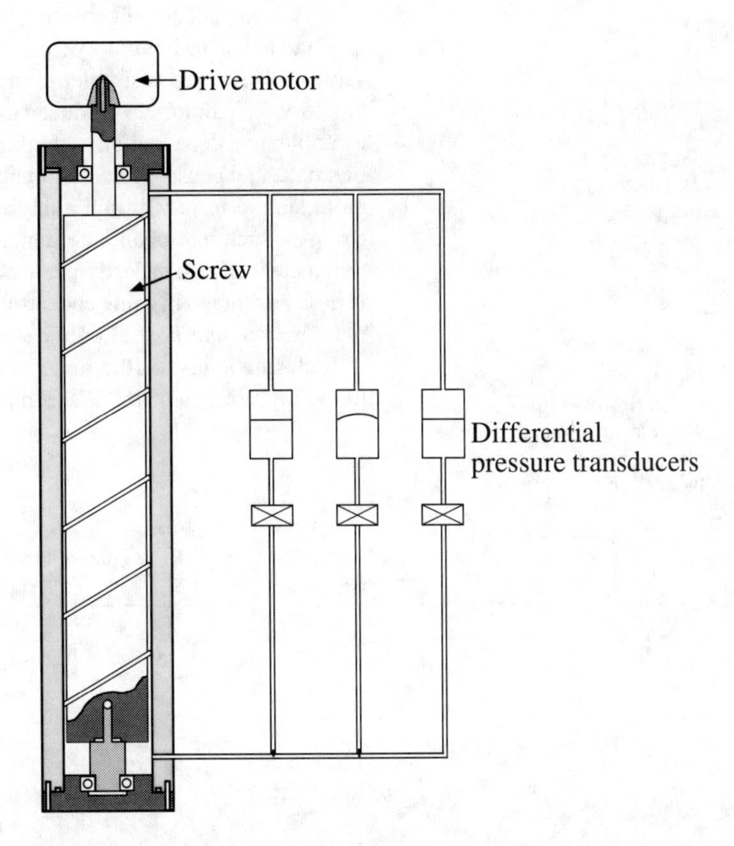

Figure 5.6.4.
Helical screw rheometer. From Kraynik (1984).

Figure 5.6.5.
Roller blades in a torque rheometer.

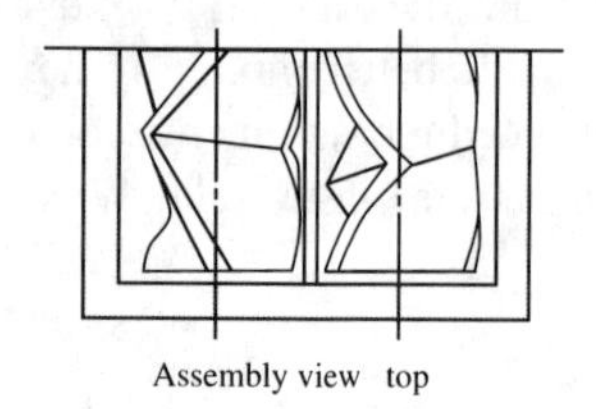

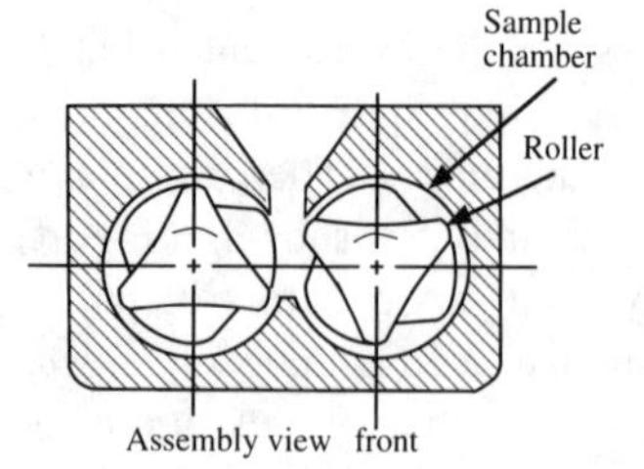

5.7 Eccentric Rotating Geometries

By taking a circularly symmetrical sample and rotating it through a deformation, it is possible to create a periodic strain in the sample somewhat similar to the sinusoidal oscillations described in Figure 3.3.3. The first use of such geometries appeared in the fatigue testing of metal rods. One end of a rod is simply inserted into a rotating holder and a weight is hung on the other end, as shown in Figure 5.7.1. Any element along the outer circumference of the rod goes through a cycle of tension and compression. The number of cycles to failure is recorded. The device, known as the Wohler fatigue tester, has been used since the late 1800s.

In 1926 Kimball and Lovell pointed out that any hysteresis or loss in the sample would result in a component of force perpendicular to the displacement because the stress distribution lags the strain. The force causes the end of the rod to displace slightly, by an amount b, as shown in Figure 5.7.1.

The other eccentric geometries described in Sections 5.7.1–5.7.3 all accomplish an oscillatory motion in a similar way by rotating the sample through a deformation fixed in laboratory space. These oscillations are not identical to sinusoidal oscillations, but in the region of linear viscoelastic response, the dynamic moduli can readily be obtained from the solution for an ideal elastic or ideal viscous material (Abbot et al., 1971; Pipkin, 1972).

The advantage of eccentric geometries is that a steady force is generated in the laboratory from an oscillatory deformation in the sample. This force is simpler to measure and analyze than the oscillatory force or torque produced by sinusoidal shearing found, for example, in a cone and plate rheometer. However, with the advent of small, inexpensive microcomputers in the late 1970s it became quite simple to record and analyze sinusoidal data (see Chapter 8). With such microcomputer control and analysis, a steady shear rheometer can also make dynamic measurements. Thus today there is little advantage in using eccentric rheometers. One area of interest, however, may be the nonlinear response of these devices. Since the deformation is not the same as sinusoidal simple shear, except in the limit of small strain, eccentric geometries could be used to

Figure 5.7.1.
Eccentric rotating rod of a viscoelastic solid. The displacement b is proportional to the loss in the sample.

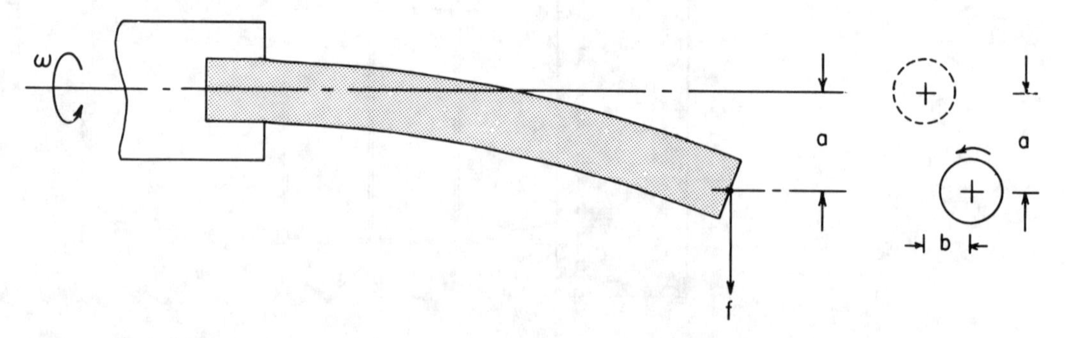

test constitutive equations (Gross and Maxwell, 1972; Jongschaap et al., 1978).

5.7.1 Rotating Cantilever Rod

The geometry of Figure 5.7.1 has been used extensively with metals by Lazan (1943, 1950) and with plastics by Maxwell (1956) and by Kaelble (1965). Instead of using a fixed load, Maxwell (1956) applied a fixed displacement to the sample and measured the forces on the end of the rod through rolling contact to a load cell. Any loss in the sample creates a component of force f_1 perpendicular to the displacement.

For small strains in the linear viscoelastic range, the measured force components can be converted to E' and E'' by using the static cantilever equations and substituting E^* for the elastic modulus E.

$$E' = \frac{64L^3 f_2}{3\pi D^4 a} \quad \text{tensile storage modulus} \tag{5.7.1}$$

$$E'' = \frac{64L^3 f_1}{3\pi D^4 a} \quad \text{tensile loss modulus} \tag{5.7.2}$$

where L is the sample length, D the diameter, and a the displacement.

Notice that in eqs. 5.7.1 and 5.7.2 the sample dimensions appear to high powers thus accurate measurement and sample uniformity are essential. Diameter variations or stresses molded into the sample can cause large oscillations in the measured force components.

5.7.2 Eccentric Rotating Disks

Figure 5.7.2a illustrates the eccentric rotating disk (ERD) geometry (recall Exercise 1.10.7 and Example 2.3.1). A sample is placed between two disks that rotate at the same angular velocity but about offset or eccentric axes. Surface tension holds the sample between the disks. The flow between these eccentric rotating disks results in a shearing motion, with material elements moving in circular paths with respect to each other. A coordinate system r, s, z that rotates with the lower disk (Figure 5.7.2b) can describe the relative motion between particles (Figure 5.7.2c). The deformation is seen to be of constant magnitude, but continually changing direction.

Gent (1960) first published an ERD experiment using vulcanized rubber. He reports that his inspiration came from Maxwell's rotating cantilever rod experiments. Mooney (1934) also tried the geometry on gum rubbers. Maxwell and Chartoff (1965) carried out the first tests on polymer melts. Maxwell's (1967) experiments caused the geometry to become known as the Maxwell orthogonal rheometer.

If an elastic rubber sample is placed between ERD$_s$, a force in the x_2 direction will be necessary to maintain the offset a. Any

Figure 5.7.2.
Eccentric rotating disks. (a) Cross section of the experimental geometry. (b) Top view showing particle paths. (c) Relative displacements of particles in a coordinate system rotating with the lower disk.

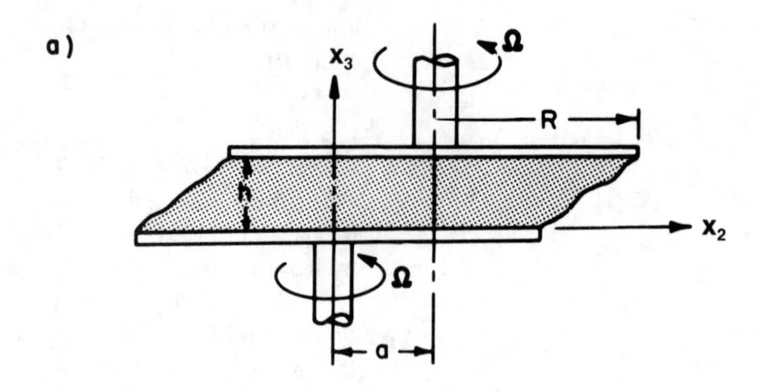

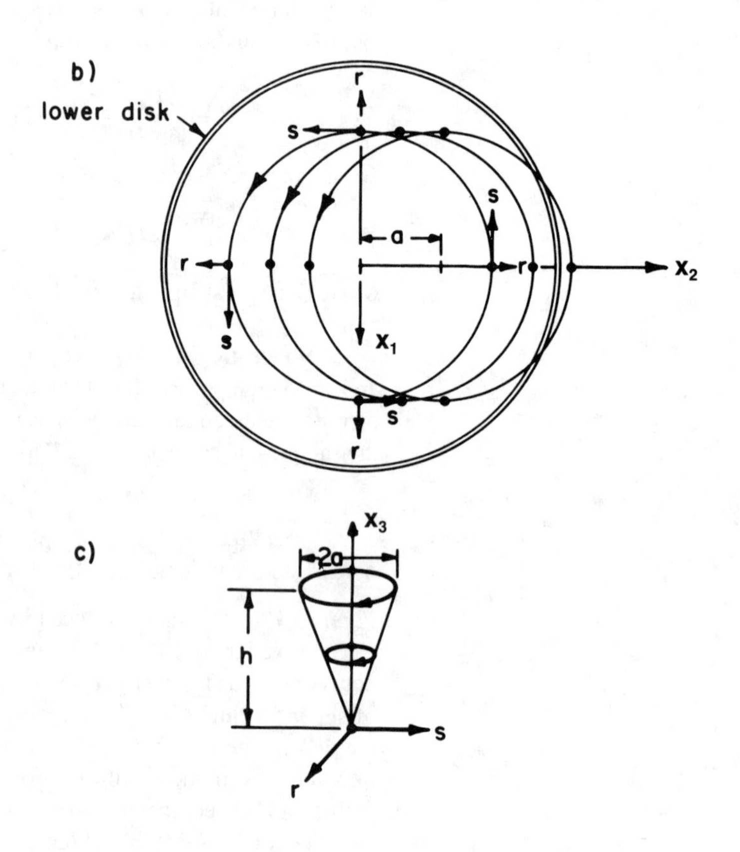

loss or hysteresis in the material will induce a force component f_1. For an ideal elastic solid, f_2 will be proportional to the deformation and shear modulus

$$f_2 = \pi R^2 \left(\frac{a}{h} \right) G \tag{5.7.3}$$

Exercise 1.10.7 derived this result from the neo-Hookean model.

For a Newtonian liquid there will be loss only in the sample. Thus $f_2 = 0$ and from Example 2.3.1

$$f_1 = \pi R^2 \, \Omega \left(\frac{a}{h} \right) \eta \tag{5.7.4}$$

For a general viscoelastic liquid we can substitute $\eta^* = \eta' + i\eta''$ for η and obtain

$$G' = \Omega \eta'' = \frac{h f_2}{\pi R^2 a} \tag{5.7.5}$$

$$G'' = \Omega \eta' = \frac{h f_1}{\pi R^2 a} \tag{5.7.6}$$

These simple equations have been verified with several different constitutive equations (Macosko and Davis, 1974; Walters, 1975; Jongschaap et al., 1978). Inertia can be included in the analysis (Abbot and Walters, 1970; Walters, 1975); however, for viscous samples the errors are negligible (Macosko and Davis, 1974).

In the typical ERD apparatus (Macosko, 1970; Payvar and Tanner, 1973) one disk is driven and the other follows, coupled through the sample. Waterman (1984) has shown that for the

Figure 5.7.3. Comparison of ERD to other rheometers for a polydimethylsiloxane melt. From Macosko and Davis (1974).

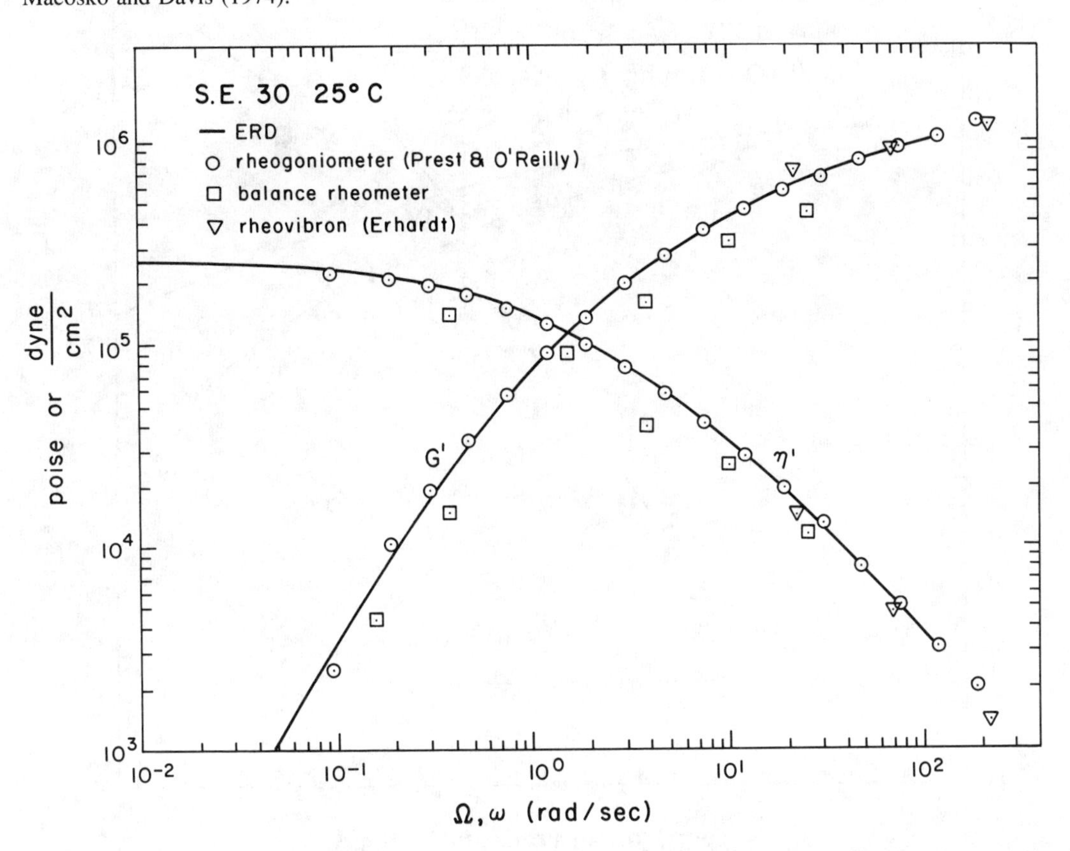

torques to balance, the freely rotating disk must always lag slightly. For a viscoelastic fluid that lag is

$$\frac{\Delta\Omega}{\Omega_1} \simeq \frac{\eta'}{\eta_0}\left(\frac{a}{R}\right)^2 + \frac{2Kh}{\pi R^4 \eta_0} \tag{5.7.7}$$

where K is the friction in the free bearing. The data of Davis and Macosko (1974) are fit well by this relation. Using an air bearing with typical operating conditions, lag is less than 1%.

Equations 5.7.5 and 5.7.6 can be tested experimentally because G' and G'' can be measured with sinusoidal shear oscillations. This has been done by Macosko and Davis (1974) for several polymeric systems. One of their results with a polydimethylsiloxane polymer is shown in Figure 5.7.3. G' and η' were measured with four different rheometers: ERD, oscillating cone and plate, oscillating sliding plates, and the tilted rotating hemispheres. The agreement between the four instruments is quite good. Note that the ERD extends to lower frequency. This enlarged range appears

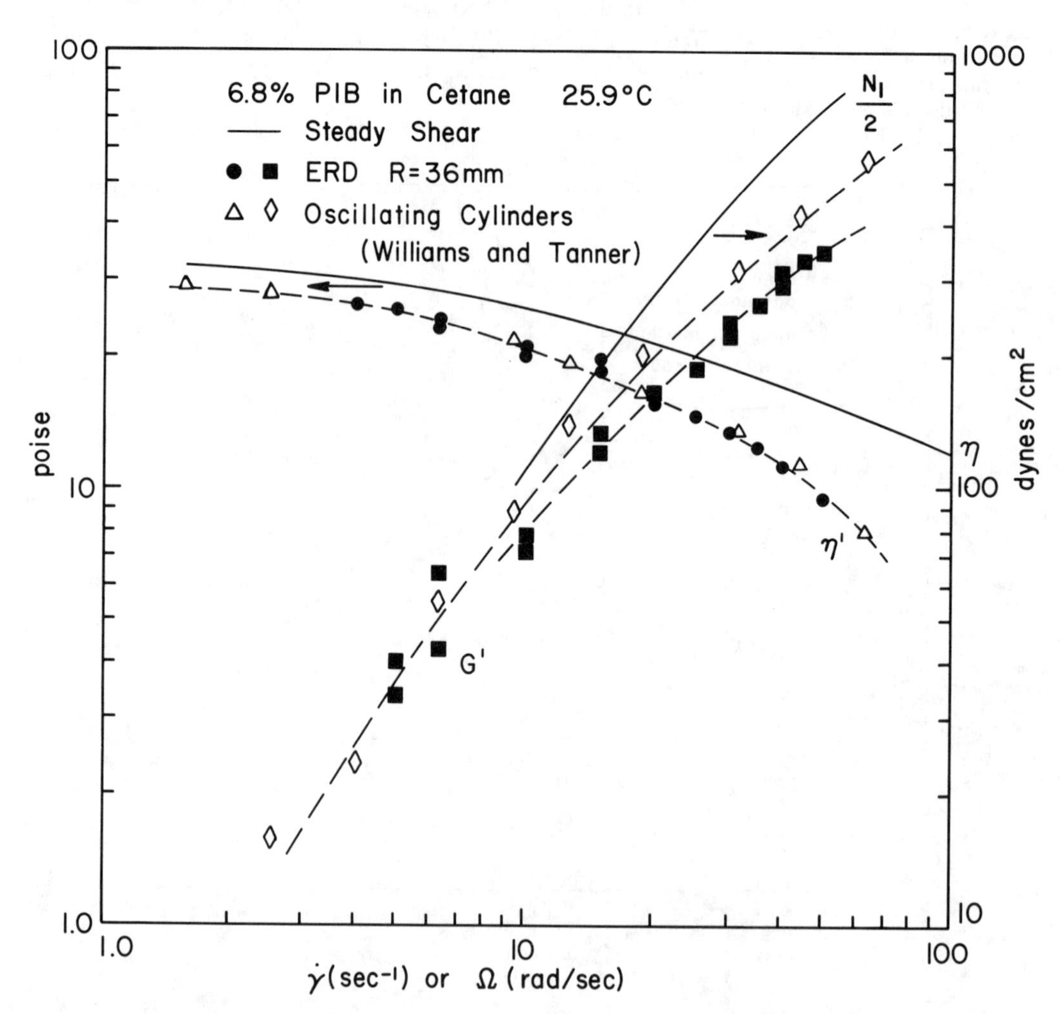

Figure 5.7.4. Comparison of ERD to other rheometers for a 6.8% polyisobutylene solution. From Macosko and Davis (1974).

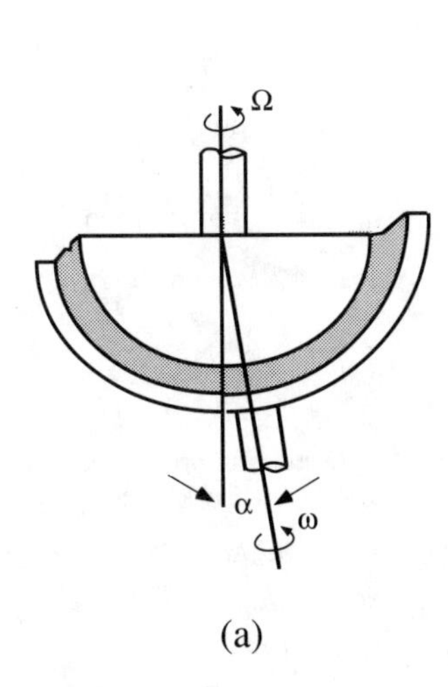

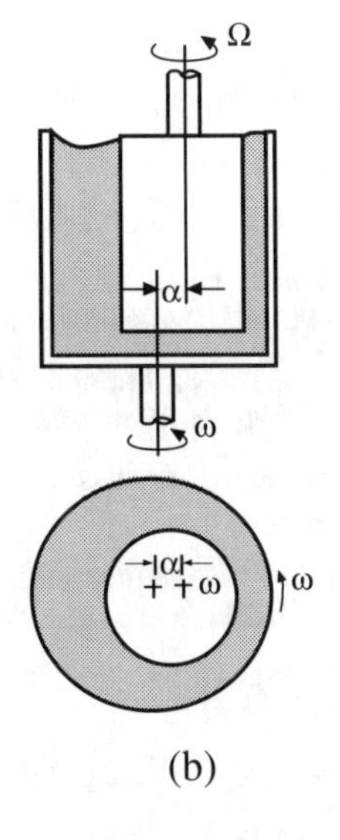

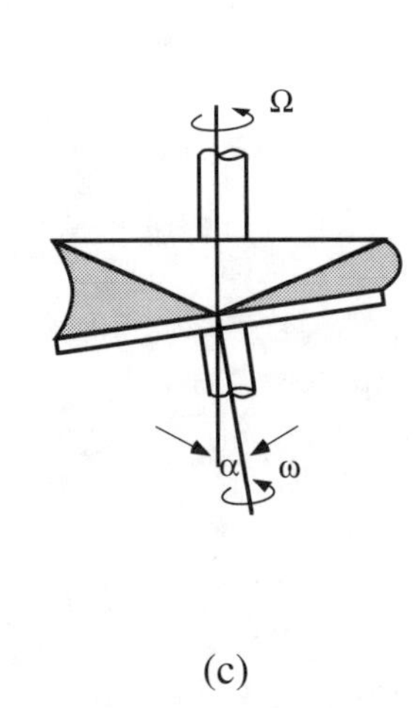

(a) (b) (c)

Figure 5.7.5.
Other eccentric rotating geometries: (a) tilted rotating hemispheres (also called balance rheometer), (b) eccentric rotating cylinders, and (c) tilted rotating cone and plate.

to be due to the ability to resolve smaller values of G'. Figure 5.7.4 shows a similar comparison between ERD and oscillating cylinders for a low viscosity polymer solution. Agreement is within the accuracy of the oscillating cylinder data (Tanner and Williams, 1971).

Macosko and Davis (1974) and Gottlieb and Macosko (1982) report relations for instrument compliance in ERD data. These corrections can become very large for G'—in the range 10^6 N/m^2 or higher—and must be accounted for.

5.7.3 Other Eccentric Geometries

Several other eccentric geometries have been described. Three of them are shown in Figure 5.7.5. Kepes (1968) and Kaelble (1969) developed tilted rotating hemispheres also known as the Kepes balance rheometer. Figure 5.7.3 shows that this rheometer can measure η' and G' data accurately .

Eccentric rotating cylinders can give accurate η' and G' data but only at rather small strains (Broadbent and Walters, 1971). At larger deformation, cavitation or extrusion of the sample from the gap can occur. This also seems to be the problem for such other eccentric geometries as the tilted rotating disks or cone and plate (Davis and Macosko, 1973; Walters, 1975).

References

Abbott, T. N. G.; Walters, K., *J. Fluid Mech.* 1970, *40*, 205.

Abbott, T. N. G.; Bowen, G. W.; Walters, K., *J. Phys. D* 1971, *4*, 190.

Abdel-Wahab, M.; Giesekus, H.; Zidan, M., *Rheol. Acta* 1990, *29*, 16.

Adams, N.; Lodge, A. S., *Phil. Trans.* 1964, *A256*, 149.

Alvarez, G. A.; Lodge, A. S.; Cantow, H. -J., *Rheol. Acta* 1985, *24*, 377.

Barnes, H. A.; Carnali, J. O., *J. Rheol.* 1990, *34*, 841.

Beavers, G. S.; Joseph, D. D., *J. Fluid Mech.* 1975, *69*, 475.

Bennett, K. E., Ph.D. thesis, University of Minnesota, 1982.

Bennett, K. E.; Hatfield, J.; Davis, H. T.; Macosko, C. W.; Scriven, L. E., in *Microemulsions*; Robb, I.D., Ed.; Plenum: New York, 1982.

Beris, A. N.; Tsamopoulos, J. A.; Armstrong, R. C.; Brown, R. A., *J. Fluid Mech.* 1985, *158*, 219.

Bird, R. B.; Curtiss, C. F.; Stewart, W. E., *Chem. Eng. Sci.* 1959 *11*, 114.

Bird, R. B.; Armstrong, R. C.; Hassager, O., *Dynamics of Polymeric Liquids,* Vol. 1: *Fluid Mechanics*, 2nd ed.; Wiley: New York, 1987.

Binding, D. M.; Walters, K., *J. Non-Newtonian Fluid Mech.* 1976, *1*, 277.

Blyler, L. L.; Daane, J. H., *Polym. Eng. Sci.* 1967, *7*, 1978.

Boger, D. V.; Rama Murthy, A. V., *Trans. Soc. Rheol.* 1969, *13*, 405.

Broadbent, J. M.; Lodge, A. S., *Rheol. Acta* 1971, *10*, 551.

Broadbent, J. M.; Walters, K., *J. Phys. D Appl. Phys.* 1971, *4*, 1863.

Broyer, E.; Macosko, C. W., *SPE Tech. Pap.* 1975, *21*, 343.

Chandrasekhar, S., *Hydrodynamic and Hydromagnetic Stability*; Oxford University Press: London, 1961.

Cheng, D. C. H., *Chem. Eng. Sci.* 1968, *23*, 895.

Cokelet, G. R., in *The Rheology of Human Blood, Symposium on Biomechanics*; Fung, Y. C.; Perrone, N.; Anliker, M.; Eds.; Prentice-Hall: Englewood Cliffs, NJ, 1972.

Coleman, B. D.; Noll, W., *Arch. Ration. Mech. Anal.* 1959, *3*, 289.

Connelly, R. W.; Greener, J., *J. Rheol.* 1985, *29*, 209.

Copley, A. L.; King, R. G., *Experimentia* 1970, *26*, 704.

Couette, M. M., *Am. Chem. Phys. Ser. VI* 1890, *21*, 433.

Davis, W. M.; Macosko, C. W., data reported by Joseph et al., 1973.

Davis, W. M.; Macosko, C. W., *AIChE J.* 1974, *20*, 600.

Dealy, J. M.; Yu, T. K. P., *J. Non-Newtonian Fluid Mech.* 1978, *3*, 127.

Dealy, J. M.; Giacomin, A. J., in *Rheological Measurement*; Collyer, A. A.; Clegg, D. W., Eds.; Elsevier: London, 1988.

Denn, M. M.; Roisum, J. J., *AIChE J.* 1969, *15*, 454.

Ehrlich, R.; Slattery, J. C., *Ind. Eng. Chem. Fundam.* 1968, *7*, 239.

Ellenberger, J.; Fortuin, J. M. H., *Chem. Eng. Sci.* 1985, *40*, 111.

Fewell, M. E.; Hellums, J. D., *Trans. Soc. Rheol.* 1977, *21*, 535.

Gavis, J.; Laurence, R. L., *Ind. Eng. Chem. Fundam.* 1968, *7*, 232; 525.

Gent, A. N., *Bull. J. Appl. Phys.* 1960, *11*, 165.

Giesekus, H., *Proceedings of the Fourth International Congress on Rheology*; Wiley- Interscience: New York, 1963; p. 249.

Geisekus, H.; Langer, G., *Rheol. Acta* 1977, *16*, 1.

Goldstein, C., *Trans. Soc. Rheol.* 1974, *18*, 357.

Gottlieb, M., *J. Non-Newtonian Fluid Mech.* 1979, *6*, 97.

Gottlieb, M.; Macosko, C. W., *Rheol. Acta* 1982, *21*, 90.

Graham, A. L.; Mondy, L. A.; Miller, J. D.; Wagner, N. J.; Cook, W. A., *J. Rheol.* 1989, *33*, 1107.

Greensmith, H. W.; Rivlin, R. S., *Phil. Trans.* 1953, *A245*, 399.

Gross, L.H.; Maxwell, B., *Trans. Soc. Rheol.* 1972, *16*, 577.

Hassager, O.; Bisgaard, C., *J. Non-Newtonian Fluid Mech.* 1983, *12*, 153.

Hoppman, W. H.; Miller, C. E., *Trans. Soc. Rheol.* 1963, *7*, 181.

Huppler, J. D.; Ashare, E.; Holmes, L. A., *Trans. Soc. Rheol.* 1967 *11*, 159.

Hutton, J. F., *Rheol. Acta* 1969, *8*, 54.

Jobling, A.; Roberts, I. E., *J. Polym. Sci.* 1959, *36*, 433.

Jongschaap, R. J. J.; Knapper, K. M.; Lopulissa, J. S., *Polym. Eng. Sci.* 1978, *18*, 788.

Joseph, D. D.; Fosdick, R. L., *Arch. Ration. Mech. Anal.* 1973, *49*, 321.

Joseph, D. D.; Beavers, G. L.; Fosdick, R. L., *Arch. Ration. Mech. Anal.* 1973, *49*, 381.

Joseph, D. D.; Beavers, G. S.; Cors, A.; Dewald, C.; Hoger, A.; Than, P. T., *J. Rheol.* 1984, *28*, 325.

Kaelble, D. H., *J. Appl. Polym. Sci.* 1965, *19*, 1201.

Kaelble, D. H., *J. Appl. Polym. Sci.* 1969, *13*, 2547.

Kaye, A.; Lodge, A. S.; Vale, D. G., *Rheol. Acta* 1968, *4*, 368.

Kepes, A., paper presented at the Fifth International Congress on Rheology, Kyoto, 1968.

Kiljanski, T., *Rheol. Acta* 1989, *28*, 61.

Kimball, A.L.; Lovell, D.E., *Trans. ASME* 1926, *48*, 479.

Kimura, S.; Osaki, K.; Kurata, M., *J. Polym. Sci., Phys. Ed.* 1981, *19*, 151.

Kotaka, T.; Kurata, M.; Tamura, M., *J. Appl. Phys.* 1959, *30*, 1705.

Kramer, J.; Uhl, J. T.; Prud'homme, R. K., *Polym. Eng. Sci.* 1987, *27*, 59.

Kraynik, A. M.; Aubert, J. H.; Chapman, R. N., in *Rheology, Vol. 4*; Astarita; G.; Marrucci, G.; Nicolias, L., Eds.; Plenum: New York, 1984; p. 77.

Krieger, I. M.; Elrod, H., *J. Appl. Phys.* 1953, *24*, 134.

Kulicke, W. M.; Kiss, G.; Porter, R. S., *Rheol. Acta* 1977, *16*, 568.

Lammiman, K. A.; Roberts, J. E., *Lab. Pract.* 1961, 816.

Larson, R. G., *Rheol. Acta* 1992, *31*, 213.

Laun, H. M.; Meissner, *J., Rheol. Acta* 1980, *19*, 60.

Lazan, B. J., *Trans. ASME* 1943, *65*, 87.

Lazan, B. J., *Trans. Am. Soc. Met.* 1950, *42*, 499.

Leary, M. L., M. S. thesis, University of Minnesota, 1975.

Liu, T. Y., Mead, D. W., Soong, D. S.; Williams, M. C., *Rheol. Acta* 1983, *22*, 81.

Macosko, C. W., Ph.D. thesis: Princeton University, 1970.

Macosko, C. W.; Davis, W. M., *Rheol. Acta* 1974, *13*, 814.

Macosko, C. W.; Morse, D. J., in *Proceedings of Seventh International Congress on Rheology*; Kalson, C.; Kubat, J., Eds.; Gothenburg, 1976; p. 376.

Markovitz, H.; Brown, D. R., *Trans. Soc. Rheol.* 1963, *7*, 137.

Maxwell, B., *J. Polym. Sci.* 1956, *20*, 551.

Maxwell, B., *Polym. Eng. Sci.* 1967, *7*, 145.

Maxwell, B.; Chartoff, R. P., *Trans. Soc. Rheol.* 1965, *9*, 41.

McKelvey, J. M., *Polymer Processing*; Wiley: New York, 1962.

McKennell, R. in *Proceedings of the Second International Congress on Rheology*; Harrison, V. G. W., Ed.; Butterworths: London, 1954; p. 350.

Middleman, S., *The Flow of High Polymers*; Wiley-Interscience: New York, 1968.

Miller, M. J.; Christiansen, E. B., *AIChE J.* 1972, *18*, 600.

Mooney, M., *J. Appl. Phys.* 1934, *35*, 23.

Mooney, M.; Ewart, R. H., *Physics* 1934, *D5*, 350.

Nakajima, N.; Collins, E. A., *Rubber Chem. Technol.* 1974, *47*, 333.

Oka, S., in *Rheology*, Vol. 4; Eirich, F. R., Ed.; Academic Press: New York, 1960.

Olabisi, O.; Williams, M. C., *Trans. Soc. Rheol.* 1972, *16*, 727.

Orafidiya, L. O., *J. Pharm. Pharmacol.* 1989, *41*, 341.

Park, N. A., Irvine, T. F., *Rev. Sci. Instrum.* 1988, *59*, 2051.

Payvar, P.; Tanner, R. I., *Trans. Soc. Rheol.* 1973, *17*, 449.

Pinkus, O.; Sternlicht, B., *Theory of Hydrodynamic Lubrication*; McGraw-Hill: New York, 1961.

Pipkin, A. C., *Lectures on Viscoelasticity Theory*; Springer-Verlag: New York, 1972.

Princen, H. M., *J. Rheol.* 1986, *30*, 271.

Quinzani, L. M.; Valles, E. M., *J. Rheol.* 1986, *30(s)*, s1.

Reiner, M., *Deformation and Flow*; Wiley-Interscience: New York, 1960.

Russell, R. J., Ph.D. thesis: London University, 1946.

Schrag, J. L., *Trans. Soc. Rheol.* 1977, *21*, 399.

Scott, C. E.; Macosko, C. W., *Polym. Eng. Sci.* 1993, *33*, 000.

Sdougos, H. P.; Bussolari, S. R.; Dewey, C. F., *J. Fluid Mech.* 1984, *138*, 379.

Siginer, A., *J. Appl. Math. Phys.* 1984, *35*, 648.

Sivashinsky, N.; Tsai, A. J.; Moon, T. J.; Soong, D. S., *J. Rheol.* 1984, *28*, 287.

Soskey, P. R.; Winter, H. H., *J. Rheol.* 1984, *28*, 625.

Sternstein, S. S., Onchin, L.; Silverman, A.; *Appl. Polym. Symp. (J. Appl. Polym. Sci.)* 1968, *7*, 175.

Stokes, G. G., *Trans. Cambr. Phil. Soc.* 1851, *9* pt II, 8.

Sukanek, P. C.; Laurence, R. L., *AIChE J.* 1974, *20*, 474.

Tadmor, Z.; Gogos, C. G., *Principles of Polymer Processing*; Wiley-Interscience: New York, 1979.

Tanner, R. I., *Trans. Soc. Rheol.* 1970, *14*, 483.

Tanner, R. I.; Keentok, M., *J. Rheol.* 1983, *27*, 47.

Tanner, R. I.; Williams, G., *Rheol. Acta* 1971, *10*, 528.

Taylor, G. I., *Phil. Trans.* 1923, *A223*, 289.

Toy, S. L.; Scriven, L .; Macosko, C.W.; Nelson, N.K.; Olmsted, R.D., *J. Rheol.* 1991, *35*, 887.

Turian, R. M., *Chem. Eng. Sci.* 1965, *20*, 771.

Turian, R. M., *Chem. Eng. Sci.* 1969, *24*, 1581.

Turian, R. M., *Ind. Eng. Chem. Fundam.* 1972, *11*, 361.

Turian, R. M.; Bird, R. B., *Chem. Eng. Sci.* 1963, *18*, 689.

Van Krevelen, D. W., *Properties of Polymers*, 2nd ed.; Elsevier: Amsterdam, 1976.

Van Wazer, J. R.; Lyons, J. W.; Lim, K. Y.; Colwell, R. E., *Viscosity and Flow Measurement*; Wiley: New York, 1963.

Vrentas, J. S.; Venerus, P. C.; Vrentas, C. M., *Chem. Eng. Sci.* 1991, *46*, 33.

Wales, J. L. S., *Rheol. Acta* 1969, *8*, 38.

Wales, J. L. S.; den Otter, J.L., *Rheol. Acta* 1970, *9*, 115.

Walters, K., *Rheometry*; Chapman & Hall: London, 1975.

Walters, K.; Waters, N. D., in *Polymer Systems Deformation and Flow*; Wetton, R. E.; Whorlow, R. W., Eds.; Macmillan: London, 1968; p. 211.

Waterman, H. A., *J. Rheol.* 1984, *28*, 273.

Whitcomb, P. W.; Macosko, C. W., *J. Rheol.* 1978, *22*, 493.

Williams, R. W., *Rheol. Acta* 1979, *18*, 345.

Yang, T. M. T.; Krieger, I. M., *J. 0Rheol.* 1978, *22*, 413.

Yoshimura, A.; Prud'homme, R. K., *J. Rheol.* 1988, *32*, 53

6

The water in the tube, when it encounters the resistance of the wall, is not able to move as a solid cylinder. The middle thread must flow much faster than the outer one.

G. H. L. Hagen (1839)

SHEAR RHEOMETRY: PRESSURE DRIVEN FLOWS

6.1 Introduction

The first measurements of viscosity were done using a small straight tube or capillary (Figure 6.1.1). Hagen (1839) in Germany and independently Poiseuille (1840) in France used small diameter capillaries to measure the viscosity of water. A key development that made their work possible was the advent of precision diameter, small bore tubing. For water, use of larger diameters usually results in turbulent flow. Precision is required, as we shall see, because the tube radius enters the viscosity equation to the fourth power.

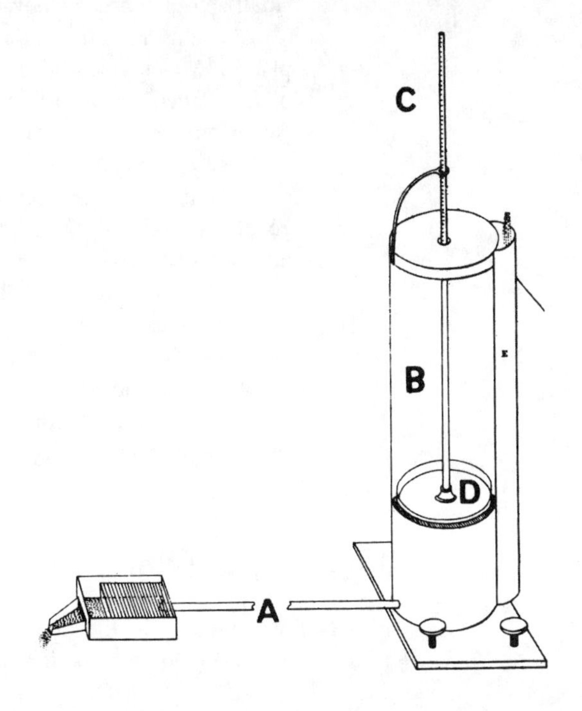

Figure 6.1.1. Hagen's capillary tube (*A*) was supplied with water from tank (*B*). The pressure head was recorded by a pointer (*C*) attached to the float (*D*) and flow rate was checked by weighing the effluent.

A capillary rheometer is a pressure-driven flow, the theme of this chapter, in contrast to the drag flows of Chapter 5. As Hagen first observed, when pressure- drives a fluid through a channel, velocity is maximum at the center. The velocity gradient or shear rate and also the shear strain will be maximum at the wall and zero in the center of the flow. Thus all pressure-driven flows are nonhomogeneous. This means that they are only used to measure *steady* shear functions: the viscosity and normal stress coefficients $\eta(\dot{\gamma})$, $\psi_1(\dot{\gamma})$, and $\psi_2(\dot{\gamma})$. Equations 5.1.1–5.1.3 define these functions, and Figure II.3 indicated how they are related to the other material functions.

If pressure-driven rheometers can measure only the steady shear functions, why are they so widely used? The first reason, of course, is that they are relatively inexpensive to build and simple to operate. Despite their simplicity, long capillaries can give the most accurate viscosity data available. A second major advantage is that closed-channel flows have no free surface in the test region. In Chapter 5, for example, we saw how edge effects in the cone and plate geometry seriously limit the maximum shear rate in rotational instruments. In fact, for viscous polymer melts, capillary or slit rheometers appear to be the only satisfactory means of obtaining data at shear rates greater than 10 s^{-1}. Capillary rheometers can also eliminate solvent evaporation and other problems that plague rotational devices with free surfaces. Because the sample flows through a capillary or slit, these rheometers can be readily adapted for on-line measurements (see Chapter 8).

Another reason capillary rheometers are so widely used is that they are very similar to process flows like pipes and extrusion dies. A capillary run is an excellent first test of processibility for a small amount of a new polymer or coating formulation.

Just as a rotational rheometer designed to produce cone and plate flow can typically be used for concentric cylinder or parallel plate geometries, a capillary rheometer usually can be adapted for slit or annular flows. This chapter focuses mainly on capillary flow but also treats these other channel geometries. Flow over a narrow channel or "pressure hole" gives ψ_1 data. Such flow can be generated by both drag and pressure, but since it is usually measured in a pressure-driven slit geometry, we discuss it here. Extrudate swell and exit pressure can also give information on normal stresses and are discussed. We also look at two important pressure-driven indexers: the melt index and squeezing flow. At the end of the chapter we compare all the shear rheometers, summarizing their advantages and limitations. Chapter 8 has a section addressing capillary rheometer design.

6.2 Capillary Rheometer

A capillary was the first rheometer, and this device remains the most common method for measuring viscosity. The basic features of the instrument are shown in Figure 6.2.1. Gravity, compressed gas, or a piston is used to generate pressure on the test fluid in a reservoir. A capillary tube of radius R and length L is connected to the bottom of the reservoir. Pressure drop and flow rate through this tube are used to determine viscosity.

In deriving the viscosity relation the important assumptions are as follows:

1. Fully developed, steady, isothermal, laminar flow
2. No velocity in the r and θ directions
3. No slip at the walls, $v_x = 0$ at R
4. The fluid is incompressible with viscosity independent of pressure

With these assumptions, the equation of motion in the x direction (Table 1.7.1) in cylindrical coordinates reduces to

$$0 = \frac{-\partial p}{\partial x} + \frac{1}{r}\frac{\partial(r\tau_{rx})}{\partial r} \tag{6.2.1}$$

Because $\partial p/\partial x$ should be constant for fully developed incompressible flow, we can integrate to obtain the shear stress distribution (see also Example 2.4.1)

$$\tau_{rx} = \frac{r}{2}\frac{p_c}{L} \tag{6.2.2}$$

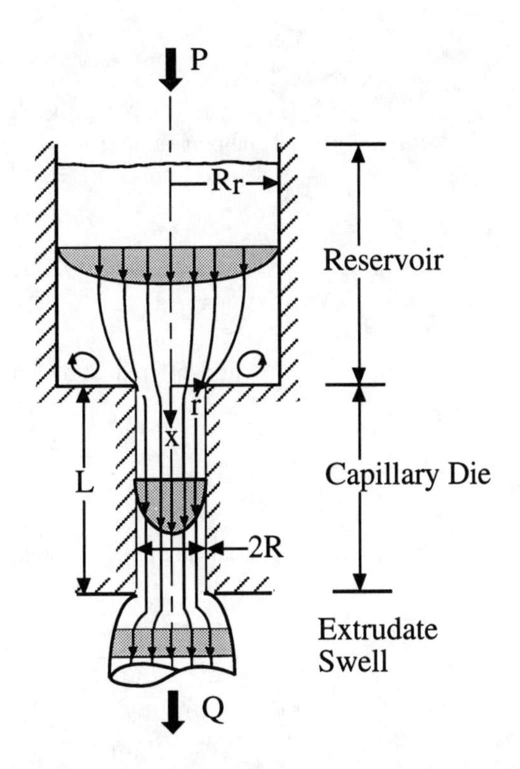

Figure 6.2.1. Schematic of capillary rheometer.

where p_c is the pressure drop over the capillary. At the wall the shear stress is

$$(\tau_{rx})_{r=R} = \tau_w = \frac{R}{2}\frac{p_c}{L} \tag{6.2.3}$$

and

$$\tau_{12} = \tau_{rx} = \frac{r}{R}\tau_w \tag{6.2.4}$$

A summary of the working equations for the capillary is presented in Table 6.2.1. Their derivations are given in Sections 6.2.1–6.2.6.

6.2.1 Shear Rate

Usually we measure the volumetric flow rate Q, but to calculate viscosity, we need the shear rate $dv_x/dr = \dot{\gamma}$. Q is related to v_x by

$$Q = 2\pi \int_0^R v_x(r) r \, dr \tag{6.2.5}$$

Integrating by parts and using the no-slip assumption gives

$$Q = -\pi \int_0^R r^2 \left(\frac{dv_x}{dr}\right) dr \tag{6.2.6}$$

We now have the shear rate dv_x/dr, but it is inside the integral. To help us remove it, we can change from the r variable to τ. Using eq. 6.2.4 gives $dr = R/\tau_w \, d\tau_{12}$. Substituting for r and dr gives

$$\frac{Q\tau_w^3}{\pi R^3} = -\int_0^{\tau_w} \tau_{12}^2 \left(\frac{dv_x}{dr}\right) d\tau_{12} \tag{6.2.7}$$

Then differentiating with respect to τ_w, we get

$$\frac{3\tau_w^2 Q}{\pi R^3} + \frac{\tau_w^3}{\pi R^3}\frac{dQ}{d\tau_w} = -\tau_w^2 \left.\frac{dv_x}{dr}\right|_{\tau_w} \tag{6.2.8}$$

Rearranging, we obtain the Weissenberg–Rabinowitsch equation

$$\left.\frac{-dv_x}{dr}\right|_{\tau_w} = \dot{\gamma}_w = \frac{1}{4}\dot{\gamma}_{aw}\left[3 + \frac{d\ln Q}{d\ln \tau_w}\right] \tag{6.2.9}$$

TABLE 6.2.1 / Working Equations for Capillary Rheometer

Wall shear stress

$$\tau_w = \frac{R}{2}\frac{p_c}{L} \qquad (6.2.3) \text{ and } (6.2.20)$$

Wall shear rate

$$\dot{\gamma}_{aw} = \frac{4Q}{\pi R^3} \qquad (6.2.10)$$

$$\dot{\gamma}_w = \frac{1}{4}\dot{\gamma}_{aw}\left[3 + \frac{d \ln Q}{d \ln p_c}\right] \qquad (6.2.9)$$

$$\dot{\gamma}_w = \frac{1}{4}\dot{\gamma}_{aw}\left[3 + \frac{1}{n}\right] \qquad \text{(for power law model)}$$

Representative shear rate

$$\eta(\dot{\gamma}) = \eta_a(\dot{\gamma}_a) \quad \pm 2\%$$

$$\text{for } \dot{\gamma} = 0.83\dot{\gamma}_a \qquad (6.2.13)$$

$$\text{and } 0.2 < \frac{d \ln Q}{d \ln p_c} < 1.3$$

First normal stress difference from extrudate swell (not rigorous)

$$(T_{11} - T_{22})^2 = 8\tau_w^2(B^6 - 1) \qquad (6.2.27)$$

$$\text{where } B = \frac{D_e}{2R} - 0.13$$

Errors

Wall slip with concentrated dispersions
Melt fracture at $\tau_c \sim 10^5$Pa
Reservoir pressure drop (6.2.21)
Entrance pressure drop
- Bagley plot
- Single die $L/R \simeq 60$
- Kinetic energy for low η, high $\dot{\gamma}$ (6.2.22)

Viscous heating—Na ≥ 1 (6.2.23)
Material compressibility
Pressure dependence of viscosity
Shear history, degradation in reservoir

Utility

Simplest rheometer, yet most accurate for steady viscosity
High $\dot{\gamma}$
Sealed system: pressurize, prevent evaporation
Process simulator
Quality control: melt index
Nonhomogeneous flow, only steady shear material functions
Entrance corrections entail more data collection

where $\dot{\gamma}_{aw}$ is the apparent or Newtonian shear rate at the wall

$$\dot{\gamma}_{aw} = \frac{4Q}{\pi R^3} \tag{6.2.10}$$

Thus we have the shear rate and the shear stress (eq. 6.2.3) at the same point, the wall. Therefore, we can calculate viscosity from the measured variables Q and p_c, if there is enough data to evaluate the derivative $1/n' = d \ln Q / d \ln p_c$

$$\eta = \frac{\tau_w}{\dot{\gamma}_w} = \frac{\pi R^4 p_c}{2QL} \left(\frac{n'}{3n' + 1} \right) \tag{6.2.11}$$

Note that n' is simply the exponent n of the power law constitutive equation (compare eq. 6.2.9 to eq. 2.4.22, Example 2.4.1). For a Newtonian fluid, $n' = 1$ and $\dot{\gamma}_w = \dot{\gamma}_{aw}$. Often we evaluate capillary data in terms of an apparent viscosity based on the Newtonian result

$$\eta_a = \frac{\pi R^4 p_c}{8QL} \tag{6.2.12}$$

For eq. 6.2.9 we see for the third time this "trick" of changing variables to extract useful rheological variables from an inhomogeneous flow. In the wide gap concentric cylinders (eqs. 5.3.21 and 5.3.24) we traded an unknown velocity gradient for the measurable gradient of torque with rotation rate. In torsional flow between parallel plates the stress distribution is unknown, but we can find it at the edge of the plate by again using the gradient of torque with rotation rate (eq. 5.5.8).

To use eq. 6.2.9 to get the true shear rate, we must have additional data near the point of interest. Furthermore, numerical differentiation of data is notoriously inaccurate. Fortunately, this correction does not greatly alter the shape of the viscosity versus shear rate function. The apparent (Newtonian) shear rate $\dot{\gamma}_{aw}$ is multiplied by $(3 + 1/n')/4$ and the viscosity is divided by it. Thus data points are shifted to the right and down along a line with slope -1 on a log–log plot as illustrated in Figure 6.2.2. The correction is similar to that for a wide gap concentric cylinders rheometer (Figure 5.3.2).

A simpler, approximate method to correct the shear rate data has been developed by Schümmer (1970, 1978) and Giesekus and Langer (1977). It uses the idea that the true and apparent shear rates must equal one another near the capillary wall. It turns out that this occurs at nearly the same point, $r_c^* = r/R = 0.83$, for a wide range of fluids. Thus the apparent viscosity equals the true viscosity evaluated at

$$\eta(\dot{\gamma}_w) = \eta_a(\dot{\gamma}_{aw}) \text{ for } \dot{\gamma}_w = 0.83\, \dot{\gamma}_{aw} \tag{6.2.13}$$

This result is derived in Example 6.2.1. Figure 6.2.2 shows how this simplified conversion shifts the data. Laun (1983, 1989) reports

Figure 6.2.2.
Effect of shear rate corrections on viscosity data for two polymer melts. Open points, assuming parabolic (Newtonian) velocity profile; solid points, assuming Weissenberg–Rabinowitsch eq. 6.2.11. Inset compares eq. 6.2.11 to the simple, approximate method of Schümmer, eq. 6.2.13.

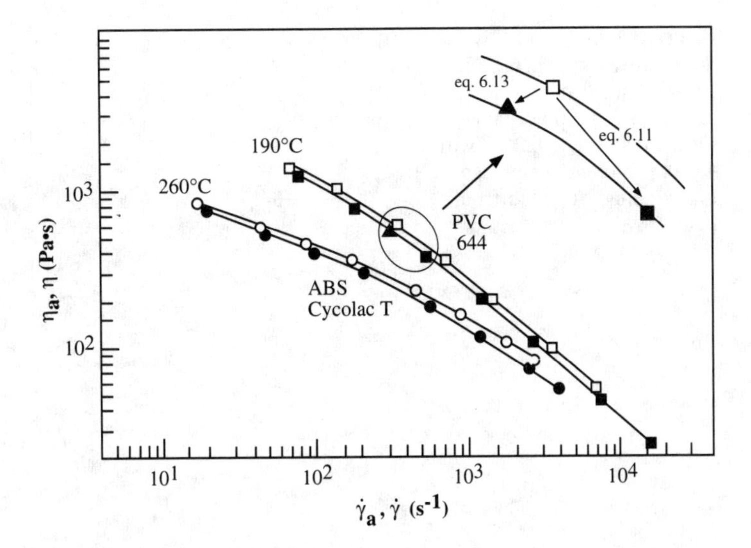

that this single point method for correcting viscosity is as accurate as the Weissenberg–Rabinowitsch method. He used both capillary and slit geometries with several polymer melts and solutions.

Example 6.2.1 Simplified Conversion from Apparent to True Viscosity

Show for a power law fluid flowing in a capillary that when the apparent (Newtonian) viscosity is evaluated at $\dot{\gamma}_{aw}$, it gives the true viscosity at $0.83\dot{\gamma}_{aw}$. Give the accuracy of this approximation.

Solution

For any fluid, the shear stress varies linearly with capillary radius (eq. 6.2.2). For a Newtonian fluid, the shear rate will also be linear in r^*, where $r^* = r/R$

$$\dot{\gamma}_a = \frac{\tau_{rz}}{\eta_o} = r^* \, \dot{\gamma}_{aw} \tag{6.2.14}$$

while for a power law it increases with $r^{*1/n}$

$$\dot{\gamma} = \left(\frac{\tau_{rz}}{m}\right)^{1/n} = r^{*1/n} \, \dot{\gamma}_w \tag{6.2.15}$$

These two functions intersect each other at r_c^* as illustrated in Figure 6.2.3. We can solve for r_c^* by equating eqs. 6.2.14 and 6.2.15 and substituting for $\dot{\gamma}_w$ from eq. 6.2.9

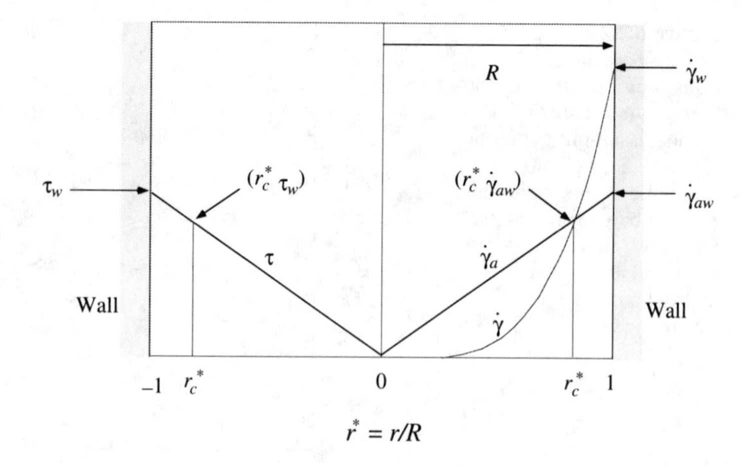

Figure 6.2.3.
Shear stress (τ_w, eq. 6.2.2), apparent shear rate ($\dot{\gamma}_a$, eq. 6.2.14), and power law shear rate ($\dot{\gamma}$, eq. 6.2.15 with $n = 0.3$) plotted versus capillary radius. Adapted from Laun (1983).

$$\dot{\gamma}_a(r_c^*) = \dot{\gamma}(r_c^*)$$

$$r_c^*\ \dot{\gamma}_{aw} = r_c^{*1/n}\ \dot{\gamma}_{aw}\left(\frac{3n+1}{4n}\right)$$

$$r_c^* = \left(\frac{4n}{3n+1}\right)^{n/(1-n)} \tag{6.2.16}$$

Figure 6.2.4 plots r_c^* versus n according to eq. 6.2.16. We see that it varies from 0.77 to 0.88 over the usual range of n. If we assume a constant value for r_c^* the viscosity error will be

$$\frac{\Delta\eta}{\eta} = (n-1)\left[r_c^*\left(\frac{3+1/n}{4}\right)^{n/(1-n)} - 1\right] \tag{6.2.17}$$

Error values are plotted for several r_c^* values. We see that for $r_c^* = 0.83$ the viscosity error will be less than 2% for $0.2 < n < 1.3$.

We have used the power law in this derivation, but this does not mean that the unknown fluid we are testing in the capillary must obey the power law over a wide range of $\dot{\gamma}$* We apply the model locally; the fluid must be power law only from $\dot{\gamma}_a$ to $0.83\dot{\gamma}_a$.

The main danger in using this approximation is for strongly shear thinning liquids (Giesekus and Langer, 1977). In Figure 6.2.4b we see that the errors are large for $n \leq 0.1$.

6.2.2 Wall Slip, Melt Fracture

A more serious error in shear rate occurs when there is wall slip. This can be particularly important for concentrated dispersions and

**Since we normally need to know only $\dot{\gamma}$ at the wall, for the rest of this chapter we will not distinguish between $\dot{\gamma}$ or $\dot{\gamma}_w$ and $\dot{\gamma}_a$ and $\dot{\gamma}_{aw}$.*

Figure 6.2.4.
(a) Variation of the intersection point r_c^* with power index. (b) Error in viscosity by using constant r_c^*. Adapted from Laun (1983).

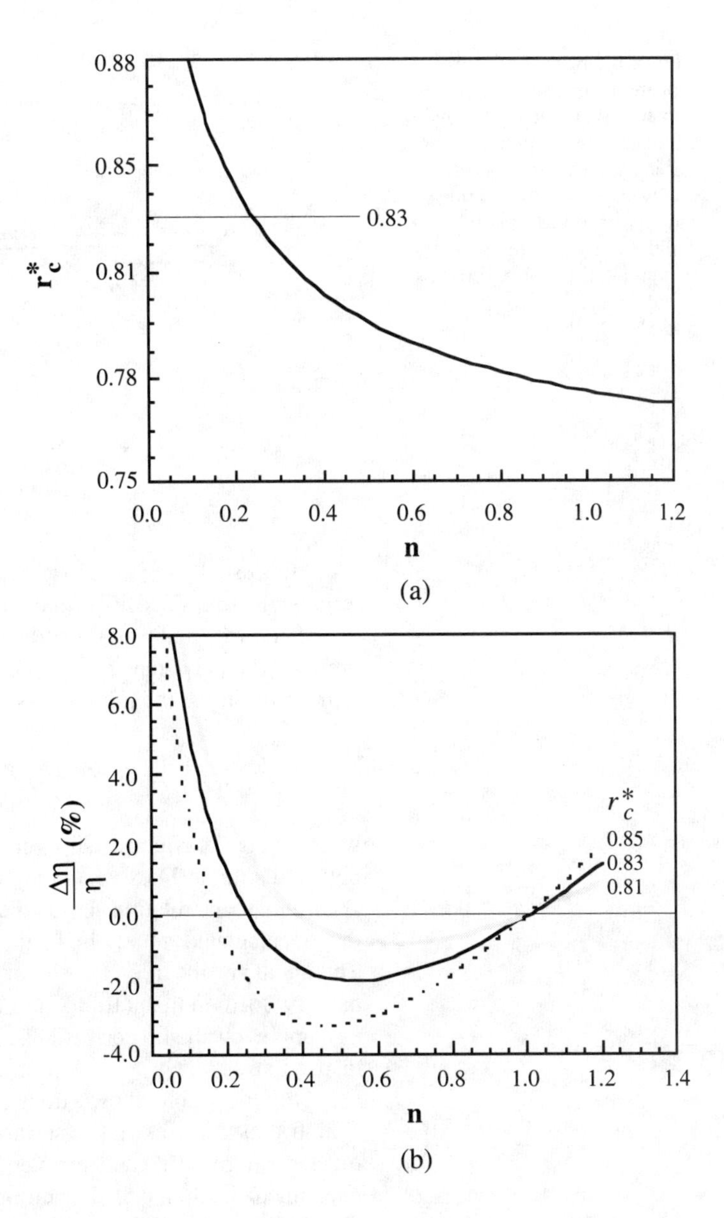

for polymer melts. Let us first consider the dispersions. Because of the impenetrability of the wall, the layer of particles next to a rheometer wall is typically more dilute than in the bulk dispersion. Furthermore, as discussed in Chapter 10, during flow the shear rate gradient causes particles to migrate away from the wall. This shear-induced migration is greater for small capillaries and higher shear rates. The thin, dilute layer near the wall will have a much lower viscosity and will act as if the bulk fluid were slipping along the wall.

The simplest way to test for this phenomenon is to compare viscosity functions determined by capillaries of similar L/R but different R. This is shown in Figure 6.2.5. We see that the smaller diameter capillaries have lower apparent viscosities.

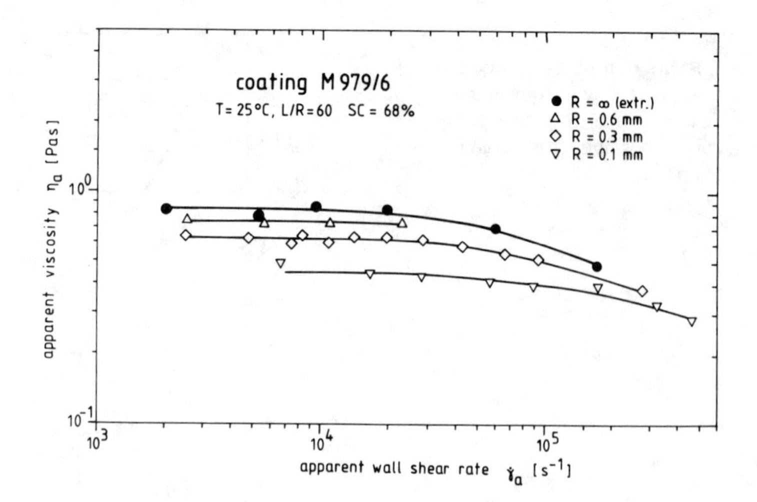

Figure 6.2.5. Apparent viscosity versus apparent wall shear rate by capillaries of various radii R but constant $L/R = 60$ for a clay paper coating formulation with 68 wt % solids. Solid circles denote values extrapolated to infinite radius using eq. 6.2.18. From Laun and Hirsch (1989).

It is usually possible to correct this apparent wall slip and determine the true viscosity of the sample by extrapolating to infinite diameter. Apparent wall shear rates measured at constant extrusion pressure (i.e., constant τ_w) for a constant L/R are plotted against $1/R$ according to the relation first developed by Mooney (1931)

$$\dot{\gamma}_a = \dot{\gamma}_{a_\infty} + \frac{4v_s}{R} \qquad (6.2.18)$$

where v_s is an effective slip velocity. Laun and Hirsch (1989) obtained straight lines when they plotted the data in Figure 6.2.5 according to the relation. Using the $\dot{\gamma}_{a_\infty}$ values from the intercepts, they recalculated η and obtained the upper curve in Figure 6.2.5. This should be the true viscosity of the sample. Gleissle and Windhab (1985), Windhab (1986), and Kurath and Larson (1990) have also applied wall slip corrections to capillary flow of concentrated dispersions.

Polymer melts show a transition from stable to unstable flow at high stress. The extrudate surface appears distorted, usually in a regular pattern at first and then very rough at higher flow rates. The phenomena is often called "melt fracture." In some cases pressure will oscillate strongly, as indicated in Figure 6.2.6 for high density polyethylene. As with the dispersions, smaller diameter dies show a greater slip effect. Uhland (1979) fit his data with eq. 6.2.18, but the $\dot{\gamma}_{a_\infty}$ were not low enough to superpose the slip data on the no-slip curve. However, he was able to model the slip region assuming a 180 μm thick film on the capillary wall with a viscosity 10% of the melt.

The melt distortion phenomenon is not well understood. Clearly it involves loss of adhesion at the die wall and slip or stick–slip flow (Uhland, 1979; Kurtz, 1984; Ramamurthy, 1986; Kalika and Denn, 1987; Lim and Schowalter, 1989). But it can also arise from unstable flow in the die entry region (White, 1973; see also Figure 7.8.5) or from rapid acceleration of the surface layer as it exits the die.

Figure 6.2.6.
Shear stress versus shear rate for a high density polyethylene melt showing flow instabilities and evidence of slip. Adapted from Uhland (1979).

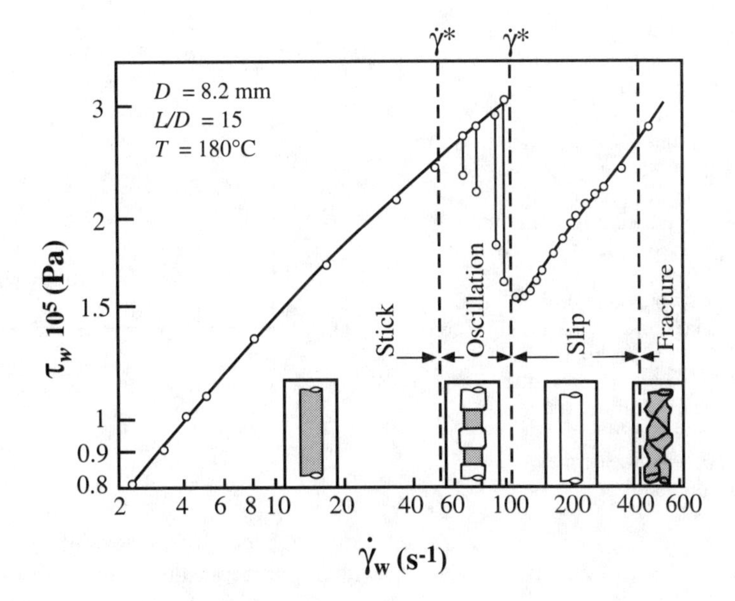

From the viewpoint of capillary rheometry, the onset of melt distortion means the end of rheological data. The onset stress should be noted, but it does not appear to be possible to correct the data to give true viscosity data at higher stresses. Although tapering the die entrance can help somewhat, distortion appears to occur at $\tau \simeq 10^5$ Pa independent of temperature, polymer type, or chain length (Tadmor and Gogos, 1979; Lim and Schowalter, 1989). This makes it particularly difficult to get data on high viscosity elastomers (Geiger, 1989; Leblanc et al., 1989).

If melt distortion limits shear stress for highly viscous (and elastic) materials, turbulent flow potentially puts an upper limit on capillary rheometry of low viscosity fluids. The classical Reynolds number criteria for the onset of turbulence in tube flow gives

$$\mathrm{Re} = \frac{\rho \dot{\gamma}_a R^2}{2\eta} = 2100 \tag{6.2.19}$$

Here we have expressed the mean velocity in terms of the apparent shear rate $\dot{\gamma}_a = 4\overline{v}/R$. For water ($\rho = 1000\ \mathrm{kg/m^3}$, $\eta = 1$ mPa·s) flowing in a 0.5 mm radius capillary, eq. 6.2.19 gives a maximum shear rate of 17,000 $\mathrm{s^{-1}}$ for laminar flow. The addition of polymer or particles will tend to raise the critical Re. Therefore, with the small diameters typically used in capillary rheometry, turbulence is rarely an issue.

6.2.3 True Shear Stress

We have presented the important corrections needed to get true shear rate from flow rate measurements. We now turn our attention to getting the correct shear stress. According to the assumptions used in deriving eq. 6.2.3, we need to measure p_c/L, the pressure

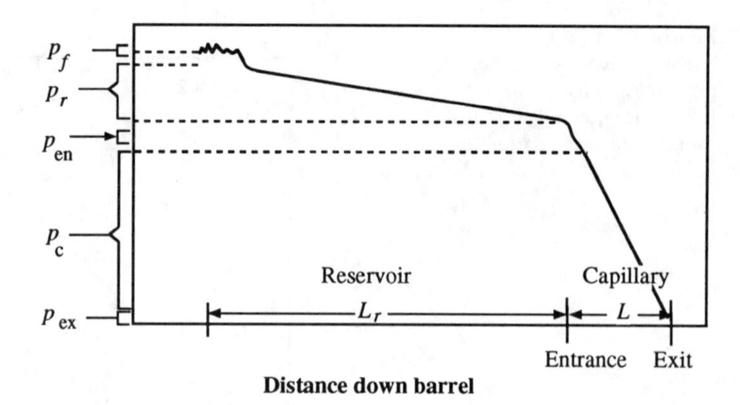

Figure 6.2.7.
Schematic of pressure profile in a capillary rheometer. Where,
p_f: pressure drop due to friction between piston & reservoir walls
p_r: pressure drop in the reservoir, due to steady, fully developed flow
p_{en}: previous drop in the capillary due to converging flow from reservoir to capillary
p_c: pressure drop due to nearly fully developed flow in capillary
p_{ex}: non-zero pressure at capillary exit due to fluid elasticity

drop over a length of steady, fully developed flow. This can be done with pressure transducers mounted on the capillary wall. The diameter of such transducers is much larger than the typical capillaries, which means that the transducers must be recessed and connected to the capillary through small holes. This can lead to plugging and cleaning problems.

As a result of these difficulties, typical capillary rheometers measure pressure in or above the reservoir as indicated in Figure 6.2.1 or from the forces on a driving piston. (Different capillary rheometer designs are discussed further in Chapter 8.) To determine the true shear stress, a number of corrections must be considered.

Figure 6.2.7 indicates the pressure profile in a typical piston-driven capillary rheometer for polymer melts. The total pressure is made up of a number of contributions:

$$p_{\text{tot}} = p_f + p_r(\dot{\gamma}_r) + p_{\text{en}}\left(\tau_w, \frac{R_r}{R}\right) + p_c + p_{\text{ex}} \qquad (6.2.20)$$

where $p_c = 2\tau_w L/R$ by eq. 6.2.3. Our goal is to evaluate τ_w from p_{tot} measurements. Thus we need to evaluate the other four terms.

Most rheometers used for higher viscosity liquids employ some type of piston and thus are subject to pressure losses due to friction between the piston and the reservoir walls. It is frequently suggested that p_f can be estimated by running the piston down an empty reservoir. However, seal friction can change significantly in the presence of fluid and back pressure. Earlier workers generally found p_f small (e.g., Marshall and Riley, 1962; Choi, 1968), but at shear rates below 1 s^{-1} the total pressure drop becomes small enough to permit frictional forces to become significant. Of course, for gas- or gravity-driven viscometers, which are used for lower viscosity fluids, p_f is not a factor. However, with the direct gas type, channeling of gas into the fluid can become a problem as viscosity increases. A ball placed on top of the reservoir liquid can prevent channeling. The mass of this ball must be added to the gravity head in the reservoir.

The pressure drop, p_r, in the reservoir shows up as a gradually falling pressure in a constant flow rate instrument or as a rising flow rate with a constant pressure rheometer. It may be eliminated by extrapolation to the capillary entrance, as indicated in Figure 6.2.8. A power law model can be used to get an estimate of p_r (Skinner, 1961; Marshall and Riley, 1962; Metzger and Knox, 1965)

$$p_r = \frac{p_{\text{cap}}\, L_r}{L\,(R/R_r)^{(n+3)/n}} \tag{6.2.21}$$

Most high temperature instruments have relatively small diameter reservoirs for fast heating and they are long, to provide enough sample. Typical dimensions are $L_r = 200$ mm and $R_r = 5$ mm. With a short, large diameter die e.g., $L = 4$ mm, $R = 1$ mm and a Newtonian fluid, $p_r = 0.08\ p_c$. For $n < 1$ the correction will be less. For $R/R_r < 0.2$ and for higher pressures (longer dies or higher shear rates), both friction and the reservoir loss can be neglected. This is the general practice, but a good alternative is to mount a pressure transducer just above the capillary entrance.

Another important aspect of the reservoir is the prehistory of the sample. The main problem is thermal *degradation* with heat-sensitive materials. Sample extruded at the end of the run may have been at high temperatures long enough to alter its rheology. Settling of dense dispersions may also occur at low flow rates. Sometimes an extruder or other pump is used to feed the capillary (see Chapter 8). Preshear in the pump can influence the capillary results. Hanson (1969) and Villemaire and Agassant (1983–1984) fed their capillaries through the gap between rotating concentric cylinders. This preshearing considerably lowered the capillary pressure drop on some polymer melts. Rauwendaal and Fernandez (1985) found lower viscosities with a slit rheometer fed by an extruder. Frayer and Huspeni (1990) found a strong preshear effect on the slit flow of liquid crystalline polymers.

Figure 6.2.1 shows a schematic representation of p_{en}, the pressure consumed in the converging flow from the large reservoir to the smaller capillary. Typically, p_{ex} is smaller and arises from normal stresses, velocity rearrangement, and perhaps surface tension at the exit. Approaches to obtain normal stresses from p_{ex} are

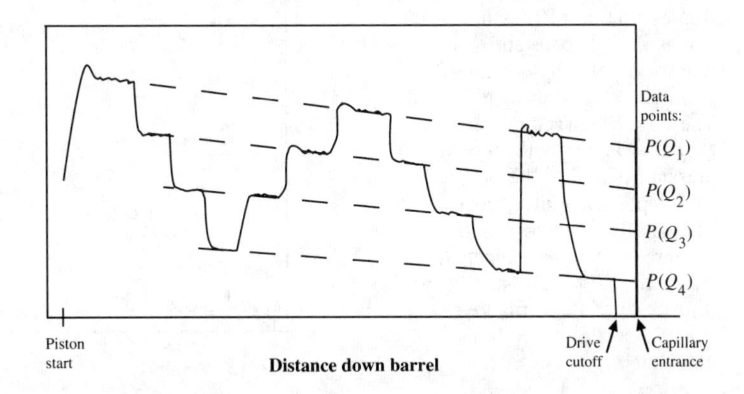

Figure 6.2.8.
Schematic of pressure on piston versus height in reservoir (barrel). To allow extrapolation to zero reservoir height, four constant flow rates (piston velocities) are repeated three times as the barrel is emptied.

discussed further in Section 6.3 on slit rheometry. Here we will lump them together as $p_e = p_{\text{en}} + p_{\text{ex}}$.

The most accurate way to correct for p_e is to take data with several capillaries. If they have the same radius and flow rate, then p_e will remain the same, but the pressure drop we need, $\Delta p = 2\tau_w L/R$, will increase with L.

If p_{tot} is plotted versus L/R at constant Q, the intercept gives p_e. This is illustrated for a polyethylene melt in Figure 6.2.9. Sometimes the intercept on the abscissa is used to express the correction as an extra die length e (Bagley, 1957). For short dies the end effects can be very significant. For example, for $L/R = 4$ in Figure 6.2.9 the intercept is about 70% of p_{tot}.

Construction of Bagley plots like Figure 6.2.9 requires considerable experimental effort. The usual practice is to use one long die ($L/R \geq 60$) and assume that *all* the corrections are negligible. This method is not always as accurate as might be expected. Notice the slight upward curvature in Figure 6.2.9. Curvature in p versus L/R data is due to compressibility and the influence of pressure on viscosity, which is discussed further in connection with slit rheometry in Section 6.3. For example, the pressure readings at $L/R =$ 66, 98, and 132 all give $\tau_w = 2.75 \times 10^5$Pa if all the corrections in eq. 6.2.20 are neglected. However, this value is 18% higher than the true value taken from the slope at small L/R.

Negative curvature rarely occurs in Bagley plots, but an example is shown in Figure 6.2.10. Laun and Hirsch (1989) attributed this to the thixotropy: that is, the decrease in viscosity with amount of shear. For this sample only dies of $L/R \leq 12$ could be used to estimate the end effects.

The most accurate method for determining p_e is to make a Bagley plot with at least two short dies of the same diameter. Another approach is to measure p_e directly with an orifice die and use it to correct data from one long die. This method usually requires using two different transducers because of the great difference in pressure.

In some cases it is possible to estimate the end corrections. For Newtonian fluids the pressure drop for a sudden contraction

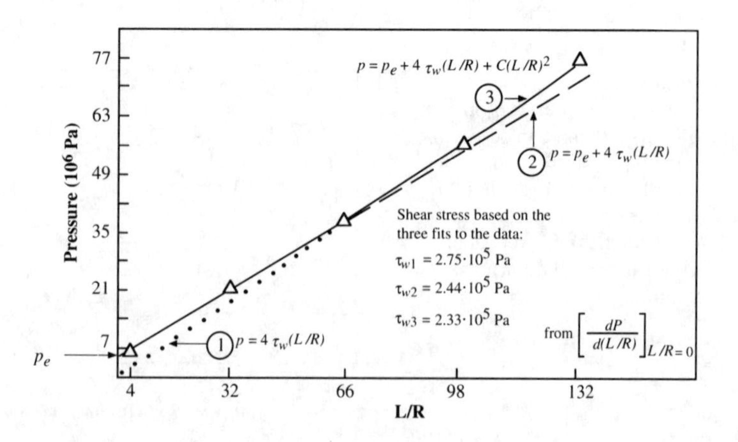

Figure 6.2.9.
Representation of pressure versus L/R at constant flow rate for a polyethylene melt, often called a "Bagley plot." The upward curvature for long dies is due both to compressibility and to the pressure dependence of viscosity. The data have been fit with the three functions indicated. Evaluating $dP/d(L/R)$ at $L/R =$ 0 for each of these fits gives $\tau_{w1} = 275$ kPa, $\tau_{w2} = 244$ kPa, and $\tau_{w3} = 233$ kPa.

Figure 6.2.10.
Pressure versus L/R for a 60% solids clay coating at 25°C, R = 1.5 mm. The negative curvature is due to strong thixotropy. From Laun and Hirsch (1989).

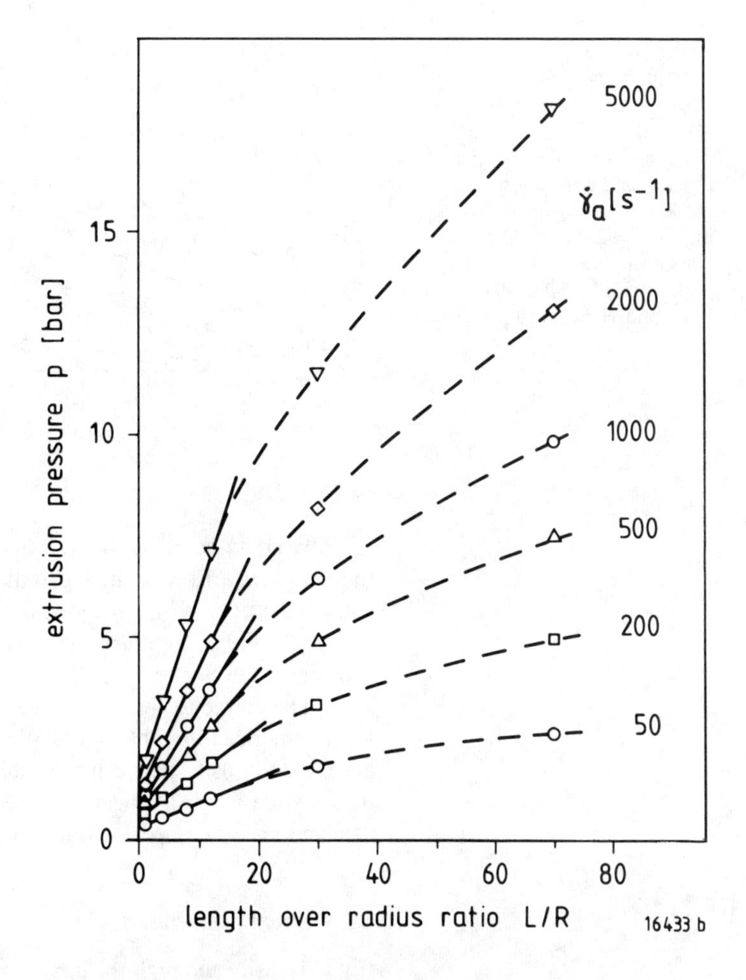

is due to the kinetic energy change (or fluid inertia) and viscous rearrangement of the velocity profile (Van Wazer et al., 1963)

$$\frac{p_{\text{en}}}{\frac{1}{2}\rho v^2} = K_H + \frac{K_c}{\text{Re}} \tag{6.2.22}$$

Here the cross-sectional area of the reservoir is assumed to be much larger than the capillary area. Sylvester and Chen (1985) have reviewed the literature and find $K_H = 2.0\text{–}2.5$ and $K_c = 30\text{–}800$ from various experimental studies. For higher Re Reynolds number (eq. 6.2.19), a best fit of their data gave $K_c = 133 \pm 30$. Boger (1987) gives the inertia correction for a power law fluid.

Figure 6.2.11 illustrates the application of the kinetic energy correction to some very high shear rate data on a low viscosity clay suspension, ~ 10 m /Pa·s. We see that in this case the Bagley plots are straight even to $L/R = 100$. Table 6.2.2 shows that at the two lowest shear rates, eq. 6.2.22 predicts p_{en} fairly well; however at higher rates it is about double the experimental values. If eq. 6.2.22 and p_{en} for only one capillary are used ($L/R = 100$), the error in viscosity is 20% at the highest $\dot{\gamma}$. Kurath and Larson (1990) also applied eq. 6.2.22 to data on similar clay coating samples.

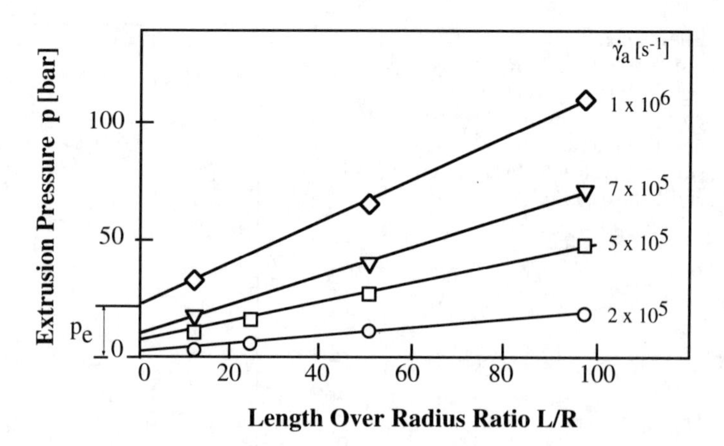

Figure 6.2.11. Pressure versus L/R for a 55.3% solids clay coating at 40°C, R = 0.205 mm. Table 6.2.2 compares these experimental end corrections to the kinetic energy term using $K_H = 2.24$ and the viscous term using $K_c = 133$ with eq. 6.2.22. Adapted from Laun and Hirsch (1989).

For polymer melts and solutions p_{en} is strongly influenced by the *extensional* flow occurring at the capillary entrance. Observed values of p_{en} can be much higher than those predicted by eq. 6.2.22. In fact, the use of capillary entrance flow as an extensional indexer is discussed in Section 7.8.

The corrections needed to get true shear rate and shear stress are summarized in Table 6.2.1. When these corrections are made, excellent viscosity data can be obtained with capillary rheometers over a wide range of shear rates. Section 6.5 gives some typical data and makes comparisons to other shear rheometers.

6.2.4 Shear Heating

High viscosity and high shear rates may generate considerable heat as a result of viscous dissipation near the capillary wall. This heat will lower the viscosity near the wall and make the fluid appear more shear thinning. The general problem of viscous dissipation was discussed in Chapter 2, and its effect on drag flow rheometers was treated in Chapter 5 (e.g., Figure 5.3.9). Shear heating in pressure-driven flows is reviewed by Winter (1975, 1977) and by Warren (1988).

The Nahme number Na is the critical parameter for estimating the importance of shear heating in rheometry. It determines how

TABLE 6.2.2 / Entrance Pressure Drop from Figure 6.2.11 Compared to Inertia Calculation

$\dot{\gamma}$ $(10^5 s^{-1})$	τ_w (bar)	p_{en} (bar)	p_{en} $eq.$ 6.2.22 (bar)	$K_H \rho v^2/2$ (bar)	$K_c \tau/16$ (bar)
10	.47	21.3	42.8	38.9	3.9
7	.32	8.8	21.8	19.1	2.7
5	.23	5.5	11.6	9.73	1.91
2	.10	2.0	2.39	1.56	.83
1	.064	0.8	.92	.389	.53

much the temperature rise will affect the viscosity (see Example 2.6.1). For capillary flow

$$\text{Na} = \frac{\beta \tau \dot{\gamma} R^2}{4k} \tag{6.2.23}$$

where k is the thermal conductivity and β is the temperature sensitivity of viscosity

$$\eta = \eta_o e^{\beta(T_o - T)} \tag{6.2.24}$$

When Na $\geq$ 1 significant errors occur in capillary measurements as a result of viscous dissipation. Figure 6.2.12 demonstrates this for flow of a polymer melt through two very different diameter capillary dies with the same L/R. If there were wall slip (recall Figure 6.2.5), the smaller diameter capillary would give a lower stress, but here it is significantly higher because there is less viscous dissipation as a result of the R^2 term in Na. Figure 6.2.12 indicates Na = 1 for each die. We see that this corresponds well to the point at which the experimental stress values depart significantly from the isothermal flow curves.

Other examples of the effect of viscous dissipation on capillary data are given by Cox and Macosko (1974b) and Warren (1988). These workers and Winter (1975, 1977) indicate how to correct data affected by shear heating to true viscosity values. This requires numerical solution of the momentum and energy equations, a capability available in many standard fluid mechanics software packages. However, note that for typical capillary dies the stress level at which viscous dissipation becomes important is near the region for polymer melt fracture, $\tau_w \sim 10^5$Pa. As already pointed out, it is not possible to get true viscosity data after the onset of melt fracture.

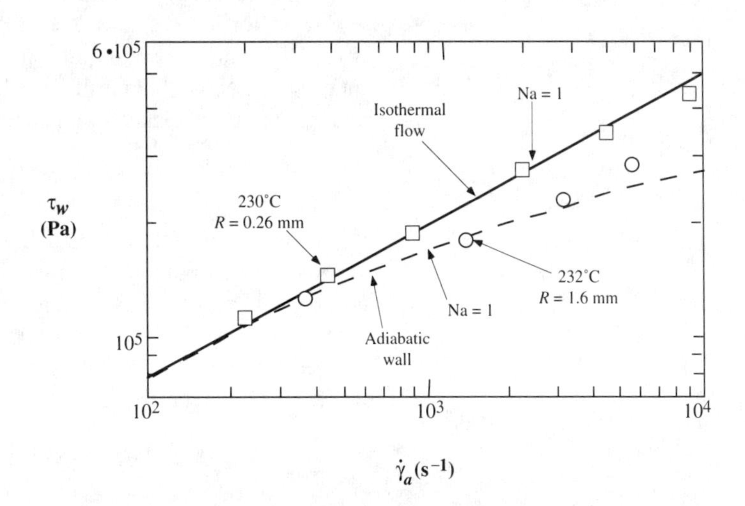

Figure 6.2.12. Shear stress versus apparent shear rate for two capillaries of different diameter but the same $L/R = 60$. ABS polymer melt: squares, R = 0.26 mm, 230°C barrel; circles, R = 1.6 mm, 232°C barrel. The solid line represents isothermal flow estimated at high $\dot{\gamma}$ from the viscosity master curve (Figure 2.6.1); the dashed line, adiabatic boundary condition, viscous dissipation with no heat transfer to the wall, for $R = 1.6$ mm. Replotted from Cox and Macosko (1974a) and Cox (1973).

6.2.5 Extrudate Swell

When an elastic liquid leaves a capillary die, it expands as sketched in Figure 6.2.1. This expansion results from tension along streamlines. The experiment is analogous to recoverable strain after steady shearing (Chapter 3), but here the confining walls evaporate at the onset of recovery. If the effects of gravity, surface tension, and air cooling are not too strong (or can be eliminated by, for example, extruding into an oil bath), then an equilibrium swell can be measured. This swelling has been related to N_1, the first normal stress difference. The derivation is model dependent and the assumptions are not rigorously obeyed, so extrudate swell must be considered to be a normal stress *index*, perhaps useful for comparing materials.

Dealy (1982) reviews the experimental methods used to measure extrudate or die swell. Typically the ratio of equilibrium extrudate diameter D_e to die diameter $2R$ is measured

$$B_{\text{ex}} = \frac{D_e}{2R} \tag{6.2.25}$$

The method that appears to give results closest to equilibrium is to extrude directly into a heated oil bath and take photographs. A simpler method is to quench a strand and then anneal it at melt temperature in an oil bath or even on a tray covered with talcum powder. Annealing time needs to be long enough to reach equilibrium swell, but short enough to prevent surface tension or gravity from distorting the shape. Diameter measurements are made at room temperature should be corrected to the melt temperature by the change in density.

Reservoir diameter and capillary L/R influence the equilibrium swell ratio. If $R_r/R < 10$ and $L/R > 40$, then swell measurements are independent of rheometer geometry (Han, 1976). Of course, the onset of melt distortion sets an upper limit on shear stress for swell measurements.

Theory and experiment show that Newtonian liquids swell at low Reynolds numbers but shrink as inertia becomes important: $B = 1.13$ for $\text{Re} \leq 2$ and $B = 0.87$ for $\text{Re} \geq 100$ (Middleman, 1977). Typically elastic liquids are extruded in the low Reynolds range, so the Newtonian result is subtracted from the experimental values to give an "elastic swell"

$$B = B_{\text{ex}} - 0.13 \tag{6.2.26}$$

Vlachopoulos (1981) and Tanner (1988a) have reviewed the various efforts to relate B to rheological material functions. The simplest and most widely used is Tanner's (1970a), which follows from Lodge's work (1964) on recoverable strain. It assumes unconstrained recovery after steady shear and applies an integral model of the BKZ type (eq. 4.2.2) with one relaxation time. The result is

$$N_1^2 = 8\tau_w^2(B^6 - 1) \tag{6.2.27}$$

Figure 6.2.13.
Viscosity and $N_1/2\tau_{12}$ versus shear rate: open points, cone and plate; solid points, capillary extrudate swell $L/R = 81$; $N_1/2\tau_{12}$ from eq. 6.2.27 with $B = B_{ex} - 0.11$. From Utracki et al. (1975).

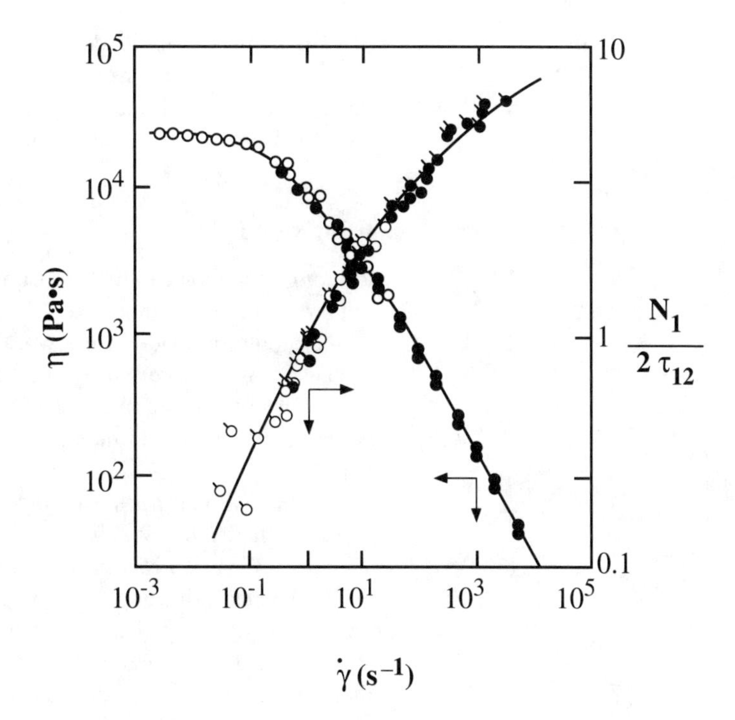

where both B and N_1 are evaluated at the same shear stress, τ_w. This result has fit some data well (Tanner, 1970b; Utracki et al., 1975), as indicated in Figure 6.2.13, but it does not always give a very accurate prediction of N_1/τ_{12}. Table 6.2.3 shows swell and normal stress data on a low density polyethylene (IUPAC A; Meissner, 1975). Here eq. 6.2.27 does a poor job of predicting B. A full numerical simulation using a viscoelastic constitutive equation does fairly well. The experimental normal stress data at 10 s^{-1} may be low because of edge failure. Vlachopoulos (1981) reports a similar underprediction of B by eq. 6.2.27 for polystyrene. Tanner (1988a) has found that in addition to N_1 the value of extensional viscosity may influence extrudate swell.

Underprediction of eq. 6.2.27 is not due to experimental errors in B_{ex} because such errors should lead to failure to reach equilibrium and smaller values.

Accurate extrudate swell measurements are not easy to make, yet because the swell ratio must be cubed $N_1 \sim B^3$, they are critical to predicting normal stresses.

TABLE 6.2.3 / Extrudate Swell of Low Density Polyethylene[2]

$\dot{\gamma}$ (s^{-1})	N_1/τ_{12} (exp)	B_{ex} $(eq.\ 6.2.27 + 0.13)$	B_{ex} $(numerical)$	B_{ex} (exp)
0.1	1.5	1.17	1.34	1.35
1.0	3.1	1.27	1.52	1.53
10	(3.8)	1.32	(1.56)	1.75

[2] Data taken with long dies at 150°C from Tanner (1988a). Parentheses indicate approximate value.

Because eq. 6.2.27 depends on both choice of constitutive equation and idealization of the deformation, it seems unwise to rely on capillary die swell data to measure normal forces. However, such data are useful as normal stress indexes for comparing materials.

6.2.6 Melt Index

The most common capillary instrument in the polymer industry is not a rheometer but an indexer. The quality of nearly every batch of thermoplastic made in the world is controlled by melt index. Because it is so widely used and has all the essential features of a capillary rheometer, and because rheologists are often asked to compare their results to melt index values, we need to examine it here.

The melt index is standardized internationally, (ISO R1133; R292) in the United States, Germany, Japan (e.g., ASTM D1238-73, DIN 53735, JIS K7210) and in other countries. Figure 6.2.14 shows the measuring apparatus. It resembles a typical capillary for polymer melts except that only one die and one driving pressure are specified to give the melt index number. The die has a rather large diameter, $2R = 2.095$ mm, and is short, $L/R = 7.637$. There are a number of different load and temperature conditions, depending on the particular polymer. The most common is "condition E": 190°C and 2.160 kg mass on the piston. Since the piston is 9.55 mm diameter, pressure on the top of the melt is 2.97×10^5 Pa and

Figure 6.2.14.
Schematic of melt index (MI) apparatus.

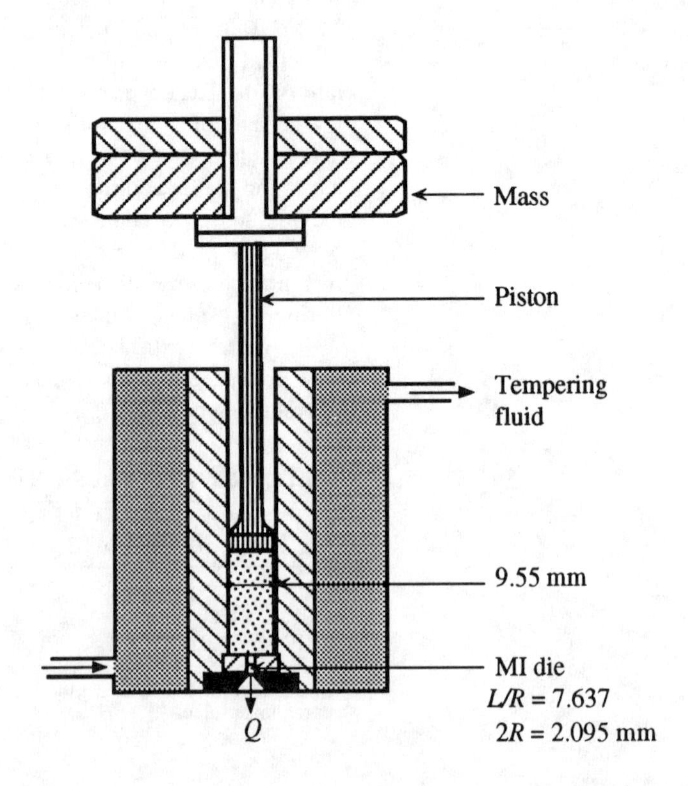

wall shear stress, assuming no end effects, is $\tau_w = 1.94 \times 10^4$ Pa, from eq. 6.2.3. This stress is relatively low for capillary rheometry. For comparison $\tau_w = 1.94 \times 10^4$ Pa would give, respectively, $\dot{\gamma} \simeq 40$ and $\dot{\gamma} < 10\ s^{-1}$ for the ABS and PVC samples shown in Figure 6.2.2.

After the sample comes to temperature, the weight is applied and flow rate is determined by cutting and weighing the extrudate obtained in 10 minutes. This weight in grams is the melt flow index (MFI) of the polymer. It resembles an inverse viscosity at moderately low stress. High MFI means low η_o and low molecular weight, hence its use to control polymerization reactors (Bremner and Rudin, 1990).

Such an instrument could measure true viscosity, but one would need to use other dies and weights. With such a short die entrance, losses can consume up to half of p_{tot} (recall Figure 6.2.9). Also the ratio of die to reservoir radius is rather large, $R/R_r = 0.219$, so reservoir losses are significant. Thus the melt index number is a combined measure of shear and extensional viscosity.

Because MFI is used so widely, there is interest in quantitatively relating it to rheological material functions. Michaeli (1984) gives a nomogram to relate MFI to melt viscosity. Such calculations generally ignore reservoir and entrance pressure losses.

Figure 6.3.1.
Schematic of slit rheometer with flush-mounted pressure transducers P_i.

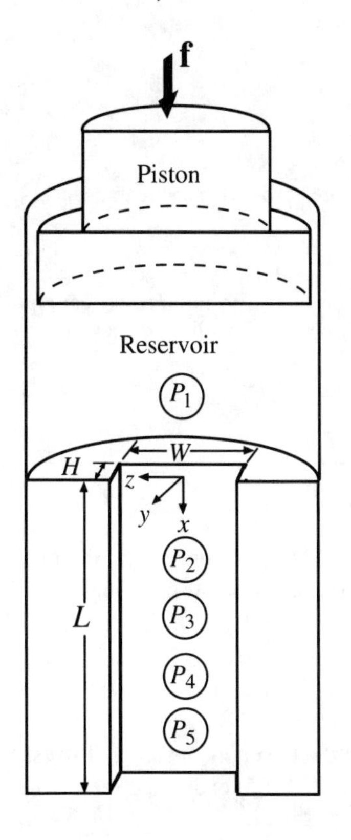

6.3 Slit Rheometry

Capillary rheometers can be modified readily to force liquid through a thin rectangular channel or slit. Figure 6.3.1 shows a typical arrangement. Derivation of the working equations follows closely that for the capillary and can be found in a number of texts (Walters, 1975; Dealy, 1982; Bird et al., 1987). The important results are summarized in Table 6.3.1, which contains eqs. 6.3.1–6.3.5, not presented separately in this section.

Except for numerical constants, the equations for calculating shear rate are the same as for the capillary. The constant 0.79 in the representative shear rate equation can be obtained following the derivation given in Example 6.2.1 (see also Laun, 1983, Appendix A). Varying slit thickness, H, can be used like capillary radius to test for wall slip. In fact, the studies by Lim and Schowalter (1989) and by Geiger (1989) referred to in Section 6.2 were done with slit dies. Slip or melt fracture in polymer melts occurs at about the same wall shear stress as for capillaries, 10^5 Pa.

Shear heating can also affect slit data. Again the Nahme number can be used to estimate when significant viscosity errors will occur. For a rectangular channel

$$\text{Na} = \frac{\beta h^2 \tau_w \dot{\gamma}_w}{k} \tag{6.3.6}$$

TABLE 6.3.1 / Working Equations for Slit Rheometer

Wall shear stress

$$\tau_w = \frac{H}{2(1+H/W)}\frac{dp}{dx} \quad (6.3.1)$$

Wall shear rate

$$\dot{\gamma}_a = \frac{6Q}{WH^2} \quad (6.3.2)$$

$$\dot{\gamma}_w = \frac{\dot{\gamma}_a}{3}\left(2 + \frac{d \ln \dot{\gamma}_a}{d \ln \tau_w}\right) \quad (6.3.3)$$

Representative shear rate

$$\eta(\dot{\gamma}) = \eta_a(\dot{\gamma}_a) = \frac{\tau_w(\dot{\gamma}_a)}{\dot{\gamma}_a} \text{ to } \pm 3\% \quad (6.3.4)$$

$$\text{for } \dot{\gamma} = 0.79\dot{\gamma}_a \quad (6.3.5)$$

$$\text{and } 0.17 < \frac{d \ln \dot{\gamma}_a}{d \ln \tau_w} < 1.2$$

First normal stress difference

By extrapolation to p_{ex}

$$N_1 = p_{ex}\left(1 + \frac{d \ln p_{ex}}{d \ln \tau_w}\right) \quad (6.3.11)$$

By pressure hole, transverse slot

$$N_1 = 2\left(\frac{d \ln p_h}{d \ln \tau_w}\right)p_h \quad (6.3.13)$$

Errors
Similar to capillary, but no Bagley plots are needed if pressure transducers on the slit wall are used

Utility (same advantages and limitations as capillary except)
Obtain dp/dx directly from pressure transducers; no entrance corrections
N_1 by p_{ex} and especially p_h more rigorous than capillary die swell, but obtaining accurate N_1 values is difficult
Effect of finite slit width
Cleaning slit corners
More complex, more expensive than capillary

Lodge and Ko (1989) tested this criteria with Newtonian oils in a miniature slit die rheometer, $0.05 \times 0.5 \times 2$ mm. At the remarkably high shear rate of $5 \times 10^6 \text{s}^{-1}$ with a 2.5 m Pa·s oil, they reached Na = 1.3 and recorded a 7% error in viscosity. Lodge and Ko (1989) and also Winter (1975, 1977) give detailed analyses of the role of viscous dissipation in slit flow.

A disadvantage of slit flow is the lower shear rate and stress obtained at the side walls. Wales et al. (1965) examined the effect of aspect ratio and found insignificant error for $B/H > 10$. For $B/H < 10$ the wall shear stress should be corrected as indicated in the denominator of eq. 6.3.1 (Table 6.3.1).

The major advantage of the slit rheometer is also indicated in eq. 6.3.1, namely direct measurement of the pressure gradient dp/dx. This is possible because pressure transducers can be mounted on the wide flat sides of the slit geometry. Even if pressure holes are used, they are easier to construct and will not disturb the flow as much as in a small diameter capillary. If pressure transducers are not used, dp/dx can be evaluated from p_{tot} as with the capillary (eq. 6.2.20).

A typical pressure profile down a slit die is shown in Figure 6.3.2. From the slope, τ_w can be determined directly using eq. 6.3.1 without the reservoir and end corrections required for accurate capillary rheometry. Thus only one flow rate is needed for each $\eta(\dot{\gamma})$ point. To ensure that accurate data are obtained, the pressure transducers must be carefully calibrated. Laun (1983) discusses the accuracy of commercial transducers and the use of a nonlinear calibration curve. When such care is taken, it is possible to get excellent viscosity data over a wide shear rate range. Figure 6.3.3 gives an example of such data with comparison to cone and plate and three sets of capillary data. There are other examples in the literature (e.g., Wales et al., 1965; Han, 1976; Hansen and Jansma, 1980).

In Figure 6.3.2 note the large drop in pressure from the reservoir to the die. Laun (1983) has shown that this entrance pressure drop is in good agreement with that measured by capillary rheometry. Notice also the curvature in the pressure versus distance data. We saw curvature with different L/R capillaries in Figure 6.2.9. It is caused primarily by the pressure dependence of viscosity. We can analyze for pressure dependence by fitting the profiles with a quadratic function

$$p(x) = a + bx + cx^2 \tag{6.3.7}$$

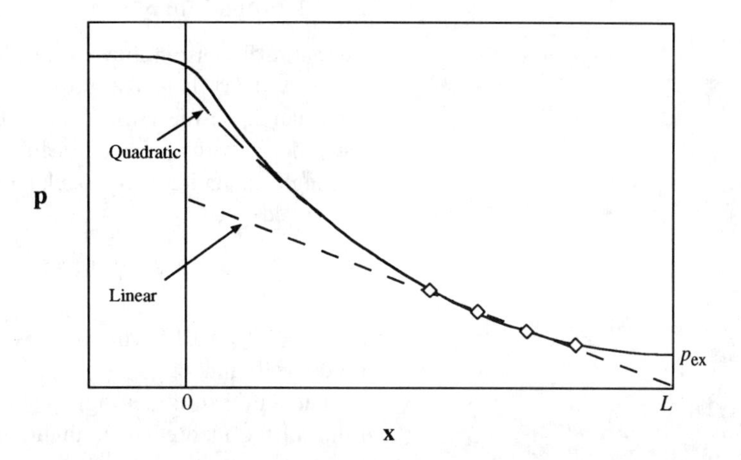

Figure 6.3.2.
Pressure versus distance down a slit die with a linear and quadratic fit through the four data points. The curves are schematic: curvature and p_{ex} have been exaggerated.

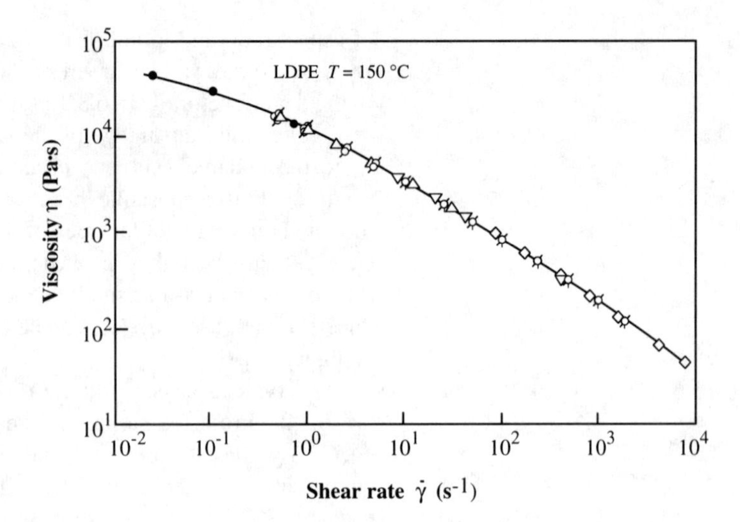

Figure 6.3.3. Comparison of viscosity versus shear rate data on a low density polyethylene at 150°C for slit (○), cone and plate (●), gas-driven capillary (△), and piston-driven capillary rheometer (▽: $R = 0.6$ mm; ◇, $R = 0.25$ mm). From Laun (1983).

Note that here $x = 0$ is at the slit *outlet*. If viscosity is assumed to depend exponentially on pressure

$$\eta_p(P, \dot{\gamma}) = \eta(\dot{\gamma}) e^{\alpha p} \tag{6.3.8}$$

then the pressure coefficient can be found from

$$\alpha = \frac{\partial \ln \eta(\dot{\gamma})}{\partial p} = \frac{2c}{b^2} \tag{6.3.9}$$

Duvdevani and Klein (1967) have used a similar approach on capillary data. This method is not highly accurate, but it yields reasonable values for α (Laun, 1983). One difficulty is that *compressibility* of the sample also can cause similar curvature in $p(x)$ data (Wales et al., 1965). One approach to separating the two effects is to apply additional hydrostatic pressure at the die exit. Shear heating also causes such quadratic curvature in the pressure profile.

6.3.1 Normal Stresses

The entire discussion above is concerned with getting accurate shear viscosity data. It is also possible to use slit geometry to obtain information on the normal stress differences in shear. As with a capillary, extrudate swell occurs as liquid leaves a slit. Again, starting from an integral model, a relation similar to eq. 6.2.27 can be derived

$$N_1^2 = 12\tau_w^2(B^4 - 1) \tag{6.3.10}$$

here $B = B_{\text{ex}} - 0.19$, where 0.19 is the correction due to Newtonian fluid swell, and $B_{\text{ex}} = H/H_o$, the ratio of the swollen extrudate thickness to that of the original slit. Planar extrudate swell suffers from all the theoretical limitations associated with capillary die swell (Tanner, 1988a). In addition, the finite width of the slit creates strong edge effects, which make experimental measurements even more difficult to interpret than with the capillary. White (1990) gives some results for extrudate swell from slits with aspect ratios of 10 and 17. As with the capillary, theory somewhat underpredicts experiment.

However, two other methods for obtaining normal stress data from slit rheometry have had reasonable success: the exit pressure and particularly the pressure hole method. These are discussed in Sections 6.3.2 and 6.3.3.

6.3.2 Exit Pressure

When pressure is plotted against distance down a capillary (eq. 6.2.7) or slit (eq. 6.3.2, Table 6.3.1) the value extrapolated to the exit typically is positive. This exit pressure comes from the same tension along streamlines that causes extrudate swell and should be related to the shear normal stresses in an elastic liquid. Such measurements are potentially quite interesting They can give N_1 on polymer melts at shear rates several orders of magnitude higher than are possible in rotational rheometers. Here we discuss p_{ex} using slit rheometry. Some work has also been done using pressure transducers mounted along a tube (Han, 1976).

If we can neglect fluid inertia and any rearrangement of the velocity profile up to the die exit, then the exit pressure can be related to the first normal stress difference

$$N_1 = p_{ex}\left(1 + \frac{d \ln p_{ex}}{d \ln \tau_w}\right) \tag{6.3.11}$$

Boger and Denn (1980) discuss problems in the use of this relation. For example, it has been shown that the velocity profile in polymer solutions rearranges near the exit (e.g., Gottlieb and Bird, 1979). However, for polymer melts evidence from experiments (Han and Drexler, 1973) and finite element analyses (Tuna and Finlayson, 1984; Vlachopoulus and Mitsoulis, 1985) indicates that velocity rearrangement is not significant enough to invalidate eq. 6.3.11. Han (1988) argues that the relation should be valid for $\tau_w > 25$ kPa.

The real problem, however, is to measure the exit pressure accurately. Exit pressure is very small, typically $p_{ex} > 2$ bar, less than 2% of the range of the pressure transducer nearest to the exit. A number of studies report negative p_{ex} using a linear extrapolation (Laun, 1983; Lodge and de Vargas, 1983; Tuna and Finlayson, 1988; Senouci and Smith, 1988). Others report wide scatter (Rauwendaal and Fernandez, 1985; Baird et al., 1986). Clearly the curvature we have already discussed makes it very difficult to evaluate p_{ex} as well as the derivative term in eq. 6.3.11. Disturbances due to pressure holes or deformation of flush-mounted diaphragms can be significant. Viscous dissipation will also complicate data analysis.

Chan, et al. (1990) have measured exit pressures for five different polymers. Pressure transducers were calibrated in situ, at temperature, with dead weights, both before and after experiments. These investigators found very little curvature in their plots of p versus x when they used three flush transducers. The introduction of a pressure hole caused some curvature. Although they report excessive scatter in their p_{ex} data and sometimes a negative slope of p_{ex} versus τ_w, Chan et al. found fair agreement between N_1 by cone and plate and predictions from eq 6.3.11. Figure 6.3.4 gives an example of their results. Tuna and Finlayson (1988) found similar agreement between cone and plate and p_{ex} data on three polyethylene melts. They used four pressure transducers and a *quadratic fit*. The derivation of eq. 6.3.11 assumes a linear pressure profile. Hence the use of p_{ex} from extrapolation of a quadratic fit needs further theoretical work.

These results indicate that exit pressure measurements can be used to obtain normal stress data on high viscosity systems. The range of measurement is 10–100 kPa (1–300 s^{-1}). Great care must be taken to get precise pressure readings, and even then the best work often shows a scatter of $\pm$100%.

6.3.3 Pressure Hole

Another method for extracting normal stress data from slit rheometry is to read the difference in pressure between a recessed and a flush transducer as indicated in Figure 6.3.5. It is rather surprising at first that the two transducers do not agree for an elastic liquid. They do agree for Newtonian fluids, and it created quite a stir when rheologists discovered that all their manometers and recessed transducers were giving errors (Broadbent et al., 1968). The good news is that these errors are proportional to something we often would like to measure, N_1 and N_2.

The source of the pressure hole effect is the slight bending of the streamlines into the hole. An elastic liquid generates tension

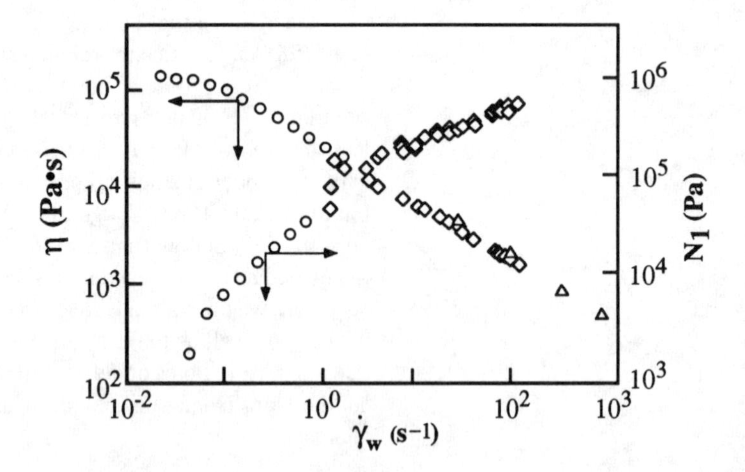

Figure 6.3.4.
Viscosity and first normal stress difference versus shear for a high density polyethylene at 200°C: cone and plate (○), slit die (◇), and capillary (△) data. Adapted from Chan et al. (1990).

Figure 6.3.5.
Schematic of pressure hole geometry showing curvature of the streamlines near the mouth of the hole or slot. The difference between the pressures measured by the flush-mounted and recessed transducers gives the hole pressure, $p_h = p_1 - p_{1,h}$.

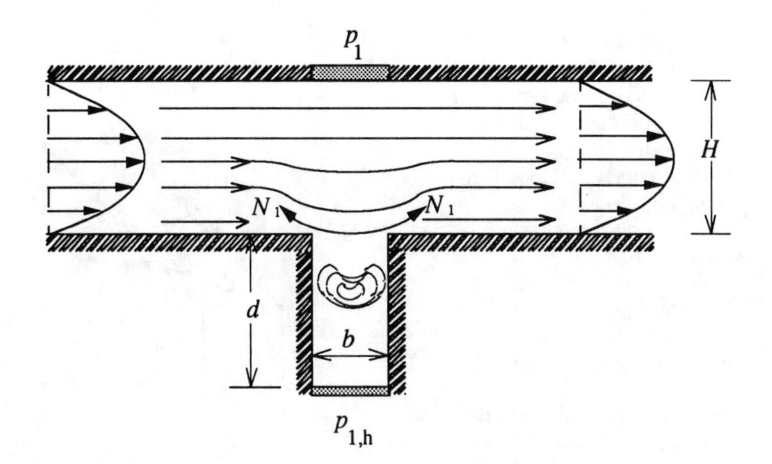

or normal stresses along these streamlines in shear flow. Where the streamlines are curved over the mouth of the hole, the normal stresses tend to lift up on the fluid in the hole as shown in Figure 6.3.5. This causes pressure that is read by a transducer at the bottom of the hole, $p_{1,h}$, to be lower than p_1, the pressure read by a flush-mounted transducer exactly opposite $p_{1,h}$. The streamline curvature has been photographed (Hou et al., 1977).

Hole pressure is defined as the difference between the two transducer readings

$$p_h = p_1 - p_{1,h} \tag{6.3.12}$$

Figure 6.3.6 shows the three different types of hole that have been used. The normal stress difference that is measured depends on how the streamlines are bent. For example, for a circular hole (Figure 6.3.6c), streamlines are bent in both directions, and thus p_h is proportional to a combination of N_1 and N_2. If flow over the hole is slow, symmetric about the hole centerline, and shear undirectional, then simple relations may be derived for the three types of hole (Lodge, 1988; Tanner, 1988b):

Slot transverse to flow

$$N_1 = 2mp_h \tag{6.3.13}$$

Or a slot parallel to flow

$$N_2 = -mp_h \tag{6.3.14}$$

Circular hole

$$N_1 - N_2 = 3mp_h \tag{6.3.15}$$

where

$$m = \frac{d \ln p_h}{d \ln \tau_w} \tag{6.3.16}$$

Figure 6.3.6.
Schematic of flow over pressure holes of three types. (a) For a long, narrow slot transverse to the flow direction, there is curvature along the streamlines, in the x_1x_2 plane. This generates a tension proportional to $\tau_{11} - \tau_{22} = N_1$ at the mouth of the slot. (b) For a slot parallel to the flow, curvature is *across* streamlines, in the x_2x_3 plane. This generates $\tau_{22} - \tau_{33} = N_2$, the second normal stress difference. (c) A circular hole bends streamlines in both directions, and thus the pressure error is a combination of N_1 and N_2.

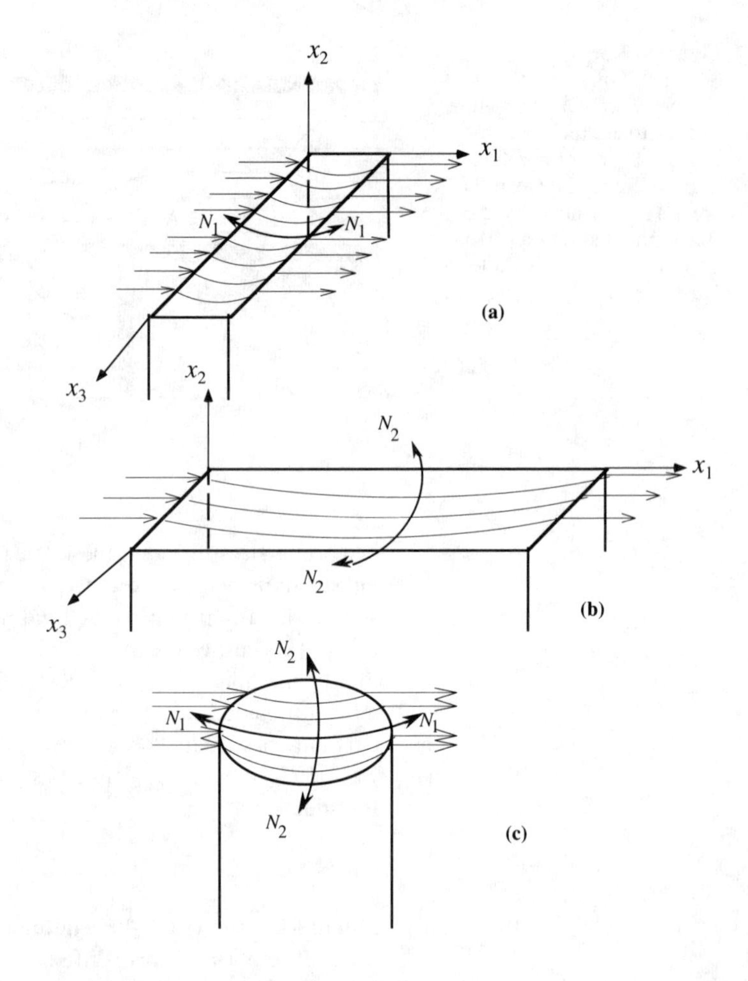

By "slot" we mean a long, rectangular hole in the channel wall. The length of this slot should be much greater than H, while its width b should be less than H and its recess depth d should exceed H. Remarkably, apart from these restrictions p_h is insensitive to specific hole dimensions.

Because N_1 is generally of greater interest than the much smaller N_2, we will concentrate on tests of eq. 6.3.13. We also will focus on hole pressure measurements in slit flow. Again more work has been done with slits, but it is of course possible to measure hole pressure on the wall of any shear flow rheometer. Lodge (1985, 1988) has extensively reviewed hole pressure results.

Lodge (1988, 1989) has shown that it is necessary to correct for misalignment of the flush-mounted transducer p_1 and for inertia. Both corrections can be made using measurements from Newtonian fluids. The misalignment correction probably comes from bending of the flush-mounted transducer diaphragm and must be measured against τ_w for each transducer in situ. The inertia correction appears to be linear in stress and Reynolds number.

$$p_h = -0.033\tau_w \mathrm{Re} \tag{6.3.17}$$

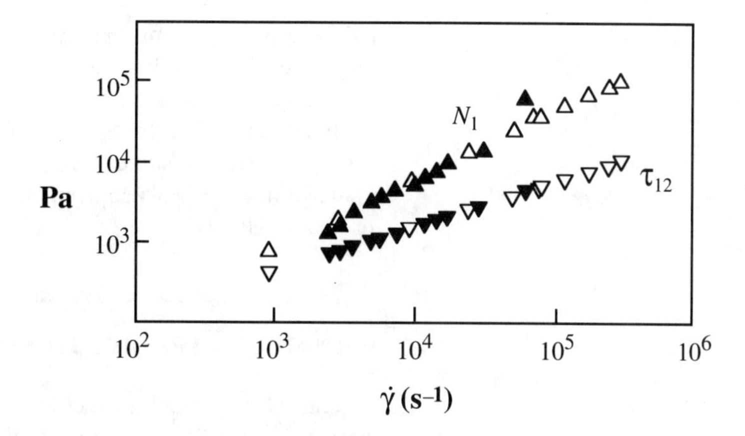

Figure 6.3.7. Comparison of shear and normal stress determined by parallel plates (solid symbols) and slit rheometer with hole pressure (open symbols) for a polyisobutylene solution at room temperature. From Lodge (1989).

When these corrections have been made, it is possible to measure N_1 over a wide shear rate range. Figure 6.3.7 compares τ_w and N_1 results from hole pressure and narrow gap parallel plates on a polymer solution. There is agreement between the two methods over nearly two decades in shear rate. Lodge (1989) has been able to measure N_1 to $\dot{\gamma} > 10^6$ s^{-1}, the highest ever reported by any method. Figure 6.3.8 compares hole pressure and cone and plate data on a polystyrene melt. Again there is significant overlap of the data, in contrast to the exit pressure data shown in Figure 6.3.4.

Actually the comparisons in Figures 6.3.7 and 6.3.8 are too good. From flow birefringence we know that at higher rates, flow over the hole is no longer unidirectional shear (Pike and Baird, 1984). This changes the normal stresses and generates new extensional stresses. Fortunately, these two contributions tend to cancel each other, leading to the wider validity of eq. 6.3.13 (Yao and Malkus, 1990).

Despite the ability of the hole pressure method to measure normal stresses reliably at high shear rates, it has not yet seen wide use. Some of the reasons may include problems with maintaining and cleaning the pressure slots for routine tests. Melt samples especially may degrade in the recess. Materials with a yield stress may

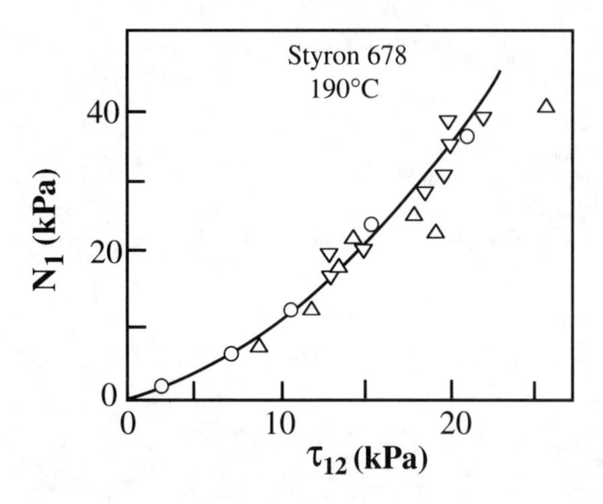

Figure 6.3.8. Normal stress versus shear stress for a polystyrene melt at 190°C: cone and plate (○) and transverse pressure holes of two different widths (△). From Baird et al. (1986).

fail to properly transmit stress in the hole. Some of these problems may be relieved by filling the hole with a viscous, stable Newtonian fluid. There is also the question of accuracy. Hole pressures are relatively small differences between large numbers. High accuracy demands careful in situ calibration of pressure transducers. Even a small amount of leakage from the hole can lead to large errors in p_h (Lodge, 1985).

6.4 Other Pressure Rheometers

Several other channel geometries have been used as rheometers, particularly annular flows. In addition in this section we describe the important indexer: squeezing flow between parallel plates.

6.4.1 Axial Annular Flow

Pressure-driven axial flow through a narrow annulus is essentially the same as flow through a slit, but without the side walls. The lack of side walls may be helpful for studying slip phenomena and for reducing residence time distribution. Furthermore, the pressure difference between the outer and inner walls of the annulus gives the second normal stress difference.

Figure 6.4.1 shows how an annular die can be installed in a capillary rheometer barrel. A screw extruder may also be used to feed a polymer melt to the annulus (Ehrmann and Winter, 1973; Ehrmann, 1976; Okubo and Hori, 1980). Using the notation given in Figure 6.4.1, in which κ is the ratio of outer to inner cylinder radius, the equations for shear stress and shear rate become

$$\tau_w = \frac{\Delta p\, R(1-\kappa)}{2L} \tag{6.4.1}$$

$$\bar{\dot{\gamma}}_w = \frac{2Q(2+1/n)}{\pi R^3(1-\kappa)^2(1+\kappa)} \tag{6.4.2}$$

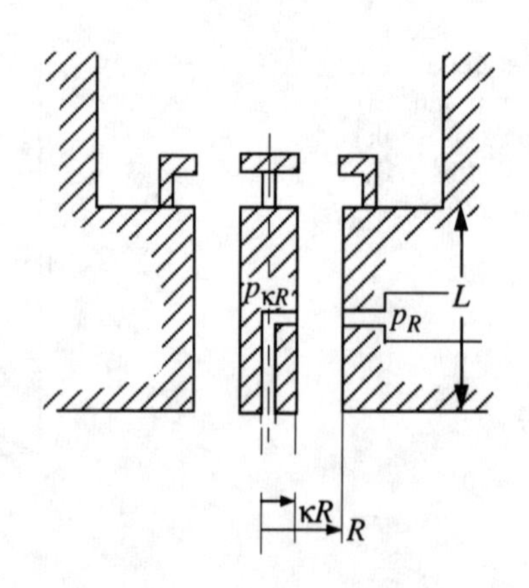

Figure 6.4.1.
Schematic of axial annular flow die with wall pressure difference measurement.

where

$$n = \frac{d \ln \Delta p}{d \ln Q}$$

The shear rate here is actually an average. Curvature causes the shear rate on the inner cylinder to be higher than on the outer wall. Equation 6.4.2 is valid for narrow annuli. Hanks and Larsen (1979) give a full solution for a power law fluid, while McEachern (1966) uses the Ellis model (Bird et al., 1987, p. 233).

In axial annular flow there is curvature *across* the flow streamlines, that is, curvature in the plane perpendicular to the flow direction x_1 (see Figure 5.1.2). This is similar to the pressure hole error for flow over a parallel slot, eq. 6.3.14. Both geometries measure the *second* normal stress difference. The pressure difference between the outer and inner cylinder gives

$$\Delta p_R = p_R - p_{\kappa R} = -\int_{\kappa R}^{R} \frac{N_2}{r}\, dr \tag{6.4.3}$$

Okubo and Hori use a change of variables and an estimate of the location of the radial position for zero shear stress to derive an explicit relation for N_2

$$N_2 \cong 2.58 \frac{\partial(\tau_w \Delta p_R)}{\partial \tau_w} \tag{6.4.4}$$

As we might expect, the pressure difference Δp_R is small. Furthermore, because of the curvature of the cylinder walls and the size of typical pressure transducers, pressure holes must be used (note Figure 6.4.1). When corrections are made for pressure hole errors, results for N_2 are in qualitative agreement with those obtained by other methods (Lobo and Osmers, 1974; Ehrmann, 1976). Figure 6.4.2 shows results for a 1% polyacrylamide solution. Here N_2 is negative and $-N_2/N_1 \simeq 0.06$. This ratio is somewhat lower than the value of 0.10 of an 0.8% solution of the same polyacrylamide (Keentok et al., 1980). This discrepancy may be due to the hole pressure error, which is somewhat lower ($0.25N_1$–$0.1N_1$) than expected by eq. 6.3.15.

6.4.2 Tangential Annular Flow

If fluid is pumped tangentially (circumferentially) around an annulus, as indicated in Figure 6.4.3, streamlines are curved. As with the transverse pressure slot (Figure 6.3.13), this curvature generates a pressure related to the first normal stress difference

$$p_R - p_{\kappa R} = \int_{\kappa R}^{R} \frac{N_1}{r}\, dr \tag{6.4.5}$$

Figure 6.4.2. Normal stress measurements versus wall shear stress for a 1% solution of polyacrylamide at 25°C in annular flows: $-N_2$ from axial annular flow, N_1 from (solid line) tangential annular flow, N_1 from (dashed line) cone and plate, and p_h the hole pressure error. Adapted from Osmers and Lobo (1976).

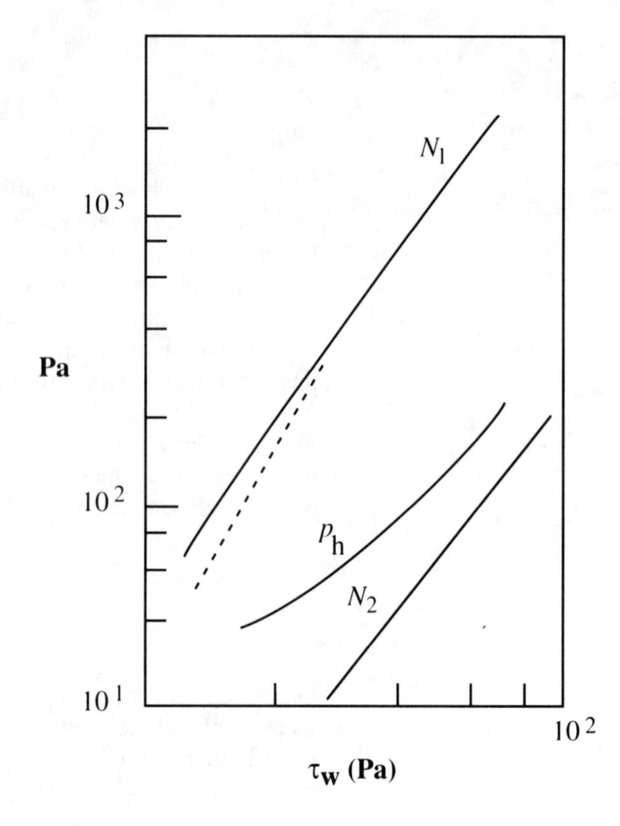

Figure 6.4.3. Schematic of tangential flow between concentric cylinders. The wall pressure *difference* gives N_1.

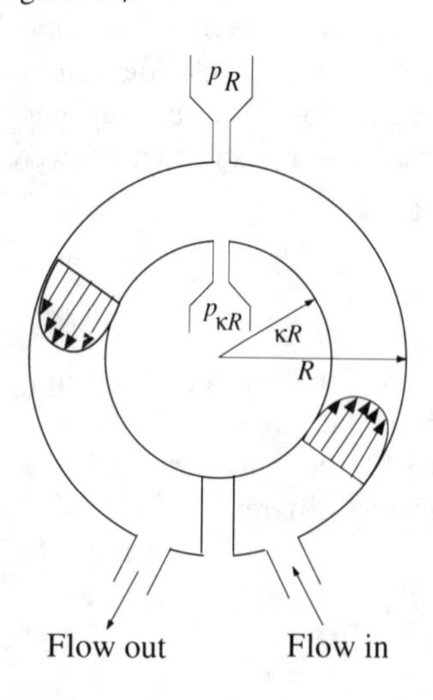

Using change of variables, it should be possible to derive an explicit expression for N_1 similar to eq. 6.4.4. Instead Osmers and Lobo (1976) have assumed an Ellis-type equation for N_1, while Geiger and Winter (1978) used a numerical method to evaluate N_1 from eq. 6.4.5. Both corrected for pressure hole errors.

As Figure 6.4.2 indicates, there is fair agreement between N_1 measured by tangential annular flow and by cone and plate. Because the tangential method involves no free surfaces, it is not limited by edge failure, as are cone and plate or parallel plate rheometers. Thus tangential annular flow permits data to be taken at significantly higher stresses. This geometry has not yet been widely used for rheometry.

6.4.3 Tilted Open Channel

When a viscoelastic fluid flows down a tilted, open channel, the free surface bulges slightly as indicated in Figure 6.4.4. This curvature is across the streamlines of the flow; thus, as with parallel pressure slot and axial annular flow, we expect the bulge to be proportional

Figure 6.4.4.
Side and cross-sectional views of flow of a viscoelastic fluid down a semicircular tube. The bulge $h(x_3)$ is an indication of negative N_2. After Tanner (1970).

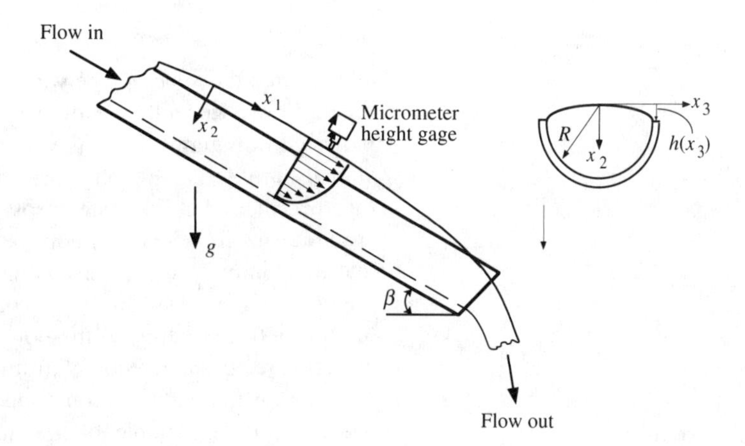

to N_2. If we include the influence of interfacial tension Γ (Keentok et al., 1980),

$$N_2|_{\tau_w} = -\rho g \cos\beta \left(h + \frac{1}{x_3} \int_0^{x_3} h \, dx_3 \right) + \Gamma \left[\frac{d^2 h}{dx_3^2} - \frac{1}{x_3} \left(\frac{dh}{dx_3} - \left. \frac{dh}{dx_3} \right|_{x_3=0} \right) \right] \quad (6.4.6)$$

where

$$\tau_w = \frac{1}{2} \rho g R \sin\beta \quad (6.4.7)$$

ρ is the fluid density, and other terms are defined in Figure 6.4.4.

Channel bulge is perhaps the simplest and most direct measurement of the second normal stress difference. If h is positive, a convex surface N_2 is negative. Since Tanner's first experiments with the tilted trough in 1970 and the discovery of the pressure hole error (Broadbent et al., 1968), nearly all studies have shown N_2 to be small and negative.

Because N_2 is relatively small and τ_w is limited by $\beta < 15°$, the channel bulge is small. However, photographing the reflection of a graduated straight edge allowed Keentok et al. (1980) to measure $h \leq \pm 10\,\mu$m. They report N_2 data on five fluids at several concentrations with an accuracy ± 0.1 Pa. The results indicate $-N_2/N_1 \simeq 0.1$ independent of shear rate. The technique is limited to lower viscosities (< 0.1 Pa·s), lower shear stresses (< 20 Pa), and room temperature.

The magnitude of the bulge can be greatly increased by floating another, immiscible fluid of similar density on top of the test fluid (Sturges and Joseph, 1980). As with rod climbing, the height rise is roughly proportional to the reciprocal of the density difference.

6.4.4 Squeezing Flow

When a liquid is squeezed between two parallel plates, a pressure-driven flow is generated. However, the flow is quite complex, as indicated in Figure 6.4.5. As with flow in a capillary or slit, the velocity profile is parabolic. But because the walls are moving together, the radial flow rate keeps increasing with r. Thus there are gradients in velocity in both z and r directions. This means that in addition to the usual inhomogeneity that accompanies all pressure-driven flows, $\partial v_r/\partial z \neq$ constant, there is also extension, $\partial v_r/\partial r > 0$. Furthermore, the flow is transient because thickness changes are normally recorded from a rest state.

Despite these limitations, squeezing flow is a popular flow indexer. It is very simple to build and operate and easy to control temperature over a wide range, and with large loads it generates high shear rates. Known as the Williams parallel plate plastometer (ASTM D926; Gent, 1960) in the rubber industry, it is particularly useful for such high viscosity materials as rubber and glass (Wilson and Poole, 1990). It has also been used to evaluate the cure of epoxy resins (Tungare et al., 1988) and flow of fiber-filled suspensions (Lee et al., 1984).

Squeezing flow is interesting to fluid mechanicians because it simulates such polymer processes as compression molding and stamping. In addition, it is a simple model for the action of a lubricant film under a bearing. Because polymers are typically added to lubricant oil, a number of studies have used squeezing flow to determine whether viscoelasticity will improve load capacity of bearing.

But the question for the rheologist is: Can squeezing flow measure rheological material functions? Strictly speaking, the answer is no. The flow is so complex that the squeezing force can be related to the gap change only through a constitutive equation. In fact, the flow has been used as a model to evaluate constitutive

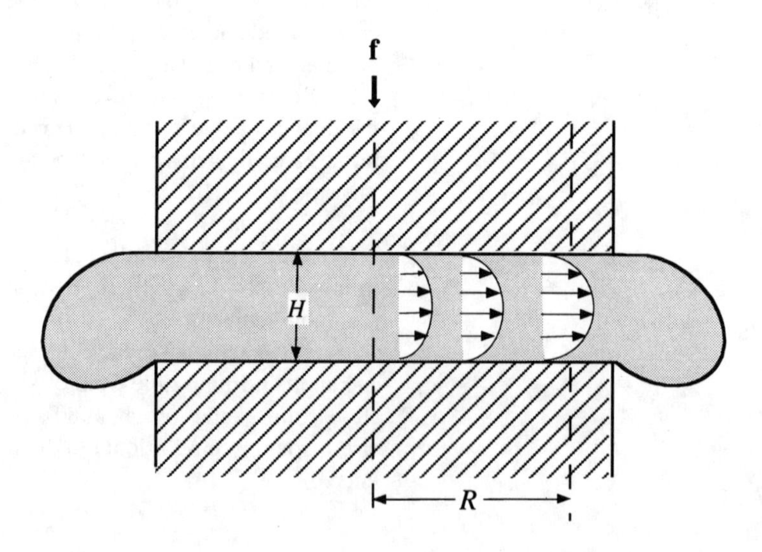

Figure 6.4.5. Schematic of squeezing flow between two parallel plates.

equations. However, if we take data at longer squeezing times, viscoelastic effects are small and useful shear viscosity measurements can be made.

For a Newtonian fluid—neglecting gravity and inertia—we obtain Stefan's simple result (1874; Bird et al., 1987, p. 21)

$$\eta_o = \frac{2fH^3}{3\pi R^4 \dot{H}} \quad \text{sample fills plates; constant sample radius} \quad (6.4.8)$$

where $\dot{H}$ is the rate of squeezing, dH/dt* Typically a weight is placed on the moving disk; thus f is constant and H is measured. However, if the plates are mounted—for example, in a tensile machine—$\dot{H}$ can be held constant and f recorded.

The parallel plate plastometer is usually run with a sample much smaller than the plates (see Figure 7.3.2). In this case the sample radius changes with time, but sample volume between the plates V is constant. This leads to (Gent, 1960)

$$\eta_o = \frac{2\pi f H^5}{3V(V + 2\pi H^3)\dot{H}} \quad \text{sample smaller; constant sample volume} \quad (6.4.9)$$

Typically $R \geq 10H_o$ and the $2\pi H^3$ term is insignificant. Most studies have been done with the plates filled with sample, the squeeze flow geometry of Figure 6.4.5. Thus we focus on this version of the indexer for the rest of our discussion.

Integrating eq. 6.4.8 for the constant force case gives

$$\left(\frac{H_o}{H}\right)^2 = 1 + \frac{4fH_o^2 t}{3\pi R^4 \eta_o} \quad (6.4.10)$$

Figure 6.4.6a shows H versus time data for a polyethylene melt. At long times, a plot of $(H_o/H)^2$ versus t will become nearly constant, as predicted by eq. 6.4.10, but at short times the data deviate from the Newtonian result. There are several reasons for this deviation. With high squeeze rates and low viscosity, inertia can become important (i.e., when $\rho H(-\dot{H})/\eta_o > 1$). When coupled with viscoelasticity, inertia can even generate oscillations in H versus t (Lee et al., 1984).

However, two more significant causes of deviation from Newtonian squeezing are typically the shear rate and time dependence of the viscosity. The importance of shear rate dependence can be estimated by applying a power law model to squeezing flow. For a power law fluid (eq. 2.4.12) with f constant (Grimm, 1978)

$$\left(\frac{H_o}{H}\right)^{(n+1)/n} = 1 + \left(1 + \frac{1}{n}\right)\left(\frac{H_o}{2}\right)^{(n+1)/n} Kt \quad (6.4.11)$$

Note that much of the squeeze flow literature uses $h = H/2$ as the gap measure.

Figure 6.4.6.
(a) Sample thickness versus time for squeezing of a low density polyethylene sample at 150°C, $R = 25$ mm. (b) Same data plotted as log $\dot{H}$ versus log H. From Laun (1992).

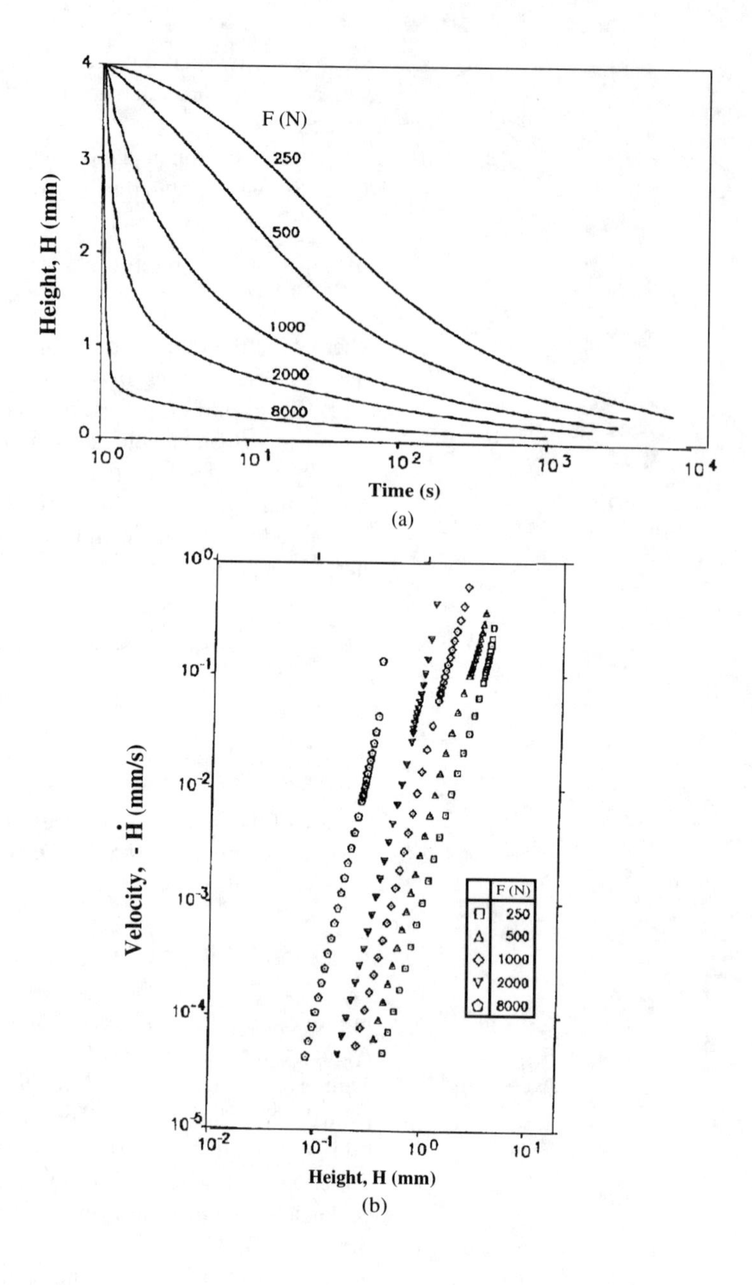

where

$$K^n = \frac{f}{\pi m R^{n+3}} \left(\frac{2n}{2n+1} \right)^n (n+3)$$

A Carreau-type model has also been applied to squeezing flow (eq. 2.4.16) (Phan Thien et al., 1987).

For any time dependent models there will still be a large deviation of H versus t at short times. This deviation is primarily due to the transient viscosity η^+, the "overshoot" in shear stress at the start-up of shear flow (recall Figure 4.2.3). Leider and Bird (1974) incorporated an empirical stress overshoot function into their analysis and found that it predicted a reduction in squeezing. Phan Thien et al. (1987) found similar results using their differential model (Table 4.4.2), which predicts shear stress overshoot on start-up. Figure 6.4.7 shows the results of their numerical calculations: reduced force is plotted versus half-time, the time to squeeze H to $H_o/2$, according to the power law model, eq. 6.4.11. Increasing the parameter ξ increases the overshoot. The ratio of the steady shear normal stresses can be used to find ξ; $\xi = 0.2$ corresponds to $-N_2/N_1 = 0.1$. It is important to note that the Maxwell model (eq. 4.3.7), which shows extensional thickening but no overshoot in shear, does not explain the short time effects in squeezing flow, nor are shear normal stresses the cause of the short time deviation (Phan Thien et al., 1987).

Laun (1992) has suggested a very simple way to analyze squeezing flow data. The flow at any location between the plates is nearly that for a slit. If we choose the edge of the plates and substitute $W = 2\pi R$ and $Q = \pi R^2 \dot{H}$ in eq. 6.3.3 (Table 6.3.1) for the shear rate, we obtain

$$\dot{\gamma}_{WR} = \frac{R(-\dot{H})}{H^2}\left(2 + \frac{1}{n}\right) \tag{6.4.12}$$

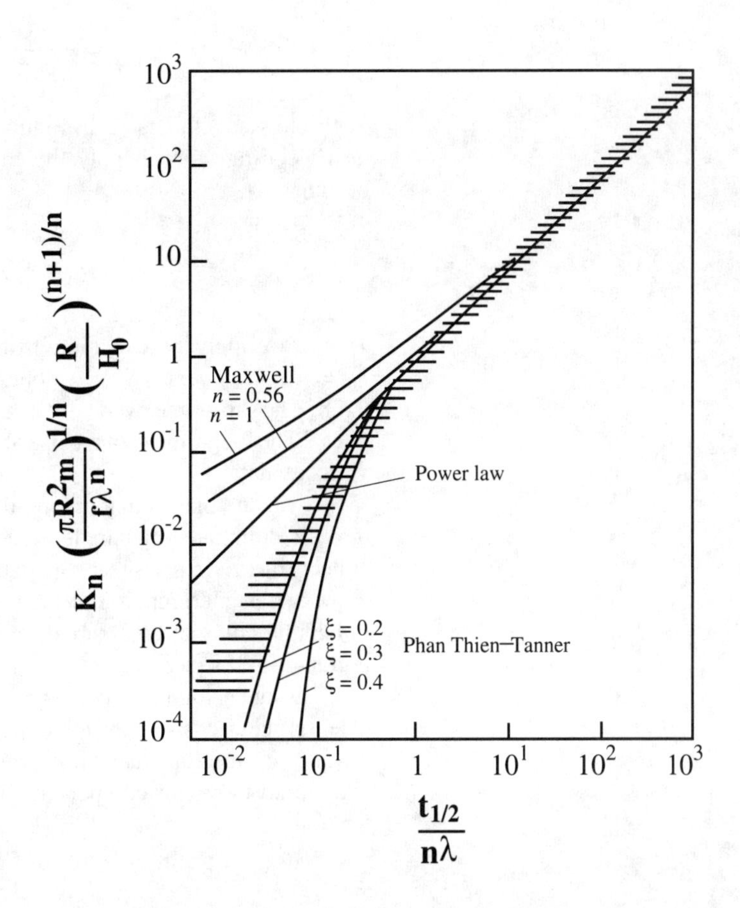

Figure 6.4.7.
Reduced reciprocal of force versus time to squeeze to half-thickness; λ is the time constant from the Carreau model, eq. 2.4.16, or the Maxwell time constant when $n = 1$. Hatched region is the range of experimental data on several polymer solutions from Leider (1974). The experiments deviate from the power law model (dashed line) at short times. A Maxwell model even with shear thinning, $n = 0.56$, gives the opposite trend, but a model with overshoot in start-up of steady shear compares well. Adapted from Phan Thien et al. (1987).

Figure 6.4.8.
Viscosity versus shear rate for a low density polyethylene: line is from capillary data, points from squeezing flow, using stress and shear rate at the perimeter. The lower curve has been shifted down one decade to show complete data for the $f = 500$ N run. The open points on this lower curve are due to transient shear behavior and have not been used to determine η versus $\dot{\gamma}$. Replotted from Laun (1992).

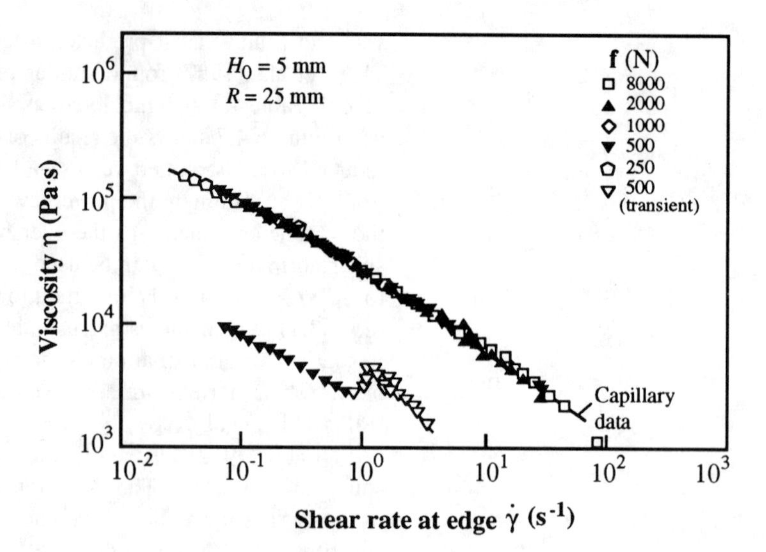

From a plot of $\log(-\dot{H})$ versus $\log H$, we can obtain the power low index directly

$$\frac{1}{n} = \frac{d \ln(-\dot{H})}{d \ln H} - 2 \tag{6.4.13}$$

Such plots are illustrated in Figure 6.4.6b. We can see that n is nearly constant for this polyethylene. The shear stress at the edge becomes

$$\tau_{WR} = (n + 3) \frac{Hf}{2\pi R^3} \tag{6.4.14}$$

If we apply these relations to the data of Figure 6.4.6a and plot $\eta_a = \tau_{WR}/\dot{\gamma}_{WR}$ versus $\dot{\gamma}_{WR}$, we obtain the curves shown in Figure 6.4.8. The short time overshoot behavior shows up at higher "shear rates." The long time data are in excellent agreement with capillary viscosity data.

Thus it appears that at long times, flow in the squeezing indexer is dominated by simple, steady shear, and it can be used to collect true viscosity shear rate data. Further analyses may even allow better interpretation of the transient regime. The influence of bulk modulus and pressure dependence of viscosity may be important in fast squeezing.

A squeezing rheometer can also be used to measure biaxial extensional viscosity if the test surfaces are lubricated with a much lower viscosity fluid than the test sample. Ideally the lubricant will take up all the shear component. This rheometer is discussed in Section 7.3.

6.5 Comparison of Shear Methods

Chapters 5 and 6 have described and evaluated the important shear rheometers and indexers. If measurements of torques, forces, and velocities or pressures and flow rates are made and interpreted properly, all these rheometers should measure the same shear material functions. To conclude these two chapters, we make a few comparisons between rheometers to emphasize this point. Table 6.5.1 highlights some of the strengths and weaknesses of each.

It is very valuable to compare different rheometers in your own laboratory. Of course comparisons are helpful in checking instrument calibration or finding such errors as a mislabled cone angle. Juxtaposition of rheometers can often identify more insidious errors, such as secondary flows, slip, or evaporation. Comparisons are also very helpful in selecting which rheometer is most useful for a particular material, deformation, or temperature range. Comparisons give confidence to operators that they are really seeing material behavior and not some instrument artifact.

We have already shown a number of rheometer comparisons. Figure 5.5.2 compares N_1 and N_2 by cone and plate, parallel disks and slit flow. Figure 2.4.1 compares viscosity data from capillary and cone and plate rheometers. Chapter 3 compares different linear viscoelastic measurements. Figure 5.7.4 compares dynamic data G' and η' by oscillating concentric cylinders with two eccentric rotating geometries. Figure 5.7.5 shows steady and dynamic data illustrating the limiting relations at low shear rate or frequency.

Figure 5.4.3 compares N_1 and N_2 as determined from total thrust in the cone and plate to values obtained from the pressure distribution. Figure 6.2.13 and Table 6.2.2 compare N_1 from cone and plate to extrudate swell. Figures 6.3.4, 6.3.6, and 6.3.7 do the same for N_1 determined by exit pressure and hole pressure. These comparisons helped us to decide which methods were reliable and over what range of measurement. The same could be said of the normal stress comparisons made for axial annular and tangential annular flow in Figure 6.4.2.

Comparisons of results from rheometers and indexers are essential in evaluating what material function dominates the indexer's response. Such comparisons can help us to determine when an indexer may give us useful rheological data, as in the case of squeezing flow, Figure 6.4.8. This ability becomes even more important in the next chapter, where we shall see that indexers are the only choice for extensional measurements on low viscosity fluids.

Figure 6.5.1 illustrates the very wide range of viscosity measurement possible on a stable suspension, eight decades in $\dot{\gamma}$, when one combines results from several rheometers. We see that concentric cylinders are typically used for the lowest shear rates, while capillaries provide high shear rate data. The increase in viscosity with shear rate is real and has been verified by tests with differ-

TABLE 6.5.1 / Comparison of Shear Rheometers[2]

Method	*Advantages*	*Disadvantages*
Sliding plates (5.2)	•Simple design •Homogeneous •Linear motion •high η, $G(t, \gamma)$ •$t \geq 10^{-3}$s	•Edges limit $\gamma < 10$ •Gap control •Loading
Falling ball (5.2)	•Very simple •Needle better •Sealed rheometer •High T, p	•Not very useful for viscoelastic fluids •Nonhomogeneous •Transparent fluid •Need ρ
Concentric cylinders (5.3) (Couette flow)	•Low η, high $\dot{\gamma}$ •Homogeneous if $R_i/R_o \geq 0.95$ •Good for suspension settling	•End correction •High η fluids are difficult to clean •N_1 impractical
Cone and plate (5.4)	•Homogeneous $\beta \leq 0.1$ rad •Best for N_1 •Best for $G(t, \gamma)$	•High η: $\dot{\gamma}$ low, edge failure, loading difficult •Low η: inertia •Evaporation •Need good alignment
Parallel disks (5.5) (torsional flow)	•Easy to load viscous samples •Best for G' and G'' of melts, curing •Vary $\dot{\gamma}$ by h and Ω •$(N_1 - N_2)(\dot{\gamma})$	•Nonhomogeneous: not good for $G(t,\gamma)$. OK for $G(t)$ and $\eta(\gamma)$ •Edge failure •Evaporation
Contained bobs (5.6) (Brabender, Mooney)	•Sealed •Process simulator	•Indexers •Friction limits range
Capillary (6.2) (Poiseuille flow)	•High $\dot{\gamma}$ •Sealed •Process simulation •η_{ext} from Δp_{ent} •Wide range with L	•Corrections for Δp_{ent} time-consuming •Nonhomogeneous no $G(t, \gamma)$ •Bad for time dependence •Extrudate swell only qualitative for N_1
Slit flow (6.3)	•No Δp_{ent} with wall-mounted pressure transients •$\eta(p)$ •p_{ex}, p_h give N_1	•Edge effects with $W / B > 5$ •Similar to capillary •Difficult to clean
Axial annular flow (6.4.1)	•Slit with no edges •Δp can give N_2	•Difficult construction and cleaning
Squeeze flow (6.4.4)	•Simple •Process simulation •$\eta(\dot{\gamma})$ at long times	•Index flow: mixed shear rates and shear transients

[2]Numbers in parentheses indicate section in chapter that describes device.

ent capillaries and a high shear rate concentric cylinders rheometer (Laun et al., 1991).

Figure 6.5.2 shows a flocculated suspension with a yield stress as well. The yield stress is best measured with a stress-controlled rotational rheometer but may be confirmed by rate-controlled and capillary measurements. In Chapter 8 we discuss

Figure 6.5.1.
Viscosity versus shear rate function using concentric cylinders (□), cone and plate (•), and capillary rheometers (Δ) for a latex suspension. Adapted from Laun (1988).

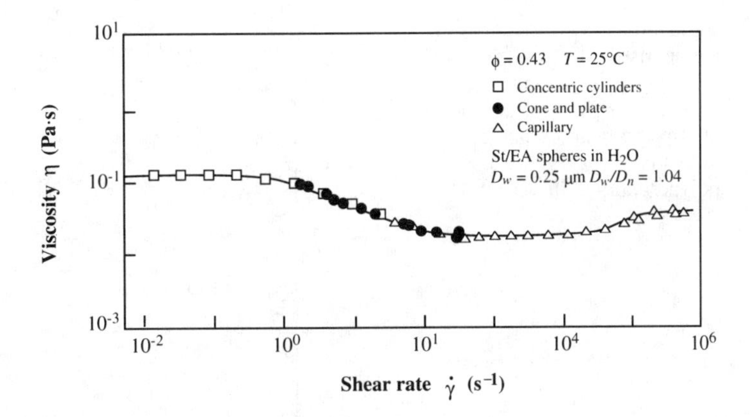

different types of rheometer control and give further data on range of deformation rate possible with current instruments.

Figures 6.5.3–6.6.5 show data collected in our lab over two days during our Rheological Measurements summer short course. No special efforts other than standard calibration procedures were made. The agreement illustrates what one can expect from normal operation of modern rheological instruments with stable, polymeric samples. Such an exercise is reassuring to the new rheologist.

Figure 6.5.3 compares data from three rheometers for a strongly shear thinning polymer solution. Note that the Cox–Merz relation, eq. 4.2.6, between steady and dynamic viscosity is very well obeyed for this fluid (see also Figure 4.2.4). Figure 6.5.4 shows viscosity and first normal stress coefficient for a polydimethylsiloxane melt. Here we compare cone and plate data and birefringence in concentric cylinders. Chapter 9 describes the birefringence method further.

Figure 6.5.2.
Viscosity versus shear stress for a latex suspension showing a yield stress. Data from concentric cylinders, cone and plate, and a capillary under stress control. Adapted from Laun et al. (1992).

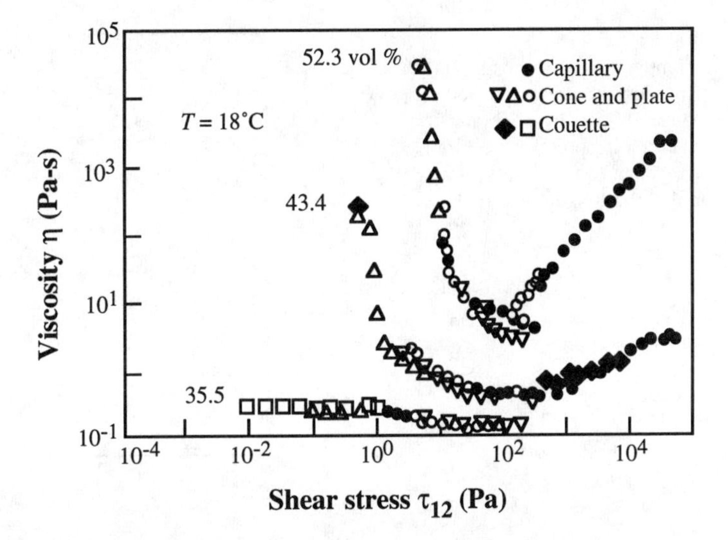

Figure 6.5.3.
Comparison of steady shear viscosity by cone and plate, concentric cylinders, and parallel plates with dynamic viscosity $|\eta^*(\omega)|$ for a 3.5% polyacrylamide in water.

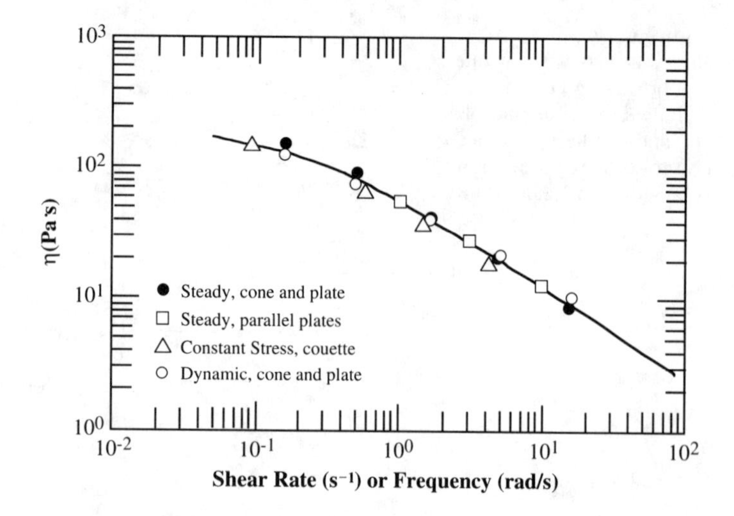

Figure 6.5.5 compares viscosity measurements on a polystyrene melt. Edge failure limits the cone and plate data to $\dot{\gamma} \leq 3\ \text{s}^{-1}$, below the range of most melt capillaries. The dynamic viscosity does a fair job of bridging the gap between the two. Dynamic data are typically easier to obtain and are recommended for a first measurement. Figure 6.5.6 shows first normal stress coefficient measured directly by cone and plate and estimated from G' by eq. 4.2.5 and from integrating the steady shear viscosity (Gleissle, 1988)

$$\psi_1(\dot{\gamma}^*) = 2 \int_0^{\dot{\gamma}^*} \frac{d\eta}{\dot{\gamma}}$$

$$\text{where } \dot{\gamma}^* = 2.7\dot{\gamma}. \qquad (6.5.1)$$

Although the cone and plate is the only true measure of Ψ_1, its range is limited on viscous polymer melts; thus the empirical

Figure 6.5.4.
Comparison of viscosity and first normal stress coefficient $\Psi_1 = N_1/\dot{\gamma}^2$ by cone and plate (solid points) and birefringence (open points) for a polydimethylsiloxane melt.

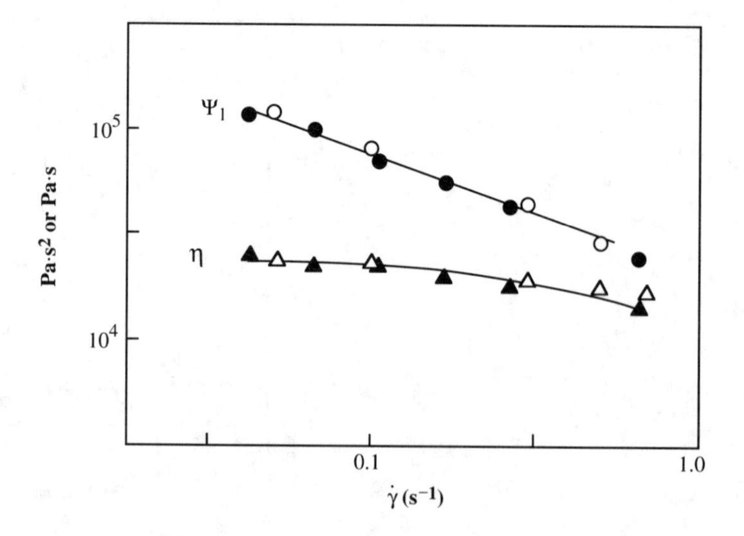

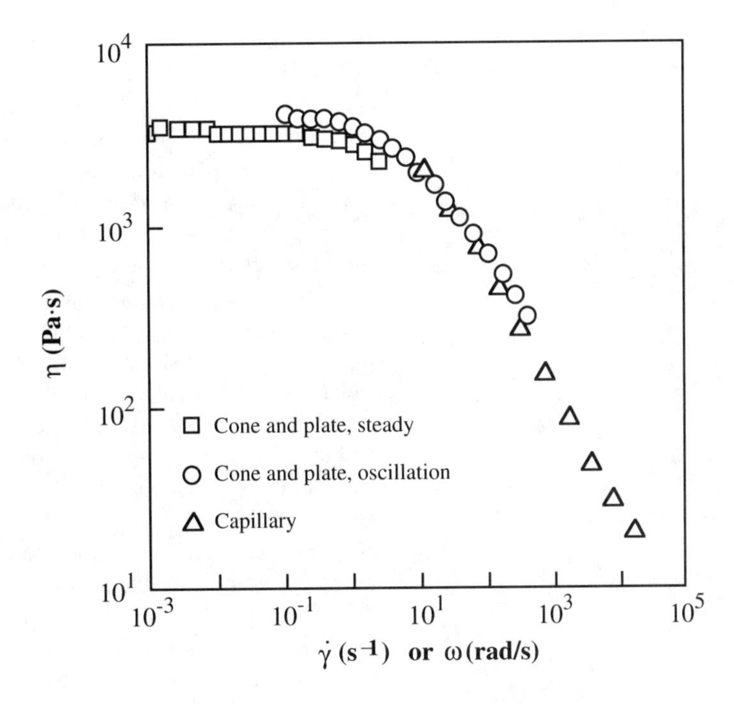

Figure 6.5.5.
Viscosity versus shear rate for polystyrene at 200°C, using capillary, cone and plate, and sinusoidal oscillations with $|\eta^*|(\omega) = \eta(\dot{\gamma})$.

correlations illustrated in Figure 6.5.6 are attractive and worthy of further investigation.

6.6 Summary

Shear rheometers are the primary tool of the experimental rheologist. The rheologist's main job is to pick the material function

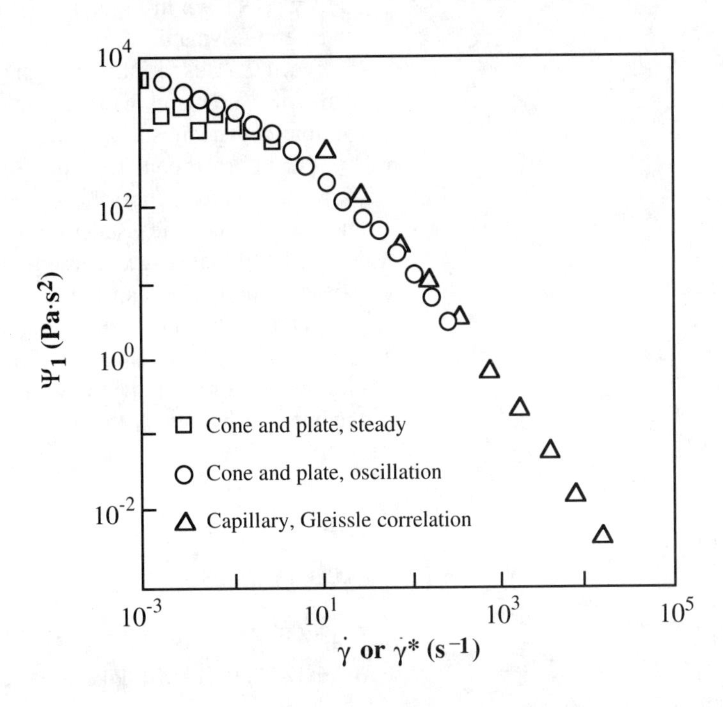

Figure 6.5.6.
First normal stress coefficient by steady cone and plate by sinusoidal oscillation, ($2G'/\omega^2$, eq. 4.2.5), and integrating capillary viscosity data using eq. 6.5.1.

needed for the particular problem and then select the best rheometer. Figure II.3 can help in connecting rheometer to material function. To assist in selecting the best rheometer for the job and to alert the operator for potential errors, Table 6.5.1 is helpful. It summarizes the major advantages and disadvantages of each rheometer.

To study *suspensions*, the first choice is a narrow gap, concentric cylinder rheometer. The outer cylinder should rotate to avoid inertia problems. If there are no settling, large particle, or sensitivity limitations, the cone and plate is a good second choice. For either geometry, stress-controlled instruments (see Figure 8.2.10) provide the lowest shear rate data and best measure of yield stress. Most of the stress-controlled instruments can also do sinusoidal oscillations that allow determination of γ_c and structure breakdown and recovery measures (see Chapter 10).

For *polymer solutions*, the rheologist should start with the cone and plate, unless the concentric cylinder sensitivity is needed. Normal stress data can be collected simultaneously, and the entire range of strain from linear to nonlinear is possible. Temperature control is typically available over a wide range, but solvent evaporation at the edge can cause problems.

For *polymer melts* one would also like to use the cone and plate, but for viscosities exceeding 10^3 Pa·s and high elasticity, edge failure is a severe limitation. Squeezing a viscous sample into the conical shape is also very time-consuming. Thus a better starting point is the parallel plate rheometer with sinusoidal oscillations. Disklike samples can be easily molded or cut from sheets. Pellets or powder can also be used. Within 15 minutes after loading the sample into a preheated instrument, one can obtain more than three decades of data on viscosity and elasticity versus frequency, as shown in Figures 6.5.5 and 6.5.6. One must take care interpreting dynamic data in terms of the steady shear functions (i.e., eq. 4.2.4), especially for highly filled or structured melts. Dynamic measurements with parallel plates are also the first choice for time-dependent studies like curing.

Capillary rheometers don't seem very "high technology," but they can give the highest shear rates, the widest range in shear rate, and the most accurate viscosity numbers. Changing capillary length is the key to that range. These rheometers offer the surest way to prevent evaporation and are the simplest extrusion and die flow simulators. Pressure drop through an orifice die can be used to estimate extensional behavior (Section 7.8). Because of the simplicity, robustness, and process simulation capabilities of capillary rheometer, it is usually the first choice for a processing lab.

References

Bagley, E. B., *J. Appl. Phys.* 1957, *28*, 624.

Baird, D. G.; Read, M. .; Pike, R. D., *Polym. Eng. Sci.* 1986, *26*, 225.

Bird, R. B.; Armstrong, R.; Hassager, O., *Dynamics of Polymeric Liquid, Vol. I: Fluid Mechanics*; 2nd ed.; Wiley: New York, 1987.

Boger, D. V., *Annu. Rev. Fluid Mech.* 1987, *19*, 157.

Boger, D. V.; Denn, M. M., *J. Non-Newtonian Fluid Mech.* 1980, *6*, 163.

Bremner, T.; Rudin, A., *J. Appl. Polym. Sci.* 1990, *41*, 1617.

Broadbent, J. M.; Kaye, A.; Lodge, A. S.; Vale, D. G., *Nature* 1968, *217*, 55.

Chan, T. W.; Pan, B.; Yuan, H., *Rheol. Acta* 1990, *29*, 60.

Choi, S. Y., *J. Polym. Sci.* 1968, *A2, 6*, 2043.

Cox, H. W., Ph.D. thesis, University of Minnesota, 1973.

Cox, H. W.; Macosko, C. W., *AIChE J.* 1974a, *20*, 785.

Cox, H. W.; Macosko, C. W., *SPE Tech. Pap.* 1974b, *20*.

Dealy, J. M., *Rheometers for Molten Plastics*; Van Nostrand Reinhold: New York, 1982.

Duvdevani, I. J.; Klein, I., *SPE J.* 1967, *23:12*, 41.

Ehrmann, G., *Rheol. Acta* 1976, *15*, 8.

Ehrmann, G.; Winter, H. H., *Kunststofftechnik* 1973, *12*, 156.

Frayer, P. D.; Huspeni, P. J., *J. Rheol.* 1990, *34*, 1199.

Geiger, K., Extensive Characterization of Rubber Compounds with Capillary and Rotational Rheometry; presentation to Society of Rheology, Montreal, October 1989.

Geiger, K.; Winter, H. H., *Rheol. Acta* 1978, *17*, 264.

Gent, A. N., *B. J. Appl. Phys.* 1960, *11*, 85.

Giesekus, H.; Langer, G., *Rheol. Acta* 1977, *16*, 1.

Gleissle, W., in *Proceedings of the 10th International Congress on Rheology*, Vol. 1; Uhlherr, P. H. T., Ed.; Sydney, 1988; p. 350.

Grimm, R. J., *AIChE J.* 1978, *24*, 427.

Gleissle, W.; Windhab, E., *Exp. Fluids* 1985, *3*, 177.

Hagen, G. H. L., *Ann. Phy.* 1839, *46*, 423.

Han, C. D., *Trans. Soc. Rheol.* 1974, *18*, 163.

Han, C. D., *Rheology in Polymer Processing*; Academic Press: New York, 1976; Chapter 5.

Han, C. D., in *Rheological Measurements*; Collyer, A. A.; Clegg, D. W., Eds.; Elsevier: London, 1988; Chapter 2.

Hanks, R. W.; Larson, K. M., *Ind. Eng. Chem.* 1979, *18*, 33.

Hanson, D. E., *Polym. Eng. Sci.* 1969, *9*, 405.

Hansen, M. G.; Jansma, J. B., in *Rheology, Proceedings of the Eighth International Congress on Rheology*, Vol 2, Astarita, G.; Marrucci, G.; Nicolais, L., Eds.; Plenum: New York, 1980; p. 193.

Hou, T. H.; Tong, P. P.; deVargas, L., *Rheol. Acta* 1977, *16*, 544.

Kalika, D. S.; Denn, M. M., *J. Rheol.* 1987, *31*, 815.

Keentok, M.; Georgescu, A. G.; Sherwood, A. A.; Tanner, R. I., *J. Non-Newtonian Fluid Mech.* 1980, *6*, 303.

Kurath, S. F.; Larson, W. S., *TAPPI Proceedings*; Boston, May 1990; p. 459.

Kurtz, S. J., in *Advances in Rheology,* Vol. 3; Mena, B.; Garcia-Regjon, A.; Rangel-Nafaile, C., Eds.; Universidal Nacional Autonoma de Mexico, 1984, p. 399.

Laun, H. M., *Rheol. Acta* 1983, *22*, 171.

Laun, H. M., in *Proceedings of the 10th International Congress on Rheology*; Uhlherr, P. H. T., Ed.; Sydney, 1988; p. 37.

Laun, H. M., *Makromol. Chem., Makromol. Symp*. 1992, 56, 55.

Laun, H. M.; Bung, R.; Hess, S.; Loose, W.; Hess, O.; Hahn, K.; Hädicke, E.; Hingmann, R.; Schmidt, F.; Lindner, P., *J. Rheol.* 1992, *36,* 743.

Laun, H. M.; Hirsch, G., *Rheol. Acta* 1989, *28*, 267.

Laun, H. M.; Bung, R.; Schmidt, F., *J. Rheol.* 1991, *35*, 999.

Leblanc, J. L.; Villemaire, J. P.; Vergnes, B.; Aggasant, J. F., *Plastics Rubber Process. Appl.* 1989, *11*, 53.

Lee, S. J.; Denn, M. M.; Crochet, M. J.; Metzner, A. B.; Riggins, G. J., *J. Non-Newtonian Fluid Mech.* 1984, *14*, 301.

Leider, P. J., *Ind. Eng. Chem. Fundam.* 1974, *13*, 342.

Leider, P. J.: Bird, R. B., *Ind. Eng. Chem. Fundam.* 1974, *13*, 336.

Lim, F. J.; Schowalter, W. R., *J. Rheol.* 1989, *33*, 1359.

Lobo, P. F.; Osmers, H. R., *Rheol. Acta* 1974, *13*, 457.

Lodge, A. S., *Elastic Liquids*; Academic Press: New York, 1964; p. 131.

Lodge, A. S., *Chem. Eng. Commun.* 1985, *32*, 1.

Lodge, A. S., *J. Rheol.* 1989, *33*, 821.

Lodge, A. S., in *Rheological Measurement*, Collyer, A. A.; Clegg, D. W., Eds.; Elsevier: London, 1988; p. 345.

Lodge, A. S.; Ko, Y. S., *Rheol. Acta* 1989, *28*, 464.

Lodge, A. S.; deVargas, L., *Rheol. Acta* 1985, *22*, 151.

Marshall, D. I.; Riley, D. W., *J. Appl. Polym. Sci.* 1962, *6*, 546.

McEachern, D. W., *AIChE J.* 1966, *12*, 328.

Meissner, J., *Pure Appl. Chem.* 1975, *42*, 553.

Metzger, A. P.; Knox, J. R., *Trans. Soc. Rheol.* 1965, *9*, 13.

Michaeli, W., *Extrusion Dies, Design and Engineering Computations*; Hanser: Munich, 1984.

Middleman, S., *Fundamentals of Polymer Processing*; McGraw-Hill: New York, 1977.

Mooney, M., *Trans. Soc. Rheol.* 1931, *2*, 210.

Okubo, S.; Hori, Y., *J. Rheol.* 1980. *24*, 275.

Osmers, H. R.; Lobo, P. F., *Trans. Soc. Rheol.* 1976, *20*, 239.

Phan Thien, N.; Sugeng, F.; Tanner, R. I., *J. Non-Newtonian Fluid Mech.* 1987, *24*, 97.

Pike, R. D.; Baird, D. G., *J. Non-Newtonian Fluid Mech.* 1984, *16*, 211.

Poiseuille, L. J., *Comptes Rendus* 1840, *11*, 961 and 1041; 1841, *12*, 112.

Ramamurthy, A. V., *J. Rheol.* 1986, *30*, 337.

Rauwendaal, C.; Fernandez, F., *Polym. Eng. Sci.* 1985, *25*, 765.

Schümmer, P., *Chem.-Ing-Tech.* 1970, *42*, 1239.

Schümmer, P.; Worthoff, R. H., *Chem. Eng. Sci.* 1978, *33*, 759.

Senouci, A.; Smith, A. C., *Rheol. Acta* 1988, *27*, 649.

Skinner, S. J., *J. Appl. Polym. Sci.* 1961, *5*, 55.

Stefan, J., *Sitzungber. K. Akad. Wiss. Math. Natur. Wien* 1874, *69 Part 2*, 713.

Sturges, L. D.; Joseph, D. D., *J. Non-Newtonian Fluid Mech.* 1980, *6*, 325.

Sylvester, N. D.; Chen, H. L., *J. Rheol.* 1985, *29*, 1027.

Tadmor, Z.; Gogos, C. G., *Principles of Polymer Processing*; Wiley: New York, 1979.

Tanner, R. I., *J. Polym. Sci.* 1970, *A-28*, 2067; *Trans. Soc. Rheol.* 1970, *14*, 483.

Tanner, R. I., in *Rheological Measurement*, Collyer, A. A.; Clegg, D. W., Eds.; Elsevier: London, 1988a; p. 93.

Tanner, R. I., *J. Non-Newtonian Fluid Mech.* 1988b, *28*, 309.

Tuna, N. Y.; Finlayson, B. A., *J. Rheol.* 1984, *32*, 285.

Tuna, N. Y.; Finlayson, B. A., *J. Rheol.* 1988, *28*, 79.

Tungare, A. V.; Martin, G. L.; Gotro, J. T., *Polym. Eng. Sci.* 1988, *28*, 1071.

Uhland, E., *Rheol. Acta* 1979, *18*, 1.

Utracki, L. A.; Bakerdjian, Z.; Kamal, M. R., *J. Appl. Polym. Sci.* 1975, *19*, 481.

Van Wazer, J. R.; Lyons, J. W.; Kim, K. Y.; Colwell, R. E., *Viscosity and Flow Measurement*; Wiley-Interscience: New York, 1963.

Villemaire, J. P.; Agassant, J. F., *Polym. Proc. Eng.* 1983–1984, *1*, 223.

Vlachopoulos, J., *Rev. Deform. Behav. Mater.* 1981, *3*, 219.

Vlachopoulos, J.; Mitsoulis, E., *J. Polym. Eng.* 1985, *5*, 173.

Wales, J. L. S.; den Otter, J. L.; Janeshitz-Kriegl, H., *Rheol. Acta* 1965, *4*, 146.

Walters, K., *Rheometry*; Wiley: New York, 1975.

Warren, R. C., in *Rheological Measurement*, Collyer, A. A., Clegg, D. W., Eds.; Elsevier: London, 1988; p. 119.

White, J. L., *J. Appl. Polym. Sci. Symp.* 1973, *20*, 155.

White, J. L., *Principles of Polymer Engineering Rheology*; Wiley: New York, 1990; p. 291.

Wilson, S. J.; Poole, D. *Mater. Res. Bull.* 1990, *25*, 113.

Windhab, E., *Untersuchungen zum rheologischen verhabten konzentrierter Suspensionen*; VDI-Verlag: Dusseldorf, 1986.

Winter, H. H., *Polym. Eng. Sci.* 1975, *15*, 84.

Winter, H. H., *Adv. Heat Transfer* 1977, *13*, 205.

Yao, M.; Malkus, D. S., *Rheol. Acta* 1990, *29*, 310.

7

EXTENSIONAL RHEOMETRY

A variety of pitch which gave by the traction method $\lambda = 4.3 \times 10^{10}$ (poise) was found by the torsion method to have a viscosity $\mu = 1.4 \times 10^{10}$ (poise).
F.T. Trouton (1906)

7.1 Introduction

In Section 2.3 we derived Trouton's important rule that the viscosity in uniaxial extension is three times the shear viscosity for a Newtonian fluid (eq. 2.3.10). Trouton carefully tested his rule using very viscous samples and slow deformation rates (Figure 7.1.1). Although these ingenious experiments with pitch, shoemaker's wax, and glass marked the beginning of extensional rheometry, the field did not advance significantly until the 1970s. Today it is an active area of research. The reason so few extension measurements had been taken in the past is that they are so hard to make. There is so much activity today because it has been recognized that extensional flows exhibit very sensitive variations in structure and are highly relevant to many process flows.

From the extensional material function data shown in Chapters 2 and 4, we know that extensional response can differ very much from shear. For example, at low extension rate tensile viscosity typically obeys Trouton's rule, but at higher extension rates it shows very little of the thinning so common for shear viscosity (Figure 2.1.3). Sometimes even thickening is observed (Figures 4.2.5–4.2.7). Such behavior is not unexpected from structural theories for rodlike suspensions (Chapter 10) and for polymers (Chapter 11).

Figure 7.1.1.
To hold his viscous pitch samples, Trouton forced a thickened end into a small metal box. A hook was attached to the box from which weights were hung. From Trouton (1906).

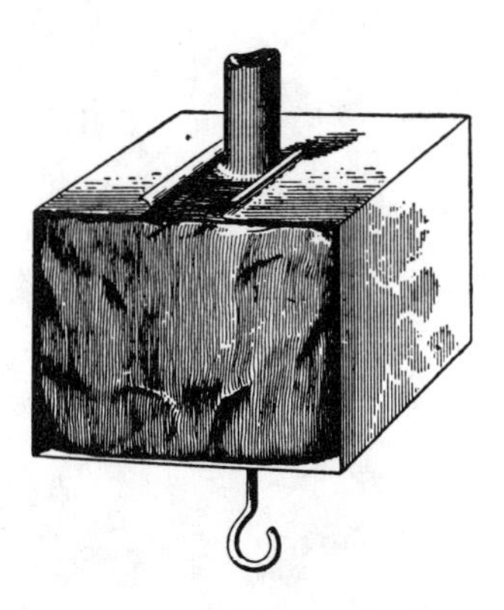

Extensional deformations also play a significant role in many processing operations. For example, fiber spinning, thermoforming, film blowing, blow molding, and foam production are all essentially extensional deformations. Flow in converging or diverging regions of dies and molds as well as flow at the moving front during mold filling can have large extensional components. Extensional material function data are needed to model these flows. Also, because extensional flows strongly orient polymer molecules and asymmetric particles, regions of extensional flow in a particular process can have a strong effect on final product properties.

Although the importance of extensional measurements is well recognized, there are relatively few data available because it is so difficult to generate homogeneous extensional flow, especially for low viscosity liquids. The basic problem is that flow over stationary boundaries results in shear stresses; but without such boundaries it is difficult to control the deformation of a low viscosity fluid. Surface tension, gravity, and inertia conspire to change the deformation.

A further problem arises from the large strains that are often required before stresses in memory fluids can reach their steady straining limit. Streamlines are parallel in shear flows, so that large strains can be achieved by going to long residence times. The streamlines in extensional flow diverge (or converge), which means that a sample must become very thin in one direction to achieve large strain. As indicated in Section 4.2, it is often not possible to reach a steady stress state before the sample ruptures or deforms nonuniformly.

Many different methods have been tried to circumvent these problems and generate purely extensional flows. In this chapter we examine the most successful, the geometries shown in Figure 7.1.2. The first three, simple extension, compression, and sheet stretching, are all, like Trouton's work, adapted from test methods used for solids, particularly those for rubber (Treloar, 1975). They can all give homogeneous, purely extensional deformations, but because their success depends on somehow holding onto the edges of a sample, they have only been used successfully with higher viscosity samples.

The remaining four geometries in Figure 7.1.2 all represent attempts to measure extensional material functions on lower viscosity liquids. Their strengths and weaknesses will be discussed in the following sections.

As with the shear rheometers, we develop the basic working equations to convert measured forces and displacements into stresses and strains. These in turn are used to calculate extensional material functions.

As indicated in Figure 7.1.2, there are several types of extensional deformation; all can be described using the convention given in Chapter 4, which defines two transient extensional viscosities (Meissner, 1985)

$$\mu_1^+(\dot{\epsilon}, t) = \frac{T_{11} - T_{33}}{2(2+m)\dot{\epsilon}} \qquad \text{(4.2.10) or (7.1.1)}$$

$$\mu_2^+(\dot{\epsilon}, t) = \frac{T_{22} - T_{33}}{2(1+2m)\dot{\epsilon}} \qquad (7.1.2)$$

To describe the deformation, a Cartesian coordinate system is chosen such that

$$\dot{\epsilon} = \dot{\epsilon}_{11} \geq \dot{\epsilon}_{22} \geq \dot{\epsilon}_{33} \qquad (7.1.3)$$

and

$$m = \frac{\dot{\epsilon}_{22}}{\dot{\epsilon}_{11}} \qquad (7.1.4)$$

Figure 7.1.2. Extensional flow geometries. As indicated in Figure II.1, only the first three geometries can be used as homogeneous rheometers. The coordinate systems indicated are chosen to give $\dot{\epsilon}_{11} \geq \dot{\epsilon}_{22} \geq \dot{\epsilon}_{33}$. Numbers in parentheses indicate section in which each geometry is discussed.

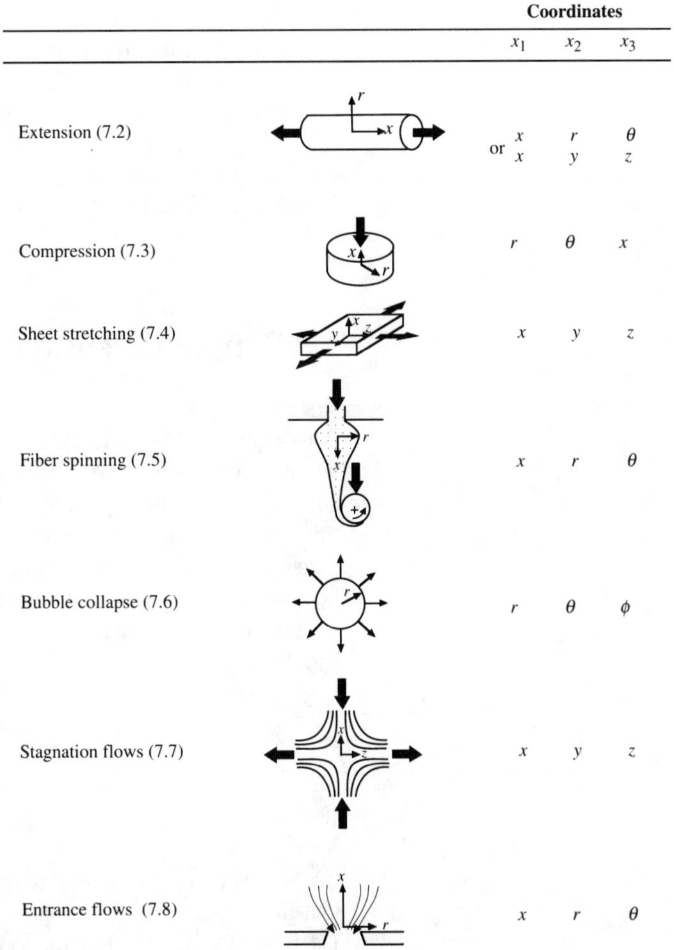

	Coordinates		
	x_1	x_2	x_3
Extension (7.2)	x	r	θ
	or x	y	z
Compression (7.3)	r	θ	x
Sheet stretching (7.4)	x	y	z
Fiber spinning (7.5)	x	r	θ
Bubble collapse (7.6)	r	θ	ϕ
Stagnation flows (7.7)	x	y	z
Entrance flows (7.8)	x	r	θ

For a general extensional deformation then

$$\mathbf{D} = \dot{\epsilon} \begin{bmatrix} 1 & 0 & 0 \\ 0 & m & 0 \\ 0 & 0 & -(1+m) \end{bmatrix} \tag{7.1.5}$$

For simple uniaxial extension $\dot{\epsilon}_{22} = -\dot{\epsilon}_{11}/2$ and from symmetry $\dot{\epsilon}_{22} = \dot{\epsilon}_{33}$ (see Example 2.2.1). Thus $m = -0.5$ and $\tau_{22} = \tau_{33}$, so $\mu_2 = 0$. For equibiaxial extension or compression $\dot{\epsilon}_{11} = \dot{\epsilon}_{22}$ and $\dot{\epsilon}_{33} = -2\dot{\epsilon}_{11}$. Thus $m = 1$ and $\tau_{11} = \tau_{22}$ and $\mu_1 = \mu_2$. For planar extension $\dot{\epsilon}_{11} = -\dot{\epsilon}_{33}$ and $\dot{\epsilon}_{22} = 0$ (see Chapter 2, Example 2.7.1). Therefore $m = 0$, and we have two extensional viscosities. Note that the extensional viscosities in eqs. 7.1.1 and 7.1.2 have been defined in such a way that all give the same linear viscoelastic shear viscosity in the limit of small strain and strain rate. Thus—for example, with uniaxial extension—the factor of 3 (often called the Trouton ratio) that arises between shear and extension (see eq. 2.3.10) is divided out when $m = -0.5$ in eq. 7.1.1.

In the following sections, in addition to giving the working equations for determining these material functions, we discuss corrections, applicability, and limitations for each of the methods depicted in Figure 7.1.2. Further information on extensional rheometry can be found in the references at the end of this chapter. Reviews by Meissner (1985, 1987), a monograph by Petrie (1979), and Dealy's book (1982) are also recommended.

7.2 Simple Extension

The simplest way we might imagine to generate uniaxial extension is to grab a rod of fluid on each end and pull on it. That is what Trouton did for his very viscous pitch and other materials. With special ways of holding the sample and supporting it, we can extend this approach down to viscosities near 10^3 Pa·s. But before we consider how to hold the sample, if we can indeed grab the ends, how should we program them? If we want to generate a *steady* uniaxial extension, then the velocity field in cylindrical coordinates should be

$$v_x = \dot{\epsilon} x \tag{7.2.1}$$

$$v_r = -\frac{1}{2}\dot{\epsilon} r \tag{7.2.2}$$

where $\dot{\epsilon}$ is a constant, the rate of extension. We should note here that the sample cross section need not be circular (see Examples 1.4.1 and 2.2.1). Rods with square or rectanglular cross section may also be used, as indicated in Figure 7.1.2.

To achieve such a velocity field, the sample ends must move with velocity (see Figure 7.2.1)

$$v_{\text{end}} = \frac{\dot{\epsilon} L}{2} \tag{7.2.3}$$

or

$$\frac{dL}{dt} = \dot{L} = \dot{\epsilon} L$$

Integrating from L_o to L, we obtain

$$L = L_o e^{\dot{\epsilon} t} \tag{7.2.4}$$

Thus the length of the sample increases exponentially, or the sample ends must move at velocity

$$v_{\text{end}} = \frac{1}{2}\dot{\epsilon} L_o e^{\dot{\epsilon} t} \tag{7.2.5}$$

Figure 7.2.1.
A cylindrical sample being pulled at each end by a force f with a velocity v_{end}.

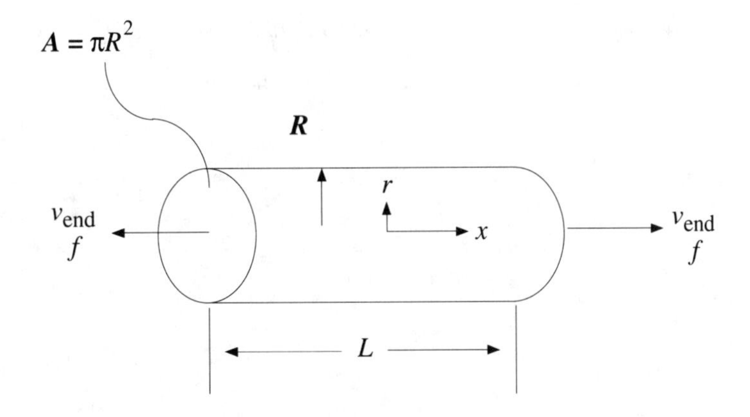

The strain in the sample is just

$$\epsilon = \dot{\epsilon}t = \ln\frac{L}{L_o} \tag{7.2.6}$$

This logarithmic strain measure is sometimes called the Hencky strain (1924). In the limit of small strain, it is the same as the usual ("engineering") strain measure $L/L_o - 1$.

The stress causing the sample to elongate is the normal stress difference $T_{xx} - T_{rr}$. If we ignore surface tension and other factors like gravity for the moment, then this stress is the force per unit area acting on the end of the sample (note Figure 7.2.1)

$$T_{xx} - T_{rr} = \frac{f}{A} \tag{7.2.7}$$

Of course the area and thus the force are changing with time. For an incompressible material, the sample volume is conserved, $\pi R_o^2 L_o = \pi R^2 L$; thus

$$A(t)L(t) = \pi R_0^2 L_o \tag{7.2.8}$$

or

$$A(t) = \pi R_o^2 e^{-\dot{\epsilon}t}$$

Thus the stress difference becomes

$$T_{xx} - T_{rr} = \tau_{11} - \tau_{22} = \frac{f e^{\dot{\epsilon}t}}{\pi R_o^2} \tag{7.2.9}$$

From the equations for stress and rate we can readily define the extensional viscosity following eq. 7.1.1 with $m = -0.5$.

$$\mu_1^+ = \frac{\tau_{11} - \tau_{22}}{3\dot{\epsilon}} = \frac{\eta_u^+}{3} \tag{7.2.10}$$

TABLE 7.2.1 / Working Equations for Uniaxial Extension

Tensile strain

$$\epsilon = \ln L/L_0 \qquad \text{(moving clamps)} \tag{7.2.6}$$

$$\epsilon = \frac{\theta R_c}{L_o} \qquad \text{(rotating clamp)} \tag{7.2.12}$$

Strain rate

$$\dot{\epsilon} = \frac{1}{L}\frac{dL}{dt} \quad \text{or} \quad \frac{2v_{\text{end}}}{L} \qquad \text{(moving clamps)} \tag{7.2.3}$$

$$\dot{\epsilon} = \frac{\Omega R_c}{L_o} \quad \Omega = \frac{d\theta}{dt} \qquad \text{(rotating clamp)} \tag{7.2.11}$$

Tensile stress

$$T_{xx} - T_{rr} = \frac{f e^{\dot{\epsilon}t}}{\pi R_o^2} \ \text{(circular)} \tag{7.2.9a}$$

$$T_{xx} - T_{yy} = \frac{f e^{\dot{\epsilon}t}}{y_o z_o} \ \text{(rectangular cross section)} \tag{7.2.9b}$$

Uniaxial extensional viscosity

$$\eta_u^+ = \frac{T_{xx} - T_{rr}}{\dot{\epsilon}} = 3\mu_1^+ \tag{7.2.10}$$

Errors
Temperature gradients
Density mismatch with surrounding fluid
Uptake of buoyancy fluid
Surface tension at low $\dot{\epsilon}$ (eq. 7.2.14)
End effects
  Moving clamps–necking at bonded clamp
  rotating clamp–slip
Sample inhomogeneity

Utility
Uniaxial extension very sensitive to macromolecular and microstructural factors relevant for fiber spinning, entrance flows
Homogeneous deformation
Highest extensional strain: $\epsilon = 3$–4 typical with careful sample preparation and 6–7 possible with rotating clamps
Buoyancy bath required for $\dot{\epsilon}\lambda > 1$ (eq. 7.2.13)
Need a solid sample to load
-$\eta \geq 10^3$ Pa·s due to gravity (density mismatch), surface tension, low forces, and drag on buoyancy fluid
Rotating clamp with horizontal bath easiest to build and use; must cut and measure length to get recovery
Movable clamps permit smaller samples and recovery versus time; but lower total strain, longer bath, usually must bond samples

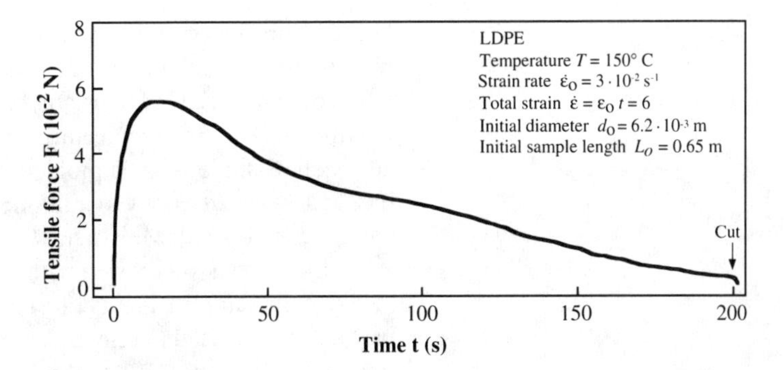

Figure 7.2.2. Recording of the tensile force as a function of time for a constant strain rate $\dot{\epsilon}_o$ test. From Laun and Munstedt (1976).

Note that the Trouton ratio of 3 is already included when μ_1 is used to define uniaxial extensional viscosity, in contrast to η_u used in Chapter 2 (eq. 2.3.9) and frequently in the literature.

As discussed in the preceding section, $\dot{\epsilon} = \dot{\epsilon}_{11}$ and $\dot{\epsilon}_{22} = \dot{\epsilon}_{33}$, and since $\tau_{22} = \tau_{33}$, there is only one viscosity function in uniaxial extension. Table 7.2.1 summarizes the working equations for uniaxial extensional rheometry.

A typical force versus time trace is shown in Figure 7.2.2. According to eq. 7.2.9, to determine tensile stress this force must be multiplied by the exponentially increasing length and divided by area. If we want to maintain constant stress on a sample, the product of force and length must be constant. Servo-controlled rheometers have been developed to increase sample length exponentially or to maintain constant stress (Munstedt, 1979; Au Yueng and Macosko, 1980). Further discussion of deformation and stress control in extensional rheometers is given in Chapter 8.

7.2.1 End Clamps

It is easy to say "make a rod of the liquid sample and pull on its ends." It is quite another matter to actually achieve this deformation in practice. A number of methods for holding the ends have been tried. As illustrated in Figure 7.1.1, Trouton (1906) thickened the ends. Connelly, Garfield, and Pearson (1979) used a blob of epoxy to thicken the ends of molten polymer rods. Other workers (Cogswell, 1968; Dealy et al., 1976) have used water-cooled collars, but these can lead to unacceptable temperature gradients. Shaw (1976) pulled ring-shaped samples like those shown in Figure 7.2.3. Mechanical clamps, like those used for rubber, are successful for rectangular samples (Muller and Froelich, 1985). Sridhar et al. (1991) have shown that for strongly extensional-thickening materials, samples can be simply confined between two plates and pulled rapidly. This method has strong end effects but is the easiest to use and has been applied to polymer solutions with shear viscosities of about 30 Pa·s.

Figure 7.2.3. Samples for uniaxial extension testing using translating clamps: (a) ring shape and (b) cylindrical sample bonded with adhesive to a metal base.

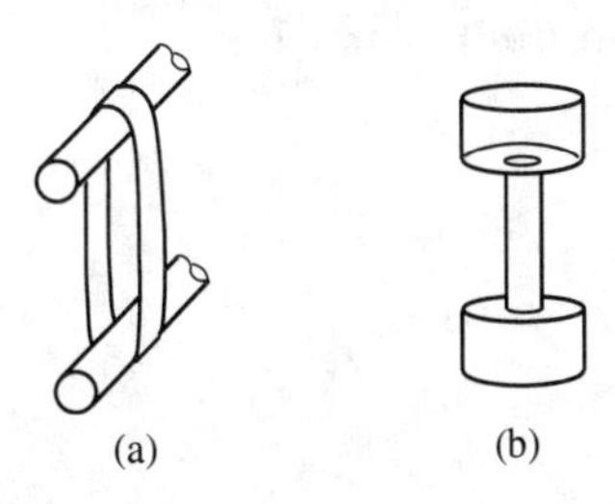

The best method for holding polymer melts over a range of deformation rates appears to be to bond the ends of the sample to a metal clip. Bonding is done in the solid state. Laun and Munstedt

(1978) found epoxy adhesives hold up to 190°C. Garritano (1982) reports success with polyimide adhesives in the 200–225°C range. For certain materials such as polyolefins, it is necessary to etch the polymer surface with concentrated sulfuric acid to achieve good adhesion. Samples can be molded under vacuum to avoid air bubbles and annealed slowly to minimize frozen-in stresses (Au Yueng and Macosko, 1980; Meissner, 1985).

End effects occur even with adhesive-bonded samples. Near the bond the sample cannot deform; see Figure 7.2.4. As a result of partially compensating effects, however, this error is not too large. Franck and Meissner (1984; Meissner, 1985) found that in creep the error in strain was only 3% at $\epsilon = 2.2$ for a cylindrical polystyrene sample approximately 6 mm × 30 mm. At larger strains the error will increase, but the control servo can be programmed to correct for it by using actual diameter (or length) measurements from a cut taken out of the control part of an earlier sample.

7.2.2 Rotating Clamps

An alternate method for pulling samples is the rotating clamp, developed by Meissner (1969) and further improved by Meissner and co-workers (1971, 1981) and Li et al. (1990). The sample is held between two pairs of rotating gears or cylinders, as indicated in Figure 7.2.5. A simpler alternative is to fix one end by gluing or other means while the free end is pulled between rotating gears or wrapped around a rotating rod (Macosko and Lorntson, 1973; Everage and Ballman, 1976; Ide and White, 1976; Laun and Munstedt, 1978; Connelly et al., 1979, Ishizuka and Koyama, 1980).

It is much easier to program a constant extension rate with the rotating clamp than with the translating clamp. Since we are always testing a constant length sample L_o from eq. 7.2.1, the velocity of

Figure 7.2.4. Test of uniformity in the drawing of a low density polyethylene sample at 150°C in creep: Initial sample length $l_o = 10$ mm; initial sample diameter $d_o = 6$ mm; stretching ratio $\lambda = l/l_o = 40$. From Laun and Munstedt (1978).

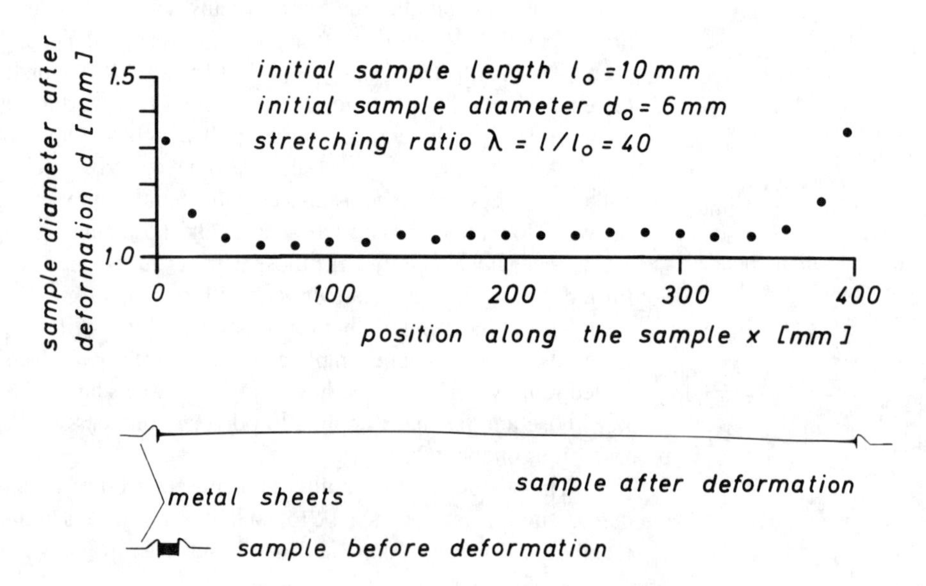

Figure 7.2.5.
Schematic drawing of the double rotating clamp apparatus (adapted from Meissner, 1971). (a) Top view. The cylindrical sample P of initial length L_o is stretched by clamps Z_1 and Z_2 rotating at Ω_1 and Ω_2. Drive motor M_1 is shown. Spring B and displacement sensor W measure force. (b) A detail of the rotating clamps. Gear teeth are used to prevent slip. (c) End view showing oil buoyancy bath with top surface O. Fourteen pairs of scissors T cut the sample into small lengths L_A for recovery measurement L_R.

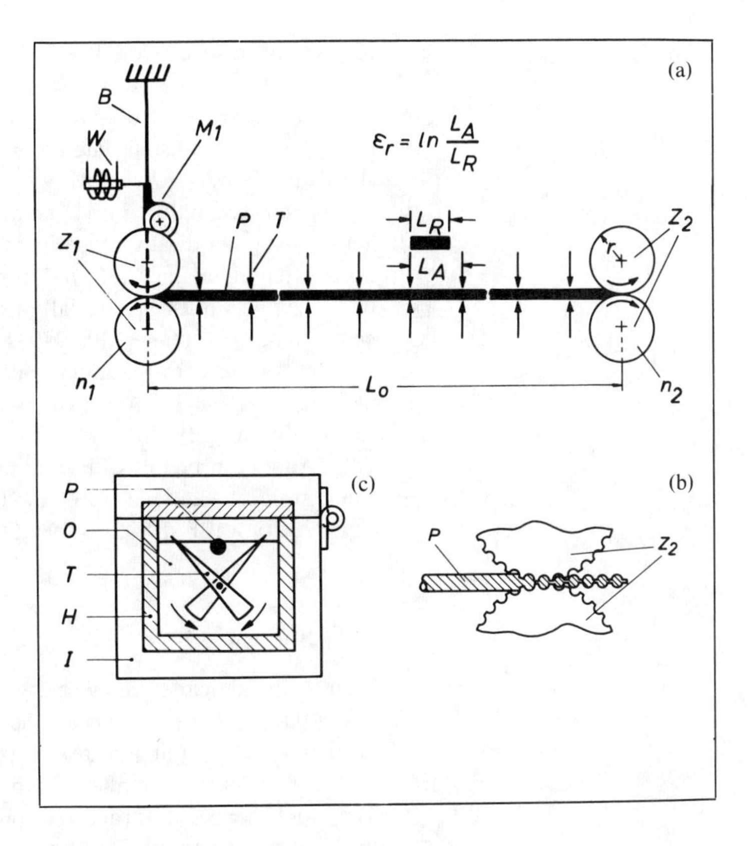

the end of the sample will be a constant, proportional to Ω, the angular velocity, and the radius, R_c, of the rotating clamp. Thus the strain rate is

$$\dot{\epsilon} = \frac{\Omega R_c}{L_o} \tag{7.2.11}$$

or

$$\dot{\epsilon} = \frac{(\Omega_1 + \Omega_2) R_c}{L_o}$$

if there are rotating clamps at each end of the sample. The strain is proportional to the angular rotation θ

$$\epsilon = \frac{\theta R_c}{L_o} \tag{7.2.12}$$

The area still decreases exponentially, eq. 7.2.8; thus eq. 7.2.9 applies for the stress.

In addition to providing a simpler control for achieving constant extension rate, another advantage of the rotating clamp is that the overall apparatus can be shorter. To reach a strain of 4, the translating clamp machine must stretch a 10 mm long sample to 550 mm. With a rotating clamp, the length remains constant. Raible et al.

(1979) have reached strains of 7, more than a 1000-fold extension, with rotating clamps (see below, Figure 7.2.7).

A disadvantage of rotating over translating clamps is that tests other than constant rate are more difficult. For example, as shown in Figure 7.2.5, recovery can be measured only by cutting the extended sample with scissors and measuring the recovered length. Another disadvantage is that longer samples are required for the rotating clamps: 70–700 mm versus 10 mm. For accurate work, samples must be specially extruded to provide a homogeneous sample (Raible et al., 1979; Meissner et al., 1981). Shorter samples for translating clamp devices can be vacuum compression molded and annealed to provide samples free from bubbles and stress (Meissner, 1985).

Another problem with rotating clamp devices is slip at the clamp surface. Meissner (1969, 1981, 1985) reports that the sample end velocity was 8–10% less than ΩR_c.

7.2.3 Buoyancy Baths

Trouton found that to achieve uniform deformation with lower viscosity materials, $\eta_o < 10^7$ Pa·s, he needed to support the sample in a low viscosity liquid (salt water) of the same density. Connelly, et al. (1979) studied samples drawn out horizontally in an air bath. They could not get uniform extension when the extension rate times the fluid's longest relaxation time was less than 1

$$\dot{\epsilon}\lambda_o < 1 \tag{7.2.13}$$

Thus for lower rate testing some sort of buoyancy bath is required. Figure 7.2.5 indicates a horizontal buoyancy bath of the type most often used with rotating clamp instruments. Both horizontal (Vinogradov et al., 1970, Franck and Meissner, 1984) and vertical baths (Munstedt, 1975, 1979) like the one shown in Figure 7.2.6 are used with translating clamp rheometers.

Matching density is less difficult with a horizontal bath. The sample merely must be less dense than the fluid. Typically, dimethyl and phenyl silicone oils and perfluorinated polyethers are used for higher density. Clearly the oil must not diffuse into the sample. A sensitive test for diffusion is actually to watch for changes with time in rheological properties like η_o. Other techniques such as infrared have been used (Munstedt, 1979).

At low stress levels the force due to interfacial tension, Γ, between the oil and sample can become appreciable. Equation 7.2.7 for the stress becomes

$$T_{11} - T_{22} = \frac{f}{A} - \frac{\Gamma}{R}\left(1 - \frac{2R}{L}\right) \tag{7.2.14}$$

Interfacial tension works against the pulling force. Typically, interfacial tension is 5 mN/m, so the correction is small in polymer

Figure 7.2.6.
Schematic of extensional rheometer with a translating clamp and vertical buoyancy control bath. The temperature control fluid is circulated through a jacket around the buoyancy fluid. An outer vacuum jacket insulates the apparatus. Redrawn from Munstedt (1979).

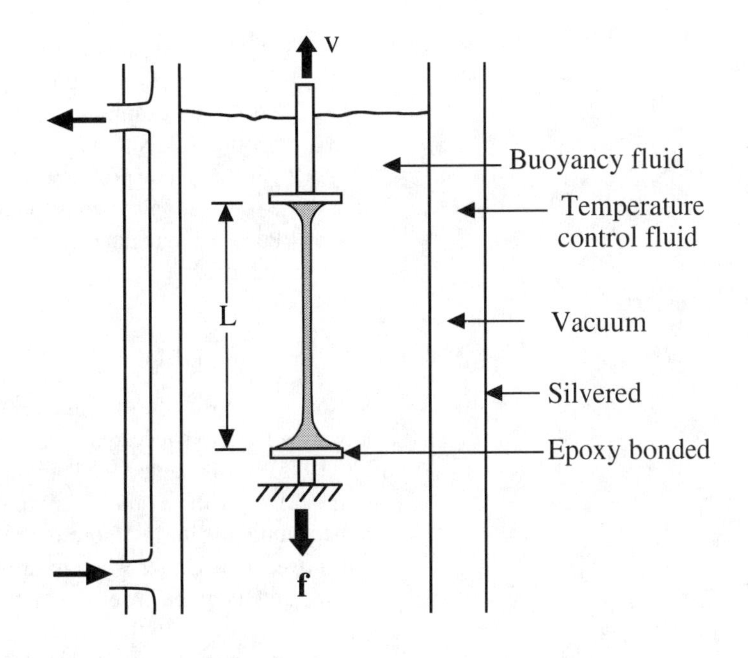

melts. Laun and Munstedt (1978) report that surface tension influenced their results on low density polyethylene for $\dot{\epsilon} < 10^{-4}\text{s}^{-1}$. Density mismatch may be a greater source of error in vertical bath instruments. With rotating clamps the $2R/L$ term in eq. 7.2.14 drops out.

With good density match, and temperature control, it is possible to get very uniform sample extension over the length of a vertical bath, as illustrated in Figure 7.2.4. Figure 7.2.7 shows that small temperature gradients in a horizontal bath can significantly affect data at extremely large strains. This figure also shows that for some polymer melts, even a strain of 7 is not sufficient to achieve steady extension.

Figure 7.2.7.
Nominal tensile stress versus strain for a low density polyethylene at 150°C: rotating clamps, $\dot{\epsilon}_o = 0.035^{-1}$. The highest curve shows the effect of small temperature gradients (> 0.1°C/10 cm) at high strain. With excellent control of temperature, data are reproducible to within the shaded region. From Meissner et al. (1981).

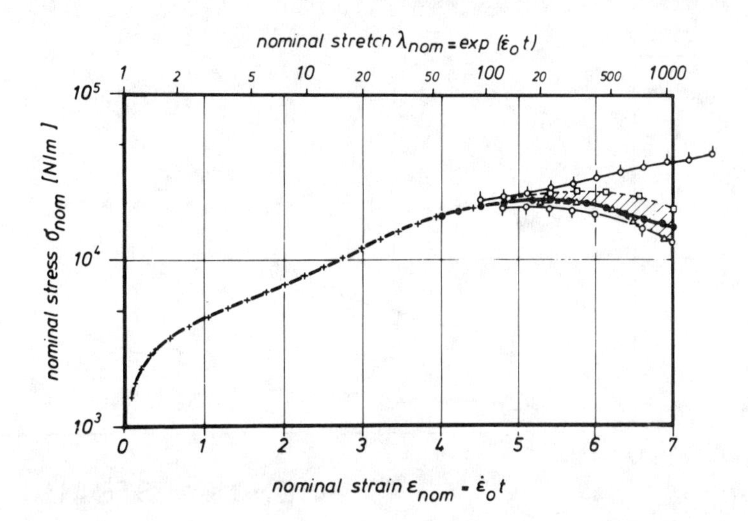

When the foregoing precautions are taken, both moving and rotating clamp devices give good agreement for η_u^+, as shown in Figure 7.2.8. Other examples of uniaxial extensional viscosity are shown in Section 4.2. Surface tension, density differences, and possibly diffusion and drag problems with the buoyancy fluid appear to limit these rod pulling methods to relatively high viscosity liquids, $\eta > 10^3$ Pa·s. The advantages and limitations of the methods are summarized in Table 7.2.1.

7.2.4 Spinning Drop

For lower viscosity fluids, an alternate to rod pulling is to use a buoyancy fluid to squeeze the sample radially. Hsu and Flummerfelt (1975) have adapted the spinning drop tensiometer (Joseph et al., 1992) shown in Figure 7.2.9 to extensional measurements. If the surrounding fluid is more dense, $\rho_2 > \rho_1$, then the test fluid will move to the center when rotation starts. It will elongate until interfacial tension balances the inertial forces.

During start-up, the tensile stress difference in the test fluid will be a balance of inertial, interfacial, and drag forces

$$T_{11} - T_{22} = \frac{(\rho_2 - \rho_1)\Omega^2 R^2}{2} + \frac{\Gamma}{R}\left(1 - \frac{R}{R_{\text{end}}}\right) + 3\eta_2 \bar{\dot{\epsilon}} \quad (7.2.15)$$

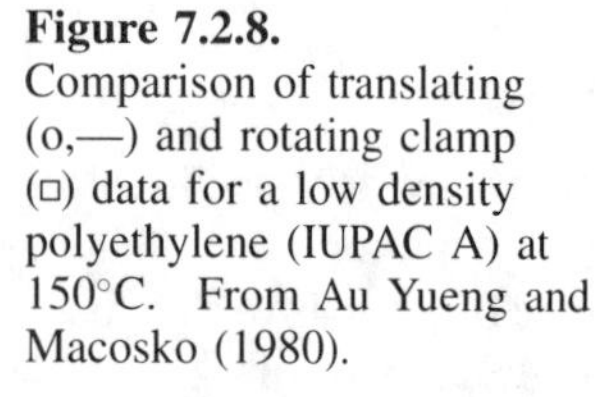

Figure 7.2.8. Comparison of translating (o,—) and rotating clamp (□) data for a low density polyethylene (IUPAC A) at 150°C. From Au Yueng and Macosko (1980).

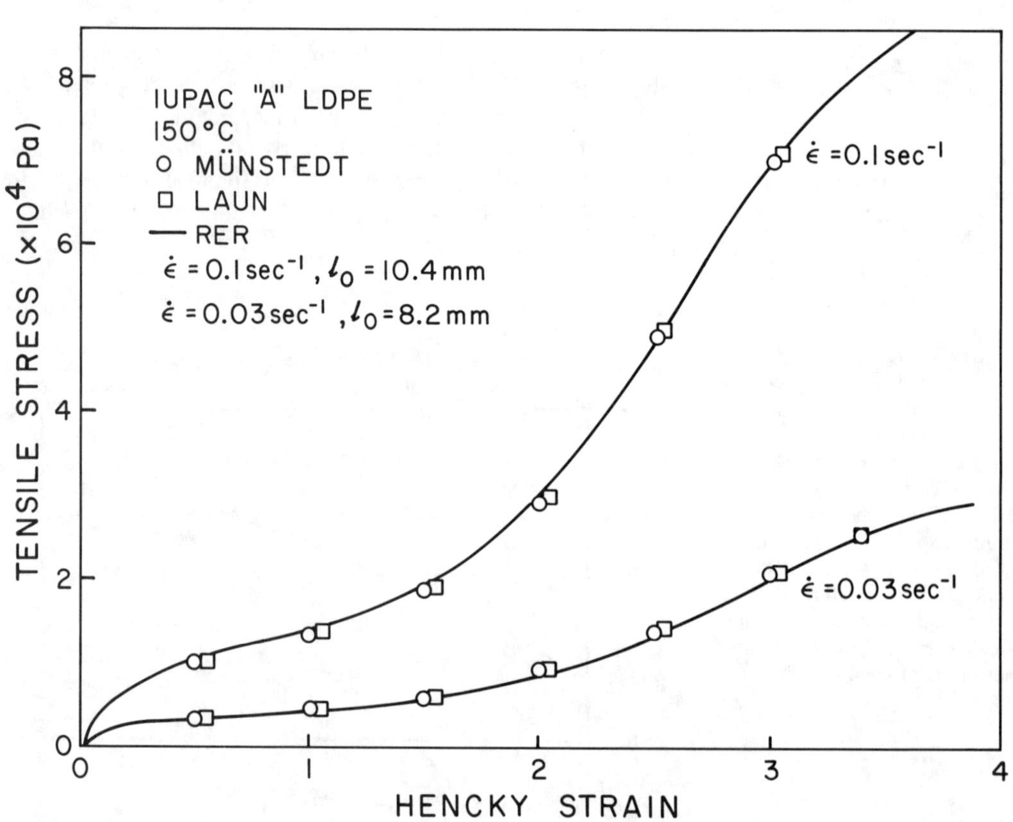

Figure 7.2.9.
Schematic of spinning drop apparatus. The shape change that occurs after rotation starts is an extensional deformation.

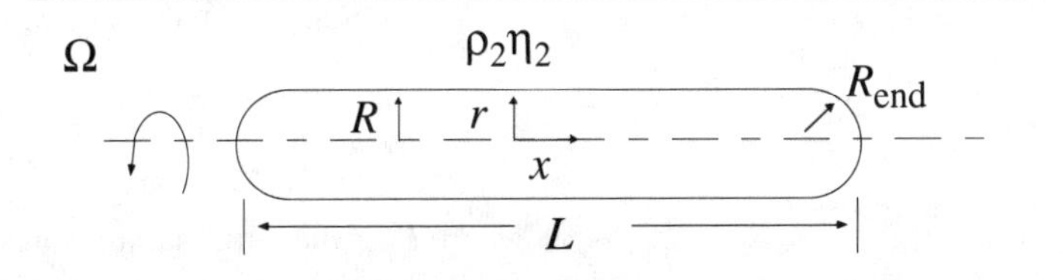

where $\bar{\dot{\epsilon}}$ represents an average extension rate in the surrounding (Newtonian) fluid at the interface, and Γ is the interfacial tension. Even when Ω is increased suddenly, the drop radius typically does not decrease exponentially; thus the deformation is not that of steady extension. However, Hsu and co-workers (1977) were able to achieve constant extension rate and constant stress tests by using a photocell system for feedback control on the drop length. Their results on two polymer solutions, $\eta_o \approx 10^3$ Pa·s, were limited to low extension rates, $\dot{\epsilon} < 0.03\ \text{s}^{-1}$, and low total strain, $\epsilon < 0.6$. Further work is needed to define the utility of this method.

7.3 Lubricated Compression

In shear if we reverse the shear direction, we obtain the same material functions. The same is not true for uniaxial extension. In general, the normal stress difference needed to generate uniaxial *compression* (or equibiaxial extension) is not merely the negative of the stresses needed to generate uniaxial *extension*. Equibiaxial extension flow is not as strongly orienting as it is uniaxial. This fact is most clearly demonstrated with dilute suspensions of rodlike particles in Chapter 10 (Figure 10.5.3). It is valuable to measure this effect, which can be used to test constitutive equations (e.g., Papanastasiou et al., 1983; Khan et al., 1987). Biaxial flow is also relevant to sheet forming of polymers and compression molding.

In principle, one way to generate uniaxial compression or equal biaxial flow would be to reverse the rod pulling experiment described in Section 7.2. However, if the initial cylinder is long and thin, it will buckle under compression. Buckling can be eliminated by going to smaller L/R, but then shear from the fixed ends will contribute significantly to the total force. One way to eliminate this shear is to lubricate both ends of the sample with a lower viscosity liquid. This approach has been used successfully to obtain biaxial viscosity data (Chatraei et al., 1981). Figure 7.3.1 shows the motion of a tracer parallel to the z axis in a polydimethylsiloxane (PDMS) melt. As the rod deforms, the tracer moves out parallel to the z axis, becoming shorter and fatter. A cylindrical weight was placed on the sample and the weight and sample put into a silicone oil bath. The same results are obtained when the lubricant is simply coated on the sample ends (Chatraei, 1981). This simpler method (see Figure 7.3.2) is the most commonly used (Soskey and Winter, 1985; Khan et al., 1987).

In compression the *maximum* strain component is

$$\epsilon = \epsilon_{rr} = -\frac{1}{2}\ln\left(\frac{L(t)}{L_o}\right) \tag{7.3.1}$$

where L_o is the initial sample thickness as indicated in Figure 7.3.2. Following the convention given in eq. 7.1.3, we must choose a coordinate system with $\mathbf{r}$ in the x_1 direction so that $\epsilon_{rr} = \epsilon_{11} = \epsilon_{22}$. This is different from the uniaxial case (see Figure 7.1.2 or compare Figures 7.3.2 and 7.2.1). It follows that

$$\dot{\epsilon} = -\frac{1}{2L}\frac{dL}{dt} \tag{7.3.2}$$

The normal stress difference that drives the flow is

$$T_{rr} - T_{xx} = T_{11} - T_{33} = \frac{f}{\pi R^2} \tag{7.3.3}$$

If the plates are only partially filled, then

$$T_{11} - T_{33} = \frac{f}{\pi r^2(t)} \tag{7.3.4}$$

Figure 7.3.1.
Lubricated squeezing of a PDMS sample, η_o = 2.7×10^4 Pa · s; lubricant viscosity = 0.45 Pa · s. A vertical, black ink line was embedded at $\approx R_o/2$. L_o = 30 mm, R_o = 28.6 mm. From Chatraei et al. (1981).

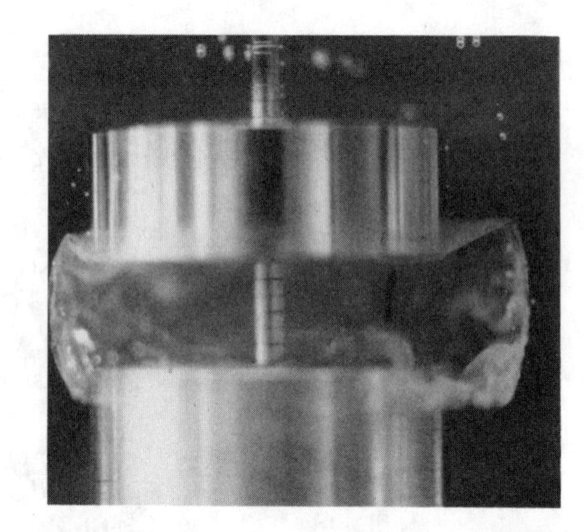

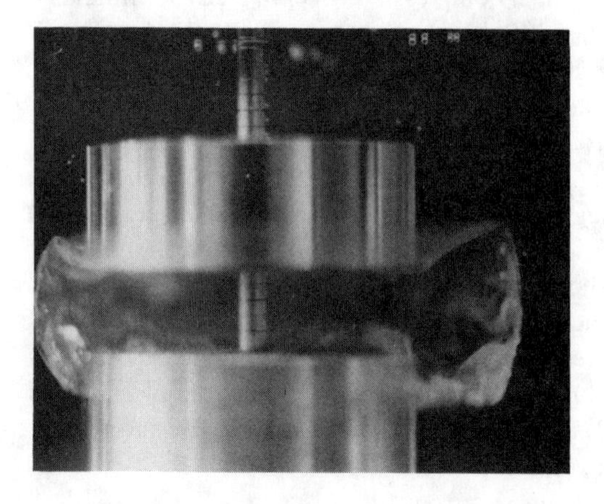

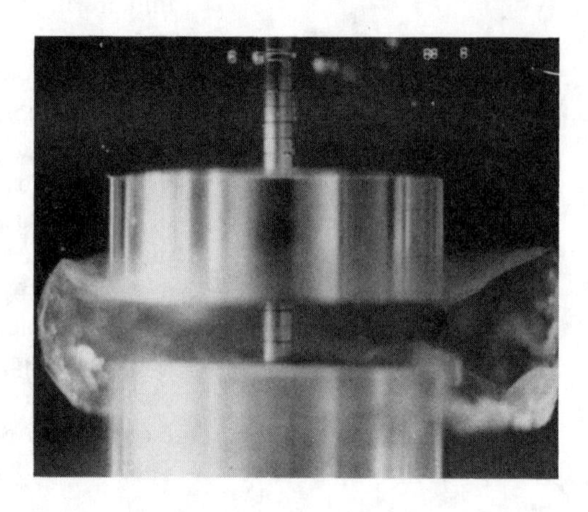

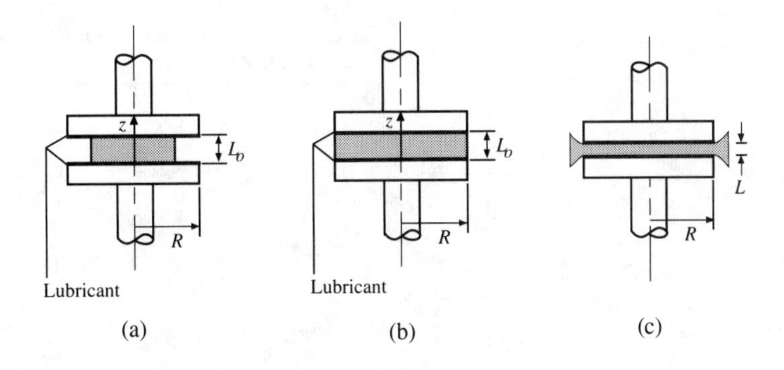

Figure 7.3.2.
Schematic of lubricated squeezing geometry: (a) the initial sample disk is smaller than plate radius; (b) sample and plate radius; are the same.

The varying sample radius can be determined from the thickness, assuming a constant volume sample

$$r(t) = r_o \sqrt{\frac{L_o}{L(t)}} \tag{7.3.5}$$

Following eq. 7.1.1, $m = 1$ and the equibiaxial viscosity becomes

$$\mu_1^+ = \frac{T_{11} - T_{33}}{6\dot{\epsilon}} = \frac{\eta_b^+}{6} \tag{7.3.6}$$

The working equations for lubricated compression are summarized in Table 7.3.1.

Figure 7.3.3 plots ϵ versus t for the sample used in Figure 7.3.1. We see that after a short transient, the lubricated sample creeps at constant rate, indicating steady equibiaxial flow. The departure at $\epsilon \sim 1$ indicates loss of lubricant. Also shown in Figure 7.3.3 is an unlubricated sample at about the same stress. It squeezes much more slowly, and the height versus time is well described by the Stefan equation (eq. 6.4.8).

Loss of lubrication at $\epsilon \sim 1$ is typical. Papanastasiou et al. (1986) analyzed lubricated squeezing of two Newtonian fluids and found limiting regimes for the flow behavior shown in Figure 7.3.4. They showed that lubricant thickness δ decreases approximately with the square root of the gap

$$\frac{\delta}{\delta_o} \approx \left(\frac{L}{L_o}\right)^{1/2} \tag{7.3.7}$$

Secor (1988) extended their work and established criteria for lubrication

$$\frac{2\delta}{L} < \frac{\eta_L R^2}{\eta \delta^2} < 20 \tag{7.3.8}$$

where η_L and η are the Newtonian shear viscosities of the lubricant and sample, respectively.

TABLE 7.3.1 / Working Equations for Compression (Equibiaxial Extension)

Strain

$$\epsilon = -\frac{1}{2}\ln\left(\frac{L(t)}{L_o}\right) \qquad (7.3.1)$$

(assuming perfect lubrication of sample ends)

Strain rate

$$\dot{\epsilon} = -\frac{1}{2L}\frac{dL}{dt} \qquad (7.3.2)$$

Stress

$$T_{rr} - T_{xx} = \frac{f}{\pi R^2}$$

$$= \frac{f}{\pi r^2(t)} \quad \text{(if partially full)} \qquad (7.3.3)$$

Equibiaxial viscosity

$$\eta_b^+ = \frac{T_{11} - T_{33}}{\dot{\epsilon}} = 6\mu_1^+ \qquad (7.3.6)$$

Errors

Lubricant viscosity must satisfy eq. 7.3.8
Always loses lubrication as δ decreases during squeezing
Stresses at edge
Influence of δ in L measurement, especially if lubricant pumped in

Utility

Biaxial extension:
  Different material function from uniaxial
  Less strongly orienting, less sensitive to structure
  Relevant for molding, foaming processes
Simple experiment: fast, small sample, easy to prepare; broad temperature range
Step strain allows exploration of nonlinear viscoelasticity
Planar flow can be done with same apparatus (see Figure 7.3.7)
Limited strain, $\epsilon < 1.5$, without external lubricant supply
Need solid sample for simple loading
-$\eta > 10^2$ Pa·s

If the lubricant viscosity is too low or the layer is too thick, it will rapidly squeeze out, according to the profile in Figure 7.3.4a. It then goes into the desired biaxial mode (Figure 7.3.4b), but eventually δ becomes thin enough (following eq. 7.3.7) that the upper limit is reached (Figure 7.3.4c). Here the flow is no longer purely equibiaxial. This point occurs at $\epsilon \sim 1$ for typical conditions: δ_o = 0.1 mm, L_o = R = 5 mm, and $\eta_L/\eta \sim 10^{-3}$.

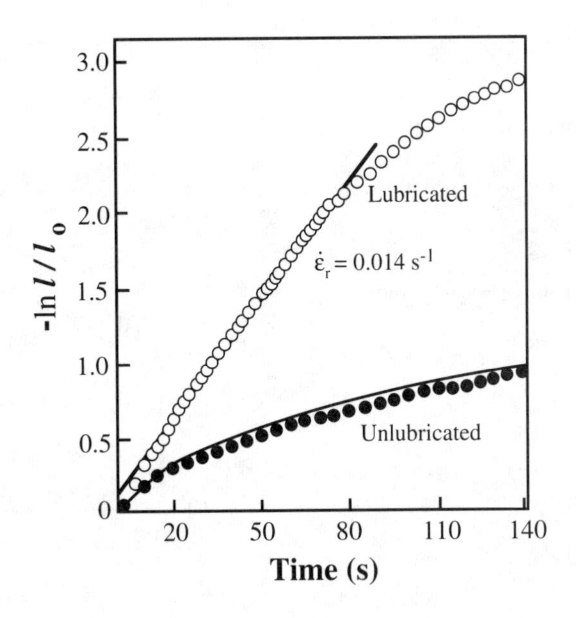

Figure 7.3.3. Typical creep, $-\frac{1}{2}\epsilon = y_2 \ln(L/L_o)$ versus time for the PDMS of Figure 7.3.1: (open points, ○) lubricated experiment f/A = 2.2. kPa, L_o = 13.13 mm; (solid points, •) unlubricated experiment f/A = 2.0 kPa, L_o = 17.01 mm; curves show Newtonian theory for unlubricated squeezing flow. Adapted from Chatraei et al. (1981).

Secor (1988) showed that by replenishing the lubricant layers from the centers of the two plates using a double-syringe pump, maximum strains could be extended to $\epsilon \sim 2.5$. Here strain becomes limited by accurate determination of the lubricant layer thickness, which must be subtracted from $L(t)$. Sample thickness becomes very small at large strain: $\epsilon = 2.5\ L/L_o = 0.007$.

Over a range of about 10^2 in stress, Chatraei et al. were able to reach steady biaxial extension ($\dot{\epsilon} = 0.003$–$1.0\ \text{s}^{-1}$). Thus a steady state biaxial viscosity η_b can be plotted versus normal stress, Figure 7.3.5. When this is compared to the data for shear viscosity versus shear stress, we see good agreement with $6\eta_o$ at low stresses. There is also reasonable agreement between the four different diameters used, indicating no strong edge effects. Note that the biaxial viscosity shows thinning but at a higher stress level than for shear.

To perform constant rate squeezing rather than constant stress requires programming the gap to close at an exponentially decreasing rate, eq. 7.2.4. Soskey and Winter (1985) have done this. They were able to get the linear viscoelastic limit, but for $\epsilon > 1$ they found it difficult to determine whether they had strain hardening or simply loss of lubrication. Isayev and Azari (1986) did the simpler constant velocity squeezing experiments. They calculated a biaxial viscosity from their force versus time curves using a differential constitutive model and found behavior very similar to Figure 7.3.5 for a polybutadiene gum ($\eta_o \sim 10^7$ Pa·s).

It is also possible to do step squeezing experiments, as indicated in Figure 7.3.6. Similar to the results of constant stress and constant rate tests, lubrication is lost at $\epsilon < 2$. Secor suggests that the ratio that controls lubrication is $\eta_L \dot{\epsilon} R^2/G\delta^2$, where $\dot{\epsilon}$ is the compression rate during the step squeezing and G is the relaxation modulus at the start of the relaxation process.

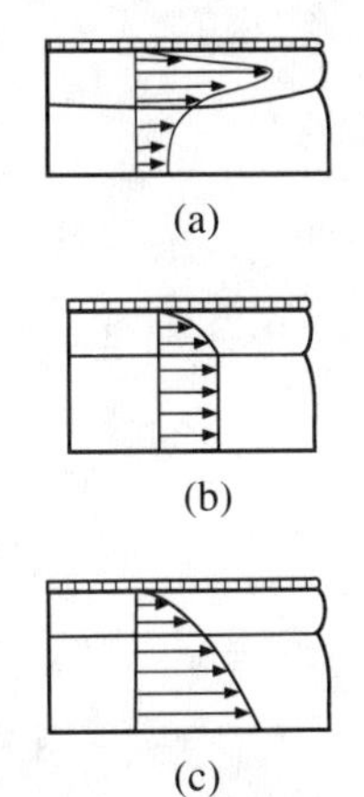

Figure 7.3.4. Velocity profile in lubricated squeezing depends on the ratio $\eta_L R^2/\eta\delta^2$ (eq. 7.3.8). (a) Ratio too small; lubricant is rapidly squeezed out. (b) The ideal situation. (c) Ratio too large; there will be shear as well as extension in the sample. From Papanastasiou et al. (1986).

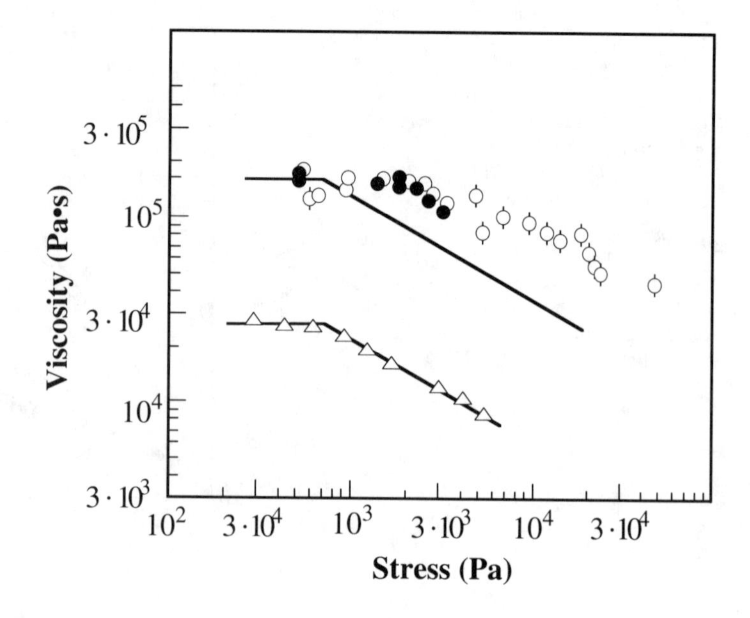

Figure 7.3.5.
Shear and biaxial viscosities versus stress at 25°C for the PDMS of Figure 7.3.1: (△), Shear viscosities (by cone and plate). Biaxial extensional viscosities (ϕ) as follows: R = 12.7 mm, (○) R = 25.4 mm, (•) R = 28.6 mm, (◇) R= 63.5 mm; (–) 6η. Replotted from Chatraei (1981).

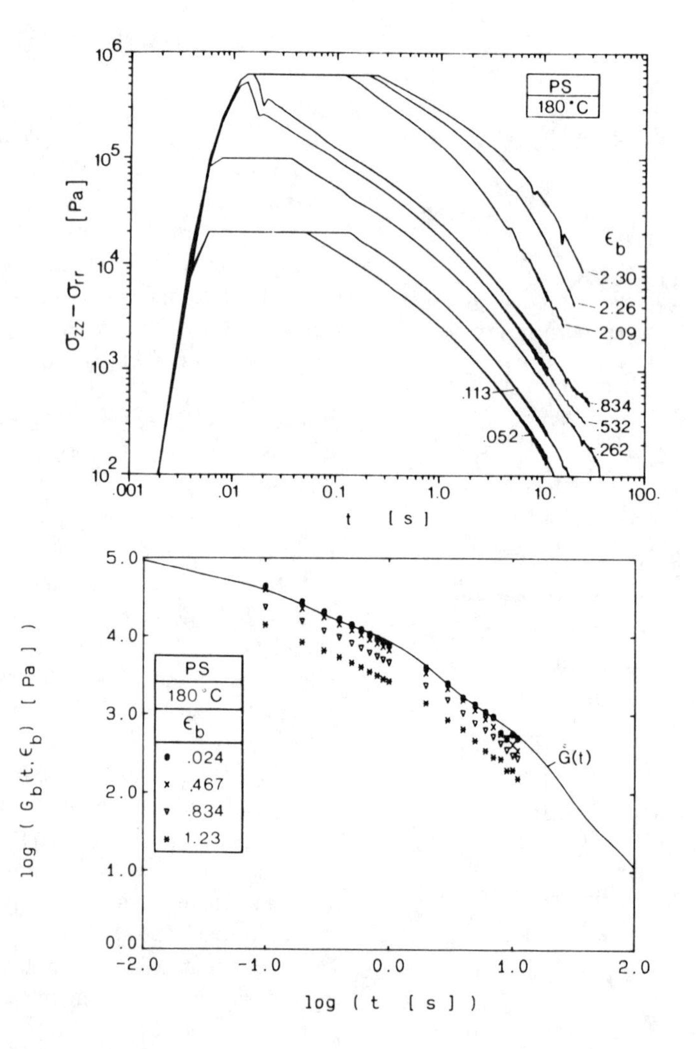

Figure 7.3.6.
Stress relaxation for step squeezing of polystyrene at 180°C. (a) Stress versus time for increasing strain steps. Stress increases at short times, 2–10 ms because the plates take a finite time to close. The horizontal stress response signifies transducer overload. The rapid drop for strains $\epsilon > 1$ indicates loss of lubricant. (b) Stress relaxation data plotted as relaxation modulis. Solid line is the linear viscoelastic relaxation modulus calculated from shear dynamic data. Adapted from Soskey and Winter (1985).

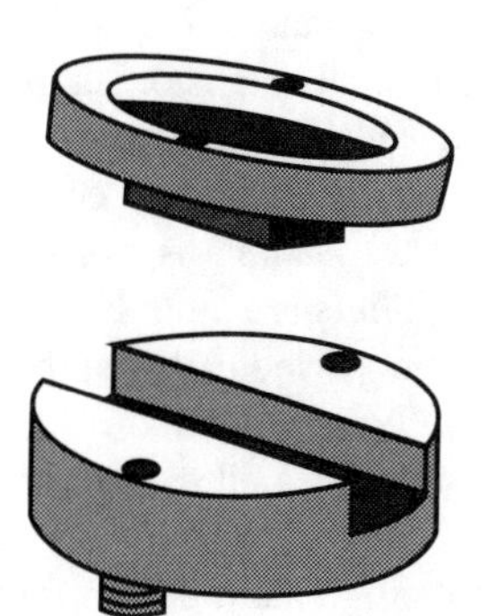

Figure 7.3.7.
Fixtures used for lubricated planar squeezing. From Khan and Larson (1991).

7.3.1 Planar Squeezing

The apparatus used for lubricated compression also can be readily adapted to give lubricated planar flow, as shown in Figure 7.3.7. Only one normal stress difference is measured, $(\tau_{11} - \tau_{33}) = F/A$. Thus only one of the two planar viscosities is available, μ_1. However, as shown elsewhere (Figures 4.2.7 and 7.4.6), this viscosity should be the more strongly thickening one.

Khan and Larson (1991) have shown that step planar squeezing gives the correct linear viscoelastic limit at small strains. Lubrication is lost at strains similar to equibiaxial, $\epsilon \leq 1.5$. Further work is needed with this geometry.

Lubricated compression is probably the simplest extensional method. If samples can be made in a solid form, they will be easy to load and test. Temperature can be readily controlled over a wide range. Strain appears to be limited to 1–1.5 by loss of lubricant, and samples with viscosity $\eta > 10^2$ Pa·s seem necessary to create enough difference between sample and lubricant. The advantages and limitations of lubricated compression are summarized in Table 7.3.1. However, the method has not been widely studied, and further work is needed to accurately determine its limitations.

7.4 Sheet Stretching, Multiaxial Extension

Nearly all the extensional tests described above were developed from experiments used to study solids, especially rubber. Another method used extensively in fundamental studies on rubber is sheet stretching. Thin rubber sheets can be pulled at their perimeter or inflated to give quite a variety of deformations (Treloar, 1975). In fact, Rivlin and Saunders (1951) showed that any state of stress could be generated by applying different combinations of stresses along the edges of an originally square rubber sheet. Figure 7.1.2 illustrated this schematically. Kawabata and Kawai (1977) have reviewed these techniques and the results for several types of rubber. A picture of their general biaxial stretching apparatus is shown in Figure 7.4.1.

A translating clamp device of similar design could be used for high viscosity polymers. Small clamps could be glued to the edges much as in the rod pulling experiments. In addition to pulling, the clamps must translate laterally to accommodate the increasing sample width. However, an apparatus like Figure 7.4.1 does not yet seem to have been applied to high viscosity liquids. Biaxial stretchers for molten polymers have been built with pneumatic clamps to simulate film tentering, but they were not designed to be used as rheometers. There appears to be too much friction in the slide mechanism to accurately measure the small forces needed to stretch polymer melts. See example 1.10.9 for another type of polymer melt orienting device.

7.4.1 Rotating Clamps

A rotating clamp device for sheets of high viscosity liquids has been developed by Meissner and co-workers (1982). Their design, illustrated in Figure 7.4.2, is analogous to the rotating clamp for uniaxial extension (Figure 7.2.5). For equibiaxial extension, eight pairs of cylindrical rollers are arranged in an octagonal pattern. Other arrangements can give uniaxial, planar, or any other combination of extensions (Figure 7.4.3). The test sample is floated on a liquid bath or on talcum powder. The rollers are mounted on leaf springs so stress can be detected. Scissors periodically cut the sheet between the rollers to permit uniform winding. Each roller is servo-controlled and a minicomputer coordinates all the rollers and scissors.

Figure 7.4.4 shows data from the rotating clamp device for the transient equibiaxial viscosity at three different extension rates. For comparison, the linear viscoelastic viscosity and the uniaxial viscosity are shown. Results for the biaxial viscosity compare well to those measured in lubricated compression on the same polyisobutylene sample as in Figure 7.4.4 (Chatraei et al., 1981). So far, only results with the rotating clamp method have been reported for this sample. Maximum strains were 2.5 in the biaxial and multiaxial tests and $\dot{\epsilon} < 0.1\ \mathrm{s}^{-1}$. Friction on the talcum powder may limit the total strain and the detectable stress values. Much larger, more homogeneous samples are required than were used in the lubricated squeezing experiments. However, because the rotating clamps can

Figure 7.4.1.
Translating clamp apparatus developed by Kawabata and Kawai (1977) for general biaxial testing of rubber sheets.

Figure 7.4.2.
Rotary clamp for multiaxial elongation of polymer melts. The two grooved cylinders RC rotate in opposite directions and draw a strip of polymeric material. After passing the nip line of the two cylinders, the strip is wound up by cylinder WU. A suspension system with eight leaf springs LS allows small elastic displacements in the length direction L of the clamp and in the perpendicular direction N, which are recorded by transducers T-L and T-N. From Meissner et al. (1982).

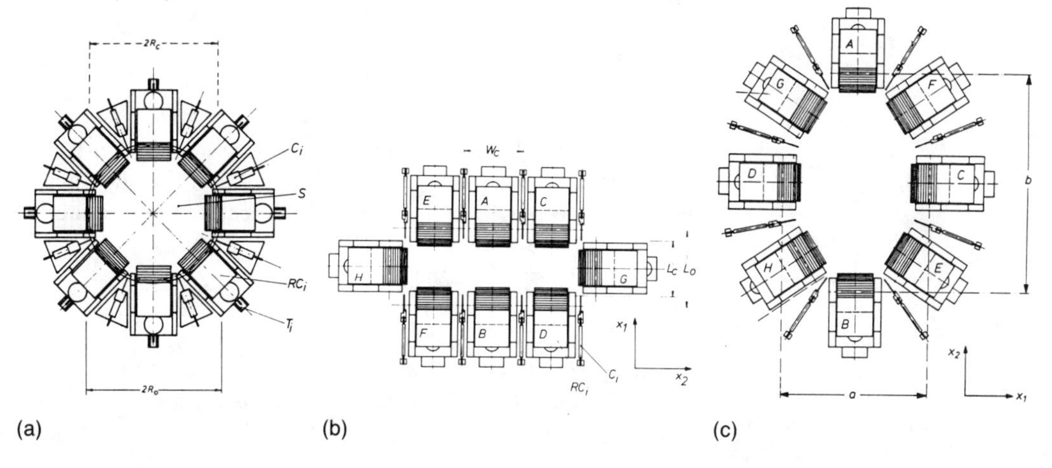

Figure 7.4.3.
Various arrangements of rotating clamps (RC_i) for multiaxial elongation. (a) Equibiaxial: S = sample; C_i = one of eight pairs of scissors; transducers T_i (T-L of Figure 7.4.2) record the forces. (b) Planar: clamps A–F rotate with constant speed, while G and H remain stationary, recording only force. L_o = 158 mm. If only G and H rotate and A–F are removed, we have uniaxial extension of a strip. (c) Multiaxial with m = 0.5 (eq. 7.1.4): ellipse axes a = 268 mm, b = 380 mm. From Meissner et al. (1982) and Demarmels and Meissner (1985).

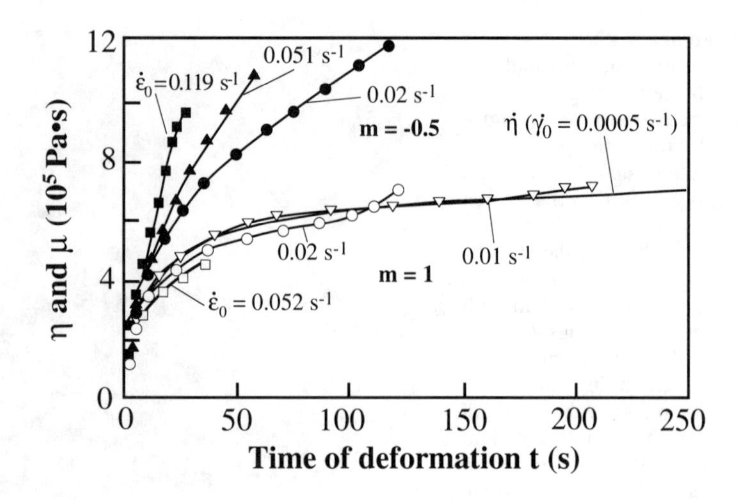

Figure 7.4.4.
Transient equibiaxial viscosity (m = 1, open symbols), uniaxial viscosity (m = −0.5, solid symbols), and linear viscoelastic shear viscosity (lines) for polyisobutylene. Replotted from Meissner et al. (1982).

be arranged in any pattern, this method permits arbitrary, multiaxial deformations.

7.4.2 Inflation Methods

To avoid the problems associated with grabbing the edge of a sheet of liquid, a number of workers have tried various inflation methods. Planar extension can be achieved by inflating a hollow cylinder of melt (Stevenson et al., 1975, Laun and Schuch, 1989). By pulling on the tube, various amounts of uniaxial extension can be superposed on the planar extension, creating any state of stress, as in the general sheet stretching (Chung and Stevenson, 1975).

The tube inflation apparatus of Laun and Schuch is shown in Figure 7.4.5. They modified the uniaxial extensional apparatus of Munstedt (Figure 7.2.6) by adding a syringe that injects oil into the tube as it is stretched. Just enough oil is injected to keep the mean radius, $\bar{R} = (R_o + R_i)/2$, constant. For constant $\dot{\epsilon}$ the tube length is extended exponentially as in uniaxial extension, eq. 7.2.4. The pressure inside the tube, P_i, measures the second planar viscosity. From eq. 7.1.1 with $m = 0$

$$\mu_2^+ = \frac{\tau_{22} - \tau_{33}}{2\dot{\epsilon}} = \frac{\eta_{p2}^+}{2} = \frac{P_i(t)}{2\dot{\epsilon}} \frac{l(t)}{L_o} \frac{\bar{R}}{\Delta R} \tag{7.4.1}$$

where $\Delta R = R_o - R_i$. The derivation assumes a thin-walled tube (see Laun and Schuch, 1989, Appendix A). The first planar viscosity is a combination of the tensile force and the pressure

$$\mu_1^+ = \frac{\tau_{11} - \tau_{33}}{4\dot{\epsilon}} = \frac{\eta_{p1}^+}{4} = \frac{F(t)}{8\dot{\epsilon}\pi\bar{R}\Delta R} \frac{L(t)}{L_o} + \frac{\eta_{p2}^+}{4} \tag{7.4.2}$$

Figure 7.4.6 shows some transient planar viscosity data obtained by tube inflation. At short times (small strains) all the ma-

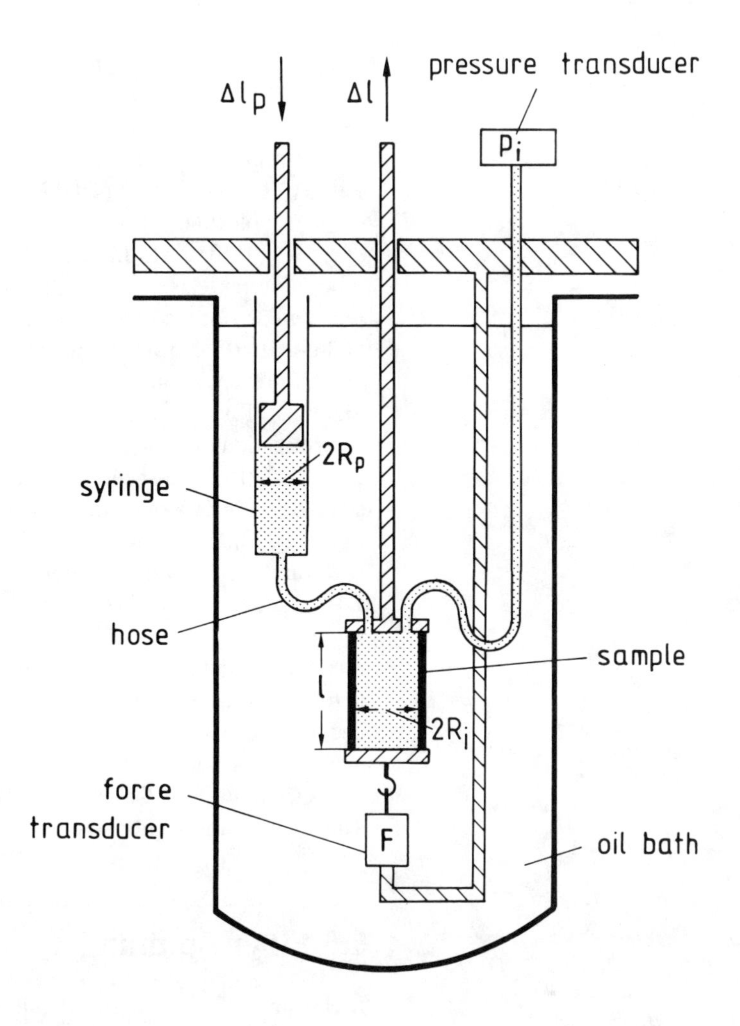

Figure 7.4.5.
Schematic of the apparatus used for planar elongation tests, in which a tube-like sample is extended and inflated. From Laun and Schuch (1989).

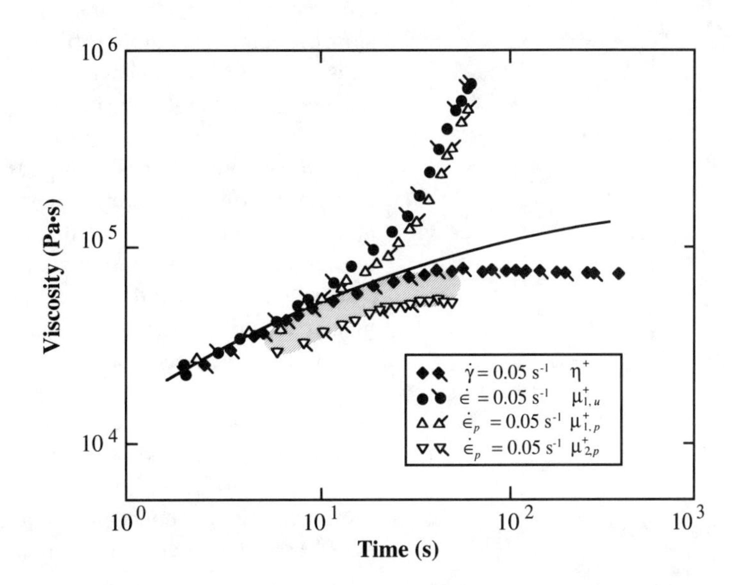

Figure 7.4.6.
Transient planar viscosities (open symbols) compared to uniaxial (•) and shear (◇)—all at constant extension or shear rate of 0.05 s^{-1}—for a low density polyethylene at 125°C. The solid line is the linear viscoeslastic limit. Ticks denote a repeat test (at half the sample length, 20 mm, in the planar case). Shaded area denotes range of μ_2 for several runs. From Laun and Schuch (1989).

terial functions should converge to the linear viscoelastic limit; μ_2 is low. The measured pressure was low, typically $P_i = 1$ mbar. The first planar viscosity does agree with linear viscoelasticity at small strain and converges toward the uniaxial result at large strain. Petrie (1990) has shown that this behavior is expected from a broad class of molecular models.

It is difficult to fabricate the hollow cylinders required for tube inflation tests. Inflating a flat rectangular sheet over a long rectangular slot also approximates planar extension (Denson and Hylton, 1980). However, the clamped edges around the rectangular slot prevent a uniform deformation over the entire sheet. Photographs of a grid in the polar region are necessary for accurate determination of strain.

Inflation of a sheet clamped over a circular hole suffers the same difficulty. More work has been done with this method (see Dealy, 1982). Inflation has been done with both liquid and gas. The stress and deformation equations are treated in Chapter 1, Exercise 1.10.8, and by Dealy (1982). The results can give the biaxial viscosity function and ϵ up to 2 has been achieved (Rhi-Sausi and Dealy, 1981; Yang and Dealy, 1987). But results so far have been limited to relatively low extension rates. The method requires photography and a somewhat complex apparatus. Inflation tests, however, are similar to the vacuum forming process for making shaped plastic items (Schmidt and Carley, 1975; DeVries and Bonnebat, 1976). Thus there is motivation to continue this work.

7.5 Fiber Spinning

In Sections 7.2–7.4, we saw that it is possible to test high viscosity samples, particularly molten polymers, in extension. The major problems are sample preparation, clamping, and buoyancy. Extensional rheometry is more difficult, and the upper limits of strain rates and strains are much lower than for shear. Nonetheless, accurate and reproducible data can be obtained, particularly in uniaxial extension.

With lower viscosity liquids, $\eta < 10^3$ Pa·s, extensional rheometry is much more problematic. It is impossible to "grab and pull" such fluids as we did with the melts. Gravity, surface tension, air drag, or confining walls all work against the desired extensional flow. Yet the effects of extensional flow can be much greater for polymer solutions than for polymer melts (see Figure 7.5.1). However, because these extensional effects are so strong, we can use extensional indexers (i.e., flows with a mixture of extension and shear or with poorly defined strain histories). These index results will give a good idea of the importance of extensional flow, but they will not give us purely extensional material functions. The rest of this chapter concentrates on these approximate extensional methods for lower viscosity samples.

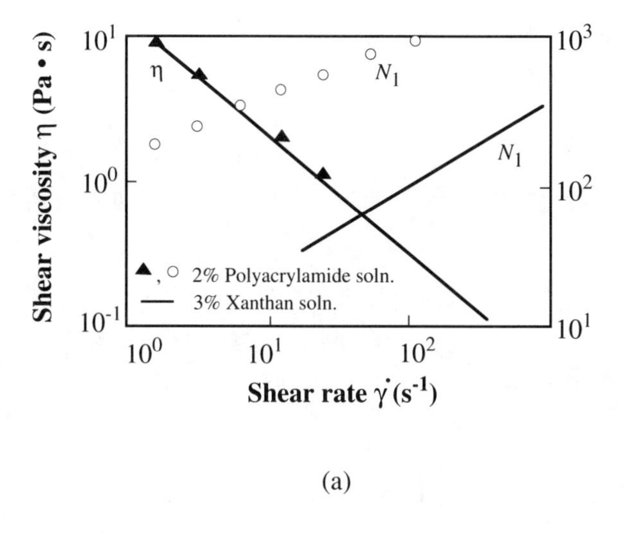

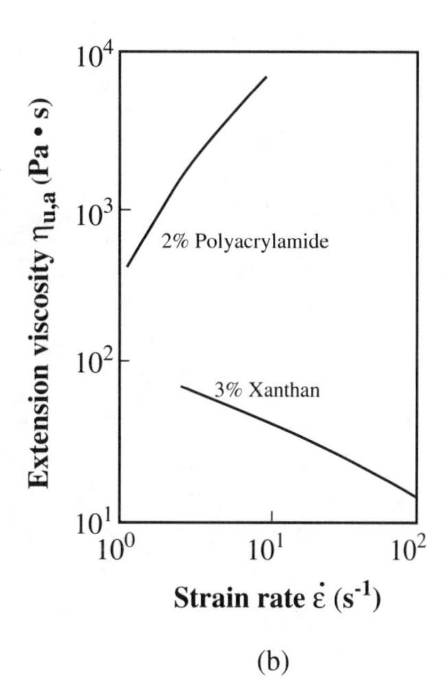

Figure 7.5.1.
Comparison of shear and extensional properties of two polymer solutions with similar shear viscosities: 2% polyacrylamide and 3% xanthan, both in water at room temperature. Apparent uniaxial extensional viscosity by fiber spinning. Replotted from Jones et al. (1987).

The most common of these methods is fiber spinning. Figure 7.5.2 shows the basic features of this method. To avoid the problem of "grabbing and pulling," the sample is continuously extruded from a tube and stretched by a rotating wheel or vacuum suction. The force on the tube or on the take-up system is measured. The fiber diameter is measured as a function of distance along the fiber, either photographically or by a video camera.

The extension rate can be determined from measurements of fiber diameter and flow rate. If we assume that the only component of velocity is v_x and that it is uniform across the radius, then the flow rate is

Figure 7.5.2
Schematic of a fiber spinning apparatus.

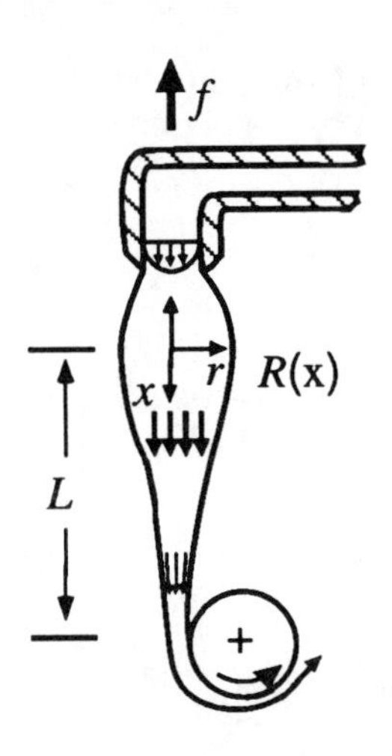

$$Q = v_x \pi R^2(x) \tag{7.5.1}$$

For constant flow rate using the definition of extension rate, we have

$$\dot{\epsilon} \equiv \frac{\partial v_x}{\partial x} = -\frac{2Q}{\pi R^3}\frac{dR}{dx} \tag{7.5.2}$$

Differentiating the $R(x)$ profile gives the extension rate down the fiber. The practice of taking derivatives of experimental data is prone to errors. Fitting the data first with a spline function can improve accuracy of dR/dx (Secor, 1988).

A simpler analysis is often used. If we assume that the extension rate is *constant* everywhere along the fiber, then

$$\dot{\epsilon}_a = \frac{\Delta v_x}{\Delta x} = \frac{v_L - v_0}{L} \tag{7.5.3}$$

and from eq. 7.5.1

$$\dot{\epsilon}_a = \frac{Q}{\pi L}\left(\frac{1}{R_L^2} - \frac{1}{R_0^2}\right) \tag{7.5.4}$$

Since the final diameter is small, a good approximation for the strain rate is

$$\dot{\epsilon}_a \cong \frac{Q}{\pi R_0^2 L} \tag{7.5.5}$$

If we equate eqs. 7.5.2 and 7.5.4, we can obtain a quadratic expression for the radius profile

$$\frac{1}{R^2} = \frac{x}{L}\left(\frac{1}{R_L^2} - \frac{1}{R_0^2}\right) + \frac{1}{R_0^2}$$

$$\simeq \frac{x}{R_L^2 L} \tag{7.5.6}$$

The total strain experienced by each material element as it is drawn down with the fiber is

$$\epsilon_a = \ln \frac{R_0^2}{R_L^2} = \ln\left(\frac{v_L}{v_0}\right) \tag{7.5.7}$$

which is equivalent to eq. 7.2.6 for rod pulling.

The measured force in the fiber spinning experiment is the sum of the extensional stress in which we are interested as well as the effects of gravity, inertia, and surface tension (Secor et al., 1989).

$$\frac{f}{\pi R^2} = (\tau_{xx} - \tau_{rr})_x + \frac{\rho g}{R^2}\int_0^x R^2 dx' - \rho v_x^2 + \frac{\Gamma}{R} \tag{7.5.8}$$

For high viscosity liquids the correction terms can often be neglected. Hence, data often are analyzed by simply dividing the force on the nozzle by the die cross section

$$\tau_{xx} - \tau_{rr} = \frac{f}{\pi R_0^2} \tag{7.5.9}$$

From this relation and eq. 7.5.5 we can define an apparent uniaxial extensional viscosity for the fiber spinning experiment

$$\eta_{u,a} = \frac{fL}{Q} \tag{7.5.10}$$

Much of the literature data is reported using this simplified analysis (e.g., Figure 7.5.1).

TABLE 7.5.1 / Working Equations for Apparent Uniaxial Extensional Viscosity from Fiber Spinning

Strain rate

$$\dot{\epsilon} = \frac{-2Q}{\pi R^3} \frac{dR}{dx} \tag{7.5.2}$$

if $\dot{\epsilon}$ is assumed to be constant down fiber

$$\dot{\epsilon}_a = \frac{Q}{\pi R_0^2 L} \tag{7.5.5}$$

Strain

$$\epsilon_a = \ln \frac{R_0^2}{R_L^2} \tag{7.5.7}$$

Tensile stress

$$[\tau_{xx} - \tau_{rr}]_x = \frac{f_o}{\pi R^2} - \frac{\rho g}{R^2} \int_0^x R^2 dx' + \rho v_x^2 - \frac{\Gamma}{R} \tag{7.5.8}$$

Apparent unixial extensional viscosity

$$\eta_{u,a} = \frac{fL}{Q} \tag{7.5.10}$$

(neglect gravity, inertia, and surface tension, and assume $\dot{\epsilon}$ constant)

Errors
Effect of shear history in feed die, extrudate swell
ϵ is not usually constant
Gravity, inertia can be significant for low η
Surface tension $\Gamma/R = 10 - 100$ Pa
Air drag, especially with vacuum take-up
Detachment of fiber on inside of die
Uncertainty in diameter due to vibration, instabilities
Nonisothermal (polymer melts)
Solvent evaporation
Unstable flow

Utility
Effects of prehistory and nonconstant $\dot{\epsilon}$ obviate interpretation of results in terms of simple extensional material function, but the method is still useful
Sensitive to changes in uniaxial extensional viscosity
Easy sample preparation, but often requires > 100 ml
Relatively simple equipment
Particularly useful for polymer solutions and suspensions
$\eta > 1$ Pa·s
Simulates fiber spinning process
$10 < \dot{\epsilon} < 10^3$ typical and usually less for a given material
$\epsilon < 3$ and typically $\epsilon < 2$; syphon may be better
Imaging system (photo or video) highly desirable, slows data analysis

These working equations, along with the limitations and utility of fiber spinning measurements, are summarized in Table 7.5.1. The major problem is that typically $\dot{\epsilon}$ is not constant, so the force, which is measured over the entire fiber, is an integration of stresses due to various strain rates and even the upstream shear history in the die. For these reasons, the fiber spinning experiment is not a true rheometer, but gives only an apparent uniaxial extensional viscosity.

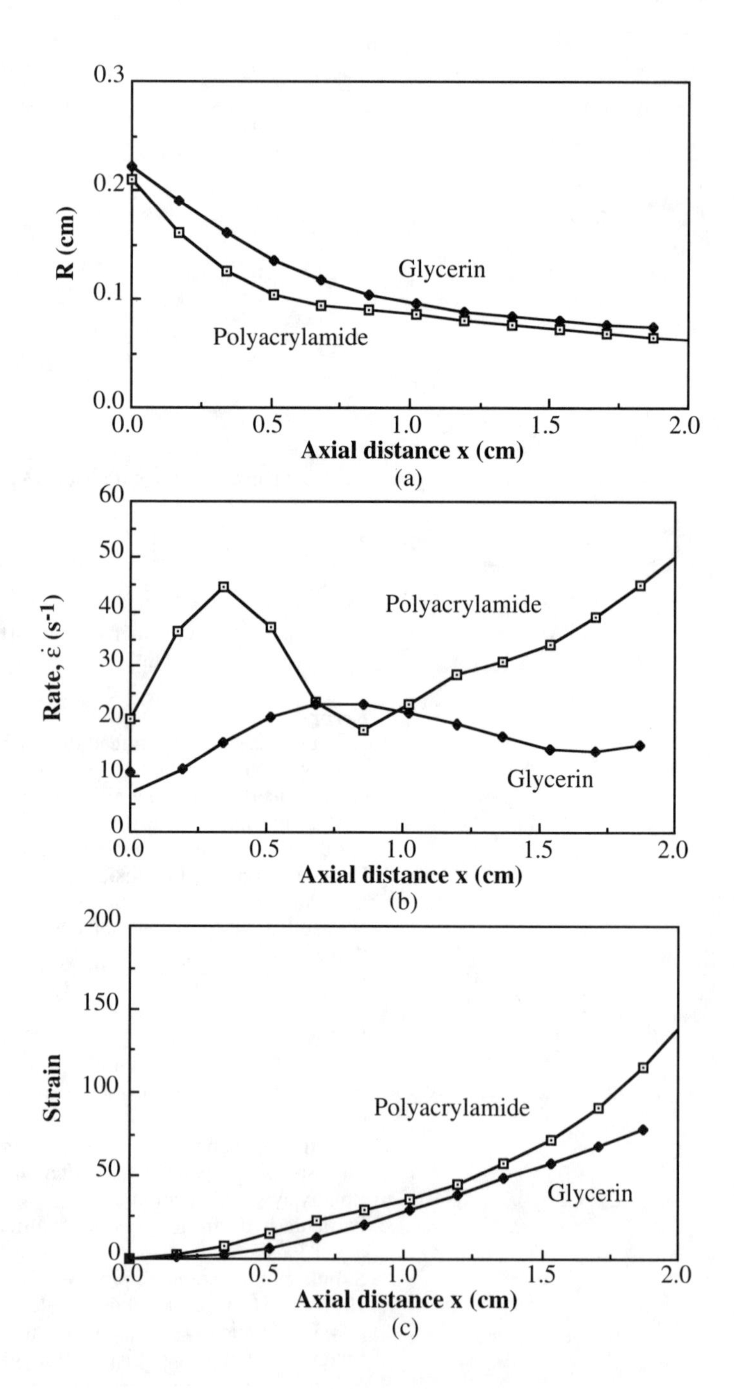

Figure 7.5.3.
Determination of extension rate in fiber spinning for a Newtonian liquid (50:50 glycerin/water mixture) and a 1.5 wt % solution of polyacrylamide in water. (a) Radius profiles used to calculate (b), the local extension rate, and (c), the cumulative strain down the fiber.

To illustrate this point, let us examine typical fiber radius profiles for a Newtonian liquid and a polymer solution (Figure 7.5.3a). For the elastic polymer solution we see that the radius actually *increases* before it draws down. This is due to extrudate swell typical for capillary flow of elastic liquids. Using eq. 7.5.2 and a five-point spline fit, extension rate was evaluated from these $R(x)$ data. Figure 7.5.3b shows that away from the nozzle $\dot{\epsilon}$ becomes constant at about 23 s^{-1} for the glycerin solution, but for the polyacrylamide $\dot{\epsilon}$ continues to increase along the entire fiber. Figure 7.5.3c shows strain versus x for the same two runs.

Figure 7.5.4 shows the tensile stress along the fiber calculated according to eq. 7.5.8. For the polyacrylamide, the correction terms are negligible, but for the lower viscosity glycerin solution, the gravity correction becomes over 90% of the measured force. With

Figure 7.5.4.
Comparison of the correction terms in the extensional stress calculation for (a) 1.5 wt% polyacrylamide in a 50:50 mixture of water and glycerin and (b) glycerine. From Secor et al. (1989).

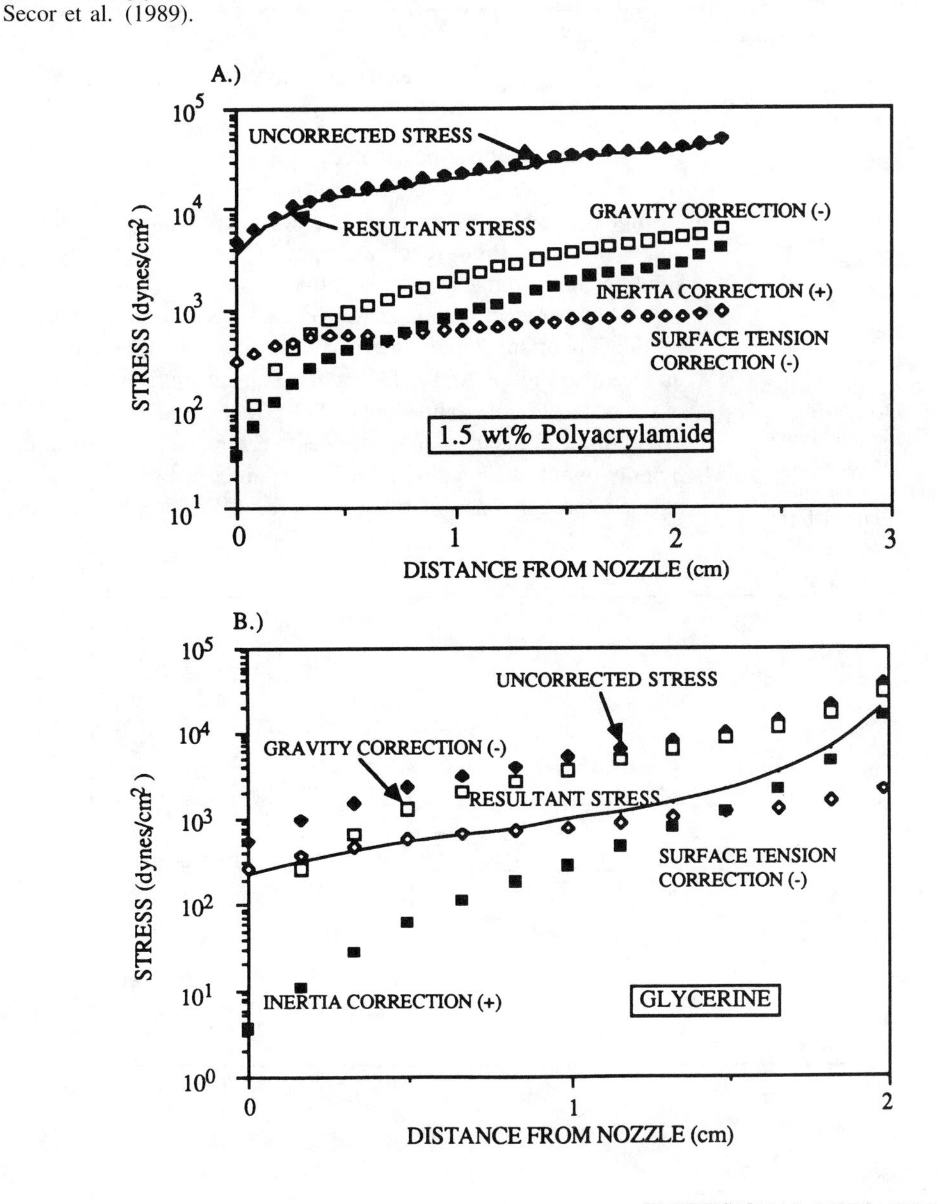

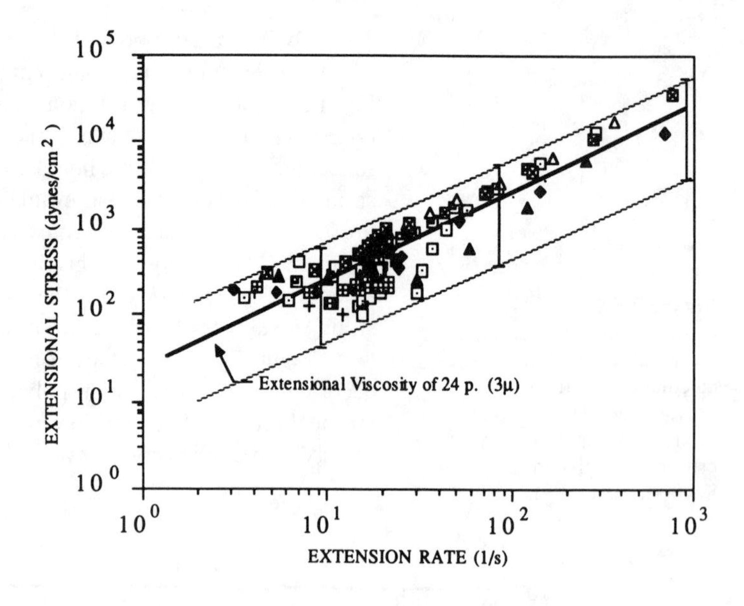

Figure 7.5.5.
Experimental data for glycerin of 0.8 Pa·s shear viscosity. The upper and lower lines represent the error bounds. The middle line corresponds to an extensional viscosity of 2.4 Pa·s, which is three times the shear viscosity. The data symbols represent measurements from different locations on 10 filaments. From Secor et al. (1989).

such large corrections, the accuracy of the results is not high. This is illustrated in Figure 7.5.5 for a wide range of extension rates. At higher rates forces are large and accuracy improved. Hudson and co-workers (1988) report accuracy of $\pm$ 10% for two glycerin solutions in the range $\dot{\epsilon} = 500 - 5000\ \text{s}^{-1}$.

Another error not included in eq. 7.5.8 is air drag, which can become important at high rates and especially with a vacuum take-up (see also Figure 7.5.8). Fibers also can detach from the inside of the nozzle (Sridhar and Gupta, 1988; Butlers and Meijer, 1990). Cavitation can occur at high stresses. The lowest rate is limited by gravity fall, while at high rates instabilities can set in. Secor et al. (1989) roughly summarize these effects to define an operating

Figure 7.5.6.
Operating diagram of extensional rheometry by fiber drawing. Inside the boundaries are ranges for which experimental measurements were made. Adapted from Secor et al. (1989).

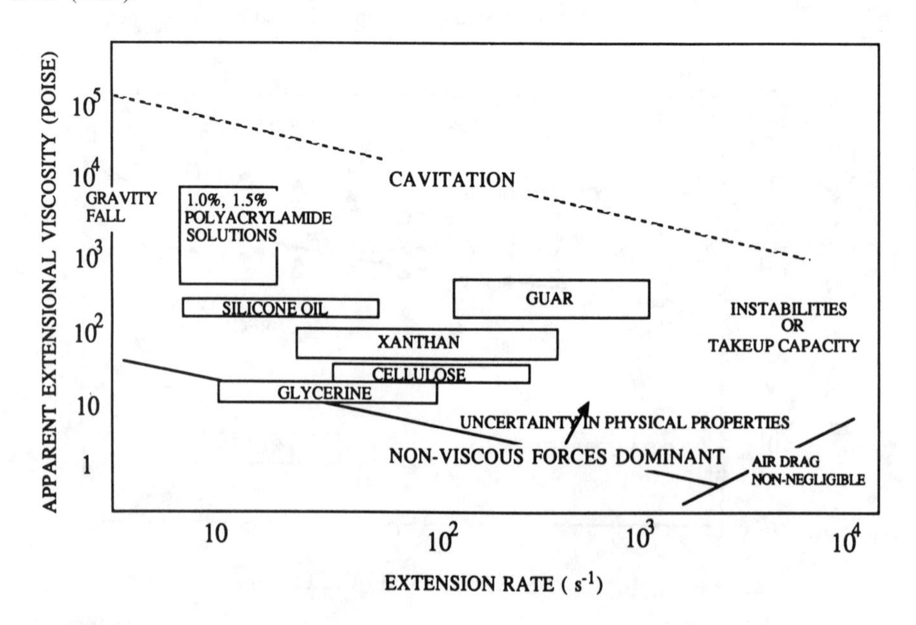

range for fiber spinning in Figure 7.5.6. As indicated, a particular fluid and apparatus combination often has a narrower range.

For polymer melts, fiber spinning results can be compared to true uniaxial extensional viscosity data measured by rod pulling. Figure 7.5.7 shows such a comparison for a low density polyethylene. We see qualitative agreement between the methods but, a very strong effect of upstream history is evident in the fiber data. Both die extrusion velocity and residence time in the die exert a big influence.

7.5.1 Tubeless Siphon

A variation on fiber spinning is the ductless or tubeless siphon, which is simply fiber spinning in reverse. A nozzle is dipped in a bath of the test fluid, a vacuum applied, and fluid sucked out of the bath. The nozzle is slowly raised, and a free-standing, rising column of fluid develops as shown in Figure I.4. Sometimes the tubeless siphon is called a Fano flow, after the physician who first reported the technique (1908).

The simplest way to use the tubeless siphon is to record the maximum height rise that can be achieved when one applies full vacuum. However, strain rates and tensile stress can be obtained with the same analysis used for fiber spinning except that the sign of the gravity term in eq. 7.5.8 is now positive. This tends to destabilize the flow and reduce the fiber length. Another problem with the

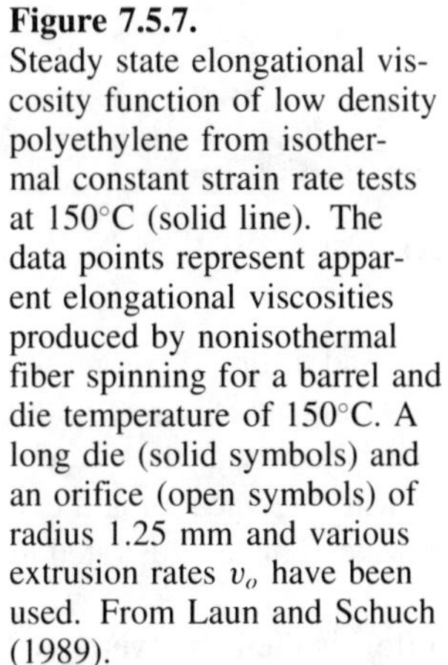

Figure 7.5.7.
Steady state elongational viscosity function of low density polyethylene from isothermal constant strain rate tests at 150°C (solid line). The data points represent apparent elongational viscosities produced by nonisothermal fiber spinning for a barrel and die temperature of 150°C. A long die (solid symbols) and an orifice (open symbols) of radius 1.25 mm and various extrusion rates v_o have been used. From Laun and Schuch (1989).

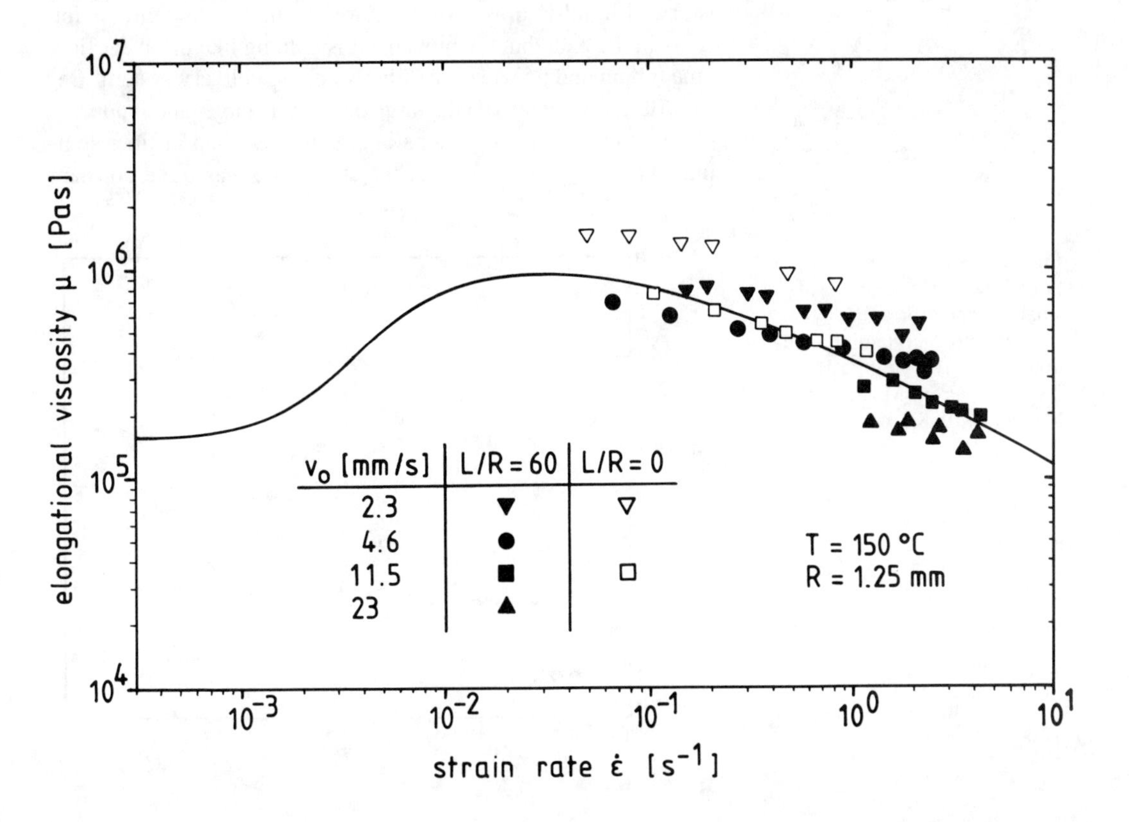

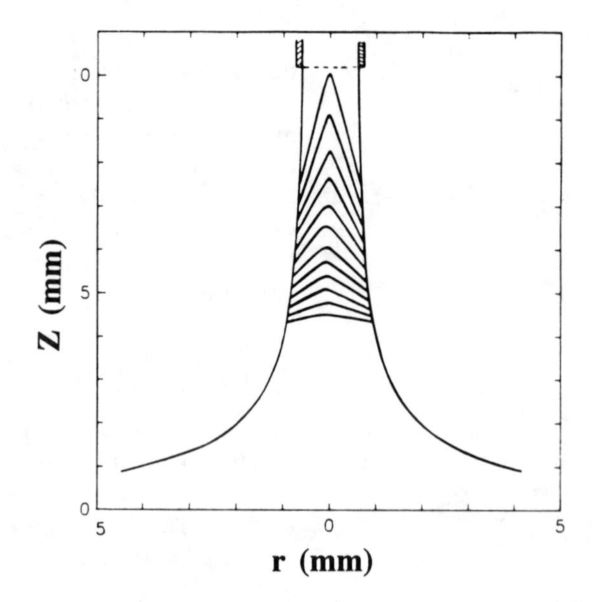

Figure 7.5.8.
Tracer line pattern obtained for a tubeless siphon experiment. The lines were drawn from a high speed movie of 2% polyisobutylene in mineral oil. Time interval between line positions, 2.5 ms; siphon length, ~ 10 mm. From Mathys (1988).

the tubeless siphon is accumulation of a bead of liquid around the nozzle tip. The siphon and some fiber spinning devices are limited in that the maximum force that can be achieved is what the vacuum can apply. Air is sometimes sucked in with the vacuum and can also alter the velocity profile near the nozzle. This is illustrated in Figure 7.5.8.

An advantage of tubeless siphon flow is the less severe prehistory. Liquid is drawn from a large bath. A slow circulation occurs in the bath, but the movement is nothing like the shear flow in the tubing and nozzle of the fiber spinning delivery system. Because fluid is drawn from the large bath, R_0 is large, and it appears possible to reach higher strains in siphon flow than in fiber spinning. Figure 7.5.9 shows that the strain rate was quite constant

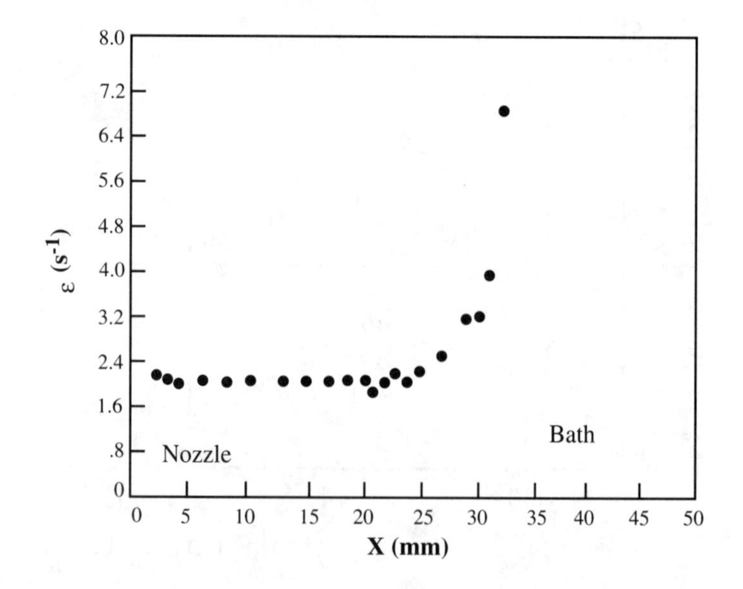

Figure 7.5.9.
Stretch rate as a function of axial distance for a 3% hydrolyzed polyacrylamide solution (3.4×10^{-4}g/cm^3, $M_w = 8 \times 10^6$). From Moan and Magueur (1988).

Figure 7.5.10.
Variation of reduced extensional viscosity with total strain for various polyacrylamide concentrations, $M_w = 8 \times 10^6$, $\dot{\epsilon} = 1 - 4\ \text{s}^{-1}$: (a) $c = 1.02 \times 10^{-3}$, (b) 7.65×10^{-4}, (c) 5.1×10^{-4}, and (d) $3.4 \times 10^{-4}\text{g/cm}^3$. From Moan and Magueur (1988).

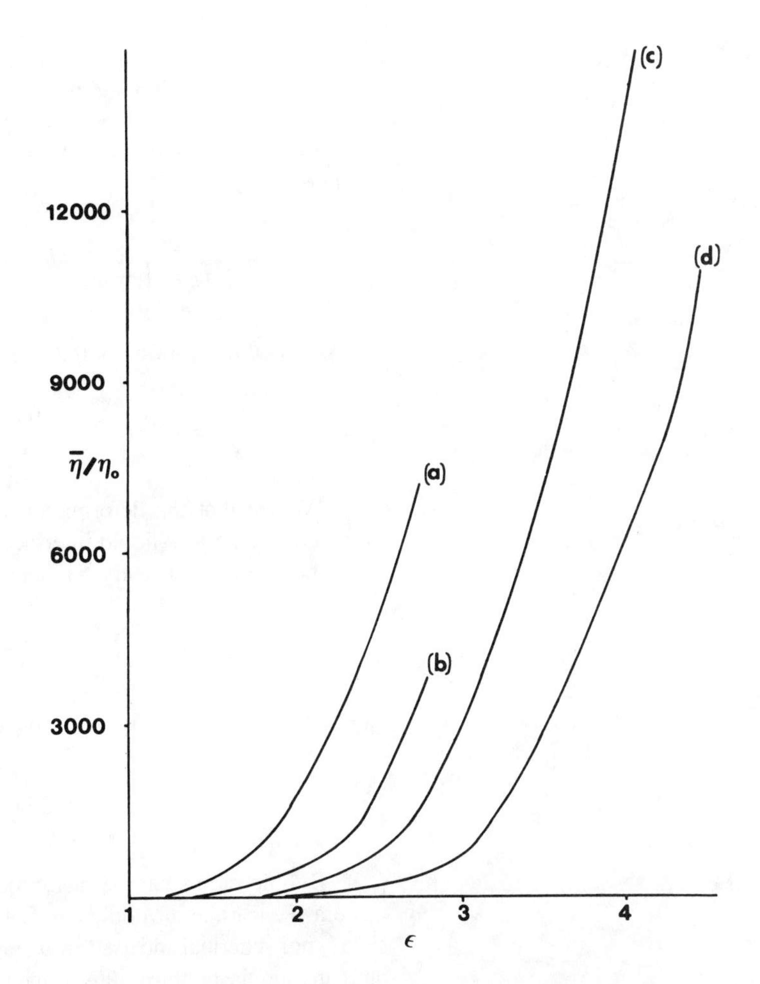

over the column for a polyacrylamide solution, and Figure 7.5.10 demonstrates that strains of nearly 4 are possible. High strain levels may be important, for example, in promoting the coil stretch transition in dilute polymer solutions. Figure 7.5.10 shows that higher strains are required to see the strong increase in apparent extensional viscosity as polymer concentration is decreased.

7.6 Bubble Collapse

Another method for measuring uniaxial extensional viscosity is by bubble collapse. A small bubble is blown at the end of a capillary tube placed in the test fluid (see Figure 7.6.1). It comes to equilibrium with the surrounding pressure and surface tension. Then at time $t = 0$ the pressure inside the bubble is suddenly lowered or the surrounding pressure increased. The decrease in bubble radius with time is recorded. If the deformation is reversed (i.e., the pressure inside the bubble is suddenly increased), the growing bubble radius can be used to give the equibiaxial viscosity. This flow appears to be less stable and has not been studied as a rheometer.

Figure 7.6.1.
Bubble of radius R on the end of a capillary tube in test fluid; p_G is the pressure inside the bubble, p_∞ in the surrounding fluid.

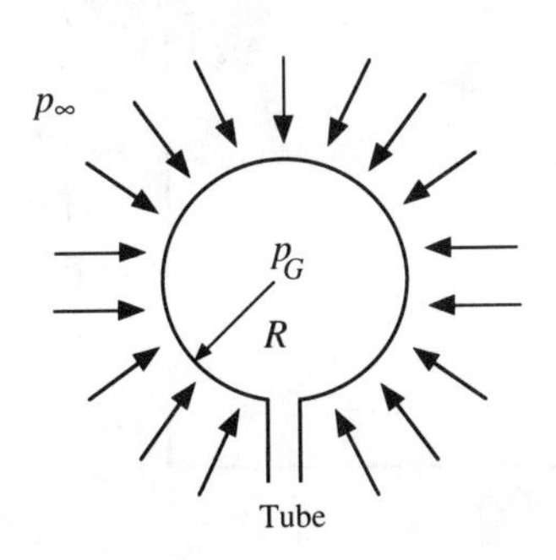

Bubble radius gives the extensional strain and strain rate. If we can ignore diffusion and the influence of the capillary and assume that the bubble collapses symmetrically in an incompressible fluid, then the continuity equation in spherical coordinates (Table 1.7.1) reduces to

$$v_r = \left(\frac{R}{r}\right)^2 \frac{dR}{dt} \qquad v_\phi = v_\theta = 0 \tag{7.6.1}$$

From the definition of the rate of uniaxial extension

$$\dot{\epsilon} = \dot{\epsilon}_{11} = \dot{\epsilon}_{rr} = \frac{\partial v_r}{\partial r} = -\frac{2R^2}{r^3}\frac{dR}{dt} \tag{7.6.2}$$

We see that the deformation is not homogeneous. It is a maximum at the gas–liquid interface, $r = R$, and decreases with r^{-3}. The rate is most easily evaluated at the interface

$$\dot{\epsilon}_R = -\frac{2}{R}\frac{dR}{dt} \tag{7.6.3}$$

and by integration, we obtain for the strain

$$\epsilon = -2 \ln \frac{R}{R_o} \tag{7.6.4}$$

Bubble radius can be measured directly by a film or video camera. Pearson and Middleman (1977) found their bubbles were sightly nonspherical and used an average radius. For opaque liquids one can infer the bubble radius by measuring the small change in gas pressure above the liquid sample caused by the collapsing bubble (Johnson and Middleman, 1978). This method is also simpler and faster to use.

Typical radius versus time data for a polymer melt are shown in Figure 7.6.2. Note that after 1.5 seconds the bubble appears to

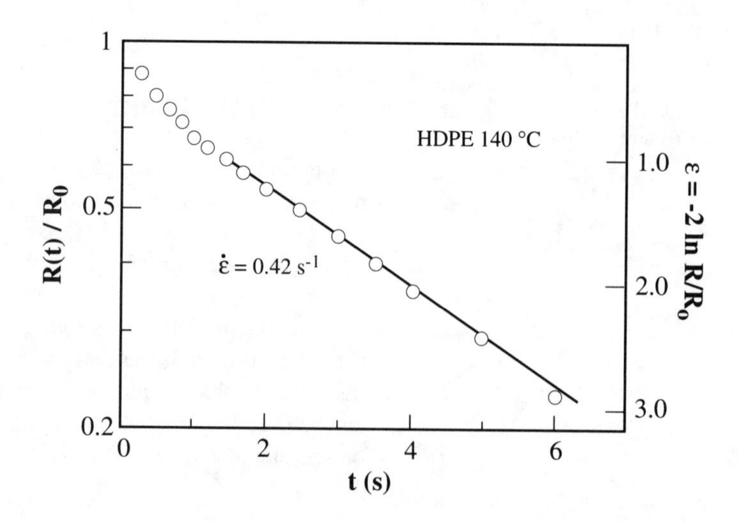

Figure 7.6.2.
Log-reduced radius versus time for the collapse of a bubble in a high density polyethylene melt at 140°C. The initial radius need not be known exactly because the ratio R/R_o comes from the decrease in pressure above the melt. Adapted from Munstedt and Middleman (1981).

collapse exponentially, giving a constant extension rate. A total strain of nearly 3 is reached in this case.

Bubbles in polymer solutions also collapse at constant rate as indicated in Figure 7.6.3. Note that for these lower viscosity materials, bubble collapse can be very rapid. When the bubble gets too small, necking occurs near the capillary and the collapse is no longer spherically symmetric (Figure 7.6.3b).

Stress can be determined from the pressure difference across the bubble. Papanastasiou et al. (1984) have shown that even for the rapid bubble collapse in polymer solutions, the unsteady and inertial terms in the momentum balance may generally be neglected. Neglecting viscosity of the gas, but considering interfacial tension, leads to

$$0 = p_G - p_\infty - \frac{2\Gamma}{R} + 2\int_R^\infty \frac{(\tau_{rr} - \tau_{\theta\theta})}{r} dr \tag{7.6.5}$$

We would like to be able to determine the extensional stress difference $\tau_{rr} - \tau_{\theta\theta}$, but in general it appears that we need a constitutive equation to evaluate the integral. This is a serious limitation on using bubble collapse as a rheometer. However, let us examine the Newtonian case for which eq. 7.6.5 becomes

$$\tau_{rr} - \tau_{\theta\theta} = \frac{-3}{2}\left(p_G - p_\infty - \frac{2\Gamma}{R}\right) \tag{7.6.6}$$

This result can be used to define an apparent uniaxial extensional viscosity

$$\eta_{u,a} = \frac{3(p_G - p_\infty - 2\Gamma/R)R}{4dR/dt} \tag{7.6.7}$$

Middleman and co-workers have shown good agreement between $\eta_{u,a}$ and 3η for Newtonian fluids. Figure 7.6.4 shows data for a mildly elastic polymer melt. There is good agreement between $\eta_{u,a}$ and both $3\eta_o$ and rod pulling measurements taken using the apparatus shown in Figure 7.2.6. Note that higher exten-

Figure 7.6.3. Log-reduced radius versus time for bubble collapse in hydroxypropylcellulose, 2% in water. (a) Low rates and (b) high rate also showing necking as the bubble shrinks to the capillary size ($R_c \simeq 0.5$ mm, $R_o \simeq 2$ mm). Data from high speed camera (Pearson and Middleman, 1977).

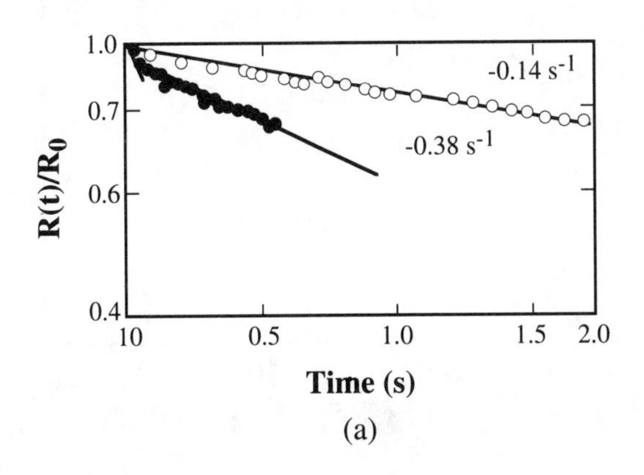

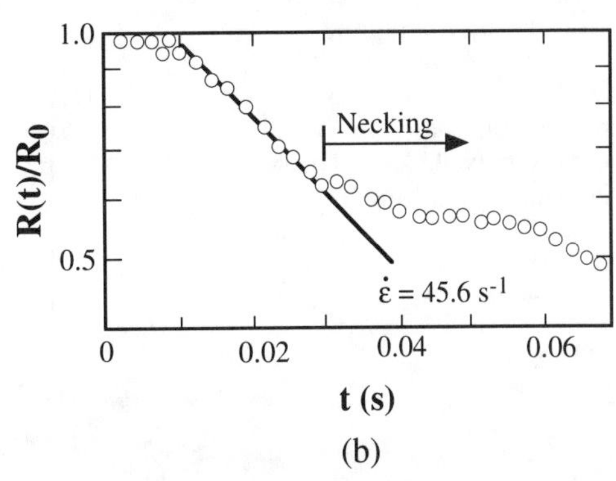

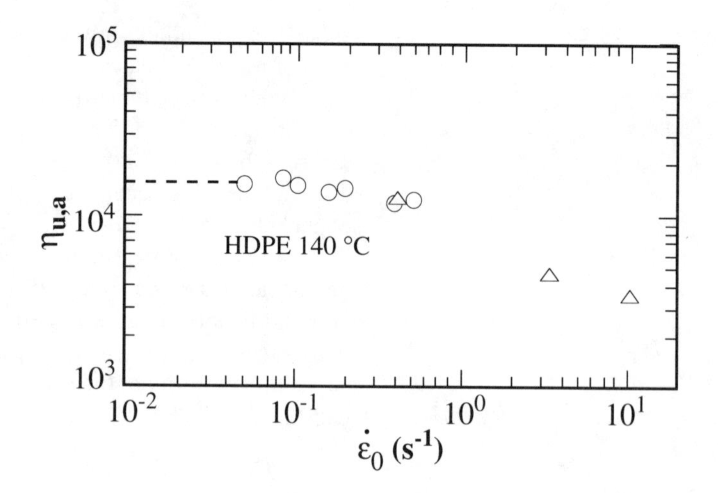

Figure 7.6.4.
Uniaxial exensional viscosity by rod pulling (○) compared to apparent extensional viscosity by bubble collapse (△). From Munstedt and Middleman (1981).

sion rates are possible with bubble collapse than with rod pulling. When these two methods are compared for a strongly extension-thickening melt, Figure 7.6.5, the apparent extensional value for bubble collapse is much too low. This indicates the importance of the form of the constitutive equation used in the integration in eq. 7.6.5.

Bubble collapse is a well-posed problem: that is, the initial condition is rest state, for analysis with viscoelastic constitutive equations (Pearson and Middleman, 1977; Papanastasiou et al., 1984). Such work would be valuable to test other extensional methods for solutions. Bubble growth and collapse measurements also have important applications for processing of foamed polymer.

7.7 Stagnation Flows

Steady extensional deformations can be created by impinging two liquid streams, creating a stagnation flow. Figure 7.7.1 illustrates both axisymmetric and planar stagnation flow. These flows are not homogeneous. A material element near the central part of the flow

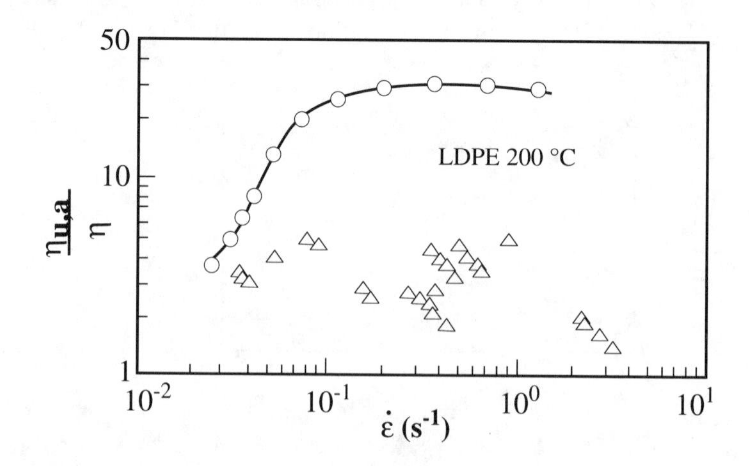

Figure 7.6.5.
The ratio of uniaxial extensional viscosity to shear viscosity by rod pulling (○) compared to apparent extensional measurements by bubble collapse (△) for a low density polyethylene. Uniaxial data shifted from 150°C to the bubble test temperature of 200°C. From Munstedt and Middleman (1981).

will experience much higher strain than one further out. Thus only the steady extensional material functions can readily be measured in these flows, and as we have seen it may take considerable strain to reach the steady state. Despite this limitation, there is a strong interest in stagnation flows, mainly because the extensional strain becomes infinite at the stagnation point. Thus at least in a portion of the flow we will get very high orientation of particles or macromolecules. Also, crude stagnation flows can be generated relatively easily for low viscosity fluids.

Fiber spinning (Figure 7.5.2) approximates one end of the axisymmetric stagnation flow. The tubeless siphon is a little closer. But neither has a stagnation point. Ideally we want to confine a fluid to flow within the stream surfaces indicated in Figure 7.7.1. For planar stagnation these surfaces are defined (Winter et al., 1979) by the relation

$$xy = \frac{hx_o}{2} \tag{7.7.1}$$

At each surface, fluid must move with velocity

$$\begin{aligned} v_x &= \dot{\epsilon}_x x \\ v_y &= -\dot{\epsilon}_x y \\ v_z &= 0 \end{aligned} \tag{7.7.2}$$

The real challenge experimentally is to hold this shape, yet keep all surfaces moving at the prescribed velocity. Probably the

Figure 7.7.1.
(a) Steady uniaxial extensional flow generated by flow into a trumpet-shaped tube (reverse impingement). The stream surface is axisymmetric. View: tilted forward by 15°. (b) The impingement of two rectangular streams generates steady planar extensional flow. From Winter et al. (1979).

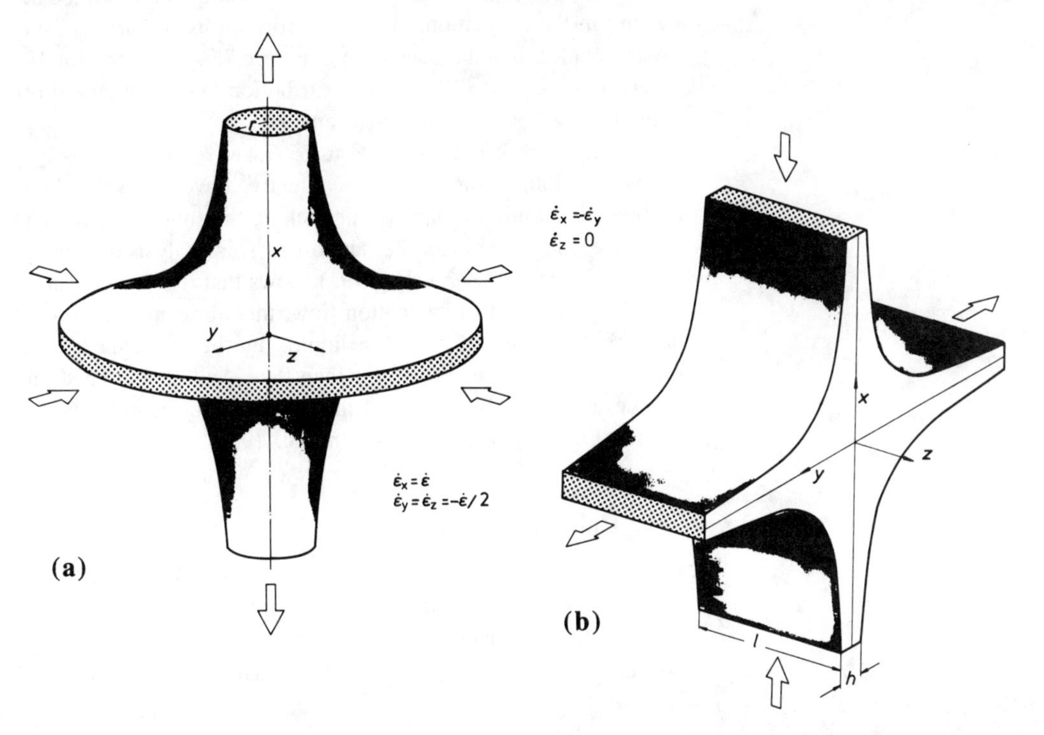

Figure 7.7.2.
The four-roller experiment of Taylor (1934) produces planar stagnation flow only near the center: the roller radius is chosen to be a radius of curvature of one of the hyperbolic streamlines of steady planar extension.

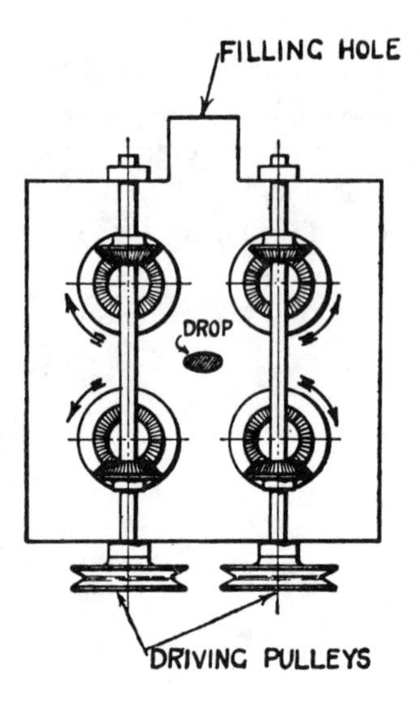

first attempt to do this involved the four-roller apparatus of Taylor (1934) shown in Figure 7.7.2. A planar stagnation flow is created near the center. Measuring pressure through the rotating rollers is very difficult, and the torque to turn the rollers is dominated by shear stresses from the surrounding fluid. However, birefringence can be used to measure the tensile stress difference directly at the center where the flow is purely extensional. With flow birefringence, the four-roller apparatus provides a convenient and accurate method for measuring planar extensional viscosities on *transparent* solutions. The method is further discussed in Chapter 9 (see Figure 9.4.6).

7.7.1 Lubricated Dies

Another attempt to generate stagnation flows involves constructing a solid die with an internal shape like Figure 7.7.1 and lubricating the die walls (Winter et al., 1979; Van Aken and Janeschitz-Kriegl, 1981). As with lubricated compression, the lubricating fluid must be Newtonian, with a significantly lower viscosity than the test fluid.

Figure 7.7.3 shows a die for creating lubricated planar stagnation flow of a polymer melt. Silicone oil is fed through four semicircular channels, from which it flows through a narrow gap and then is directed along the die walls. Pressure required to drive the polymer melt through the diverging then converging flow is recorded by a transducer. As Figure 7.7.4 shows, pressure in the unlubricated flow can be modeled by a simple viscous, shear thinning constitutive equation. The good fit implies that shear effects at the wall dominate any extensional thickening that might be occurring in the stagnation. This is confirmed by the dramatically lower data for the lubricated case in Figure 7.7.4. Lubrication has essentially eliminated the shear contribution to the pressure drop, leaving only that due to planar extension.

The lower set of curves in Figure 7.7.4 were calculated assuming various lubricating conditions. The two lowest curves assume perfect slip, while the upper pair include the lubricant flow and thus give a slightly higher pressure drop. This analysis of Secor et al. (1987; see also Zahorski, 1992) shows that it is impossible to achieve perfect planar stagnation flow; the lubricant fluid always produces some shear in the test liquid. Furthermore, the experimental pressures are even lower than the calculations. This may be due to effects of gravity on the exiting sheet causing the top lubricant layer to be thicker (Secor et al. 1987).

7.7.2 Unlubricated Dies

Lubricated stagnation is a difficult, messy experiment and given the results reported so far, does not seem to be worth the trouble. Unlubricated stagnation is much easier to achieve and, although wall pressure measurements are dominated by shear as Figure 7.7.4

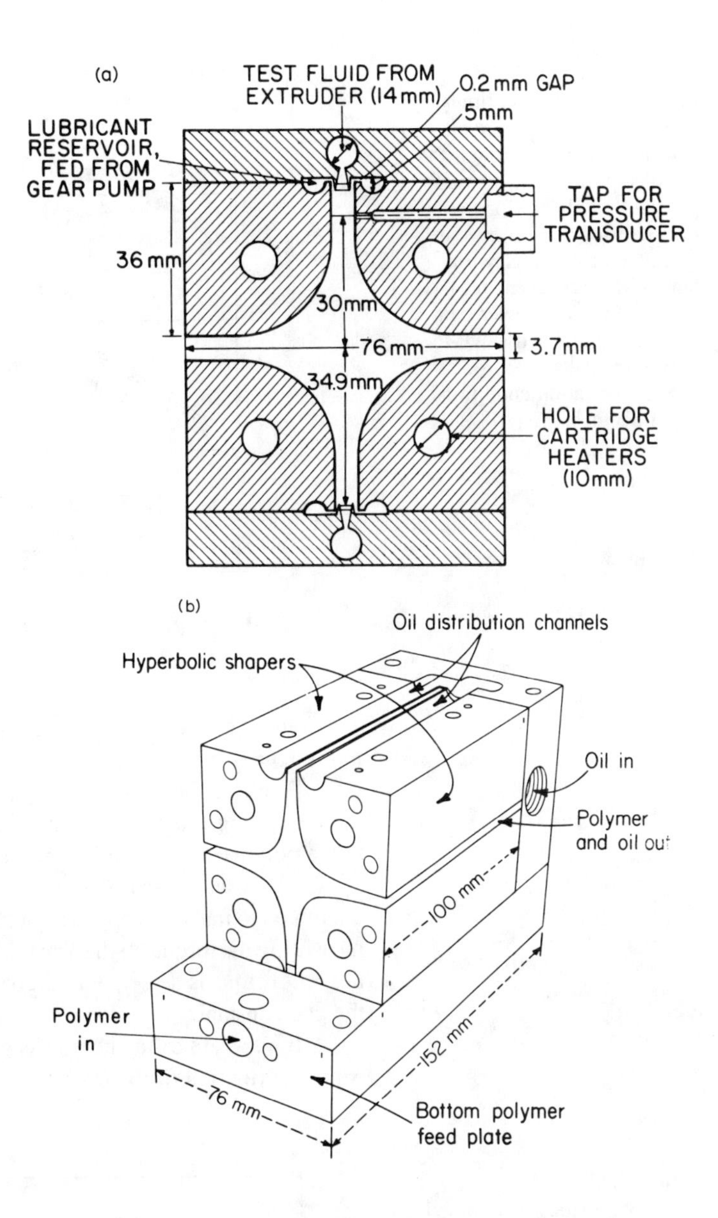

Figure 7.7.3.
(a) Cross section of a die designed to create lubricated, planar stagnation flow of a polymer melt. (b) Projection of the same die with top and end melt feed plates removed. From Macosko et al. (1982).

shows, birefringence can be used to probe stresses near the center, where the stagnation flow assumption is very good. Keller (1975) showed dramatic birefringence in polystyrene solutions impinging in a simple crossed-slot die. Figure 7.7.5 shows planar viscosity measurements calculated from birefringence measurements along the y and z axes of unlubricated flow through the same die shown in Figure 7.4.3.

7.7.3 Opposed Nozzles

If one is going to use an unlubricated stagnation flow, it is best to remove the walls as far as possible. Keller and co-workers (1987) have placed two tubes (or nozzles) in close opposition (see Figure

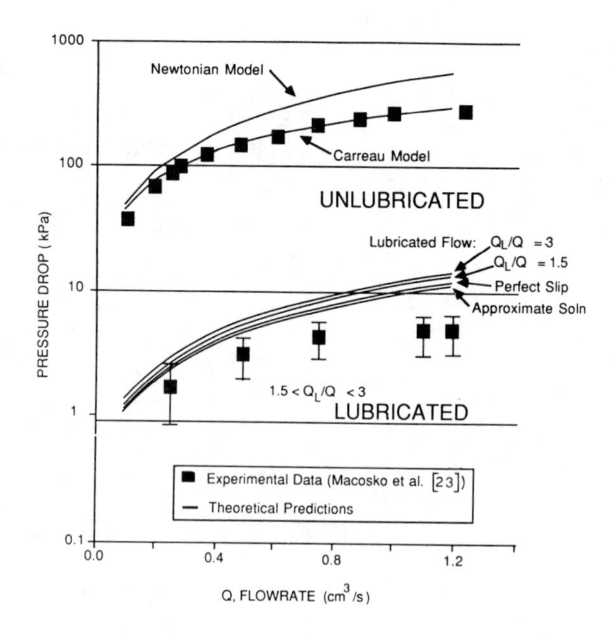

Figure 7.7.4.
Pressure drop versus flow rate for a polystyrene melt flowing at 200°C through the die of Figure 7.7.3. The upper data points are for unlubricated flow; the curves are calculated using a Newtonian and a Carreau model (see eq. 2.4.16). The lower data set is for lubricated flow; the curves are calculated for different lubrication conditions. From Secor et al. (1987).

7.7.6) in a "sea" of high polymer solution. When fluid is sucked into the nozzle from the surrounding sea, a stagnation flow is created. Birefringence and pressure drop measurements show strong increase at high suction rate, indicating extensional thickening due to the polymer coils rapidly stretching out.

Fuller and co-workers (1987; Mikkelsen et al., 1988) have attempted to make the opposed-nozzle device more quantitative by measuring the torque on the arm that holds the nozzle down into the beaker. Torque is detected by a rebalance transducer that maintains the gap constant.

If we assume that the flow is purely extensional and that there is no contribution from pressure in the surrounding fluid, the force

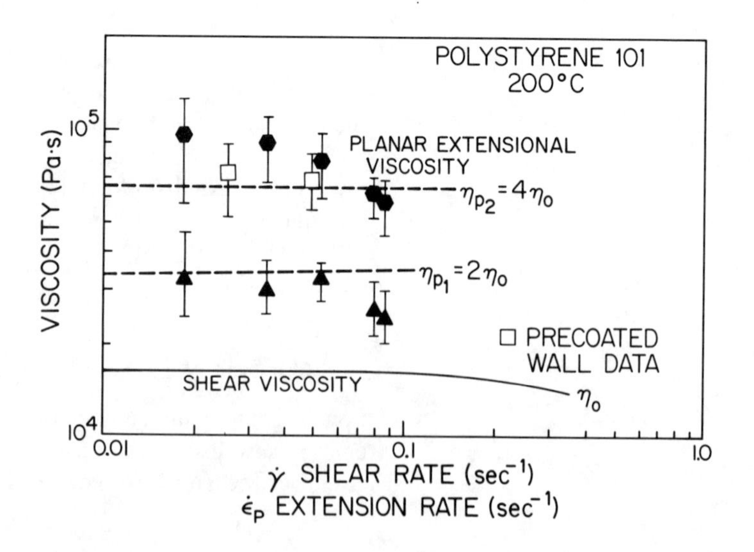

Figure 7.7.5.
Planar viscosities determined by measuring birefringence in unlubricated stagnation flow in both directions. The same polystyrene melt as Figure 7.4.4 was used in the same die. From Macosko et al. (1980).

Figure 7.7.6.
Opposed-nozzle (or jet) device creates a uniaxial extensional flow by sucking in surrounding fluid. Flow creates a force that tends to pull the nozzles together and is measured by the torque on one of the flow tubes.

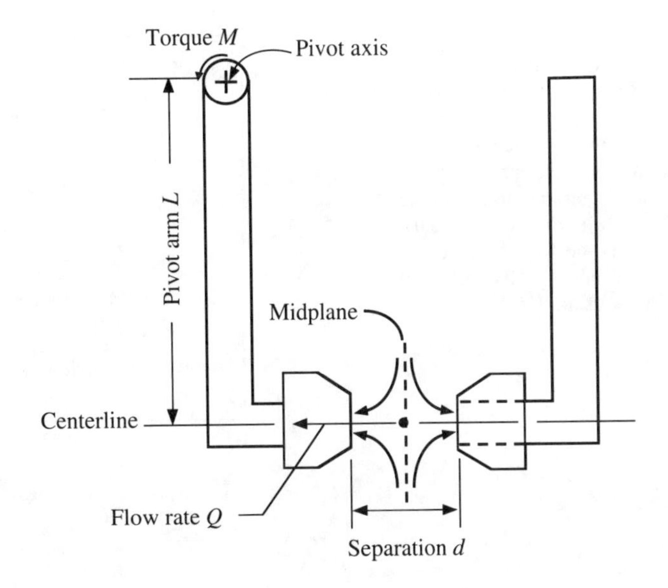

on the nozzles (M/L) divided by the nozzle sectional area is the tensile stress difference on the fluid.

$$\tau_{11} - \tau_{22} = \tau_{11} - \tau_{33} = \frac{M}{L\pi R^2} \tag{7.7.3}$$

As mentioned with respect to Figure 7.7.1, the strain and strain rates in stagnation flow, as in capillary flow, are not constant. We can define an apparent strain and strain rate in the cylindrical test region

$$\epsilon_a = < \dot{\epsilon} t > \cong 1 + 2 \ln\left(\frac{d}{R}\right) \tag{7.7.4}$$

$$\dot{\epsilon}_a = \frac{2Q}{\pi R^2 d} \tag{7.7.5}$$

Apparent extensional viscosity is calculated by dividing eqs. 7.7.3 and 7.7.5.

Figure 7.7.7 shows data collected in suction (uniaxial extension) and expulsion (biaxial or compression) plotted as apparent viscosity versus apparent extension rate. The liquid was Newtonian: a glycerin–water mixture with shear viscosity of 1.6 poise. The dashed line gives $3\eta_o$, the value we would expect for pulling of the sample in air. We see that the $\eta_{u,a}$ values are closer to $4\eta_o$ because of the departure of the flow from ideal extensional flow. The solid lines represent calculations by Schunk et al. (1990), who have carried out a fairly complete analysis of this flow, solving the Navier-Stokes equations via the finite element method.

Figure 7.7.8 shows this plot of the streamlines and the relative amount of extension in the flow. We see that only near the stagnation point is the flow nearly pure uniaxial extension. Shear flow

Figure 7.7.7.
Measured extensional viscosities versus rate for glycerin–water compared to predictions from flow of a Newtonian liquid between opposed nozzles. The upturn in the expulsion data is due to secondary flows as the Reynolds number exceeds 1. Adapted from Schunk et al. (1990).

Predictions
Suction
Expulsion
Trouton line (rod pulling)
$3\eta_0$
Data from Mikkelsen et al. (1988)
Suction
Expulsion
$\eta_{u,a} = \frac{Md}{2QL}$ (Pa•s)
1
0.1
10
100
1000
10,000
Nominal extension rate $\frac{2Q}{\pi R^2 d}$ (s^{-1})

dominates along the nozzle walls and "pollutes" the pure extension around the nozzle tip. However, this shear flow does not contribute as much to the measured torque as does the pressure from the surrounding liquid. Schunk et al. are able to calculate the total torque reasonably well, including the effect of fluid inertia at higher flow rates (Figure 7.7.7).

The opposed-nozzle device is an extensional viscosity indexer. The flow is not homogeneous, nor is it purely extensional. However, it has a strong extensional component. Figure 7.7.9 demonstrates that it can detect strong thickening in a dilute polymer solution where shear shows slight thinning and biaxial extension only very slight thickening. All these results are expected from molecular theory. Cathey and Fuller (1988) find good agreement between $\eta_{u,a}$ and theory for rodlike suspensions (eq. 10.3.19) on solutions of collagen. They also have measured the coil–stretch transition for dilute, high molecular weight polystyrene solutions with the device. The opposed-nozzle device is particularly useful for low viscosity liquids, for which few methods are available. It is convenient to use and has a significantly lower viscosity and wider rate range than fiber spinning (Cai et al., 1992). One problem with interpreting results from the device may be flow instability as polymer molecules become highly oriented (Keller et al., 1987).

Figure 7.7.8.
Predictions showing one-quarter of the flow into a nozzle: (a) Giesekus criteria, namely +1 in pure extension, 0 in simple shear, and minus 1 in pure rotation, and (b) streamlines. Glycerin–water solution η_o = 0, 17 Pa·s, Q = 0.1 ml/s, d = 1 mm, $2R$ = 0.98 mm. Adapted from Schunk et al. (1990).

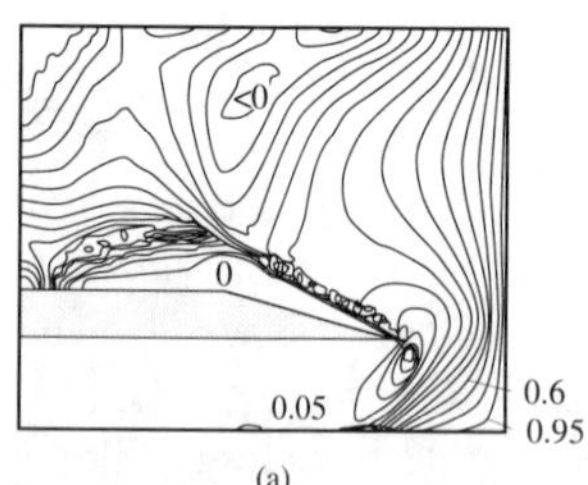

(a)

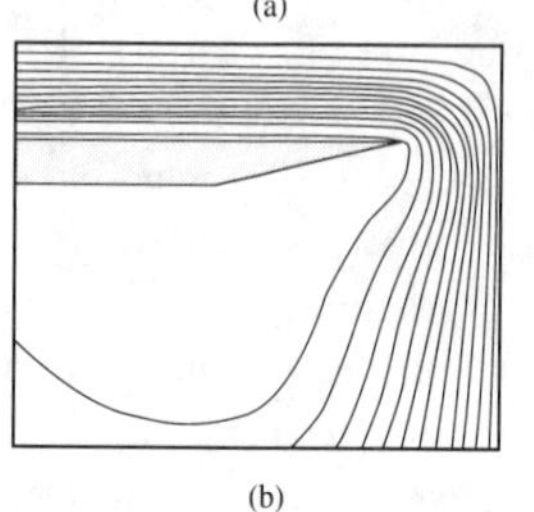
(b)

7.8 Entrance Flows

Entrance flows can be perceived as a part of the stagnation flows shown in Figure 7.7.1. For example, flow into an axisymmetric contraction is similar to the stagnation flow starting from the y–z plane and proceeding along the x direction (Figure 7.7.1a). For a planar contraction, which involves the flow from a rectangular reservoir into a slit or rectangular orifice, the similarity is with the planar stagnation flow starting from the x–z plane and going along

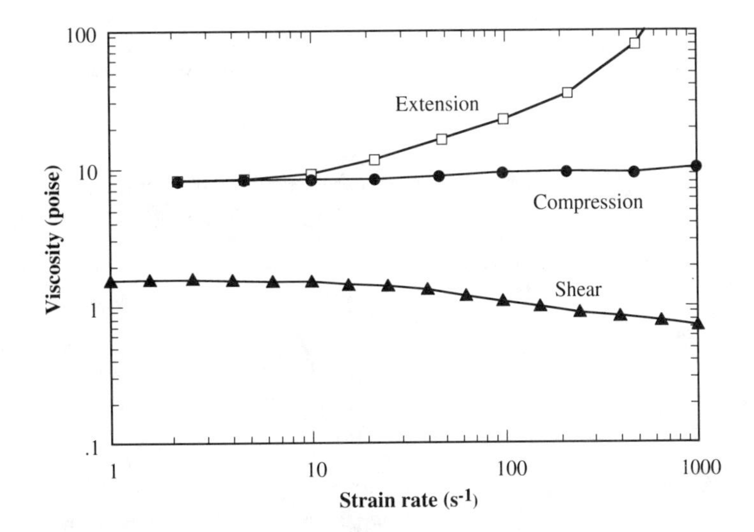

Figure 7.7.9.
Apparent uniaxial and biaxial (compression) viscosity versus apparent extension rate for 1% polyacrylamide in glycerin–water.

the y direction (Figure 7.7.1b). As the fluid flows from a large cross section tube into a small cross section tube, the streamlines converge (Figure 7.8.1). To overcome this reduction in the cross-sectional area and continue to flow, the fluid dissipates extra energy, which is expressed as the entrance pressure drop Δp_{en}. The converging streamlines indicate the existence of the extensional flow. However, the presence of the walls along the contraction imparts a shear component to the flow. Attempts to minimize the shear component were made by Everage and Ballman (1978) and Winter et al. (1979) by lubricating the tube walls. However, just as with the lubricated stagnation dies, the increased difficulty involved in lubricating the walls is not really worth the trouble. The major advantage of the entrance flow is that it is the easiest extensional flow to generate and measure because it primarily involves forcing the fluid through an orifice and measuring the pressure drop.

Several different analyses have been presented to estimate the extensional viscosity from Δp_{en} measurements. The three major approaches are discussed below: sink flow (Metzner and Metzner, 1970), Cogswell's analysis (1972), and Binding's analysis (1988).

The sink flow analysis, which assumes a purely extensional flow (i.e., no shear component), was presented by Metzner and Metzner (1970) to evaluate the extensional viscosity from orifice Δp_{en} measurements. For an axisymmetric contraction, the flow into the orifice is analogous to a point sink; for a planar contraction flow, the analogy is with a line sink (Batchelor, 1967). In the case of axisymmetric contraction (Figure 7.8.1), the use of spherical coordinates and continuity gives the velocity components

Figure 7.8.1.
Streamlines showing the entrance flow into an orifice.

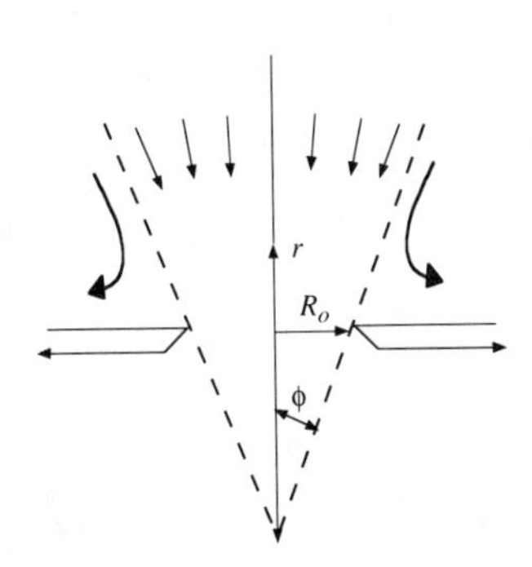

$$v_r = \frac{Q}{A}; \qquad v_\theta = v_\phi \tag{7.8.1}$$

where A is the cross-sectional area in the converging region. By considering the isosceles triangle with the included angle ϕ (where

ϕ is the half-angle of convergence) and employing the law of cosines (from trigonometry), the area is given by

$$A = \pi 2r^2(1 - \cos\phi) \quad (7.8.2)$$

where r is the radial coordinate distance (Figure 7.8.1). From eqs. 7.8.1 and 7.8.2, the radial velocity is

$$v_r = \frac{Q}{2\pi r^2(1 - \cos\phi)} \quad (7.8.3)$$

The extension rate at the orifice is

$$\dot{\epsilon} = \frac{dv_r}{dr} = \frac{Q \sin^3\phi}{\pi R_0^3(1 - \cos\phi)} \quad (7.8.4)$$

where $r = R_0/\sin\phi$ has been used, R_0 being the radius of the orifice (Figure 7.8.1). The normal stress difference is assumed to be equal to the entrance pressure drop

$$\tau_{11} - \tau_{22} \approx \Delta p_{\text{en}} \quad (7.8.5)$$

The apparent extensional viscosity is given by

$$\eta_{u,a} = \frac{\Delta p_{\text{en}}}{\dot{\epsilon}} \quad (7.8.6)$$

The major problem in using eq. 7.8.4 to evaluate the extension rate is the need to know ϕ, which is typically determined by flow visualization. The approximations involved in the analysis and the difficulty in obtaining ϕ, which varies with flow rate, result in the following approximate equation for the extension rate:

$$\dot{\epsilon} = \frac{\dot{\gamma}_a}{8} \quad (7.8.7)$$

where $\phi \approx 15^\circ$ has been used and $\dot{\gamma}_a$ is the apparent shear rate evaluated at the orifice.

Other kinematical problems associated with the sink flow analysis are discussed by Denn (1977) and Cogswell (1978).

For the planar contraction, the cylindrical coordinate system will be convenient, and the velocity and extension rate equations respectively are given by

$$v_r = -\frac{Q}{2Wr \sin\phi} \quad (7.8.8)$$

and

$$\dot{\epsilon} = \frac{Q \sin\phi}{2Wh^2} \quad (7.8.9)$$

where W is the slit width, h is half of slit height, and $r = h/\sin\phi$ has been used to eliminate r.

Because the sink flow analysis is based on the assumption of purely extensional flow, it requires flow situations in which the shear components are negligible. Cogswell (1972) considers both shear and extensional components in the contraction. By assuming that the pressure drop Δp_{en} can be separated into shear and extensional components, Cogswell calculates these pressure drop components from the shear and normal stresses in an elemental cone (or wedge) by applying simple force balances. The shear and normal stresses are then replaced by the shear and extensional viscosities. The overall Δp_{en} is then calculated by minimizing an infinite sum of the elemental pressure drops along the contraction. For axisymmetric contraction, the orifice extension rate is

$$\dot{\epsilon} = \frac{\tau_w \dot{\gamma}_a}{2(\tau_{11} - \tau_{22})} \tag{7.8.10}$$

where τ_w is the wall shear stress, and the normal stress difference is calculated from

$$\tau_{11} - \tau_{22} = \frac{3}{8}(n+1)\Delta p_{\mathrm{en}} \tag{7.8.11}$$

The corresponding equations for a planar contraction are

$$\dot{\epsilon} = \frac{\tau_w \dot{\gamma}_a}{3(\tau_{11} - \tau_{22})} \tag{7.8.12}$$

and

$$\tau_{11} - \tau_{22} = \frac{1}{2}(n+1)\Delta p_{\mathrm{en}} \tag{7.8.13}$$

respectively. The application of Cogswell's analysis requires shear viscosity data. With a knowledge of the shear viscosity parameters, Cogswell's equations permit the calculation of the extensional viscosity from a single Δp_{en} and Q measurement with an orifice die—a procedure very convenient for quickly ranking fluids.

A number of recent studies have attempted to evaluate extensional viscosity from Δp_{en} using Cogswell's analysis (Laun and Schuch, 1989; Tremblay, 1989; Boger and Binnington, 1990). Laun and Schuch (1989) compared extensional viscosities from rod pulling tests with that from circular orifice pressure drop measurements for several polymer melts. Their data for different polyethylene melts are shown in Figure 7.8.2. The entrance flow $\eta_{u,a}$ deviates from the rod pulling data at low $\dot{\epsilon}$ but overlaps at higher $\dot{\epsilon}$.

Binding (1988) analyzed the entrance flow by applying variational principles to minimize the overall energy consumption for

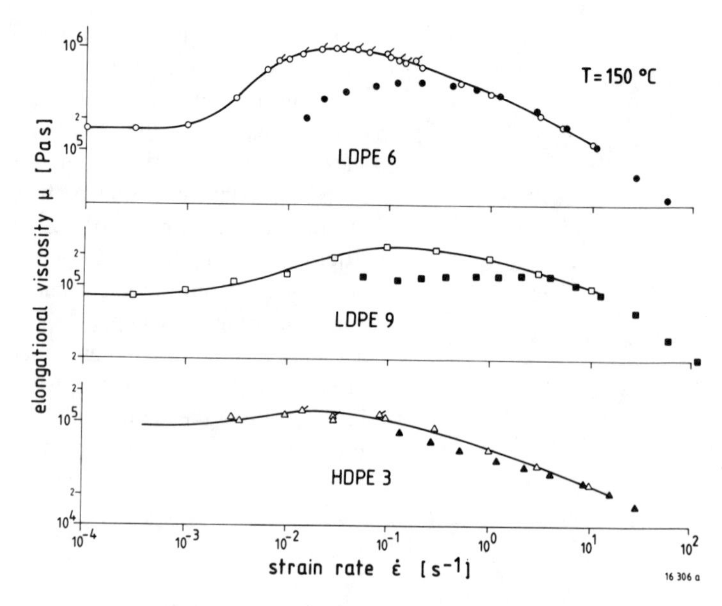

Figure 7.8.2.
Comparison of uniaxial extensional viscosities from rod pulling (open symbols), entrance flow using Cogswell's analysis (solid symbols), and tensile creep (solid symbols with ticks) measurements. From Laun and Schuch (1989).

axisymmetric and planar contractions. His analysis also considers both shear and extensional components in the contraction. In addition, the extensional viscosity is assumed to be a power law function of $\dot{\epsilon}$

$$\eta_u = s\dot{\epsilon}^{t-1} \tag{7.8.14}$$

with $t > 1$ for an extension-thickening fluid and $t < 1$ for an extension-thinning fluid. By minimizing the overall energy dissipated in the entrance region, an expression relating Δp_{en} and the power law parameters of the shear and extensional viscosities, shear rate, and the contraction ratio is obtained. The expressions and the calculations of Binding's analysis are more elaborate than those of the sink flow and Cogswell's analyses. Interestingly, both Cogswell's and Binding's analyses predict the same value for t when η_e is plotted versus $\dot{\epsilon}$; this can also be shown theoretically (Tremblay, 1989).

Binding and co-workers (1988, 1990) used his analysis to estimate the extensional viscosity of polymer solutions. For an aqueous polyacrylamide solution, Binding and Walters (1988) found that the entrance flow η_u agreed within the experimental scatter of the fiber spinning data (Figure 7.8.3); however, for a Boger fluid the predictions were considerably apart. Tremblay (1989) compared sink flow, Cogswell's, and Binding's analyses for polyethylene melts. For LLDPE he found that the sink flow and Binding predictions were reasonably close, while the Cogswell prediction was considerably larger in magnitude (Figure 7.8.4).

Figure 7.8.5 shows the flow field for a Boger fluid obtained by Binding and Walters (1988). The pictures show that as the flow rate increases, vortices generated in the corners of the contraction increases in size as a result of the extension-thickening nature of the fluid.

Figure 7.8.3.
Data from fiber spinning (symbols, with different symbols corresponding to different flow rates) and orifice entrance flow using Binding's (1988) analysis (line) measurements for a 1% aqueous solution of polyacrylamide. From Binding and Walters (1988).

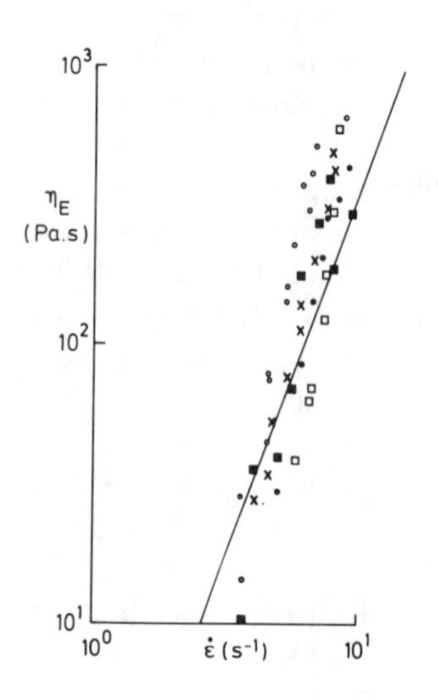

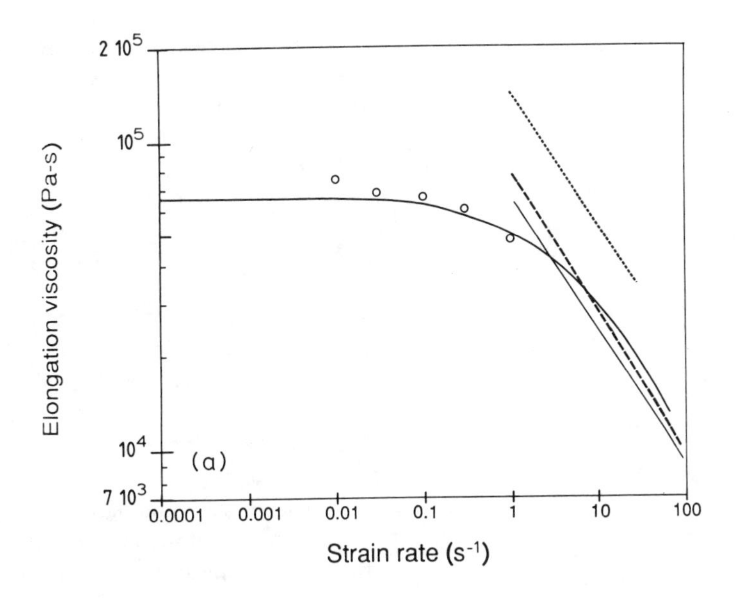

Figure 7.8.4.
Comparison between Cogswell's (dotted line), Binding's (dashed line), and Tremblay's sink flow (solid straight line) predictions of uniaxial extensional viscosity for LLDPE at 155°C. The symbols are data from rod pulling, and the continuous curve is a prediction based on Larson's constitutive equation (see Table 4.4.2). From Tremblay (1989).

The complexities of the entrance flow are clearly evident from Figure 7.8.5, and the analyses above are simplified interpretations. The reviews of Boger (1987), White et al. (1987), and Binding (1991) provide an excellent overview of the entrance flow problem. The greatest advantage of the entrance flow is that it is the easiest method to obtain extensional data. It can be especially useful in quickly ranking the extensional effects. For example, Laun and Schuch (1989) made single point Δp_{en} measurements with an orifice die, and using Cogswell's analysis they were able to

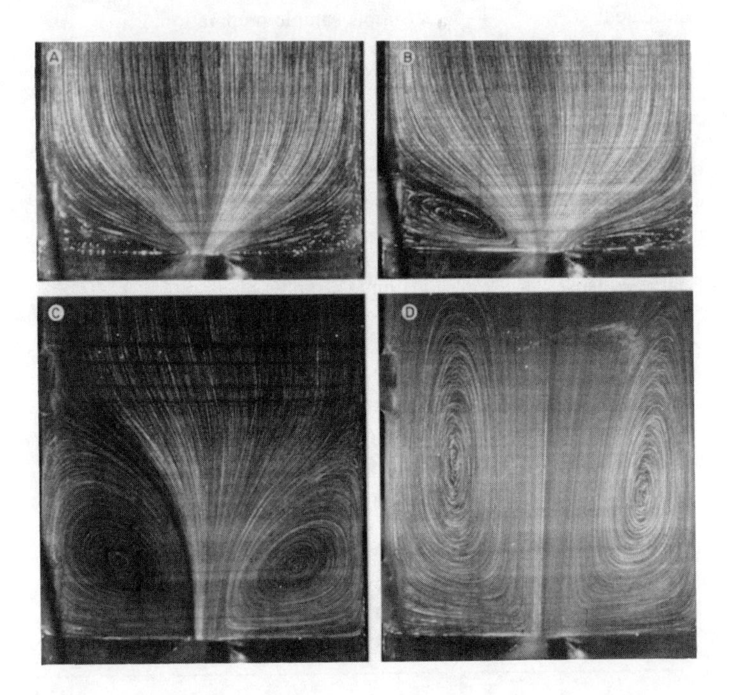

Figure 7.8.5.
Flow fields of a Boger fluid at various flow rates in an axisymmetric contraction with a contraction ratio of 14.375: (A) $Q = 0.08$ mL/s, (B) $Q = 0.18$ mL/s, (C) $Q = 0.5$ mL/s, and (D) $Q = 0.76$ mL/s. From Binding and Walters (1988).

differentiate the extensional flow behavior of various polyethylene melts. Thus, experimental evidence seems to indicate that entrance flows give valuable information that increases our understanding and use of extensional rheology.

7.9 Summary

Table 7.9.1 summarizes the main flow geometries that have been tested as extensional rheometers. They are listed from top to bottom in the order of this chapter, but also from rheometer to indexer, and generally from use with more viscous to less viscous test samples. The key advantages and disadvantages of each method are noted. The types of material function that these rheometers and indexers can measure were summarized in Figure II.3.

Extensional rheometry will continue to be an area of active research for some time. The critical problems in extensional rheom-

TABLE 7.9.1 / Comparison of Extensional Methods

Method	*Advantages*	*Disadvantages*
Tension	• Homogeneous • "Clean" data • Windup is easiest to generate	• Requires high viscosity • Sample gripping • Low $\dot{\epsilon}$ • Sample history, preparation • Need bath
Lubricated compression	• Simple sample preparation, grip • Easy to generate small displacement	• Need lubricant • High η • $\eta_b < \eta_e$ • $\epsilon \leq 2$
Fiber spinning, ductless siphon	• Low η • Process simulator • Sample prep easy	• Entrance condition • Corrections; g, F_D • Photo
Bubble collapse	• Simple sample preparation • Process simululation	• Transparent for photo • $\eta_b < \eta_e$ • Not homogeneous
Stagnation	• Large strain center but nonhomogeneous • Birefringence • Opposed nozzles low η, convenience wide range of $\dot{\epsilon}$	• Wall effects (e.g., shear) • Lubricant • Stability of flow
Entrance flows	• Simplest • Wide η range • Process simululation	• Complex flow

etry are elimination of shear due to confining walls and the difficulty in determining the role of strain in nonhomogeneous flows. No extensional device can compare with such shear rheometers as concentric cylinders, cone and plate, or capillary for generating a pure deformation to large strains over a wide range of rates and test liquids. However, we are able to recommend a few methods for families of materials.

For high viscosity systems like polymer melts, where a solid sample can be prepared, grabbed, melted, and then tested, the first choice is tension of a cylindrical or rectangular sample. This is a homogeneous deformation that provides true η_u^+ data. The rotating clamp or simple windup is the easiest gripping method. High extension rates are still difficult with tension devices.

By far the easiest extensional method is the last listed in Table 7.9.1, entrance pressure drop. Entrance pressure drop data can be readily obtained with a capillary rheometer over a very wide range of rates and viscosities. More work is needed to determine the utility of data from this attractive extensional indexer. Can it give information on extensional response of new materials? Can we guess extensional parameters for a constitutive equation and use entrance flow to test them (eg. Schunk and Scriven, 1990)?

In the low viscosity range, the opposed-nozzle device seems most attractive. Like entrance pressure drop, it is an indexer, but the flow from opposed nozzles is more nearly extensional.

Lubricated squeezing is also attractive because of its simple operation and sample preparation. The flow is homogeneous, but loss of lubricant at relatively low strains is discouraging. The fact that biaxial extension is not as strong a flow as uniaxial reduces the interest in making many biaxial experiments. Results to date show no surprises in biaxial response. Current constitutive equations can predict biaxial behavior from shear and uniaxial measures.

References

Au Yueng, V.; Macosko, C. W., in *Proceedings of the Eighth International Congress on Rheology, Vol. III*; Astarita, G. Marrucci, G., Nicolais, L., Eds., Plenum: New York, 1980; p. 717; *Mod. Plast.* 1981.

Batchelor, G. K., *An Introduction to Fluid Dynamics*; Cambridge University Press: Cambridge, 1967.

Binding, D. M., *J. Non-Newtonian Fluid Mech.* 1988, *27*, 193.

Binding, D. M., *J. Non-Newtonian Fluid Mech.* 1991, *41*, 27.

Binding, D. M.; Walters, K., *J. Non-Newtonian Fluid Mech.* 1988, *30*, 233.

Binding, D. M.; Jones, D. M.; Walters, K., *J. Non-Newtonian Fluid Mech.* 1990, *35*, 121.

Boger, D. V., *Annu. Rev. Fluid Mech.* 1987, *19*, 157.

Boger, D. V.; Binnington, R. J., *J. Non-Newtonian Fluid Mech.* 1990, *35*, 339.

Bulters, M. J. H.; Meijer, H. E. H., *J. Non-Newtonian Fluid Mech.* 1990, *38*, 43.

Cai, J.; Souza Mendes, P. R.; Macosko, C. W.; Scriven, L. E.; Secor, R. B., in *Theoretical and Applied Rheology*; Moldenaers, P., Keunings, R., Eds.; Elsevier: 1992; p. 1012.

Cathey, C. A.; Fuller, G. G., *J. Non-Newtonian Fluid Mech.* 1988, *30*, 303.

Cathey, C. A.; Fuller, G. G., *J. Non-Newtonian Fluid Mech.* 1990, *34*, 63.

Chatraei, S., M. S. thesis, University of Minnesota, 1981.

Chatraei, S.; Macosko, C. W.; Winter, H. H., *J. Rheol.* 1981, *25*, 433.

Chung, S. C. -K., Stevenson, J. F., *Rheol. Acta* 1975, *14*, 832.

Cogswell, F. N., *Plast. Polym.* 1968, *36*, 109.

Cogswell, F. N., *Polym. Eng. Sci.* 1972, *12*, 64.

Cogswell, F. N., *J. Non-Newtonian Fluid Mech.* 1978, *4*, 23.

Connelly, R. W.; Garfield, L. J.; Pearson, G. H., *J. Rheol.* 1979, *23*, 651.

Dealy, J. M., *Rheometers for Molten Plastics*; Van Nostrand Reinhold: New York, 1982.

Dealy, J. M.; Farber, R.; Rhi-Sansi, J.; Utracki, L., *Trans. Soc. Rheol.* 1976, *20*, 455.

Demarmels, A.; Meissner, J., 1985, *Rheol. Acta* 1985, *24*, 253.

Denn, M. M., in *The Mechanics of Viscoelastic Fluids*; Rivlin, R. S. Ed.; ASME: New York, 1977.

Denson, C. D.; Hylton, D. C., *Polym. Eng. Sci.* 1980, *20*, 535,.

DeVries, A. J.; Bonnebat, C., *Polym. Eng. Sci.* 1976, *16*, 93.

Everage, A. E.; Ballman, R. L., *J. Appl. Polym. Sci.* 1976, *20*, 1137.

Everage, A. E.; Ballman, R. L., *Nature* 1978, *273*, 213.

Fano, G., *Arch. Fisiol.* 1908, *5*, 365.

Franck, A.; Meissner, J., *Rheol. Acta* 1984, *23*, 117.

Fuller, G. G.; Cathey, C. A.; Hubbard, B.; Zebrowski, B. E., *J. Rheol.* 1987, *31*, 235.

Garritano, R. E., Rheometrics Inc., personal communication, 1982.

Hencky, H., *Z. Angew. Math. Mech.* 1924, *4*, 323.

Hsu, J. C.; Flummerfelt, R. W., *Trans. Soc. Rheol.* 1975, *19*, 523.

Hsu, J. C.; Flummerfelt; R. W., Schneider; W. P.; Everett, R. L., Extensional Flow Measurements with an Elongating Drop Technique, I. The Method, Instrumentation and Technique for Controlled Testing, presented at AIChE annual meeting, 1977.

Hudson, N. E.; Ferguson, J.; Warren, B. C. H., *J. Non-Newtonian Fluid Mech.* 1988, *30*, 251.

Ide, Y.; White, J. L., *J. Appl. Polym. Sci.* 1976, *20*, 2511; 1978, *22*, 1061.

Isayev, A. I.; Azari, A. D., *Rubber Chem. Technol.*, 1986, *5*, 868.

Ishizuka, O.; Koyama, K., *Polymer* 1980, *21*, 164.

Johnson, E. D.; Middleman, S., *Polym. Eng. Sci.* 1978, *18*, 963.

Jones, D. M.; Walters, K.; Williams, R. W., *Rheol. Acta* 1987, *26*, 20.

Joseph, D. D.; Arney, M. S.; Gillberg, G.; Hu, H.; Hultman, D.; Verdier, C.; Vinagre, T. M. *J. Rheol.* 1992, *36*, 621.

Kawabata, S.; Kawai, T., *Adv. Polym. Sci.* 1977, *24*, 89.

Keller, A; Muller, A. J.; Odell, J. A., *Progr. Colloid Polym. Sci.* 1987, *75* 179.

Khan, S. A.; Larson, R. G., *Rheol. Acta* 1991, *30*, 1.

Khan, S. A.; Prudhomme, R. K.; Larson, R. G., *Rheol. Acta* 1987, *26*, 144.

Laun, H. M.; Munstedt, H., *Rheol. Acta* 1976, *15*, 517.

Laun, H. M.; Munstedt, H., *Rheol. Acta* 1978, *17*, 415.

Laun, H. M.; Schuch, H., *J. Rheol.* 1989, *33*, 119.

Li, L.; Masuda, T.; Takahashi, M., *J. Rheol.* 1990, *34*, 103.

Mackley, M. R.; Keller, A., *Phil. Trans. Roy. Soc.* 1975, *278A*, 1276.

Macosko, C. W.; Lorntson, J. M., *SPE Tech. Pap.* 1973, *19*, 461.

Macosko, C. W.; Ocansey, M. A.; Winter, H. H., in *Proceedings of the Eighth International Congress on Rheology*; Astarita, G., Marrucci, G., Nicolais, L., Eds.; Plenum: New York, 1980; p. 723.

Macosko, C. W.; Ocansey, M. A.; Winter, H. H., *J. Non-Newtonian Fluid Mech.* 1982, *11*, 301.

Mathys, E. F., *J. Rheol.* 1988, *32*, 773.

Meissner, J., *Rheol. Acta* 1969, *8*, 78.

Meissner, J., *Rheol Acta* 1971, *10*, 230.

Meissner, J., *Chem. Eng. Commun.* 1985, *33*, 159.

Meissner, J., *Polym. Eng. Sci.* 1987, *27*, 537.

Meissner, J., Raible, T. Stephenson, S. E., *J. Rheol.* 1981, *25*, 1; 673.

Meissner, J.; Stephenson, S. E.; Demarmels, A.; Portman, P., *J. Non-Newtonian Fluid Mech.* 1982, *11* , 221.

Metzner, A. B.; Metzner, A. P., *Rheol. Acta* 1970, *9*, 174.

Mikkelsen, K. J.; Macosko, C. W.; Fuller, G. G., in *Proceedings of the 10th International Congress on Rheology, Vol. 2*; Uhlherr; P. H. T., Ed., Sydney, 1988; p. 125.

Moan, M.; Magueur, A., *J. Non-Newtonian Fluid Mech.* 1988, *30*, 343.

Muller, R.; Froelich, D., *Polymer* 1985, *26*, 1477.

Munstedt, H., *Rheol. Acta* 1975, *14*, 1077.

Munstedt, H., *J. Rheol.* 1979, *23*, 421.

Munstedt, H.; Laun, H. M., *Rheol. Acta* 1981, *20*, 117.

Munstedt, H.; Middleman, S., *J. Rheol.* 1981, *25*, 29.

Papanastasiou, A. C.; Scriven, L. E.; Macosko, C. W., *J. Rheol.* 1983, *27*, 387.

Papanastasiou, A. C.; Scriven, L. E.; Macosko, C. W., *J. Non-Newtonian Fluid Mech.* 1984, *16*, 53.

Papanastasiou, A. C.; Macosko, C. W.; Scriven L. E., *J. Numer. Methods Fluids* 1986, *6*, 819.

Pearson, G. H.; Middleman, S., *AIChE J.* 1977, *23*, 714.

Petrie, C. J. S., *Elongational Flows*; Pittman: London, 1979.

Petrie, C. J. S., *J. Non-Newtonian Fluid Mech.* 1990, 37, 37.

Raible, T; Demarmels, A.; Meissner, *J., Polym. Bull.* 1979, *1*, 397.

Rhi-Sausi, J.; Dealy, J. M., *Polym. Eng. Sci.* 1981, *21*, 227.

Rivlin, R. S.; Saunders, D. W., *Phil. Trans.* 1951, *A243*, 251.

Secor, R. B., Ph.D. thesis, University of Minnesota, 1988.

Secor, R. B.; Macosko, C. W.; Scriven, L. E., *J. Non-Newtonian Fluid Mech.* 1987, *23*, 355.

Secor, R. B.; Schunk, P. R.; Hunter, T. B.; Stitt, T. F.; Macosko, C. W.; Scriven, L. E., *J. Rheol.* 1989, *33*, 1329.

Schmidt, L. R.; Carley, J. F., *Polym. Eng. Sci.* 1975, *15*, 51; *Int. J. Eng. Sci.* 1975, *13*, 563.

Schunk, P. R.; deSantos, J. M.; Scriven, L. E., *J. Rheol.* 1990, *34*, 387.

Schunk, P. R.; Scriven, L. E., *J. Rheol.* 1990, *34*, 1085.

Shaw, M. T., in *Proceedings of the Seventh International Congress on Rheology*, Gothenburg, 1976; p. 304.

Soskey, P. R.; Winter, H. H., *J. Rheol.* 1985, *29*, 493.

Sridhar, T.; Gupta, R. K., *J. Non-Newtonian Fluid Mech.* 1988, *30*, 285.

Sridhar, T; Tirtaetmadja, V.; Nguyen, D. A.; Gupta, R. K., *J. Non-Newtonian Fluid Mech.* 1991, *40*, 271.

Stevenson, J. F.; Chung, S. C. -K.; Jenkins, J. T., *Trans. Soc. Rheol.* 1975, *19*, 397.

Taylor, G. I., *Proc. R. Soc.* 1934, *A146*, 501.

Treloar, L. R. G., *The Physics of Rubber Elasticity*, 3rd ed.; Clarendon Press: Oxford, 1975.

Tremblay, B., *J. Non-Newtonian Fluid Mech.* 1989, *33*, 137.

Trouton, F. T., *Proc. R. Soc.* 1906, *77*, 426.

Van Aken, J.; Janeschitz-Kriegl, H., *Rheol. Acta* 1981, *20*, 419.

Vinogradov, G. V.; Fikman, V. D.; Radushkevich, B. V.; Malkin, A. Ya., *J. Polym. Sci. A-2* 1970, *8*, 657.

White, S. A.; Gotsis, A. D.; Baird, D. G., *J. Non-Newtonian Fluid Mech.* 1987, *24*, 121.

Williams, P. R.; Williams, R. W., *J. Non-Newtonian Fluid Mech.* 1985, *19*, 53.

Winter, H. H.; Macosko, C. W.; Bennett, K. E., *Rheol. Acta* 1979, *18*, 323.

Yang, M. -C.; Dealy, J. M., *J. Rheol.* 1987, *31*, 113.

Zahorski, S., *J. Non-Newtonian Fluid Mech.* 1992, *41*, 309.

8

RHEOMETER DESIGN

More than 40 years after Weissenberg's pioneering proposal [Figures 4.1.1 and 8.1.1] for the measurement of shear and normal stresses during shear flow of elastic liquids, the determination of the rheological material functions of polymers in simple shear flow is still a nightmare for the experimenter.

J. Meissner et al. (1989)

8.1 Introduction

Chapters 5–7, which describe shear and extensional rheometry, give the most important deformation geometries and derive the working equations for each. These equations permit conversion of measured quantities like force, torque, pressure, and angular velocity to stress and strain on the sample. Such stress and strain data allow us to determine rheological material functions, which are needed to evaluate the parameters in particular constitutive equations.

In Chapters 5–7 we discussed difficulties in getting true stress and strain values as a result of sample-related problems like inertia, flow instabilities, shear heating, and evaporation of matrix solvent.

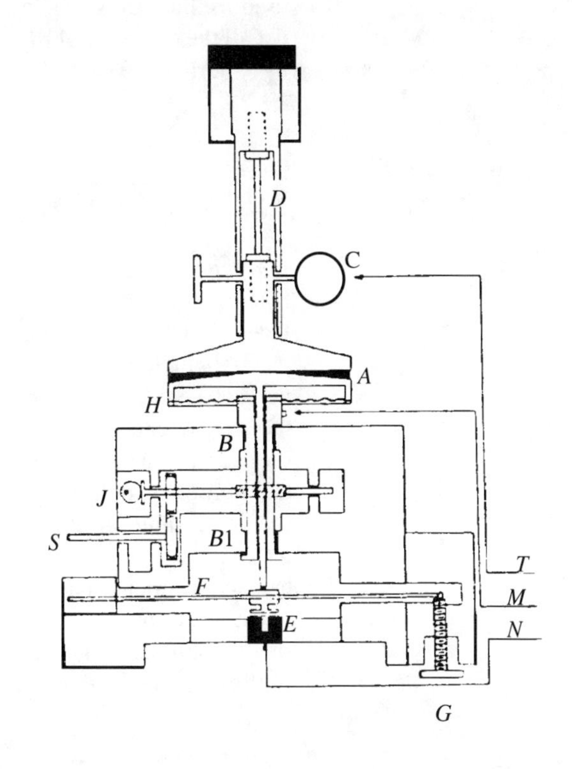

Figure 8.1.1. The Weissenberg Rheogoniometer, the first apparatus used to measure normal stresses in shear flows. The lower cone H is driven by a motor through shaft S and a gear train to a hollow spindle which is supported by bearings B and $B1$. The sample A generates a torque that twists a torsion bar D. The torque is measured by capacitance gage C. Normal thrust on the cone deflects the leaf spring F detected by the gage at E. From Jobling and Roberts (1958).

We assumed that the instrument was perfect. In this chapter we try to uncover the most important instrument imperfections. In the process we hope to show that with proper design and operation, rheological experiments need not be the nightmare experienced by Meissner and co-workers, even with polymer melts.

The organization of this chapter parallels that of its predecessors in Part II, describing first the design of drag flow rheometers, then pressure-driven ones. Separate sections are devoted to analysis of data, particularly sinusoidal oscillations, and to special designs for process-line rheological measurements. The section on extensional rheometry is brief because most of the design issues were discussed in Chapter 7.

The focus of this chapter is on how we can actually generate the deformations already described, particularly those given in Chapters 3 and 4. For example, even if the sample can follow a step strain without inertia problems, we still need an apparatus to generate that motion. No rheometer can do it instantly. We describe the major methods of rheometer control and measurement and discuss instrument compliance, imperfect temperature distribution, and similar problems. The final section gives some criteria for instrument and test selection, as well as guidelines for detecting instrument errors.

Many rheometer designs have been published. A number have been reviewed by Van Wazer et al. (1963), Whorlow (1980, 1992), Dealy (1982), and Collyer and Clegg (1988). In this chapter we concentrate on describing features that are common to commercially available instruments. Particular commercial instruments are used to illustrate specific designs; however we do not attempt to list all the commercial rheometers. Such lists have been compiled by Whorlow (1980, 1992) and Dealy (1982).

8.2 Drag Flow Rheometers

In a typical drag flow rheometer we measure the velocity or displacement of the moving surface and the force on one of the surfaces. As Chapter 5 indicates, most drag flow rheometers are based on rotary motion and use one of three geometries: concentric cylinder, cone and plate, and parallel disks. In most cases the same rotary instrument can use all three of these flow geometries. Thus we concentrate on the design of rotary instruments. The most important nonrotary, drag flow rheometers are the sliding plate devices shown in Figures 5.2.1 and 5.2.2. To generate the needed linear motion, they typically use actuators like the hydraulic pistons and ball screws found in standard tensile testing machines for solids. Solenoids or other electromechanical actuators are often used for small amplitudes and low forces. Most of the same problems of control and measurement found in the rotary devices occur in these linear ones. We will point out some features that are unique to sliding plate rheometers.

There are two basic designs of drag flow rheometers: controlled strain with stress measurement and controlled stress with strain measurement. Below we first discuss strain control and torque measurement (Section 8.2.2) followed by instrument alignment problems (Section 8.2.3) and normal stress measurement (Section 8.2.4). Then we treat special design issues for stress control. Both designs use the same type of environmental control system, as discussed in Section 8.2.6.

8.2.1 Controlled Strain

The first drag flow rheometer, Couette's concentric cylinders shown in Figure 5.1.1, was a controlled strain device. Couette fixed the angular velocity of the outer cup and measured the torque on his inner cylinder by the deflection of a suspending wire.

The Weissenberg Rheogoniometer (Figure 8.1.1) also controlled the strain and measured the generated torque and normal force. Like Couette's device, the rheometer described by Jobling and Roberts (1958) drove the outer cup. Using a rotating outer cylinder eliminates the problem of flow instabilities discussed in Chapter 5.3, eq. 5.3.42, but can present some difficulties in attaching temperature control baths.

Jobling and Roberts used an ac synchronous motor coupled to a gear box to generate a wide range of constant rotation rates. This method is still used in some rheometers: for example, the Brookfield Synchro-Electric (Whorlow, 1992, pp. 140–141). However, most rotary rheometers now use dc motors with closed-loop servo control. This permits infinite speed variation and programming of the deformation. Several instruments use feedback from a tachometer: for example, the Bohlin VOR (Bohlin, 1988) and the Rotovisco RV20 (Haake, 1986). Both these devices drive the rotating shaft *indirectly* through a gear train (e.g., Figure 8.2.1). This gives some flexibility in motor size and location, but backlash in the gears prevents smooth reversing of the motion and thus limits the deformations that can be programmed.

The most versatile controlled strain design is *direct* coupling of a dc motor to the rotating shaft. As illustrated in Figure 8.2.2, a tachometer is used to control angular velocity and a capacitance transducer is used to control angular position. Combined with a servo control system, this design has a fast response and can deliver a wide range of shear deformations to the test sample. Typical rotation rates are from 0.001 to 100 rad/s. The position control servo can generate oscillation frequencies from 10^{-5} to 500 rad/s, with an amplitude of angular motion from 50 μrad to 0.5 rad. Angular accelerations typically exceed 10^3 rad/s^2, which means a 0.1 rad rotation can be accomplished in ≤ 0.01 s. This fast transient response, illustrated in Figure 8.2.3, is important for measuring short relaxation times in stress relaxation tests (recall Figure 3.1.2a and see Figure 8.3.5) and start up of steady straining.

Sinusoidal oscillations can also be accomplished mechanically. The rotating cam (J in Figure 8.1.1) and the oscillating, eccentric solenoid arm (D in Figure 8.2.1) are examples. These mechanical devices can sometimes generate smaller amplitude oscillations than a position servo. However, they generally cannot cover as wide a range nor generate as smooth a sine wave.

Optical encoders have been used to provide feedback control of both rate and position to direct current motors (Michel, 1988; Amari et al., 1992). In contrast to capacitance transducers, which are typically limited to a $\pm$ 0.5 rad window, encoders can control position over 360°. This permits, for example, the superposition of sinusoidal oscillations on steady rotation. However, encoders are digital devices with a minimum step size around 1 mrad. With interpolation circuits encoders can resolve about 10 μrad. Thus at very low rates and low frequency there may be a long time between pulses, which can make control difficult. At high rates, accurately reading all the pulses sets an upper limit. Amari et al. report a steady rotation range of 0.001 to 3 rad/s with their stepper motor

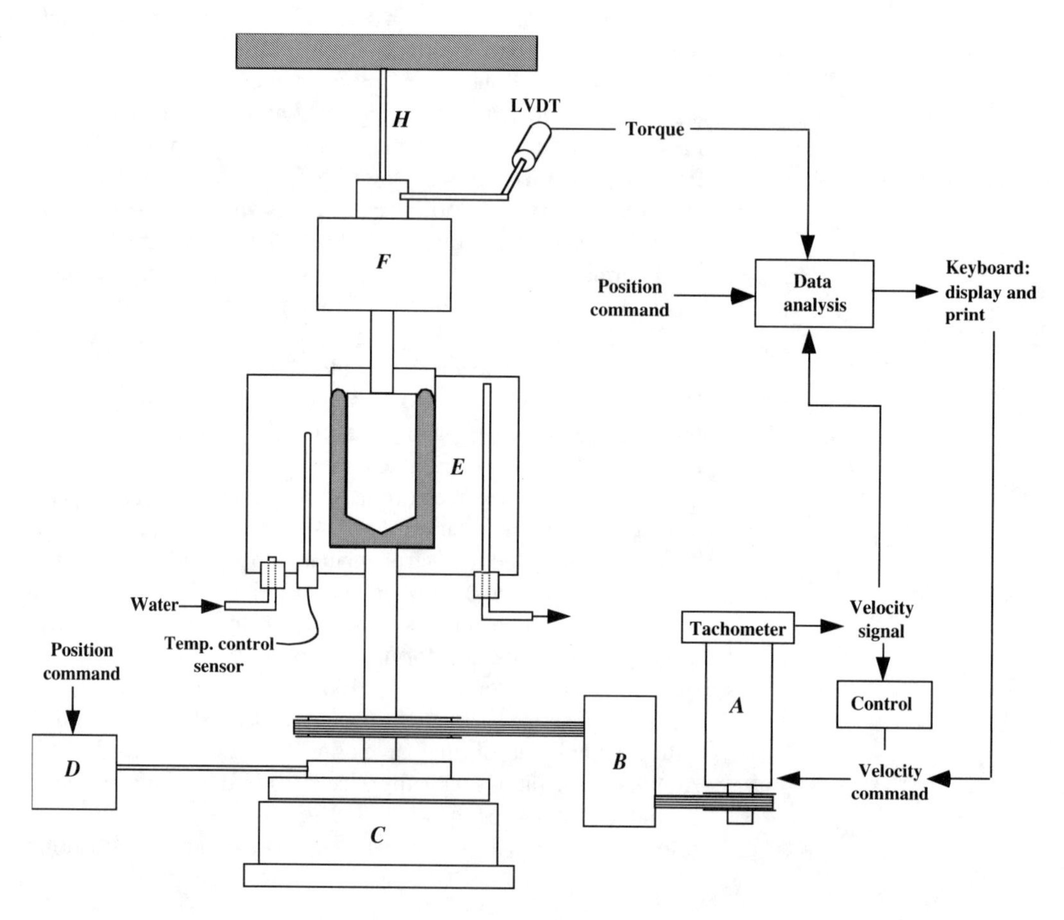

Figure 8.2.1.
Schematic diagram of the Bohlin VOR rheometer. A dc motor with tachometer A drives gear box B, which drives the timing belt to provide steady rotation of the shaft. When the electromagnetic clutch C, is activated, the eccentric arm D can oscillate the shaft sinusoidally. The sample cell is surrounded by a temperature control bath E. An air bearing F centers a torsion bar H, whose rotation is sensed by linear variable differential transformer. Adapted from Bohlin (1988).

Figure 8.2.2.
Schematic diagram of the Rheometrics RFSII. A dc motor C is controlled in steady rotation by a tachometer B and in sinusoidal or other angular position changes by a rotary variable capacitance transformer (RVCT) A. The sample is surrounded by a temperature control bath D. The upper fixture is attached to a rebalance transducer E, described further in Figure 8.2.5. Adapted from Rheometrics (1992).

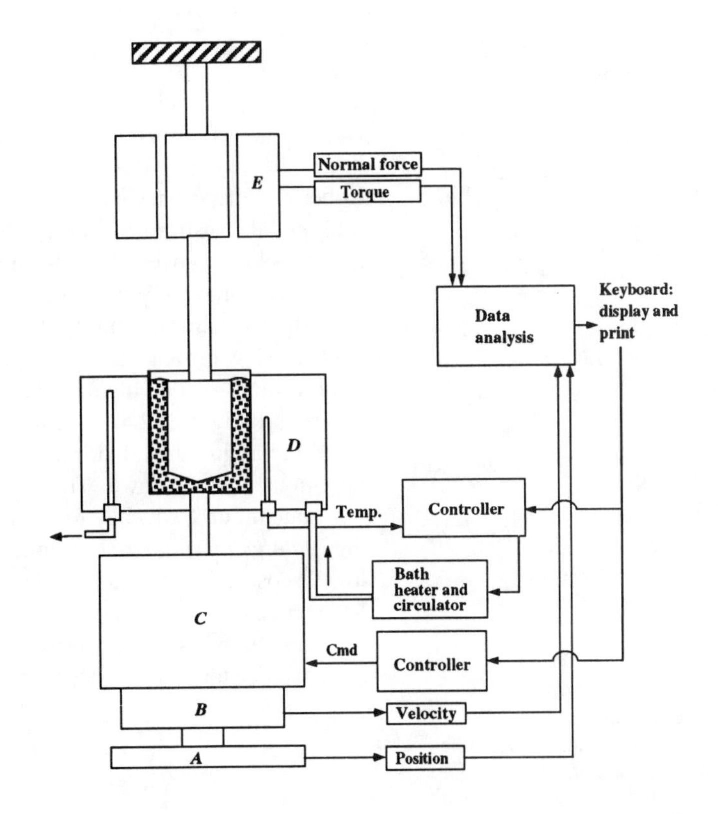

system. Inductive or capacitive coupled devices are similar to an encoder but generate a continuous series of sine waves. Using such a device for feedback gives control of velocity from 2×10^{-6} to 1 rad/s with 0.01% accuracy (Starita, 1980; Rheometrics, 1990; Whorlow, 1992).

Figure 8.2.3.
Normalized angular displacement versus time for a dc motor with RVCT feedback. The 0.1 rad step is achieved in less than 5 ms but shows some overshoot. From Starita (1980).

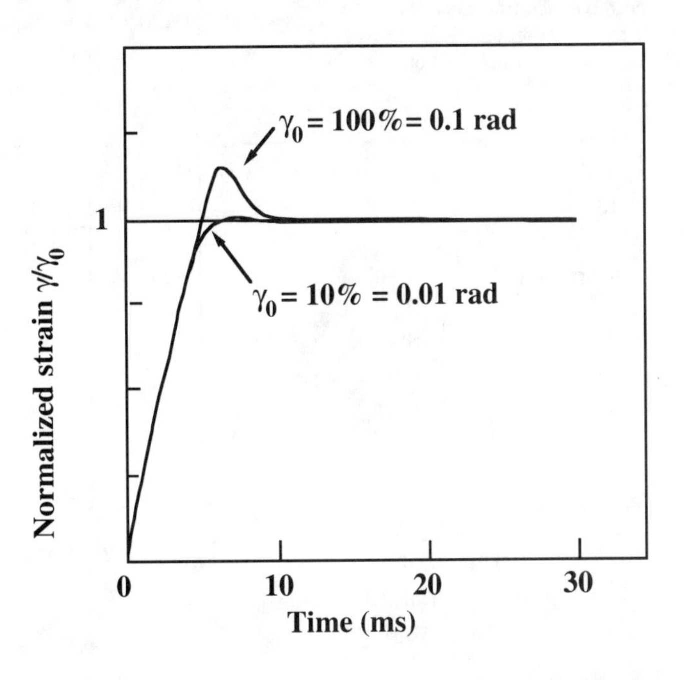

8.2.2 Torque Measurement

The measured variable in a controlled strain rheometer is the torque. Couette's original method of measuring the twist in a torsion bar has been followed up to modern times. He used a mirror; Jobling and Roberts used a capacitance gage, and Bohlin uses a linear variable differential transformer (LVDT). By using different diameter bars, it is possible to cover a very wide range of torque. Centering the bar with a radial air bearing permits measurement of torques down to below 10^{-8}Nm (Bohlin, 1988). However, such very thin torsion bars have low resonant frequency, which reduces the range of transient testing. This problem is illustrated for relaxation after steady shearing in Figure 8.2.4a. The torque in the 1 Pa · s Newtonian oil should relax instantly, but it takes about 3 seconds for the inertial vibrations to damp down. Transducer inertia and compliance also limit maximum frequency for accurate sinusoidal oscillation testing. *Sample inertia*, too, can limit the range of transient testing of low viscosity liquids, but as discussed in Chapter 5, the range can be extended by simply decreasing the sample thickness.

Figure 8.2.4a was constructed from the equation of motion for a damped torsion oscillator (Walters, 1975; Mackay et al. 1992)

$$\frac{d^2\theta}{dt^{*2}} + (g + C)\frac{d\theta}{dt^*} + \theta = 0 \qquad (8.2.1)$$

where $g = U/I\omega_n$, $C = \frac{2\pi R^3\omega_n\eta}{3K\beta}$, $t^* = t\omega_n$; θ is the angular motion of the transducer, g is the transducer damping element, and C is the test fluid damping. I is the transducer inertia, U its damping coefficient (typically due to the air bearing), ω_n resonant frequency, and K torsional spring constant ($1/K$ is the compliance). R is the radius of the cone, β its angle, and $\dot{\gamma}$ the shear rate. This equation gives good agreement with experimental transducer response. In

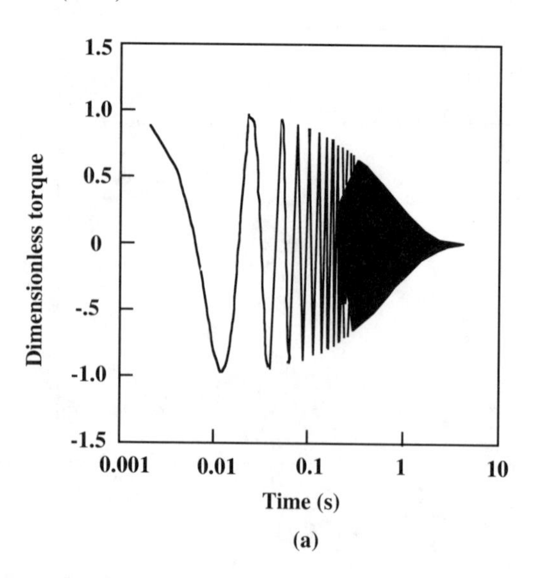

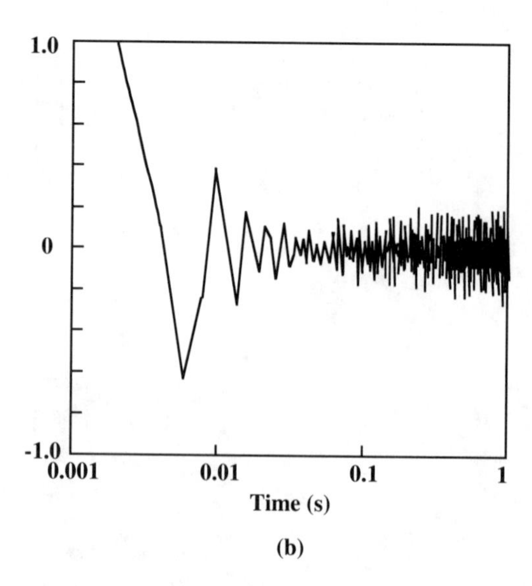

Figure 8.2.4. Reduced torque versus time after steady shear of a 1 Pa·s Newtonian oil at 10 s^{-1} with a 25 mm diameter, 0.1 rad cone. (a) Strong spring: response of a torsion bar transducer with a spring constant, K = 10 N·m/rad and 250 rad/s resonant frequency, simulated using eq. 8.2.1. (b) Force rebalance transducer with resonant frequency of 960 rad/s. From Mackay et al. (1992).

Figure 8.2.4a a stiff spring was used; a more sensitive one would oscillate more slowly and take even longer to damp out.

In one design variation the torsion bar is mounted on the rotating shaft and deflection of the bar is detected through slip rings, by a capacitive device, or even optically (Haake, 1986; van Wazer et al., 1963; Whorlow, 1992). This design puts everything on one shaft of the rheometer, reducing cost and making temperature control design simpler. However, to detect the torque through a rotating shaft, these torsion springs are typically made even more compliant than the simple torsion bars.

Even very stiff torsion bars must twist slightly to record the torque. With high viscosity polymer melts, this small twist can lead to significant errors in the strain or transient strain rate imposed on a sample (Gottlieb and Macosko, 1982). For example, consider stress relaxation after a step strain of 100% on a polymer melt with $G_0 = 10^5$ Pa (see Figure 3.3.6). Using a 0.1 rad, 25 mm diameter cone, to achieve a strain of 100% requires an initial torque of 0.4 N·m. However, if the transducer stiffness is 10 N·m/rad then it will twist 0.04 rad, and the true strain in the sample will only be 60%. There are also transient errors. For very viscous samples, the parameter C in eq. 8.2.1 dominates the damping of the transducer, and it can be used to estimate the time constant of the error:

$$\lambda = \frac{2\pi R^3 \eta}{3K\beta} \tag{8.2.2}$$

Meissner (1972) shows how this small transient deflection of the torque transducer can affect short time start-up data (see Figure 8.3.7). Gottlieb and Macosko give relations to correct G' and G'' for compliance. These corrections are made in the software in some rheometers. Corrections are reliable only if the spring constant K is known accurately, and then typically only when the measured strain is 10% or more of the commanded value.

Stiffer transducers have been made with strain gages (Macosko and Starita, 1971; Drislane et al., 1974). These devices can be very sensitive to temperature changes (Franck, 1985a). Very stiff and more thermally stable are piezoelectric transducers. Laun and Hirsch (1989) were able to measure stress overshoot and stress relaxation in less than 10 ms with a torque and normal force quartz load cell. However, it is difficult to get low torque levels with piezoelectric transducers. Furthermore, they are capacitance devices with a slowly decaying signal, making it difficult to accurately measure long relaxation processes.

One solution to transducer deflection is to eliminate it with a feedback control servo; see Figure 8.2.5 (Franck, 1985a). The transducer is essentially a dc motor in which torque is measured by the current needed to prevent any deflection. No servo system is instantaneous, and a combination of high frequency and torque can lead to transducer compliance (Mackay and Halley, 1991). The upper frequency is about 100 rad/s, whereas stiff torsion bar systems have resonant frequencies above 10^3 rad/s and piezoelectric sys-

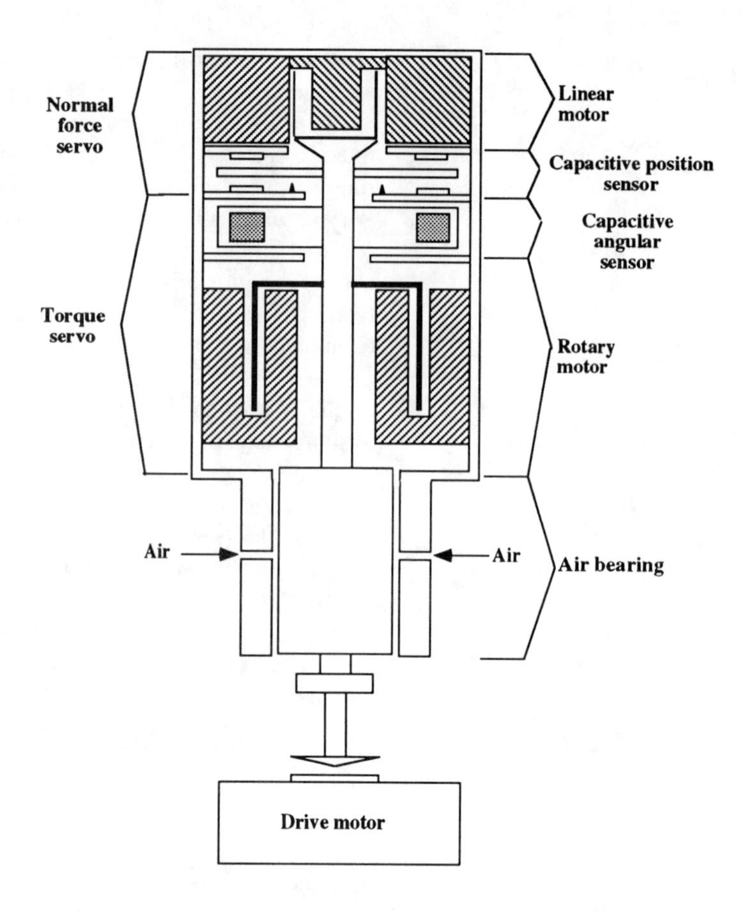

Figure 8.2.5. Force rebalance transducer. From Franck (1985a).

tems even higher. However, as Figure 8.2.4b indicates, the active servo system results in tremendously improved time response for low viscosity liquids, where more compliant torsion springs must be used. A key advantage of force rebalance transducers is their wide torque range, 1 to 10^4 or 10^5, with good low torque resolution, 10^{-7}N · m or less (Whorlow, 1992; Rheometrics, 1992).

A very different approach is to measure shear stress directly with a transducer mounted in the wall (Giacomin et al., 1989). The deflection of a small disk flush with the wall is proportional to the shear force acting on its surface. Such a device can be mounted in rheometers of many different types. Figure 8.2.6 illustrates a wall shear stress transducer in a sliding plate rheometer. The wall transducer is not affected by edge failure (e.g., Figure 5.4.7), sample area, or degradation and drying of exposed surfaces. Like a torsion bar, the cantilever design is subject to errors due to transducer inertia and compliance. Wall stress transducers have been designed mainly for high viscosity samples. It is necessary to periodically remove test material, which leaks into the clearance around the disk. This material can also damp transient measurements (Dealy, 1992).

Figure 8.2.6 also illustrates a sliding plate shear rheometer design. If a wall shear stress transducer is not used, then the total force on the fixed plate is usually measured with a strain gage load cell. The linear actuators used have time constants in the

Figure 8.2.6.
Wall shear stress transducer mounted in a sliding plate rheometer. Shear between the sample and the wall causes the cantilever to deflect slightly, which motion is detected by the capacitance probe. Adapted from Interlaken (1992).

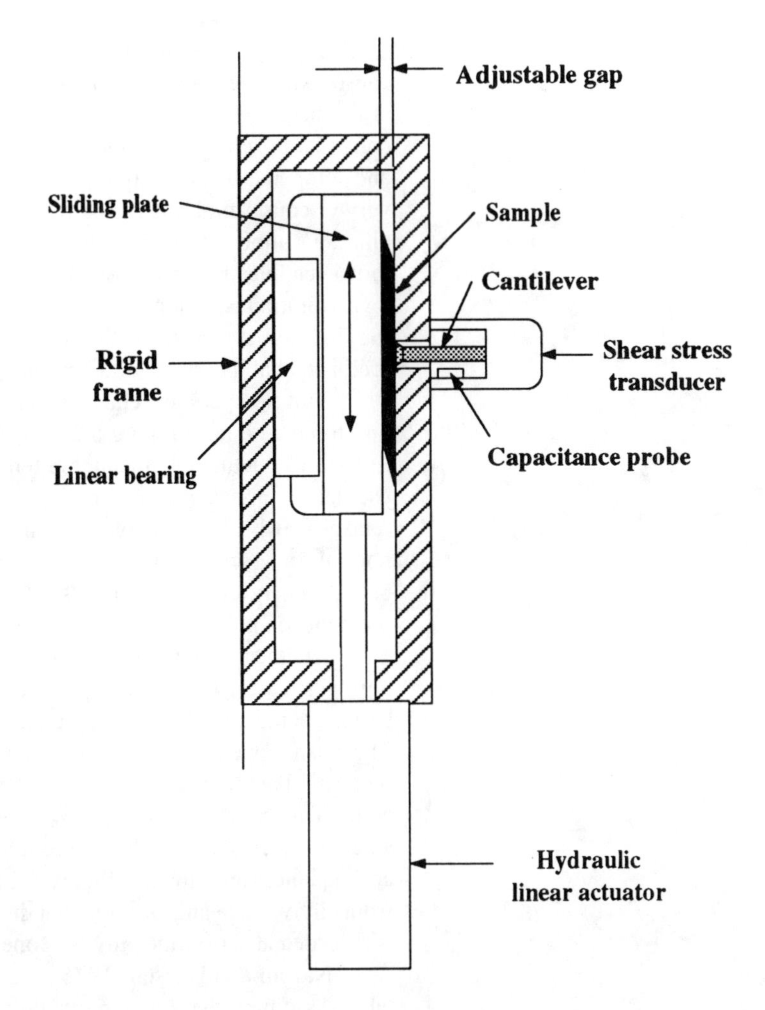

5–10 ms range. They are servo controlled similar to the rotary actuators. Hydraulic drivers can deliver much more force than a typical dc motor. The flat design of sliding plate rheometers can be an advantage for combining optical or scattering experiments with rheometry (Kannan and Kornfield, 1992; Koppi et al., 1993; see also Chapter 9).

8.2.3 Normal Stresses

Typically only more sophisticated rotary rheometers measure normal stresses. When measurements are made, they usually entail total thrust in the cone and plate or parallel disk geometries. The Weissenberg Rheogoniometer transferred the thrust through the driven shaft to a leaf spring (Figure 8.1.1). Other rheometers have used multiaxis strain gage (Macosko and Weissert, 1974; Drislane et al., 1974) or piezoelectric load cells (Laun and Hirsch, 1989). A rebalance transducer design for both torque and normal force was shown in Figure 8.2.5.

Trying to make normal force measurements on polymer melts caused Meissner and co-workers (1989) the nightmares referred to at the beginning of this chapter. One source of such nightmares consists of oscillations in the thrust readings due to bearing runout and temperature fluctuations. Because of their slow relaxation, very viscous samples in the narrow cone and plate gap transmit oscillations almost directly to the thrust transducer. Smooth, normal thrust readings become possible when the instrument is mounted on a vibration isolation table, gear noise eliminated, and the rotating shaft directly driven with a dc motor rotating on high precision, preloaded ball bearings, or a stiff air bearing.

Tiny temperature changes can also wreak havoc with normal force readings. Figure 8.2.7 illustrates the potential effect of $\pm$ 0.02°C fluctuations in sample temperature. The normal force oscillations are caused by sample expansion or contraction, which relaxes slowly as a result of the high sample viscosity. A rebalance normal force servo can greatly reduce this noise (Franck, 1985a), but ultimately temperature needs to be very stable for reliable measurements on polymer melts (Meissner et al., 1989).

Another nightmare is the accurate measurement of *transient* normal stresses. In the start-up of steady shear, as the normal force builds up, the transducer must deflect slightly to measure the force. This pushes open the gap between cone and plate (or parallel disks) by a small amount, and the sample flows toward the center of the cone. This cross flow reduces the true normal force reading. The smaller the cone angle, the longer the material takes to flow. The consequences are shown in Figure 8.2.8. After 15 seconds, the data from all five cone angles agree; but the true normal force overshoot is recorded at short time only for cone angles of 0.1 rad or larger.

Nazem and Hansen (1976) analyzed the gap opening problem. Assuming that the gap opening is infinitesimally small and the liquid is Newtonian, the time for the transducer to reach 63% of the total travel when a step normal force is applied is

$$\lambda_{\text{melt}} = \frac{6\pi R\eta}{K_z \beta^3} \tag{8.2.3}$$

Applying this equation to the data in Figure 8.2.8 gives 2.6 seconds for the 0.035 rad (2°) cone, 0.3 seconds for the 0.07 rad cone, and

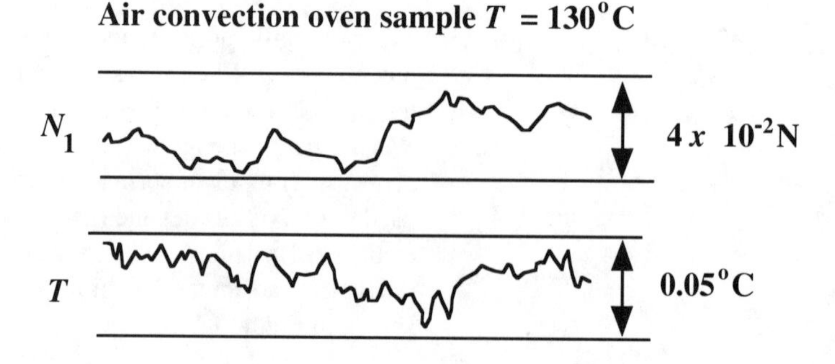

Figure 8.2.7. Oscillation in normal force due to small temperature fluctuations in a viscous polyethylene sample (Franck, 1985a). Tiny vertical oscillations in the rotating shaft (bearing runout) can lead to similar oscillations in the normal force.

< 0.1 seconds for all the larger angles, in reasonable agreement with the time for the normal stress to reach 10^5 Pa. The normal force transducer used to collect the data was specially stiffened. Greater stiffness will just accentuate the temperature and bearing runout problems. Increasing the cone angle or decreasing the radius for samples with large normal forces leads to accurate data (and more pleasant rheological dreams). A normal force rebalance transducer can virtually eliminate this problem (Franck, 1985a).

Although nearly all shear normal stress data reported in the literature are measured by total thrust methods, some work has been done with the distribution of pressure across the plate with small flush transducers (Christiansen and Leppard, 1974; Magda et al., 1991; see Figure 5.4.2). Meissner and co-workers (1989) have used the total thrust on a central disk of smaller radius than the sample. Both techniques permit measurement of N_2 as well as N_1. Birefringence methods, discussed in the next chapter, can also give normal stress distributions.

8.2.4 Alignment

The bearing runout problems discussed above highlight the importance of mechanical construction in rheometers. Clearly a small departure from 90° between the axis of a small angle cone and the surface of its plate will cause the shear rate to be higher on one side than on the other. The same is true for eccentricity in the axes of concentric cylinders (eq. 5.3.43); but with typical gap sizes (≥ 0.5 mm), alignment is less critical than with the cone and plate.

To load and unload samples, vertical motion is required. This motion must be precise for cone and plate rheometers. Typically

Figure 8.2.8.
Normal stress versus time upon start-up of steady shear, $\dot{\gamma} = 10\ s^{-1}$, for a polyethylene melt, $\eta_0 = 5 \times 10^4$ Pa · s. Cone diameter, 24 mm and five different cone angles. Normal force spring stiffness, $K_z = 10^5$ N/m. Replotted from Meissner (1972).

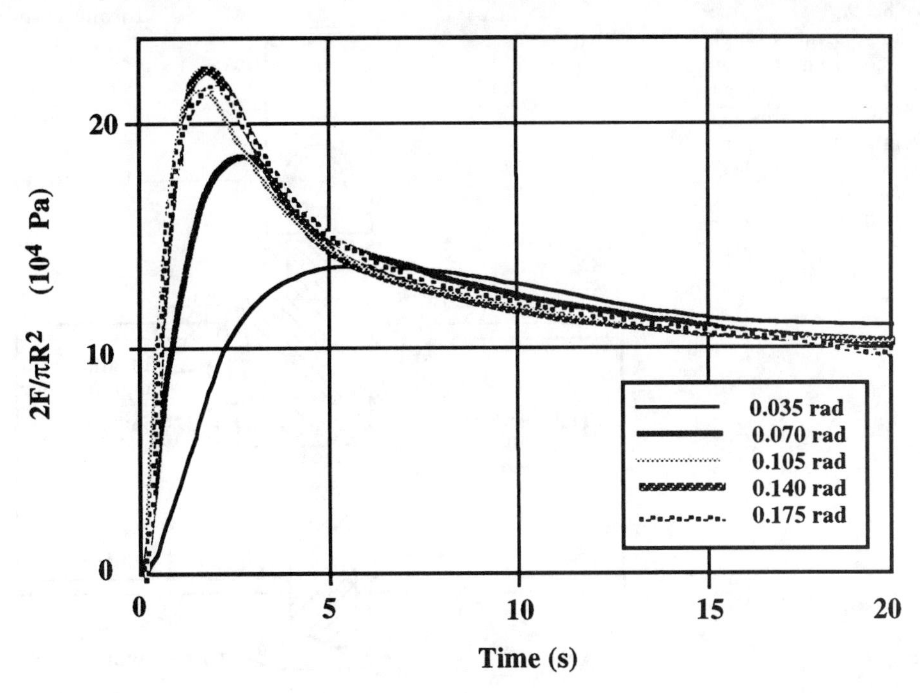

the tip of the cone, about 50 μm, is removed to avoid unwanted normal forces due to particles in the sample. Thus the gap between the truncated cone and plate must be set at this value. This is usually done without a sample by noting the vertical position at which the normal force is first sensed and then resetting that position with the sample in place. For rheometers without normal force capability, the null position is determined by electrical contact or the onset of torque under slow rotation. Increasing the sample test temperature causes the test fixtures to expand, thereby closing the gap, typically on the order of 1 μm/°C.

Precise alignment and vertical motion are critical for obtaining high shear rate data with parallel disk rheometers. As discussed in Chapter 5, several workers have reported that for gaps below 300 μm they must add as much as 40 μm to their apparent gap readings (see eq. 5.5.11). This is surprising because these investigators used rheometers that were able to set the gap and parallelism between the disks within 2 μm over the 25 mm diameter.

High shear rate data also have been obtained using very narrow gap concentric cylinder rheometers. Rather than trying to align the axes perfectly, a universal joint is used on one shaft. It is believed that the flow will be self-centering (Taylor, 1992). The inner surfaces of the cylinders are tapered inward slightly, so that lowering the inner cylinder will decrease the gap (see Figure 8.2.9). Gaps down to 3 μm and shear rates up to 10^6 s^{-1} have been reported using this method (Whorlow, 1992).

Figure 8.2.9 also illustrates a design for quick but precise mounting and removal of test fixtures. There are many such de-

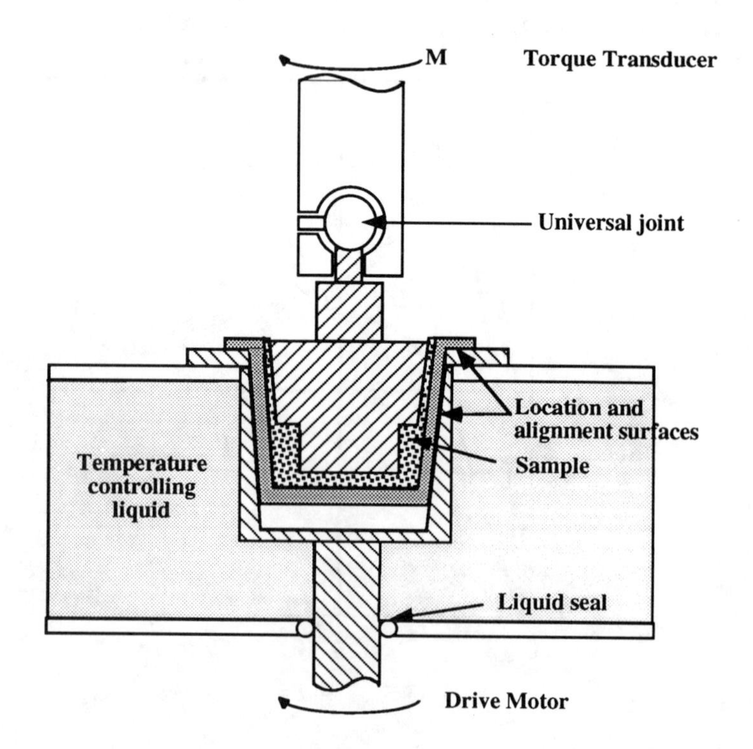

Figure 8.2.9.
Schematic diagram of higher shear rate concentric cylinder test fixtures. The tapered inner cylinder surfaces and universal joint allow operation at very narrow gaps. The outer cup is aligned to the tapered shaft and can be removed quickly from the rotating shaft and bath.

signs in commercial rheometers. Each should include two precision locating surfaces, one parallel and one perpendicular to the axis of rotation.

8.2.5 Controlled Stress

Controlling torque and measuring the angular motion has also been done since the earliest days of rotational rheometry (van Wazer et al., 1963). A string wrapped around the shaft attached over a weight to a pulley provided the constant torque. This design is simple but limited by bearing friction, total strain (how far the weight can fall), and maximum strain rate.

Torque may also be maintained in a controlled strain rheometer such as shown in Figure 8.2.2 by using feedback from the torque sensor to adjust the motor velocity or position. However, control is typically difficult because of the sample response. Ideally one should include the viscosity of the sample in the control algorithm. Performance can be improved significantly by avoiding the sample and closing the feedback loop around a torque sensor on the motor, such as motor current to a dc motor (Michel, 1988). However, brush friction in the motor limits the lowest torque levels to 10^{-3} to 10^{-4} N·m.

Zimm and Crothers (1962) used a rapidly rotating magnet to produce a constant torque on a steel pellet contained in an inner glass cylinder of a concentric cylinder viscometer. Changing the distance between the magnet and cylinder changes the torque. Very low torque levels can be maintained (10^{-9} N·m), and this design has been used to study dilute biopolymer solutions. Van den Brule and Kadijk (1992) used a concentric cylinder filled with a Newtonian reference liquid to convert a constant angular velocity to a constant driving torque. Because of the large difference between driving velocity and velocity at the sample, neither this nor the Zimm–Carothers design can be used for transient stress control.

The most versatile design and the one used in commercial controlled stress rheometers is shown in Figure 8.2.10. It consists of a magnetic field, which rotates around a copper or aluminum cup, which is supported by an air bearing. The field induces eddy currents in the cup, which tries to follow the rotating field. The torque on the cup is proportional to the square of the voltage on the stators. The device is called a drag cup motor (Davis et al., 1968; Plazek, 1968, 1980; Franck, 1985b; Berry et al. 1989).

Drag cup motors are not very efficient. It takes a high voltage to generate high torques. This high voltage can cause heating of the motor and a transient decrease in the resulting torque. Cooling and compensating circuits can extend the upper torque range to about 50 mNm. The lowest torque is limited by imperfections in the air bearing and eccentricities in the rotating shaft to about 10^{-6} N·m. A magnetically supported shaft has considerably lower residual torque but is much less stiff to axial loads than an air bearing (Plazek, 1968, 1980; Franck, 1985b).

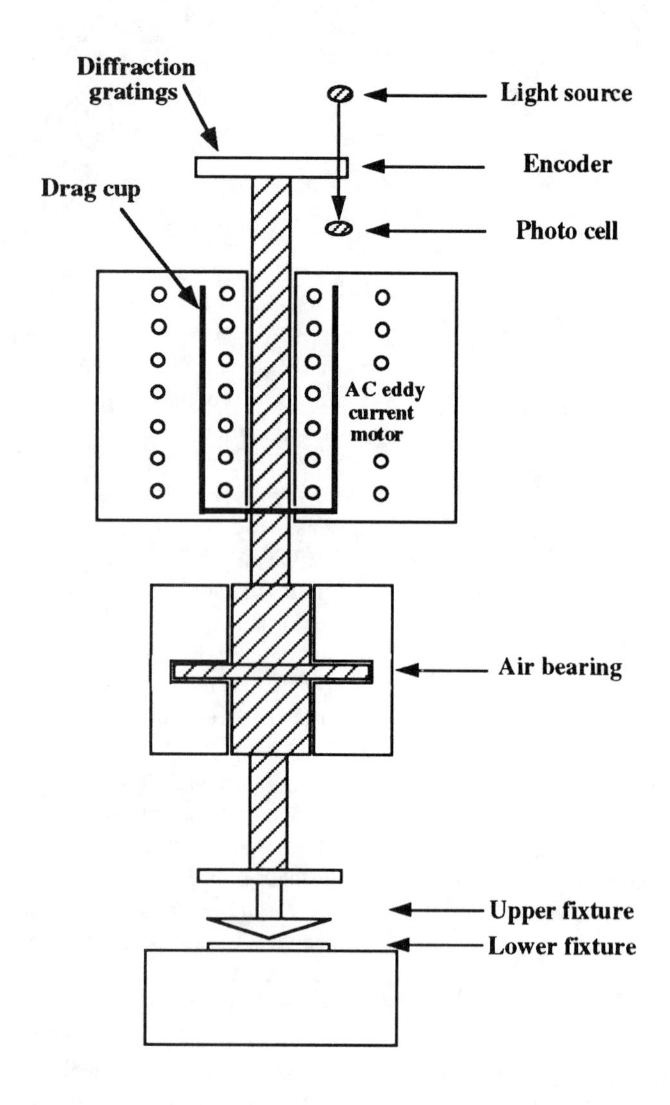

Figure 8.2.10.
Schematic diagram of an open loop, controlled stress rheometer. Torque is provided by a drag cup motor, and motion is measured with an optical encoder. Designs similar to this are used by Bohlin, Carri-Med, Haake, and Rheometrics.

Angular displacement of the shaft can be measured by a variety of methods. The optical lever similar to Couette's is still used (Magnin and Piau, 1991). Capacitance transducers can be very sensitive to angular position (Franck, 1985b) but are linear over only a limited angular range (Giles and Denn, 1990). Optical encoders have a great advantage because they divide 360° into uniform steps, as small as 5 μrad/step with recent models. This step size translates to less than 0.01% strain with a 0.1 rad cone. Yet with high speed counters it is possible to use such encoders to measure velocity up to 100 rad/s.

A major limitation in controlled stress rheometers is instrument inertia. It is very similar to the transducer inertia described with eq. 8.2.1. The torque imposed on the drag cup must overcome reluctance torque, air bearing friction, and rotor inertia as well as the sample viscosity. The inertia portion dominates with low viscosity samples (Kreiger, 1990). Figure 8.2.11a illustrates the problem. Stress was commanded to increase linearly from 0 to

Figure 8.2.11.
Errors introduced in transient viscosity measurements (broken lines) as a result of instrument inertia can be eliminated by an active control loop (—). (a) Stress ramp 0–90 seconds for a 5 mPa·s Newtonian standard. Adapted from Franck (1992). (b) Sinusoidal oscillations on a 100 mPa·s standard.

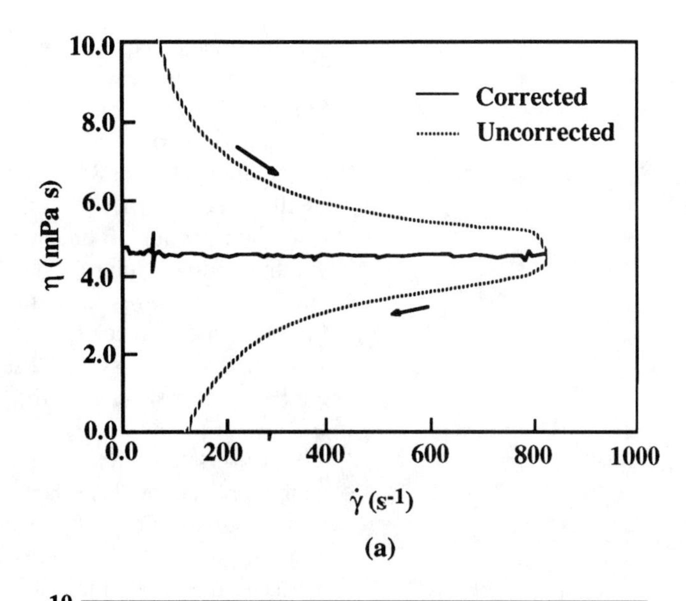

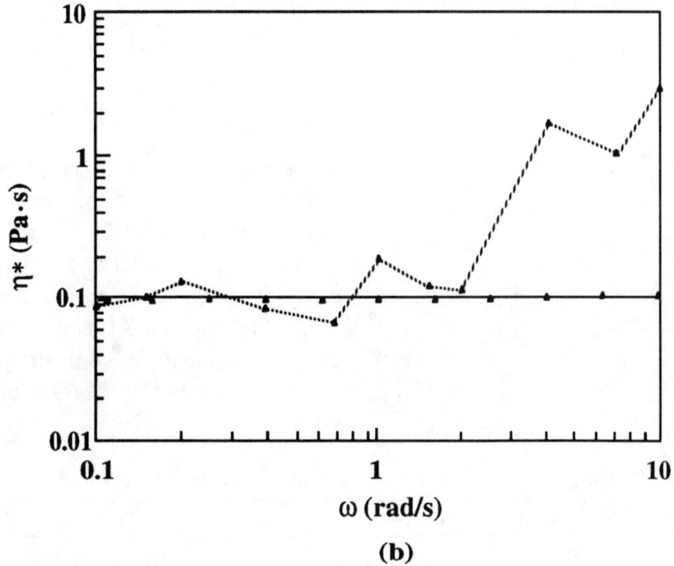

0.5 Pa in 90 seconds. The response time of the system with parallel disk geometry is

$$\lambda_{\text{inert}} = \frac{2hI}{\pi R^4 \eta} \tag{8.2.4}$$

where h is the sample gap and I is the inertia of rotor and the upper test fixture. For the sample and instrument used in Figure 8.2.11a, $\lambda_{\text{inert}} = 7.5$ seconds. Thus the sample never sees the commanded torque (although it gets closer at long times), and the Newtonian standard looks like a strongly thixotropic material. If the stress is ramped at a slower rate, the effect of course is reduced. However, Kreiger (1990) has shown that it is relatively straight-forward to use the equations of motion to correct the data. This can be done automatically in the instrument software, with the results illustrated in Figure 8.2.11a (Franck, 1992).

Rotor inertia affects all transient measurements on low viscosity fluids. Figure 8.2.11b illustrates the problem in sinusoidal oscillation testing. Even for this relatively high viscosity standard, it was not possible to obtain accurate data above about 0.2 rad/s without compensating for inertia. The most difficult corrections are for step changes in torque like start-up and recovery. Even for these cases, Franck (1992) reports that his correction software yields true values less than 2 seconds after the step for a very low 5 mPa · s, viscosity standard.

With proper inertia correction controlled stress rheometers are very versatile and may replace controlled strain instruments for many applications. The fact that stress and strain are measured on the same shaft in controlled stress instruments allows lower cost and simpler temperature control but also is the source of the inertia limitations. Table 8.2.1 summarizes the advantages and disadvantages of each control mode.

8.2.6 Environmental Control

The two types of rotational rheometer, controlled strain and controlled stress, can use the same environmental control. It can be as important to control the temperature, pressure, or humidity of a sample as it is to control the shear stress. Yet these factors, especially temperature, seem to be less exciting to the engineers who design rheometers. At least there are often large temperature gradients even in popular commercial instruments. Fortunately, with a few thermocouples it is easy to check for such problems, and often

TABLE 8.2.1 / Comparison of Typical Controlled Stress and Controlled Strain Rheometers

s *Controlled Stress*	s *Controlled Strain*
Torque: $10^{-6} - 5 \times 10^{-2}\ N \cdot m$	$10^{-7} - 10^{-2}\ N \cdot m$
Angular velocity: $10^{-7} - 10^{2}$ rad/s	$10^{-3} - 10^{2}\ (10^{3})$ rad/s
Oscillation frequency: $10^{-5} - 10^{2}$ rad/s	$10^{-5} - 500$ rad/s
Angle: $5 \times 10^{-6} - \infty$ rad	$5 \times 10^{-5} - 0.5$ rad[a]
Stress and strain measured on the same shaft	Typically stress and strain on different shafts.
Inertia limits short time response at low viscosity.	Better short time response at low viscosity (see Fig. 8.2.4b).
No limit to long time creep data or long relaxation times.	Low torque signals, transducer drift may limit.
Constant stress is often a natural loading more sensitive to material changes. $\tau < \tau_y$	Natural way to do step strain experiments, get $G(t, \gamma)$.
doesn't break structure. May reach steady state η more quickly.	Normal force available on these instruments.

[a] For 5×10^{-6}–0.02 rad (see Bohlin, 1988).

the situation can be improved. Here we mainly focus on temperature control but also point out some ways of maintaining desired pressure and humidity in rotary rheometers.

There are three fundamental approaches to temperature control: radiation, convection, and conduction. Radiation has been used for some very high temperature designs. For example, Sopra et al. (1988) describe a wide gap concentric cylinder viscometer with a radiation furnace for studying silicate magna to 1300°C. Normally radiation is combined with a gas convection oven. Figure 8.2.12 shows a typical design. Air or inert gas is forced to flow over a resistance wire heater into an insulated chamber surrounding the sample. The sample is heated by radiation from the oven walls, as well as by the circulating gas. Baffles are used to ensure that the hot gas does not hit the sample directly and to create a more uniform temperature distribution. The gas can be recirculated (Macosko and Weissert, 1974), but usually it is simply allowed to escape around the test fixture bases. These bases are typically hollow stainless steel cylinders with windows cut out near the sample and an insulating plug. Both features reduce conduction loss from the sample. Sample temperature is recorded by a thermocouple either very close to the sample in the gas or mounted in the fixture base at the center.

A platinum resistance thermometer or thermocouple in the gas stream is used to control the temperature. Placement of this sensor is important. Close to the edge of the sample yields better agreement between the controller set point and sample temperature. Close to the gas inlet gives less overshoot.

This overshoot is illustrated in Figure 8.2.13, curve 2a. Large overshoot can lead to degradation of the sample free surface. A control strategy which resets the integration constant can eliminate overshoot, as illustrated in curve 2b. However, as curve 1b illus-

Figure 8.2.12. Temperature distribution in a typical gas convection oven. Each gray tone represents ~ 1°C. From Mora and Macosko (1991).

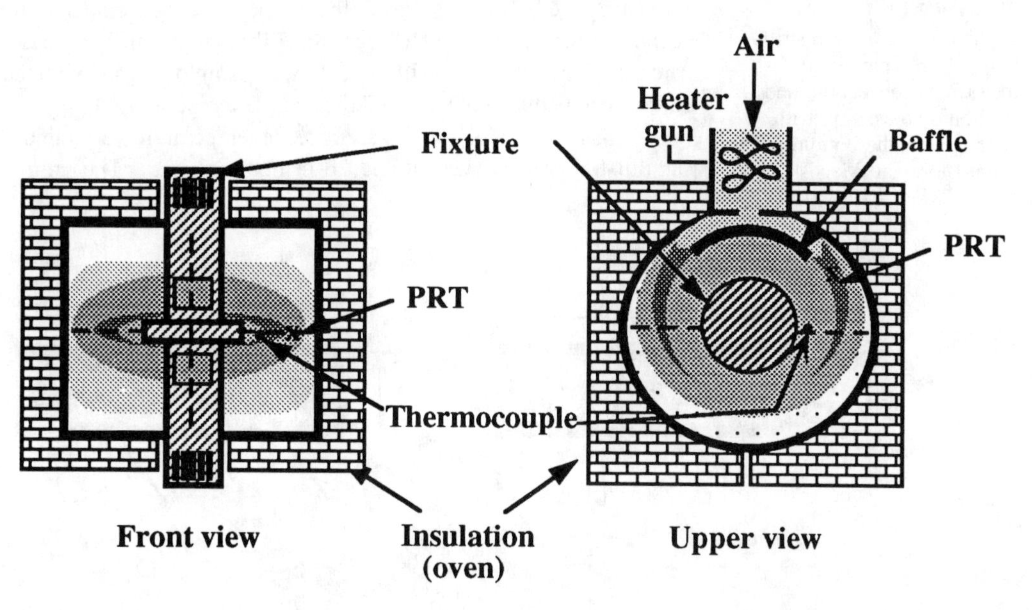

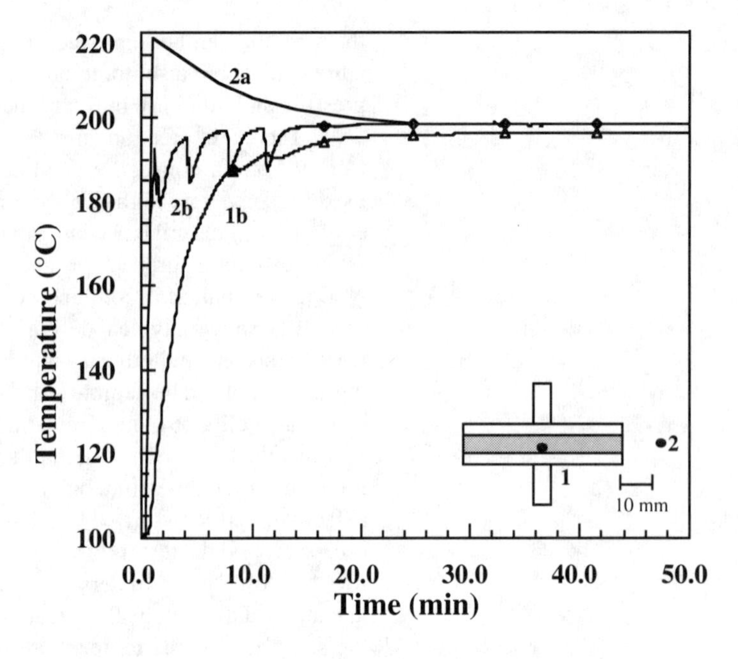

Figure 8.2.13.
Temperature versus time in parallel disks after a step change in the set point from 100 to 200°C. From Mora (1991).

trates, such a strategy means longer time to *sample* equilibration. In this example the final sample center point temperature reached was 197°C. For curve 1b, 196°C was reached in about 20 minutes. Allowing the gas temperature to overshoot (curve 2a) reduced the time to reach within 1°C of the final temperature to about 10 minutes.

Note in Figure 8.2.13 that even at steady state there is a 1.5°C difference between the gas and sample center temperatures. This means that there will be temperature gradients inside the sample. Figure 8.2.14a illustrates these gradients. In this example the upper plate is cooler than the lower one. This gradient probably could be reduced by lowering the fixture with respect to the oven or by adjusting the baffles. However, the temperature gradient from the center to the sample edge (Figure 8.2.14b) is more difficult to decrease. These gradients will increase with sample size and with the difference between the test and the sample temperatures.

Gas convection ovens can increase temperature very rapidly as illustrated by curves 2a and 2b in Figure 8.2.13. The sample

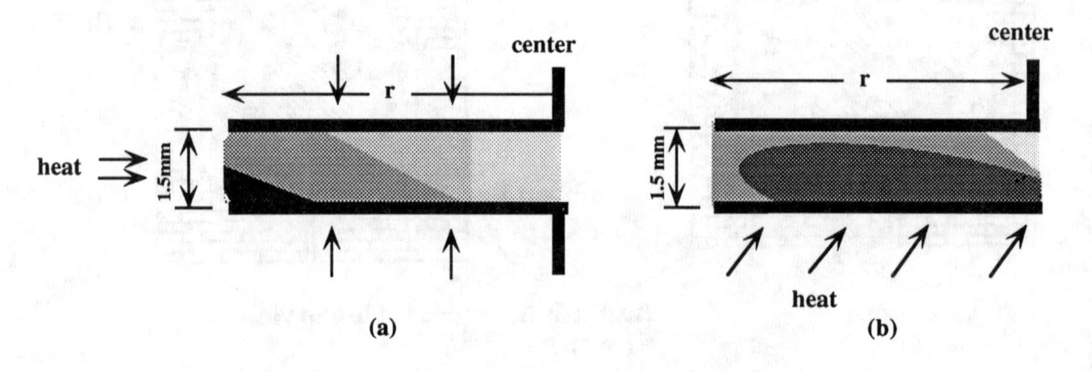

Figure 8.2.14.
Typical temperature distribution between parallel disks for (a) gas convection, and (b) liquid convection bath, heating only the bottom plate. From Mora and Macosko (1991); Mora, (1991).

temperature rises more slowly, $\sim$ 40°C/min using the initial slope of curve 1. Under *temperature-programmed* testing, gradients in the sample increase. Mora (1991) reports that the temperature difference between center and sample edge for a 1.1 mm thick, 50 mm diameter parallel disk sample nearly doubled from its steady state value when programmed at 4°C/min. The difference increased from 0.65 to 1.1°C at 200°C.

Temperature programming is frequently used to study the viscoelasticity of solid polymers. Typically rectangular samples about 40 mm x 10 mm x 20 mm are tested in sinusoidal torsion. Temperature gradients in such samples, particularly if they are thick and programmed rapidly, can cause significant errors. For example, Mora and Macosko (1991) found that the temperature in the central 20 mm region of a 30 mm x 12 mm x 3 mm sample was constant to within 1°C, but the ends were 3°C colder at 100°C and about 5°C colder when temperature was programmed at 3°C/min. These temperature gradients can cause a significant error in transition temperatures. For example, T_g, based on tan δ maximum using a 4 mm thick sample was 5°C too high, −39°C rather than −45°. When Mora and Macosko reduced the sample thickness to 0.5 mm, the error was eliminated. Similar temperature gradient problems in thick samples " *buried* " the T_g of a rubber additive under the β transition of the epoxy matrix (Gerard et al., 1990).

Another error associated with temperature programming is sample and fixture expansion. As mentioned above, a typical change in the gap is 1 μm/°C due to fixture expansion. This can cause significant errors for thin cone and plate or parallel disk samples. With solids it can lead to high normal forces at low temperatures and sample buckling at high temperatures. Some instruments automatically adjust the gap based on the normal force (Rheometrics, 1990).

The testing of both liquids and solids discussed above illustrates a major advantage of gas convection ovens: flexibility. They can accommodate a wide variety of test fixtures from three-point bending to torsion rectangular to concentric cylinders. Gas convection ovens also offer the widest temperature range. Using the boil-off from liquid nitrogen, it is possible to control temperature from −150 to 600°C (van der Wal et al., 1969). Such a liquid nitrogen system is particularly useful for rapid cooling—for example, to study rheological changes during crystallization or phase separation.

The main disadvantages of gas convection ovens are temperature gradients and sample evaporation. Gas flow can also disturb sensitive torque transducers. Control using liquid convection is much better for reducing evaporation but temperature gradients can still be a problem. This is illustrated in Figure 8.2.14b, where the top and edges of the sample are cooler because the sample is heated only from the bottom plate. Gradients are less for a concentric cylinder geometry, which can be immersed in the liquid bath as indicated in Figures 8.2.1 and 8.2.2. Around 80°C the temperature difference across a 1 mm gap is typically $< 0.5°C$; over a length of 30 mm it is $\sim 1°C$ (Mora, 1991).

Liquid convection baths typically use water as the circulating liquid and thus are limited to $5 - 80°C$. With oils and a refrigerating/heating circulator this range can be extended from -50 to 200°C. Temperature in these circulators can be programmed but typically slowly ($\leq 3°C/min$). Rheometer designs like Figures 8.2.1 and 8.2.2 require liquid-tight rotating seals between the bath and driven shaft. Devices in which stress and strain are measured on the same shaft (Figure 8.2.10) make the design of such baths simpler.

Liquid convection baths are best suited for the concentric cylinder geometry and for temperatures between -10 and 150°C. Temperature gradients with cone and plate and parallel disk geometries can be a problem because of heat loss from the upper fixture. Insulated covers and thin samples help. For other geometries and wider temperature ranges, gas convection is preferred.

Direct electrical heating is less common in commercial rheometers, but it can provide rapid programming and precise control for parallel disk or cone and plate geometries (Meissner et al., 1989; Mani et al., 1991). Heaters can be readily controlled on rotating shafts using slip rings, so precise temperature can be maintained on both surfaces. Typically an inert gas chamber surrounds the sample for temperatures of 200°C or more. With normal resistance heaters, controlled cooling is not possible. However, semiconductors known as Peltier elements can heat and cool over a limited temperature range: 0 to 80°C. A Peltier device is incorporated into the lower plate of some controlled stress rheometers (Carri-med, 1988; Rheometrics, 1993) while others use electrical heating (Bohlin, 1990; Haake, 1992; Rheometrics, 1993). Induction heating of the test fixtures is also used (Carri-med, 1992). Electrically heated concentric cylinders geometry is available commercially (Physica, 1989).

Above we discussed the problem of sample evaporation. Nearly all rotational rheometers have free surfaces, where the sample is exposed to the environment. To prevent evaporation, the usual method is to surround the sample with porous pads filled with the sample or an appropriate solvent, to try to maintain a proper humidity in the environmental chamber. Some workers coat the free surface with a low viscosity oil to prevent evaporation.

An alternative approach to preventing evaporation is to seal the entire rheometer. This can be easily done with the falling or rolling ball indexer (Figure 5.2.5). Contained rotors are used in the rubber industry (see Section 5.6). The viscous rubber forms a seal in the tight clearance around the rotating shaft. This design can contain a few atmospheres of pressure, which prevents oil in the rubber from vaporizing.

Figure 8.2.15 shows a design for a pressurized rheometer. The key is the thin-walled tube, which acts as a torsion spring and also permits the motion to be transferred out of the chamber.

Figure 8.2.15.
Schematic diagram of a torque transducer for a pressurized rheometer. The thin-walled tube twists slightly when torque is exerted by the sample on the inner cylinder. This twist is recorded by the rod and capacitance gage.

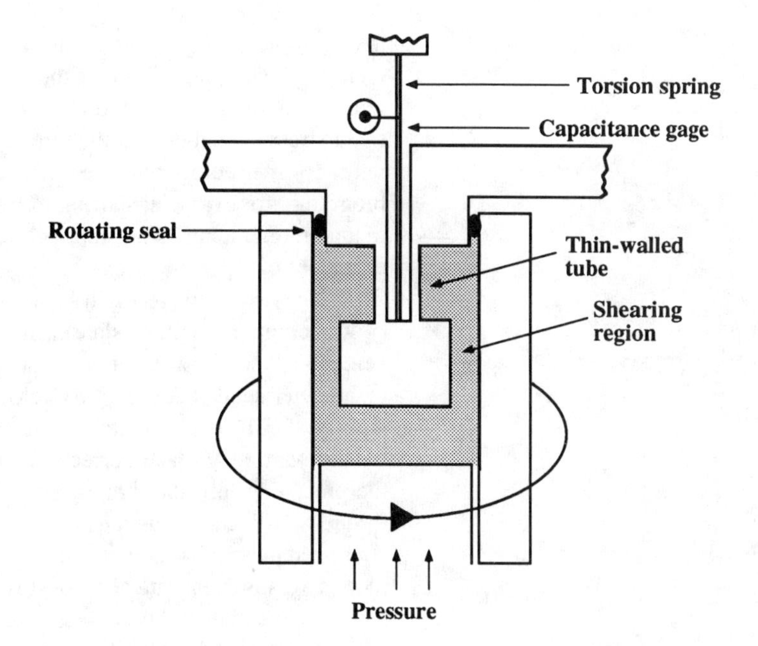

Richards and Prud'homme (1987) have described such a design for testing polymer solutions up to 70 atm. One application is in polymerization reactors; another is oil recovery with polymer pusher fluids. Christmann and Knappe (1976) describe a similar design that can test polymer melts at 500 atm. These levels are necessary to probe the pressure dependence of material functions.

8.3 Data Analysis

Section 8.2 described how different rotary rheometers are designed to control and to measure rotation rate, angular position, torque, temperature, and other variables. Equally important is the analysis of these measurements, conversion of the raw millivolts to material functions. Twenty years ago this was all done by hand, but today commercial rheometers spit out materials functions like G' and G'' in real time. Data analysis software is becoming a more and more important part of rheometer design. We have already seen that the inertia correction algorithms illustrated in Figure 8.2.11 can significantly extend the performance of controlled stress rheometers.

This section first considers the general features of data analysis software and future trends in this area. Then we focus on analysis of transient strain or stress tests, particularly sinusoidal oscillations. We will apply this analysis to data from rotational rheometers, but some of the strategies are also applicable to pressure-driven shear rheometers and extensional rheometers described in the following sections.

Figure 8.2.2 is a good starting point for considering the design of data analysis software. Let us say that the operator wants to determine the $\eta(\dot{\gamma})$ function for her sample. Through the instrument keyboard she will enter the concentric cylinder dimensions into the analysis/control software which will calculate the range of shear stress available with the torque transducer and the shear rate range for the motor-controller. She will then select the shear rate program—for example, starting from the minimum rate and going to maximum rate in 3 or 4 steps per decade, holding for 30 seconds at each rate.

The rest of the test will be automatic. Every 30 seconds the control computer will translate each desired shear rate to an angular velocity set point for the circuit that controls the motor speed. The tachometer sends back the true velocity, which the software uses with eq. 5.3.11 to calculate the shear rate. In some instruments the software may make corrections for wide gap using data from the preceding low shear rates (eq. 5.3.24). The voltage from the torque transducer is typically averaged over several seconds, then converted to shear stress with eq. 5.3.8. Viscosity is calculated and plotted versus shear rate as the test is running.

To ensure that the viscosity data are true steady state values, the operator can command longer measuring times at lower shear rates and request that, for example, torque be averaged over only the last 5 seconds of the measuring time. The software can even be asked to keep measuring until the torque has reached steady state. Readings from clockwise and counter-clockwise rotation can be averaged for each shear rate. After the desired $\eta(\dot{\gamma})$ data have been collected, the program can automatically change temperature and repeat the process.

Many higher level control functions are starting to appear in rheometers. In Section 8.2.5 on environmental control we mentioned gap control based on torsion when solid samples are tested as a function of temperature. Luckenbach (1992) describes a novel gap control procedure for testing asphalt and soft solids between parallel disks. At low temperatures torques are very high, so an 8 mm diameter disk can be used. At higher temperatures, when torque falls below a certain level, the instrument brings a 25 mm disk in contact with the sample. This procedure permits measurement of G^* from 10^9 to 10^3 Pa allowing testing from the solid to liquid state on one sample. The data obtained can then be used to construct the time–temperature master curve and fit to a relaxation spectra using the analysis software.

Software can be used to control the test range—for example, taking more data when something rheologically interesting is happening. A temperature scan can be made more slowly, or in smaller steps when $\tan\delta$ starts to change. This helps to accurately identify T_g. We usually don't need much $\dot{\gamma}$ data in the Newtonian region. Thus in the $\eta(\dot{\gamma})$ example given above, we can reduce the number of steps per decade or even cut off the shear rate sweep when η_0 is reached.

Such types of analysis and control will probably be the greatest area for innovation in future rheometer design. Fitting of constitutive equations will be done on line. After the general type of sample has been identified, the software will decide which type of test to run and over what ranges. It will compare the results to a library of relaxation spectra and even use molecular models to calculate polymer molecular weight distribution.

8.3.1 Sinusoidal Oscillations

The most commonly measured viscoelastic material function is $G^*(\omega, T)$. It is so popular because sinusoidal oscillations can be used to follow viscoelastic changes with time, such as during curing and crystallization. As discussed below, cross-correlation analysis of the signal can provide accurate G' and G'' values over a wide range of frequency and signal levels.

There have been many designs for measuring the dynamic moduli (Ferry, 1980). Most of them fall into the three basic categories illustrated in Figure 8.3.1. The simplest is shear wave propagation—how fast a pulsed deformation travels through a sample. If the damping of the probe is measured, both G'' and G' can be calculated. However, usually the damping per wave is small

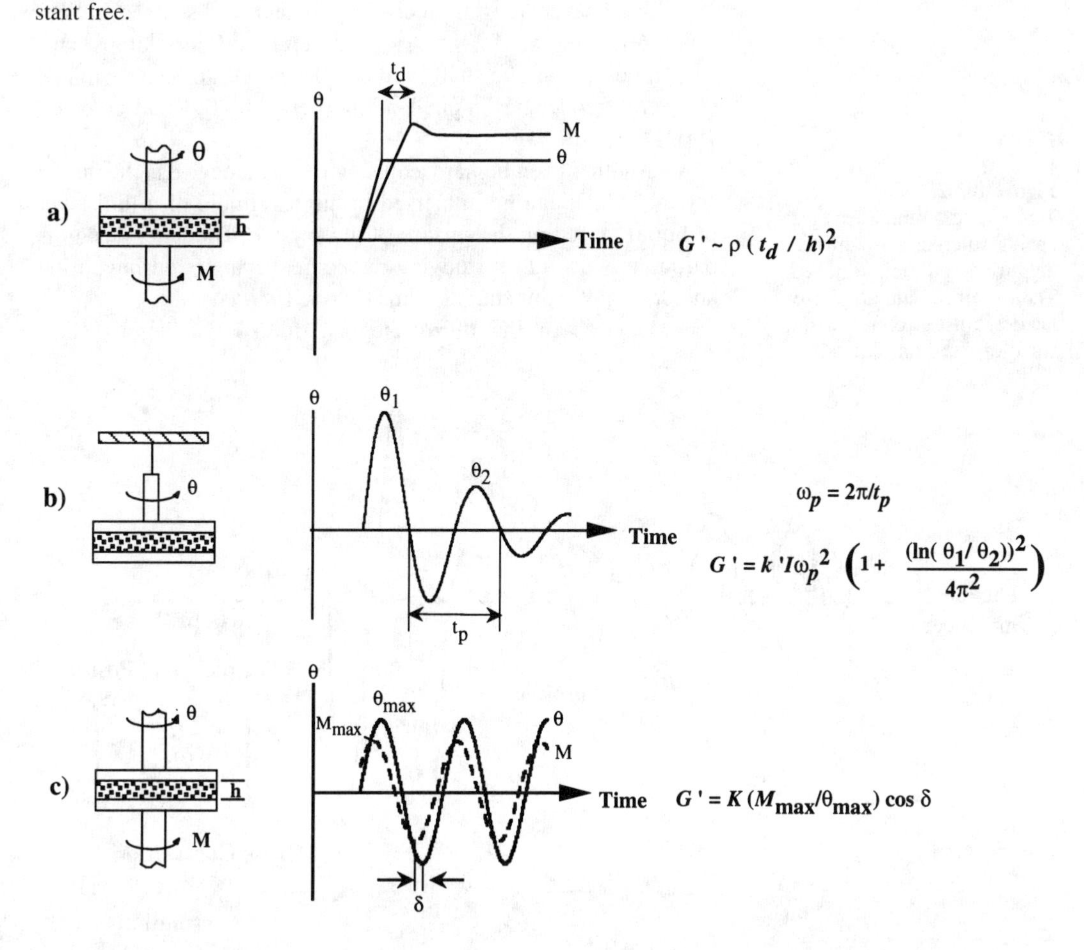

Figure 8.3.1.
Methods for measuring G': (a) wave speed, (b) resonance, and (c) forced oscillations. K is a geometry constant free.

and G' is just the sample density times wave speed squared. G' is measured at one frequency in the 10–1000 Hz range. Van Olphen (1956) and Buscall et al. (1982) have developed such a device for studying flocculation of colloidal systems (Rank Pulse Shearometer, 1992). Joseph (1990) has also developed a wave speed meter and used it to study many liquids.

The second major design type also operates at resonance and thus is limited to one frequency. Shown in Figure 8.3.1b is a freely decaying torsion pendulum. It is used primarily on elastic samples; that is, the damping per cycle is small enough that the decay can be measured over several cycles. The frequency range, about 0.01–25 Hz, can be controlled by varying I, the moment of inertia. Free torsion pendula have been widely used to measure the viscoelastic properties of solid polymers (Nielsen, 1977, Ferry, 1980).

Forced resonance devices are better suited for lower elasticity samples like dilute polymer solutions. A mass is oscillated sinusoidally to the resonant frequency of the sample–apparatus combination. The resonant frequency with and without the sample and the energy used to drive at resonance allow calculation of G' and G''. These devices operate at around 1000 Hz. Since both measurements are made on the oscillator, it can be conveniently immersed into a lab beaker or even process equipment as discussed in Section 8.6 (Nametre, 1990). Several different stiffness lumps have been mounted on one shaft to allow G' and G'' to be determined at several frequencies in one loading (Ferry, 1980; Knudsen et al., 1992).

Another even higher frequency resonant device is the quartz crystal, which can be immersed in the test liquid or a thin layer of liquid placed on the surface of the crystal. The latter design is particularly useful for studying rheological changes during curing and drying of paint and ink films (Ferry, 1980).

Devices in the third category, forced oscillations (Fig-

Figure 8.3.2.
Torque and strain waves are analog filtered, zero shifted, amplified and then digitized. The digitized data are correlated against a reference sine wave of fixed amplitude.

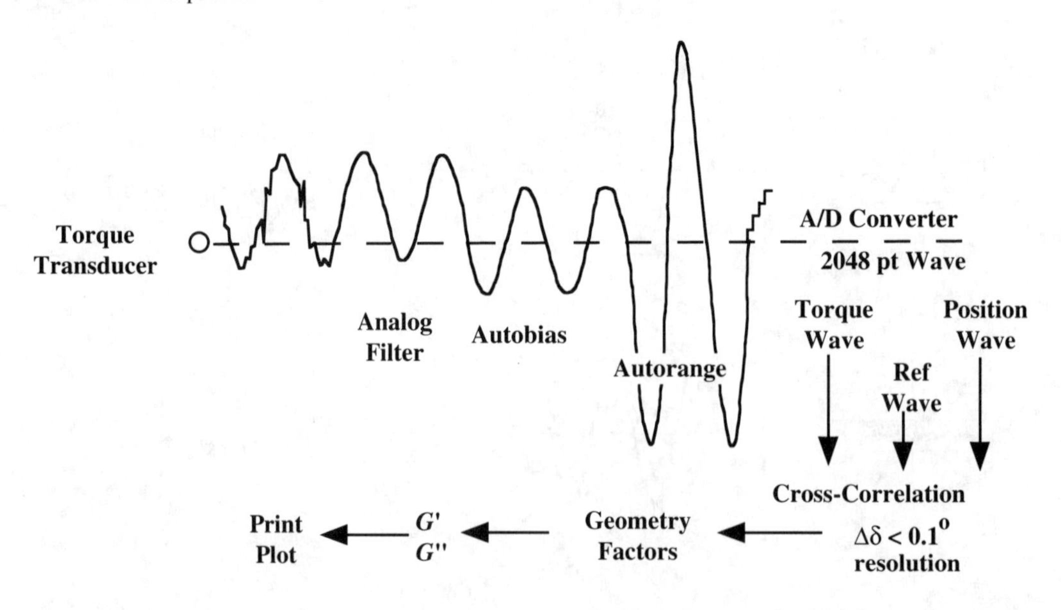

ure 8.3.1c), are by far the most common and versatile. They are generally run at lower frequencies (0.001 – 500 rad/s). This is mainly because of limits of mechanical oscillators but also reflects the desire to maintain gap loading conditions (Schrag, 1977, see also Section 5.3).

Figure 8.3.2 ilustrates how data are treated. Cross-correlation is the heart of the analysis. Each signal can be considered as a combination of

$$R = R_o(\sin \omega t + \phi) + \text{noise} + \text{harmonics} \tag{8.3.1}$$

where $\sin \omega t$ is the reference wave. An analog filter removes the high frequency noise. The cross-correlating "beats" the signal against the reference wave.

$$\frac{2\omega}{N}\int_0^N \omega R \sin \omega t \; dt = R_o \cos \phi \tag{8.3.2}$$

and

$$\frac{2\omega}{N}\int_0^N \omega R \cos \omega t \; dt = R_o \sin \phi \tag{8.3.3}$$

where N = number of cycles. As Figure 8.3.3 shows, noise rejection improves significantly with number of cycles.

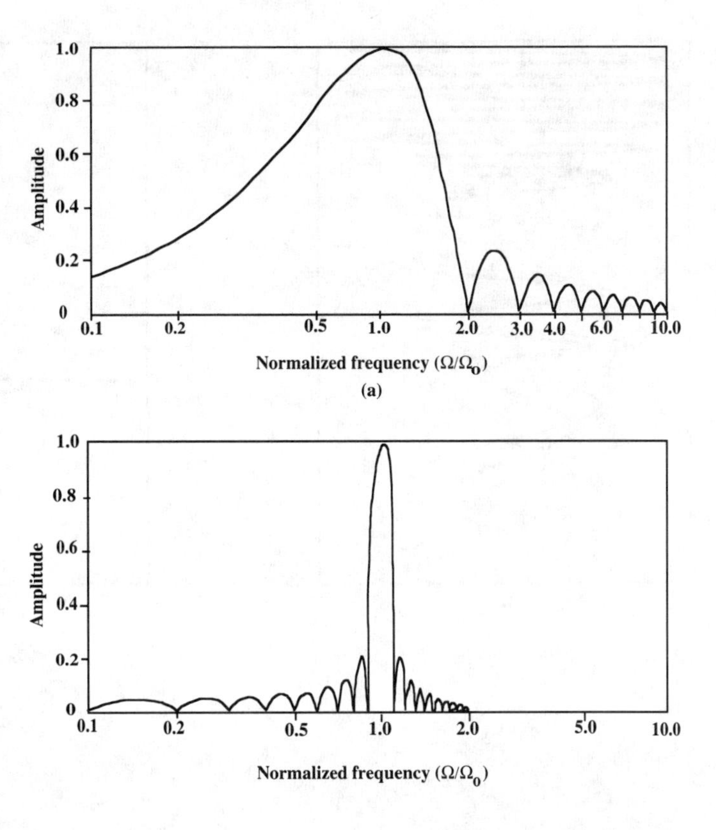

Figure 8.3.3.
Filter characteristics of Fourier integral cross-correlation: (a) one-wave integration and (b) ten-wave integration.

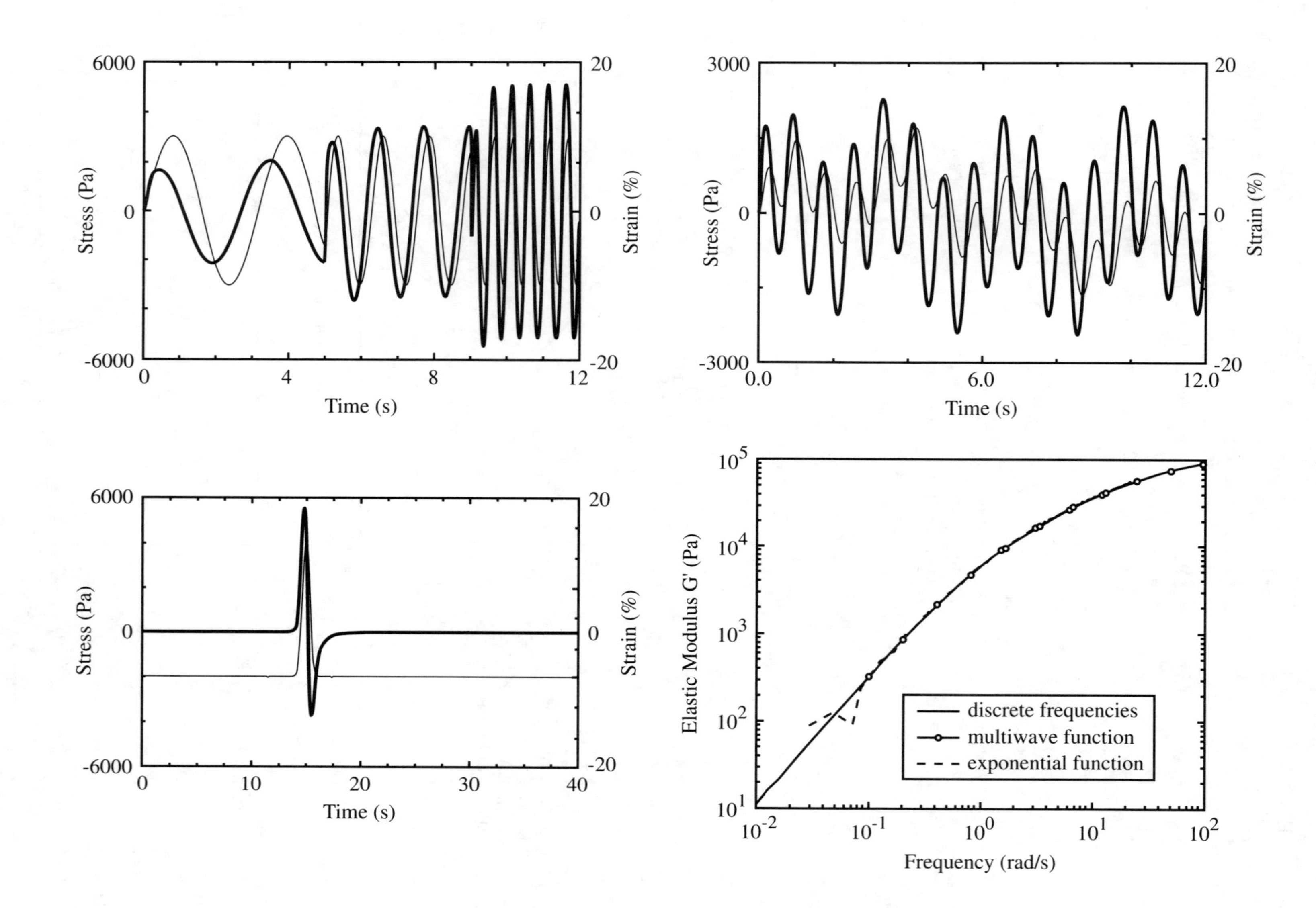

Stress (Pa)
Strain (%)
Time (s)
6000
0
-6000
20
0
-20
0
4
8
12
Stress (Pa)
Strain (%)
Time (s)
3000
0
-3000
20
0
-20
0.0
6.0
12.0
Stress (Pa)
Strain (%)
Time (s)
6000
0
-6000
20
0
-20
0
10
20
30
40
Elastic Modulus G' (Pa)
Frequency (rad/s)
discrete frequencies
multiwave function
exponential function

Figure 8.3.4
Comparison of three methods for measuring elastic modulus versus frequency $G'(\omega)$: (a) time sequence of discrete sine waves, (b) three sine waves superposed on each other, (c) exponential pulse (thick line represents stress and the thin line represents strain), and (d) comparison of G' determined by each of the methods. From Berting et al., (1990)

Often it is desirable to collect G' and G'' rapidly, for example if a sample is curing or crystallizing. When discrete sine waves (Figure 8.3.4a) are employed to collect the necessary data, the test duration can be rather long. To speed up the experiment, one can use features such as running the test with a multi-wave function which is a superposition of waves of several frequencies (Holly, et al. 1988; Figure 8.3.4b), or an exponential function (Figure 8.3.4c). A comparison of the discrete frequency data obtained from each of the three different experiments are shown in Figure 8.3.4d. In terms of the test time, the discrete frequency test took approximately 50 min, the multi-wave function test took 16 min for 10 frequencies, and the exponential function took less than 10 min (Berting et al., 1990).

8.3.2 Transient

Besides the sinusoidal oscillations, transient tests such as stress relaxation, start-up of steady shear flow, and cessation of steady shear flow are also important in the rheological characterization of polymeric liquids. The instrument limitations and features, and their effect on the data obtained in some types of transient tests, are examined in this section. Also, we shall illustrate the use of linear viscoelastic transformation to obtain, for example, stress relaxation data from tests such as start-up of steady shear flow and sinusoidal oscillations.

In the stress relaxation test, the material is subjected to a step strain at time zero (inset, Figure 8.3.5), and the decay in the stress (or modulus) is monitored as a function of time (Figure 8.3.5). At short time, limitations due to electronic hardware can affect the data collection speed and the transducer inertia can affect the torque values, thereby affecting the quality of the data. At long time, the low torque signal and hysteresis of the torque transducer affects the data quality.

Figure 8.3.6 illustrates the danger of using an electronic analog filter in eliminating high frequency noise from the torque signal for a low concentration xanthan solution in a Newtonian solvent (Mackay et al., 1992). The experiment employed was a stress decay test after the cessation of steady shear flow. Note that the steady shear flow was ceased at 100s. We expect an instantaneous relaxation of the Newtonian solvent followed by the polymer relaxation. The analog filter hides the Newtonian relaxation.

From Figure 8.3.5 we saw the limitation of the stress relaxation test in obtaining long time data. Also, in some cases the limitations may be due to the type of instrument that is available. But, for long chain molecules the long time data becomes important. The use of linear viscoelastic transformations (Ferry, 1980) can be valuable in obtaining, for example, the long time stress relaxation data based on experiments such as the start-up of steady shear or sinusoidal oscillations. Figure 8.3.7 compares the experimentally obtained stress relaxation data with that obtained by the transformation of the stress growth and sinusoidal oscillation tests (Meissner, 1972). The good agreement between the different tests is noteworthy. Deviation of the η^+ transformed data (from start up of shearing) at short times is due to torsional compliance and small cone angles as predicted by eq. 8.2.2. In addition, it is worthwhile to note that the transformed data allow the calculation of the relaxation modulus for longer times.

8.4 Pressure-Driven Rheometers

As with drag flow rheometers, there are two basic design types: one features controlled drive pressure and measurement of flow rate, and the other uses controlled flow rate and measures pressure drop. Pressure is controlled by a hydrostatic head, external gas or hydraulic pressure, or even a weight. Flow rate can be controlled by motion of a driving piston. Whorlow (1992) has an extensive review of pressure driven rheometer design.

In the first capillary rheometer, Hagen (1839; Figure 6.1.1) controlled pressure by a gravity head. This is still the case with the common glass capillary viscometers. As indicated in Figure

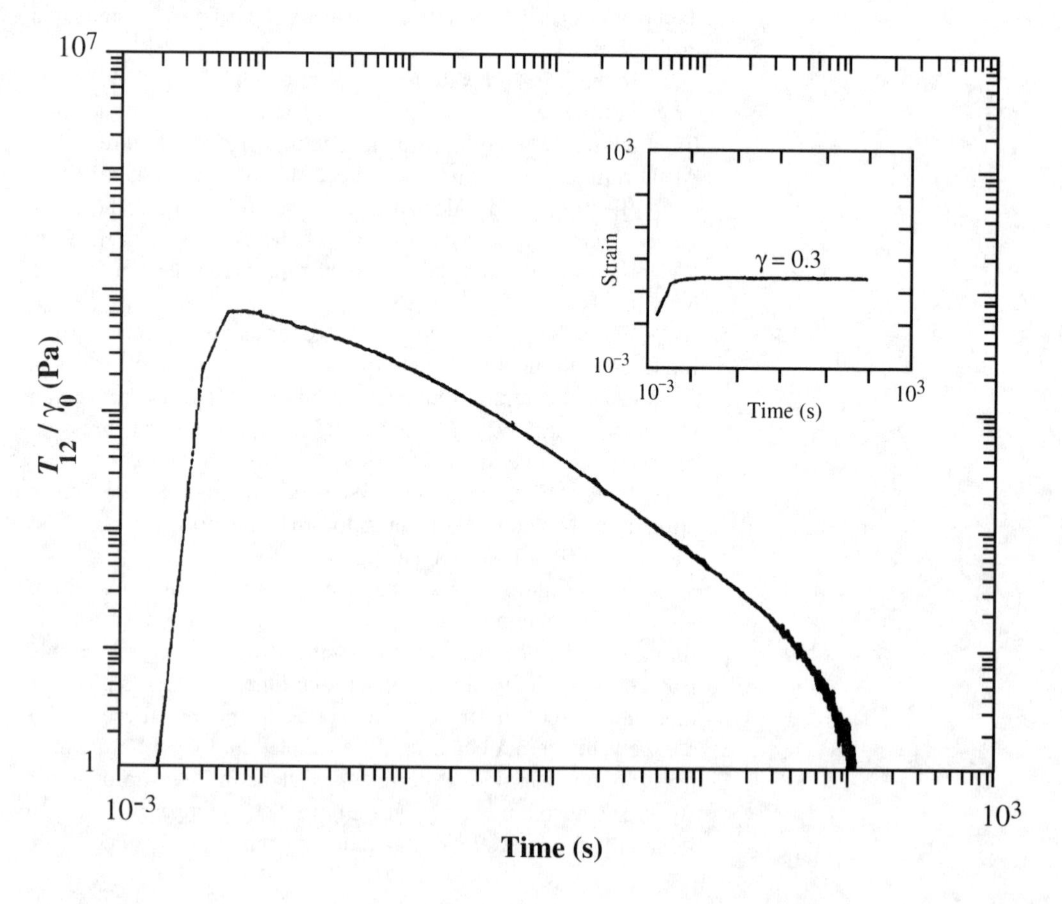

Figure 8.3.5.
Torque versus time after a step strain for a polydimethylsiloxane. The inset graph shows that the commanded strain is reached in ~3ms. Stress reaches a maximum in ~6ms due to transducer inertia. Low torque signal and transducer hysteresis are limiting at long time.

Figure 8.3.6.
Effect of analog filter on torque after cessation of steady shear for a dilute xanthan solution. From Mackay et al. (1992).

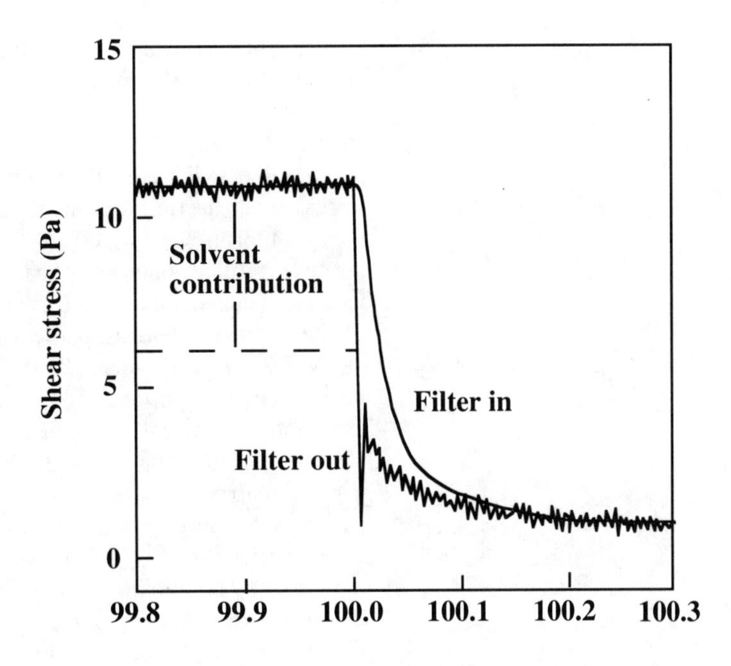

Figure 8.3.7.
Comparison of $G(t)$ by different methods. Adapted from Meissner (1972).

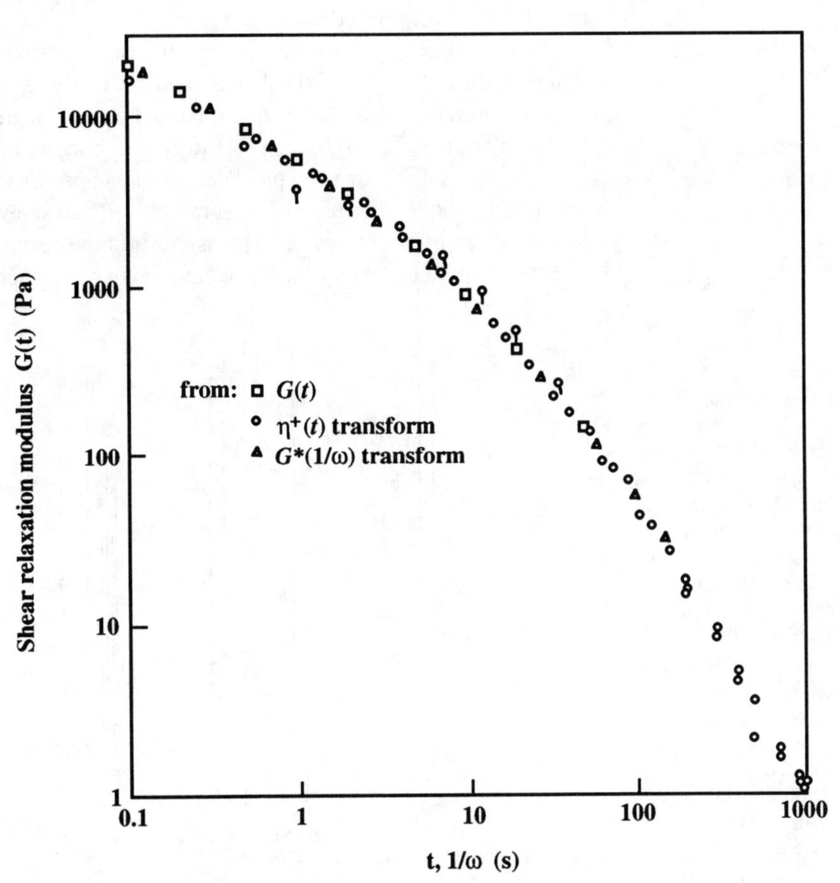

8.4.1, the gravity head is provided by liquid in the bulb above the capillary. Since the height change in emptying the bulb is relatively small, pressure changes are small during the test. Glass capillaries are used for lower viscosity liquids: $\eta < 10$ Pa·s and typically $\eta \sim 10\,\text{mPa}\cdot\text{s}$. Different diameters are used for different viscosity ranges to keep the flow time in the range of 1–5 minutes. Wall shear rates are typically about 100 s^{-1}. Since gravity is usually used as the driving force, only one shear stress is available for a given diameter. Glass capillaries are used almost exclusively to measure relative viscosity changes, particularly of polymer solutions (e.g., ASTM D2857; Rodriguez, 1989). The liquids are assumed to be independent of shear rate, and thus single point data are sufficient. Temperature is controlled by immersing the capillary in a bath. Van Wazer et al. (1963, pp. 215–230) have an extensive discussion of glass capillaries.

Another single point capillary instrument is the melt indexer shown earlier (Figure 6.2.14). It consists of a single capillary of rather low L/D ratio and a plunger on which a weight is placed. This drives the molten polymer through the capillary at constant pressure and the extrudate is timed, collected, and weighed. Although the melt index apparatus can be used as a rheometer by employing two different length capillaries and different weights to vary the stress, typically it is used as an indexer (ASTM D1238; see Chapter 6 for more details).

High pressure air or nitrogen can be used to provide a wide range of shear stresses. Extrudate can be cut and weighed or extruded onto a recording balance. Figure 8.4.2 show such an automated instrument for high shear rate testing of coating liquids.

Figure 8.4.3 shows a typical design of a capillary rheometer for polymer melts. The piston is controlled by a ball screw drive or in some cases by gas or hydraulic pressure. In some designs the ball screws are driven by a constant speed motor. In more sophisticated

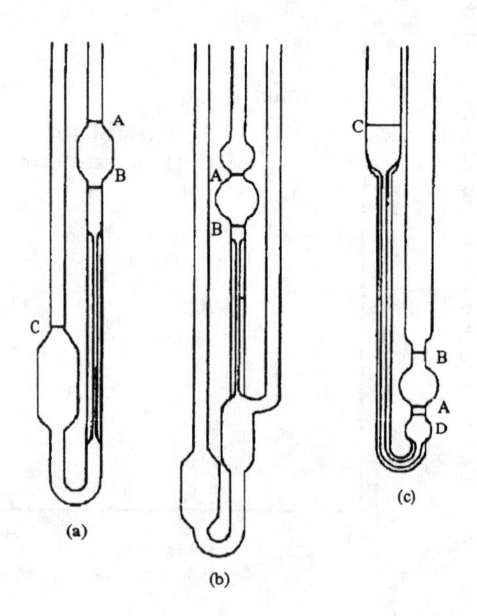

Figure 8.4.1. Hydrostatic head viscometers. (a) In the Ostwald design, liquid is filled exactly to mark C then sucked through the narrow capillary section to above mark A. The time for the meniscus to fall from A to B is proportional to viscosity. (b) The side arm of the Ubbelehde design eliminates the need to fill with a precise volume. (c) The Cannon–Fenske design reverses the flow from (a) and is used for opaque fluids. The dark meniscus rises from A to B during the timing. (from Van Wazer et al., 1963).

Figure 8.4.2.
An automated gas-driven capillary viscometer. A pressure transducer (not shown) records the head. Valve *A* opens the head to sample tank *B*. Jacket *C* controls temperature. *D* is a ball valve, which opens flow to capillary *E*. Balance *F* (not shown) records the flow rate. From Grankvist and Sandas (1990; Oy Gradek Ab, Kauniainen, Finland).

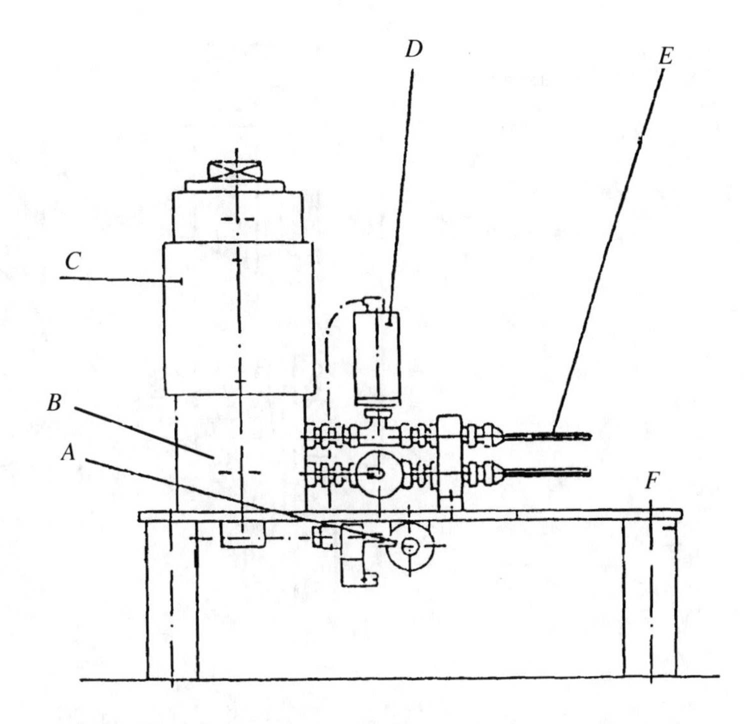

designs the drive feeds back on plunger force or a transducer placed at the die entrance. If a slit die is used, pressure distribution can be measured with flush-mounted pressure transducers as shown earlier (Figure 6.3.1). Pressure transducer calibration and thermal stability are critical because accurate pressure differences are needed.

Figure 8.4.3.
Typical melt capillary rheometer (similar to commercial designs of Göttfert, Instron; Kayeness, Monsanto; Rosand, Seiko).

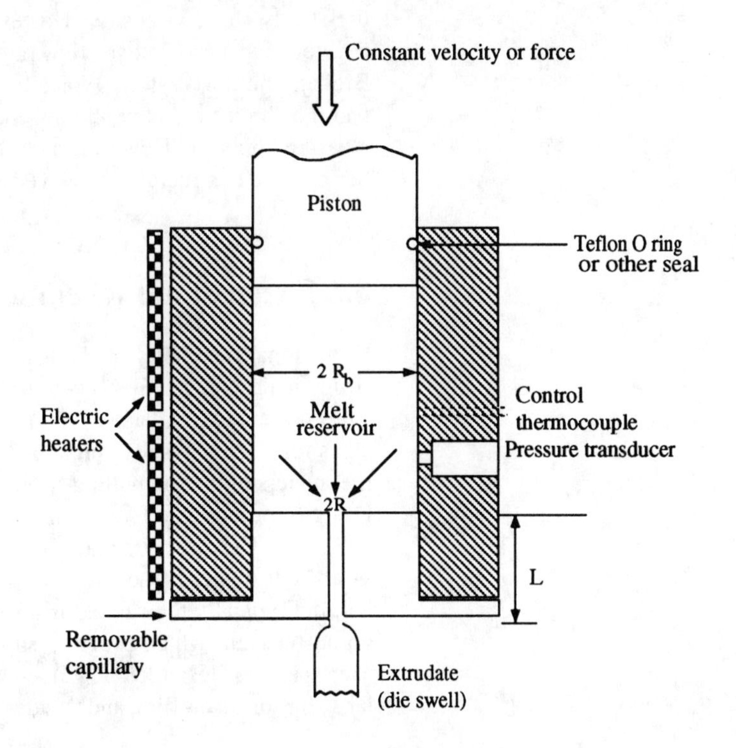

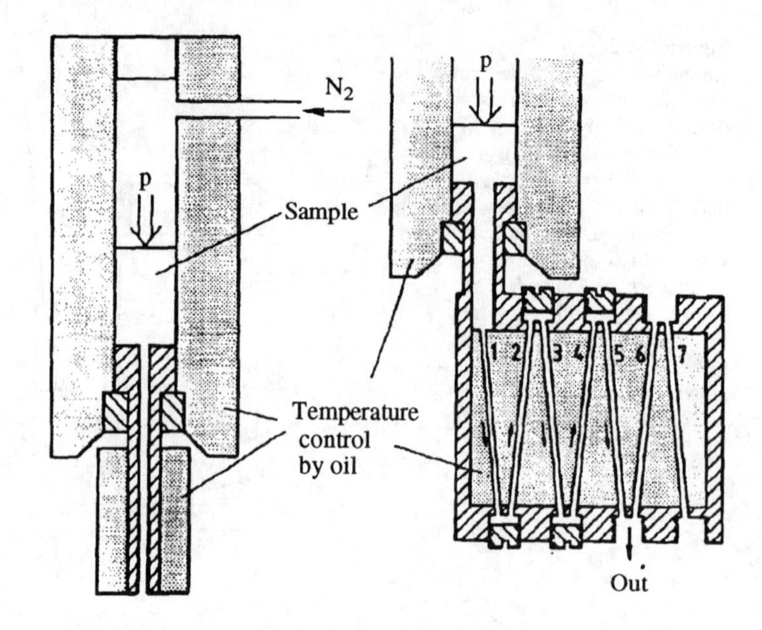

Figure 8.4.4.
Methods for temperature control of long capillary dies. From Laun and Hirsch (1989).

Temperature is controlled by band heaters on the barrel. Typically a separate band is placed at the die end to reduce losses. Figure 8.4.4 illustrates two methods for designing and heating long capillaries. At high shear rate, viscous dissipation can make temperature control very difficult. Small diameter capillaries minimize the problem (see Figure 6.2.12).

Normally pressure-driven rheometers are used only to measure steady shear viscosity. However, several devices have been developed that oscillate the flow rate sinusoidally (Thurston, 1961; Brokate and Gast, 1992). Typically oscillations are large amplitude and the strain field is nonhomogeneous, so G' and G'' cannot be measured directly. However, such rheometers have been shown to be sensitive to structure in low viscosity liquids (Vilastic, 1992).

8.5 Extensional Rheometers

Extensional rheometry methods are strongly connected to the instrument design. The important designs for high viscosity samples were presented in Chapter 7. Meissner (1992) has reviewed designs for polymer melts. Figures 7.2.6 and 7.4.5 show designs of rheometers of the rod pulling type, where the sample is immersed. Designs for lubricated squeezing are shown in Figures 7.3.1, 7.3.2, and 7.3.6. The most popular extensional rheometer design for high viscosity samples is the windup method (Figures 7.2.5, 7.4.2, and 7.4.3). Typically the sample is a rod or rectangular strip floating on an oil bath, which provides support and temperature control. Meissner and Hostettler (1992) report that they can float rectangular sheets on an air film, and this design is offered commercially by

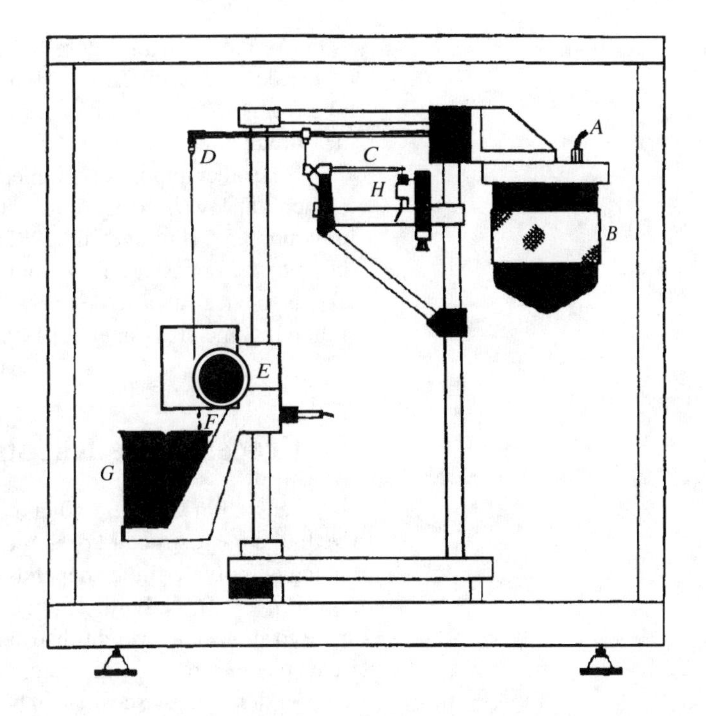

Figure 8.5.1. Schematic diagram of the Carri-med Spin Line Rheometer (Jones et al., 1987). Air pressure, A forces test liquid from the reservoir B through tube C to the spinning nozzle D, where it is drawn down by the rotating drum E. Liquid is scraped from the drum by F and falls into the beaker G. The feed tube C also acts as a force spring, whose deflection is measured by LVDT, H. Fiber diameter is recorded with a video camera near J. The whole apparatus is contained in an environmental chamber.

Rheometrics. The importance of temperature uniformity in extensional testing of polymer melts is highlighted in Figure 7.2.7.

In Chapter 7 we discussed the fiber spinning rheometer and its limitations. A specific design is pictured in Figure 8.5.1. In this design the sample is drawn by a rotating wheel. The same method is used to draw polymer melts extruding from a capillary rheometer (Göttfert, 1989; see also Figure 7.5.2). Another method for pulling the sample is vacuum (Secor et al., 1989).

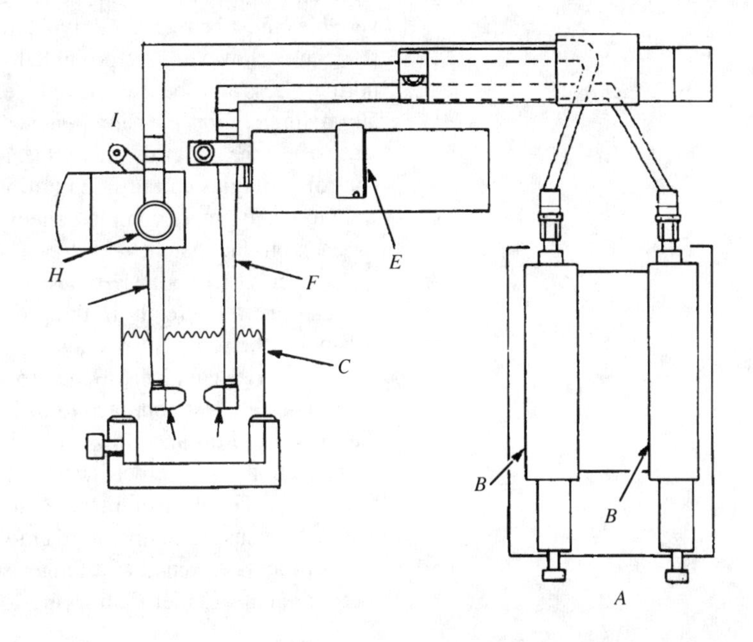

Figure 8.5.2 Schematic of opposed-nozzle extensional rheometer. Stepper motor A moves a pair of syringes B, which suck (or blow) the test liquid from the beaker C through a matched pair of nozzles D. Screw E can translate arm F to change the gap between the nozzles. Arm G pivots around the inlet tube at H. Torque on the tube is measured by a rebalance transducer I. From Rheometrics (1991).

As indicated in Figure 8.5.1, the tensile force is measured by the deflection of the feed tube. Secor et al. suspended their tube from an analytical balance. In the melt drawing, tension is measured on the rotating wheel.

The other important commercial design for extensional measurements on low viscosity fluids is the opposed nozzle device shown in Figure 8.5.2 (Fuller et al., 1987; Mikkelsen et al., 1988). In addition to the opposed-nozzle configuration, if the arm G is turned $90°$, the device can also be operated as fiber spinning and tubeless siphon rheometers (Cai et al., 1992).

8.6 Process Line Rheometers

As discussed in Chapters 10 and 11, rheology can be very sensitive to the microstructure of liquids. For example, the viscosity of entangled polymer melts depends on molecular weight to the 3.4 power, $\eta_0 \sim M_w^{3.4}$. Equilibrium creep compliance J_e^0 is very sensitive to molecular weight distribution. The yield stress and low frequency G' are good indicators of the flocculation state of colloids. Extensional viscosity can be an important indicator of bread dough quality (Padmanabhan, 1993).

It is natural to try to use these sensitive indicators for process control, but there are some important design challenges in going from a lab measurement to the process line (Dealy, 1990). To be part of a continuous feedback control system, the sensor must be totally automatic and extremely reliable. Very rugged designs which have few moving parts are favored. The rheological sensor should not interrupt the process flow, causing dead zones where material may accumulate and degrade. Simultaneous temperature measurement and control around the sensor are required. Two rheological variables must be measured, typically pressure drop or torque and flow rate. However, it is possible to use a pair of stresses such as normal stress and shear stress or extensional and shear stress. Usually a single point measurement will be used for control such as viscosity at the melt flow index shear stress, but more sophisticated control strategies can utilize multiple points. The user may also want to make a more complete rheological characterization on-line for each product batch.

There are two basic types of process line rheometer: those that can operate directly in the process and those that pull a side stream off the process for analysis. The side stream may be either dumped or recycled back to the process.

The simplest approach to in-line measurement is to monitor the pressure drop and flow rate over a flow channel of constant cross section in the process. This is more difficult than it might first appear. Getting a uniform section with a reasonable Δp can be difficult because of temperature gradients. A low pressure gradient in the process means that a longer section is needed for accurate measurement. Rather than trying to infer the shear stress by the pressure drop between two transducers, this variable can be directly measured by a wall shear stress transducer like the one shown in Figure 8.2.6.

Flow rate is not trivial either. Invasive devices like turbines or gear pumps can produce excessive pressure drop and lead to fouling. They also are nonlinear at low viscosity. Such noninvasive devices as coriolis, ultrasonic meters, and magnetic meters can be used for particular liquids, usually in the lower viscosity range (Mattingly, 1983). The process flow rate normally fixes the test shear rate. At this rate the viscosity (or other variable) may not be sensitive to the process changes. To get around this limitation, Pabedinskas et al. (1991) have developed an in-line, tapered slit with several transducers along its length that can measure viscosity over more than a decade in $\dot{\gamma}$ at one process flow rate.

It is also possible to mount an oscillating probe in the process line. As discussed in Section 8.4, these probes operate at relatively high frequency, which may not be sensitive to the process variable. These surface loading devices typically measure material within a few micrometers of the surface. Thus they should be located to have good flow over the surface. However, such high flow rates may damage the probe.

An alternative approach is to measure two stresses. Lodge (1988) has shown how the pressure hole method can be used as an in-line rheometer. N_1 is measured in slit flow by two opposed pressure transducers as shown earlier (Figure 6.3.5). A third, recessed transducer downstream is used to get the shear stress. Together this gives N_1 (τ), which is very sensitive to polymer concentration and molecular weight. Padmanabhan and Bhattacharya (1993) have proposed a combination of entrance pressure drop and wall shear stress.

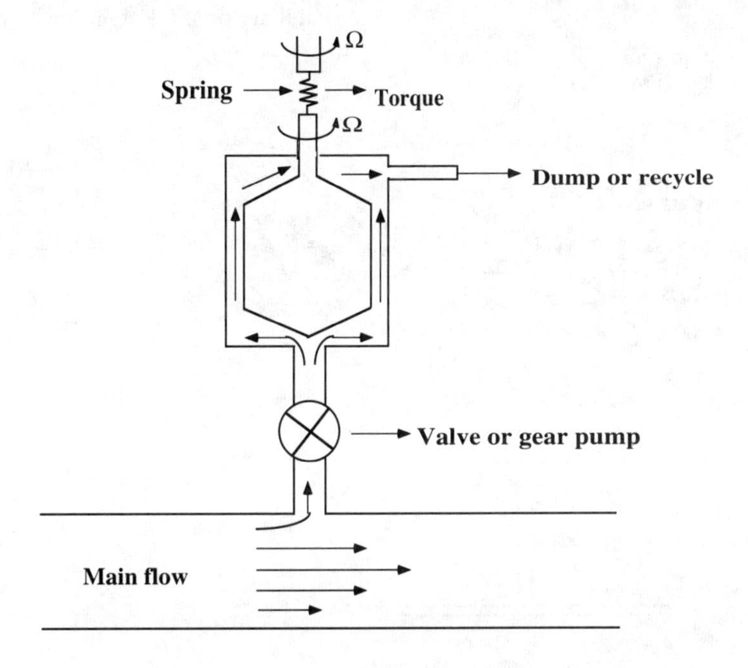

Figure 8.6.1.
On-line concentric cylinder rheometer with torque sensor on the rotating bob. Similar to commercial designs from Brookfield, Haake, and Mettler.

Figure 8.6.2.
On-line concentric cylinder rheometer with torque sensor on fixed bob. Designed for sinusoidal oscillations. From Zeichner and Macosko (1982).

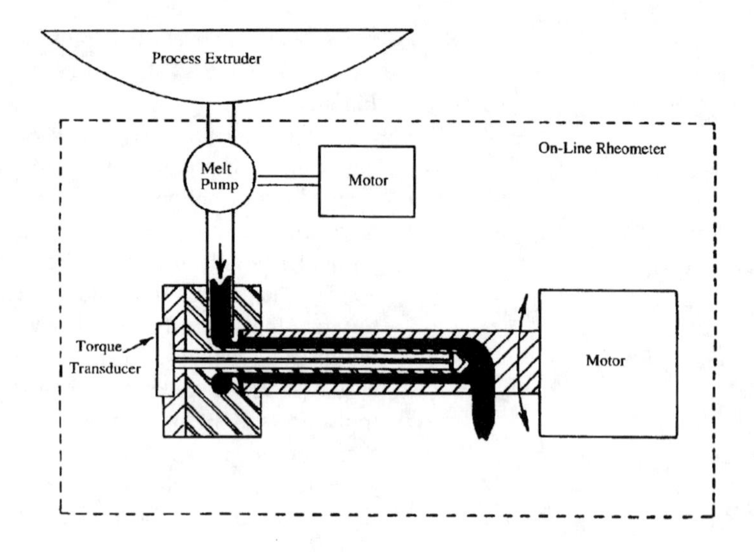

Side stream or on-line methods provide greater flexibility in the type and temperature of rheological measurement. This is the area that has seen the most commercial development. Figure 8.6.1 shows a common design for on-line viscosity measurement of slurries and polymer solutions. Often there is a holding tank or reservoir in the process where the viscometer may be mounted. A strong superposed cross flow will typically lead to reduction in the measured viscosity (recall Figure 2.4.4).

Figure 8.6.2 shows a concentric cylinder design for polymer melts. The outer cylinder rotates or oscillates, while torque is measured on the inner cylinder. To get a fresh representative sample and to avoid cross flow, the gear pump is operated intermittently and measurements taken intermittently (Zeichner and Macosko, 1982). The capillary or slit design shown in Figure 8.6.3 can be operated in the same way.

Figure 8.6.3.
On-line capillary rheometer, flange mounted on a single-screw extruder: *a*, melt stream from extruder; *b*, gear pump; *c*, pressure transducer; *d*, thermocouple; and *e*, capillary. From Göttfert (1991).

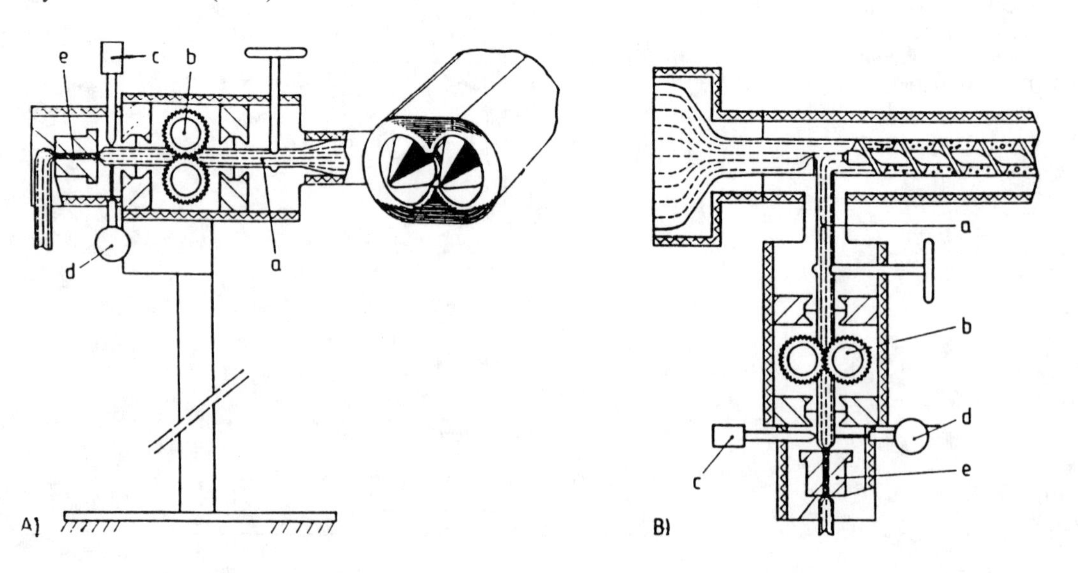

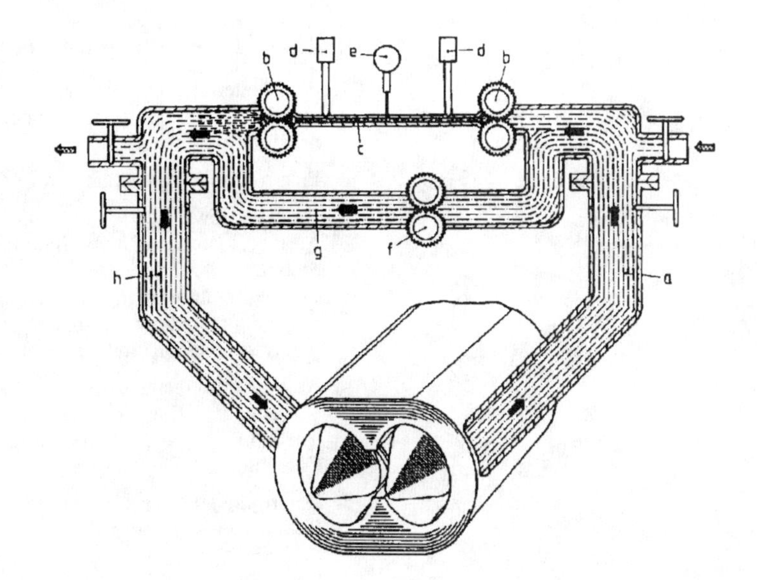

Figure 8.6.4.
On-line slit rheometer with recycle. The conveying gear pump permits rapid sampling and flushing: *a* melt stream from extruder; *b*, gear pumps in front of and behind *c*, the measuring capillary; *d*, pressure transducers; *e*, thermocouple; *f*, additional gear pump; *g*, circulating melt stream; and *h*, melt back flow. From Göttfert (1991).

The on-line designs in Figures 8.6.1–8.6.3 typically dump the bypass stream. This can lead to plugging the exit port in addition to the problem of waste. Recycle schemes like the one shown in Figure 8.6.4 have been developed. With this design, a sample can be taken in less than 5 minutes from the process and the temperature can be more than 20°C different from the process as a result of the efficient heat transfer. This slit design also lends itself to on-line optical measurements such as infrared for composition or light scattering for gel particles.

8.7 Summary

The rheology literature is rich with homemade instruments. However, today at least in the area of shear rheometry, there are a large number of excellent commercial instruments to choose from. Even if the user has an especially difficult material or measurement, a rheometer manufacturer will often help to modify an existing design.

The real challenge usually lies in wisely using the available instruments. Table 8.7.1 summarizes the experimental philosophy suggested in Figure II.3 and throughout Chapters 5–8. The table outlines a check list to follow when dealing with new instruments, new types of material, and even new ranges of testing. One problem that is difficult to cover adequately is item 8 in Table 8.7.1: sample changes. These are typically the biggest single cause of "bad data." The operator's eyes and brains are the most important option on any rheometer.

As indicated in Section 8.3, perhaps the most rapidly developing area of rheometer design is interactive software control. Significant advances have also been made recently in extensional

TABLE 8.7.1 / Typical Approach to Rheological Measurements

1. Pick material function of interest. Then pick range of variables desired and test temperature (and see if it is possible to achieve)
2. Pick a simple, controllable test geometry: a rheometer
3. Analyze flow[a]:
   Assume material is a continuum (i.e., particle size << smallest rheometer dimensions)
   Check literature on this flow geometry
   Solve equations of motion, energy balance
   Apply simple constitutive equations (especially for indexers)
4. Assume homogeneous or simple stress field:
   No secondary flow or slip
   Check by birefringence, pressure profiles for new geometries
   Compare to predicted stress
5. Determine stress from:
   Boundary force, shape
   Pressure
   Birefringence
6. Determine deformation from:
   Boundary motion
   Bulk flow rate
   Check this by flow visualization, strain distribution, changing sample geometry
   Compare to predicted velocities, strains
7. Watch out for systematic errors:
   Pressure hole errors
   Secondary flow
   Instrument compliance
   Gap changes with temperature
8. Watch for sampling changes:
   Prehistory before loading
   Air bubbles
   Slip, buckling
   Evaporation, settling, phase separation, crystallization, chemical reaction
9. Report material function data!
   Indicate uncertainty (error bars) and limits of measurements

[a]For simple geometries this is done in the instrument software. Check a few points manually.

rheometry. Rheo-optics is also a fast changing area of rheometry and is the subject of our Chapter 9, the last in Part II.

References

Amari, T.; Wei, X.; Watanabe, K., in *Theoretical and Applied Rheology*; Moldenaers, P.; Keunings, R., Eds.; Elsevier: Amsterdam, 1992; p. 573.

Berry, G. C.; Park, J. O.; Meitz, D. W.; Birnboim, M. H.; Plazek, D. J., *J. Polym. Sci. Phys.* 1989, *27*, 273.

Berting, J. P.; O'Connor, J. J.; Grehlinger, M., Use of Arbitrary Functions to Determine the Rheological Properties of Viscoelastic Systems, presented to Society of Plastics Engineers, May 1990. Manuscript available from Rheometrics, Inc., Piscataway, NJ.

Bohlin, L., in *Progress and Trends in Rheology*, Vol. II; Giesekus, H., Hibberd, M. F., Eds.; Steinkopf: Darmstadt, 1988; p. 151.

Bohlin Reologi AB, CSM product literature, Lund, Sweden, 1990.

Brokate, T.; Gast, T., in *Theoretical and Applied Rheology*, Moldenaers, P.; Keunings, R., Eds.; Elsevier: Amsterdam, 1992; p. 1992.

Brookfield Engineering Laboratories Inc, product literature, Stoughton MA, 1992.

Buscall, R.; Goodwin, J. W.; Hawkins, M. W.; Cotterill, R. M., *J. Chem. Soc., Faraday Trans. 1* 1982 *78*, 2873.

Cai, J. J.; Souza Mendes, P. R.; Macosko, C. M.; Scriven, L. E.; Secor, R. B., in *Theoretical and Applied Rheology*, Moldenaers, P.; Keunings, R., Eds.; Elsevier: Amsterdam, 1992; p. 1012.

Carri-med Ltd., CSL product literature, Dorking, U.K., 1988.

Carri-med Americas Inc, Extended Temperature Module product literature, Valley View, Ohio, 1992.

Christiansen, E. B.; Leppard, W. R., *Trans. Soc. Rheol.* 1974, *18*, 65.

Christmann, L.; Knappe, W., *Rheol. Acta* 1976 *15*, 296.

Collyer, A. A.; Clegg, D. W., Eds., *Rheological Measurement*; Elsevier: London, 1988.

Davis, S.; Deer, J. J.; Warburton, B., *J. Phys. E* 1968, *1*, 933.

Dealy, J. M., *Rheometers for Molten Plastics*; Van Nostrand-Reinhold: New York, 1982.

Dealy, J. M., *Rheol. Acta* 1990, *29*, 519.

Dealy, J. M., in *Theoretical and Applied Rheology*; Moldenaers, P.; Keunings, R.; Eds.; Elsevier: Amsterdam, 1992; p. 39.

Drislane, C. J.; DeNicola, J. P.; Wareham, W. M.; Tanner, R. I., *Rheol. Acta* 1974, *13*, 4.

Ferry, J. D., *Viscoelastic Properties of Polymers*, 3rd ed.; Wiley-Interscience: New York, 1980.

Franck, A. J. P., Quasi-Infinite Stiff Transducer for Measuring Torque and Normal Force, presented at conference on New Techniques in Experimental Rheology, University of Reading, U.K., September 1985a. Manuscript available from Rheometrics, Inc., Piscataway, NJ.

Franck, A. J. P., *J. Rheol.* 1985b, *29*, 833.

Franck, A. J. P., in *Theoretical and Applied Rheology*; Moldenaers, P.; Keunings, R., Eds.; Elsevier: Amsterdam, 1992; p. 982.

Fuller, G. G.; Cathey, C. A.; Hubbard, B.; Zebrowski, B. E., *J. Rheol.* 1987, *31*, 235.

Gerard, J. F.; Andrews, S. J.; Macosko, C. W., *Polym. Composites* 1990, *11*, 90.

Giacomin, A. J.; Samurkas, T.; Dealy, J. M., *Polym. Eng. Sci.* 1989, *29*, 499.

Giles, D. W.; Denn, M. M., *J. Rheol.* 1990, *34*, 603.

Göttfert Werkstadt-Prüfmaschinen, product literature, Buchen, Germany, 1989.

Göttfert, A., *SPE Tech. Pap.* 1991, 37, 2299.

Gottlieb, M.; Macosko, C. W., *Rheol. Acta* 1982, *21*, 90.

Grankvist, T. O., Sandas, S. E., in *Proceedings of the TAPPI Coating* Conference, Boston, May 1990.

Haake AG, Rotovisco RV20, product literature, Karlsruhe, Germany, 1986.

Haake Mess-Technik GmbH, RS100, product literature, Karlsruhe, Germany, 1992.

Hagen, G. H. L., *Ann. Phys.* 1839, *46*, 423.

Holly, E. E.; Venkataraman, S. K.; Chambon, F.; Winter, H. H., *J. Non-Newtonian Fluid Mech.* 1988, *27*, 17.

Interlaken Technology Corporation, True Shear Rheometer product literature, Eden Prairie, MN 55346, 1992.

Jobling, A.; Roberts, J. E., in *Rheology: Theory and Applications*, Vol. 2; Eirich, F.R., Ed.; Academic Press: New York, 1958; p. 503.

Jones, D. M.; Walters, K.; Williams, P. R., *Rheol. Acta* 1987 *26* 20.

Joseph, D. D., *Fluid Dynamics of Viscoelastic Liquids*, Springer-Verlag, New York, 1990, Appendix F.

Kannan, R. M.; Kornfield, J. A., *Rheol. Acta* 1992, *31*, 535.

Knudsen, K. D.; Mikkelsen, A.; Elgsaeter, A., *Rheol. Acta* 1992, *31*, 431.

Koppi, K.; Tirrell, M.; Bates, F. S., *Phys. Rev. Lett.* 1993, *70*, 1494.

Krieger, I. M., *J. Rheol.* 1990, *34*, 471.

Laun, H. M.; Hirsch, G., *Rheol. Acta* 1989, *28*, 267.

Lodge, A. S., in *Rheological Measurements*, Collyer, A. A.; Clegg, D. W., Eds.; Elsevier: Amsterdam, 1988; p. 345.

Luckenbach, T. A., Rheological Testing of Asphalt Using a Cup and Plate Geometry, presented at the Petersen Asphalt Research Conferences, Laramie, WY, 1992. Manuscript available from Rheometrics, Inc., Piscataway, NJ.

Mackay, M. E.; Halley, P. J., *J. Rheol.* 1991, *35* 1609.

Mackay, M. E.; Liang, C.-H.; Halley, P. J., *Rheol. Acta* 1992, *31,* 481.

Macosko, C. W.; Starita, J. M., *SPE J.* 1971, *27* (*11*), 38.

Macosko, C. W.; Weissert, F. L, in *Rubber and Related Products: New Methods for Testing and Analyzing,* ASTM STP 553, American Society for Testing and Materials, Philadelphia, 1974; p. 127.

Magda, J. J.; Lou, J.; Baek, S. G.; DeVries, K. L., *Polymer* 1991, *32*, 2000.

Magnin, A.; Piau, J. M., *J. Rheol.* 1991, *35*, 1465.

Mani, S.; Malone, M. F.; Winter, H. H.; Halary, J. L.; Monnerie, L., *Macromolecules* 1991, *24*, 5451.

Mattingly, G. E., Volume Flow Measurement, in *Fluid Mechanics Measurements*; Goldstein, R.J., Ed.; Hemisphere: Washington, DC, 1983; Chapter 6.

Meissner, J., *J. Appl. Polym. Sci.* 1972, *16*, 2877.

Meissner, J., *Makromol. Chem., Makromol. Symp.* 1992, *56*, 25.

Meissner, J.; Hostettler, J., in *Theoretical and Applied Rheology*, Moldenaers, P.; Keunings, R., Eds.; Elsevier: Amsterdam, 1992; p. 938.

Meissner, J.; Garbella, R. W.; Hostettler, J., *J. Rheol.* 1989, *33*, 843.

Michel, H., A Universal Sensor for Rotational Rheometers, presented at Belgian Society of Rheology, Leuven, 1988. Manuscript available from Physica GmbH, Stuttgart, Germany.

Mikkelsen, K. J.; Macosko, C. W.; Fuller, G. G., in *Proceedings of the 10th International Congress on Rheology*, Vol. 2; Uhlherr, P. H. T., Ed.; Monaash University: Clayton, 1988; p. 125.

Mora, E., MS thesis, University of Minnesota, September 1991.

Mora, E.; Macosko, C. W., *Proc. North Am. Therm. Analysis Soc.* 1991; p. 506.

Nametre Viscometer, product literature, Nametre Company, Metuchen, NJ, 1990.

Nazem, F.; Hansen, M. G., *J. Appl. Polym. Sci.* 1976, *20*, 1355.

Nielsen, L. E., *Polymer Rheology*; Marcel Dekker: New York, 1977.

Pabendinskas, A.; Cluett, W. R.; Balke, S. T., *Polym. Eng. Sci.* 1991, *31*, 365.

Padmanabhan, M., *J. Food Eng.* 1993, to appear.

Padmanabhan, M.; Bhattacharya, M., *Rheol. Acta* 1993, submitted.

Physica Mess Technik GmbH, Rheolab product literature, Stuttgart, Germany, 1989.

Plazek, D. J., *J. Polym. Sci.* 1968, *6*, 621.

Plazek, D. J., Viscoelastic and Steady-State Rheological Response, in *Methods of Experimental Physics*, Vol. 16; Fava, R. A., Ed.; Academic Press: New York, 1980.

Rank Pulse Shearometer, product literature, PenKem, Inc., Bedford Hills, NY, 1992.

Rheometrics Inc., RMS-800 Operation Manual, Piscataway, NJ, 1990.

Rheometrics Inc., RFX product literature, Piscataway, NJ, 1991.

Rheometrics Inc., RFS II operation manual, Piscataway, NJ, 1992.

Rheometrics Inc., DSR product literature, Piscataway, NJ, 1993.

Richards, W. D.; Prud'homme, R. K., *Polym. Eng. Sci.* 1987, *27*, 294.

Rodriguez, F., *Principles of Polymer Systems*, 3rd ed., McGraw-Hill: New York, 1989.

Schrag, J. L., *Trans. Soc. Rheol.* 1977, *21*, 399.

Secor, R. B.; Schunk, P. R.; Hunter, T. B.; Stitt, T. F.; Macosko, C. W.; Scriven, L. E., *J. Rheol.* 1989, *33*, 1329.

Sopra, F. S.; Borgia, A.; Strimle, J.; Feigenson, M., *J. Geophys. Res.* 1988, *93*, 10273.

Starita, J. M., in *Rheology*; Astarita, G., Marrucci, G., Nicolais, L., Eds.; Plenum: New York, 1980; p. 229.

Taylor, J. C., as cited on p. 161, in Whorlow (1992).

Thurston, G. B., *J. Acoust. Soc. Am.* 1961, *33*, 1091.

Van den Brule, B. H. A. A.; Kadijk, S. E., *J. Non-Newtonian Fluid Mech.* 1992, *43*, 127.

Van der Wal, C. W.; Nederveen, C. J.; Schwippert, G. A., *Rheol. Acta* 1969, *8*, 130.

Van Olphen, H., *Clays and Clay Materials* 1956 *4*, 204.

Van Wazer, J. R.; Lyons, J. W.; Lim, K. Y.; Colwell, R. E., *Viscosity and Flow Measurement*; Wiley: New York, 1963.

Vilastic 3, product literature, PenKem, Inc., Bedford Hills, NY, 1992.

Walters, K., *Rheometry*; Chapman & Hall: London, 1975.

Whorlow, R. W., *Rheological Techniques*; Wiley: New York, 1980.

Whorlow, R. W., *Rheological Techniques*, 2 ed.; Ellis Horwood: London, 1992.

Zeichner, G. R.; Macosko, C. W., *SPE Tech. Pap.* 1982, *28*, 79.

Zimm, B. H.; Crothers, D. M., *Proc. Natl. Acad. Sci. U.S.* 1962, *48*, 905.

9

RHEO-OPTICS: FLOW BIREFRINGENCE

In 1866 I made some attempts to ascertain whether the state of strain in a viscous fluid in motion could be detected by its action on polarized light. . . . It is easy . . . to observe the effect in Canada balsam. . . .
James C. Maxwell (1873)

In flowing liquids in which the same mechanism . . . gives rise to birefringence and to stress, it is reasonable to expect these to vary together when factors such as rate of shear and concentration of solute are altered.
Arthur S. Lodge (1955)

Timothy P. Lodge

9.1 Introduction

Optical experiments of various kinds are being used increasingly across the spectrum of macromolecular science, and polymer rheology is certainly no exception. In Figure 9.1.1, for example, the birefringence of a polymer melt is measured as a function of position in the entrance region of a slit flow (Han, 1981). The light intensity patterns are essentially contour lines of the stress field and provide a direct picture of the spatial evolution of stress in the liquid. In the general case, a rheological experiment entails the

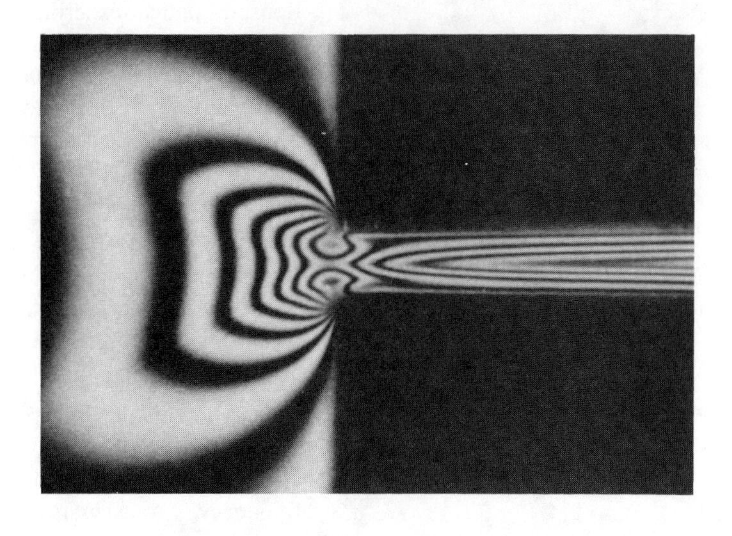

Figure 9.1.1. Representative stress–birefringence pattern (isochromatics) for a high density polyethylene melt flowing into a slit die. From Han (1981).

measurement of a force (related to the stress) and either a displacement (related to the strain) or a velocity (related to the strain rate). In a rheo-optical experiment, the measurement of force is replaced by the measurement of some optical property of the sample. The results obtained from the two experiments usually are very closely related, but differences in both principle and practice render the two approaches complementary rather than redundant.

Some of the important features of the rheo-optical approach are as follows

1. The measured quantity is a direct reflection of molecular orientation and shape; in contrast, a mechanical experiment senses the dissipation and/or storage of energy.
2. The measured quantity is decoupled from the applied field. In other words, the sample is subjected to a flow while an optical signal is monitored. This property has important implications in terms of sensitivity. In a mechanical experiment, the measured quantity—for example, the force—results from the displacement of both sample and apparatus, and the two contributions must be resolved. If the sample is a polymer solution, the solvent contribution must also be considered. In either case, if the polymer contribution is relatively small, it may be very difficult to extract with precision. In contrast, optical experiments often can be designed such that the polymer contribution dominates the measured signal.
3. Spatial resolution within the sample is possible. This is obtained either by appropriate positioning of the light beam or by the use of finite area detectors, as in Figure 9.1.1. In contrast, a mechanical experiment inherently integrates the response over the entire sample.
4. Molecular labeling is possible. By appropriate manipulations, an optical experiment can sense the behavior of one selected component in a multicomponent fluid. For example, molecules can be labeled by refractive index, polarizability anisotropy, isotopic substitution, or chromophore attachment. This kind of specificity is impossible to achieve in a mechanical experiment.
5. Optical experiments are very sensitive. Given the ready availability of lasers, emitting approximately 10^{20} photons per second, and photomultiplier detectors, which can respond to single photons, it is possible to make measurements of very weak signals with good precision.
6. The inherent speed of optical detectors, in combination with various modulation schemes, can provide information on shorter time scales (e.g., submilliseconds) than most mechanical techniques, permitting examination of either high frequency or transient responses.
7. Optical experiments provide an independent means to assess the utility and applicability of rheological constitutive rela-

tions. For example, two models that give similar expressions for the stress tensor may differ markedly in their predictions for the birefringence properties.

As is to be expected, the rheo-optical approach is not without significant restrictions. The two most prominent are:

1. The sample must have suitable optical properties, particularly in terms of transparency and freedom from macroscopic contaminants. This problem can be severe for commercial products.
2. Additional characterization is often required. Although the direct measurement of a modulus or a viscosity is often of immediate utility, in the optical case some additional measurements may be required to establish the relationship between the optical properties and the mechanical ones. For example, the stress–optical relation, to be discussed in Section 9.4, predicts that the shear stress and first normal stress difference may be obtained from birefringence measurements, but only after a quantity called the stress–optic coefficient is determined.

These disadvantages notwithstanding, in many situations the appropriate rheo-optical method would be the ideal tool for rheological characterization. A wide variety of experimental techniques have been developed and will be used increasingly as they become less mysterious to polymer scientists and engineers. In this introductory chapter, the intention is to provide a sense of the breadth of opportunity in rheo-optics in addition to the necessary background information. In the next two sections, birefringence measurements are placed in the context of optical experiments in general, and the basics of light polarization are reviewed. In Section 9.4, the general principles of flow birefringence measurements are outlined, while nine illustrative examples of rheo-optical applications are discussed in Section 9.5. The subject cannot possibly be considered in detail in a limited space, so appropriate references are provided.

9.2 Review of Optical Phenomena

When incident electromagnetic radiation interacts with matter, three broad classes of phenomena are of interest. First, energy can be *absorbed*, with the possible subsequent emission of some or all of the energy. Second, the radiation can be *scattered* (i.e., change direction), with either no change in energy (elastic scattering) or a measurable change in energy (inelastic scattering). Third, the light can *propagate* through the material with no change in direction or energy, but *with a change in its state of polarization*. This last possibility is the basis for birefringence, which is the most relevant process from the rheo-optical perspective. Scattering experiments are also of very great importance in polymer science, while absorption and emission spectroscopy are used for the chemical analysis of polymers in entirely the same manner as for small molecule systems.

9.2.1 Absorption and Emission Spectroscopies

The probability of absorption of electromagnetic energy by a given molecule depends on whether the incident photon energy matches that of a quantum mechanically allowed transition between stationary states of the molecule. Thus, the wavelength dependence of absorption efficiency is a diagnostic of chemical structure; infrared photons absorbed correspond to vibrational and rotational transitions, whereas ultraviolet and visible photons absorbed correspond to changes between electronic energy levels. The subsequent emission of radiation (i.e., by fluorescence or phosphorescence) is also used in quantitative and qualitative chemical analysis. However, with the exception of linear dichroism to be discussed subsequently, neither absorption nor emission is of immediate relevance to rheo-optical methods, and these subjects are not discussed further here.

9.2.2 Scattering Techniques

In a scattering experiment, incident radiation is redirected when it encounters a change in the impedance of the medium. Thus, for light, x-rays, and neutrons, the scattered intensity depends on changes in refractive index, electron density, and nuclear scattering cross section, respectively. For all three types of radiation, the physical description of the scattering process is identical, and all three are of central importance to polymer science. However, the measured quantities reflect the size, shape, and spatial arrangements of polymer molecules rather than their rheological properties directly. Thus, although simultaneous scattering measurements are now being performed in rheological apparatuses (a difficult experimental undertaking), the discussion of these techniques will be rather brief.

Static Light, X-Ray, and Small-Angle Neutron Scattering

In these experiments the time-averaged scattered intensity I_s is measured as a function of the scattering vector $\mathbf{q}$. The net detected intensity can be computed as the superposition of the signals from each scattering center (e.g., each monomer unit). According to the spatial arrangement of the scatterers, the individual scattered waves may interfere constructively or destructively at the detector. Thus, $I_s(\mathbf{q})$ is proportional to the so-called static structure factor $S(\mathbf{q})$, which sums the waves with different phases from different locations. $S(\mathbf{q})$, in fact, reflects the spatial Fourier transform of the distribution of scatterers (i.e., the pair correlation function), and thus in a polymer system it is possible to determine the monomer distribution function. More commonly, one obtains the second moment of this distribution, known as the mean square radius of gyration. Furthermore, if the absolute scattered intensity is measured in the dual limits of zero scattering angle and zero concentration, the absolute weight-average molecular weight of the polymer sample can be determined. This is the classical approach developed by Debye, Zimm, and others (Hiemenz, 1984). Although light and x-ray scattering apparatuses are routinely available in polymer laboratories, neutron scattering experiments can be performed at only a handful of locations around the world. However, the particular advantage of neutron scattering is that the nuclear scattering cross sections of hydrogen and deuterium are very different, and thus the size and shape of isotopically labeled molecules in bulk systems can be determined.

Dynamic Light Scattering

Also known as quasi-elastic light scattering, this technique monitors the temporal fluctuations in $I_s(\mathbf{q})$ (Berne and Pecora, 1976; Chu, 1990). These fluctuations result from random thermal motions, which change the instantaneous spatial arrangement of molecules and thus the net scattered intensity. As these random motions result in microscopic concentration fluctuations, a mutual diffusion coefficient can be determined from the time constant of the decay of the time autocorrelation function of $I_s(\mathbf{q}, t)$. Rapid advances in laser and autocorrelator technology during the last two decades have made this experiment a routine characterization and research tool.

Other Scattering Experiments

Three other scattering experiments used in polymer science deserve mention for the sake of completeness; these are Raman scattering, Brillouin scattering, and forced Rayleigh scattering. The first two are inelastic; that is, the scattered radiation is of a frequency different from that of the incident wave. Raman frequency shifts correspond to changes in vibrational or rotational energy states of the molecule and thus reflect the molecular chemical structure, much like infrared absorption spectroscopy. Brillouin scattering arises from propagating density fluctuations, or phonons, in the liquid. Thus the Brillouin spectrum contains information about the modulus of the material. Forced Rayleigh scattering is a relatively new technique that follows the decay by mass diffusion of a transient chemical diffraction grating created in a sample. It is a very powerful experiment, as tracer diffusion coefficients can be determined over a very wide range (from $10^{-5}\text{cm}^2/\text{s}$ to $10^{-16}\text{cm}^2/\text{s}$), and it has been applied to many different polymer systems (see e.g., Huang et al., 1987).

9.2.3 Birefringence and Dichroism

The phenomenon of birefringence is the one of most importance to rheo-optics and is introduced in this section. Section 9.3 on polarization of light, explains the primary methods for measuring birefringence. Dichroism, although in fact an absorption process, is very closely coupled to birefringence and therefore is also defined here.

A light wave propagates through a nonabsorbing medium with a speed v, which is reduced from the speed in vacuo, c (Born and Wolf, 1980; Hecht, 1987). The ratio c/v defines the refractive index of the material, n. The wavelength of the radiation is also changed, from λ_o in vacuo to λ in the medium, with $\lambda = \lambda_o/n$, but the frequency ν remains constant. The change in speed results in a change in propagation direction at an interface between two materials of different n, as given by Snell's law (see Figure 9.2.1).

$$n_1 \sin \psi_1 = n_2 \sin \psi_2 \tag{9.2.1}$$

This phenomenon is called refraction. The actual value of n, which is always greater than or equal to one, reflects the polarizability of the constituent molecules α, or, loosely speaking, the ability of the electric field of the light wave to distort the electron distribution of the molecules. In general, the more polarizable the molecules, the larger n will be. Typical values for n for transparent liquids and glasses fall in the range 1.3–1.7. Most molecules actually possess an anisotropic polarizability, which can be represented as a tensor quantity $\boldsymbol{\alpha}$. This is the direct consequence of anisotropy in the chemical structure (see Denbigh, 1940); thus, nitrobenzene, chloroform, and biphenyl are anisotropic, whereas carbon tetrachloride and methane are isotropic (see Figure 9.2.2). By virtue of their chain structure, almost all polymers exhibit some degree of opti-

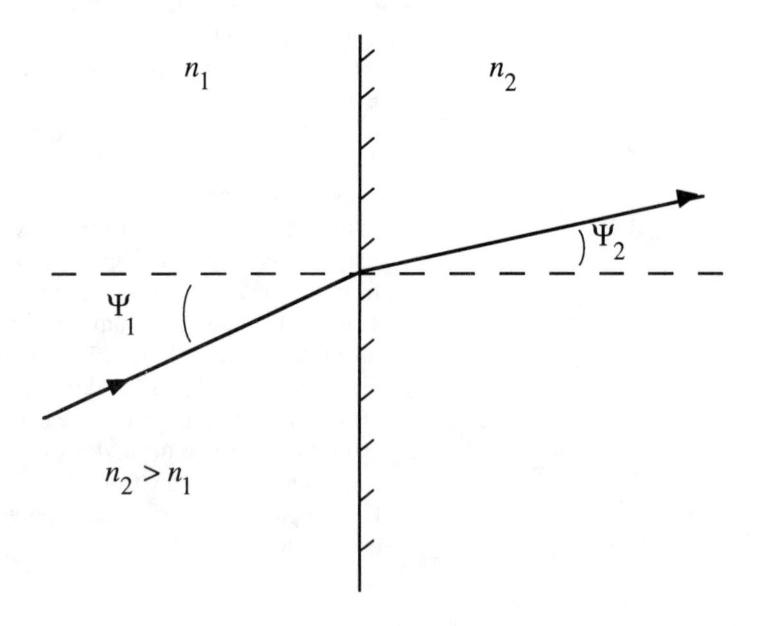

Figure 9.2.1.
Illustration of Snell's law.

cal anisotropy. If anisotropic molecules in a material are oriented preferentially in one direction, the value of n will depend on the relative orientation of the electric field of the light (i.e., its polarization) to the molecular axes. This is the origin of birefringence or, as it is also known, double refraction. For when a beam of light containing two orthogonal polarizations enters a birefringent material, the two values of n experienced lead to two different angles of refraction. The birefringence Δn can be defined simply as $n_1 - n_2$, where n_1 and n_2 are the refractive indices of the material along two appropriately selected orthogonal axes.

The crucial point from the preceding discussion is the necessity of producing a net orientation of anisotropic molecules in order to generate a measurable birefringence. In a quiescent liquid, all molecular orientations are equally likely (except in the particular case of liquid crystals), and therefore anisotropy in the molecular property $\boldsymbol{\alpha}$ does not lead to anisotropy in the material property n. However, in a polymer liquid under flow, a nonspherically symmetric distribution of macromolecular orientation is induced. Even a slight perturbation to the equilibrium (random) orientation distribution can produce a measurable birefringence. Birefringence in liquids can be generated by the application of various orienting fields; historically, when the field is hydrodynamic, electric, or magnetic, the resulting birefringence is referred to as the Maxwell, Kerr, or Couton–Mouton effect, respectively. The phenomenon of dichroism occurs when the wavelength of light entering a macroscopically anisotropic material is such that it will be absorbed to

Figure 9.2.2.
Examples of molecules possessing (a) isotropic polarizability and (b) anisotropic polarizability.

Isotropic polarizability: CH_4 and CCl_4

Anisotropic polarizability: $C_6H_5NO_2$, biphenyl, $CHCl_3$, polystyrene, $(\ldots)_n$ with Cl, $(\ldots)_n$ with CH^3

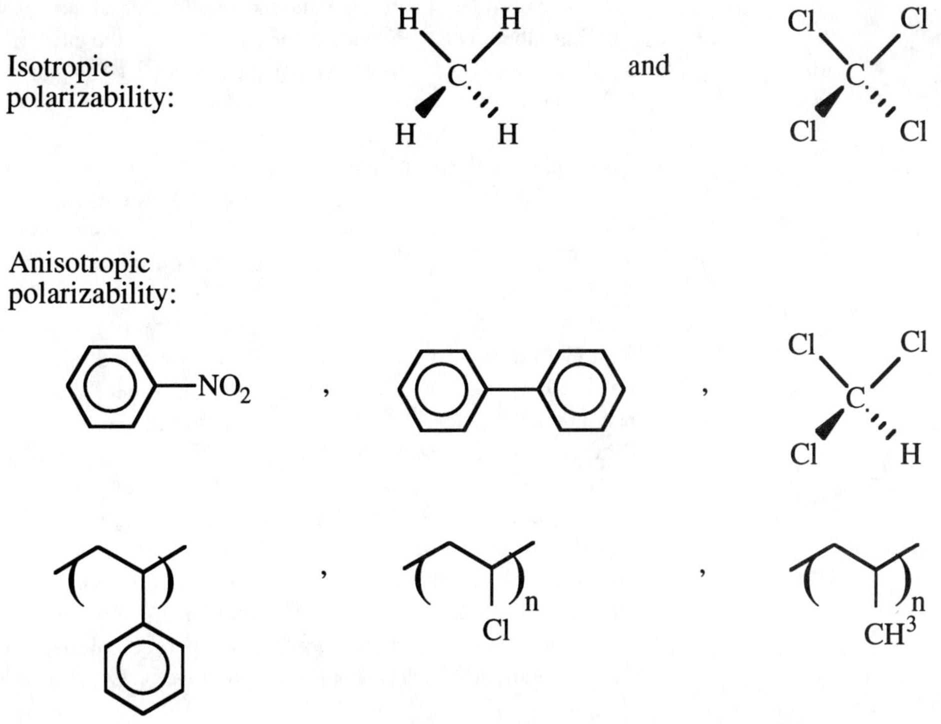

some extent. The value of the molar absorptivity also depends on the relative orientation of the light polarization and the so-called transition dipole of the molecules. The resulting variation in absorption with polarization orientation is referred to as dichroism.

9.3 Polarized Light

According to Maxwell's equations, light is a traveling electromagnetic wave; for our purposes, only the electric field component is of importance. For a beam propagating in the z direction, as will be assumed henceforth, the electric field amplitude $|\mathbf{E}|$ oscillates in the x–y plane, and the wave may be expressed in either of two equivalent ways:

$$\mathbf{E}(z,t) = \mathbf{E}_m \cos(2\pi\nu t - kz) \tag{9.3.1}$$

$$\mathbf{E}(z,t) = R_e\{\mathbf{E}_m e^{j2\pi\nu t} e^{-jkz}\} \tag{9.3.2}$$

where $j = (-1)^{0.5}$, ν is the frequency, and the wave vector k is given by $2\pi/\lambda$. The use of complex notation in eq. 9.3.2 is for arithmetical convenience only; the observable quantity is the intensity of the beam, which is proportional to the electric field squared. Note that use of complex notation is potentially unreliable for nonlinear equations. In the complex notation, $I \sim |\mathbf{E}^*\mathbf{E}|$, where $\mathbf{E}^*$ is the complex conjugate of $\mathbf{E}$ (obtained by replacing j with $-j$). The product of a number and its complex conjugate is always a real number. The frequency is the rate at which the electric field amplitude oscillates in time and is on the order of $10^{14}\ \mathrm{s}^{-1}$ for visible light. (This oscillation is much too rapid to be detected directly and is not of further concern.) The term kz determines how the wave amplitude oscillates in space at a given instant of time; in other words, it keeps track of the spatial phase of the wave and is crucial to the subsequent development. Note that k depends on the value of n through the appearance of λ, which is the wavelength of the light *in the material.*

A light beam can be thought of as containing components with electric field orientations randomly distributed in the $x - y$ plane. However, in a "plane-polarized" or "linearly polarized" beam, the electric field lies along one line in the $x - y$ plane (Shurcliff, 1962; Azzam and Bashara, 1977; Born and Wolf, 1980; Hecht, 1987). Figure 9.3.1 illustrates schematically unpolarized, x-polarized, and y-polarized light. In the case of unpolarized light, it is also important that the phases (kz) of the components be randomly distributed throughout one cycle. To illustrate the importance of the relative phases of the components, consider the situation of a pure x-polarized wave and a pure y-polarized wave, each with an amplitude E_o but with a relative phase angle ϕ (that is therefore independent of z). We can write the two components as follows:

$$E_x = E_o \cos(2\pi \nu t) \tag{9.3.3}$$

$$E_y = E_o \cos(2\pi \nu t - \phi) \tag{9.3.4}$$

Light waves obey the principle of superposition, and thus the net effect of the two waves can be computed simply by adding eqs. 9.3.3 and 9.3.4 together. (Conversely, any given wave can be viewed as an appropriate sum of components along selected axes.) This sum depends critically on ϕ, as illustrated by the following specific cases.

Figure 9.3.1.
Schematic illustration of (a) y-polarized, (b) x-polarized, and (c) un-polarized light.

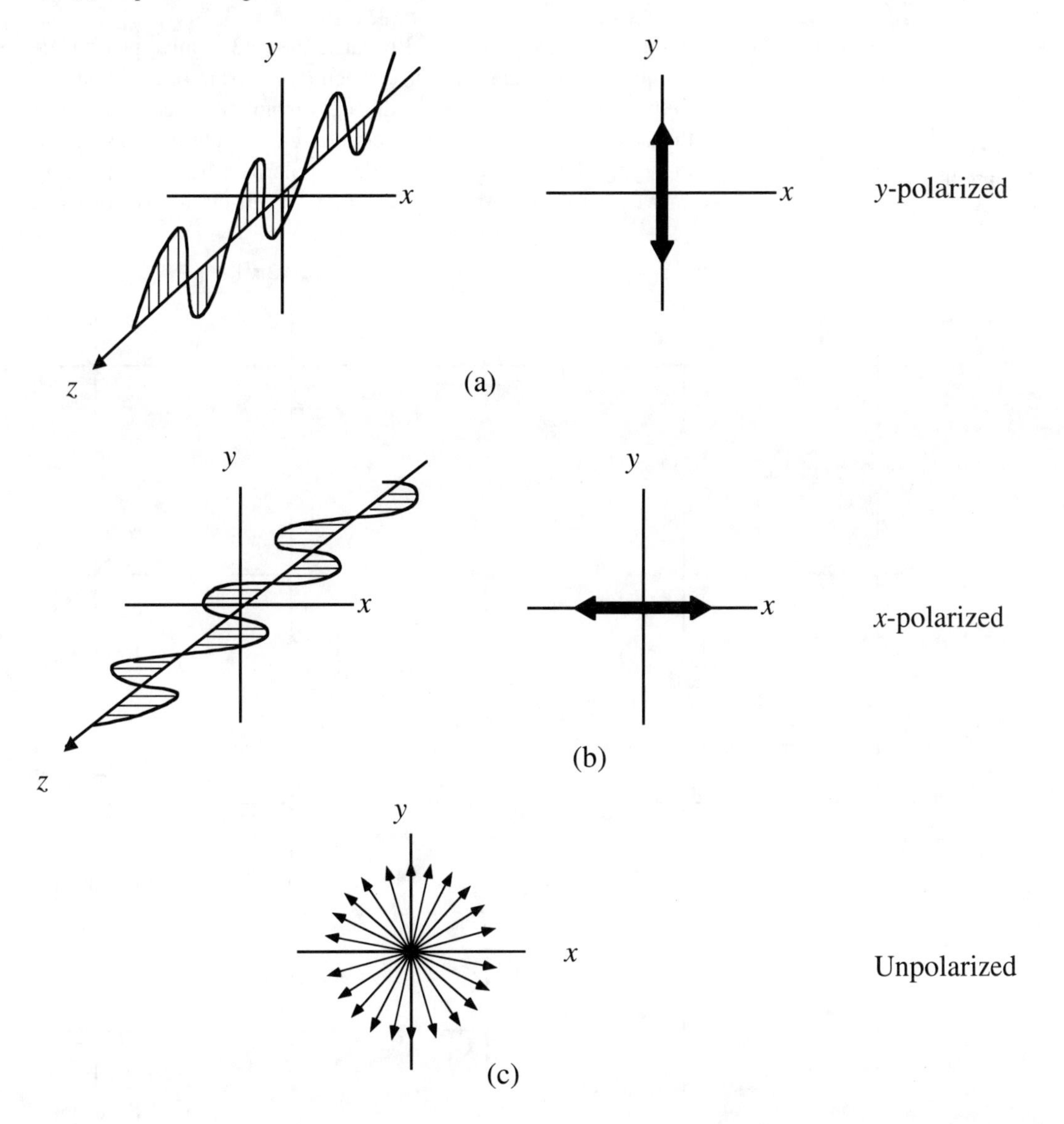

$\phi = 0^\circ$

The two components are in phase. That means that whenever E_x is at its positive maximum ($E_x = E_o$), so too is $E_y = E_o$. The resultant is a wave of amplitude $(2)^{0.5} E_o$ oscillating along the line at 45° to both the x and y axes. Thus, the resultant is also linearly polarized light, as illustrated in Figure 9.3.2a.

$\phi = 180^\circ$ (π rad)

Now E_x achieves its positive maximum when E_y is at its most negative and vice versa. The resultant is still linearly polarized light, as shown in Figure 9.3.2b, albeit oriented at 90° in the x–y plane from the resultant for $\phi = 0$.

$\phi = 90^\circ$ ($\pi/2$ rad)

Here E_x has its positive and negative maxima when $E_y = 0$, and vice versa. Note what happens when $|E_x| = E_o/(2)^{0.5}$, which occurs whenever $2\pi\nu t = i\pi/4$, with $i = 1, 3, 5, 7, \ldots$. At that point, $|E_y|$ also $= E_o/(2)^{0.5}$ because the $\pi/2$ relative phase shift is equal to the change in phase when i goes from one odd integer to the next. The resultant in this case is not confined to a line in the x–y plane; rather, it maps out a circle of radius E_o, as shown in Figure 9.3.2c. This is circularly polarized light. Depending on whether ϕ is -90° or $+90^\circ$, right- or left-circularly polarized

Figure 9.3.2.
(a) E_y and E_x are in phase; the resultant is always 45° to the x and y axes; $\omega = 2\pi\nu$. (b) E_y and E_x are 180° out of phase; the resultant is always 45° to the x and y axes. (c) E_y and E_x are 90° out of phase; the resultant sweeps out a circle in the x–y plane. (d) E_y and E_x have an arbitrary phase shift; the resultant sweeps out an ellipse in the x–y plane.

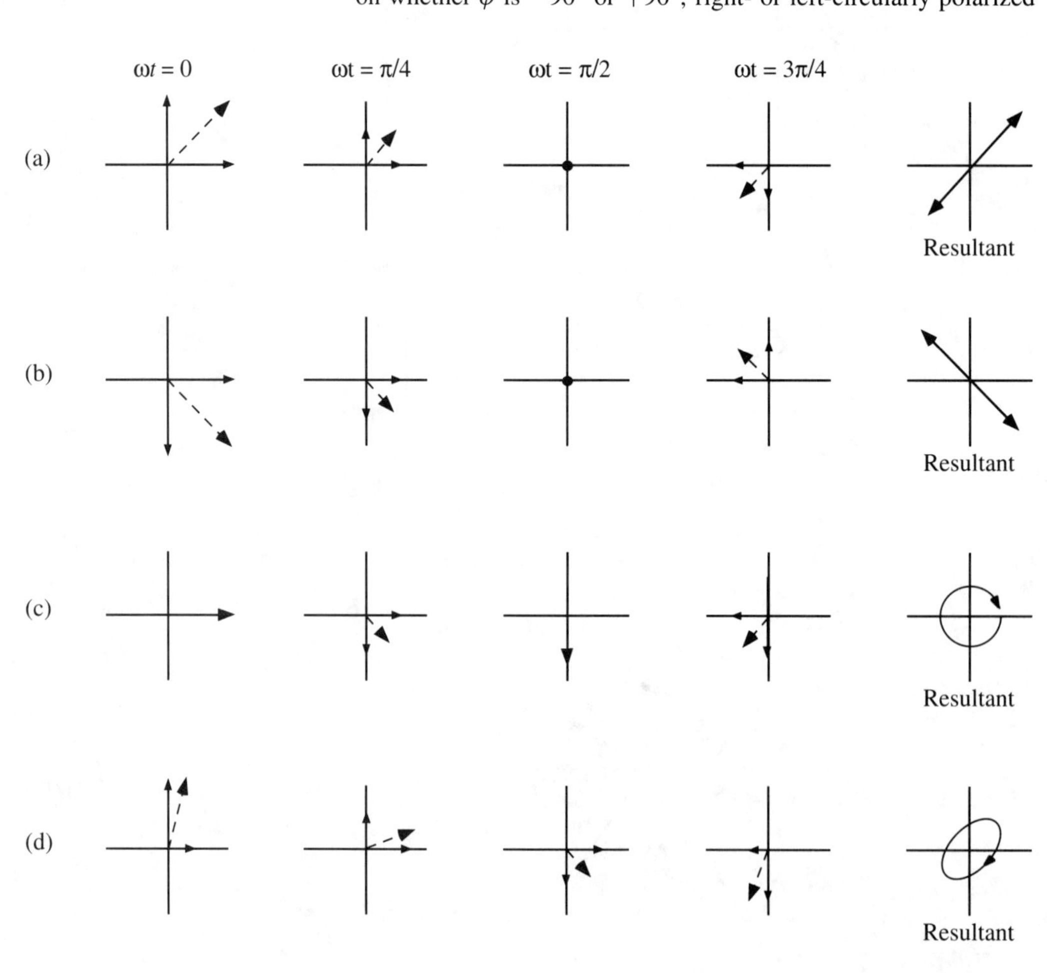

light is generated, respectively. Thus, circularly polarized light is exactly equivalent to two copropagating beams of orthogonally linearly polarized light of equal amplitude, which are 90° out of phase with each other. The resultant vector sweeps out 2π radians in the x–y plane every 10^{-14}s, which, as mentioned previously, is too rapid for optical detectors to sense.

ϕ = anything

In general, the resultant sweeps out an ellipse in the x–y plane; linearly polarized and circularly polarized light are just special cases of an ellipse when the minor axis is zero or equal to the major axis, respectively. An example is given in Figure 9.3.2d. Note also that when the amplitudes of E_x and E_y are not equal, it is impossible to generate circularly polarized light.

In practice, circularly polarized light is produced by use of an element called a quarter-wave plate, which is nothing more than a carefully selected piece of birefringent material, usually mica. As stated earlier, birefringence means a difference in n along two axes in a material: $\Delta n \equiv n_1 - n_2$. Which axes are chosen as 1 and 2? For light propagating along z, the value of n can be represented as an ellipse in the x–y plane, and the 1 and 2 axes can be chosen as the major and minor axes, thus giving Δn its largest value. (For any ellipse, there is a pair of orthogonal axes that would give $\Delta n = 0$ although the material is birefringent. In three dimensions, a refractive index tensor, **n**, can be defined, and an associated refractive index ellipsoid. However, in most rheo-optical experiments the light beam propagates along one of the principal axes of this ellipsoid, thus reducing the problem to two dimensions.) In a flow birefringence experiment, the 1 and 2 axes will be determined by the flow geometry and the beam propagation direction; note the possibility that Δn can be positive or negative. Consider two mutually orthogonal, linearly polarized beams, of equal amplitude and relative phase, entering the mica quarter-wave plate such that one beam is polarized along axis 1 and the other along axis 2. They now propagate at different speeds, and the phase kz changes at a different rate for the two components. The phase difference between the two components after passing through a thickness L of the mica is easy to compute:

$$\begin{aligned}\delta &= k_1 L - k_2 L \\ &= \frac{2\pi n_1 L}{\lambda_o} - \frac{2\pi n_2 L}{\lambda_o} \\ &= \frac{2\pi \Delta n L}{\lambda_o}\end{aligned} \tag{9.3.5}$$

Given that the birefringence Δn is a property of the material, it remains only to cut a thickness L such that $\delta = \pi/2$. The quantity δ is called the *retardation*, for obvious reasons; a quarter-wave plate retards one component by $\pi/2$ rad, or one-quarter of a wave, relative to the other. Note that in a birefringence experiment, it is always a retardation that is measured directly, not the birefringence itself.

One other important point from this discussion of the quarter-wave plate: the orientation of the axes of the birefringent material relative to the incoming polarized beams is crucial. If the "fast" axis (i.e., the axis of lower n) of the quarter-wave plate is along the y axis, the incoming beams must be x- and y- polarized exactly to produce circularly polarized light. This is exactly equivalent to one incident linearly polarized beam oriented at 45° to the x and y axes. Left- or right-circularly polarized light can be generated, depending on whether the incoming linearly polarized beam is 45° to the left or to the right of the quarter-wave plate fast axis.

9.3.1 Transmission Through a Series of Optical Elements

Consider a light beam propagating along the z axis through a series of polarizers, wave plates, and arbitrary birefringent elements. Because of the necessity of specifying the orientation angles in the x–y plane as well as the nature of each element, the general solution to the problem of identifying the polarization properties of the emerging beam quickly becomes very complicated when there are three or more elements. Some simplifying mathematical schemes have been developed, particularly the Jones and Mueller calculus (Shurcliff, 1962; Azzam and Bashara, 1977) and the scalar product mechanism set out by Thurston (1964). In the former, each element is replaced by a matrix that encodes its retardation and orientation, and the net result is obtained by simple matrix multiplication. In the latter, the projections along selected orthogonal axes are computed at each stage. By either method the problem can be solved directly, if not necessarily simply; however, for the purposes of this brief introductory chapter, only five particular situations are considered and without recourse to these more elaborate methods. The general scheme is shown in Figure 9.3.3, where up to four optical elements are placed in series. In all cases, the first element is a linear polarizer oriented along the y axis, and the last element is another linear polarizer, the "analyzer," with variable orientation. The intensity emerging from the first polarizer will be designated I_o, and the transmitted intensity emerging from the analyzer I_t.

The five important arrangements of the optical train are listed next.

- *No intermediate elements, variable analyzer orientation.*

What is I_t as a function of θ, the angle of rotation of the analyzer axis away from the y axis? The answer is found by taking the projection of the electric field amplitude E_o onto the analyzer axis, which is just $E_o \cos\theta$. Thus, the emerging intensity is proportional to $(E_o \cos\theta)^2$ or $(I_o/2)(1 + \cos 2\theta)$. This function is plotted as a function of θ in Figure 9.3.4a. As expected, for crossed polarizers ($\theta = \pi/2$ or $3\pi/2$), the emerging intensity is zero. Note also that it is important to deal with electric field amplitudes until the last step, at which point the intensity should be computed.

• *An intermediate linear polarizer P, of variable orientation, and the analyzer oriented along the x-axis.*

In this case, the polarizer and analyzer are crossed, and so in the absence of P, $I_t = 0$. Clearly, also, when P is oriented with either the x or y axis, it will have no effect at all. However, the situation is different at all other angles. The electric field amplitude emerging from P is $E_o \cos\theta$, just as in the preceding arrangement (no intermediate elements). This becomes the amplitude incident on the analyzer; thus, the amplitude emerging from the analyzer is $(E_o \cos\theta)\cos(\theta - \pi/2)$. Using trigonometric identities, it can be seen that $I_t = (I_o/4)(1 - \cos^2 2\theta)$, as displayed in Figure 9.3.4b. Note the somewhat counterintuitive result that light can be obtained from crossed polarizers simply by insertion of another polarizer between them.

• *An intermediate quarter-wave plate, variable analyzer orientation.*

Here we assume that the quarter-wave plate is correctly aligned to give circularly polarized light. This means that the fast and slow axes are oriented at $\pm 45°$ to the y axis; the incident y-polarized beam is equivalent to two equal-amplitude, equal-phase beams oriented at $\pm 45°$ to the y axis. In this case, it should be

Figure 9.3.3.
Four optical elements in series. The first element is a linear polarizer oriented along the y axis. The last element is a linear polarizer oriented at an angle θ to the y axis.

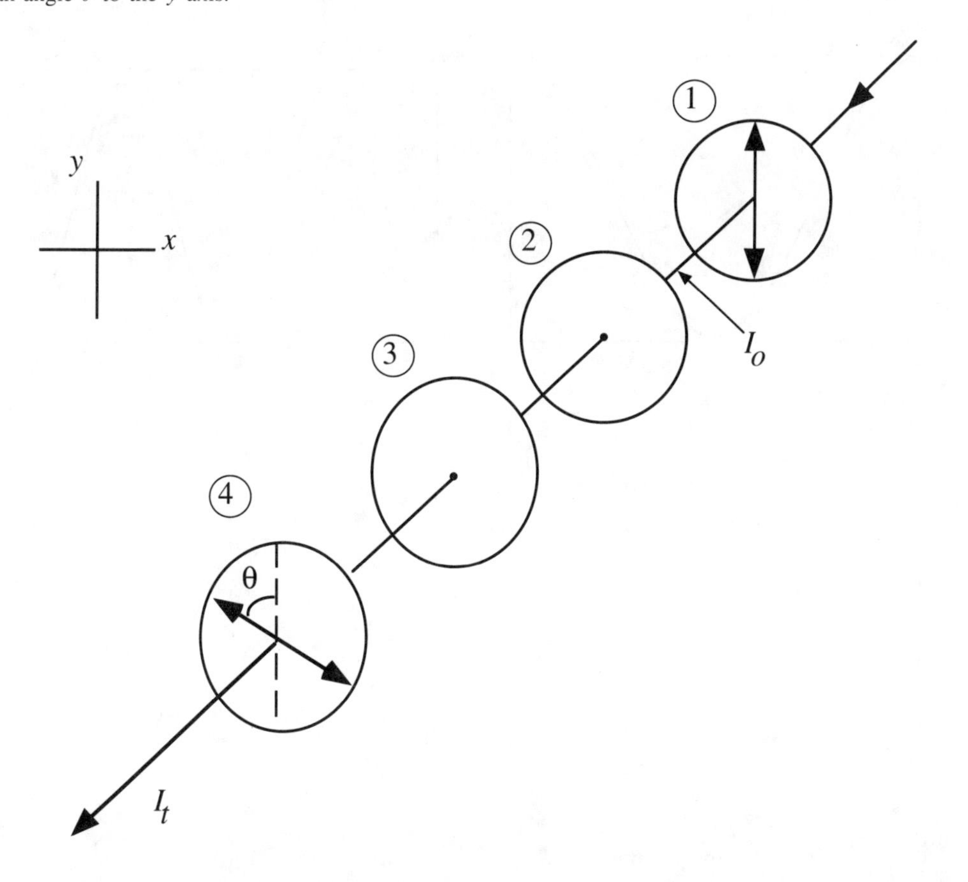

obvious that I_t is independent of analyzer orientation and (slightly less obviously) equal to $I_o/2$.

- *An intermediate element of arbitrary retardation δ, with principal axes at $\pm 45°$ to the y axis, and the analyzer along the x axis.*

This is the first model in this series that could describe an actual birefringence experiment, in that δ could be due to the orientation of polymer molecules. In this case the answer, which is not as straightforward to compute, is simply stated: $I_t = (I_o/2)(1 - \cos\delta)$. This function is represented in Figure 9.3.4c. Just to establish the plausibility of this result, consider some special cases. If $\delta = 0$, or no birefringence, such as in a quiescent liquid, $I_t = 0$, as expected. If $\delta = \pi/2$, the material acts as a quarter-wave plate, and the result is the same as the case of an intermediate quarter-wave plate. If $\delta = \pi$, a so-called half-wave plate, the element acts to rotate the plane of polarization by 90°; thus, $I_t = I_o$.

- *A quarter-wave plate aligned to generate circularly polarized light, followed by an element of unknown retardation δ_x and the analyzer oriented along the x axis.*

This is the typical experimental configuration for birefringence measurements on liquids. It is, in fact, a special case of Fig-

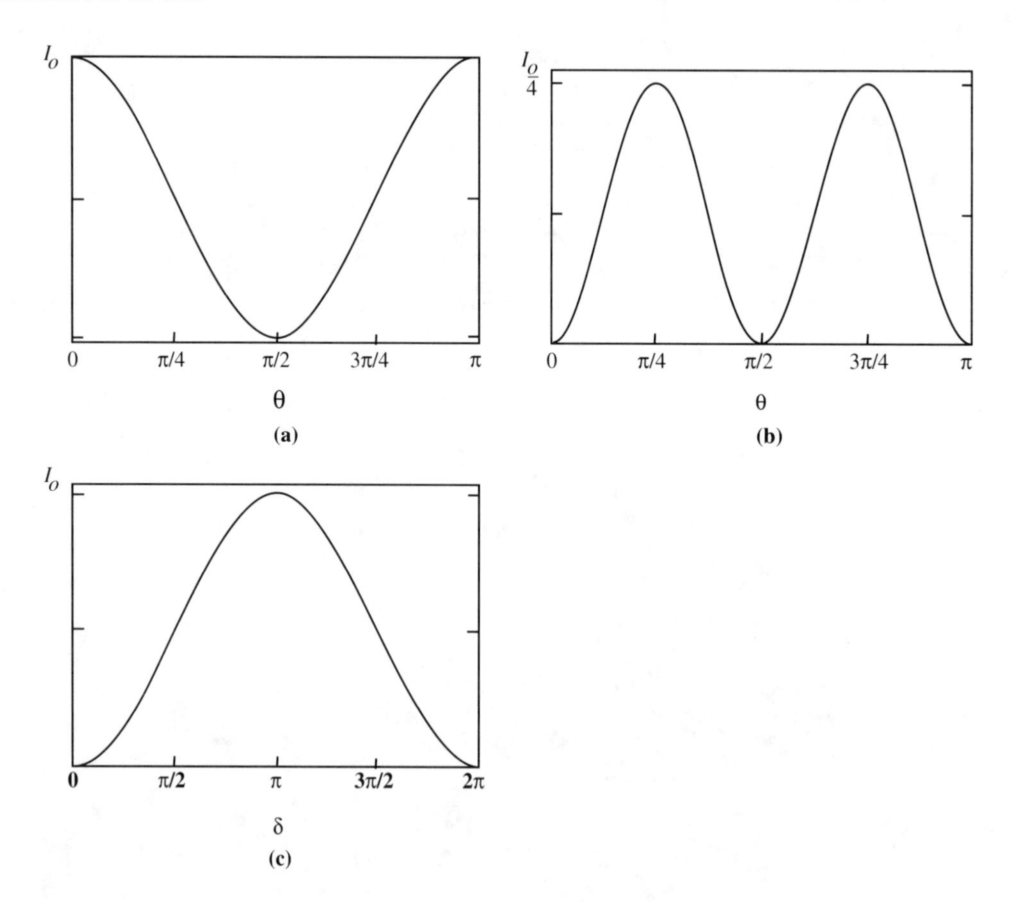

Figure 9.3.4.
Transmitted intensities from a series of optical elements, as described in the text.

ure 9.3.4c. For when the optical axes of two successive elements are aligned, their retardations simply add; note that in general it is not correct to add the retardations of two successive elements. Thus the net retardation in this case is $\delta_x + \pi/2$, which we can consider to be δ in Figure 9.3.4c. Now the advantage of the quarter-wave plate in this experiment can be explained. The measured intensity I_t changes most rapidly with changing δ_x in the vicinity of $\delta = \pi/2$. In most situations, $\delta_x \ll \pi/2$, so it is important to achieve as high a sensitivity as possible. As a rough estimate, Δn is often about 10^{-6} to 10^{-8}, which for a 1 cm path length cell and visible light gives δ_x in the range of 10^{-1} to 10^{-3} radian. Values of Δn as small as 10^{-13} can be determined (Morris and Lodge, 1986), an indication of the potential sensitivity of rheo-optical techniques. At the other extreme, values of Δn sufficiently large to make $\delta_x > \pi$ can also occur (Figure 9.1.1 is an example). In this case, I_t can go through maxima and minima as δ_x cycles through integer multiples of π.

9.4 Flow Birefringence: Principles and Practice

9.4.1 The Stress–Optical Relation

The stress-optical relation (SOR) lies at the very heart of the use of flow birefringence in rheology (Janeschitz-Kriegl, 1969, 1983; A.S. Lodge, 1955; Tsvetkov, 1964; Fuller, 1990). Given a polymer liquid undergoing flow, both a stress tensor $\boldsymbol{\tau}$ and an index of refraction tensor $\mathbf{n}$ can be defined. The SOR comprises two statements about these tensors:

- The principal axes of the stress tensor and the refractive index tensor are collinear.
- The differences in principal values of the stress tensor and the refractive index tensor are proportional; the constant of proportionality is called the stress–optic coefficient, C.

The SOR may be written as follows:

$$\mathbf{n} - \langle n \rangle \mathbf{I} = C\{\boldsymbol{\tau} + p\mathbf{I}\} \quad (9.4.1)$$

where $\langle n \rangle$ is the mean refractive index, p is the hydrostatic pressure, and $\mathbf{I}$ is the unit tensor. This expression becomes much more direct when we consider a shear flow, with x the flow direction and y the perpendicular to the shear planes. In this case, the z direction is neutral and the stress components can be represented as an ellipse in the $x-y$ plane (recall Figure 1.3.1). Designating the principal values of the stress σ_1 and σ_2 and the angle ($< 45°$) between the flow direction and a principal axis, χ', as shown in Figure 9.4.1, the following relations hold:

$$2\tau_{12} = \Delta\sigma \sin 2\chi' \quad (9.4.2)$$

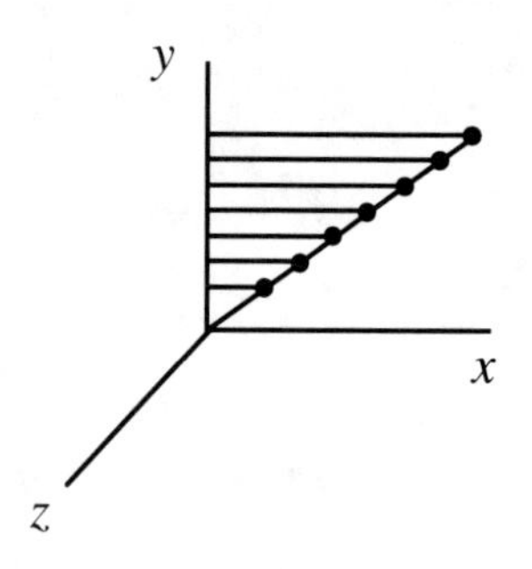

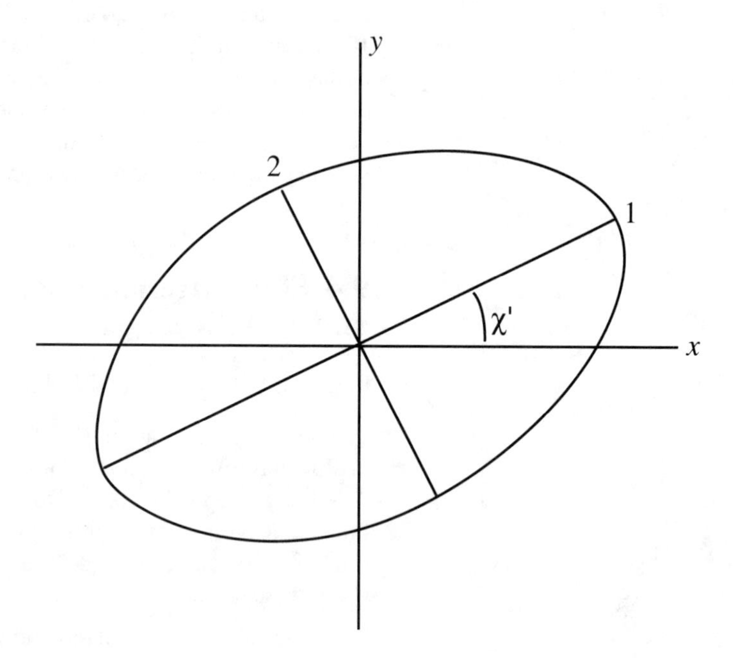

Figure 9.4.1.
Orientation of the stress ellipsoid in the $x - y$ plane.

$$\tau_{11} - \tau_{22} = \Delta\sigma \cos 2\chi' \tag{9.4.3}$$

where $\Delta\sigma = \sigma_1 - \sigma_2$. These relations are illustrated in Example 1.3.1 and can be obtained directly by recognizing that the stress matrix $\boldsymbol{\tau}$ in the x–y coordinate system is related to the diagonalized matrix $\boldsymbol{\Sigma}$ by a rotation matrix $\mathbf{U}$:

$$\mathbf{U}^T \cdot \boldsymbol{\Sigma} \cdot \mathbf{U} = \boldsymbol{\tau} \tag{9.4.4}$$

where (recall Example 1.4.1c) §

$$\mathbf{U} = \begin{pmatrix} \cos\chi' & \sin\chi' \\ -\sin\chi' & \cos\chi' \end{pmatrix}$$

The terms τ_{12} and $\tau_{11} - \tau_{22}$ correspond to the shear stress and the first normal stress difference, respectively. The SOR may now be rewritten

$$\tau_{12} = \left(\frac{1}{2C}\right)\Delta n_{12} \sin 2\chi \tag{9.4.5}$$

$$\tau_{11} - \tau_{22} = \left(\frac{1}{C}\right) \Delta n_{12} \cos 2\chi \tag{9.4.6}$$

where χ, the orientation angle of the refractive index ellipsoid, is referred to as the extinction angle and is equal to χ' by the SOR. Thus, eqs. 9.4.5 and 9.4.6 indicate that measurements of Δn_{12} and χ in a flow birefringence experiment are equivalent to measurements of the shear stress and the first normal stress difference, once the SOR has been established for the particular system and C is known.

It is important to emphasize that the SOR is not the inevitable consequence of fundamental physical principles; rather, it is a very plausible hypothesis, which has extensive experimental support for polymer solutions and melts. In other words, there is no reason to assume that the SOR is valid under all possible flow conditions or for all possible polymer liquids. Some situations under which the SOR is expected to fail are mentioned in the next section. Many constitutive relations for solutions and melts predict that the SOR will hold, but even this apparent generality is somewhat misleading. The derivation of an SOR starts at a measurable molecular property, the optical polarizability of an isolated molecule $\boldsymbol{\alpha}$, and leads to a macroscopic refractive index tensor $\mathbf{n}$, in a nontrivial way; several substantial assumptions are necessary. Most rheological models (for flexible chains) that proceed to an SOR assume the derivation of Kuhn and Grün (1942) for the polarizability anisotropy of a Gaussian subchain and thus in a sense make the same assumptions for the optical half of the SOR (Larson, 1988). Therefore differences between constitutive relations and their predictions for an SOR usually stem from differences in the calculation of $\boldsymbol{\tau}$.

For flexible homopolymer melts, the SOR is almost universally valid. Examples are shown in Figure 9.4.2 for polystyrene, polyethylene, and a silicone oil (Janeschitz-Kriegl, 1969, 1983; White, 1990). Values of C are found to be independent of shear rate (even in the shear thinning regime), molecular weight, and molecular weight distribution; they do, however, depend on the identity of the monomer unit and also slightly on temperature and optical wavelength. Furthermore, C is independent of time following the onset of shear, as shown in Figure 9.4.2d for polystyrene (Janeschitz-Kriegl and Gortemaker, 1974). In solutions, C is independent of polymer concentration, but depends slightly on the solvent employed. As discussed in the next section, however, the SOR for solutions applies only to the *polymer contributions* to $\mathbf{n}$ and $\boldsymbol{\tau}$. These dependences for C can be understood in rather general terms. The magnitude of the measured birefringence is the product of two factors. The first is the degree of molecular orientation, and the second is the actual difference in refractive index along the two axes per degree of orientation. The first factor is proportional to the stress and is the basis for the SOR. The second factor reflects entirely the polarizability anisotropy of the monomer unit, and it determines the sign and magnitude of C. Thus, aligning relatively isotropically polarizable monomers, such as dimethylsiloxane, will

Figure 9.4.2.
Steady shear viscosity (open symbols) and (absolute value of the) stress–optical coefficient (solid symbols): (a) For melts of polystyrenes of narrow molecular mass distribution as functions of shear rate,

have a much smaller effect on the refractive index than aligning more anisotropic monomers, such as styrene. The independence of C of molecular weight and concentration is therefore quite plausible. The dependence on solvent is a little more subtle and is thought to reflect differences in the way solvent molecules solvate the polymer chains and modify the local electric fields (Frisman et al., 1963; Tsvetkov, 1957). Some typical values of C are listed

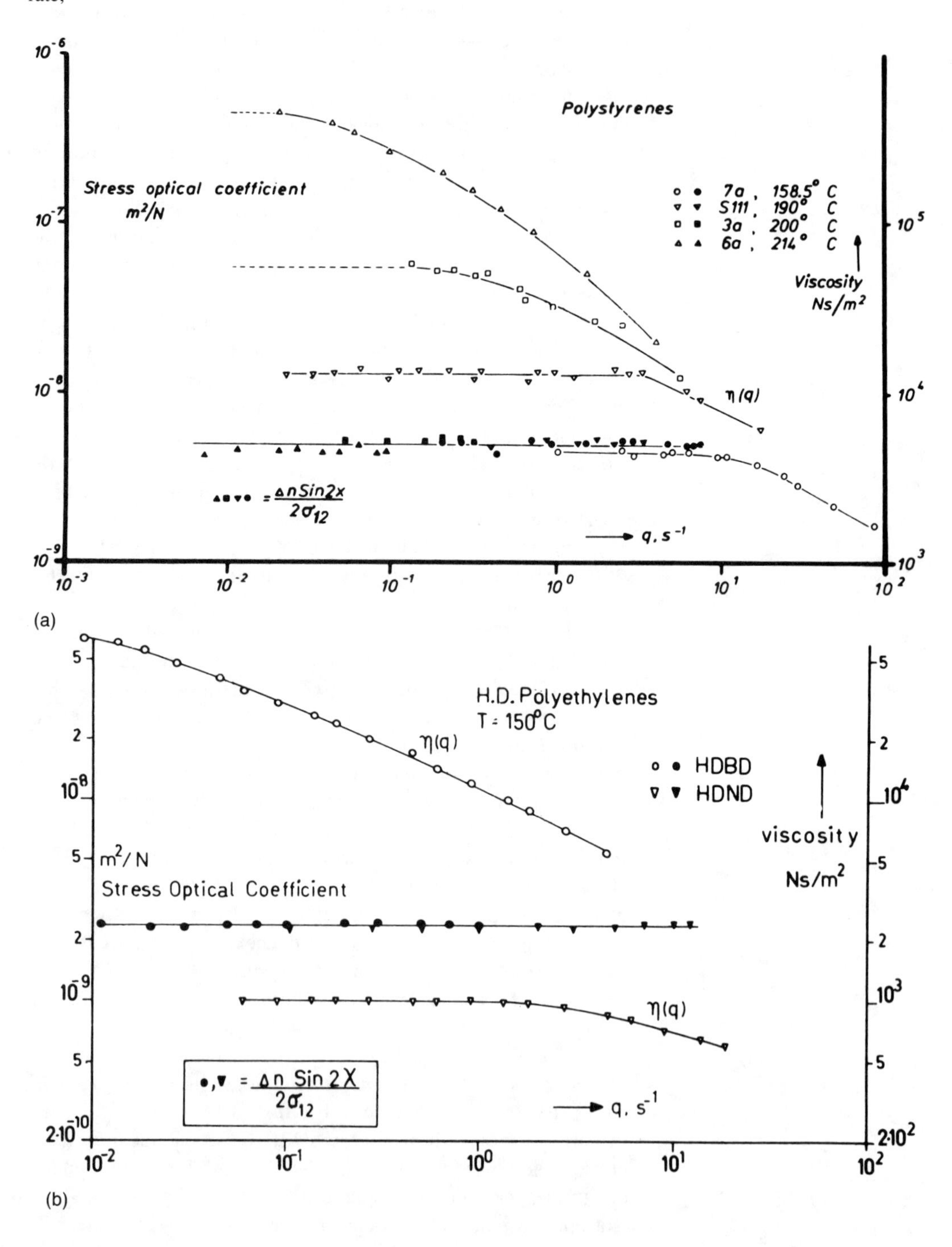

in Table 9.4.1 (Janeschitz-Kriegl, 1969, 1983; White, 1990). Note the very important point that C, and thus Δn_{12}, can be negative. This simply reflects the fact that the refractive index can be greater along either the x or the y direction; polystyrene has a negative C because when the chain backbone is aligned, the more polarizable side groups are aligned on the average perpendicularly to the chain axis.

Figure 9.4.2. (*Continued*) (b) for two high density polyethylenes of widely differing molecular mass distributions, and (c) for a silicone oil; onset of melt fracture (M.F.) in capillary flow at $\tau_w = 0.39. \times 10^5$ Pa. From Janeschitz-Kriegl (1983). (d) absolute value of the stress–optical coefficient for a polystyrene melt after the onset of steady shear; Δn, χ, and τ_{12} reached their steady state values after about 100–200 seconds. From Janeschitz-Kriegl and Gortemaker (1974).

9.4.2 Range of Applicability of the Stress–Optical Relation

As mentioned earlier, the applicability of the SOR does not rest on fundamental physical laws. However, there are two important statements to emphasize. First, the observation of a breakdown of the SOR can be very informative about the molecular level contributions to the stress tensor, and second, in principle rheo-optical measurements can be at least as informative in regions where the SOR does not apply. Apparently, the SOR is successful when the

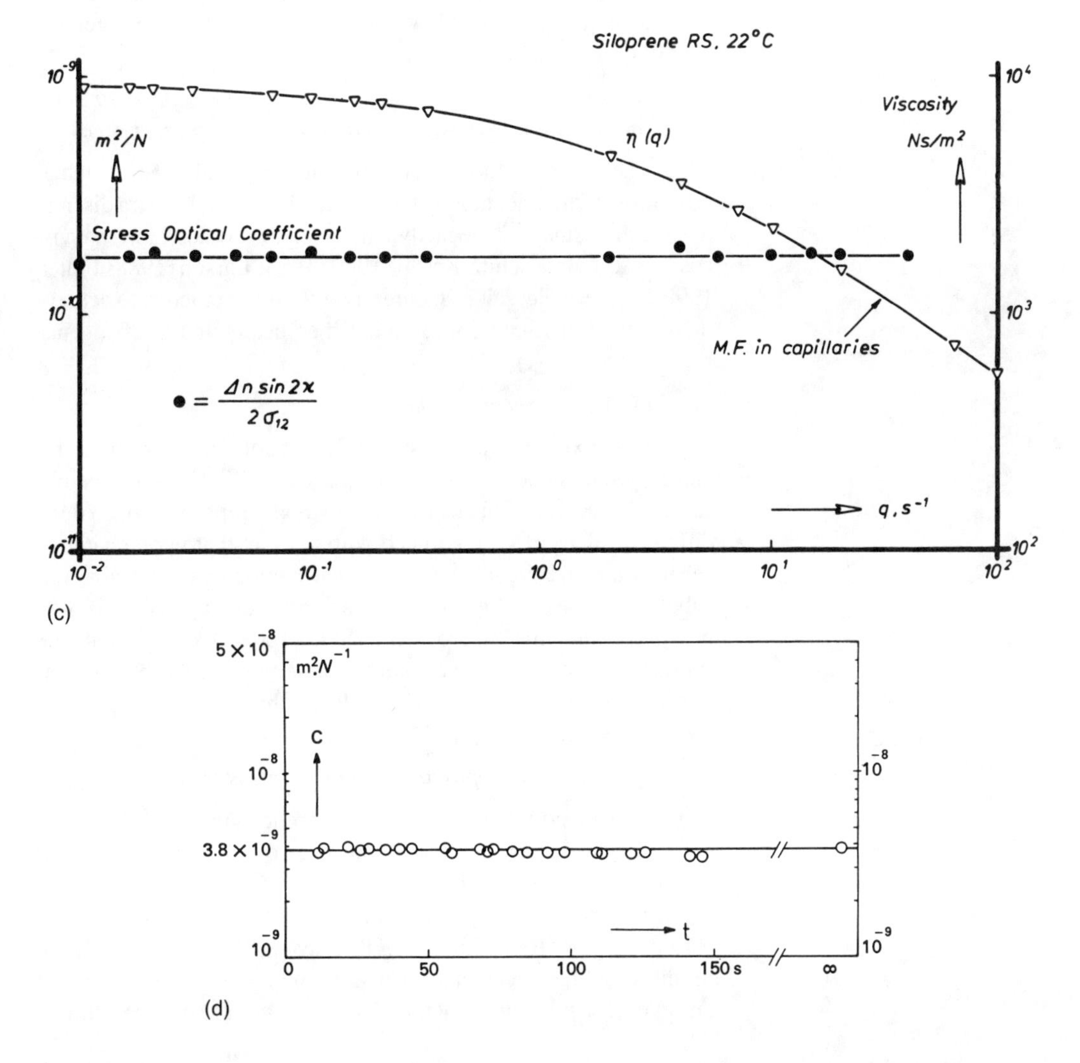

TABLE 9.4.1 / Stress–Optic Coefficients

Polymer	*Typical Values for Polymer Melts* $(10^{-9} m^2/N)$
Polybutadiene	+2.2
Polydimethylsiloxane	+0.14 to +0.26
Polyethylene	+1.2 to +2.4
Polyisobutylene	+1.5
Polyisoprene	+1.9
Polypropylene	+0.6 to +0.9
Polystyrene	-4 to -6
Polyvinyl choride	-0.5

Adapted from Janeschitz-Kriegl (1983) and White (1990).

polymer contributions to $\boldsymbol{\tau}$ and $\mathbf{n}$ are each proportional to the average dyadics $< \mathbf{rr} >$ or $< \mathbf{uu} >$, where $\mathbf{r}$ is the end-to-end vector of a flexible subchain and $\mathbf{u}$ is the orientation vector for a rigid link (Janeschitz-Kriegl, 1969, 1983; Larson, 1988; Fuller, 1990). One can expect the breakdown of the SOR in at least four general situations:

- *Homopolymer melts and solutions at very high shear rates (or very high frequencies in an oscillatory experiment).*

The failure of the SOR can be due to several effects, such as saturation of the orientation at very high shear rates or extra dissipative mechanisms in the local dynamics (chain stiffness, internal viscosity, etc.) (Janeschitz-Kriegl, 1969, 1983; Larson, 1988; Fuller, 1990). Nevertheless, it is apparently difficult in practice to achieve these conditions with conventional rheometers and birefringence apparatuses.

- *Multicomponent systems.*

For mixtures of chemically different polymers, for solutions, and for block copolymers, a "macroscopic" SOR will not be observed in general. This can be understood simply in terms of the different values of C, combined with the fact that upon changing shear rate or frequency, the individual components will not contribute to the total stress in constant proportions. However, it is possible to imagine "microscopic" SORs for each component, and if C is much greater for one component, it may be possible to monitor its behavior selectively. Examples of this strategy are given in the next section.

- *Systems near, or passing through, a glass transition.*

A given polymer will exhibit different values of C above and below its glass transition, which may even differ in sign (Rudd and Gurnee, 1957); this is a consequence of the different conformational rearrangements available in the two regimes. Furthermore, below the glass transition the SOR may well not hold for all but infinitesimal displacements. Thus, by the arguments given above one expects a breakdown of the SOR in the vicinity of such a transition. Similarly, the onset of crystallization will also cause the SOR to fail.

• *Systems in which form birefringence is significant.*

Form birefringence is a phenomenon that arises from a spatially anisotropic arrangement of domains with different mean refractive indices (Bullough, 1960; Onuki and Doi, 1986). It can occur in polymer solutions, block copolymer melts, and polymer blends, but presumably not in homopolymer melts. The effect may also incorporate contributions from anisotropic pair correlations on a very local scale, which can modify the internal electric field (Bullough, 1962). It can be thought of as a result of multiple scattering, in which the superposition of scattered waves from the various domains modifies the polarization state of the transmitted beam. In other words, even a suspension of perfect spheres can give rise to form birefringence if the spheres are arranged in an anisotropic manner. This phenomenon is generally very difficult to treat rigorously, but its contribution to the measured signal is always positive in sign and can be minimized by matching the refractive indices of the components (i.e., refractive index increment $\partial n/\partial c = 0$). Form birefringence increases with the size (molecular weight) of the dissolved molecules. There are two situations in which the form effect is likely to be a major factor: in dilute polymer solutions and suspensions and in microphase-separated block copolymer liquids. An example of the former is shown in Figure 9.4.3 for polystyrene solutions in steady shear (Frisman and

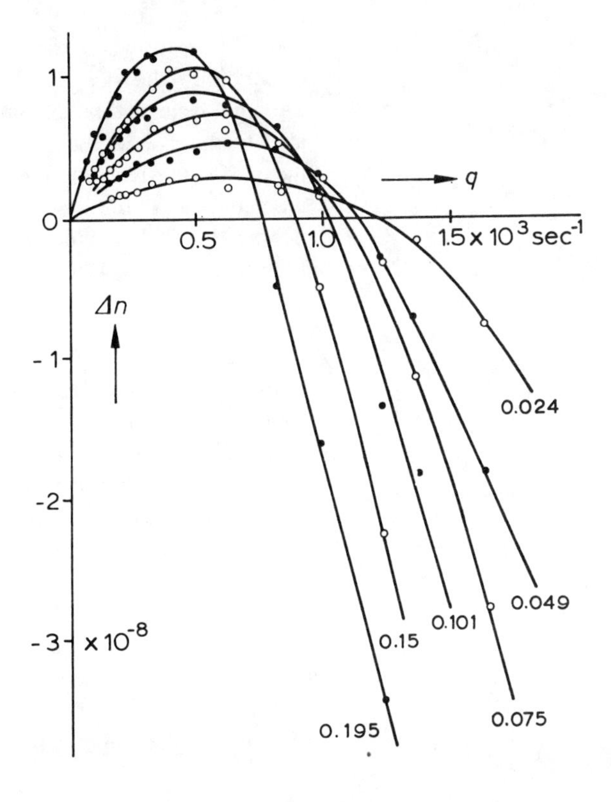

Figure 9.4.3. Flow birefringence Δn versus shear rate q for solutions of a polystyrene fraction of high molecular weight ($M = 1.7 \times 10^7$) in dioxane. Concentrations (g/100 cm^3) are given near the curves. From Frisman and Mao (1964).

Mao, 1964). Because polystyrene has a negative C, the *intrinsic birefringence* (i.e., that due to molecular polarizability anisotropy, and the primary subject of this chapter) should be negative. At low shear rates, however, the signal is positive, due to the form effect. At higher shear rate the intrinsic birefringence dominates, and the net response becomes negative. For block copolymers, the microdomains formed by microphase separation can give rise to substantial form effects, which could also be exploited as a means to examine such ordering transitions (Fredrickson, 1987).

9.4.3 Geometries for Measuring Flow Birefringence

This section provides a brief discussion of the experimental geometries that have been used for flow birefringence measurements on polymer liquids. These techniques may be classified into three groups according to whether they determine birefringence in the x–y plane in shear, in the x–z plane in shear, or in elongation. The first category has been by far the most extensively employed.

The discussion leading to eqs 9.4.5 and 9.4.6 was based on the assumption of a shear flow with the light beam propagating along the neutral z direction. There are three standard experimental realizations of this situation.

Couette Flow

In Couette flow, one of a pair of concentric cylinders is made to rotate. The liquid is confined in the narrow gap between the cylinders, and in the limit where the gap width is much less than the radius of curvature, the ideal case of parallel plane shear is approached. The light beam is directed through the gap along the direction parallel to the axis of the cylinders. This approach has been used the most extensively of all geometries in flow birefringence measurements and is discussed in greater detail below (Thurston and Schrag, 1968: Janeschitz-Kriegl, 1969, 1983; Osaki et al., 1979; Frattini and Fuller, 1984).

Cone and Plate

In the cone and plate or Rheogoniometer geometry, the liquid is confined between a flat disk and a cone that rotates around its axis (Janeschitz-Kriegl, 1969, 1983). The angle between the cone and the disk is typically less than 10°. For birefringence measurements to be effected, the rheological apparatus requires some modification. Rather than have the cone meet the plate at its apex, the center of the cone and plate is removed and replaced with a cylindrical annulus. Thus, the liquid is confined to an annular region, the bottom surface of which is the flat plate and the upper surface is part of the rotating conical section. The light beam comes up vertically into the center of the apparatus, where it is redirected by a prism to be horizontal and thus parallel to the flat surface. Windows in the cylindrical annulus and outer containing wall permit the beam to pass through the sample. A schematic of the apparatus appears in Figure 9.4.4.

Thin Fluid Layer Transducer

The geometry used for flow birefringence measurements in an oscillatory shear (Miller and Schrag, 1975) consists of two flat, parallel surfaces, one of which is fixed; the other is made to oscillate in the plane parallel to the fixed surface. The liquid is confined to the thin layer between the surfaces. The light beam propagates through the layer in a direction perpendicular to the flow direction and parallel to the surfaces. This experimental arrangement and its application are discussed in detail below.

In the case of an x–y shear flow, it is also possible to direct the light beam along either the x or the y axis, thus enabling determination of Δn_{23} or Δn_{13}, respectively. According to the SOR, optical measurements of $\tau_{22} - \tau_{33}$ or $\tau_{11} - \tau_{33}$ are then possible. When either quantity is combined with the first normal stress difference obtained from the more common measurement of Δn_{12} and χ, the second normal stress difference can be determined. Measurement of Δn_{13} in the x–y plane has been achieved by at least two geometries.

Parallel Glass Plates

In this apparatus—one of the first to be used for flow birefringence measurements of polymer liquids—one glass plane is translated parallel to another with the liquid contained between them (Dexter et al., 1961). Thus, the light beam can propagate directly through the apparatus along the y direction.

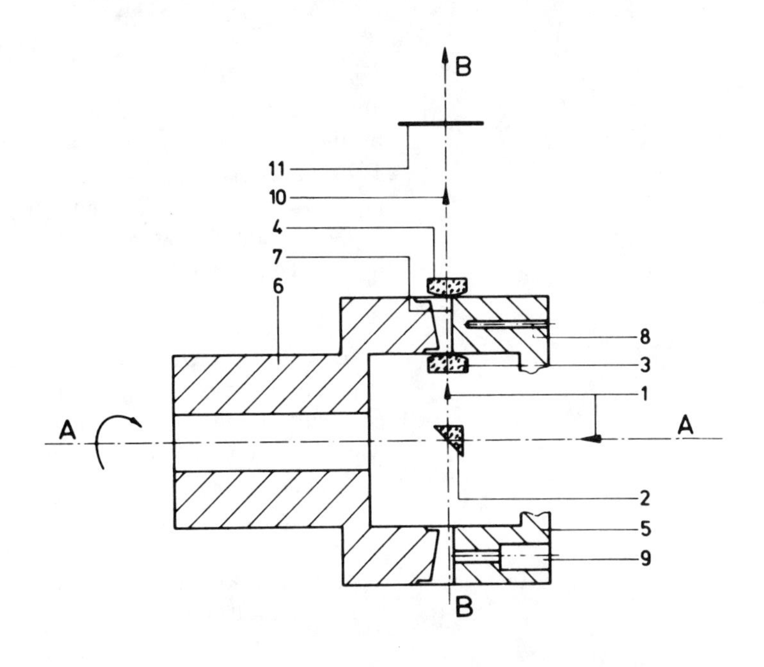

Figure 9.4.4.
Cross section through the most recent version of the cone and plate unit for measuring flow birefringence according to Van Aken et al. (1980): 1 linearly polarized light beam entering the ring-shaped gap; 2, reflection prism; 3, inner window; 4, outer window; 5, stationary plate; 6, rotor with conical front surface; 7, ring-shaped gap (gap angle 1°8′, causing a maximum gap width of $\simeq$ 0.4 mm); 8, blind hole for the thermocouple; 9, sample injection hole; 10, elliptically polarized light beam emerging from the gap; and 11, analyzer. From Janeschitz-Kreigl (1983).

Figure 9.4.5.
Slit apparatus for the measurement of flow birefringence (and velocity profiles); A, outer body; V, conical inner member containing the slit; W_1, W_2, windows; E, entrance to slit; and H_1, H_2, H_3, band heaters. From Janeschitz-Kreigl (1983).

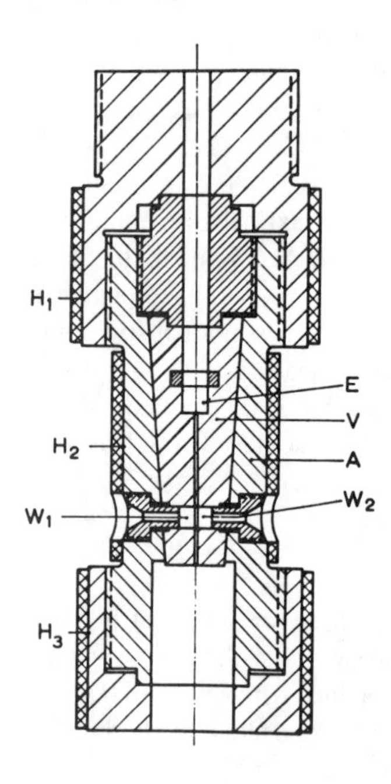

Capillary Slit
In this case, the liquid is directed through a slit of rectangular cross section with the long side of the rectangle much greater than the short side (Janeschitz-Kriegl, 1969, 1983). The light beam again passes through the long side walls, which are transparent. Unlike the case of parallel glass plates, the shear rate is not constant along the path length through the liquid; however, the flow geometry is well defined and the inherently integrating measurement can still be employed to evaluate the desired normal stress difference. This geometry is shown in Figure 9.4.5.

Capillary Slit and the y–z Plane
In the same apparatus as in Figure 9.4.5, the light beam is directed along the capillary slit axis. This rather delicate experiment has not been used extensively since the pioneering work in the 1960s (Janeschitz-Kriegl, 1969, 1983).

Extensional rheometry is a very important subset of rheology, and flow birefringence in extensional flow has been measured using a variety of geometries. General observations that apply to measurements of extensional viscosity relative to shear viscosity are reflected also in optical measurements. In particular, it is usually much harder to generate a well-defined extensional flow, either uniaxial or biaxial, than a shear flow; but on the other hand, chain deformations in extensional flow are usually much greater, giving rise to large birefringence signals. The analysis can also be somewhat simpler than in shear flow, since the principal directions of the stress and refractive index tensors are always known. Three of the methods used to measure flow birefringence in extensional flows are now mentioned very briefly; some examples of their application are presented in Section 9.5.

Four-Roll Mill
Four cylindrical units that can rotate independently around their parallel axes are arranged as shown in Figure 9.4.6a (Frank et al., 1976; Fuller and Leal, 1981; see also Fig. 7.7.2). When the sense of rotation is as shown, a stagnation point is generated in the center of the apparatus, and an elongational flow is created along the indicated direction.

Cross-Slots Apparatus
A piece of transparent material is machined to provide four perpendicular, intersecting coplanar channels. The liquid is pumped with equal force in opposite directions toward the center along two collinear channels. The exiting streams along the two other channels generate an extensional flow along the axis of the exiting streams, as indicated in Figure 9.4.6b. Alternatively, two opposing jets can provide a well-defined extensional flow in an equivalent fashion (Farrell et al., 1980).

Spinning Apparatus
In this case, a thin strip of liquid is drawn up between two counter rotating cylinders, and the light beam passes directly through the strip, permitting determination of Δn.

9.4.4 Birefringence in Steady and Transient Couette Flow

The Couette geometry is by far the most commonly employed for flow birefringence measurements, as mentioned above. Among the advantages of this geometry are the relative ease with which the optical beam can be introduced into the experiment, the wide range of shear rates available (by varying both rotation rate and gap width), and the very well-characterized flow profile that is developed. We do not address any of the mechanical aspects of this approach but make some remarks about the various types of optical configuration that have been used.

There are several issues at stake in the measurement of flow birefringence in a Couette cell. First, there are two quantities to be determined, Δn_{12} and χ. For transient flows in particular, it is highly desirable to measure these two quantities simultaneously. Second, Δn_{12} can have a positive or negative sign, and not all optical arrangements can make this distinction. Third, for large birefringence signals, the retardation can exceed 2π; not all measurements schemes can sense this or count the "order" of the retardation (i.e., the number of cycles of 2π involved). All these factors need to be considered when selecting an appropriate optical detection method.

It is instructive to consider the simplest method of measuring birefringence: using crossed polarizers surrounding the sample. The linearly polarized light strikes the sample and can be resolved into components along the principal axes of the sample's refractive index ellipsoid in the x–y plane. These two components propagate at different speeds, developing a relative phase difference. As discussed in Section 9.3, this results in elliptically polarized light. With the analyzer at a fixed orientation, an intensity is measured

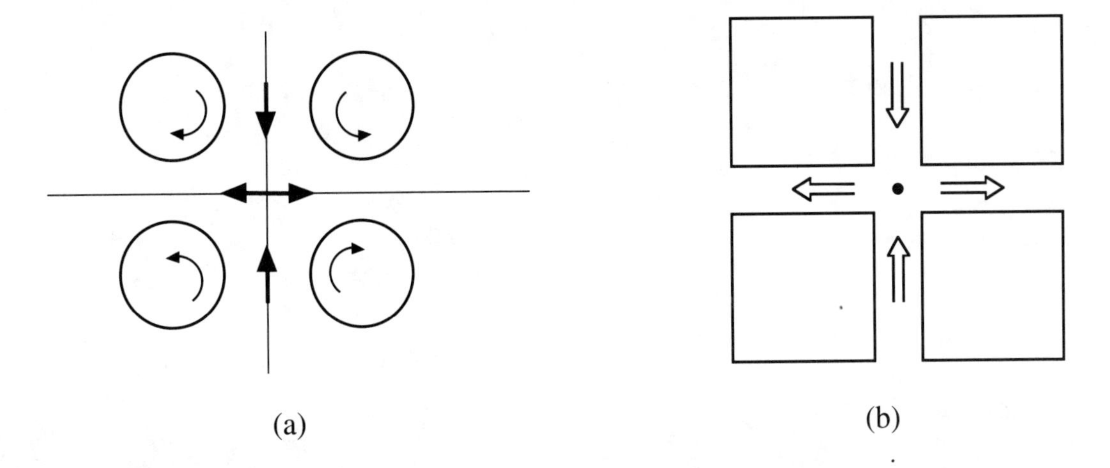

Figure 9.4.6.
Apparatuses for generating extensional flow: (a) four-roll mill and (b) cross-slots.

that depends on both Δn_{12} and χ. Then the analyzer and the polarizer are rotated together through a measured angle. When the analyzer becomes parallel to either the fast axis or the slow axis of the sample, the polarizer is parallel to the other axis. Thus, the incident light is resolved into orthogonal components along axes in the sample with equal n, and no light will emerge from the analyzer. This results in the condition of extinction, with the rotation angle necessary to achieve extinction equal to χ. The net intensity that was measured originally can now be interpreted in terms of Δn_{12} and the known χ.

It should be apparent that the foregoing approach is rather limited: it cannot be used to determine the sign or the order of the birefringence, and two consecutive measurements must be made. A variety of improvements on it have been developed, generally involving the use of a quarter-wave plate. Although the quarter-wave plate can improve the sensitivity considerably and permit resolution of the sign of Δn_{12}, the problem of requiring two separate measurements remains. A method to overcome this difficulty has been developed (Frattini and Fuller, 1984). In this case, a polarization rotator is incorporated into the optical train, as shown in Figure 9.4.7. The polarizer rotates at frequency ω (with $\omega \gg \tau^{-1}$, where τ corresponds to the shortest time constant of interest in the sample), and thus the polarization incident on the sample rotates at frequency 2ω. After the sample, a standard quarter-wave plate and

Figure 9.4.7.
Optical train incorporating a polarization rotator.

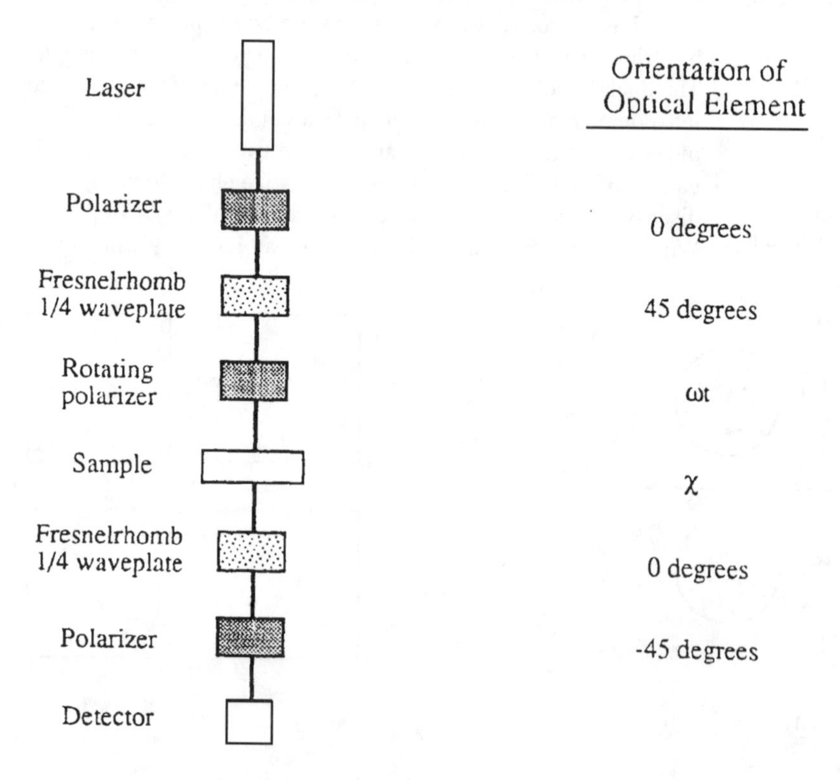

analyzer combination is used. The analysis of the system reveals that the net transmitted intensity is given by:

$$I = \frac{I_o}{4}\{1+(\cos 2\chi \sin\delta)\sin 2\omega t+(\sin 2\chi \sin\delta)\cos 2\omega t\} \quad (9.4.7)$$

and thus phase-sensitive detection at frequency 2ω permits simultaneous determination of χ and the retardation δ. The signs of both Δn_{12} and χ can be determined, and the order of the birefringence emerges in the relaxation that follows the cessation of a shear flow. If the order is greater than one, the net intensity will oscillate as the signal decays through several orders.

9.4.5 Birefringence in Oscillatory Shear Flow

A cross-sectional view of the thin fluid layer (TFL) transducer is shown in Figure 9.4.8 (Miller and Schrag, 1975). The moving

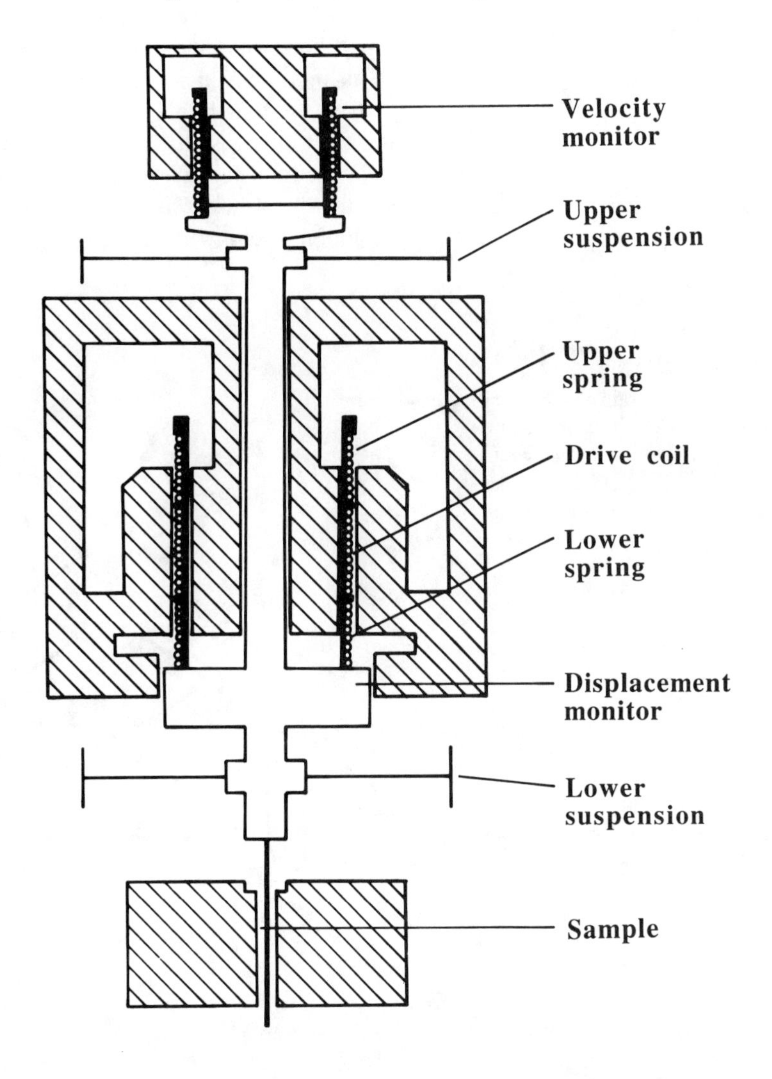

Figure 9.4.8.
Cross-sectional view of the Miller-Schrag thin fluid layer (TFL) transducer, used to generate a precise sinusoidally time-varying shear flow for the oscillatory flow birefringence experiment (see Section 9.4.6).

surface is a black glass plane rigidly attached to a vertical steel shaft. The shaft is attached to a solenoid that sits in a permament magnetic field; application of a sinusoidal current to the solenoid induces a sinusoidal vertical motion of the shaft. A second solenoid at the top of the shaft in an independent magnetic field develops an induced current proportional to the shaft velocity. This induced current is therefore directly related to the shear rate. The frequency range of the TFL is approximately 0.01–2500 Hz, with a maximum vertical displacement of the plane of about 1 mm. The gap width is usually between 0.1 and 1 mm, to maintain gap loading conditions (i.e., to keep the ratio of the shear wavelength to the gap width of order 50 or greater, which guarantees that the shear rate is constant across the gap: see Chapter 8).

The optical train is illustrated in Figure 9.4.9 and follows the standard polarizer, quarter-wave plate, sample, analyzer arrangement discussed in Section 9.3. However there is one important modification: the rotating analyzer is replaced by a beam-splitting prism and two detectors, allowing the simultaneous detection of both horizontal and vertical polarizations (Amelar et al., 1991). This modification is important for the following reason. The light emerging from the cell is elliptically polarized, with the magnitude of the ellipse oscillating at the motional frequency of the plane. Thus, both detectors sense an oscillating intensity, but the two detector signals are 180° out of phase with each other. Because the solution layer is so thin, the optical beam fills the gap. Any lateral

Figure 9.4.9.
Optical train for the oscillatory flow birefringence experiment. From Amelar et al. (1991).

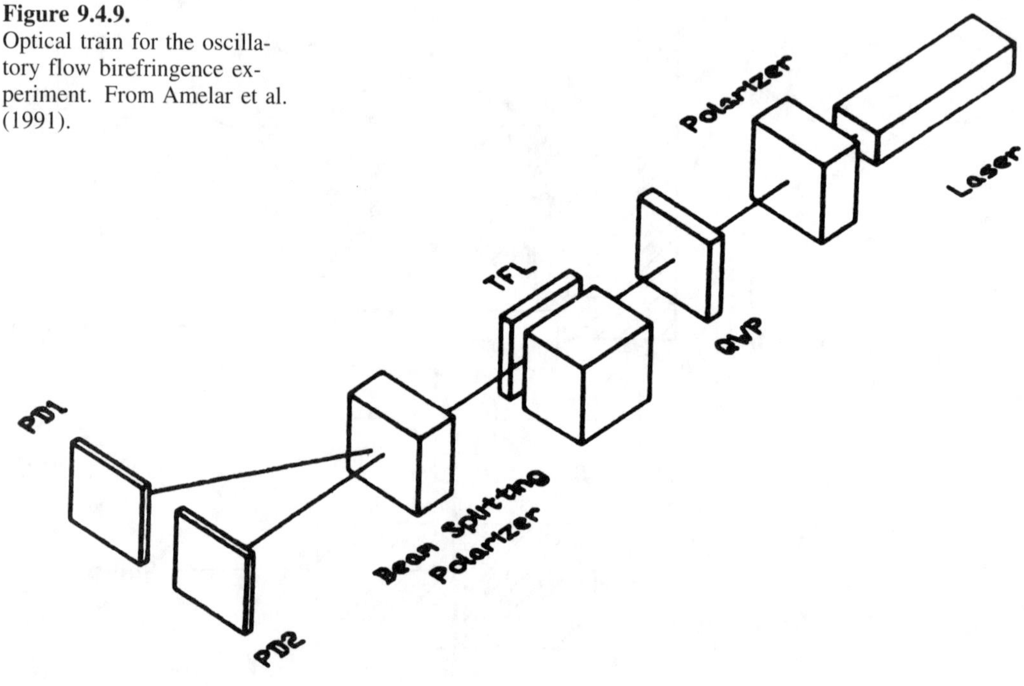

or rotational component to the motion of the plane will therefore produce a modulation of the net transmitted intensity at the driving frequency. This contamination of the signal can be comparable in magnitude to the birefringence. However, this modulation is independent of polarization and therefore is in phase between the two detectors. Thus, subtraction of the two detector signals gives the birefringence signal alone.

The detection of the two optical signals and the velocity of the moving plane is accomplished by a computerized data acquisition and processing system, which has been described elsewhere in detail (Morris and Lodge, 1986). Essentially, it acts as a digital lock-in amplifier, which utilizes the precisely known frequency of the desired signal component to reject contaminating random and coherent noise selectively. As a result, extremely small signals can be measured accurately, and the instrument is usually employed to examine very dilute polymer solutions and to run analyses at very low shear rates. In the latter limit, the extinction angle χ is always 45°, which is assumed in the fixed analyzer orientation scheme employed. However, in principle it is possible to monitor χ directly and obtain information regarding the first normal stress difference in oscillatory shear as well. This can be achieved by rotating the axes of the analyzing prism through 45° and detecting the response at the second harmonic of the driving frequency.

9.4.6 Experimental Considerations

This section makes a few points concerning potential sources of error in flow birefringence measurements.

Temperature Control

Clearly, temperature control is always important in rheological measurements, as the material properties can be quite strong functions of temperature. Two particular situations deserve special mention. In the oscillatory flow birefringence (OFB) experiment, very viscous solvents such as chlorinated biphenyls are often used to bring the characteristic relaxation times of flexible chains into the frequency range of the apparatus. The viscosity of such solvents can change by several orders of magnitude over a temperature interval of 20–30°C. Thus in OFB instrumentation the temperature is routinely controlled to within $\pm 0.005^\circ$C. The other special case occurs in Couette flow at relatively high shear rates, when viscous heating can become significant (see Section 5.3). In this event, temperature gradients in the sample liquid can lead to thermal lensing or other alterations in the profile of the transmitted beam, which appear at the detector as spurious changes in intensity. This difficulty has been dealt with in detail elsewhere (Janeschitz-Kriegl, 1960).

Edge Effects
In either the Couette geometry or the TFL, the beam must pass through a region of sample that is near the edge of the moving surface and therefore is experiencing a different flow pattern. The only safe way to assess the magnitude of this problem is to vary the path length through the sample by changing the relevant dimension of the moving part. If the resulting retardation increases linearly with increasing path length, with an intercept of zero retardation when extrapolated to zero path length, it is reasonable to assume that edge effects are negligible. Another constraint imposed on the experimental configuration by optical detection is the required absence of a free liquid surface along the light path. The combination of lensing and scattering at the air–liquid interface is usually sufficient to prevent accurate evaluation of the polarization state of the transmitted beam.

Window Birefringence
Any rheo-optical instrument must have transparent windows for the entrance and exit of the light beam. All glass materials are birefringent at some level. The first step, therefore, is to hand-select window material of very low birefringence. Then it is important to design the mounting of the windows carefully. Most rigid cements, for example, can introduce large stresses into glass when they harden, thus generating a substantial window birefringence. None of this would present a significant problem except that the principal axes of the window refractive index tensor are not known, and therefore it is not possible to perform a direct subtraction of the window contribution. (Recall that the retardations of optical elements in series can be added only when the principal axes are parallel.) In the case of concentrated solutions or melts, where the signals are relatively large, the window contribution is of course less significant.

9.5 Flow Birefringence: Applications

Sections 9.5.1–9.5.9 briefly discuss applications of rheo-optics that have been selected to illustrate the main points made in the preceding text and to demonstrate the breadth and power of rheo-optical experiments.

9.5.1 Stress Field Visualization

Rheo-optical methods provide a unique opportunity to examine the spatial and temporal evolution of the stresses in a liquid; an example was given in Figure 9.1.1 (Han, 1981). A second example is shown in Figure 9.5.1 (Fuller, 1990). Here the birefringence was monitored as a function of time at various distances from the outer, rotating wall in a Couette cell. The rotation rate was stepped from

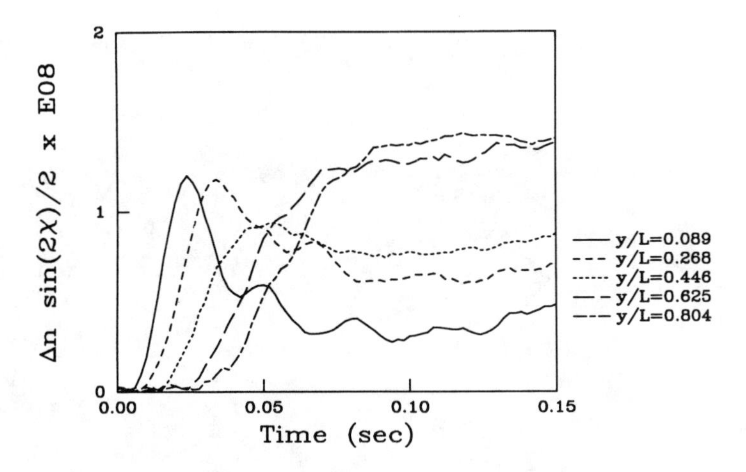

Figure 9.5.1. Optically measured shear stress as a function of time for various distances from the outer, rotating wall of a Couette cell. The solution was a 0.5 wt % mixture of polystyrene (molecular weight 2×10^7) in tricresyl phosphate. The other wall of the Couette cell was impulsively started to a speed of 5cm/s. The ratio y/L measures the distance from the outer cylinder, normalized by the gap separation. From Fuller (1990).

0 to 5 cm/s at $t = 0$. The different curves illustrate the propagation of the shear wave through the liquid, as the effect is first apparent near the moving surface. The wave speeds deduced from these measurements were in good agreement with theoretical analysis (Joseph et al., 1986).

9.5.2 Extensional Flow

Extensional flows are ubiquitous in processing operations and are well known to be "stronger" than shear flows in the sense of the extent of deformation of individual chains (see Chapter 7). For dilute solutions, it has long been predicted that there is a "coil–stretch" transition at some critical value of the strain rate. Birefringence offers a sensitive means to detect such a transition, as shown, for example, in a series of papers (Farrell et al., 1980). Figure 9.5.2a illustrates this effect by means of a bright line, due to birefringence, in the region of greatest elongation rate between opposing jets; the sample is a dilute polystyrene solution. Figure 9.5.2b displays the measured retardation as a function of strain rate, for two solutions that differ in concentration, clearly revealing the onset of a substantial deformation of the chain. The development of a plateau in the retardation with increasing strain rate after the transition, combined with numerical estimates of the birefringence, implies that the deformed state corresponds to a nearly completely extended and oriented chain. However, recent light scattering measurements of coil dimensions in elongational flow challenge this interpretation (Menasveta and Hoagland, 1991).

9.5.3 Dynamics of Isolated, Flexible Homopolymers

The oscillatory flow birefringence (OFB) experiment yields a measurement of a quantity S^*, which is directly related to the dynamic viscosity η^* and dynamic modulus G^* via the SOR

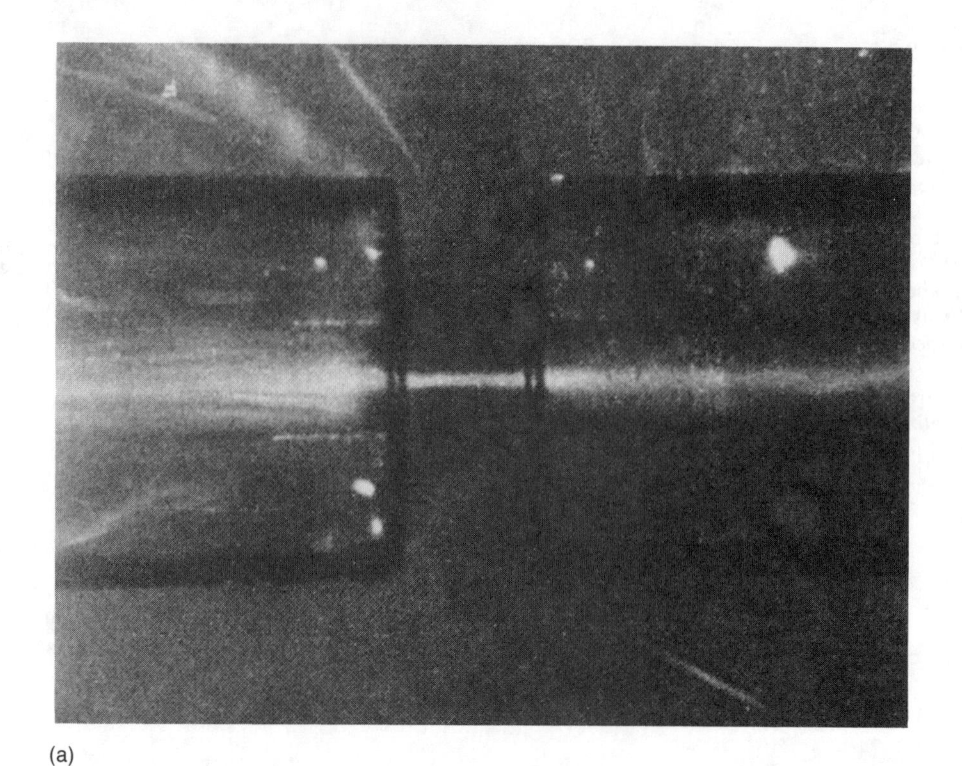

(a)

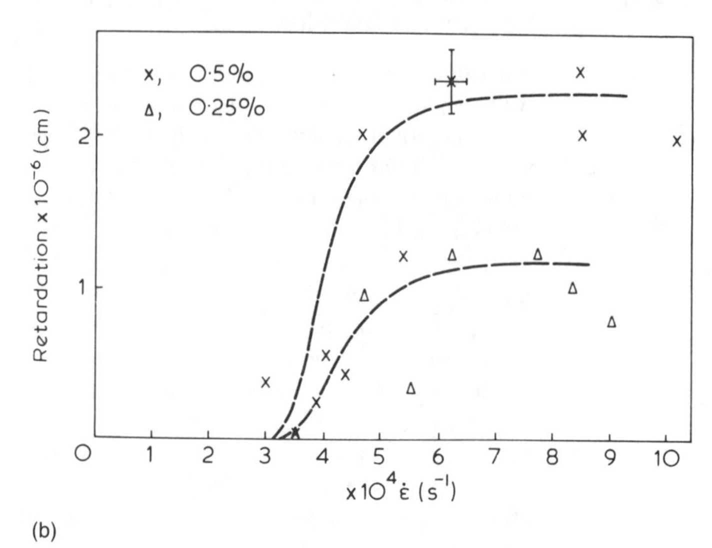

(b)

Figure 9.5.2.
(a) Birefringent line between jets for a 0.1% solution of polystyrene observed with monochromatic light and crossed polarizer and analyzer at 45° to jet axis. (b) Retardation plotted as a function of strain rate $\dot{\epsilon}$ for concentrations of 0.5 and 0.25%. Error bars, shown for one point only, are representative of the rest. From Farrell et al. (1980).

$$S^* \equiv \frac{\Delta n^*}{\dot{\gamma}^*} = S' + jS'' = S_m e^{j\theta} \tag{9.5.1}$$

$$= 2C\eta^* = 2C\eta' - j2C\eta''$$

$$= \frac{2CG^*}{j\omega}$$

where in all cases the asterisk indicates a sinusoidally time-varying quantity represented in complex notation; $\dot{\gamma}^*$ is the shear rate, and ω is the drive frequency ($= 2\pi\nu$). The frequency dependence of S^*, like that of η^* or G^*, can be used to examine the relaxation time spectrum of the polymer molecules in the sample. Extensive measurements at finite concentrations, and extrapolations to infinite dilution (Martel et al., 1983; Amelar et al., 1991), have revealed that this frequency dependence can be described extremely well using the bead–spring model of Rouse and Zimm (see Chapter 11), with exact eigenvalue calculations (Rouse, 1953; Zimm, 1956; Sammler and Schrag, 1988). Figure 9.5.3 displays an example of the infinite dilution OFB properties of a 390,000 molecular weight polystyrene in a chlorinated biphenyl solvent (Lodge et al., 1982). Note that measurements have been made at several temperatures and reduced to a master curve via time–temperature superposition (see Chapter 11). In Figure 9.5.4 the longest relaxation time at infinite dilution, τ_1°, is plotted as a function of molecular weight, along with the corresponding bead–spring model prediction (Amelar et al., 1991). Again, the agreement with theory is remarkably good, even down to very low molecular weights ($< 10^4$).

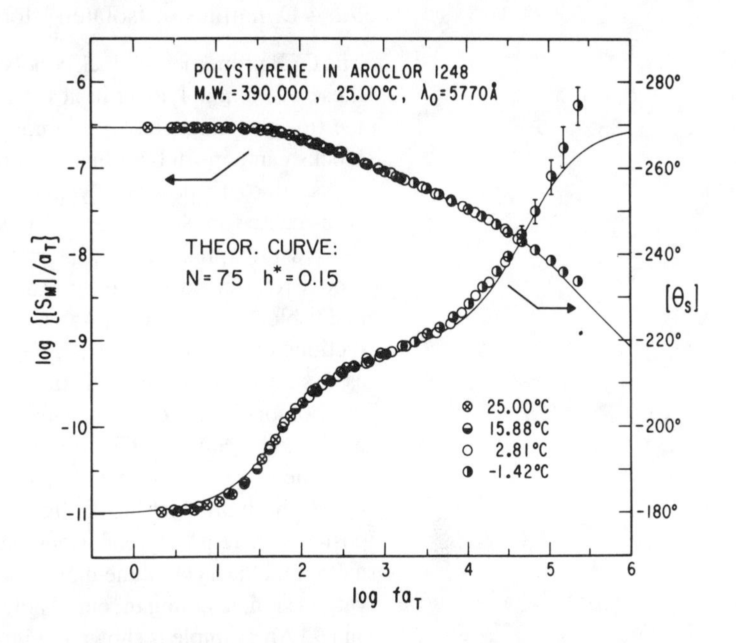

Figure 9.5.3. Infinite dilution OFB properties for polystyrene in Aroclor 1248. From Lodge et al. (1982).

Figure 9.5.4.
Infinite dilution longest relaxation time for polystyrene in Aroclor 1248. From Amelar et al. (1991).

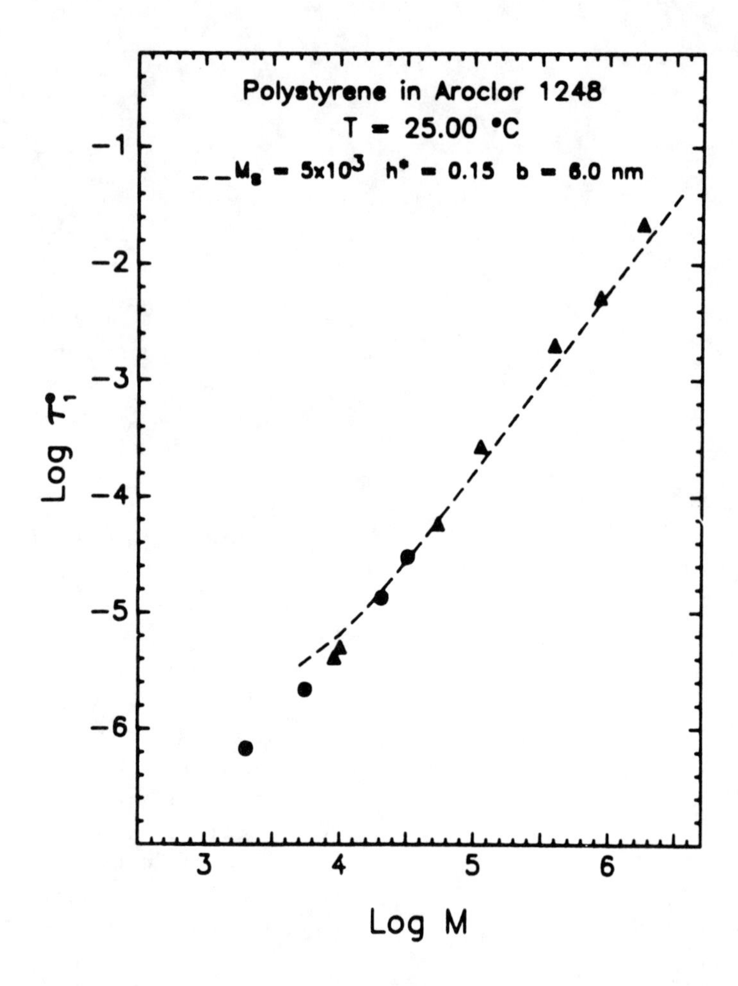

9.5.4 Dynamics of Isolated Block Copolymers

The OFB properties of block copolymers in solution can differ from those of homopolymers in at least two important respects. First, the (usually unfavorable) thermodynamic interactions among the blocks can perturb both the structure and dynamics of the chains, and second, the polymer dynamics will not follow a simple SOR, as discussed in Section 9.4. However, the latter issue has been dealt with explicitly, with interesting results. The Rouse–Zimm model referred to in Section 9.5.3 has been extended to diblock and triblock copolymers (Wang, 1978; Man et al., 1991). The predictions of this model for η^*, G^*, and S^* may all be expressed as a sum over discrete relaxation modes. The difference from the homopolymer case appears only in S^*, where each normal mode now acquires its own optical weighting factor, which is a complicated function of the model parameters. For polymers in which the individual blocks have stress–optic coefficients of opposite sign (e.g., styrene–isoprene, styrene–butadiene, styrene–methyl methacrylate), the mode optical weighting factors can also vary in sign, resulting in remarkable frequency dependences for S_m and θ. An example is shown in Figure 9.5.5, where numerical pre-

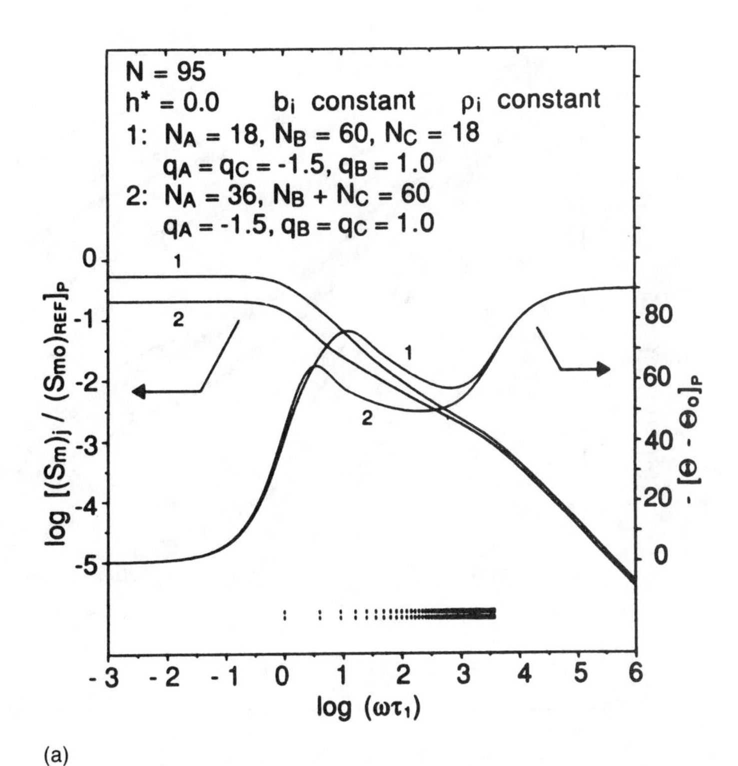

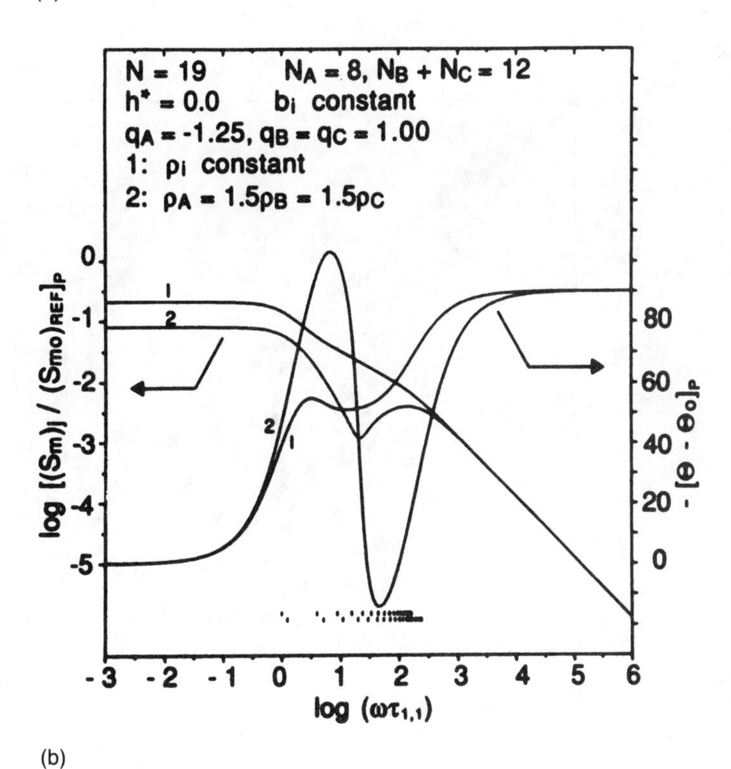

Figure 9.5.5.
(a) Predictions of the Wang–Man bead–spring model for the OFB properties of two block copolymers: curve 1 is for an ABA (18-60-18) triblock, curve 2 is for a composition-matched AB (36-60) diblock. The stress–optic coefficients for each block differ in sign and magnitude. The mechanical properties (b, ζ) of the blocks are identical, and therefore the relaxation spectra for the two model chains are identical.
(b) Predictions of the Wang–Man BSM for the OFB properties of two AB (8-12) block copolymers. The difference between the two curves is that in chain 1, $\zeta_A = \zeta_B$, while in chain 2, $\zeta_A = 1.5\zeta_B$. The stress–optic coefficients for the two blocks differ in sign and magnitude. From Man et al. (1991).

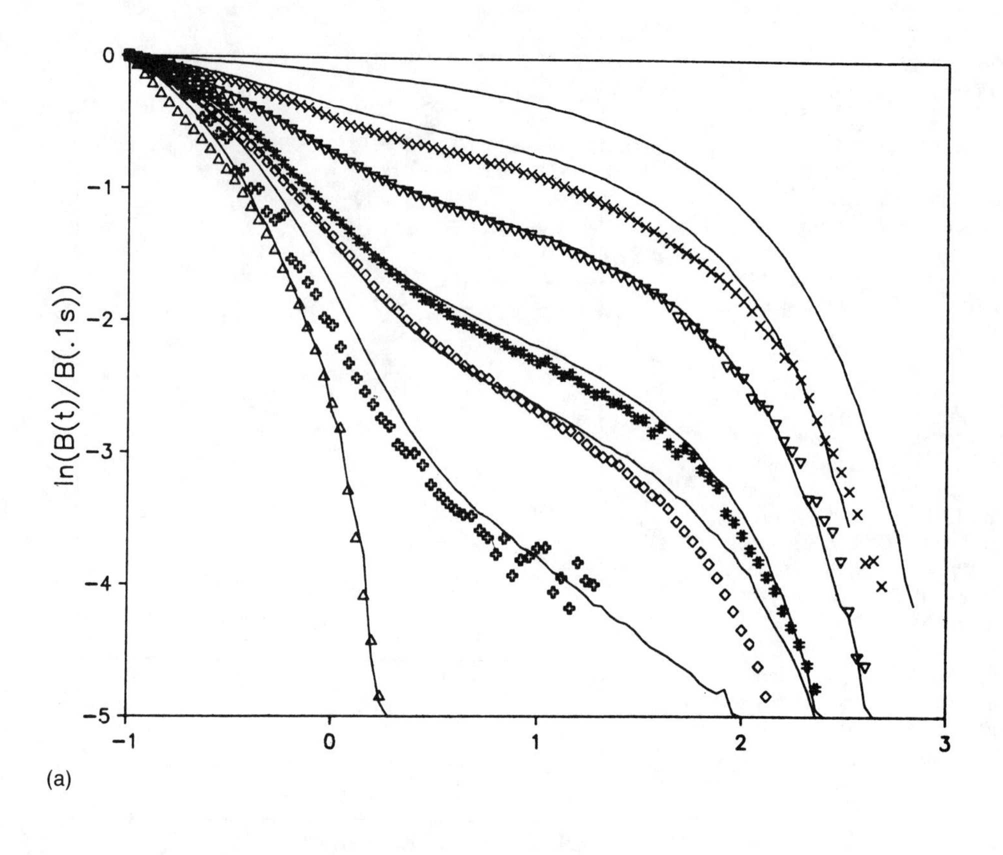

(a)

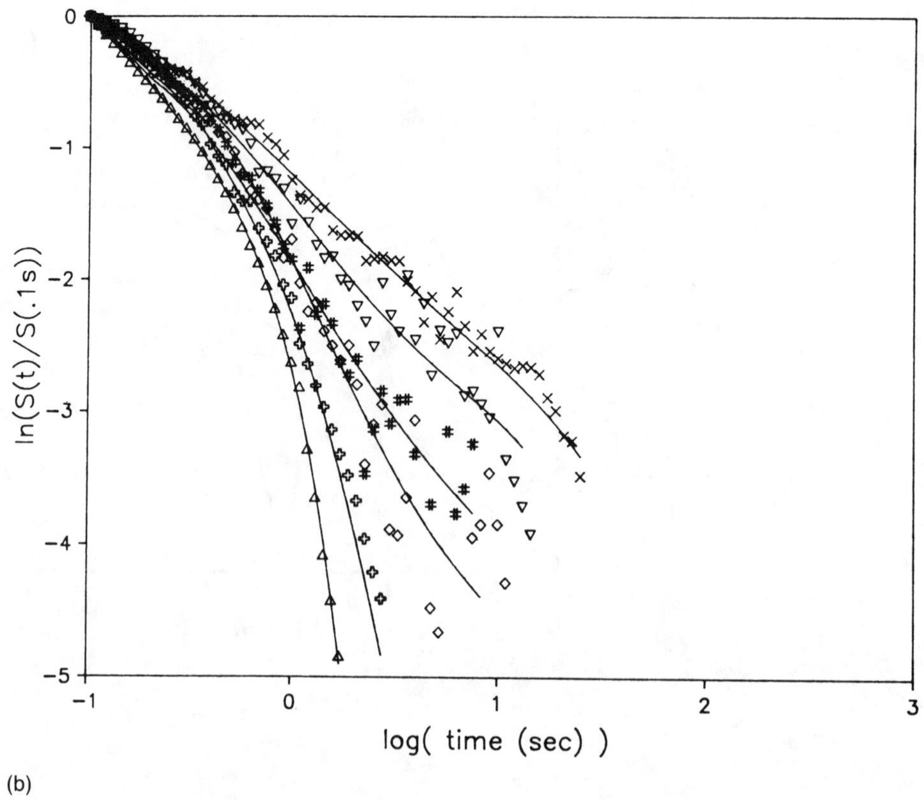

(b)

Figure 9.5.6.
(a) Bulk relaxation following a step strain as measured by stress and birefringence. The solid curves show the normalized relaxation modulus for blends containing 0, 10, 20, 30, 50, 75, and 100 vol% long polymer. The symbols show the normalized birefringence relaxation for blends containing 0 (Δ), 10 (+), 20 ($\diamond$), 30 (#), 50 (∇), and 75 ($\times$) vol% long polymer. Note that the two independent measurements show close agreement. (b) Short chain dichroism relaxation for blends containing 0 (Δ), 10 (+), 20 ($\diamond$), 30(#), 50 (∇), and 75 ($\times$) vol% long polymer. The solid curves are arbitrary smooth curves through the data. From Kornfield et al. (1989).

dictions are shown for two particular cases (Man et al., 1991). The first is a comparison between a diblock and a symmetric triblock with the same total composition, and the second is between two diblocks that differ slightly in only one parameter (the local friction coefficient ζ). These results illustrate the unique potential of OFB for characterizing block copolymer architecture because in each case the two model chains differ only slightly, but the predicted responses are strikingly different. Preliminary results on styrene–diene block copolymers are in at least qualitative accordance with these predictions (Soli, 1978; Man, 1984).

9.5.5 Dynamics of Block Copolymer Melts

This rapidly developing field involves a fascinating array of physical phenomena. The dominant feature is the so-called order–disorder transition or microphase–separation transition in which the interblock repulsion overcomes the entropic drive for molecular mixing, and the system spontaneously adopts a highly structured morphology (Bates and Fredrickson, 1990). In surprising contrast to conventional theories, rheological measurements even 50°C above such a transition suggest the existence of incipient domains (Bates et al., 1990a,b). Recently, Fredrickson proposed a model for the form birefringence of such block copolymer liquids in which the difference in refractive index between blocks should lead to a form contribution that could easily dominate the intrinsic birefringence (Fredrickson, 1987). As the transition is approached (e.g., by lowering the temperature), the form birefringence is predicted to diverge; as yet, no measurements in this regime exist.

9.5.6 Dynamics of a Binary Blend

The combination of flow birefringence and flow dichroism has recently been exploited by Kornfield et al. (1989) to examine the dynamics of binary blends of low and high molecular weight polymers, specifically hydrogenated and deuteriated polyisoprenes. Because the isotopic substitution has a negligible effect on the stress–optic coefficient, the flow birefringence measurements reflect the total stress; in this case, relaxation after an imposed step shear strain was examined. The dichroism measurements were made with an infrared light source tuned to the carbon–deuterium stretch at 2180 cm^{-1}, where the hydrogenated polymer does not absorb significantly. Thus, the dichroism measurements reflect solely the orientation of the labeled component (either the short or the long chains). Figure 9.5.6a shows the birefringence relaxation for a series of blends having different volume fractions of the longer polymer, compared to smooth curves representing the measured relaxation modulus; the SOR is reasonably well obeyed. Figure 9.5.6b shows the dichroism relaxation from the labeled shorter chains in the same blends. The most interesting result is that the longest relaxation time of the shorter chains is a strongly increasing function of the volume fraction of longer chains. This contrasts with the predictions of the basic reptation model, for example, in which the short chain relaxation should be essentially independent of the presence of longer chains (see Chapter 11). These results have been interpreted in terms of a model of orientational coupling whereby the presence of oriented long chains biases (and therefore retards) the reptational relaxation of the shorter chains.

9.5.7 Birefringence in Transient Flows

An example of birefringence in a transient flow was given in Section 9.5.1. Two additional examples are shown in Figures 9.5.7 and 9.5.8. Both cases reveal a distinct stress overshoot often observed in mechanical shear measurements under similar conditions. In Figure 9.5.7, the birefringence is plotted as a function of time after the inception of flow for different shear rates in a Couette geometry. The sample was a high molecular weight polystyrene in chlorinated biphenyl. In Figure 9.5.8, the overshoot and subsequent "ringing" were observed after start-up of an extensional flow in a four-roll mill for different strain rates. The sample was a very dilute solution of high molecular weight polystyrene in tricresyl phosphate. Neither the ringing, nor the overshoot in extension itself, is predicted by many constitutive equations.

9.5.8 Rheo-Optics of Suspensions

The application of rheo-optics is by no means restricted to polymer solutions and melts. As an example, Figure 9.5.9 shows the birefringence of a dilute colloidal suspension (ferric hydrous oxide) in glycerin as a function of shear rate times time (Johnson et al., 1985). The data reveal the transient response both upon start-up of the shear (in a Couette cell) and upon reversal of the flow di-

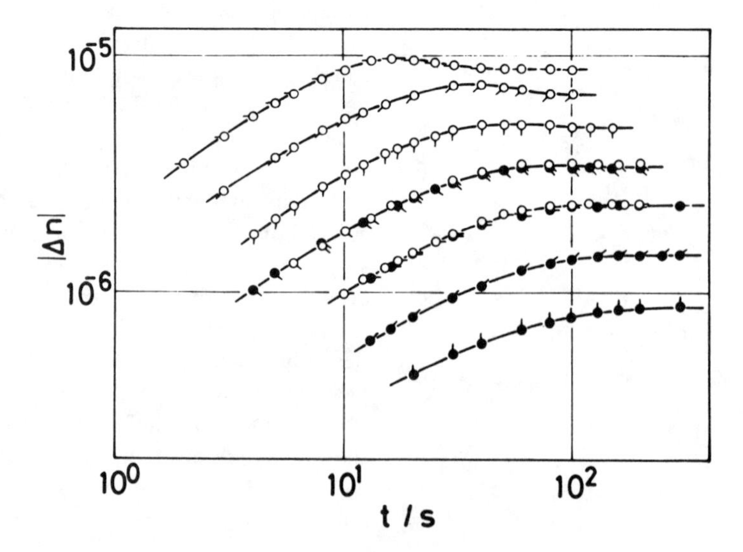

Figure 9.5.7.
Amount of birefringence Δn at start of steady shear flow. Directions of pips represent rates of shear; pip up, 0.0066 s^{-1} with successive 45° rotations clockwise representing 0.0118, 0.0216, 0.038, 0.066, 0.118, and 0.214 s^{-1}, respectively. Open and solid circles, respectively, indicate data obtained with 2 and 3 mm widths of cylindrical gap. From Osaki et al. (1979).

Figure 9.5.8.
Light intensity due to birefringence versus time after the start-up of purely extensional flow for the polystyrene–tricresyl phosphate solution. From Fuller and Leal (1981).

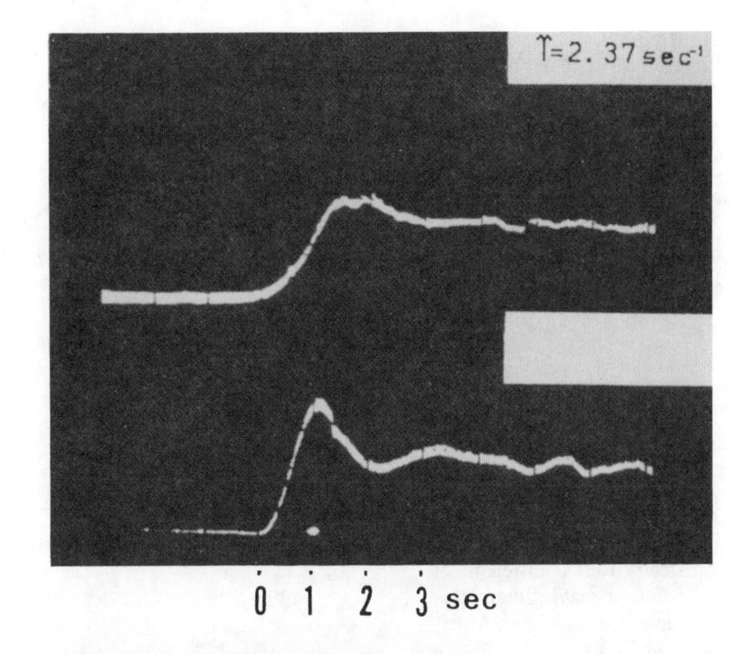

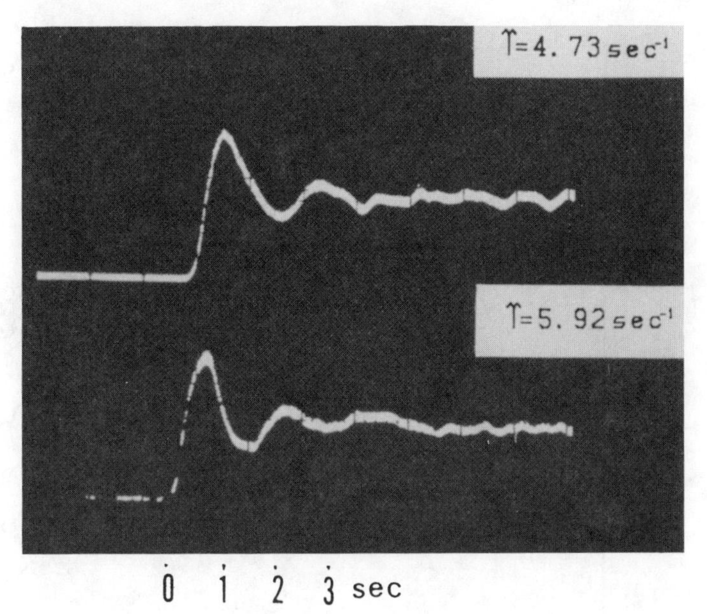

rection. The oscillation is due primarily to the rotational period of the anisotropic particles, and the flow reversal produces a response that is nearly the mirror image of the earlier time behavior as the particles retrace their paths. The lack of perfect reversal is due predominantly to Brownian motion.

9.5.9 Rotational Dynamics of Rigid Rods

In this example (Mori et al., 1982), the experimental method is oscillatory electric birefringence (Morris and Lodge, 1986), in which

Figure 9.5.9.
Linear birefringence for a 400 ppm βFeOOH suspension in 97:3 glycerin–water. The experiment consisted of a flow reversal sequence with a shear rate of 4 s^{-1}. Abscissa units are dimensionless time, shear rate times t. From Johnson et al. (1985).

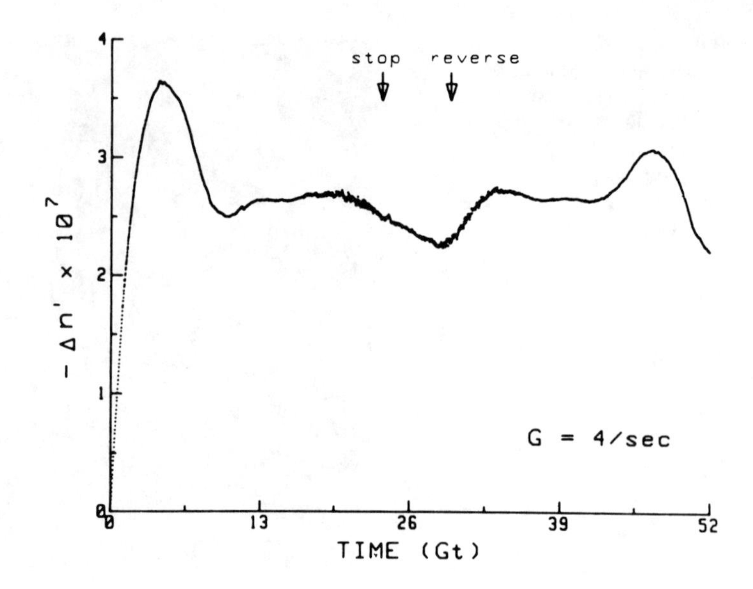

Figure 9.5.10.
Frequency dependence of the steady Kerr coefficient for poly(γ-benzyl glutamate) rods in m-cresol. From Mori et al. (1982).

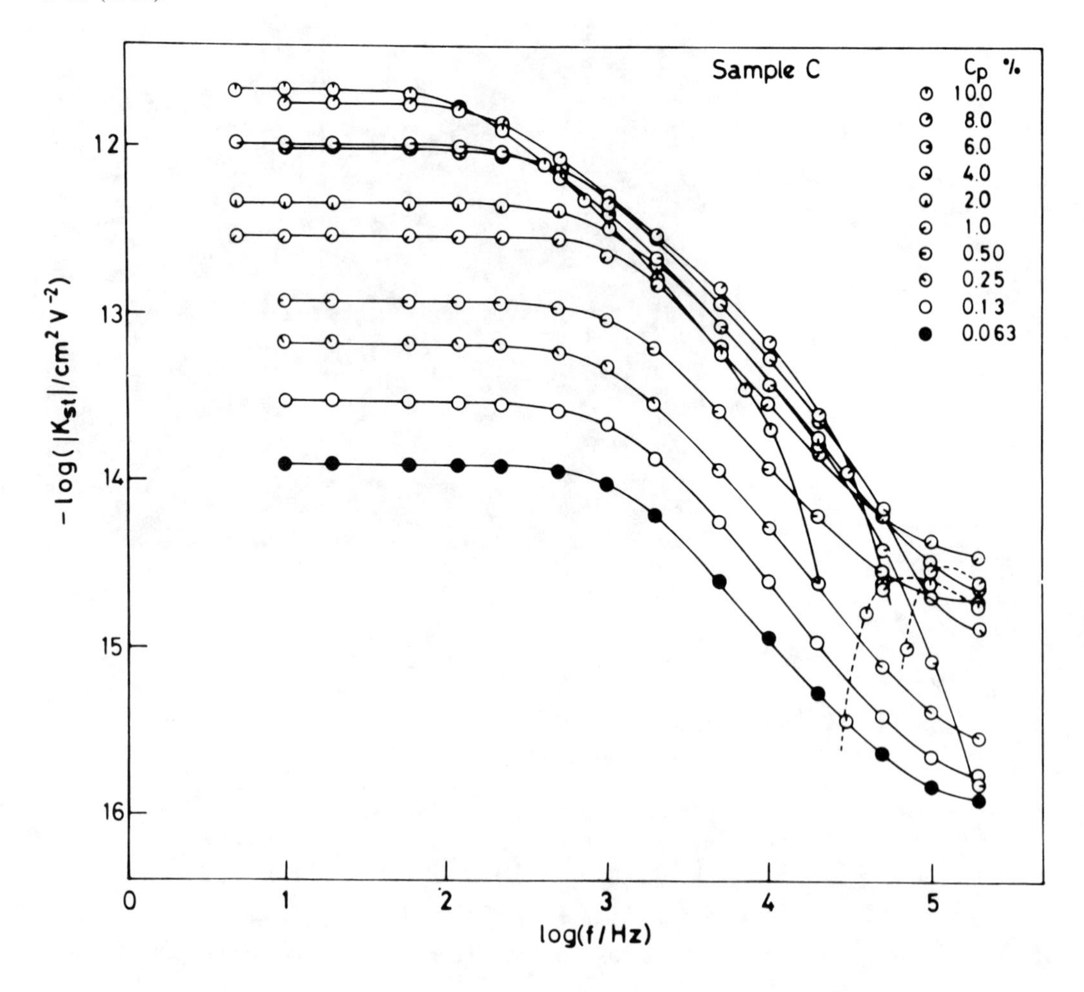

the sample is subjected to a sinusoidally time-varying electric field. Thus, this measurement of the Kerr effect is not a rheo-optical measurement in the strict sense and is more directly analogous to a dielectric relaxation experiment. Nevertheless, the basic principles are similar, and the dynamic response is interpretable by a hydrodynamic theory. The sample, poly(γ-benzyl glutamate), is a synthetic polypeptide with a large persistence length. The molecule has a permanent dipole moment oriented along its major axis, and thus the applied electric field drives the end-over-end tumbling mode. The frequency dependence of the steady Kerr coefficient K_{st}, defined as the birefringence normalized by the squared amplitude of the applied electric field, is displayed in Figure 9.5.10 for a range of concentrations (Mori et al., 1982). As the concentration is increased, the magnitude of K_{st} increases significantly, and the onset of the frequency dispersion shifts to lower frequencies, indicating a monotonic increase in the rotational relaxation time of the rods.

9.6 Summary

To conclude this introduction to rheo-optics in general, and flow birefringence in particular, it is appropriate to summarize the main points of the discussion. Birefringence in a polymer liquid results from the orientation of optically anisotropic monomer units upon imposition of a suitable flow. A postulated, intimate connection between the aspect ratios and orientations of the stress and refractive index tensor ellipsoids, known as the stress–optical relation, allows one to extract shear and normal stress data from optical measurements. The stress–optical relation does not hold under all circumstances; but even when it does not apply, rheo-optical measurements can provide useful information about the system. The two primary experimental geometries for measurements of flow birefringence are the Couette geometry, which gives information about molecular response to steady and transient shear flows, and the thin fluid layer transducer, which permits measurement of dynamic birefringence in a sinusoidally time-varying shear. However, a range of other geometries, including some for extensional flows, have been developed. A series of examples of applications of rheo-optics have been described very briefly. The systems of interest include suspensions, solutions, and melts, involving flexible homopolymers and copolymers, and rigid rods. Related techniques, such as flow dichroism and electric birefringence, have also been mentioned briefly.

References

Amelar, S.; Eastman, C. E.; Morris, R. L.; Smeltzly, M. A.; Lodge, T. P.; von Meerwall, E. D., *Macromolecules* 1991, *24*, 3505.

Azzam, R. M. A.; Bashara, N. M., *Ellipsometry and Polarized Light*; North-Holland: Amsterdam, 1977.

Bates, F. S.; Fredrickson, G. H., *Annu. Rev. Phys. Chem.* 1990a, *41*, 525.

Bates, F. S.; Rosedale, J. L.; Fredrickson, G. H., *J. Chem. Phys.* 1990b, *92*, 6255.

Berne, B. J.; Pecora, R., *Dynamic Light Scattering*; Wiley: New York, 1976.

Born M.; Wolf, E., *Principles of Optics*; 6th ed.; Pergamon: Oxford, 1980.

Bullough, R. K., *J. Polym. Sci.* 1960, *46*, 517.

Bullough, R. K., *Phil. Trans. R. Soc. A* 1962, *254*, 397.

Chu, B., *Laser Light Scattering*, 2nd ed.; Academic Press: New York, 1990.

Denbigh, K. G. *Trans. Faraday Soc.* 1940, *36*, 936.

Dexter, F. D.; Miller, J. C.; Philippoff, W., *Trans. Soc. Rheol.* 1961, *5*, 193.

Farrell, C. J.; Keller, A.; Miles, M. J.; Pope, D. P., *Polymer* 1980, *21*, 1292, and references therein.

Frank, F. C.; Keller, A.; Mackley, H. H., *J. Polym. Sci., Polym. Phys. Ed.* 1976, *14*, 1121.

Frattini, P. L.; Fuller, G. G., *J. Rheol.* 1984, *28*, 61.

Fredrickson, G. H., *Macromolecules* 1987, *20*, 3017.

Frisman, E. V.; Mao, S., *Vysokomol. Soedin.* 1964, *6*, 34.

Frisman, E. V.; Dadivanyan, A. K.; Dynzhev, G. K., *Dokl. Akad. Nauk SSSR* 1963, *153*, 1062.

Fuller, G. G., *Annu. Rev. Fluid Mech.* 1990, *22*, 387.

Fuller, G. G.; Leal, L. G., *J. Polym. Sci., Polym. Phys. Ed.* 1981, *19*, 557.

Han, C. D., *J. Rheol.* 1981, *25*, 139.

Hecht, E., *Optics*, 2nd ed.; Addison Wesley: Reading, MA, 1987.

Hiemenz, P. C., *Polymer Chemistry*; Marcel Dekker: New York, 1984; Chapter 10.

Huang, W. J.; Frick, T. S.; Landry, M. R.; Lee, J. A.; Lodge, T. P.; Tirrell, M., *AIChE J.* 1987, *33*, 573; and references therein.

Janeschitz-Kriegl, H., *Rev. Sci. Instrum.* 1960, *31*, 119.

Janeschitz-Kriegl, H., *Adv. Polym. Sci.* 1969, *6*, 170;

Janeschitz-Kriegl, H., *Polymer Melt Rheology and Flow Birefringence*; Springer-Verlag: Berlin, 1983.

Janeschitz-Kriegl, H.; Gortemaker, F. H., *Delft Progress Report, Ser. A* 1974 1, 73.

Johnson, S. J.; Frattini, P. L.; Fuller, G. G., *J. Colloid Interface Sci.* 1985, *104*, 440.

Joseph, D. D.; Riccius, O.; Arney, M., *J. Fluid Mech.* 1986, *171*, 309.

Kornfield, J. A.; Fuller, G. G.; Pearson, D. S., *Macromolecules* 1989, *22*, 1334.

Kuhn, W.; Grün, F., *Kolloid. Z.* 1942, *101*, 248.

Larson, R. G., *Constitutive Equations for Polymer Melts and Solutions*; Butterworths: Boston, 1988.

Lodge, A. S., *Nature* 1955, *176*, 838.

Lodge, T. P.; Miller, J. W.; Schrag, J. L., *J. Polym. Sci., Polym. Phys. Ed.* 1982, *20*, 1409.

Man, V. F., Ph.D. thesis, University of Wisconsin, 1984.

Man, V. F.; Schrag, J. L.; Lodge, T. P., *Macromolecules*, 1991, *24*, 3666.

Martel, C. J. T.; Lodge, T. P.; Dibbs, M. G.; Stokich, T. M.; Sammler, R. L.; Carriere, C. J.; Schrag, J. L., *J. Chem. Soc. Faraday Soc.* 1983, *18*, 173.

Maxwell, J. C., *Proc. R. Soc. A* 1873, *22*, 46.

Menasveta, M. J.; Hoagland, D. A., *Macromolecules* 1991, *24*, 3427.

Miller, J. W.; Schrag, J. L., *Macromolecules* 1975, *8*, 361.

Mori, Y.; Ookubo, N.; Hayakawa, R.; Wada, Y., *J. Polym. Sci., Polym. Phys. Ed.* 1982, *20*, 2111.

Morris, R. L.; Lodge, T. P., *Anal. Chim. Acta* 1986, *189*, 183.

Onuki, A.; Doi, M., *J. Chem. Phys.* 1986, *85*, 1190.

Osaki, K.; Bessho, N.; Kojimoto, T.; Kurata, M., *J. Rheol.* 1979, *23*, 457.

Rouse, P. E., Jr., *J. Chem. Phys.* 1953, *21*, 1272.

Rudd, J. F.; Gurnee, E. F., *J. Appl. Phys.* 1957, *28*, 1096.

Sammler, R. L.; Schrag, J. L., *Macromolecules* 1988, *2*, 1132.

Shurcliff, W. A., *Polarized Light*; Harvard University Press: Cambridge, MA, 1962;

Soli, A. L., Ph.D. thesis, University of Wisconsin, 1978.

Thurston, G. B., *Appl. Opt.* 1964, *3*, 755.

Thurston, G. B.; Schrag, J. L., *J. Polym. Sci.* 1968, *A-2 6*, 1331.

Tsvetkov, V. N.; *J. Polym. Sci.* 1957, *23*, 151.

Tsvetkov, V. N., *Polym. Rev.* 1964, *6*, 563. Ed.

Van Aken, J. A.; Gortemaker, F. H.; Janeschitz-Kriegl, H.; Laun, H. M.; *Rheol. Acta* 1980, *19*, 159.

Wang, F. W., *Macromolecules* 1975, *8*, 364; 1978, *11*, 1198.

White, J. L., *Principles of Polymer Engineering Rheology*, Wiley: New York, 1990.

Zimm, B. H., *J. Chem. Phys.* 1956, *24*, 269.

PART III

APPLICATIONS

The major motivation for rheology research has been the application of rheological measurements to solving material characterization and processing problems. The theoretical foundations of rheology described in Chapters 1 through 4 have been pursued so vigorosly in recent years precisely because they are important to making and understanding measurements on a large group of materials, particularly polymers.

We can divide the applications of rheology into three categories: characterization, processing, and design. Rheology can be a powerful tool for identifying certain structural features. For example, zero shear rate viscosity of polymer melts increases with weight average molecular weight to the 3.4 power, i.e. $M_w^{3.4}$. Normal stresses can be even more sensitive to molecular weight, and recoverable compliance can measure molecular weight distribution.

In these application chapters we will concentrate on rheologically interesting materials. Thus, for example, we will not deal with relations between molecular structure and Newtonian viscosity for small molecule liquids. The two most rheologically important groups of materials are suspensions and polymers. Chapter 10 describes relations between suspension rheology and particle shape, concentration and size interparticle forces. The flow of concentrated suspensions is an important industrial field that has become an active area for rheological research. Chapter 11 applies rheology to the concentration of polymeric liquids. In the area of polymer rheology in particular, whole books are available; several are cited in the chapter. Here the emphasis is on the major physical concepts relation structure to rheology.

10

If rigid spheres are added to a fluid, its viscosity coefficient increases by a fraction which equals 2.5 times the volume fraction of the suspended spheres.
Albert Einstein (1906, 1911)

SUSPENSION RHEOLOGY

Jan Mewis and Christopher W. Macosko

10.1 Introduction

A suspension, or perhaps more broadly a dispersion, consists of discrete particles randomly distributed in a fluid medium. Generally we divide suspensions into three categories: solid particles in a liquid medium (often the word *suspension* is restricted to this meaning), liquid droplets in a liquid medium (or an *emulsion*), and gas in a liquid (or a *foam*). All these categories have great practical importance, from biological materials like milk and blood to paint, ink, ceramics, and many other industrial dispersions.

The addition of a rigid sphere to a liquid alters the flow field, as indicated in Figure 10.1.1. This hydrodynamic disturbance, first calculated by Einstein, has a small effect on viscosity, shown by the lower curve in Figure 10.1.2. However, if the spheres are small ($< 1\ \mu$m), colloidal forces between particles can become enormous. This is illustrated in Figure 10.1.2 for suspensions of TiO_2 and carbon black. Even at low concentration, viscosity can be increased more than an order of magnitude.

Adding particles does not simply change the magnitude of the viscosity it also can introduce all the known deviations from Newtonian behavior. This is illustrated in Figure 10.1.3 with data from Laun (1984, 1988) on various concentrations of a polymer latex. The occurrence of shear thinning and shear thickening is very obvious in this figure. At high concentrations and low shear stress, the Newtonian plateau seems to disappear and to develop into a yield stress. At shear rates exceeding $10^3 s^{-1}$, the high concentration samples display shear thickening. Not visible in the equilibrium viscosities are eventual time effects. By adjusting the interaction forces between the particles, the shear can induce gradual changes in aggregate structure. These give rise to gradual (i.e., time-dependent) changes in viscosity or *thixotropy*. The interaction forces are potential forces, which are therefore elastic in nature.

Figure 10.1.1.
Creeping flow around a sphere. From Taneda (1979).

This can be seen in the viscoelastic response during the oscillatory testing of colloidal suspensions.

In this chapter we deal first with very dilute suspensions, in which particles never interact. For dilute suspensions many rigorous theoretical results are available and a number of them have

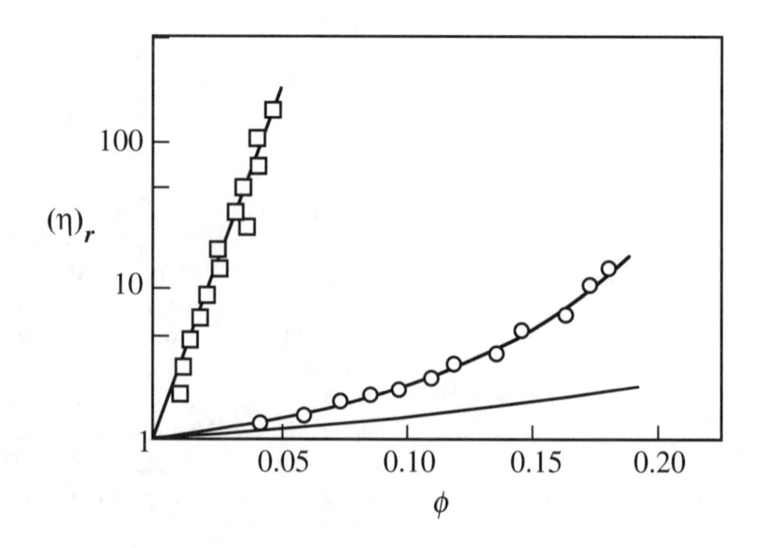

Figure 10.1.2.
Log of relative viscosity (η/η_s) versus volume fraction of particles for TiO_2, $0.1\mu m$ in diameter, in linseed oil (circles) and for carbon black in mineral oil (squares) compared to the ideal dilute sphere result, eq. 10.2.4 (solid line). Adapted from Mewis and Spaull (1976).

Figure 10.1.3.
Viscosity versus shear stress for a polystyrene ethylacrylate latex at different volume fractions. From Laun (1984, 1988).

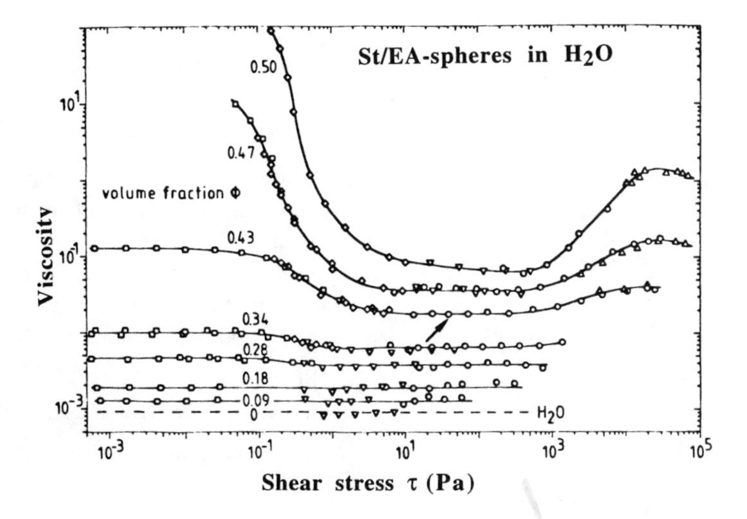

been confirmed experimentally. We start with rigid spheres, then treat liquid droplets, and finally, rigid axisymmetric particles. In each case, the mutual interference between the flow field and the particle motion is discussed. It is shown how interfacial tension for emulsions, and rotary Brownian motion for nonspherical particles, lead even in dilute systems to non-Newtonian effects such as shear thinning and viscoelasticity. Interference from sedimentation, migration, and inertia is pointed out.

Nearly all systems of practical interest are nondilute. Theories for such materials are less developed and can become extremely complicated. Nevertheless, useful semiquantitative and even quantitative information is becoming available from new theoretical approaches and from systematic experimental studies on well-characterized model suspensions.

The treatment of nondilute systems starts in Section 10.4 with a review of the various interaction forces between particles. Hydrodynamic effects, Brownian motion, electrostatic and polymeric repulsion, and van der Waals attraction are considered. The balance between these forces can be expressed by means of dimensionless groups, which are used to scale rheological experiments. Three limiting cases are covered systematically. First, attention is focused on the case in which only hydrodynamic forces and Brownian motion are present (i.e., Brownian hard particles). This section includes brief discussions of particle size distribution, non-Newtonian media, and nonspherical particles. The second case includes the effects of repulsion forces, both electrostatic and polymeric. Finally, we consider the role of attractive forces, which cause particles to aggregate either reversibly or irreversibly. Very divergent rheological responses can be obtained in this manner, including yielding and thixotropy. The first results of applying some new concepts in analyzing complex microstructures on these materials are reviewed.

10.2 Dilute Suspensions of Spheres

10.2.1 Hard Spheres

As shown in Figure 10.1.1., the presence of a particle will modify the velocity distribution in a flowing liquid. Extra energy dissipation will arise because of this and will be reflected in a proportional increase in viscosity. Einstein followed such reasoning in his classic papers of 1906 and 1911.

The important assumptions for the analysis are as follows.

1. Surrounding fluid or solvent is incompressible and Newtonian and can be treated as a continuum.
2. Creeping flow (i.e., negligible body forces, torques, and inertia).
3. Neutral density, $\rho_s = \rho_p$ (i.e., no settling).
4. No slip between the particle and the fluid.
5. Rigid, spherical particles.
6. Dilute (noninteracting) particles.
7. No influence of walls.
8. No particle migration.
9. Velocity perturbations due to a particle are local; the average velocity field in the surrounding fluid is the same as if the particles were not present.

Assumptions 1, 8, and 9, imply that there are three widely different length scales for suspension rheology (Brenner, 1972). The first assumption (continuous fluid) implies that the size of fluid molecules is much less than the suspended particle size:

$$\text{fluid molecules } (\leq 1 \text{ nm}) \ll a, \text{ radius of particles} \tag{10.2.1}$$

To use the average velocity profile (assumption 9), the length scale of the velocity perturbations due to the particles must be small compared to the distance over which the velocity varies appreciably. In other words, the particle size must be much smaller than the scale of motion.

$$a \ll \begin{matrix}\text{scale of the main} \\ \text{velocity variation}\end{matrix} = \frac{dv/dx}{d^2v/dx^2} \tag{10.2.2}$$

In addition, to avoid wall effects on the velocity field around the particle (assumption 7), suspended particles must be much smaller than the viscometer gap.

$$a \ll h, \text{ viscometer gap} \tag{10.2.3}$$

With the assumptions given above and the velocity profile around a sphere, Einstein found the extra energy dissipated by adding one particle to the fluid per unit volume. Then, assuming no interactions, he multiplied by ϕ, the volume fraction of particles, to obtain the very simple result for the viscosity of the suspension:*

$$\eta = \eta_s(1 + \frac{5}{2}\phi) \qquad (10.2.4)$$

To get a complete constitutive equation for dilute spheres, we must solve for the stress tensor directly rather than for the energy dissipation. If the particles add only small perturbations to the main velocity field, the stress tensor can be divided into two parts: one due to the main flow and the other due to the particles (Batchelor, 1970; Schowalter, 1978).

$$\boldsymbol{\tau} = \boldsymbol{\tau}_s + \boldsymbol{\tau}_p \qquad (10.2.5)$$

The stress from the main flow is determined by the Newtonian viscosity of the suspending fluid and the *average* rate of deformation tensor (recall assumption 9)

$$\boldsymbol{\tau}_s = 2\eta_s \overline{\mathbf{D}} \qquad (10.2.6)$$

$\overline{\mathbf{D}}$ is averaged over a volume large compared to the particle separation, $V \gg a^3/\phi$.

$$\overline{\mathbf{D}} = \frac{1}{V}\int_V \mathbf{D}\, dV \qquad (10.2.7)$$

For subsequent work with dilute suspensions we will drop the overbar and assume that $\mathbf{D}$ is averaged according to eq. 10.2.7.

The particle stress is a volume average of the stresses contributed by one particle.

$$\boldsymbol{\tau}_p = \underbrace{\tfrac{1}{V}\sum_i}_{\text{sum overall } i \text{ particles}} \int_{S_i} \Big[\underbrace{\mathbf{n}\cdot(\mathbf{Tr})}_{\text{surface stress}} - \underbrace{\eta_s(\mathbf{vn} + \mathbf{nv})}_{\text{velocity perturbation}}\Big] dS_i \qquad (10.2.8)$$

Here S is the particle surface area, $\mathbf{n}$ is the vector normal to the particle surface, and $\mathbf{r}$ is the particle position. Equation 10.2.8 is actually valid for particles of any shape, provided inertia and external couples can be neglected. The real problem is to solve this equation. The velocity field $\mathbf{v}$ can be quite complex. Happel and Brenner (1965) and Schowalter (1978) show how to evaluate

The first paper (1906) reported the front factor of the volume fraction to be 1. Einstein corrected this in 1911.

eq. 10.2.8 with the velocity field around a spherical particle. With this result and eq. 10.2.6, the stress becomes

$$\boldsymbol{\tau} = 2\eta_s\left(1 + \frac{5}{2}\phi\right)\mathbf{D} \tag{10.2.9}$$

which agrees with the Einstein result for the shear viscosity. This constitutive equation is surprisingly simple. It states that particle size does not affect a suspension of dilute spheres. The spheres may be protein globules or basketballs as long as they are much smaller than the flow gap in our rheometer. We would need a very large rheometer to verify eq. 10.2.9 for a suspension of basketballs, but it should still hold. The only effect of temperature is through η_s. There should be no shear rate dependence; the constitutive equation is Newtonian.

10.2.2 Particle Migration

Einstein's result can be verified experimentally in the limit as $\phi \to 0$. However, doing so is not a trivial matter. Large spherical particles can be made rather easily, however, settling, migration, wall effects, and particle inertia can cause serious problems with such particles. A criterion for neglecting the effect of settling can be obtained from Stokes' law (recall the falling ball viscometer in Chapter 5). The time required for a sphere to migrate 10% of the rheometer gap, h, under the influence of gravity, g, acting perpendicular to the gap is

$$t_{0.1h} = \frac{0.45\,\eta_s h}{|\rho_p - \rho_s| a^2 g} \tag{10.2.10}$$

For 100 μm diameter spheres in water and a 10 mm rheometer gap, the particle density must be within 2% of that of water to prevent settling during typical measurements (1000 s). As pointed out in Chapter 5, gravitational settling will affect torque readings in a concentric cylinder rheometer much less than in a cone and plate or parallel plate rheometer because in the concentric cylinders gravity normally acts parallel rather than perpendicular to the narrow gap of the rheometer. Thus, typically for concentric cylinders, $h \simeq$ 50 mm, versus $h = r \sin\beta < 3$ mm for cone and plate.

However, inertia can still create difficulties. Lin et al. (1970) show that when the particle Reynolds number becomes large, inertia can alter the velocity field around the sphere, causing deviations in viscosity measurements:

$$\eta = \eta_0\left(1 + \frac{5}{2}\phi + 1.34\,\mathrm{Re}_p^{3/2}\right) \tag{10.2.11}$$

Thus a criterion for neglecting inertia in shear flow is

$$\mathrm{Re}_p = \frac{\rho_s \dot{\gamma} a}{\eta_s} \ll 0.1 \tag{10.2.12}$$

Even if Re_p is small, the stress field around a sphere can interact with the wall, causing it to migrate inward. Ho and Leal (1974) have analyzed this problem extensively for both drag (Couette) and pressure-driven (plane Poiseuille) flow. They show that if the spheres are small enough and the flow rate is low, Brownian motion can keep the particles uniformly distributed. Their criteria for neglecting migration are

$$K = \frac{\rho_p \overline{v}^2 a^4}{hkT} \begin{array}{l} < 0.1 \text{ for Couette flow} \\ < 0.01 \text{ for plane Poiseuille flow} \end{array} \tag{10.2.13}$$

where h is the narrow gap of the flow, k is Boltzmann's constant, and $\overline{v}$ is $\frac{1}{2}v_{\max}$ in Couette and $\frac{2}{3}v_{\max}$ in Poiseuille flow. Figure 10.2.1 shows results of their calculations. In a concentric cylinder

Figure 10.2.1. Concentration distribution across the shear gap for various K values (eq. 10.2.13) in (a) Couette flow and (b) planar Poiseuille (slit) flow. From Ho and Leal (1974).

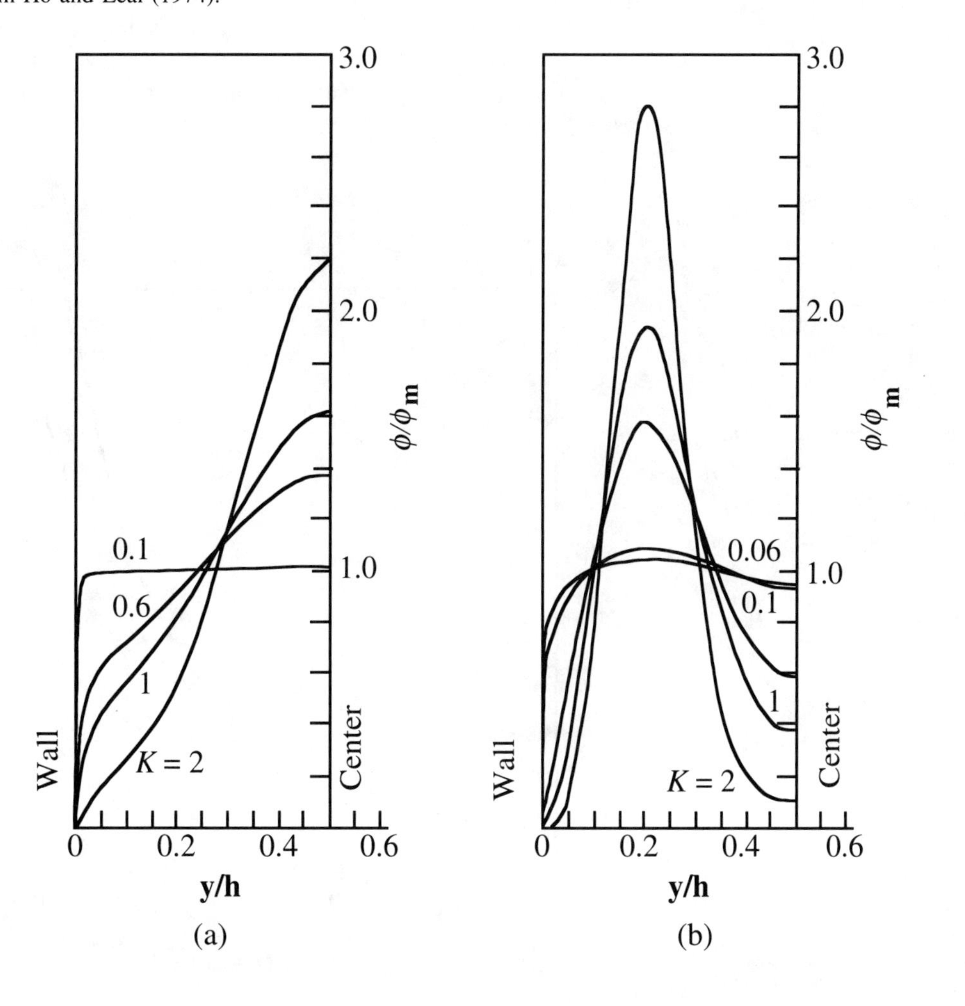

rheometer, spheres move toward the centerline of the flow, while in a slit rheometer particles will concentrate about midway between the wall and centerline because of the nonuniform shear rate. Migration in tube flow will be similar. These results are in good agreement with experiments on large particles (Segre and Silberberg, 1962; Halow and Willis, 1970).

Looking at each of the foregoing criteria, we can see that as the sphere radius a decreases, the effects of settling and migration are smaller. Inorganic particles as well as polymer latices with very uniform size in the 0.01–10 μm range can be made (Woods et al., 1968; van Helden et al., 1981). As we shall see in Section 10.4, with such small spheres colloidal forces can become significant. These also act to prevent migration, but they can lead to strong interparticle interactions. If interparticle repulsion forces are suppressed, particle agglomeration can occur. With care, all these problems can be controlled, and for polymer latices good agreement with Einstein's result for shear viscosity has been obtained, as shown in Figure 10.2.2. However, we see that even neutrally buoyant spheres with low interparticle forces show deviations from eq. 10.2.9 at very low volume fraction.

A more sensitive way to plot the data to test for the 5/2 coefficient of ϕ and to examine the higher ϕ terms is shown in Figure 10.2.3. Frequently, the viscosity relation is expressed as a power series in ϕ

$$\eta = \eta_s\left(1 + \frac{5}{2}\phi + k\phi^2 + \cdots\right) \tag{10.2.14}$$

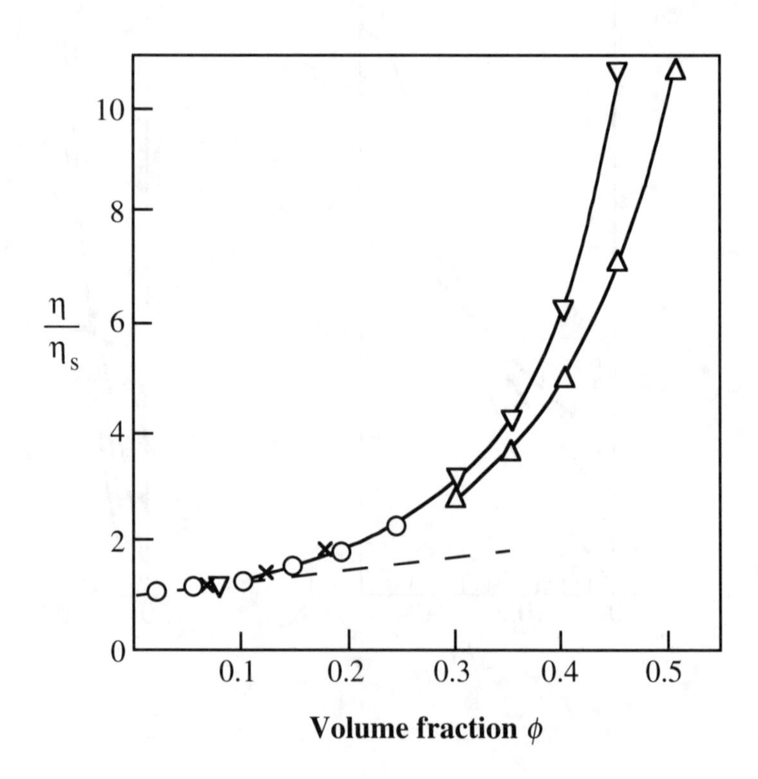

Figure 10.2.2. Some experimental viscosity results on dilute suspensions of rigid spheres: $\times$, glass $\sim$ 5 μm in zinc iodide glycerin (Manley and Mason, 1954); ○, polystyrene aqueous latices, 0.42, 0.87 μm (Saunders, 1961); $\triangledown$, low shear rate, and $\triangle$, high shear rate limits for nonaqueous polystyrene latices, 0.16–0.43 μm (Krieger, 1972).

Figure 10.2.3.
Data of Saunders (1961) from Figure 10.2.2 replotted as intrinsic viscosity according to eq. 10.2.15. The slope of $k = 6.2$ is the theoretical result (eq. 10.5.2) for two sphere interactions.

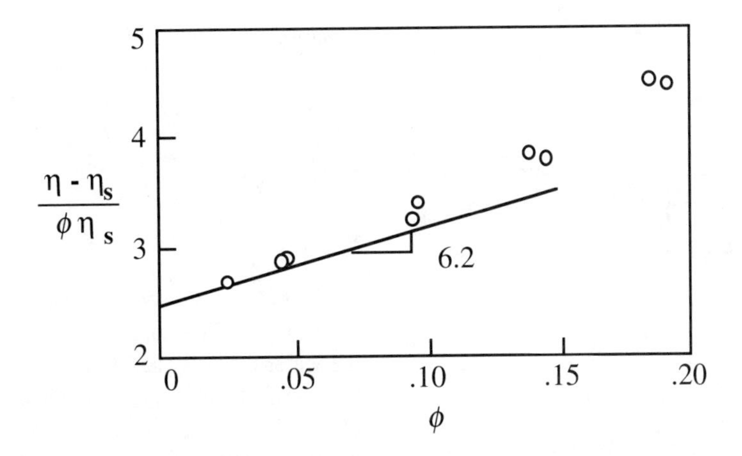

and rearranged to give

$$\frac{\eta - \eta_s}{\eta_s \phi} = \frac{5}{2} + k\phi + \cdots \tag{10.2.15}$$

These equations are used to define some common terminology in suspension rheology:

$$\frac{\eta}{\eta_s} = \eta_r, \quad \text{the relative or reduced viscosity} \tag{10.2.16}$$

$$\frac{\eta - \eta_s}{\eta_s} = \eta_{sp}, \quad \text{the specific viscosity} \tag{10.2.17}$$

and

$$\lim_{\phi \to 0} \frac{\eta_{sp}}{\phi} = [\eta], \quad \text{the intrinsic viscosity} \tag{10.2.18}$$

The intrinsic viscosity is 5/2 for the ideal rigid sphere case. The specific viscosity per unit volume, η_{sp}/ϕ (without the limit), is often called the inherent viscosity.

The higher order dependence on ϕ for spheres, k in eq. 10.2.15, is shown in Figure 10.2.3. This departure from the Einstein equation is due to hydrodynamic interactions between spheres and to other interparticle forces. We will examine these effects in Section 10.4, but first we look at the influence of particle shape on the rheology of dilute suspensions.

Example 10.2.1 Latex Suspension

A polystyrene latex of monodisperse 1.0 μm diameter spheres is diluted to a volume fraction of 0.05 in water. Specific gravity of the particles is 1.05; $\rho_p = 1.05\rho_s$. (a) Using the constitutive equation for dilute, noninteracting spheres, estimate its shear and extensional viscosity. (b) Determine whether settling, particle inertia, or migration will present measurement problems in a narrow gap Couette rheometer: $L = 50$ mm, $R_i = 25$ mm, $R_o = 26$ mm, and $\Omega < 100$ rad/s (see Figure 5.3.1 for Couette geometry).

Solution

(a) From Einstein's equation, eq. 10.2.4, $\eta/\eta_s = 1 + (2.5)(0.05) = 1.13$. Thus, a 5% suspension of spheres increases the viscosity of the water by 13%. Einstein's equation was derived only for shear, but the full three-dimensional constitutive relation in eq. 10.2.9 shows that dilute suspensions of spheres are Newtonian; thus the extensional viscosity will be simply $\eta_e = 3\eta = 3.38\eta_s$. We should note that even for a 5% suspension of spheres, particle interactions are measurable. This can be seen in Figure 10.2.3. These results indicate that because of hydrodynamic interactions, the reduced viscosity would be 1.15. See eq. 10.5.2 for further discussion.
(b) From eq. 10.2.10, the settling time of this suspension, assuming gravity is in the direction of L, will be very long

$$t_{0.1h} = \frac{0.45\eta_s L}{(\rho_p - \rho_s)(d^2/4)g} = \frac{0.45(10^{-3}\ \text{kg/ms})(0.05\ \text{m})}{(50\ \text{kg/m}^3)(25 \times 10^{-14}\ \text{m}^2)(9.8\ \text{m/s}^2)}$$

$$t_{0.1h} = 1.8 \times 10^5 \text{ seconds}$$

From the definition of the particle Reynolds number (eq. 10.2.12) and $\dot{\gamma}$ in narrow gap Couette flow (eq. 5.3.11)

$$\text{Re}_p = \frac{\rho_p R_i \Omega (d/2)^2}{\eta_s (R_o - R_i)} = \frac{(1.05 \times 10^3\ \text{kg/m}^3)(0.025\ \text{m/s})(5 \times 10^{-7}\ \text{m})^2 \Omega}{(10^{-3}\ \text{kg/ms})(0.001\ \text{m})}$$

$$\text{Re}_p = 6.6 \times 10^{-6}\Omega$$

Thus, inertial effects are negligible. Following eq. 10.2.13, the migration criterion becomes

$$K = \frac{\rho_p (R_i\Omega)^2 (d/2)^4}{(R_0 - R_i)kT} = \frac{(1.05 \times 10^3\ \text{kg/m}^3)(6.25 \times 10^{-4}/\text{s}^2)(6.25 \times 10^{-26}\ \text{m}^4)\Omega^2}{(10^{-3}\ \text{m})\left(1.38 \times 10^{-23} \frac{\text{kg m}^2}{\text{s}^2\text{K}}\right)(298\ \text{K})}$$

$$K = 0.01\Omega^2$$

By eq. 10.2.13, particle migration effects will be negligible for $\Omega < 3$ rad/s.

10.2.3 Emulsions

Many heterogeneous systems of practical importance consist of droplets of one liquid dispersed in another. Soaps, some paints, cosmetics, creams, mayonnaise, milk, and butter are all examples

Figure 10.2.4.
External streamlines around a fluid sphere in shear flow. Solid lines for $\eta_{dr} = 0$ (a gas bubble); dotted line for $\eta_{dr} = \infty$ (a rigid sphere) Note that the flow is less disturbed for a gas bubble. From Bartok and Mason (1958).

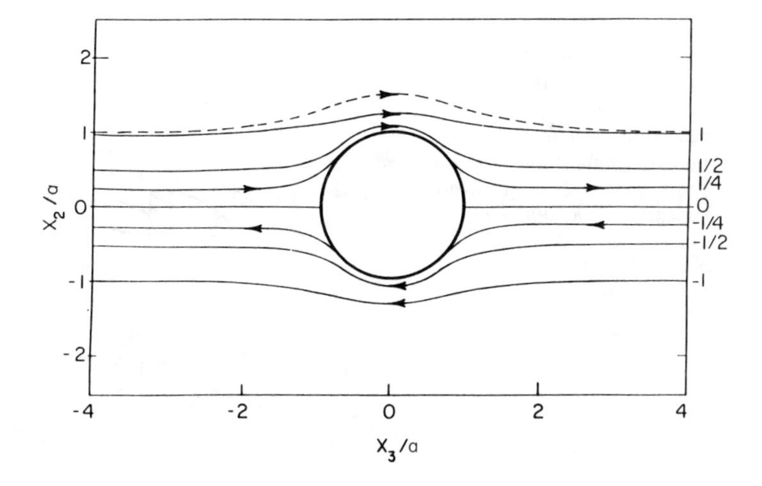

of emulsions. When these emulsions flow—for example through a pump or during spreading—the liquid inside the droplets also flows, rolling around in a "tank-tread-like" motion. This circulation causes extra dissipation over that in the suspending fluid, though not as much as in the case of rigid spheres, where particles disturb the flow more profoundly. We can see this difference in the streamlines shown in Figure 10.2.4. The higher the viscosity ratio (the ratio of viscosity of the droplet liquid to that of the suspending medium),

$$\eta_{dr} = \eta_d / \eta_s \tag{10.2.19}$$

the more the streamlines will be deformed, approaching the rigid sphere case in the limit of high droplet viscosity.

If the flow is strong enough, the hydrodynamic forces can overcome interfacial tension and cause the drop to change shape and even to break up. Mason and co-workers have studied droplet flow and deformation in detail (e.g., Goldsmith and Mason, 1967). Some drop configurations sketched from their movies taken at high deformation rates are shown in Figure 10.2.5. We see that when a drop is viewed along the principal axes of the deformation, up to the breakup region, its shape is the same for both extension and shear. But drop breakup depends on both the type of flow and the viscosity ratio.

For small deformations the drop assumes an elliptical form as indicated in Figure 10.2.5. Taylor (1932) showed that the difference between major and minor axes, $2a$* and $2b$, of the ellipsoid depends on

$$D = \frac{a-b}{a+b} = 2\lambda_d II_{2D}^{1/2}\left(\frac{1 + 19\eta_{dr}/16}{1+\eta_{dr}}\right) \tag{10.2.20}$$

Note that a is the radius for a sphere **and** *the semimajor axis of an ellipsoid.*

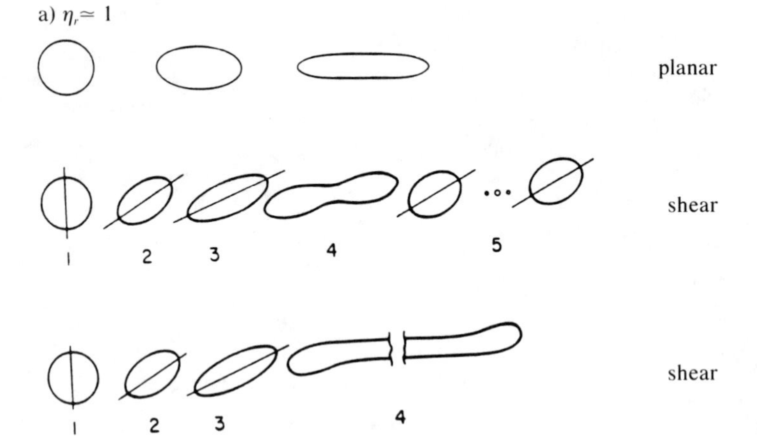

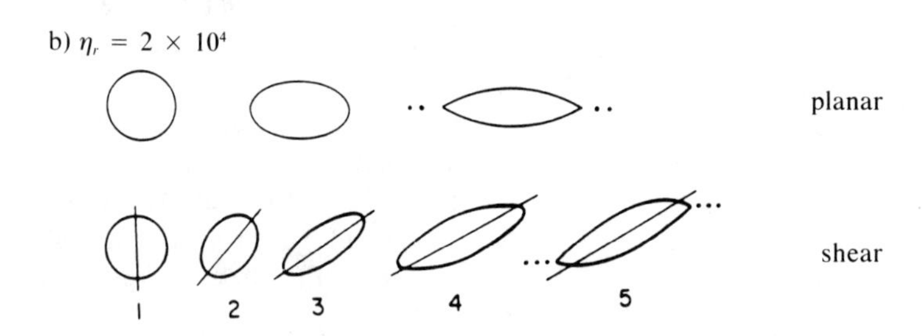

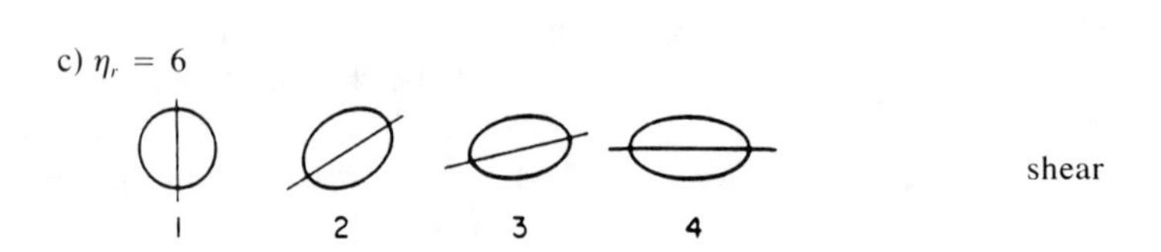

Figure 10.2.5.
The influence of deformation rate, type of flow, and viscosity ratio on the deformation of droplets in shear and extensional flow. At high deformation rates (right-hand shapes) drop breakup can occur. Adapted from Rumscheidt and Mason (1961).

where II_{2D} is the second invariant of the rate of deformation, eq. 2.2.16. The drop relaxation time λ_d is given by

$$\lambda_d = \frac{a\eta_s}{\Gamma} \tag{10.2.21}$$

where Γ is the interfacial tension. Interfacial tension tends to pull the drop back to a sphere of radius a, while the viscosity of the surrounding fluid η_s slows down the motion. For typical oil–water emulsions $\eta_s = 1\,\text{mPa·s}$, $\eta_{dr} \simeq 3$, $\Gamma = 0.02\,\text{N/m}$, and $a \simeq 1\,\mu\text{m}$. Thus the droplet relaxation time is very short, $\lambda_d \simeq 5 \times 10^{-8}\text{s}$ and even for relatively high shear rates, the deformation of a drop will be quite small.

If the viscosity of the drop is very large, $\eta_{\text{dr}} \gg 1$, interfacial tension will have a negligible effect in shear flows. At high deformation rates the drop will align with the flow and assume a limiting shape (Goldsmith and Mason, 1967)

$$\lim_{\dot{\gamma}\to\infty} D = \frac{5}{2(2\eta_{dr} + 3)} \tag{10.2.22}$$

This is shown for the shear case in Figure 10.2.5c. For lower viscosity drops, breakup can occur. Taylor (1932) argued that drops will burst when the hydrodynamic force exceeds the interfacial tension or

$$D_{\text{burst}} > 1/2 \tag{10.2.23}$$

A more detailed theoretical study by Acrivos and Lo (1978) on droplet breakup supports this result.

10.2.4 Deformable Spheres

For small deformations it is possible to solve for the velocity field around a drop. However, in contrast to the case for rigid spheres, the circulation inside the drop must be considered. This results in less distortion of streamlines than with rigid spheres, as shown in Figure 10.2.4. With the velocity field, eq. 10.2.8 can be solved for the particle stress. Schowalter (1978) and Frankel and Acrivos (1970) show how to carry out this solution. The latter obtain a constitutive relation of the form

$$\boldsymbol{\tau} = 2\eta_s \mathbf{D}\left[1 + \left(\frac{1 + \frac{5}{2}\eta_{\text{dr}}}{1 + \eta_{\text{dr}}}\right)\phi + F\right] \tag{10.2.24}$$

where F is a function of η_{dr}, λ_d, and the history of the deformation through the convected derivative of $\mathbf{D}$. For steady simple shear, eq. 10.2.24 reduces considerably to give shear rate independent material functions. The steady shear viscosity becomes

$$\eta = \eta_s\left[1 + \left(\frac{1 + \frac{5}{2}\eta_{\text{dr}}}{1 + \eta_{\text{dr}}}\right)\phi\right] \tag{10.2.25}$$

Note that for large η_{dr} the drop becomes "rigid" and the coefficient of ϕ goes to 5/2, the Einstein result. In the other limit, a gas bubble, the coefficient is 1.

For steady shear, eq. 10.2.24 gives a positive first normal stress coefficient

$$\frac{\tau_{11} - \tau_{22}}{\dot{\gamma}^2} = \psi_1 = 2f(\eta_{\text{dr}})\eta_s\lambda_d\phi \tag{10.2.26}$$

while the second is negative

$$\frac{\tau_{22} - \tau_{33}}{\dot{\gamma}^2} = \psi_2 = \left[\frac{g}{2}(\eta_{dr}) - f(\eta_{dr})\right]\eta_s \lambda_d \phi \qquad (10.2.27)$$

where

$$f(\eta_{dr}) = \frac{1}{80}\left(\frac{19\eta_{dr} + 16}{\eta_{dr} + 1}\right)^2 \quad \text{and} \quad g(\eta_{dr}) = \frac{3(19\eta_{dr} + 16)(25\eta_{dr}^2 + 41\eta_{dr} + 4)}{560(\eta_{dr} + 1)^3}$$

For steady uniaxial extension Frankel and Acrivos (1970) report

$$\eta_e = 3\eta_s\left[1 + \left(\frac{1 + \frac{5}{2}\eta_{dr}}{1 + \eta_{dr}}\right)\phi + g(\eta_{dr})\lambda_d\dot{\epsilon}\right] \qquad (10.2.28)$$

We see that just going from rigid to deformable spheres leads to great rheological complexity: time dependence (through F in eq. 10.2.24), normal stresses, and an extensional viscosity that increases with rate. The new physical effect, interfacial tension, is responsible. Surface tension tends to restore the spherical shape that is distorted by the viscous forces of the surrounding fluid during flow. Hence, the former shape serves as a "memory" of the equilibrium condition and consequently as a source of viscoelastic effects like normal force differences and storage moduli.

Some experimental results are available on model emulsions to test the viscosity relation. Nawab and Mason (1958) prepared fairly monodisperse butyl benzoate oil droplets, $a \simeq 3$ μm, by electrical atomization. The droplets were suspended in various water solutions and stabilized with a nonionic surfactant. Viscosities were measured in a capillary viscometer at shear rates of 200 to 900 s^{-1}. Their results are plotted in Figure 10.2.6. The agreement between theory and experiment for intrinsic viscosity ($\phi = 0$ limit) is excellent.

In Figure 10.2.6 all the intrinsic viscosities for dilute emulsions lie between the solution for rigid spheres, $\eta_{dr} \to \infty$, on the high side and that for gaseous bubbles or foam, $\eta_{dr} \to 0$, on the low end. It is interesting to note that adding "nothing" to a liquid, in the form of small gas bubbles, actually raises the viscosity. In many emulsions the surfactant tends to form a stiff film around the droplet, making it more like a rigid sphere. In these cases the rigid sphere results may be closer to the real behavior. Another approach is to model the particle as an elastic ball (Roscoe, 1967; Goddard and Miller, 1967). Here non-Newtonian effects arise from the modulus of the sphere G rather than from interfacial tension. The form of the equations is very similar to eq. 10.2.24 with $\lambda_d = \eta_s/G$, and one obtains similar results.

What about normal stresses? Because the viscosity is so low for the oil-in-water emulsions of Figure 10.2.6, eq. 10.2.26 predicts that the first normal stress coefficient will be $\sim 10^{-4}$ Pa·s^2. This is below the detection limit of typical measurements. It is interesting

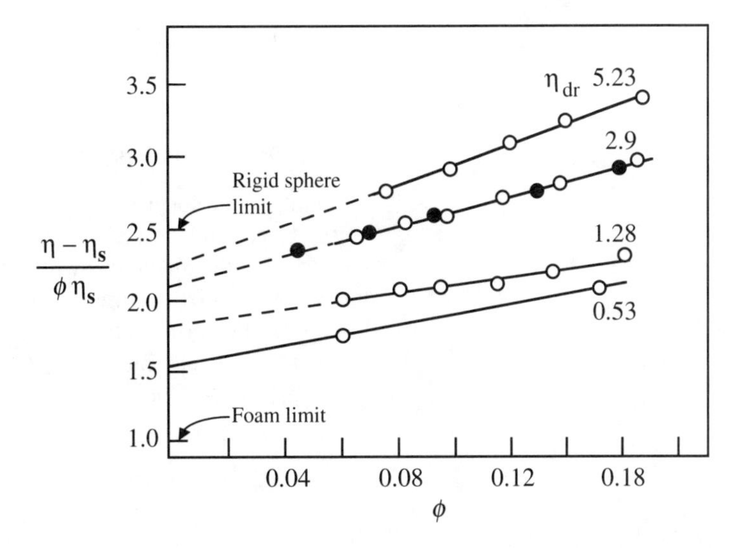

Figure 10.2.6.
Reduced viscosity versus volume fraction for several emulsions with varying viscosity ratios. Adapted from Nawab and Mason (1958).

to note that ψ_2, the second normal stress coefficient, is negative and less than $|\psi_1|/2$. This finding agrees with experimental results on polymer solutions and melts. Oscillatory experiments at high frequencies offer a better opportunity to test elasticity of emulsions, and some progress has been made in this direction (Oosterbroek et al., 1980). At high concentration, emulsions droplets (or gas bubbles in foam) interact strongly to form a network, which behaves like a weak solid with modulus proportional to Γ/a (Otsubo and Prud'homme, 1994).

10.3 Particle–Fluid Interactions: Dilute Spheroids

Look back at the assumptions we used in developing the constitutive relation for a suspension of dilute, rigid spheres. We now want to change the fifth one to consider *axisymmetric* particles. We generally divide such particles into two groups: prolate spheroids or rodlike particles, such as glass and graphite fibers, viruses, proteins, and even very stiff polymer molecules, and oblate spheroids or disklike shapes, such as red blood cells and mica flakes. Some of these are sketched in Figure 10.3.1. Most real suspension particles can be approximated fairly accurately by a spheroidal shape. We usually define a spheroid in terms of its axis ratio, $r_p = a/b$, where a is the axis of symmetry.

In comparison to suspensions of rigid spheres, the overwhelming additional effect with axisymmetric particles is orientation. Obviously the orientation of a nonspherical particle with respect to the flow will greatly affect the velocity field around it and thus the particle stress, $\boldsymbol{\tau}_p$ in eq. 10.2.8. For example, if the particle is a rod with its long axis aligned in the flow direction, the alteration of the streamlines will be much less than if the axis were perpendicular to the flow.

10.3.1 Orientation Distribution

Orientation of a spheroid is determined by the balance of hydrodynamic forces and rotary Brownian motion. Hydrodynamic forces tend to align the major axis with the flow, while Brownian motion tends to randomize the orientation. The relative importance of each is expressed in terms of the Peclet number Pe, the ratio of the time scales for Brownian motion ($1/D_r$) to that for convective motion ($1/II_D^{1/2}$).

$$\begin{aligned} \mathrm{Pe} &= \frac{II_{2D}^{1/2}}{D_r} \\ &= \frac{\dot{\gamma}}{D_r} \quad \text{for shear flow} \end{aligned} \tag{10.3.1}$$

For rigid spheres the rotary Brownian diffusion coefficient is

$$D_r = \frac{kT}{8\pi\eta_s a^3} \tag{10.3.2}$$

for circular disks

$$D_r = \frac{3kT}{32\eta_s b^3} \tag{10.3.3}$$

and for long thin prolate spheroids, $r_p \gg 1$,

$$D_r = \frac{3kT(\ln 2r_p - 1/2)}{8\pi\eta_s a^3} \tag{10.3.4}$$

Brenner (1974) gives D_r for many other shapes.

If the particles are small, shear rate and viscosity of the suspending fluid are low, so Brownian motion randomizes orientations completely and $\mathrm{Pe} = 0$. With high viscosity and shear rates or with large particles, the disperse phase will orient with the flow as $\mathrm{Pe} \to \infty$.

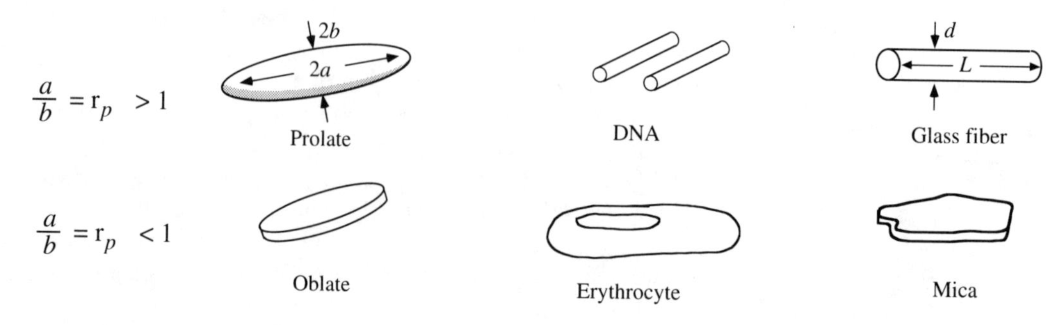

Figure 10.3.1.
Prolate and oblate spheroids and related shapes of typical suspension particles.

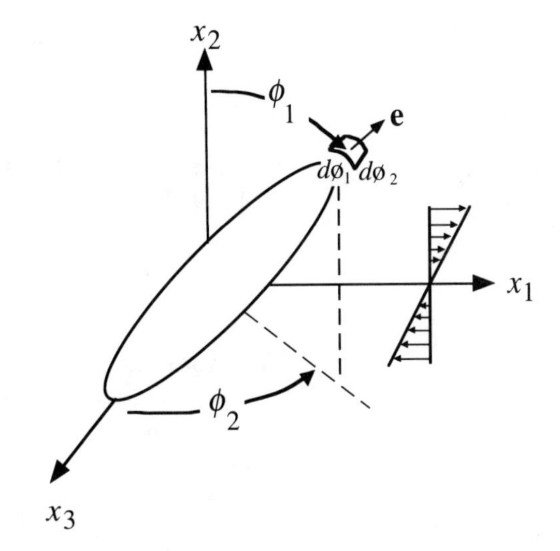

Figure 10.3.2.
Prolate spheroid with orientation vector **e** along its major axis. The spherical coordinate system to describe **e** is shown with the Cartesian system of the shear flow of surrounding fluid. Both coordinate systems are fixed to the center of the particle.

However, particles are never fully aligned. To calculate particle stresses, we have to add contributions of the various orientations. Hence we have to keep track of the orientation distribution. The orientation of a particle can be described by a unit vector **e** parallel to its symmetry axis. Figure 10.3.2 shows **e** for a prolate spheroid with its azimuthal and polar angles ϕ_1 and ϕ_2. If all vectors **e** for particles occupying a given part of space are drawn from a single origin, their ends are points on a sphere with radius unity. The function $f(\phi_1, \phi_2)$ gives the number of points per unit area or the "point density" for a particular orientation. Figure 10.3.3 shows a schematic representation of f.

The probability that **e** will lie within some solid angle $d\Omega = \sin\phi_1 d\phi_1 d\phi_2$ is $f(\phi_1, \phi_2)d\Omega$. The probability that **e** will lie on the sphere is 1, so $f(\mathbf{e})$ must obey

$$\int_{\mathrm{e}} f(\mathbf{e})d\mathbf{e} = 1 \tag{10.3.5}$$

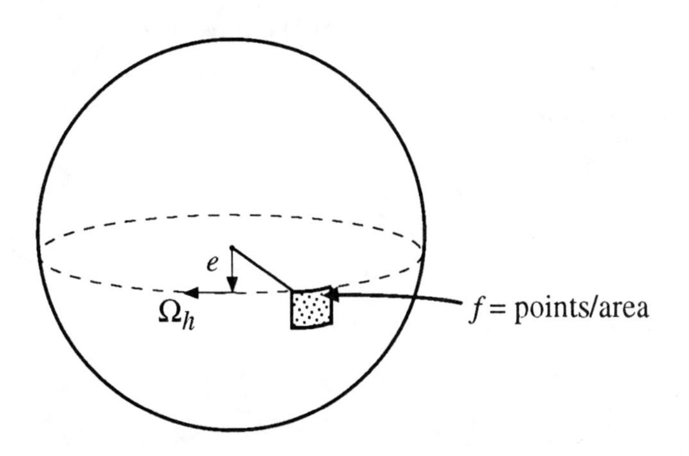

Figure 10.3.3.
Representation of the distribution function as points on the surface of a sphere.

The function f is called the orientation distribution function. Changes in this function are governed by a dynamic conservation equation. It takes into account contributions to the rotational flux from Brownian motion ($\mathbf{j}_d$) and from the hydrodynamic convection ($\mathbf{j}_h$) (Brenner 1972; Hinch and Leal 1972, Schowalter 1978; Bird et al., 1987)

$$\frac{\partial f(\phi_1, \phi_2)}{\partial t} + \nabla \cdot (\mathbf{j}_h + \mathbf{j}_d) = 0 \qquad (10.3.6)$$

The hydrodynamic convection can be expressed in terms of the angular velocity $\mathbf{\Omega}_h \times \mathbf{e}$, whereas the Brownian term is determined by the product of the rotary diffusion coefficient Dr and the gradient ∇f:

$$\frac{\partial f}{\partial t} + \nabla \cdot (f \mathbf{\Omega}_h \times \mathbf{e}) - D_r \nabla^2 f = 0 \qquad (10.3.7)$$

The solution of the conservation relation for the distribution function must be done for each flow field of interest. This often necessitates numerical methods. Fortunately, the results for many important cases are available from Brenner (1974) or Kim and Karrila (1991) for a range of spheroidal and other axisymmetric shapes.

Mason and co-workers have measured the distribution function for a number of axisymmetric particles. Figure 10.3.4 is typical of their results for rodlike particles. For visualization purposes, their particles were always large enough so that Brownian motion was negligible $D_r \ll 1$, and thus the Peclet number Pe $\gg 1$. For this case, Jeffery (1922) calculated the trajectories ϕ, (t)

$$\frac{d\phi_1}{dt} = \frac{\dot{\gamma}}{r_p^2 + 1} (r_p^2 \cos \phi_1 + \sin^2 \phi_1) \qquad (10.3.8)$$

Integrating this gives

$$\tan \phi_1 = r_p \tan \left(\frac{\dot{\gamma} t}{r_p + 1/r_p} \right) \qquad (10.3.9)$$

and the period of rotation T as

$$T = \frac{2\pi}{\dot{\gamma}} \left(r_p + \frac{1}{r_p} \right) \qquad (10.3.10)$$

Figure 10.3.4. Sequences of a rod, r_p = 10, rotating in shear flow. Adapted from Goldsmith and Mason (1967).

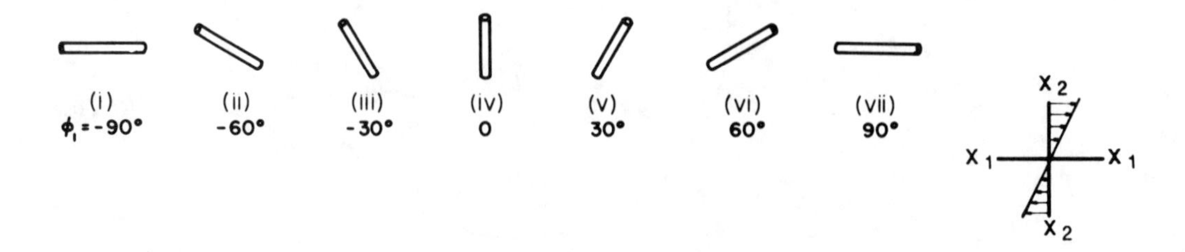

The excellent agreement between the calculated distribution function and experiments such as those shown in Figure 10.3.4, has been obtained (Karnis et al., 1966; Anczurowski and Mason, 1967).

10.3.2 Constitutive Relations for Spheroids

The same basic approach used to calculate the constitutive equation for dilute suspensions of spheres can be applied to spheroids. The difficulty lies in calculating the particle stress $\boldsymbol{\tau}_p$ in eq. 10.2.8. Not only is the velocity field more complex, but $\boldsymbol{\tau}_p$ depends on the orientation. Thus, to get the bulk value of the stress contribution of the particles, we need to integrate over all orientations, weighting by the distribution function

$$< \boldsymbol{\tau}_p >= \int_{\mathrm{e}} \boldsymbol{\tau}_p f(\mathbf{e}) d\mathbf{e} \qquad (10.3.11)$$

Rotary Brownian motion enters in two ways: first, it alters $f(\mathbf{e})$ according to eq. 10.3.7, and second, it contributes to the particle's angular velocity, which in turn alters $\boldsymbol{\tau}_p$ (Brenner, 1972). The final solution, except for limiting cases, must be numerical, so the results for various material functions can be given only in tabular or graphical form. A number of solutions for specific geometries exist. Brenner has unified and extended these solutions in an extensive paper (1974). We reproduce some of his results in Figure 10.3.5. For other geometries and material functions, the reader is referred to his work.

Figure 10.3.5 shows calculations of the intrinsic viscosity (see eq. 10.2.18) versus Peclet number for both oblate and prolate spheroids. We see that they both show considerable shear thinning. Qualitatively, as Pe increases, the hydrodynamic forces become strong enough to align the particles more with the shear flow direction, reducing their contribution to the viscosity. Good agreement has been found between these theoretical results and measurements of $[\eta]$ versus $\dot{\gamma}$ on rigid rod polymers and biological macromolecules (e.g., Whitcomb and Macosko, 1978; Bird et al., 1987, p. 124).

In the low Peclet limit (i.e., small particles and/or low shear rates), Brownian motion randomizes the orientation totally. For prolate spheroids the intrinsic viscosity becomes

$$\lim_{\mathrm{Pe}\to 0}[\eta] = [\eta]_0 = \frac{r_p^2}{15}\left[\frac{3}{\ln 2r_p - 0.5} + \frac{1}{\ln 2r_p - 1.5}\right] + \frac{8}{5} \qquad (10.3.12)$$

This result is accurate for $r_p > 10$ ($< 3\%$ error). In the high Pe limit for prolate spheroids

$$\lim_{\mathrm{Pe}\to\infty}[\eta] = [\eta]_\infty = \frac{0.315 r_p}{\ln 2r_p - 1.5} \qquad (10.3.13)$$

Figure 10.3.5
The variation of intrinsic viscosity of (a) oblate and (b) prolate spheroids with reduced shear rate. Adapted from Brenner (1974).

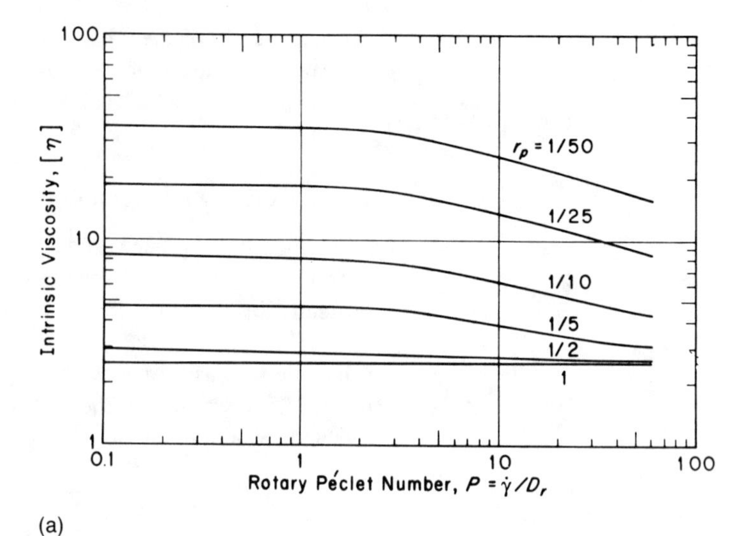

(a)

(b)

The aspect ratio must be much larger, $r_p > 100$, for good accuracy here. Also, as indicated in Figure 10.3.5, this limit is approached slowly.

The Brownian motion that gives rise to shear thinning in axisymmetric suspensions also results in normal forces. Particles, forced away from their equilibrium configurations by the flow, generate Brownian torques that are manifest as normal forces. "Intrinsic normal stress differences" are defined as

$$[\tau_{11} - \tau_{33}] = \frac{\tau_{11} - \tau_{33}}{\phi \eta_s D_r} \tag{10.3.14}$$

$$[\tau_{22} - \tau_{33}] = \frac{\tau_{22} - \tau_{33}}{\phi \eta_s D_r} \tag{10.3.15}$$

Note that Brenner (1974) uses $\tau_{11} - \tau_{33} = N_1 + N_2$ rather than the usual $\tau_{11} - \tau_{22} = N_1$. Figure 10.3.6 plots $[\tau_{11} - \tau_{33}]$ and $[\tau_{22} - \tau_{33}]$ for various aspect ratios. We see that the normal stresses increase proportional to $(\mathrm{Pe})^2$ and then level off at high Pe where full particle alignment is achieved. Note in Figure 10.3.6b that the second difference is about 0.1 of the first and opposite in sign. Brenner gives limiting relations for the normal stresses at high Pe that are accurate within 10% at $r_p = 10$ and improve with increasing r_p

$$[\tau_{11} - \tau_{33}]_\infty = \frac{r_p^4}{4(\ln 2r_p - 1.5)} \tag{10.3.16}$$

$$[\tau_{22} - \tau_{33}]_\infty = \frac{-[\tau_{11} - \tau_{33}]_\infty}{r_p^2} \tag{10.3.17}$$

Figure 10.3.7 gives the intrinsic viscosity of prolate spheroids in uniaxial extensional flow. We define the intrinsic viscosity in extensional flow as

$$[\eta]_u = \frac{\eta_u - 3\eta_s}{3\eta_s \phi} \tag{10.3.18}$$

For $\eta_u = 3\eta$, the Trouton limit, we see by comparing Figures 10.3.5 and 10.3.7 that $[\eta]_u = [\eta]$. In fact, at low Pe, the intrinsic viscosity is the same in extension as in shear. This is as expected because at low Pe there is no preferred orientation and extension does not differ from shear. However, at large positive Pe these cigar-shaped particles line up with the flow and increase the drag, thus producing extensional *thickening*.

Why do we see the opposite effect with negative Pe? As discussed in Chapter 7, uniaxial compression is the same as equal biaxial extension. Unlike changing direction in shear, changing direction in extension can cause different a response with a non-Newtonian material. Prolate spheroids are a good example of this point. Particles in biaxial extension can take any orientation in the plane of stagnation, whereas in uniaxial extension particles tend to align along the symmetry axis. As sketched in Figure 10.3.8, uniaxial flow will align particles with the streamlines converging toward the particle ends. In biaxial flow particles align such that streamlines diverge from their ends. This produces less drag, and as Figure 10.3.7 indicates, results in compressional *thinning*.

Figure 10.3.6.
The variation of intrinsic normal stress functions (a) $[\tau_{11} - \tau_{33}]$ and (b) $-[\tau_{22} - \tau_{33}]$ with reduced shear rate for prolate spheroids of various aspect ratios. Adapted from Brenner (1974).

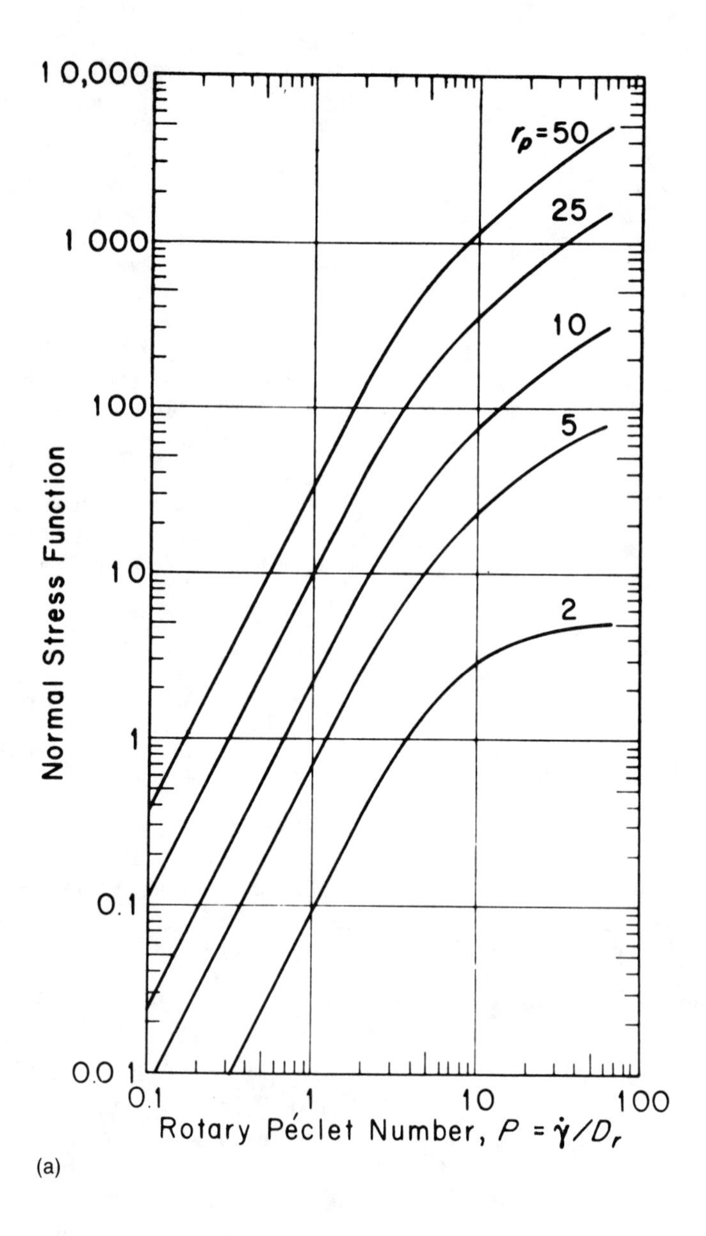

(a)

Brenner (1974) gives limiting relations for the extensional viscosities. The last term can be dropped for $r_p > 5$.

$$[\eta]_{u+\infty} = \frac{r_p^2}{3(\ln 2r_p - 1.5)} - \frac{6 \ln 2r_p}{r_p^2} \tag{10.3.19}$$

$$[\eta]_{u-\infty} = \frac{r_p^2}{12(\ln 2r_p - 1.5)} + \frac{5}{2} - \frac{6 \ln 2r_p}{r_p^2} \tag{10.3.20}$$

Spheroids will also show time-dependent rheological behavior through the time dependence of the orientation distribution function in eq. 10.3.7. Including the $\partial f/\partial t$ gives rise to a convected derivative in the constitutive equations. Brenner (1974) gives results for the problem of stress relaxation after steady shear of long

Figure 10.3.6
Continued.

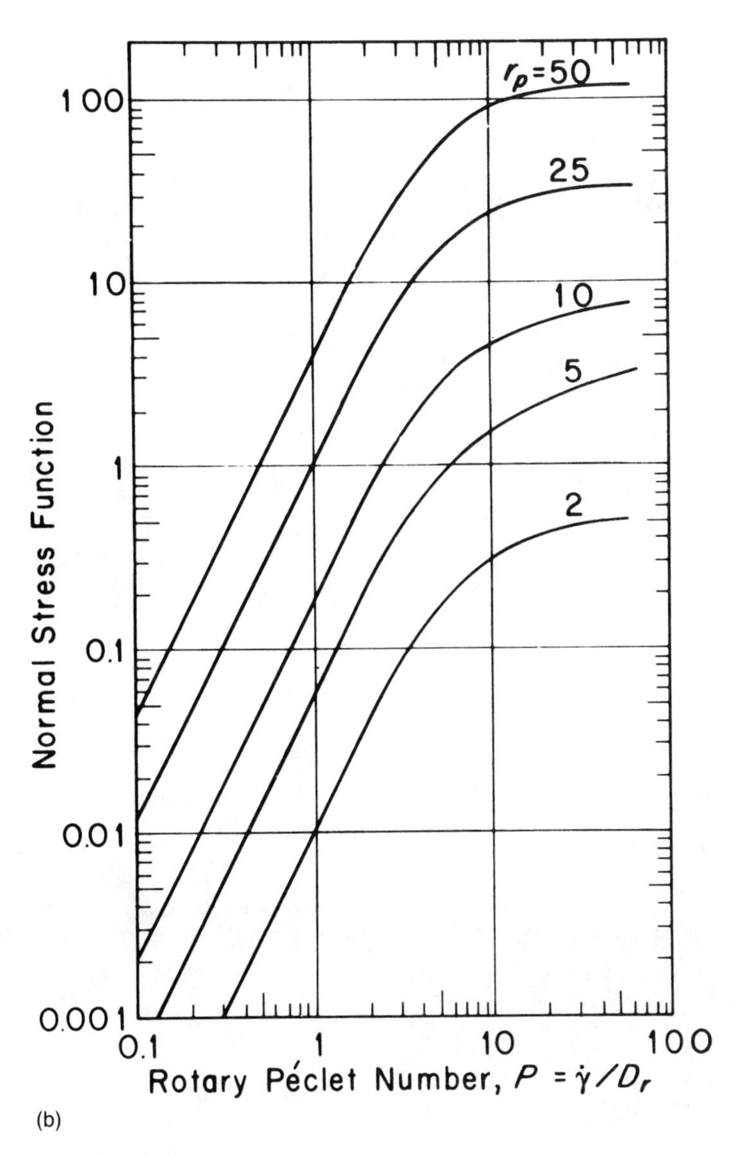

Figure 10.3.7.
Intrinsic viscosity of prolate spheroids in uniaxial extensional flow. From Brenner (1974).

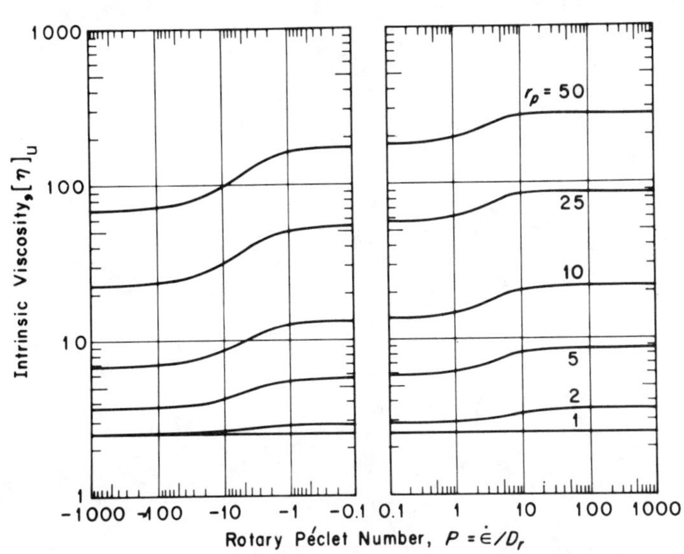

thin spheroids. He shows that the shear and normal stresses drop instantaneously to less than 10% of the steady state values. This stress relaxation effect would be difficult to measure.

With sinusoidal oscillations it is easier to measure short relaxation times. Cerf (1951) and Scheraga (1955) solved the time-dependent constitutive relation for the case of sinusoidal shear and $r_p \gg 1$. They used polymer solution notation: M = molecular weight, which for an ellipsoid is $4/3\pi\ ab^2\rho N_{AV}$, and c, mass concentration, rather than ϕ. They find for the reduced, intrinsic dynamic moduli

$$[G']_R = \frac{\omega^2\lambda^2}{1+\omega^2\lambda^2} \tag{10.3.21}$$

$$[G'']_R = \omega\lambda\left(\frac{2}{5} + \frac{1}{1+\omega^2\lambda^2}\right) \tag{10.3.22}$$

where

$$[G'']_R = \frac{5M}{3RT}\lim_{c\to 0}\frac{G'}{c}$$

and

$$[G'']_R = \frac{5M}{3RT}\lim_{c\to 0} G'' - \omega\eta_{\frac{s}{c}} \tag{10.3.23}$$

The relaxation time λ is inversely proportional to D_r, the rotary Brownian diffusion coefficient, eq. 10.3.4.

$$\lambda = \frac{\pi}{48 D_r} \tag{10.3.24}$$

or in terms of M

$$\lambda = \frac{25}{21}\frac{[\eta]\eta_s M}{RT} \tag{10.3.25}$$

Figure 10.3.8.
Different orientations of a prolate spheroid at the stagnation plane in strong (a) uniaxial extension and (b) uniaxial compression or equal biaxial extension. In biaxial flow the particle can take any orientation in the stagnation plane, resulting in a lower drag force.

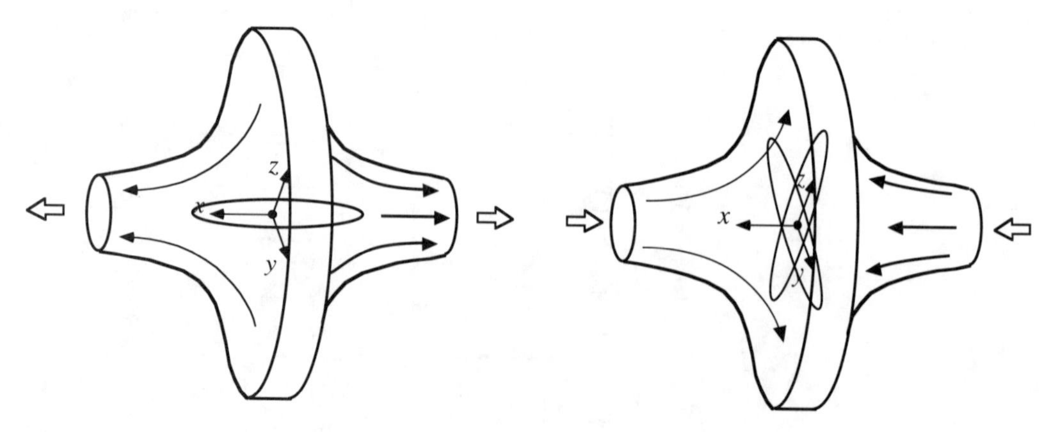

Figure 10.3.9.
Reduced dynamic moduli for the ellipsoid models, eqs. 10.3.21 and 10.3.22 compared to experimental results on tobacco mosaic virus. From Nemoto et al. (1975).

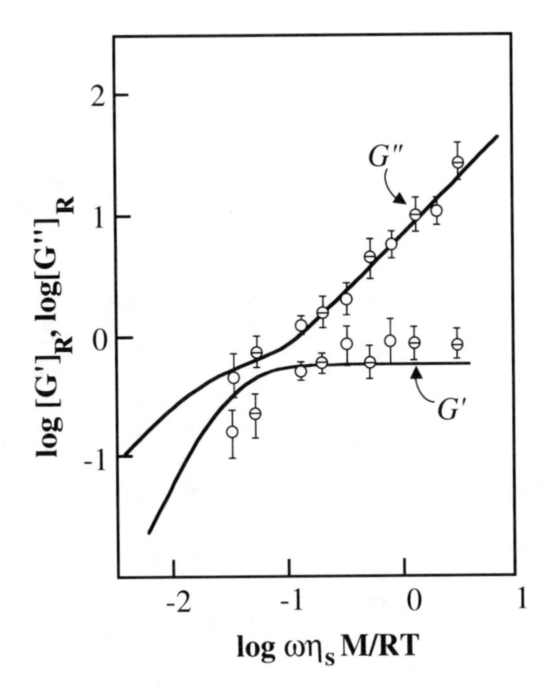

Note that the form for G' is the same as that for a Maxwell model, eq. 3.3.31, while that for G'' shows liquid rather than solid-like behavior at high frequency. Figure 10.3.9 shows good agreement between $[G'_R]$ and $[G''_R]$ and measurements on tobacco mosaic virus. DuPauw (1968) gives the virus dimensions as $2a = 300$ nm and $2b = 18$ nm (DuPauw, 1968).

10.4 Particle–Particle Interactions

When the volume fraction grows larger than 0.01, particles increasingly enter the neighborhood of other particles. The resulting disturbance of the flow increases the viscosity. At relatively small concentrations only binary interactions are likely to occur. With increasing concentration, more than two particles can interact simultaneously. This causes the viscosity to grow at an increasing rate with concentration. Different kinds of force (described below) are active in particle interactions. Depending on their relative magnitude, the microstructure and the rheology of the suspension can vary widely. A suspension with 1 vol % of dispersed phase can be a fluid with approximately the viscosity of water, or it can respond as a solid, depending on the relative value of the various interaction forces. In this section, key scaling groups, which express the balance between interparticle forces, are identified and used to classify the different types of suspension, which are discussed in Sections 10.5–10.7.

The relative motion of neighboring particles causes a hydrodynamic interaction force, which is present in all flowing nondilute suspensions. Most of the other interaction forces are relevant only for small particles. If the particle size is smaller than roughly 1 μm, Brownian motion of the particles becomes noticeable. As we saw in Section 10.3 for dilute systems, rotary Brownian motion of nonspherical particles results in elasticity and shear rate effects. For nondilute systems, translational diffusivity plays a similar role, even for spherical particles. Interparticle distance and its distribution (the so-called radial distribution function) must now substitute for what average orientation and orientation distribution meant for dilute suspensions of spheroids. The radial distribution function constitutes the simplest measure of microstructure in nondilute suspensions of spherical particles. Its dependence on flow causes shear thinning and viscoelasticity in such suspensions (Russel and Gast, 1986).

Additional interparticle forces exist in colloidal systems. They can be derived from a potential because they depend only on interparticle distance. Hence they act as springs and as such can cause pronounced elastic effects. However, the spring force depends on interparticle distance, and the springs even "rupture" when the particles move too far apart. This results in a highly nonlinear material response. During flow, the potential forces will affect the interparticle distances and consequently the frictional forces and the viscosity. We review each of these forces briefly in Sections 10.4.1–10.4.4. More thorough discussion can be found in Russel et al. (1989).

10.4.1 Dispersion Forces

Several potential interparticle forces can be distinguished. Dispersion (i.e., London–van der Waals) forces always exist because of interactions between induced dipoles in the molecules of neighboring particles (Mahanty and Ninham, 1976). Integration of the interactions between all induced dipoles in two bodies results in an expression for the total attraction force. It is characterized by the Hamaker constant A, which depends on the material composition of the particle. In principle, it can be calculated from the frequency-dependent polarizability of the material, and it determines the attraction forces in vacuum. For particles with Hamaker constant A_p in a medium with Hamaker constant A_i, an effective A_m can be calculated from

$$A_m = (A_p^2 - A_i^2)^{1/2} \qquad (10.4.1)$$

The potential V_D, from which the dispersion forces between two particles can be derived, is of the general form

$$V_D = -A_m H_g \qquad (10.4.2)$$

where H_g is a function of particle geometry and of interparticle distance H. For identical spheres, the potential is proportional to $1/H$ at close contact and to $1/H^6$ at large distances (Figure 10.4.1).

Figure 10.4.1.
Interaction potential for dispersion forces (dipole attraction) and electrostatic repulsion (V_D: attraction from dispersion forces; V_E: electrostatic repulsion; V_t: total interaction potential).

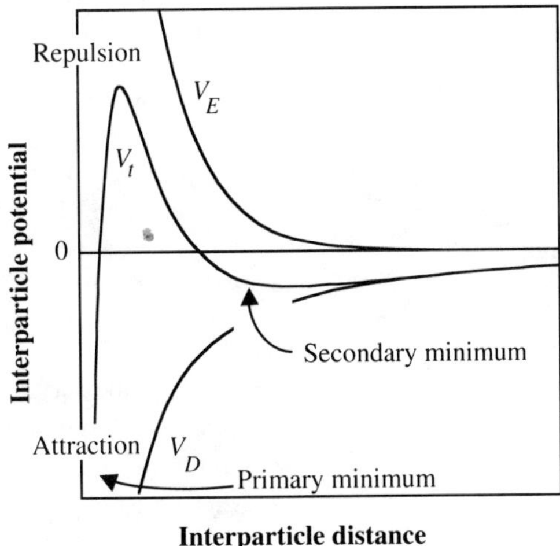

10.4.2 Electrostatic Forces

In aqueous media, particles can be electrically charged (e.g., by transfer of ions between the particles and the water phase). As a result, the ions in the water that have sign opposed to that of the particle charge (the counter ions) will be drawn toward the particle. Immediately near the surface, a monolayer composed of such ions develops, forming a narrow double layer with the particle surface (the Stern layer). Outside this layer the concentration of counterions gradually decreases toward the bulk concentration in the water; these counterions make up the diffuse double layer, as shown schematically in Figure 10.4.2. Equally charged particles repel

Figure 10.4.2.
Distribution of ions around a charged particle (dashed line separates Stern layer from the diffuse double layer).

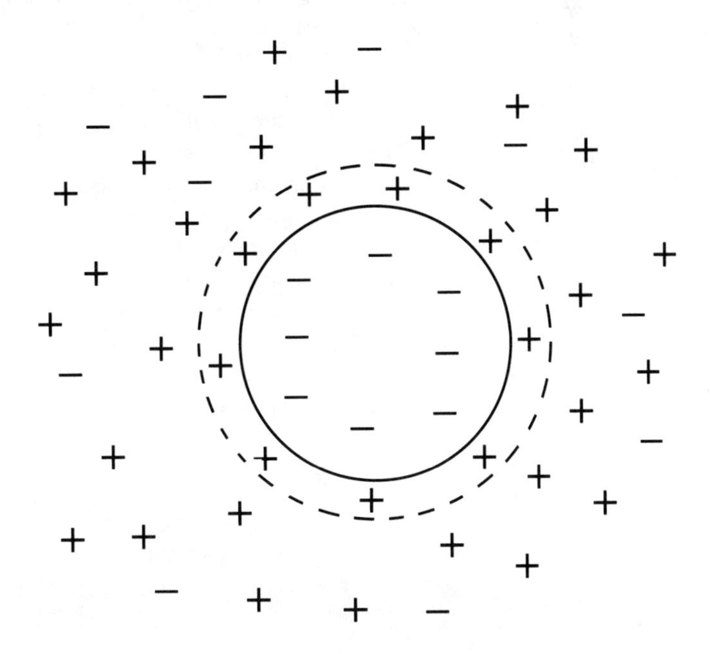

each other electrostatically. The net repulsion force will depend on the surface charge and the screening effect of the double layer. As for the dispersion force, the electrostatic repulsion force can be derived from a potential (V_E). For small surface potentials (ψ_o) and thick double layers, the potential energy for identical spheres is approximated by (e.g., Hunter, 1989)

$$V_E = 4\pi\epsilon \frac{a}{2 + H/a} \psi_o^2 \exp(-\kappa H) \tag{10.4.3}$$

where ψ_o is the surface potential. The Debye–Huckel constant κ is given by

$$\kappa = \left(\frac{e^2 \Sigma n_{oi} z_i^2}{\epsilon k T} \right)^{1/2} \tag{10.4.4}$$

where e is the charge of an electron, n_{oi} the concentration of ions of type i far from the particle, z_i the valence of ions of type i, and ϵ the dielectric constant of the medium.

The factor κ has the dimension of 1/length. Its inverse measures the thickness of the double layer, which is determined by the concentration of ions in the water phase (eq. 10.4.4). An electrostatic repulsion becomes noticeable when particles approach close enough for the double layers to overlap ($\kappa H \simeq 2$).

The total interaction potential for electrostatically stabilized systems is the sum of the attraction and repulsion potentials V_D and V_E. A possible net result, V_t, is shown in Figure 10.4.1. The attraction forces dominate at both small and large distances. Close to the surface is a deep minimum (primary minimum) caused by the very steep rise of the attraction forces near the surface and by the existence of a supplementary repulsion of a different nature very close to the surface. The deep minimum means that if another particle can approach to such a distance, it will be in a very stable position, resulting in a permanent aggregate (sometimes called agglomerate). If the potential barrier outside this potential well is sufficiently large, particles are kept from falling in the primary minimum. Outside the barrier there can be another shallow minimum (secondary minimum). Particles are then only weakly kept in position, and aggregation there is reversible. The energy in this secondary minimum is less than a few kT; hence Brownian motion is strong enough to eventually deflocculate the particles again.

10.4.3 Polymeric (Steric) Forces

When a polymer layer is present at the surface of the particles (either adsorbed or chemically grafted), a repulsion force can be created when the layers on two neighboring particles overlap. This happens whenever the polymer molecules would rather become more compact than mix as the two layers are squeezed together

(Figure 10.4.3). The simplified theories predict a potential V_S (e.g., Napper, 1983)

$$V_S = \frac{akT(1/2 - \chi)\delta^2}{H_s v_s} \tag{10.4.5}$$

where χ is the interaction parameter according to the Flory–Huggins theory (see Flory, 1953), H_s is a function of interparticle distance, which also contains polymer and system characteristics, δ the thickness of the stabilized layer, and v_s the volume of medium or solvent molecule.

The interaction parameter should be smaller than 1/2 to provide a repulsion. This requires the medium to be a good solvent for the free dangling polymer. Polymeric repulsion occurs only when polymeric stabilizer layers overlap. The thickness of these layers is often of the order of 10 nm. In contrast, electrostatic double layers can be much thicker if the ion concentration of the medium is low (eq. 10.4.4). Also, the polymer repulsion potential is quite steep. As a result, the total potential for polymerically stabilized systems (Figure 10.4.3) shows no deep primary minimum. A shallow minimum, similar to the secondary minimum for electrostatically stabilized systems, is possible.

If polymer molecules are dissolved and moving freely in the medium, they can favor flocculation rather than stabilization. Whenever the interparticle distance becomes smaller than the dimensions of the polymer molecules, the gap is depleted of these molecules for thermodynamic reasons. As a result, the particles are forced even closer together and become flocculated (depletion

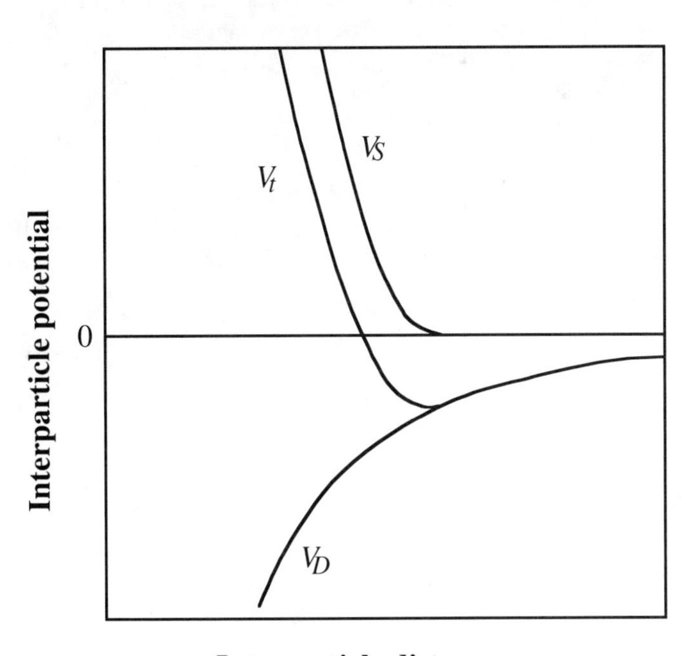

Figure 10.4.3.
Interaction potential for polymerically stabilized suspension (V_D: attraction from dispersion forces; V_S: polymeric repulsion; V_t: total interaction potential).

flocculation). Polymers can also flocculate particles by other mechanisms, which are discussed further by Napper (1983).

The forces described above are responsible for most colloidal effects. Other forces could be present and are briefly mentioned here. The most relevant ones result from external fields (e.g., gravity, electrical, or magnetic fields). While acting on the particles, they can cause dramatic changes in structure and behavior, including particle migration and special types of aggregation. They offer the possibility of increasing the viscosity by applying an external field (e.g., electrorheological or magnetorheological effects described for example in Block and Kelly, 1988).

10.4.4 Scaling

The magnitudes of the colloidal forces range widely. Depending on which one dominates, the resulting structure and the rheology can change drastically. The relative importance of the various contributions can be expressed by means of the dimensionless groups used in scaling arguments (Krieger, 1972; Russel, 1980).

The hydrodynamic forces are always proportional to the viscosity of the medium. Therefore, suspension viscosities are scaled with the viscosity of the suspending medium, meaning that relative viscosities are used. As for dilute systems, the balance between Brownian motion and flow can be expressed by a Peclet number. Here the translational diffusivity D_t has to be used, but that does not change the functionality (for spheres, D_r is proportional to D_t). A dimensionless number is obtained by taking the ratio of the time scales for diffusion (D_t) and convective motion ($\dot{\gamma}$). This is again a Peclet number:

$$\mathrm{Pe} = \frac{\eta_s \dot{\gamma} a^3}{kT} \text{ for shear flow} \tag{10.4.6}$$

As with dilute suspensions, inertia effects can be estimated from a particle Reynolds number, eq. 10.2.12.

The other colloidal forces can be compared by taking ratios of characteristic energies, that is, A for the dispersion force, $\epsilon \psi_o^2 a$ for electrostatic repulsion, and $a\delta^2(1/2 - \chi)kT/v_s$ for polymer stabilization. A comparison with Brownian motion (Russel, 1980) leads to the following dimensionless groups:

For dispersion forces:

$$\frac{A}{kT} \tag{10.4.7}$$

For electrostatic repulsion forces:

$$\frac{\epsilon \psi_o^2 a}{kT} \tag{10.4.8}$$

For the polymeric repulsion (note kT cancels out):

$$\frac{(1/2-\chi)\,a\delta^2}{v_s} \tag{10.4.9}$$

Sections 10.5–10.7 discuss the rheology of the major classes of suspensions. The classification is based on the relative magnitude of the various interaction forces. First we consider the case in which only Brownian motion interferes with the always-present hydrodynamic forces during flow. Such "Brownian hard spheres" are not easy to produce. The repulsion forces can easily be made small, but then the attractive dispersion forces usually start to dominate. Krieger and co-workers (Krieger, 1972) have systematically reduced the electrostatic repulsion and found a region in which Brownian motion dominates the residual attraction forces. De Kruif et al. (1985) have produced similar results by nearly matching the Hamaker constant of particles and medium, thus requiring only a minor steric stabilization.

In Section 10.6 we consider strongly *stable colloidal suspensions*, which occur when electrostatic or polymeric repulsion forces become larger than dispersion or Brownian forces. Finally, in Section 10.7, we treat the case of dominating attraction forces that result in aggregating particles and a *flocculated suspension.*

10.5 Brownian Hard Particles

10.5.1 Monodisperse Hard Spheres

In this section we discuss systems for which potential interaction forces can be ignored, leaving only hydrodynamic and Brownian forces. Large particles are a limiting case of this group, namely when hydrodynamic forces totally dominate other forces, including the thermal motion. For Brownian hard spheres the potential force between particles is zero whenever the particles are not touching. Once contact has been made it suddenly becomes an infinite repulsion because the particles are assumed to be rigid. Among the dimensionless groups of Section 10.4, only the relative viscosity, the Peclet number, and eventually the particle Reynolds number are relevant in this case. If we neglect particle inertia, the Peclet number is the only relevant flow parameter. General viscosity curves of nondilute suspensions were shown in Figure 10.1.3. As indicated in Figure 10.2.2, two Newtonian regions can be detected, separated by a shear thinning region. At higher concentrations shear thickening might appear at a certain level of shear. In this latter condition, more complex structural changes occur, and other phenomena should be taken into account (Barnes, 1989; Laun et al., 1991).

In the two Newtonian regions, no length or time scales are left in the scaling. Hence, the corresponding relative viscosities should be universal for monodisperse hard spheres: they should not depend on shear rate, medium viscosity, temperature, or particle size. In between these regions, the Peclet number provides a suitable scaling factor for the shear rate. At a given volume fraction, the η_r versus Pe curve should be universal for suspensions of monodisperse hard spheres, assuming that effects like particle inertia can be ignored. On the low shear Newtonian plateau, Brownian motion totally dominates the structure. This means that the relative positions of the particles during flow are identical to those at rest. From model experiments (Krieger, 1972; de Kruif et al., 1985) that approximate Brownian hard spheres, the curves of the plateau viscosities versus volume fraction are known (Figure 10.2.2). They can be used for all suspensions of Brownian hard spheres. The Krieger–Dougherty relation (Krieger, 1972) offers an adequate expression for the concentration dependence of the viscosity:

$$\eta_{rx} = \left(1 - \frac{\phi}{\phi_{\max}}\right)^{-2.5\phi_{\max}} \tag{10.5.1}$$

where $x = 0$ (low shear plateau) or ∞ (high shear plateau), $\phi_{mo} = 0.63$ (maximum packing at low shear rates), and $\phi_{m\infty} = 0.71$ (maximum packing at high shear rates).

Ab initio calculations for nondilute systems become very complicated. Einstein derived the linear term in the concentration law for the viscosity in 1906 and 1911. The quadratic term, which is the first interaction term, was published in 1977 by Batchelor:

$$\eta_r = 1 + 2.5\phi + 6.2\phi^2 + 0(\phi^3) \tag{10.5.2}$$

Equation 10.5.2 fits available data (see Figure 10.2.3 and de Kruif et al., 1985) within measurement accuracy. Higher order expansions do not seem to be useful because they are applicable over increasingly small concentration regions. Various approaches are being used to compute viscosities at higher concentrations. The hydrodynamics for multiple particle interactions become very involved. They have been studied mainly by simulation (e.g., Brady and Bossis, 1988; Phillips et al., 1988). Other workers have used an approach based on nonequilibrium thermodynamics (Russel and Gast, 1986). Finally, Woodcock (e.g., 1984) uses molecular dynamics simulations, ignoring the medium viscosity, to calculate the flow-induced structure and then the viscosity.

The Peclet number of eq. 10.4.6 is based on the diffusivity of an isolated sphere (i.e., Stokes' law for the hydrodynamic effect). This is obviously incorrect in a concentrated suspension, where the presence of other particles can have an enormous effect on the mobility. The viscous resistance a particle encounters will be the sum of all hydrodynamic interactions with all neighboring particles. This resistance is much higher than that given by Stokes' law and should be comparable with the global viscosity of the suspension.

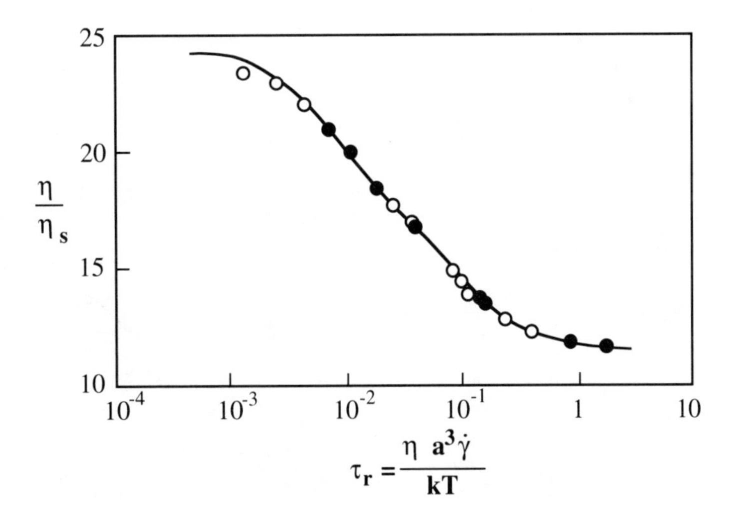

Figure 10.5.1.
Reduced viscosity versus reduced shear stress for suspensions of polystyrene spheres, $\phi = 0.50$, solid line, in water; open circles, in benzyl alcohol; solid circles, *m*-cresol. Replotted from Krieger (1972).

On this basis Krieger (1972) has suggested substituting suspension viscosity for the medium viscosity in eq. 10.4.6. The resulting correction of Pe is really a reduced shear stress τ_r, as can be seen in eq. 10.5.3:

$$\tau_r = \frac{\eta \dot{\gamma} a^3}{kT} = \frac{\tau a^3}{kT} \tag{10.5.3}$$

As shown in Figure 10.5.1, the viscosity versus reduced shear stress curves can be described rather well by the semi-empirical relation (Krieger, 1972)

$$\frac{\eta - \eta_\infty}{\eta_o - \eta_\infty} = \frac{1}{1 + \tau_r/\tau_c} \tag{10.5.4}$$

The location of the shear thinning region on the reduced stress axis can be characterized by the value of a critical reduced stress τ_c (i.e., the reduced stress for which the viscosity reaches the average value between the two Newtonian values). This critical stress increases with concentration up to volume fractions of 0.50 and then decreases to zero at maximum packing (Figure 10.5.2).

Woodcock (1984) has argued that scaling is of limited use because in all real flows of concentrated suspensions, particle inertia or kinetic energy is important (see also Woodcock and Edwards, 1984; Barnes et al., 1987). Using molecular dynamics simulations, Woodcock has generated results for the changes in structure during flow. The complex effect of shear on structure has been demonstrated experimentally (e.g., Ackerson, 1986). Depending on concentration and shear rate, a colloidal suspension can have the structure of a gas, a fluid, a glass, a mesomorphic system, or a crystal. The structure of flowing suspensions is currently studied intensively with scattering techniques using light, X-rays, and neutrons (Markovic et al., 1986; Johnson et al., 1988; Ackerson et al.,

Figure 10.5.2.
Critical shear stress for shear thinning, Brownian hard spheres. Data from de Kruif et al. (1985).

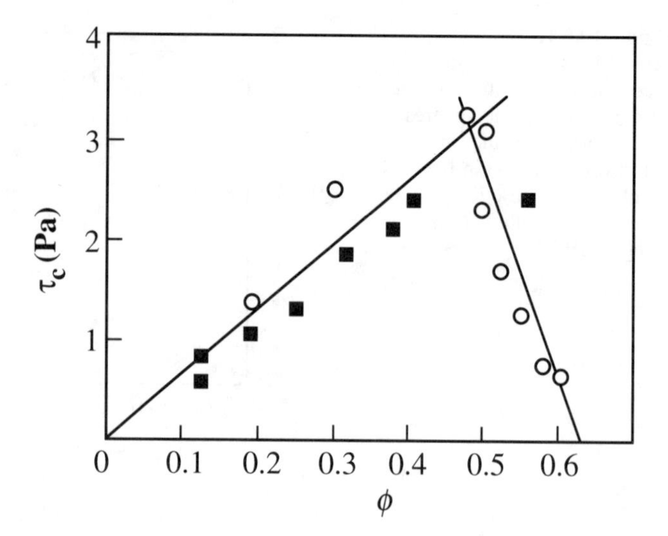

1990) and via nuclear magnetic resonance imaging (Graham et al., 1991).

The reduced shear stress is proportional to a^3 (eq. 10.5.3). When the particle radius becomes larger than 1 μm, the critical shear stress and consequently the shear thinning region shift rapidly to extremely low values and essentially become experimentally inaccessible. The lack of Brownian motion for these large particles also eliminates the reversibility in structural changes, which complicates experiments. At the same time, particle inertia becomes more important. The combination of these effects gives rise to a number of complex phenomena: for example, migration toward areas of low shear stress and wall slip, especially at higher concentrations (Leighton and Acrivos, 1987). Measurements also become less consistent (Woodcock, 1984; Cheng, 1984): reproducibility is poorer and the results depend on measurement geometry. It should also be kept in mind that the hydrodynamic interactions, and consequently the viscosity, depend on the type of flow. The Brownian motion provides a driving force to bring the structure back to its equilibrium condition. Hence it serves as a "memory" or source for elastic phenomena. The effect is usually too small to generate noticeable normal stress differences, but it gives rise to storage moduli in oscillatory flow (van der Werff et al., 1989).

10.5.2 Particle Size Distribution

In the dilute concentration region, where particles hardly ever feel each other, the particle size distribution is not important. For moderate concentrations experimental evidence suggests a minor effect of this parameter. Close to the maximum packing, however, the effect becomes very large (Figure 10.5.3). This can be understood on the basis of the drastic increase in maximum packing when bimodal or multimodal particle size distributions are used. The effect can be predicted if the maximum packing, calculated or measured,

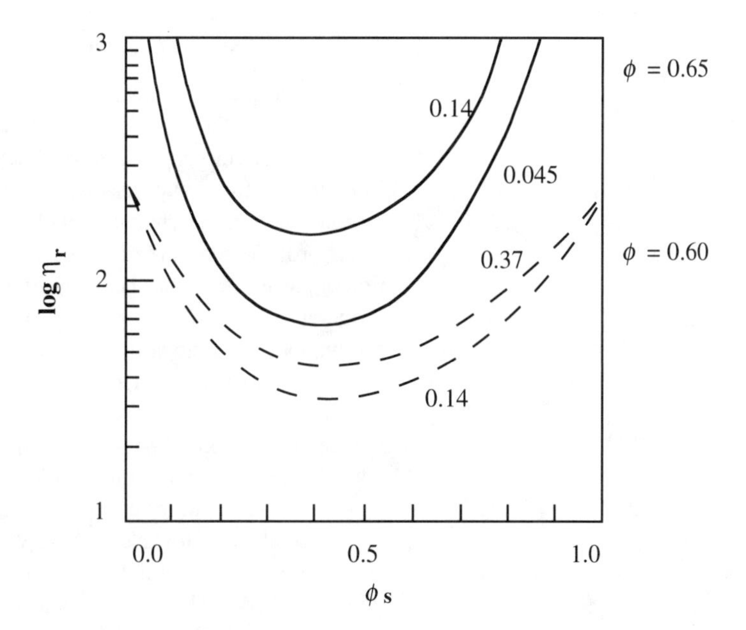

Figure 10.5.3.
Relative viscosity for bimodal distributions of hard spheres as function of the proportion of small particles (ϕ_s). The parameter is the ratio of particle radii. Note that increasing total volume fraction ϕ from 0.6 (dashed line) to 0.65 has a large effect. After Chong et al. (1971).

is introduced in the Krieger–Dougherty equation (eq. 10.5.1) for monodisperse systems. Commercially bimodal or trimodal distributions are used to minimize the viscosity in highly concentrated suspensions.

10.5.3 Nonspherical Particles

In principle, some types of nonspherical particles could be packed more tightly than spheres, although they would start to interact at lower concentrations. In reality, higher viscosities are normally found with nonspherical particles. The concentration law is approximately exponential at low to moderate concentrations, but equations similar to eq. 10.5.1 can still be used as well. The empirical value of ϕ_m can be much smaller than that for spherical particles (e.g., 0.44 for rough crystals with aspect ratios close to unity: Kitano et al., 1981). If fibers are used, this value drops even further, down to 0.18 for an aspect ratio of 27 (see also Metzner, 1985). The decrease with aspect ratio seems to be roughly linear. Homogeneous suspensions of fibers with large aspect ratios are difficult to prepare and handle. As in dilute systems, the type of flow will determine the extent of the shape effect. Extensional flows are discussed below.

Doi and Edwards (1986) have used a tube model to describe flow of semidilute suspensions of rods. Predicted behavior is qualitatively similar to their theory for entangled, flexible polymer chains (see Chapter 11). This approach has also been extended to describe the rheology of nematic liquid crystalline polymers (Doi and Edwards, 1986; Larson, 1988).

10.5.4 Non-Newtonian Media

In industrial applications, particles are often suspended in non-Newtonian media, especially in polymer fluids. There are basic differences between suspensions in Newtonian and viscoelastic media. The latter seem to suppress particle rotation, and the radial distribution function can change as well because of the altered flow and stress profiles near and in between particles. In addition, segregation of particle sizes has been reported (Michele et al., 1977). Up to moderate concentrations, the power law index of the shear thinning region does not change with concentration (Nicodemo et al., 1974; Mewis and de Bleyser, 1975). A slight decrease in power law can sometimes be detected, for which there is also some theoretical evidence (Jarszebski, 1981). Hence, to a first approximation, adding particles to a shear thinning fluid does not alter the general shape of the viscosity curve, except for the additional colloidal effects discussed above. However, a purely vertical shifting of the curves, assuming a constant relative viscosity, independent of shear rate, does not work. The shifting is always less in the shear thinning region than in the Newtonian region. In addition, the onset of shear thinning shifts to smaller shear rates with increasing concentration.

Both phenomena can be understood if the local shear rate in the fluid around the particles is considered. Local rates increase more with concentration than the average shear rate does and thus reduce the apparent medium viscosity. The resulting shift in the curves can be partially compensated if the relative viscosities are calculated from medium and suspension viscosities at equal values of the shear stress, rather than at equal shear rates.

The elastic properties of a viscoelastic medium also change when particles are added. These changes are qualitatively similar to those of the viscosity (Mewis and de Bleyser, 1975). In the non-Newtonian region, elasticity increases less than viscosity. Hence filled polymers are always less elastic than the suspending polymers under processing conditions. Reviews available on suspensions in non-Newtonian media include Metzner (1985) and Kamal and Mutel (1985).

10.5.5 Extensional Flow of Ellipsoids

Relative particle motion, and consequently particle interaction and viscosity, depend in principle on the type of flow. For nearly spherical particles the effect seems to be limited, except perhaps near the maximum packing. Long slender particles, especially long fibers, provide a totally different picture. In extensional flow they orient more or less in the flow direction, depending on Brownian motion, as discussed in Section 10.3 for dilute suspensions. The presence of a fiber will affect the flow of the part of the fluid that is near the particles, actually within a sphere having the fiber length as its diameter. With large aspect ratios, a small volume fraction of fibers can affect a large volume of medium. When the affected volumes of neighboring particles overlap, the increase in viscosity becomes

Figure 10.5.4.
Stress–strain rate curves for extensional flow of glass fibers suspended in polybutene ($\phi \simeq 0.1$ to 1%, $r_p = 282, 586$, and 1259). Adapted from Mewis and Metzner (1974).

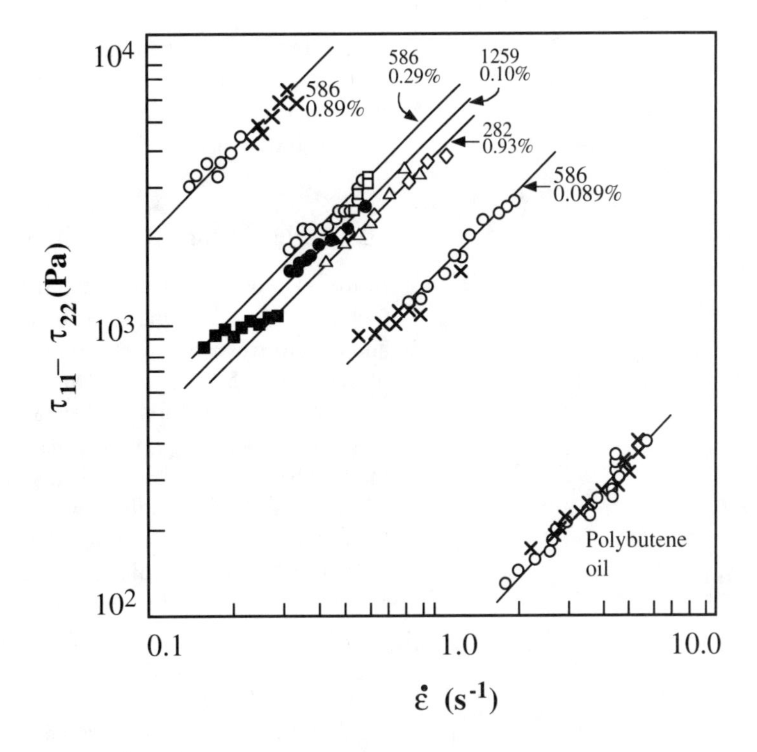

very large. This constitutes one of the very few cases in which an adequate ab initio theory is available for strong interaction effects in suspensions. Batchelor (1971) calculated for the uniaxial extensional viscosity η_u of a suspension with interacting fibers

$$\eta_u = \eta_s \left[\frac{3 + 4\phi r_p^2}{3 \ln(\pi/\phi)} \right] \quad \text{assuming } L \gg H \gg D \qquad (10.5.5)$$

where $r_p = L/D$, the aspect ratio, and H is the distance between the parallel fibers.

Equation 10.5.5 predicts large contributions from the particles to the stresses within the range of its validity. This has been verified experimentally (Mewis and Metzner, 1974), as is shown in Figure 10.5.4. Polymeric media themselves can give rise to high stresses in extensional flow. Possibly the introduction of particles then tends to transform local stretching in shearing motion. In the latter mode the medium is shear thinning. As a result, the suspension viscosities could be smaller than expected from eq. 10.5.5 (Chan et al., 1978; Goddard, 1978).

10.6 Stable Colloidal Suspensions

In Section 10.5 we considered systems in which colloidal forces, with the exception of Brownian motion, did not play a role. Dispersion forces are always present in real materials; a repulsion force is required to keep small particles from aggregating together. This section discusses systems in which the repulsion forces dominate the attraction forces, creating stable colloidal dispersions. Electrostatic and polymeric stabilization are treated separately.

10.6.1 Electrostatic Stabilization

In the case of electrostatic stabilization, the total interaction potential is the sum of the contributions from the dispersion forces and the electrostatic forces. To ascertain stability, the electrostatic forces should be larger than the dispersion forces at long range. As discussed in Section 10.4, aggregation nonetheless can occur in such a material if a particle has sufficient energy to overcome the repulsive energy barrier. This is possible at high shear rates (flow-induced flocculation), whereas at intermediate shear rates the hydrodynamic forces might be capable of deflocculating suspensions that are flocculated in the secondary minimum (van de Ven and Mason, 1976; Zeichner and Schowalter, 1977). In this section the suspension is always assumed to be stable.

The presence of electric charges affects the rheology in different ways. In dilute systems flow will distort the charge cloud around the particles and thus produce additional stresses. This so-called primary electroviscous effect has been modeled over a wide range of conditions by Sherwood (1980). In nondilute systems, charged particles can interact. When electrostatically stabilized particles approach each other, the repulsion force increases gradually, contrary to the case of Brownian hard spheres, in which the repulsion suddenly jumps from zero to infinity upon contact ("soft" versus "hard" repulsion). During flow, the repulsion forces keep particles farther apart than in the neutrally stable systems discussed in Section 10.5. As a result, the energy dissipation—and consequently the viscosity—become larger. This phenomenon is called the second electroviscous effect. The case of relatively dilute systems has been studied well. The viscosity–concentration relation then contains the linear term from the dilute systems and a quadratic term for binary particle interactions. In eq. 10.5.2 the numerical coefficient for the quadratic term was given as 6.2 for Brownian hard spheres. In the case of electrostatic stabilization, the corresponding coefficient can be expressed as a function of the length H_o, which measures the distance of closest contact between two particles for the case of electrostatic repulsion in the limit of low shear rates (Russel, 1978)

$$H_o = \left(\frac{1}{\kappa}\right) \ln\left[\frac{\alpha}{\ln(\alpha/\ln\alpha)}\right] \tag{10.6.1}$$

where $\alpha = 4\pi\epsilon\psi_o a^2 \kappa \exp(2a\kappa)/kT$

Figure 10.6.1 shows that the effect of electrostatic stabilization on viscosity can be very pronounced for low ionic strengths.

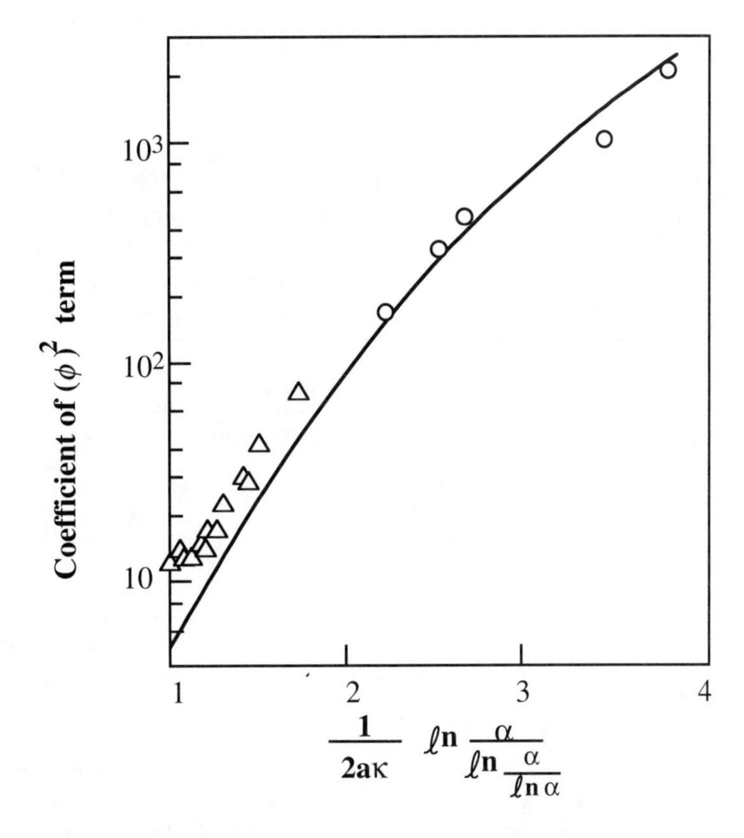

Figure 10.6.1.
Effect of electrostatic repulsion on the coefficient of the ϕ^2 term in the viscosity–concentration law. After Russel (1980).

This result also provides a qualitative indication about the effect of electrostatic stabilization and its parameters in more concentrated systems. The importance of the Debye–Huckel constant κ is obvious. At higher concentrations and higher shear rates, the excluded volume around a particle is no longer given exactly by eq. 10.6.1. However, the equation is still suitable as a scaling factor. Electrostatic forces are not a function of shear rate; therefore their effect is larger at low shear rates than at high ones. This leads to shear thinning and eventually to the appearance of yield stresses (Krieger and Eguluz, 1976).

If the repulsion is strong enough, the particles position themselves as far apart as possible. This can lead to a lattice structure and the formation of colloidal crystals. These solidlike materials are characterized by a frequency-independent storage modulus (e.g., Russel and Benzing, 1981), which again depends strongly on the repulsion forces. In more dilute systems, or when the repulsion forces are smaller, the behavior under oscillatory flow can be that of a viscoelastic fluid, displaying a nearly Maxwellian behavior. Hence, a characteristic relaxation frequency can be determined, as well as a limiting high frequency modulus, both of which depend again strongly on the Debye–Huckel constant (Goodwin et al., 1984). The relation between stability parameters and rheology is quite well understood now for electrostatically stabilized dispersions (Goodwin et al., 1982). It is even possible to calculate the interparticle potential from the modulus-concentration curves (Buscall et al., 1982).

10.6.2 Polymeric (Steric) Stabilization

The polymeric repulsion force is relatively "hard"; that is, it increases quite rapidly with decreasing interparticle distance. It is also a short-range force, as it drops to zero where the polymer layers do not overlap. If these layers do not deform much, the rheological behavior resembles that of Brownian hard spheres. The results for hard spheres can be used as a first approximation, substituting for the volume fraction an effective volume fraction that includes the stabilizer layer δ:

$$\phi_{\text{eff}} = \phi\left(1 + \frac{\delta}{a}\right)^3 \tag{10.6.2}$$

This corresponds roughly to the case of electrostatic stabilization if H_o is replaced by the thickness of the stabilizer layer. For small particles this layer can be a substantial part of the effective volume. In that case the deformability of the stabilizer layer can become significant, especially at high shear rates and at high volume fractions. The viscosity curves will still resemble those for hard spheres, but the viscosity will drop systematically below that for the hard spheres (Figure 10.6.2). This is especially clear near the maximum packing (Mewis et al., 1989). Viscosity-concentration curves, like those of Figure 10.6.2. can be superimposed if the

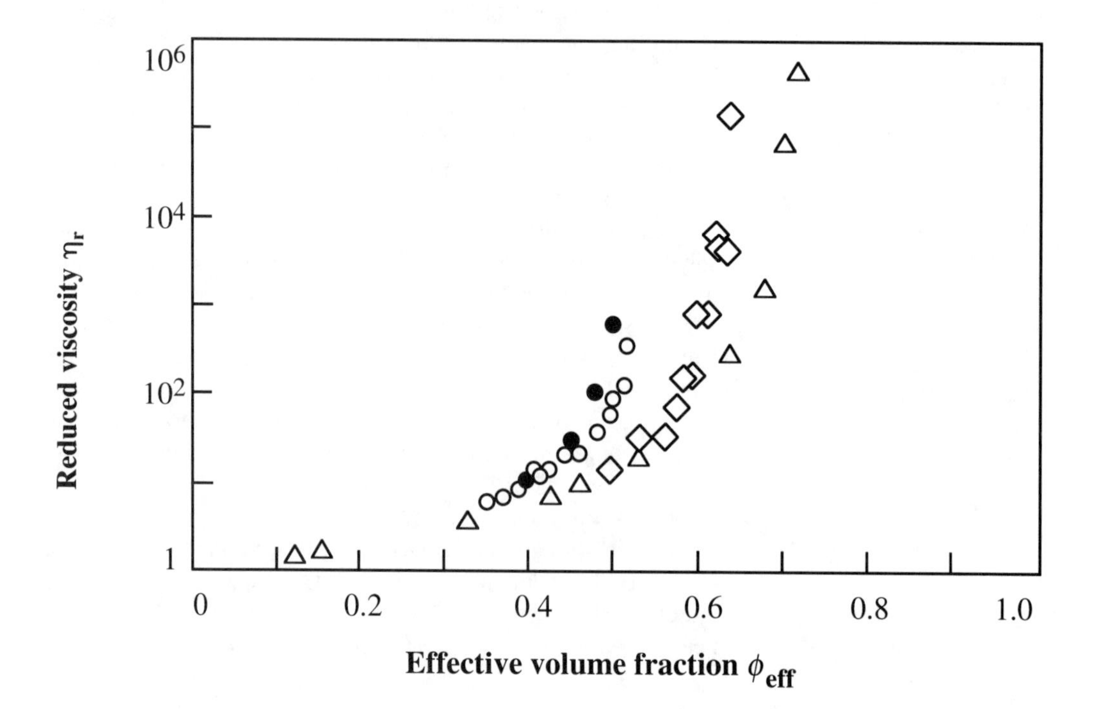

Figure 10.6.2. Effect of particle size (i.e., softness) on the low shear rate viscosity of suspensions containing sterically stabilized particles. Polymethyl methacrylate latex in decalin, stabilizer layer thickness, 9 nm; particle diameters: •, 475 nm; ○, 376 nm; ◇, 129 nm; and △, 84 nm. From D'Haene (1991).

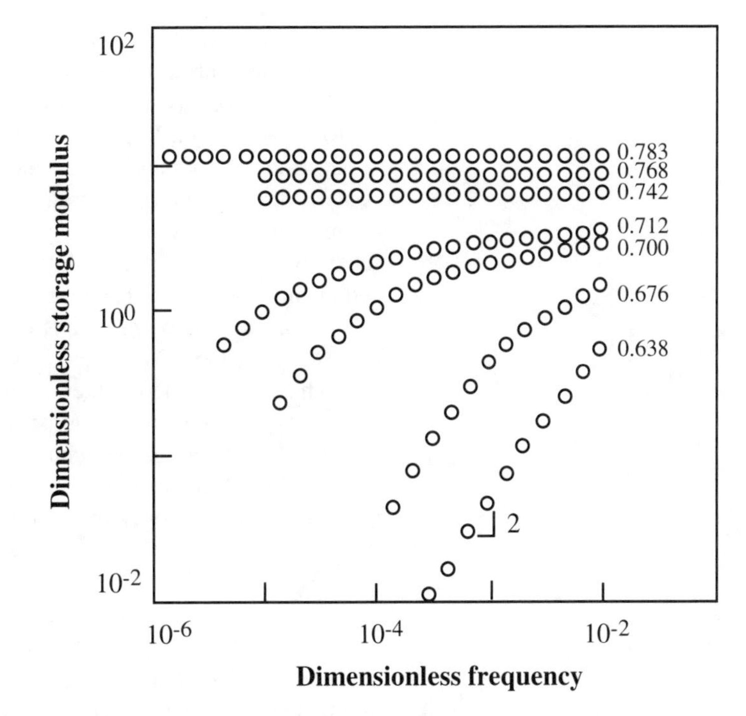

Figure 10.6.3. Storage modulus versus frequency for the same 84 nm polymethyl methacrylate dispersions as shown in Figure 10.6.2 at several ϕ_{eff}. After Frith et al. (1990).

ratio $\phi_{\text{eff}}/\phi_{\text{max}}$ is used instead of ϕ_{eff}. The scaling for hard spheres can still be used to describe the effect of temperature and medium viscosity (Willey and Macosko, 1978).

As was the case for electrostatic repulsion, polymer repulsion shows up directly in elasticity measurements (Figure 10.6.3). At intermediate concentrations Maxwellian-like behavior is encountered again: the slope in Figure 10.6.3 changes from 2 below the relaxation frequency to zero above this frequency. The average relaxation time becomes extremely sensitive to the volume fraction when the maximum packing is approached (Frith et al., 1990). With relaxation frequencies below the measuring range, a solidlike response is recorded. Only a plateau modulus can be measured. The latter is much smaller than the values measured for hard spheres at the same effective volume fraction (Frith et al., 1990). The plateau modulus–concentration curve reflects the increasing repulsion when the particles come closer together. Such a curve can be used to calculate the interparticle potential, as was the case for electrostatically stabilized dispersion (Mewis and D'Haene, 1993).

10.7 Flocculated Systems

10.7.1 Structure in Flocculated Dispersions

Once the attraction forces have become larger than the repulsion, and also larger than Brownian motion, particles can remain together when they collide, lying in the primary or secondary minimum discussed in Section 10.4. The resulting aggregates or flocs have a very complex structure that has long evaded theoretical treatment and experimental analysis. Theories have been developed on an intuitive basis, especially by Hunter and co-workers (Firth and Hunter, 1976). They used a Bingham model to correlate rheological parameters to interparticle forces and floc structure. For aqueous systems of moderate concentration, they could reproduce empirical relations like the dependence of Bingham yield stress on the square of the surface potential, the square of the volume fraction, and the inverse of particle radius.

Considerable progress has recently been made in this area (e.g., Meakin, 1983; Weitz and Oliveria, 1984). Most of the flocs do not have homogeneous internal structures. The center is usually more dense than the outer regions; hence the mass does not change with the third power of the radius r as in normal objects with constant density. Still they are often self-similar in the sense that their mass m, or the number N of particles in a floc, grows as

$$m \sim N \sim r^D \tag{10.7.1}$$

where D is smaller than 3, the Euclidean dimension. Structures that obey eq. 10.7.1 are called fractal objects. Substructures of different sizes taken from a given fractal object look similar if they are observed under a magnification that has been adjusted to give them the same size.

Fractal aggregates have been extensively investigated by computer simulation. Various assumptions can be made, each of which leads to a specific fractal dimension D. Witten and Sander (1983) simulated the convective diffusion of single particles toward a central floc, assuming aggregation at each collision. This situation gave a fractal dimension of 2.5. Taking into account the fact that several flocs exist which can themselves collide and aggregate (Meakin, 1983), a more open structure is obtained ($D = 1.8$). If the particles or clusters do not always stick on first contact, a sticking probability must be introduced. This leads to more compact flocs as particles or flocs can penetrate further into other flocs. For cluster–cluster aggregation, this results in a value of 2.1 for D. Reversible flocculation, in which particles or clusters can detach and connect again, also should give rise to more dense flocs.

The fractal dimension of flocs can be deduced experimentally from measurements such as electron micrographs or from scattering measurements (e.g., Weitz and Oliveria, 1984; Pusey and Rarity, 1987). Measured values lie within the range of the theoretical predictions. When the flocs are formed during flow, they are expected to generate more dense and more complex structures. Flow may cause breakdown of the outer regions of the flocs and the formation of more dense units. Experimentally, higher D-values are indeed measured for flow-induced flocculation (Sonntag and Russel, 1986). Eventually the floc might also lose its fractal nature.

Ultimately growing flocs can touch, thereby forming a space-filling network of particles. This phenomenon is studied by means of percolation theory (de Gennes, 1976; Feng and Sen, 1984). Depending on the assumed structure and on the physical property under consideration, one finds a lower concentration limit for network formation, the percolation threshold ϕ_c, and a concentration law for properties such as conductivity, elastic modulus, and yield stress. As long as the network is not broken down, the system will react as a solid. Often rupture is initiated at extremely small strains (e.g., 10^{-4}). Beyond this strain the material gradually weakens to become ultimately fluidlike.

10.7.2 Static Properties

Space-filling networks of particles normally display a frequency-independent modulus. Near the percolation threshold, theory predicts a relation (de Gennes, 1976; Feng and Sen, 1984)

$$G \propto (\phi - \phi_c)^n \tag{10.7.2}$$

with n between 2 and 4.5.

This relation is often valid only on a very narrow concentration region. Over wider ranges either an exponential or a power law relation between G and ϕ is found theoretically (de Gennes, 1976; Patel and Russel, 1988) and experimentally (Sonntag and Russel, 1986; Buscall et al., 1987; Navarrete, 1991). For the power law index, experimental values of 2.4 to 4.4 have been reported. Figure 10.7.1 shows an example.

Figure 10.7.1. Storage modulus versus volume fraction relation for a flocculated silica–methyl laurate system. From Van der Aerschot and Mewis (1992).

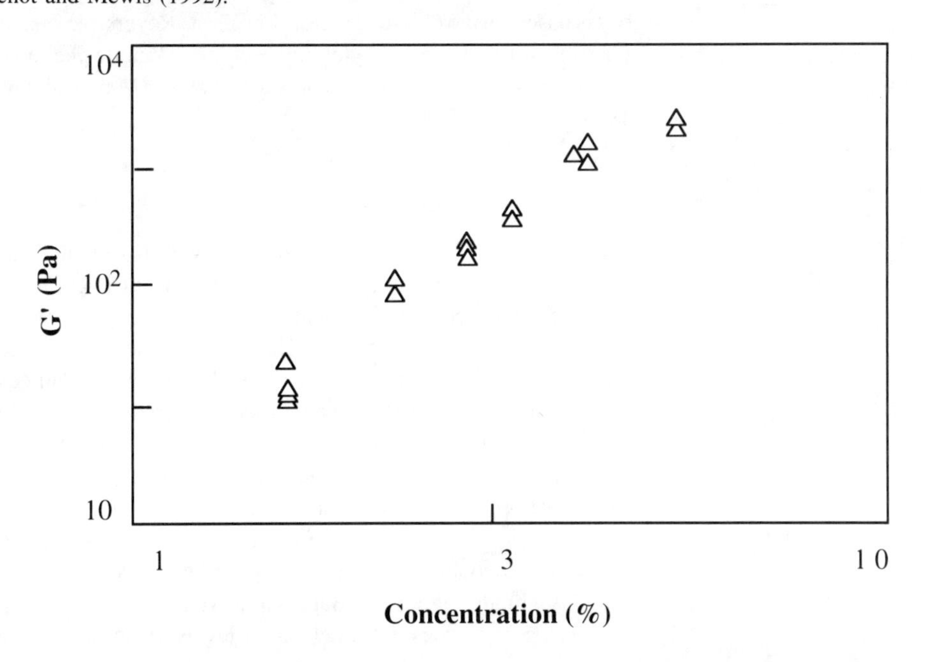

Figure 10.7.2. Relation between the power law indices of the concentration laws for modulus and yield stress: comparison between experiment (hatched rectangles) (van der Aerschot, 1989) and theory (lines). From Patel and Russel (1988).

Yielding is more difficult to measure and to model. For strongly flocculated systems, Buscall et al. (1987) measured the yield stress under compression and found a concentration law quite similar to that for the shear modulus. This relation differed, however, from that for the yield stress in shear. Patel and Russel (1988) predicted nearly identical power law indices for modulus and shear yield stress. This prediction has been confirmed experimentally (Figure 10.7.2), albeit for reversibly flocculated systems. The theory is based on the classical yielding criteria. Reversible systems do not follow these criteria as yielding becomes a kinetic phenomenon. The yield stress then depends on shear history (Mewis and Meire, 1984).

10.7.3 Flow Behavior

There are major experimental difficulties with flocculated suspensions. Evidence exists that flow between concentric cylinders can be quite heterogeneous (Toy et al., 1991). For many systems structural changes are not reversible. Here we consider only systems in which the structural changes are reversible; otherwise an equilibrium viscosity curve cannot be defined. Reversibility is caused by attractive forces and Brownian motion. In such systems the flocs gradually break down when shear rate or shear stress are increased, thus causing pronounced shear thinning. At very low stress levels, all or most of the rest structure persists and the response is either elastic or viscoelastic. Whether it ever becomes really solidlike remains a matter of debate (Barnes and Walters, 1985). In any case the viscosity reaches extremely high levels, even at low volume

Figure 10.7.3.
Viscosity versus shear stress for flocculated 2.5% silica particles in methyl laurate: open triangles, stress-controlled cone and plate instrument; others symbols, shear rate controlled. Replotted from Van der Aerschot and Mewis (1992).

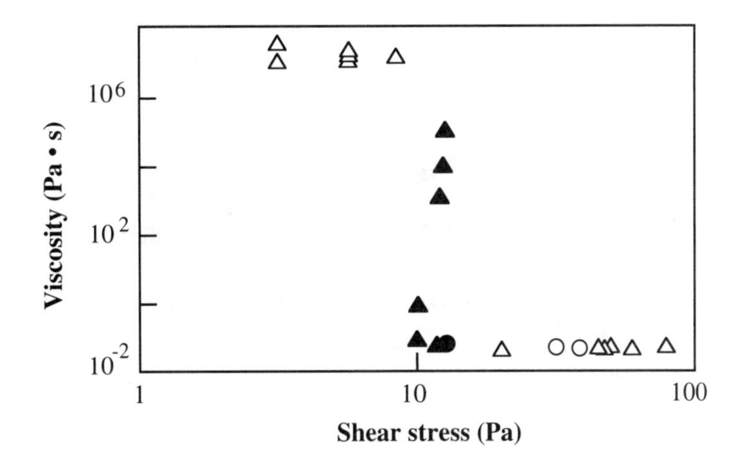

fractions (Figure 10.7.3). Then the mechanism of motion is creep rather than flow, meaning that a particle network exists at all times and that motion is based on subsequent local rearrangements of the structure. The pioneering work of Rehbinder and co-workers in this area should be mentioned (Fedotova et al., 1967). These authors measured systematic reductions in elasticity and viscosity at critical conditions of strain or stress. They were the first to measure dramatic drops in viscosity. A remarkable 10^9 decrease for less than 10% change in shear stress is shown in Figure 10.7.3! At the highest shear rates, a Newtonian region is regained when the flocs cannot be broken down further. Another example of viscosity data on flocculated systems is shown for iron oxide in mineral oil in Chapter 2 (Figures 2.5.3 and 2.5.4). Recently some progress has been made in relating interaction forces to the viscosity curve. Especially for relatively weak interactions a square well approximation for the potential can be used to correlate the data. (Woutersen and de Kruif, 1991). A somewhat different approach was followed by Buscall et al., (1990).

A reversible change in floc structure requires a finite amount of time, resulting in time-dependent viscosities. This phenomenon is called thixotropy (Mewis, 1979). Figure 10.7.4 demonstrates the response of a thixotropic system to stepwise changes in shear rate. Under a sudden increase in shear rate, the viscosity gradually decreases while the structure breaks down to smaller flocs. If a sudden decrease is applied, the initial structural units are below the new equilibrium size and therefore they gradually grow, causing a gradual increase in viscosity. Qualitatively this behavior can be described by structural models, including a kinetic equation for the changes in structure. However, no adequate quantitative models are generally valid. Thixotropy can be understood on the basis of the underlying structure. Shear initially reduces the network to individual flocs, which are then gradually further reduced in size. The flow around and through an aggregate has been discussed by Adler (1978), who also attempted to calculate the resulting rheology for aggregating particles. Gradual erosion seems to be the dominating structural change. If the flow stops, the aggregates will grow again,

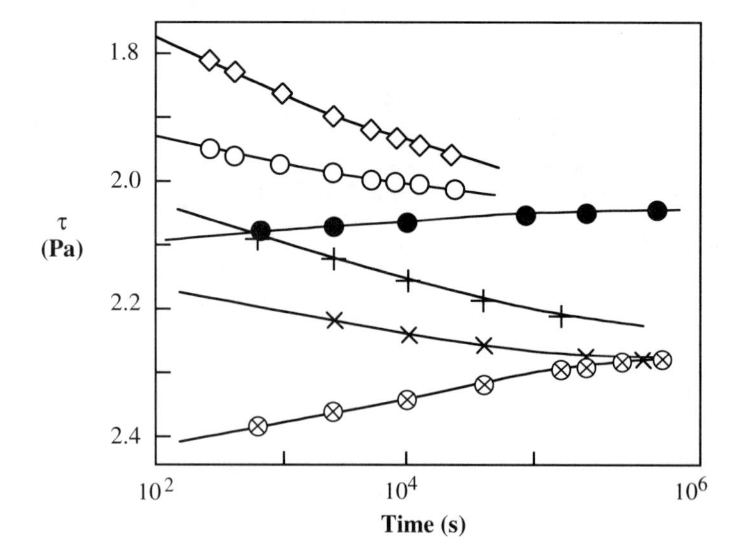

Figure 10.7.4.
Shear stress transients for stepwise changes in shear rate (s^{-1}) for treated sepiolite (a claylike mineral): ◇, 0.4 → 20; ○, 4 → 20; •, 40 → 20; + 1 → 10; ×, 4 → 10; ⊗, 40 → 10. From Van der Aerschot (1989).

but by sticking together rather than by a reverse process of erosion. Hence breakdown and recovery of structure do not proceed through the same intermediate states, giving rise to a complex dependency of structure on shear history. This complexity can be detected by various techniques, including rheological (Mewis et al., 1975) and dielectrical (Mewis et al., 1987).

Weakly flocculated systems are the most common encountered in industry and, as we have seen, the most complex rheologically because of the flow-dependent floc size and long time constants. Such systems continue to be an active area of research.

10.8 Summary

Suspensions, especially colloidal suspensions, can display all known rheological phenomena from shear thinning or thickening to time-dependent normal stresses and strong extensional effects. Particle shape, interparticle forces, and the resulting microstructure are responsible for this behavior. For dilute suspensions our theoretical understanding is very good. Theories are emerging for both the structure and rheology of colloidally stable, concentrated suspensions. In addition, scaling principles, that can reduce and interpret many of the data are available. This progress has been confirmed with systematic measurements on well-characterized model systems. Measuring problems arise with very concentrated systems and with large particles (migration, wall effects, nonlinearities).

For flocculated systems the state of the art is less satisfactory. The role of the interaction forces is understood qualitatively, but quantitative models are still in their initial stages of development. Also, experimentally flocculated systems are more difficult to handle. They can cause inhomogeneities and instabilities because of extreme shear thinning. The complex effect of shear history and the slow rate at which equilibrium is reached often make low shear data difficult to reproduce. It is also more difficult to test theories because of the difficulty in generating model systems whose properties as well as structure are well characterized. The emergence of fractal and percolation theories provided new opportunities which are still being explored. Elasticity has been closely linked with percolating structures. Attempts have been made to link viscosity with fractal structures. Suitable approaches to manipulate complex rheological properties, such as yield stress and thixotropy, are essentially still lacking. Even measuring these properties is not a totally unambiguous process.

References

Ackerson, B. J., *J. Rheol.* 1990, *34*, 553.

Ackerson, B. J.; Hayter, J. B.; Cotter, L., *J. Chem. Phys.* 1986, *84*, 2344.

Acrivos, A.; Lo, T. S., *J. Fluid Mech.* 1978, *86*, 641.

Adler, P. M., *Rheol. Acta* 1978, *17*, 288.

Anczurowski, E.; Mason, S. G., *J. Colloid Interface Sci.* 1967, *23*, 522.

Barnes, H., *J. Rheol.* 1989, *33*, 329.

Barnes, H.; Walters, K., *Rheol. Acta* 1985, *24*, 323.

Barnes, H. A.; Edward, M. F.; Woodcock, L. V., *Chem. Eng. Sci.* 1987, *42*, 591.

Bartok, W.; Mason, S. G., *J. Colloid Sci.* 1958, *13*, 293.

Batchelor, G. K., *J. Fluid Mech.* 1970, *41*, 545.

Batchelor, G. K., *J. Fluid Mech.* 1971, *46*, 813.

Batchelor, G. K., *J. Fluid Mech.* 1977, *83*, 97.

Bird, R. B.; Curtiss, C. F.; Armstrong, R.; Hassager, O., *Macromolecular Hydrodynamics,* Vol. 2*: Kinetic Theory*, 2nd ed.; Wiley: New York, 1987.

Block, H.; Kelly, J. P., *J. Phys. D.* 1988, *21* 1661.

Brady, J. F.; Bossis, G., *Annu. Rev. Fluid Mech.* 1988, *20*, 111.

Brenner, H., in *Progress in Heat and Mass Transfer*, Vol. 5; Schowalter, W. R., et al. Eds., Pergamon: Oxford, 1972; p. 89.

Brenner, H., *Int. J. Multiphase Flow* 1974, *1*, 195.

Buscall, R.; Goodwin, J. W.; Hawkins, M. W.; Ottewill, R. H., *J. Chem. Soc. Faraday Trans.* 1982, *78*, 2889.

Buscall, R.; McGowan, I. J.; Mills, P. D. A.; Stewart, R. F.; Sutton, D.; White, L. R.; Yates, G. E., *J. Non-Newtonian Fluid Mech.* 1987, *24*, 183.

Buscall, R.; McGowan, I. J.; Mumme-Young, C. A., *Faraday Discuss. Chem. Soc.* 1990, *90*, 155.

Cerf, R.; *J. Chim. Phys.* 1951, *48*, 59.

Chan, Y.; White, J. L.; Oyanagi, Y., *J. Rheol.* 1978, *22*, 507.

Cheng, D. C. -H., *Powder Technol.* 1984, *37*, 255.

Chong, J. S.; Christiansen, E. B.; Baer, A. D., *J. Appl. Polym. Sci.* 1971, *15*, 2001.

de Gennes, P. G., *J. Phys. (Paris), Lett.* 1976, *37*, L1.

de Kruif, C. G.; van Iersel, E. M. F.; Vrij, A.; Russel, W. B., *J. Chem. Phys.* 1985 *83*, 4717.

D'Haene, P., Ph.D. thesis, Katholieke Universiteit, Leuven, Belgium, 1991.

Doi, M.; Edwards, S. F., *The Theory of Polymer Dynamics*; Clarendon Press: Oxford, 1986.

DuPauw, E. J., *Cell and Molecular Biology*; Academic Press: New York, 1968; p. 326.

Einstein, A. *Ann. Phy.* 1906, *19*, 289; 1911, *34*, 591.

Fedotova, V. A.; Zhodzhaeva, Kh.; Rehbinder, P. A., *Dokl. Akad. Nauk SSSR* 1967, *177*, 155.

Feng, S.; Sen, P., *Phys. Rev. Lett.* 1984, *52*, 216.

Flory, P. J., *Principles of Polymer Chemistry*; Cornell University Press: Ithaca, NY, 1953.

Frankel, N. A; Acrivos, A., *J. Fluid Mech.* 1970, *44*, 65.

Firth B. A.; Hunter R. J., *J. Colloid Interface Sci.* 1976, *57*, 266.

Frith, W. J.; Strivens, T. A.; Mewis, J., *J. Colloid Interface Sci.* 1990, *139*, 559.

Goddard, J.; Miller, C., *J. Fluid Mech.* 1967, *28*, 57.

Goddard, J., *J. Rheol.* 1978, *22*, 615.

Goldsmith, H. L.; Mason, S. G., Microrheology of Dispersions, in *Rheology, Vol. 4*; Eirich, F. R., Ed.; Pergamon: New York, 1967; p. 85.

Goodwin, J. W.; Gregory, T.; Stile, J. A., *Adv. Colloid Interface Sci.* 1982, *17*, 185.

Goodwin, J. W.; Gregory, T.; Miles, J. A.; Warren, B. C. H., *J. Colloid Interface Sci.* 1984, *97*, 488.

Graham, A. L.; Atobelli, S. A.; Fukushima, E.; Mondy, L. A.; Stephens, T. S., *J. Rheol.* 1991, *35*, 191; 721; 773

Halow J. S.; Willis G. B., *AIChE J.* 1970, *16*, 281.

Happel, J.; Brenner, H., *Low Reynolds Number Hydronamics*; Prentice-Hall: Englewood Cliffs: NJ, 1965.

Hinch, E. J.; Leal, L. G., *J. Fluid Mech.* 1972, *52*, 683.

Ho, B. P.; Leal, L. G., *J. Fluid Mech.* 1974, *65*, 365.

Hunter, R. J., *Foundations of Colloid Science*, Vol. 1; Clarendon Press: Oxford, 1989.

Jarzebski, G. J., *Rheol. Acta* 1981, *20*, 280.

Jeffery, G. B., *Proc. R. Soc.* 1922, *A 102*, 161.

Johnson, S. J.; de Kruif, C. G.; May, R. P., *J. Chem. Phys.* 1988, *89*, 5909.

Kamal, M. R.; Mutel, A., *J. Polym. Sci.* 1985, *5*, 293.

Karnis, A; Goldsmith, H. L.; Mason, S. G., *Can. J. Chem. Eng.* 1966, *44*, 181.

Kim, S.; Karrila, S. J., *Microhydrodynamics: Principles and Selected Applications*; Butterworths: Boston, 1991.

Kitano, T.; Kataoka, T.; Shirota, T., *Rheol. Acta* 1981, *20*, 207.

Krieger, I. M., *Adv. Colloid Interface Sci.* 1972, *3*, 111.

Krieger, I. M.; Eguluz, M., *Trans. Soc. Rheol.* 1976, *20*, 29.

Larson, R. G., *Constitutive Equations for Polymer Melts and Solutions*; Butterworths: Boston, 1988.

Laun, H. M., *Angew. Makromol. Chem.* 1984, *124-125*, 335.

Laun, H. M., in *Proceedings of the 10th International Congress on Rheology*, Vol. 1, Uhlherr, P.H.T., Ed.; Sydney, 1988; p. 37.

Laun, H. M.; Bung, R.; Schmidt, F., *J. Rheol.* 1991, *35*, 999.

Leighton, D.; Acrivos, A., *J. Fluid Mech.* 1987, *181*, 415.

Lin, C. J.; Perry, J. H.; Schowalter, W. R., *J. Fluid Mech.* 1970, *44*, 1.

Mahanty, J.; Ninham, B. W., *Dispersion Forces*; Academic Press: London, 1976.

Manley, R. St. J.; Mason, S. G., *Can. J. Chem.* 1954, *32*, 763.

Markovic, I.; Ottewil, R. H.; Underwood, S. M.; Tadros, T. F., *Langmuir* 1986, *2*, 625.

Meakin, P., *Phys. Rev. Lett.* 1983, *51*, 1119.

Metzner, A. B., *J. Rheol.* 1985, *29*, 739.

Mewis, J., *J. Non-Newtonian Fluid Mech.* 1979, *6*, 1.

Mewis, J.; D'Haene, P., *Macromol. Chem. Macromol. Symp.* 1993, *68*, 213.

Mewis, J.; de Bleyser, R., *Rheol. Acta* 1975, *14*, 721.

Mewis, J.; De Groot, L. M.; Helsen, J., *Colloids Surfaces* 1987, *22*, 271.

Mewis, J.; Frith, W. J.; Strivens, T. A.; Russel, W. B., *AIChE J.* 1989, *35*, 415.

Mewis, J.; Meire C., in : *Advances in Rheology, Proceedings of Ninth International Congress on Rheology,* Vol. 2; Mena, B.; Garcia-Rejon, A.; Rangel-Nafaile, C.; Eds.; Acapulco, 1984, p. 591.

Mewis, J.; Metzner, A. B., *J. Fluid Mech.* 1974, *62*, 593.

Mewis, J.; Spaull, A. J. B., *Adv. Colloid Interface Sci.* 1976, *6*, 173.

Mewis, J.; Spaull, A. J. B.; Helsen, J., *Nature* 1975, *253*, 618.

Michele, J.; Patzold, R.; Donis, R., *Rheol. Acta* 1977, *16*, 317.

Napper, D. H., *Polymeric Stabilization of Colloidal Dispersions*; Academic Press: London, 1983.

Navarrete, R. C., Ph.D. thesis; University of Minnesota, 1991.

Nawab, M. A.; Mason S. G., *J. Colloid Sci.* 1958, *13*, 179.

Nemoto, N.; Schrag, J. L.; Ferry J. D.; Fulton, R. W., *Biopolymers* 1975, *14*, 407.

Nicodemo, L.; Nicolais, L.; Landel, R. F., *Chem. Eng. Sci.* 1974, *29*, 729.

Oosterbroek, M.; Lopulissa, J. S.; Mellema, J., in *Rheology*, Vol. 2; Astarita, G.; Marrucci, G.; Nicolais, L.; Eds.; Plenum: New York, 1980.

Otsubo, Y.; Prudhomme, R., 1994, in press.

Patel, P. D.; Russel, W. B., *Colloids Surf.* 1988, *31*, 355.

Phillips, R. J.; Brady, J. F.; Bossis, G. *Phys. Fluids* 1988, *11*, 3462.

Pusey, P. N.; Rarity, J. -G., *Mol. Phys.* 1987, *62*, 411.

Roscoe, R., *J. Fluid Mech.* 1967, *28*, 273.

Rumscheidt, F. D.; Mason, S. G., *J. Colloid Sci.* 1961, *16*, 238.

Russel, W. B., *J. Rheol.* 1978, *85*, 209.

Russel, W. B., *J. Fluid Mech.* 1980, *24*, 287.

Russel, W. B.; Benzing, D. W., *J. Colloid Interface Sci.* 1981, *83*, 163.

Russel, W. B.; Gast, A. P., *J. Chem. Phys.* 1986, *84*, 1815.

Russel, W. B.; Saville, D. A.; Schowalter, W. R., *Colloidal Dispersions*; Cambridge University Press: Cambridge, 1989.

Saunders, F. L., *J. Colloid Sci.* 1961, *16*, 13.

Scheraga, H. A., *J. Chem. Phys.* 1955, *23*, 1526.

Schowalter, W. R., *Mechanics of Non-Newtonian Fluids*; Pergamon: Oxford, 1978.

Segre, G.; Silberberg, A., *J. Fluid Mech.* 1962, *14*, 115; *J. Colloid Sci.* 1963, *18*, 312.

Sherwood, J. D., *J. Fluid Mech.* 1980, *101*, 3.

Sonntag, R. C.; Russel, W. B., *J. Colloid Interface Sci.* 1986, *113*, 339.

Taneda, S., *J. Phys. Soc. Jpn.* 1979, *30*, 262.

Taylor, G. I., *Proc. R. Soc.* 1932, *A 138*, 41.

Toy, M. L.; Scriven, L. E.; Macosko, C. W., *J. Rheol.* 1991, *35*, 887.

Van der Aerschot, E., Ph.D. thesis, Katholicke Universiteit Leuven, Belgium, 1989.

Van der Aerschot, E.; Mewis, J., *Colloids and Surfaces*, 1992, *69*, 15.

Van der Werff, J. C.; de Kruif, C. G.; Blom, C.; Mellema J., *Phys. Rev. A* 1989, *39*, 795.

Van de Ven, T. G. M.; Mason, S. G., *J. Colloid Interface Sci.,* 1976, *57*, 505, 517.

Van Helden, A. K.; Jansen, J. W.; Vrij, A., *J. Colloid Interface Sci.* 1981, *26*, 62.

Weitz, D.; Oliveria, M., *Phys. Rev. Lett.* 1984, *52*, 1433.

Whitcomb, P.; Macosko, C. W., *J. Rheol.* 1978, *22*, 493.

Willey, S. J.; Macosko, C. W., *J. Rheol.* 1978, *22*, 525.

Witten, T. A.; Sander, L. M., *Phys. Rev. Lett.* 1983, *51*, 1123.

Woodcock, L. V., *Chem. Phys. Lett.* 1984, *111*, 455.

Woodcock, L. V.; Edwards, M. F., *Inst. Chem. Eng. Symp. Ser.* 1985, *91*, 107.

Woods, M. E.; Dodge, J. S.; Krieger, I. M.; Percy, E., *J. Paint Technol.* 1968, *40, 541.*

Woutersen, A. T. J. M., de Kruif, C. G., *J. Chem. Phys.* 1991, *94*, 5739.

Zeichner, G. R.; Schowalter, W. R., *AIChE J.* 1977, *23*, 243.

11

Given the complexity of polymer molecules, the theories are astonishingly simple, . . . it is reasonable to attack the problem of interacting macromolecules with a confidence which would not be justified for say liquid benzene, let alone water.

Masao Doi and
Sam F. Edwards (1986)

RHEOLOGY OF POLYMERIC LIQUIDS

Matthew Tirrell

11.1 Introduction

At one time or another, virtually everyone who has tried to synthesize, fabricate, or utilize polymeric materials has encountered the peculiar flow properties of polymeric liquids. Depending on the nature of the encounter, it may have been amusing, intellectually challenging, or totally exasperating. These adjectives more or less describe the relationship that has developed between polymeric liquids (Figure 11.1.1) and the science of rheology. The study of the flow properties of polymeric liquids (polymer rheology or, more elegantly, macromolecular hydrodynamics) has been and is

Figure 11.1.1.
Model suggested by Kimmich et al. (1988) for a polyethylene melt.

an extremely incisive tool in elucidating various levels of macromolecular structure. By the same token, accurate description, measurement, and classification of the flow behavior of macromolecular liquids is directly useful in polymer processing and has been a tremendous stimulus to the science of rheology. In this chapter we connect *macromolecular structure to rheology*. Our treatment is necessarily brief. For deeper treatments of this important and challenging topic, see Bird et al. (1987), Doi and Edwards (1986), or des Cloizeaux and Jannink (1990).

11.2 Polymer Chain Conformation

We begin by considering the liquid state conformation of a typical synthetic polymer, such as a polymer chain consisting of n bonds of identical length l joined at fixed valence angles θ. Linear polyethylene or vinyl polymers $(-CH_2 - C_Y^X-)_n$ fit this model rather well. Polymers with heterobackbone atoms, such as nylons and polyesters, can also be described by this model if n is taken as the number of repeat units, l the length of a repeat unit, and θ some average angle between repeat units (see Flory, 1969). The end-to-end distance vector $\mathbf{r}$ may be expressed as the sum of the n bond vectors l.

$$\mathbf{r} = \sum_{i=1}^{n} l_i \tag{11.2.1}$$

Then

$$r^2 = \mathbf{r} \cdot \mathbf{r} = \sum_{i=1}^{n} \sum_{j=1}^{n} l_i \cdot l_j \tag{11.2.2}$$

Now we must realize that the polymer chain conformation is dynamic, constantly changing as a result of Brownian movement. If we wish to deal with some equilibrium conformational property of a macromolecule, we must define average properties such as

$$\overline{r^2} = \sum_{i=1}^{n} \sum_{j=1}^{n} \overline{l_i \cdot l_j} \tag{11.2.3}$$

If the chain were completely freely jointed (θ' unrestricted), then the average projection of one bond vector on any other, given by the average dot product $\overline{l_i \cdot l_j}$, would be zero. Then

$$\overline{l_i \cdot l_j} = 0 \quad \text{for} \quad i \neq j$$

so

$$\overline{r^2}_{\text{freely jointed}} = \sum_{i=1}^{n} \overline{l_i \cdot l_i} = nl^2 \tag{11.2.4}$$

The important result here is that the mean-square, end-to-end vector $\overline{r^2}$ is proportional to chain length (molecular weight), n. This result also can be derived from consideration of a random walk of n steps of length l (Flory, 1953). If fixed valence angle θ is considered, the result is

$$\overline{r^2} = n\ell^2 \left(\frac{1 - \cos\theta}{1 + \cos\theta} \right) \tag{11.2.5}$$

For the tetrahedral angle $\theta = 109.5°$, we get

$$\overline{r^2} = 2nl^2 \tag{11.2.6}$$

and $\overline{r^2}$ is still proportional to molecular weight. For a complete discussion of conformational statistics of macromolecules, see Flory (1953), Volkenshtein (1963), Birshtein and Ptitsyn (1966), Flory (1969), and Yamakawa (1971).

Equations 11.2.4–11.2.6 describe what are termed the "unperturbed" dimensions of polymer molecules, in reference to the conformation obtained in the absence of interactions with solvent or some imposed field, such as velocity gradient. Unperturbed dimensions exist in bulk amorphous or molten polymers or in solutions at the so-called theta temperature, T_θ. For most synthetic polymers a characteristic ratio can be defined as

$$C = \frac{\overline{r_o^2}}{nl^2} \tag{11.2.7}$$

where the subscript o refers to unperturbed dimensions. Experimentally, C is found to be in the range 4–10 for most synthetic polymers (Flory, 1969), C increasing with chain stiffness. Note $\overline{r_o^2}$ is still proportional to n, now with an empirical coefficient.

Up to this point we have discussed only the effects of the chemical structure—that is, bond angles, bond lengths, and steric factors—on the size of a polymer molecule in a solution or melt. These are the factors that go into $\overline{r_o^2}$. It is also possible to describe the same factors in terms of an "equivalent freely jointed chain." By this we mean the following. Suppose we write

$$\overline{r_o^2} = NL^2 \tag{11.2.8}$$

where N and L refer to the number and length of what we call "statistical segments." These segments may be made up of more than one monomer unit. Comparing eqs. 11.2.7 and 11.2.8 we see:

$$NL^2 = Cnl^2$$

We also know that NL must equal nl; thus we obtain the relations $L = Cl$ and $N = n/C$. Therefore, the stiffer chain (higher C) may be said to have longer (and consequently fewer) statistical segments. This statistical segment idea is represented by the dashed lines in Figure 11.2.1b.

Before turning to the flow properties, let us briefly discuss the effect of solvent on the conformations of polymers because conformations also affect the rheology.

In addition to the effects of chain stiffness on $\overline{r_o^2}$, if a macromolecule is placed in a good solvent, its average dimensions expand, that is,

$$\overline{r^2} = \alpha^2 \overline{r_o^2} \tag{11.2.9}$$

Polymer solution thermodynamics (Flory, 1953) also can be used to show that the expansion factor α is given by

$$\alpha^5 - \alpha^3 = A\left(1 - \frac{T_\theta}{T}\right) M^{1/2} \tag{11.2.10}$$

where A is a combination of constants and thermodynamic parameters, T is absolute temperature, and M is molecular weight. Two points here are relevant to our future discussion: (1) at $T = T_\theta$, $\alpha^5 - \alpha^3 = 0$ and α must be unity; therefore, $\overline{r^2} = \overline{r_o^2}$, and unperturbed dimensions are obtained; (2) above T_θ, α is greater than 1 and depends on molecular weight, at most to the 1/10 power. Nonetheless, for high molecular weight polymers, the hydrodynamic volume of a polymer can be increased by a factor of 5 or more on dissolution in a good solvent. These points will be important in our discussion of the molecular weight dependence of polymer intrinsic viscosity.

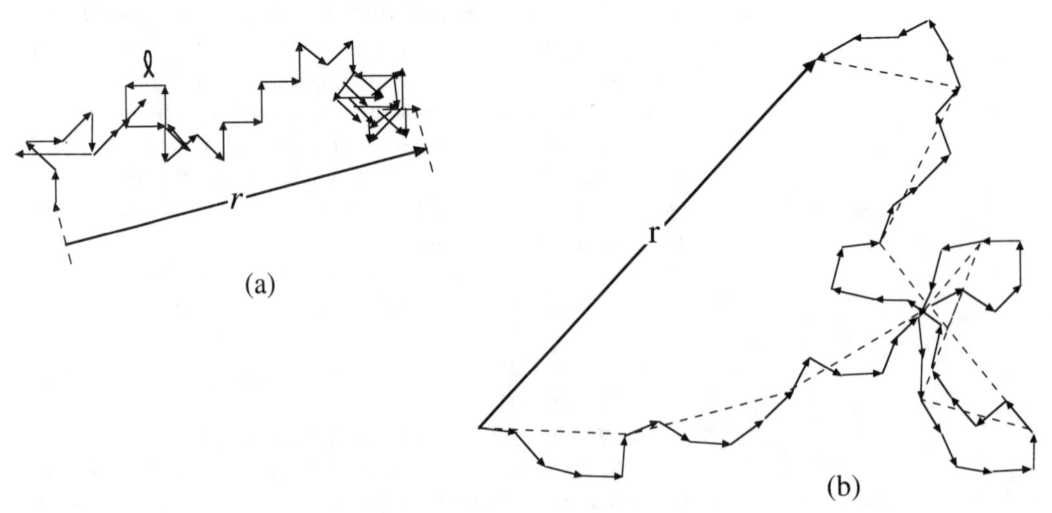

Figure 11.2.1. Vectorial representation in two dimensions of a freely jointed chain. A random walk of 50 steps. (a) Two-dimensional random coil represented as a random walk. (b) Representation of a hindered chain in two dimensions. A random walk of 50 steps with series between successive bonds limited to $\pm\pi/2$. The scale in (b) is identical to that in (a) for an unrestricted random walk of the same number of steps. From Flory (1953).

11.3 Zero Shear Viscosity

11.3.1 Dilute Solutions

We can construct a useful and reasonably accurate theory of intrinsic viscosity of dilute polymer solutions, building directly on the Einstein result for viscosities of dilute suspensions of spheres (Chapter 10). Recalling this result in a slightly different form:

$$\frac{\eta - \eta_s}{\eta_s} = 2.5\,\phi = 2.5\left(\frac{n_2}{V}\right)V_{\text{sph}} \qquad (11.3.1)$$

where n_2/V is number of molecules (n_2) per unit volume of solution (V), and V_{sph} is the volume of the molecular sphere. Clearly, $n_2/V = c\,N_{\text{Av}}/M$, where c is concentration, M molecular weight, and N_{Av} Avogadro's number. Now we need to decide on the appropriate V_{sph} for a randomly coiling macromolecule. It should be proportional to the cube of the root-mean-square end-to-end distance:

$$(V_{\text{sph}})_{\text{random coil}} = k_1(\overline{r^2})^{3/2} \qquad (11.3.2)$$

Combining, we get

$$\frac{\eta - \eta_s}{\eta_s c} = [\eta] = \frac{2.5\,k_1\,N_{\text{Av}}(\overline{r^2})^{3/2}}{M} \qquad (11.3.3)$$

The constants $2.5\,k_1$ and N_{Av} are usually lumped into one constant called Φ to give what is known as the Flory–Fox equation (Flory and Fox, 1951)

$$[\eta] = \Phi\frac{(\overline{r^2})^{3/2}}{M} \qquad (11.3.4)$$

Experimentally, Φ is found to be approximately constant for all synthetic polymers with a value of 2.0 to 2.6×10^{23} mol^{-1}. The best theory (Yamakawa, 1971) predicts Φ to be asymptotically constant at 2.25×10^{23} mol^{-1}. From eq. 11.2.7, $\overline{r_o^2}$ is proportional to M, so we have

$$[\eta] \propto M^{1/2} \qquad (11.3.5)$$

at the theta condition. Experimentally, intrinsic viscosity appears to depend on molecular weight, as the Mark–Houwink relation suggests:

$$[\eta] = KM^a \qquad (11.3.6)$$

The constant a varies between 0.5 and 0.8 for different polymer–

solvent combinations. The theoretical justification for this variation can be seen by combining eqs. 11.2.9 and 11.3.4:

$$[\eta] = \Phi\alpha^3 \frac{(\overline{r_o^2})^{3/2}}{M} \tag{11.3.7}$$

Since from eq. 11.2.10, $\alpha \propto M^{0.0 \text{ to } 0.1}$ and $(\overline{r_o^2})^{3/2}/M \propto M^{0.5}$, we can write

$$[\eta] \propto M^{0.5 \text{ to } 0.8}$$

These results form the basis of utilization of intrinsic viscosity as a molecular weight measuring technique. Values of K and a from eq. 11.3.6 are extensively tabulated in the *Polymer Handbook* of Brandrup and Immergut (1989).

It is worthwhile to keep in mind that this development treats the hydrodynamic (but not the thermodynamic) interactions between polymer molecule and solvent as if the polymer were an impenetrable sphere. This seems to provide an accurate picture for dilute solutions, but certainly not the only imaginable one. The degree to which the medium permeates and flows through the domain of the polymer molecule (the nature of the hydrodynamic interaction) is a sophisticated and much-discussed problem in theoretical macromolecular rheology (Bird et al., 1987).

Up to this point, we have neglected the effect of shear rate on the measured viscosity. More will be said about this later, but it is important to realize that we have been discussing *zero shear rate intrinsic viscosity*. Hydrodynamic forces are capable of perturbing the average dimensions of a polymer molecule and will qualitatively alter the observed rheological behavior. As noted in Chapters 6 and 8, intrinsic viscosities are most often measured in gravity-driven capillary viscometers. Typical shear rates are of the order of 100 s^{-1}. When planning an intrinsic viscosity measurement for molecular weight determination, care should be taken to ensure that data representative of zero shear rate behavior are obtained. It may be necessary to obtain values of $[\eta]$ at several shear rates and extrapolate to zero shear rate.

11.3.2 Nondilute Polymeric Liquids

The intrinsic viscosity discussed in Section 11.3.1 is, strictly speaking, an infinite dilution value. To account for increasing solution viscosity with increasing concentration, an expansion in powers of concentration is usually used, as is also done to account for concentration effects in suspension rheology:

$$\eta = \eta_s(1 + [\eta]c + k'[\eta]^2c^2 + \cdots) \tag{11.3.8}$$

We see that the proper definition of $[\eta]$ for polymers is

$$[\eta] = \lim_{\dot{\gamma}, c \to 0} \left(\frac{\eta - \eta_s}{\eta_s c} \right)$$

Plots of $(\eta - \eta_s)/\eta_s c$ versus c are usually very linear at low concentration (justifying truncating the series), and the slopes of these plots are found to vary with solvent quality; $[\eta]^2 k'$ is known as the Huggins constant (Huggins, 1942) after the person who made this observation. Experimentally, k' is independent of molecular weight for long chains, with values of roughly 0.30 to 0.40 in good solvents and 0.50 to 0.80 in theta solvents.

The literature on viscosity of polymer solutions at low and high concentrations of polymer suggests very clearly that *two fundamentally different types of intermolecular interaction need to be considered* (Graessley, 1974). We have seen that the viscosity behavior of low concentration solutions in many ways resembles that of a suspension of discrete rigid particles. As concentration is increased from infinite dilution, intermolecular effects are introduced by the interactions of the flow fields in the neighborhood of each molecule, as in the case of rigid particles (Section 10.5; see also Happel and Brenner, 1965). This interaction seems to depend on the volume occupied by the molecules, or alternatively, on the degree of overlap of the individual molecular domains. If we have n_2/V *inter*penetrable polymer molecules per unit volume, then $n_2 V_{\text{sph}}/V$ loses its *artificial* interpretation of eq. 11.3.1 as an "equivalent hard sphere" volume fraction and now in a more concentrated regime represents the average number of other molecules with centers lying within the pervaded volume of any one molecule. From eqs. 11.3.2 and 11.3.3

$$V_{\text{sph}} = \frac{[\eta]M}{2.5\ N_{\text{Av}}}$$

and using $n_2/V = cN_{\text{Av}}/M$, we get

$$\frac{n_2}{V} V_{\text{sph}} = 0.4\ c[\eta] \qquad (11.3.9)$$

So we see that a good rule of thumb for predicting when concentration effects will become important is when the coil overlap parameter $c[\eta]$ (or cM^a) is near unity.

A relation that fits viscosity data well in the coil overlap region is the Martin equation:

$$\frac{\eta - \eta_s}{\eta_s c} = [\eta] e^{kc[\eta]} \qquad (11.3.10)$$

Note that if $k = k'$, eq. 11.3.8 may be viewed as a power series expansion of eq. 11.3.10. Experimentally, k is often quite close to k'.

11.3.3 Coil Overlap

A second source of intermolecular interactions arises from segment–segment contacts between molecules. This mode of interaction has become popularly known as entanglement.

We have more to say about entanglement later. Here we note that in practice, the coil overlap considerations leading to eq. 11.3.9 seem to be the dominant type of intermolecular interaction in the concentration range $1 < c[\eta] < 10$. Figure 11.3.1 shows some data correlated with the use of the Martin equation (eq. 11.3.10). Above $c[\eta] \approx 10$, this second mode of interaction known as entanglement begins to dominate, and other viscosity correlations are more effective.

It is useful to think about the various concentration regimes as illustrated in Figure 11.3.2 for a thermodynamically good solvent. We refer to the concentration regimes, following deGennes and coworkers (1979) as dilute ($c < c^*$), semidilute ($c^* < c < c^{\ddagger}$), and concentrated ($c > c^{\ddagger}$). We can estimate where these boundaries occur and gain much useful information about polymer conformational properties through the use of a sort of dimensional analysis which physicists call scaling laws. Equations 11.2.9 and 11.2.10 can be combined to show that, in very good solvents ($\alpha \gg 1$):

$$\overline{r_o^2} \sim M^{1.2} \tag{11.3.11}$$

The symbol $\sim$ signifies "depends on" and is meant to give a functional dependence, not an equality. Thus eq. 11.3.11 states that the

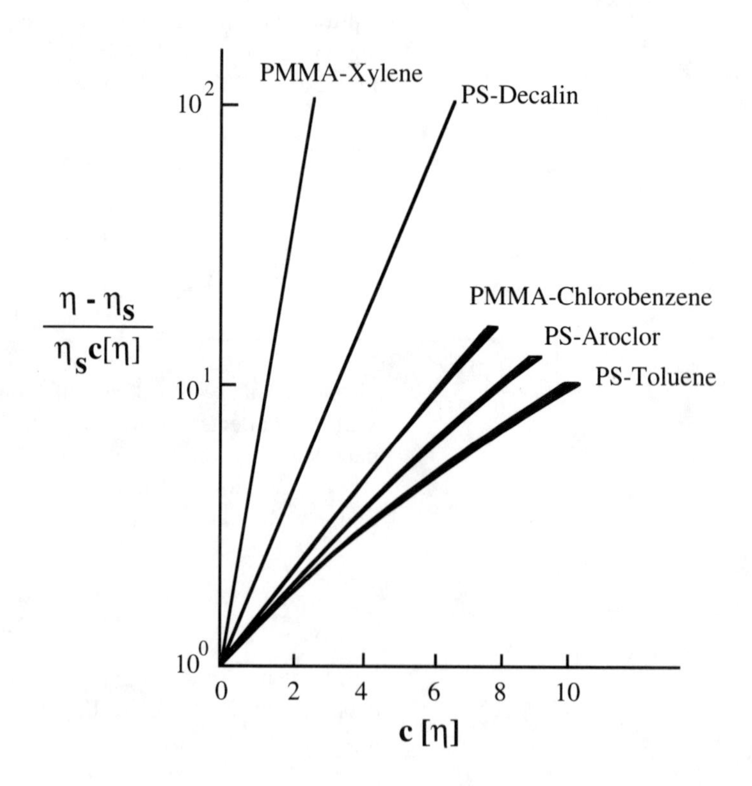

Figure 11.3.1. Viscosity at various concentrations and molecular weights in the low to moderate concentration range. Polystrene–decalin and polymethyl methacrylate–xylene are theta or near-theta systems; the remainder are good solvent systems. Note that the $c[\eta]$ reduction is somewhat better in theta solvents, and that the Martin equation, which would give a straight line in the figure, is a somewhat better representation for theta solvents. Adapted from Graessley (1974).

mean-square, end-to-end distance depends on molecular weight (chain length) to the 1.2 power. This is an example of a scaling law derived from fundamental physical arguments. What we hope to do now is derive further functional dependences from this law combined with some plausible physical insights.

A simple illustration is an estimation of the overlap threshold c^*. We expect c^* to be comparable with the local concentration within a single chain, thus

$$c^* = \frac{\text{no. of monomer units/chain}}{\text{pervaded volume/chain}} \sim \frac{M}{(\overline{r^2})^{1/2}} \tag{11.3.12}$$

or using eq. 11.3.11

$$c^* \sim \frac{M}{(M^{1.2})^{1/2}} \sim M^{-4/5} \sim \frac{1}{[\eta]} \text{in good solvents} \tag{11.3.13}$$

(The readers should convince themselves that this is essentially the same argument that led to eq. 11.3.9.) We can go further to obtain information about the semidilute regime. A quantity of interest here is the correlation length ξ, that is, a mean distance between monomer units (Figure 11.3.3) on separate chains.

We can determine how ξ varies with concentration from scaling arguments and a few simple physical considerations. First, we expect that for $c > c^*$, the solution structure on the length scale will not depend on molecular weight because we are looking only at rather small sections of the molecule. Second, at $c = c^*$, we expect ξ to be approximately the same as the isolated coil size $\xi \simeq \sqrt{(\overline{r^2})^{1/2}}$. Thus we are led to the scaling form:

$$\xi \simeq (\overline{r^2})^{1/2} \left(\frac{c^*}{c}\right)^m \tag{11.3.14}$$

$$\xi(c) \simeq M^{3/5} \left(\frac{M^{-4/5}}{M^0}\right)^m \sim M^0 \tag{11.3.15}$$

Figure 11.3.2. Concentration regimes in good solvents.

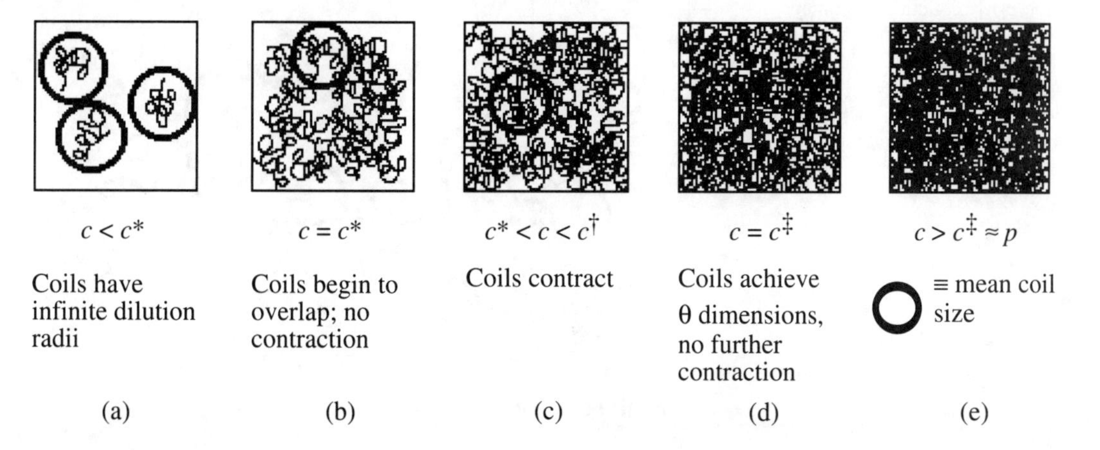

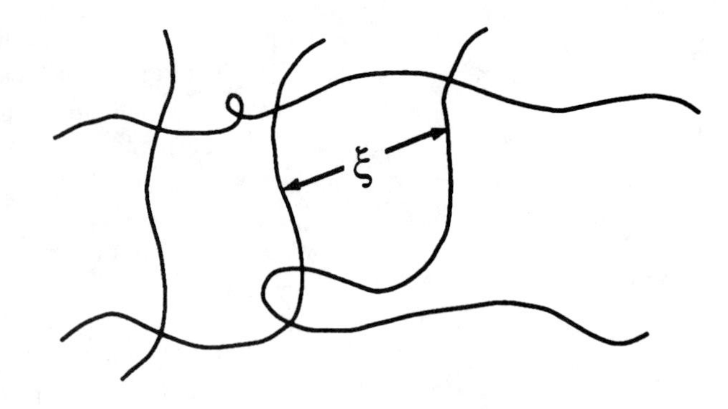

Figure 11.3.3.
Magnified view of Figure 11.3.2a–c.

By our first supposition that ξ is independent of M, we must then choose m so that this is so:

$$m = \frac{3}{4} \text{ and } \xi \sim c^{-3/4} \tag{11.3.16}$$

Thus the average distance between monomers decreases like $c^{-3/4}$ above c^*. This can be measured experimentally by scattering, and data support this scaling law (Wiltzius et al., 1983).

We can also use these techniques to determine how the radius of an individual coil changes with concentration. Look at Figure 11.3.4. We visualize an individual chain as a succession of "blobs", each of molecular weight M_b. Inside each blob the coil is swollen so that the contour length of the coil (and therefore M_b) within each blob is related to the size of the blob ξ by

$$\xi \sim l N_b^{3/5} \tag{11.3.17}$$

where N_b is the number of monomer units per blob ($N_b = M_b/M_o$). Using eq. 11.3.16, we find

$$N_b \sim c^{-5/4} \tag{11.3.18}$$

At distance $x >> \xi$, the chain is ideal; thus, we can use the

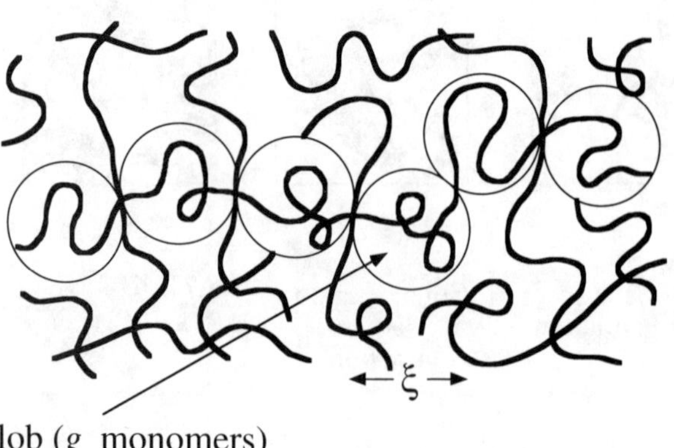

Figure 11.3.4.
Representation of polymer chains as a series of blobs of molecular weight M_b.

statistical segment idea of eq. 11.2.7 to express the mean-square, end-to-end distance

$$\overline{r^2}(c) = \frac{n}{N_b(c)}\xi^2(c) \tag{11.3.19}$$

where n is the total number of monomer units per chain; hence n/N_b is the number of blobs or statistical segments, and the length of a blob is the length of a statistical segment. Combining the results of eqs. 11.3.16, 11.3.18, and 11.3.19, we find

$$\overline{r^2}(c) \sim \frac{c^0}{c^{-5/4}}(c^{-3/4})^2 \sim c^{-1/4} \tag{11.3.20}$$

Thus, in the semidilute regime the coils shrink with concentration according to $c^{-1/4}$. This shrinking has been observed experimentally by Daoud et al. (1975): see Figure 11.3.5.

Shrinking does not continue indefinitely; the chain reaches its unperturbed minimum (θ) dimensions at concentration $c^{\ddagger}$. From eq. 11.3.20 we expect

$$\overline{r^2}(c) = \overline{r^2}(0)\left(\frac{c^*}{c}\right)^{1/4} \tag{11.3.21}$$

Thus we further expect $\overline{r^2}(c)$ to reach its θ value at $c = c^{\ddagger}$, so

$$\overline{r^2}(\theta) = \overline{r^2}(0)\left(\frac{c^*}{c^{\ddagger}}\right)^{1/4} \tag{11.3.22}$$

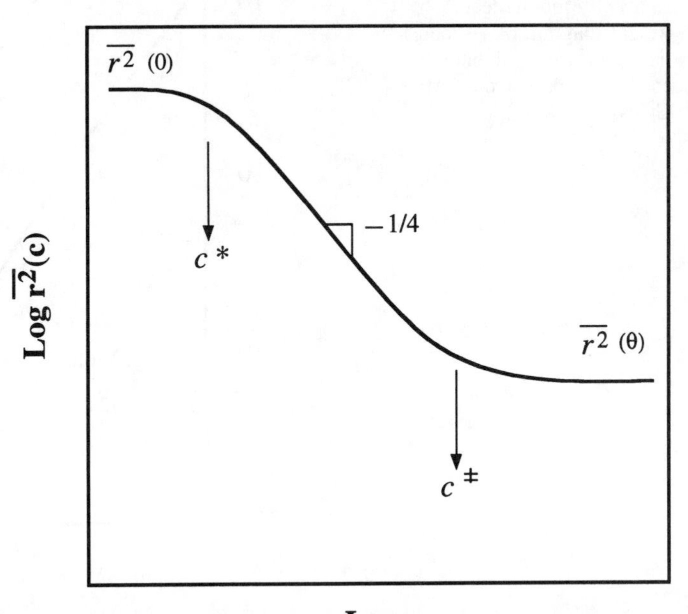

Figure 11.3.5.
Chain dimensions versus concentration according to the proposal of Daoud et al. (1975) and Graessley (1980).

or

$$c^{\ddagger} = c^* \left(\frac{\overline{r^2}(0)}{\overline{r^2}(\theta)} \right)^4 \tag{11.3.23}$$

or

$$c^{\ddagger} = c^* \alpha^8(0) \tag{11.3.24}$$

So the molecular weight dependence of $c^{\ddagger}$ is

$$c^{\ddagger} \sim M^{-4/5}(M^{1/10})^8 \sim M^o \tag{11.3.25}$$

That is, $c^{\ddagger}$ is independent of molecular weight. Its magnitude can be estimated by combining eqs. 11.3.12, 11.2.10, and 11.3.24. Figure 11.3.5 plots $\overline{r^2}(c)$ versus log (c).

As we shall see, the transition from nonentangled to entangled rheological behavior requires a minimum molecular weight independent of concentration and, for any given molecular weight entanglement may occur before or after $c^{\ddagger}$, the semidilute to concentrated transition. The various regimes are illustrated in Figure 11.3.6.

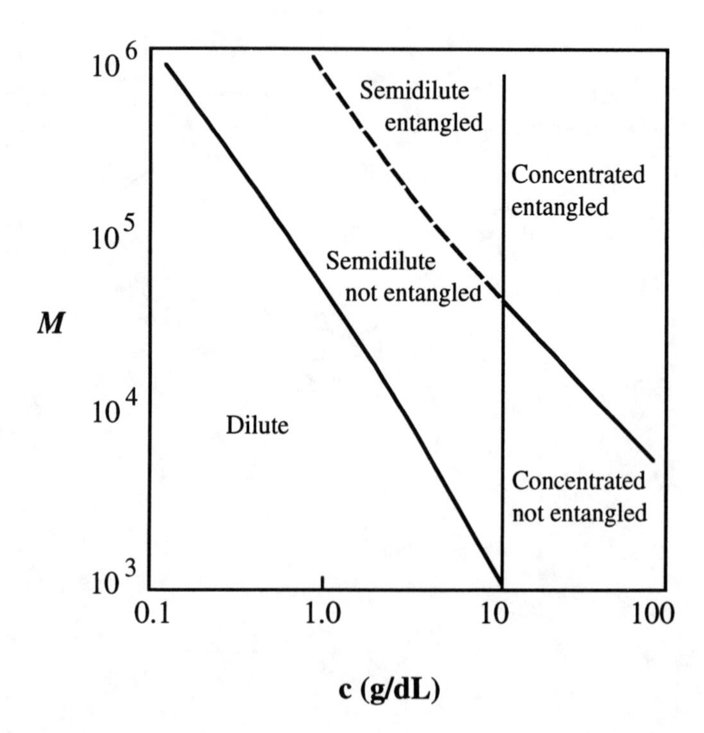

Figure 11.3.6. Concentration–molecular weight diagram of viscoelastic regimes for polybutadiene in a good solvent. After Graessley (1980).

11.4 Rheology of Dilute Polymer Solutions *

To obtain more insight into factors governing macromolecular hydrodynamics, it is useful to study the dynamics of some simplified model macromolecules, which come in varying degrees of complexity. We discuss only the simplest to see what kind of useful information can be obtained.

What we are trying to do is examine how macromolecular structural features are altered by hydrodynamic forces and how this in turn affects the macroscopic rheology of the bulk fluid containing them. What macromolecular features are we trying to include? At the least, we should examine the dynamics of a model that stretches and aligns like a random coil, has viscous drag interactions with the solvent, and has the springiness or entropy elasticity associated with macromolecular conformation. The tools we use to formulate our model will be those of statistical mechanics, but they should be readily comprehensible.

11.4.1 Elastic Dumbbell

We analyze the *rheology of elastic dumbbell suspensions.* This is the simplest kinetic theory system that can be studied and serves as an introduction to the flexible chain molecule theories of Rouse, Zimm, and others. We consider a dumbbell consisting of two "beads," labelled "1" and "2," joined by a "connector," which we will take to be elastic (see Figure 11.4.1). The position and orientation of the dumbbell are specified by the position vectors of the centers of the two beads with respect to a laboratory-fixed coordinate system; these are designated $\mathbf{r}_1$ and $\mathbf{r}_2$, respectively. In place of $\mathbf{r}_1$ and $\mathbf{r}_2$ it is usually convenient to use a position vector $\mathbf{r}_o = (1/2)(\mathbf{r}_1 + \mathbf{r}_2)$, which gives the location of the center of mass of the dumbbell, and a vector $\mathbf{R} = \mathbf{r}_2 - \mathbf{r}_1$, which gives the interbead separation and orientation of the dumbbell. The vector $\mathbf{R}$ has components x_1, x_2, and x_3 in Cartesian coordinates.

Next we summarize the parameters used to characterize the dumbbell suspension:

η_s = the viscosity of the Newtonian solvent in which the dumbbells are suspended

n_o = the number of dumbbells per unit volume

m = the mass of one bead of the dumbbell

ξ = the friction coefficient of a bead as it moves through the solvent; this is defined as the force acting on the bead divided by the velocity of the bead with respect to the solvent; if Stokes' law is used, then $\xi = 6\pi\eta_s \mathrm{r}$, where r is the radius of the bead

For a more complete treatment of this subject, see Bird et al. (1987).

Figure 11.4.1.
Elastic dumbbell.

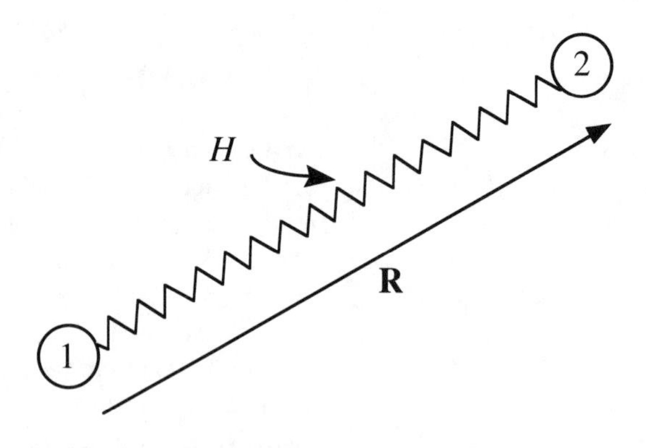

H = the spring constant of the Hookean spring, which we assume joins the two beads; in other words, we are assuming that the tension in the connector $\mathbf{F}^c$ is given by a linear spring law $\mathbf{F}^c = \mathrm{H}\mathbf{R}$, where force is pro— portional to separation

So what do we have so far? We have a model macromolecule that: (1) has viscous drag interactions with the surrounding solvent, (2) has variable **R**, which describes how the macromolecule stretches and aligns when subjected to flow, (3) accounts for elasticity with a spring connector, and (4) includes the possibility for concentration and molecular weight effects through n_o.

The dumbbells are to be suspended in a Newtonian solvent that is flowing with some kind of velocity distribution. For example, for shear flow the velocity distribution is $v_1 = \dot{\gamma} x_2$; for uniaxial extension flow $v_1 = 2\dot{\epsilon} x_1$, $v_2 = -\dot{\epsilon} x_2$, $v_3 = -\dot{\epsilon} x_3$. Both kinds of flow are *homogeneous flows* for which the velocity components are linear combinations of the Cartesian coordinates [or to put it differently, the rate of deformation tensor $\mathbf{D} = (\nabla \mathbf{v} + \nabla \mathbf{v}^T)$ is independent of position]. Hence, if we wish to formulate a kinetic theory valid for shear flows, elongational flows, and some other flows as well, we might as well set up the theory to handle all homogeneous flows. This means, then, that we consider a solvent velocity distribution given by $\mathbf{v} = \mathbf{K} \cdot \mathbf{r}$, where **K** is a traceless tensor (**K** has to be traceless because we are considering only incompressible fluids for which $\nabla \cdot \mathbf{v} = 0$). Written out in component form the solvent velocity field is

$$v_1 = K_{11} x_1 + K_{12} x_2 + K_{13} x_3$$

$$v_2 = K_{21} x_1 + K_{22} x_2 + K_{23} x_3$$

$$v_3 = K_{31} x_1 + K_{32} x_2 + K_{33} x_3$$

Specification of the K's then gives the flow field.

To develop a theory for dilute solutions, we can consider that the dumbbells all move independently. We further assume that the effect of the suspended dumbbells on the rheological properties of the solution can be obtained by finding the statistical contribution of one dumbbell. The dumbbell on which we focus our attention will be located with its center of gravity at the origin of the coordinate system $\mathbf{r} = 0$; there $\mathbf{v} = 0$, and it is assumed that the center of gravity of the dumbbell does not drift away from $\mathbf{r} = 0$.

If the dumbbell is in a solvent with $\mathbf{v} = 0$ everywhere (a fluid at rest), the dumbbell nonetheless will rotate continually because of the Brownian forces acting on it. However, all angular orientations are equally likely. On the other hand, if the solution is being sheared, then the dumbbells will tend to be aligned and all orientations are not equally likely. This idea leads to the notion of a distribution function, which we will call $\Psi(\mathbf{R},t)$. This function has the meaning:

> $\Psi(\mathbf{R},t)d\mathbf{R}$ is the probability that at time t, the dumbbell will have a configuration (a bead–bead separation vector) in the range between $\mathbf{R}$ and $\mathbf{R} + d\mathbf{R}$.

This formulation is adequate if the molecule does not experience spatially varying external fields.

In other words, if we have a collection of similar systems [i.e., systems consisting of one dumbbell with its center of gravity at $\mathbf{r} = 0$ in the flow field $\mathbf{v} = (\mathbf{K}\cdot\mathbf{r})$], then $\Psi(x_1, x_2, x_3, t)$ $dx_1dx_2dx_3$ gives the percentage of systems in the collection whose bead–bead separation vector has components in the ranges x_1 to $x_1 + dx_1, x_2$ to $x_2 + dx_2$, and x_3 to $x_3 + dx_3$. Clearly, for every specified flow situation (steady shear flow, oscillatory shear motion, stress relaxation after cessation of steady shear flow, elongational stress growth, etc.), $\Psi(\mathbf{R},t)$ will be different. Once $\Psi(\mathbf{R},t)$ is known, the stresses can be found. Hence, we have to know how the stresses are related to the distribution function $\Psi(\mathbf{R},t)$, and we have to know how to get $\Psi(\mathbf{R},t)$.

For dumbbell suspensions the main outline of this story can be told rather simply although no attempt will be made to give detailed derivations since they may be found elsewhere (Kirkwood, 1967; Bird et al., 1971, 1987). The basis of the kinetic theory involves just three equations, which we now summarize for dumbbells with any kind of connector (rigid or elastic).

Equation of Motion (or Force Balance) for the Beads

We can write an equation of motion for *each bead* in the form

mass × acceleration	=	viscous drag + Brownian motion force + force of one bead on another through the connector

When these two equations are written down, the equation for bead "2" may be subtracted from that for bead "1" to get an equation in terms of the bead–bead separation vector:

$$m\ddot{\mathbf{R}} = \xi(\dot{\mathbf{R}} - [\mathbf{K}\cdot\mathbf{R}]) - 2kT\frac{\partial}{\partial\mathbf{R}}\ln\Psi - 2\mathbf{F}^c \qquad (11.4.1)$$

inertial or acceleration	viscous drag	Brownian or fluctuating	connector tension

The Brownian motion term involves the distribution function $\Psi(\mathbf{R},t)$. Whereas the flow field tends to orient the particles, the Brownian forces tend to randomize the orientations. The term in $\ddot{\mathbf{R}}$ may be omitted because accelerations are usually quite small. The quantity $\mathbf{F}^c$ cannot be specified until the force law for the connector is known. For our case with Hookean springs, $\mathbf{F}^c = \mathrm{H}\mathbf{R}$. The notation $\partial/\partial\mathbf{R}$ means the vector having components $(\partial/\partial x_1,\ \partial/\partial x_2,\ \partial/\partial x_3)$, that is, the gradient operator in $\mathbf{R}$ space (the configuration space of the molecule).

Equation of Continuity for $\Psi(\mathbf{R},t)$

When a dumbbell leaves one orientation, it has to end up in another orientation. This simple idea enables us to write an equation of continuity for the distribution function. The derivation is quite similar to that of the equation of continuity in hydrodynamics. The result is

$$\frac{\partial\Psi}{\partial t} = -\left(\frac{\partial}{\partial\mathbf{R}}\cdot\dot{\mathbf{R}}\Psi\right) \tag{11.4.2}$$

When eq. 11.4.1 is solved for $\dot{\mathbf{R}}$ and the result substituted into eq. 11.4.2, we get a partial differential equation for Ψ. This "diffusion equation," as it is often called, must be solved after the flow field (the K's) has been specified.

$$\frac{\partial\Psi}{\partial t} = -\frac{\partial}{\partial\mathbf{R}}\cdot\left\{[\mathbf{K}\cdot\mathbf{R}]\Psi - \frac{2kT}{\xi}\frac{\partial}{\partial\mathbf{R}}\Psi - \frac{2}{\zeta}\mathbf{F}^c\Psi\right\} \tag{11.4.3}$$

Expression for the Stress Tensor

The stress tensor $\boldsymbol{\tau}$ of the dilute suspension will be made up of two parts: $\boldsymbol{\tau}_s$, the solvent contribution (which is just $\eta_s(2\mathbf{D})$ for a Newtonian solvent), and $\boldsymbol{\tau}_p$, the particle (or polymer) contribution. The particle (i.e., the dumbbell in this case) in turn will contribute in two ways to the stress transmitted across an arbitrary plane in the fluid:

- Any dumbbell that straddles the plane will contribute to the stress because of the tension in the connector (Figure 11.4.2a).
- Any bead that flies across the plane will contribute to the stress because of the momentum transported by the bead itself (Figure 11.4.2b).

When these two ideas are translated into mathematical terms (Bird et al., 1987), the final result is:

$$\boldsymbol{\tau}_p \;=\; n_o < \mathbf{F}^c\mathbf{R} > \;+\; n_o kT\,\mathbf{I} \tag{11.4.4}$$

particle contribution to stress	connector tension contribution	momentum transported by beads

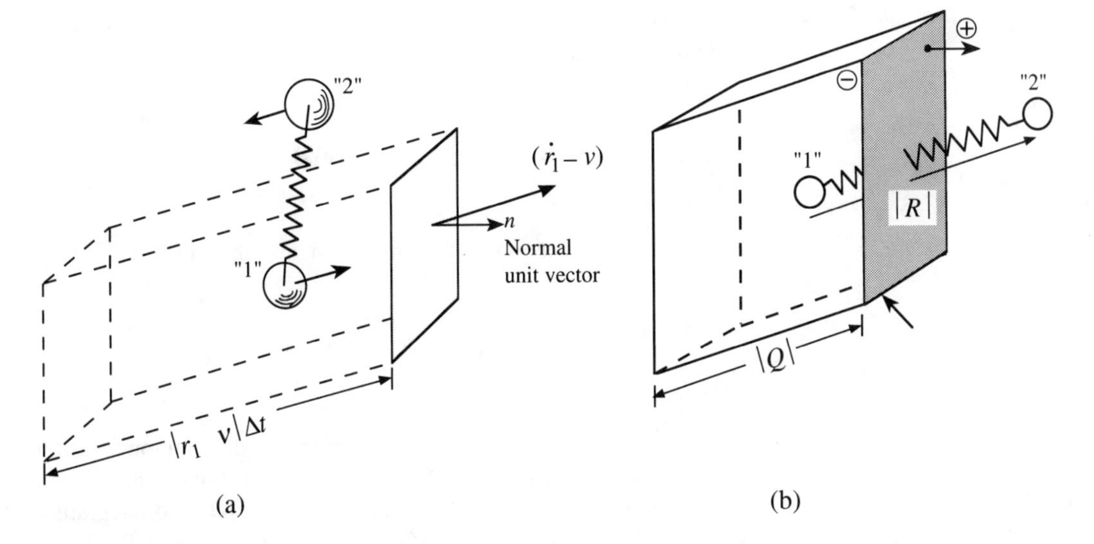

Figure 11.4.2.
(a) Stress contribution by connector tension. A dumbbell of orientation **R** intersects an arbitrary plane in a suspension of dumbbells; the plane is moving with the local fluid velocity v. The magnitude of the vector **R** is equal to the distance between the centers of the beads, and **n** is a unit vector normal to the plane. (b) Stress contribution by bead convection. A bead of a dumbbell crosses an arbitrary surface and contributes to the stress tensor τ (From Bird et al., 1971).

We see that the second term on the right-hand side contributes only an isotropic pressure to the total stress. We also see that if the Hookean spring law is used, we get

$$\boldsymbol{\tau}_p = n_o H < \mathbf{RR} > + n_o kT\ \mathbf{I} \tag{11.4.5}$$

where **RR** is a tensor formed by taking the "dyadic" product of **R** with itself. The "dyadic" has the following components:

$$\begin{pmatrix} x_1x_1 & x_1x_2 & x_1x_3 \\ x_2x_1 & x_2x_2 & x_2x_3 \\ x_3x_1 & x_3x_2 & x_3x_3 \end{pmatrix}$$

So we see that the stress contribution is related to the average dimensions of the molecule in the three directions. Specifically, for simple shear flow, $v_1 = \dot{\gamma}x_2$, the shear stress component $\tau_{p_{12}}$ is given by the average product of the molecular dimensions in direction of flow (1) and direction of the velocity gradient (2), respectively:

$$\tau_{p_{12}} = n_o H < x_1 x_2 > \tag{11.4.6}$$

Thus a positive first normal stress difference reflects an expansion of the molecular dimensions in the flow direction (1) relative to those in the direction of the velocity gradient (2):

$$\tau_{11} - \tau_{22} = n_o H[< x_1^2 > - < x_2^2 >] \tag{11.4.7}$$

and a negative second normal stress difference implies an effective contraction of the dimensions in the (2) direction relative to those in the (3) direction:

$$< \mathbf{RR} >= \int \int_{-\infty}^{\infty} \int \mathbf{RR} \, \Psi(\mathbf{R}, t) dx_1 dx_2 dx_3$$

This can be taken as the *definition* of an average value, that is,*

$$< X > \equiv \int \int_{-\infty}^{\infty} \int X \, \Psi(\mathbf{R}, t) \, dx_1 dx_2 dx_3$$

This value seems to require knowledge of the distribution function Ψ by solving the partial differential eq. 11.4.3. We can get around this requirement by multiplying eq. 11.4.3 by **RR** and integrating both sides. This gives a differential equation for $< \mathbf{RR} >$, not containing Ψ:

$$\frac{\partial}{\partial t} < \mathbf{RR} > -\{\mathbf{K}^{\mathrm{T}} \cdot < \mathbf{RR} >\} - \{< \mathbf{RR} > \cdot \mathbf{K}\} = \frac{4kT}{\xi}\mathbf{I} - \frac{4}{\xi} < \mathbf{RF}^c > \quad (11.4.9)$$

Since **K** is homogeneous, $\nabla \mathbf{K} = 0$. Thus, the left-hand side is the upper-convected time derivative referred to in Chapter 4 (eq. 4.3.2), which is often designated by ∇ over the tensor

$$< \overset{\nabla}{\mathbf{RR}} >= \frac{4kT}{\xi}\mathbf{I} - \frac{4}{\xi} < \mathbf{RF}^c > \quad (11.4.10)$$

or for a Hookean spring:

$$< \overset{\nabla}{\mathbf{RR}} >= \frac{4kT}{\xi}\mathbf{I} - \frac{4H}{\xi} < \mathbf{RR} > \quad (11.4.11)$$

Combining this with eq. 11.4.5, we can get an equation for the stress

$$\boldsymbol{\tau}_p + \frac{\xi}{4H}\overset{\nabla}{\boldsymbol{\tau}}_p = -\frac{n_o kT\xi}{4H}2\mathbf{D} \quad (11.4.12)$$

(Note that $\overset{\nabla}{\mathbf{I}} = -2\mathbf{D}$.) $\xi/4H$ has the units of time and is referred to as the relaxation time for the dumbbell, λ.

$$\boldsymbol{\tau}_{\mathrm{p}} + \lambda\overset{\nabla}{\boldsymbol{\tau}}_p = - n_o kT\lambda 2\mathbf{D} \quad (11.4.13)$$

This is the constitutive equation or rheological equation of state for the elastic dumbbell suspensions. It is identical to the upper-convected *Maxwell* model, eq. 4.3.7. The molecular dynamics have led to a proper (frame-indifferent) time derivative and to a definition

**Note that this definition of the average is equivalent to that given in eq. 11.2.3.*

of the relaxation time in terms of the molecular parameters ξ and H. It is easy to show that $H = 3kT/ < r_o^2 >$, so that the relaxation time is given by

$$\lambda = \frac{\xi < r_o^2 >}{12kT} \qquad (11.4.14)$$

The rheological predictions that derive from this simple molecular model are very similar to the upper-convected Maxwell model; see example 4.3.3. Recall that $\boldsymbol{\tau} = \boldsymbol{\tau}_s + \boldsymbol{\tau}_p = \eta_s\, 2\mathbf{D} + \boldsymbol{\tau}_p$. We obtain:

Steady shear viscosity

$$\eta = \frac{\tau_{12}}{\dot{\gamma}} = \eta_s + n_o kT\lambda \qquad (11.4.15)$$

First normal stress coefficient

$$\psi_1 = \frac{\tau_{11} - \tau_{22}}{\dot{\gamma}^2} = 2n_o kT\lambda^2 \qquad (11.4.16)$$

Second normal stress coefficient

$$\psi_2 = \frac{\tau_{22} - \tau_{33}}{\dot{\gamma}^2} = 0 \qquad (11.4.17)$$

We see first of all that this model predicts η independent of shear rate. This serious deficiency results from the oversimplified molecular model. There are four classes of intramolecular effects (we are still saving the intermolecular effects until later) we have not included; when they are included, however, they will give a shear rate dependent viscosity (Williams, 1975; Larson, 1988) incorporating: (a) hydrodynamic interaction, (b) reduced excluded volume effects, (c) nonlinear spring force law, and (d) internal viscosity. Effect a refers to the perturbation of the velocity field in the vicinity of a bead by the presence of a nearby bead. These effects have been incorporated into the Zimm model version of the bead–spring theory (Zimm, 1956). Zimm uses the Oseen tensor (Happel and Brenner, 1965) to model these hydrodynamic interactions between beads. Effect b refers to the expansion of the coil by deformation and the consequent reduction of intramolecular segment–segment contacts. Effect c results from the fact that a real coil is not infinitely extendable and frequently is embodied theoretically by the Finite Extensibility Nonlinear Elastic connector force, which is a connector of variable stiffness from the elastic dumbbell ($b = \infty$) to rigid rod ($b = 0$), where $b = HR_0^2/kT$ (Bird et al., 1987). Effect d refers to the polymer coil sluggishness of the response of the polymer coil to the deforming forces. Inclusion of any of these effects leads to shear thinning viscosity prediction. This behavior is, of course, experimentally observed (Figure 11.4.3).

This simple model does predict a finite positive first normal stress coefficient in qualitative agreement with observations.

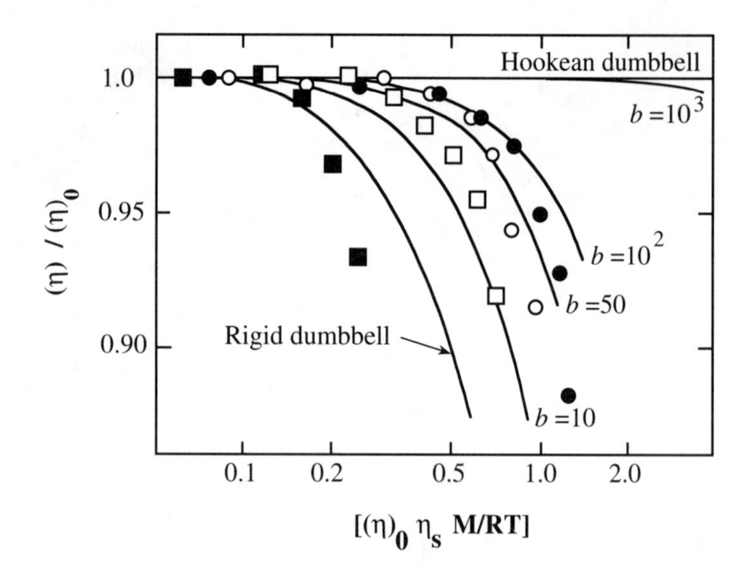

Figure 11.4.3. Illustration of shear thinning. Experimental data of intrinsic viscosity of a series of monodisperse poly (α-methylstyrene) samples in toluene: (filled in square $M = 694,000$; square $M = 1,240,000$; circle $M = 1,460,000$; filled in circle $M = 1,820,000$, the data being taken from Noda et al. (1968). The theoretical results for Hookean and rigid dumbbells also are shown; the other four curves are the theoretical results for the FENE dumbbells for several values of the dimensionless stiffness parameter b. From Bird et al. (1976).

Second normal stress coefficients have not yet been reliably determined, and thus eq. 11.3.17 is difficult to evaluate at this time. They do seem to be finite, small ($\sim$ 0.1 N_1), and negative.

The elongational viscosity prediction is

$$\eta_e = 3\eta_s + \frac{3n_o kT\lambda}{(1 + \lambda\dot{\epsilon})(1 - 2\lambda\dot{\epsilon})} \tag{11.4.18}$$

giving an η_e that increases with $\dot{\epsilon}$, going to ∞ as $\dot{\epsilon} \to 1/2\lambda$. This is also presently difficult to evaluate experimentally because a *steady constant strain rate* elongational flow is difficult to achieve.

The dynamic properties are

$$\eta' = \eta_s + \frac{n_o kT\lambda}{1 + \lambda^2\omega^2} = \frac{G''}{\omega} \tag{11.4.19}$$

$$\eta'' = \frac{n_o kT\lambda^2\omega}{1 + \lambda^2\omega^2} = \frac{G'}{\omega} \tag{11.4.20}$$

Plots of some of these functions are shown in Figure 11.4.4. The qualitative behavior is quite reasonable.

Figure 11.4.4 shows that experimentally $\eta(\dot{\gamma})$ and $\eta'(\omega)$ and $|\eta^*(\omega)|$ are all very similar functions. Other predictions for some of the linear viscoelastic functions are recoverable steady state shear compliance

$$J_e^0 = \frac{\gamma_r}{\tau_o} = \frac{n_o kT\lambda^2}{(\eta_s + n_o kT\lambda)^2} \tag{11.4.21}$$

and stress relaxation modulus

$$G(t) = \frac{1}{3} n_o kT e^{-t/\lambda} \tag{11.4.22}$$

The final prediction that comes from the model of the Hookean elastic dumbbell concerns the molecular conformation changes induced by flow. In shear flow the prediction is

$$\frac{\bar{x}_1}{\bar{x}^2_{1_o}} = 1 + 2\dot{\gamma}^2\lambda^2 \tag{11.4.23}$$

$$\frac{\bar{x}^2_2}{\bar{x}^2_{2_o}} = \frac{\bar{x}^2_3}{\bar{x}^2_{3_o}} = 1 \tag{11.4.24}$$

Thus, in contrast to the constitutive relations, which have their origins in continuum mechanics, molecular theories give results on the conformational properties as well as the macroscopic rheological properties.

11.4.2 Rouse and Other Multibead Models

The results of the simple Hookean elastic dumbbell lack realism because they have only a single relaxation time. This is the chief virtue of the Rouse model (Rouse, 1953; Ferry, 1980). It consists of N beads connected by $N-1$ linear springs (refer back to Figure 11.4.1). Its rheological predictions are *exactly* the same as those in eqs. 11.4.15–11.4.22 if λ is replaced by λ_i and $\sum_{i=1}^{N}$ is inserted before all terms involving λ_i. This introduces the idea of a relaxation time distribution where

$$\lambda_i = \frac{\lambda}{2\sin^2(i\pi/2N)} \approx \lambda\left(\frac{2N^2}{i^2\pi^2}\right) \quad \text{for large } N \tag{11.4.25}$$

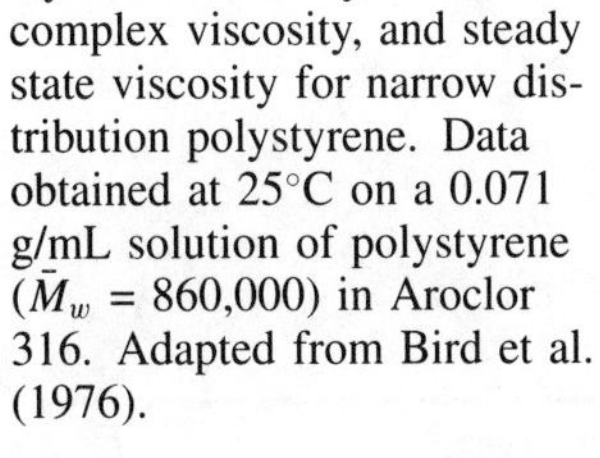

Figure 11.4.4. Dynamic viscosity, absolute complex viscosity, and steady state viscosity for narrow distribution polystyrene. Data obtained at 25°C on a 0.071 g/mL solution of polystyrene ($\bar{M}_w$ = 860,000) in Aroclor 316. Adapted from Bird et al. (1976).

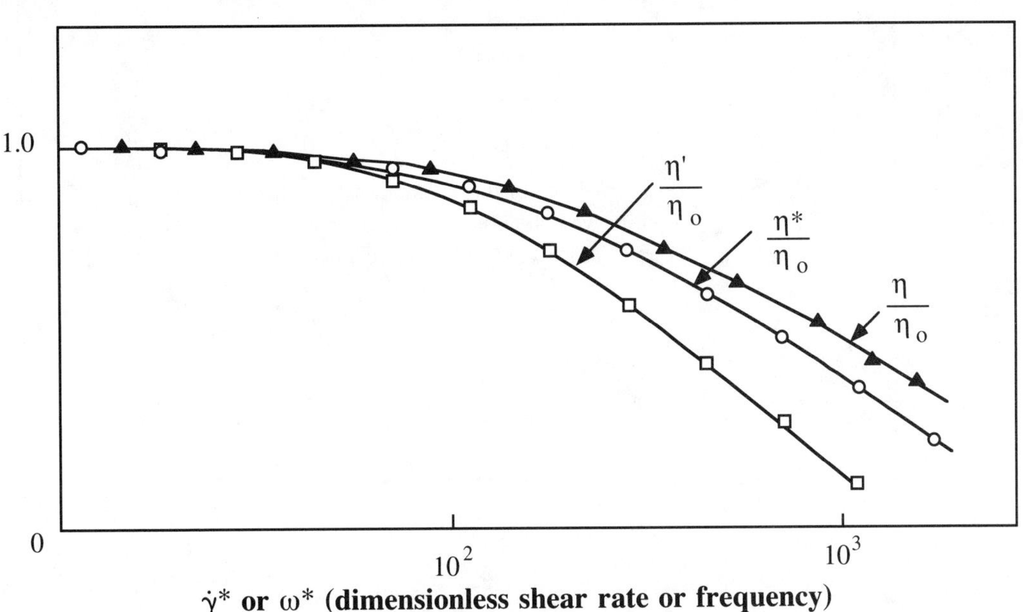

Equations for J_e^0 at $G(t)$ now are

$$J_e^0 = \frac{1}{n_o kT}\frac{(\eta - \eta_s)^2}{\eta^2}\left[\sum_{i=1}^{N}\lambda_i^2 \Big/ \left(\sum_{i=1}^{N}\lambda_i\right)^2\right] \tag{11.4.26}$$

$$G(t) = \frac{1}{3}n_o kT \sum_{i=1}^{N} e^{-t/\lambda_i} \quad \text{or} \quad \frac{\rho RT}{M}\sum_{i=1}^{N} e^{-t/\lambda_i} \tag{11.4.27}$$

This is an improvement in fitting the data on linear viscoelastic properties (see Figure 11.4.5) but does not help in the lack of nonlinear prediction. We see that the recoverable shear compliance is governed by the breadth or relative spacings of the relaxation time distribution. It is important to realize that the idea of a relaxation time distribution is independent from the idea of a molecular weight distribution. Even a monodisperse polymer sample will have a distribution of relaxation times as a result of the various internal degrees of freedom or modes of motion of the chain molecule. For the bead–spring models all the relaxation times are of the form

$$\lambda \sim \frac{\xi}{4H}$$

which are readily related to measurable molecular properties (Ferry, 1980)

$$\lambda = \frac{\bar{r}_o^2 \xi}{6\pi^2 kT} \quad \text{or} \quad = \frac{6[\eta]\eta_s M_w}{\pi^2 RT} \quad \text{or} \quad \frac{6\eta_o M}{\pi^2 cRT} \tag{11.4.28}$$

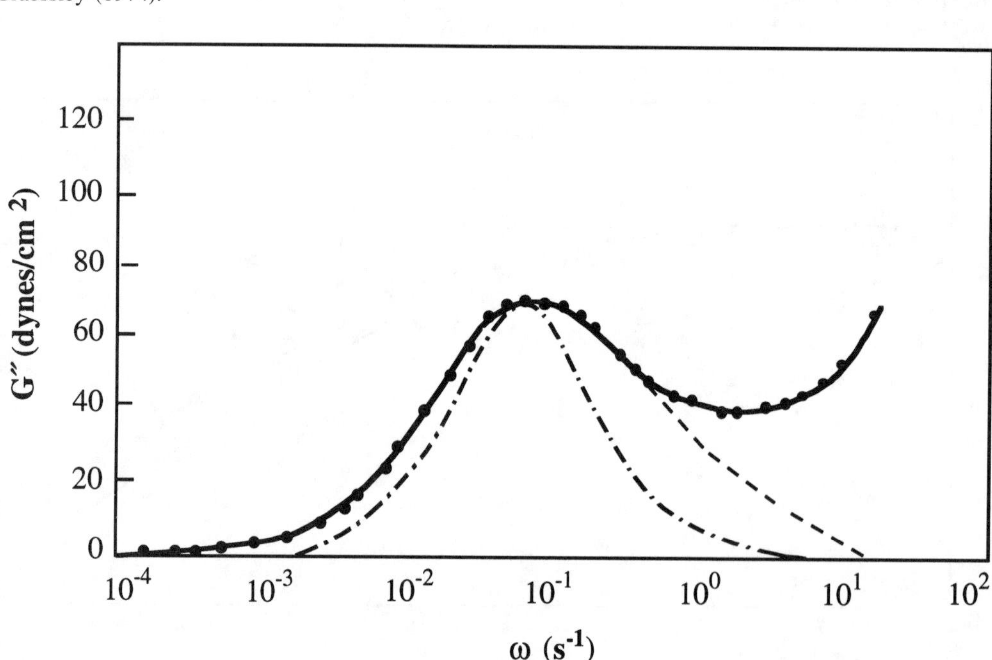

Figure 11.4.5. "Breadth" of relaxation process even for a monodisperse polymer: loss modulus versus log frequency for a narrow distribution polystyrene melt reduced to 160°C ($\bar{M}_w = 21,500$). The dashed line is an approximate resolution of the terminal relaxation peak. The constructed line (— · — · —) is G'' versus ω calculated from the Rouse model. From Graessley (1974).

where η_o is the zero shear viscosity. Note that λ depends basically on molecular weight and temperature.

11.5 Concentrated Solutions and Melts*

As we move away from dilute solutions to more concentrated systems, we can no longer look at isolated molecules, and a significant theoretical problem is: How does one model intermolecular entanglement? Before attempting to answer this question, let us look at some experimental evidence indicative of entanglement.

11.5.1 Entanglements

Look at Figure 11.5.1. The plateau between log $E = 6$ and 8 is due to entanglements. The Rouse theory (surprisingly!) is able to model the terminal (long time) portion of this type of curve rather well, but shows no plateau. This is due to entanglements. They are very pronounced in the G' data of Figure 11.5.2. These entanglements also show up as an additional loss peak (Figure 11.4.5) in tan δ or G''. There is a new relaxation appearing for polymer liquids above a critical molecular weight and/or concentration. The Rouse theory is unable to account for this because it considers intramolecular effects only.

Figure 11.5.3 shows another important feature of macromolecular rheology in the entanglement regime. That is, an abrupt change in slope of the zero shear viscosity–cM curve. This break

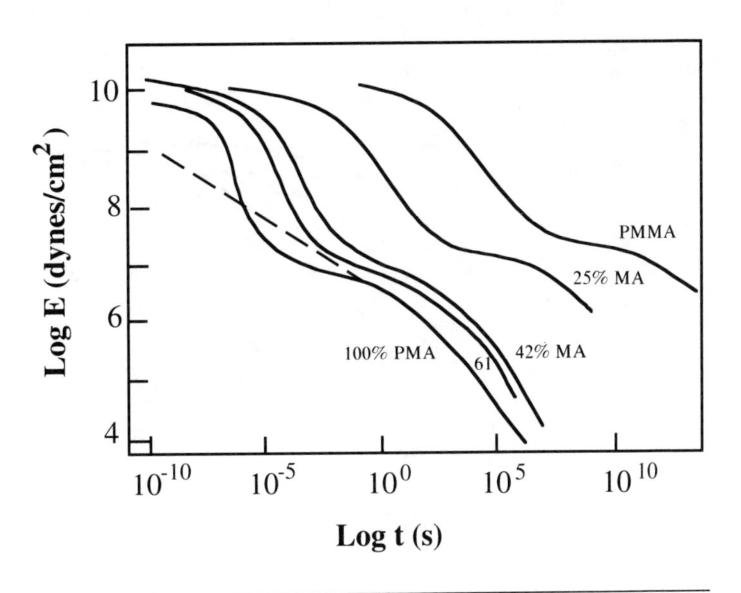

Figure 11.5.1. Stress relaxation modulus for methyl acrylate–methyl methacrylate copolymers. Dashed line illustrates a typical Rouse theory prediction. Adapted from Fujino et al. (1961).

This section relies heavily on reviews by Graessley (1974, 1982).

also occurs in the zero shear viscosity–molecular weight curve for polymer melts (see Figure 11.5.4). It can be seen that

$$\eta_o = K\ M \qquad M < M_c$$

$$\eta_o = K\ M^{3.4} \quad M > M_c$$

where M_c is the molecular weight at the break point. M_c has been interpreted as being a critical molecular weight for the formation of effective entanglement couples. Other methods exist for estimating the spacings between entanglement couples from linear viscoelastic measurements (Ferry, 1980).

Entanglements have become part of the folklore of polymer science; the idea they represent is a particular type of intermolecular interaction, to be distinguished from the coil overlap type of interaction mentioned earlier. However their exact topological character is quite difficult to define.

Entanglements arise from segment–segment contacts between molecules. The number of intermolecular contacts per unit volume is proportional to c^2. Since the number of polymer molecules per unit volume is proportional to c/M, the number of intermolecular contacts per molecule is proportional to cM. Thus we see that because $c[\eta]$ or cM^a (coil overlap) and cM (entanglement) involve different combinations of c and M, in principle it should be possible to experimentally distinguish between the two types of interactions. Figure 11.5.5 compares both data reduction schemes

Figure 11.5.2.
Storage modulus versus frequency for narrow distribution polystyrene melts of increasing molecular weight, reduced to 160°C by temperature–frequency superposition. Molecular weight ranges from $\bar{M}_w$ = 8900 (L9) to $\bar{M}_w$ = 581,000 (L18). From Onogi et al. 1970).

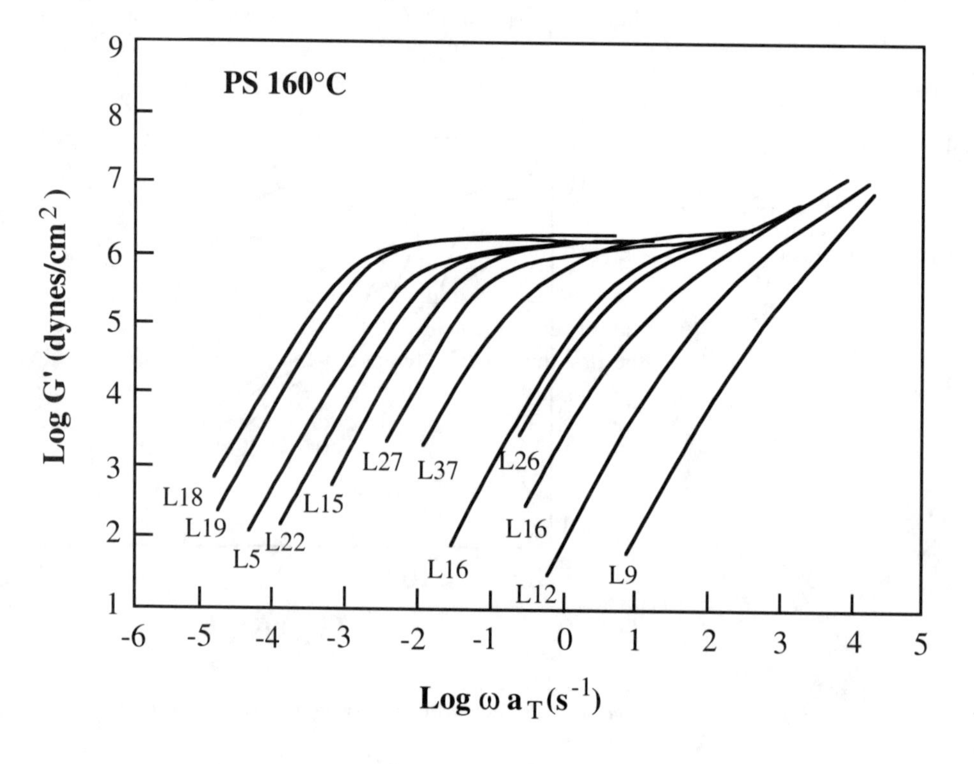

Figure 11.5.3.
Viscosity versus the product $c\bar{M}_w$ for polystyrenes at concentrations between 25 and 100%. Data at the various concentrations have been shifted vertically to avoid overlap. From Graessley (1974).

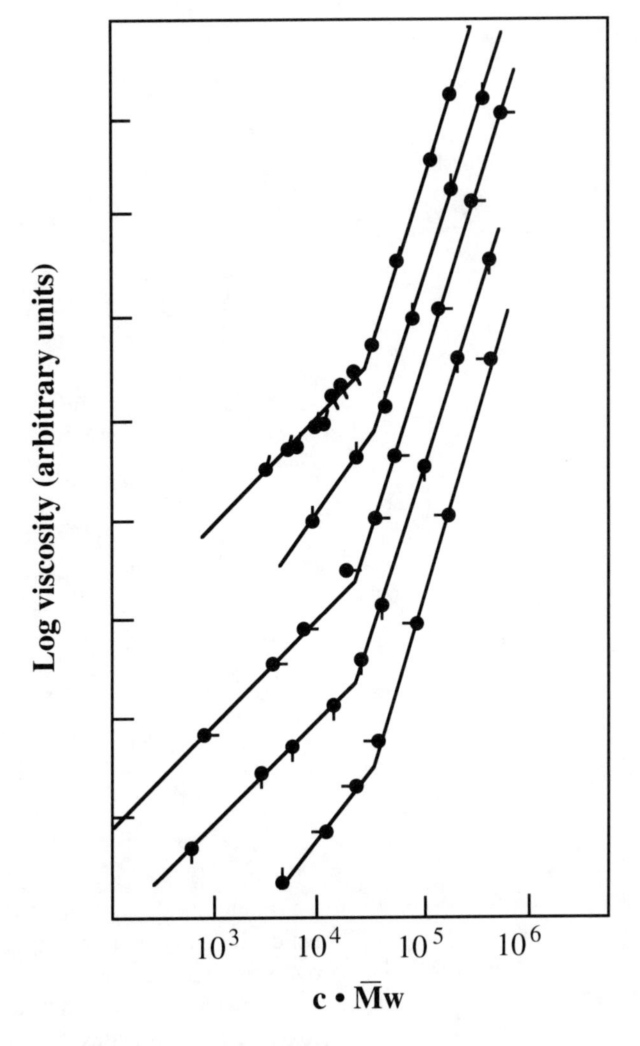

for concentrated polymer solutions ($c[\eta] > 10$); the cM correlation deduced from entanglement considerations is clearly superior in the concentrated regime.

We will mention only the barest essential features of the numerous entanglement theories (Graessley, 1974). We begin with the first theoretical attempt to capture the entanglement effect. It is necessary to know what factors control the number of entanglements per unit volume, the entanglement density, in steady flow. Relative motion is imposed on the system as molecules, which had been close enough to entangle extensively, separate. The net number of entanglements for any particular molecule, therefore, must depend on its rate of entanglement formation with approaching molecules. Thus, the entanglement process may be conceptualized as a kinetic process. If the entanglement process is modeled as a first-order kinetic process (Lodge, 1956), then some of the elastic character associated with entanglements is well-mimicked, but a shear rate independent viscosity function is obtained. However,

Figure 11.5.4.
Typical viscosity–molecular weight dependence for molten polymers. x_w is proportional to the number of backbone atoms and M_w. From Berry and Fox (1968).

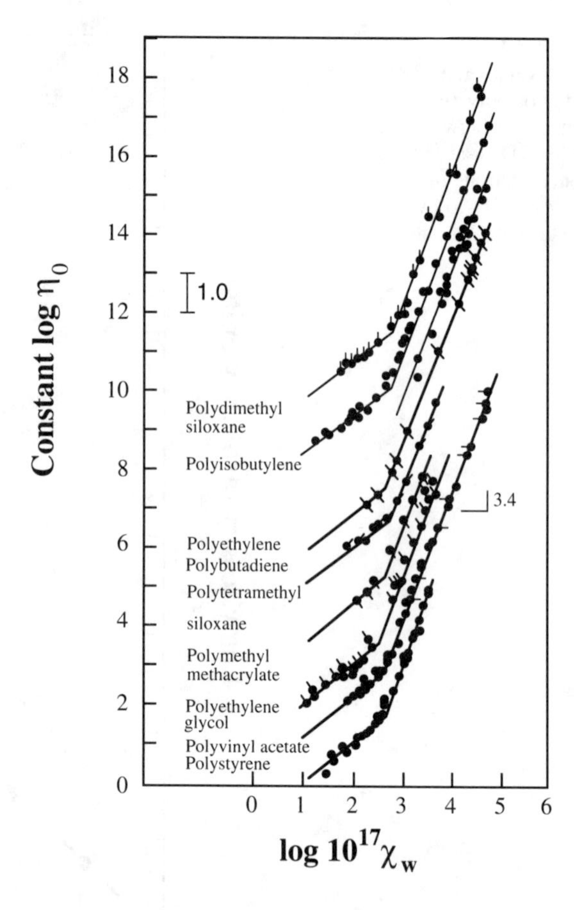

we should realize from our earlier discussion that entanglement kinetics should not even be expected to be a first-order process, since by nature, entanglement is a bimolecular process.

Imagine two initially widely separated chains, rapidly brought together and maintained at some fixed (center-of-mass) separation. Although considerable overlap in domains occurs, one can imagine the chains being relatively undisturbed in a gross sense by each other's presence. To a casual observer both chains exhibit random thermal motions, and nothing essential changes with time. However, a change *is* occurring in that the chains are becoming increasingly entangled. This might be observed (hypothetically) by attempting to separate the chains after some passage of time and noting the initial amount of force resisting the separation. (See Figure 11.5.6a.)

This process may be understood by noting that for any pair of chains close enough to entangle extensively, the majority of conformations available to the two chains furnishes a high entanglement density. However, at the instant the chains are brought together, the conformation is necessarily an improbable one because it involves *no* entanglement. Thus the passage of the pair to a high entanglement density involves a succession of diffusive rearrangements. The chains enter progressively more probable states with respect to entanglement density, approaching some equilibrium entanglement density as the conformations become increasingly independent of the initial conformation.

The formation of entanglements between chains requires the thermal motion of rather sizable sections of the chains. Furthermore, at least two or three sequential movements should be necessary to produce loops for entanglement. Thus, entanglement kinetics may be sluggish initially; on the other hand, once entanglement has begun to occur, a relatively rapid growth to the equilibrium amount of entanglement is to be expected. The series of random motions required to form the first entanglement loop leads to other entanglement loops nearly simultaneously. Furthermore, other sections of the same chains would be diffusing also and would be expected to begin producing entanglements at approximately the same time. Figure 11.5.6b shows qualitatively the presumed behavior of the entanglement density with time, averaged over many pairs of chains. This type of process is known in physics as a *cooperative* process.

Returning to the system of molecules in steady flow, we can see that the entanglement density between any two passing molecules will depend on the characteristic time necessary for entanglement compared to the "contact" time between the two

Figure 11.5.5. Viscosity versus (a) $CM_w^{0.68}$ and (b) $c\bar{M}_w$ for polystyrenes at concentrations between 25 and 100%. The data have been shifted vertically to produce superposition at high molecular weights. From Graessley (1974).

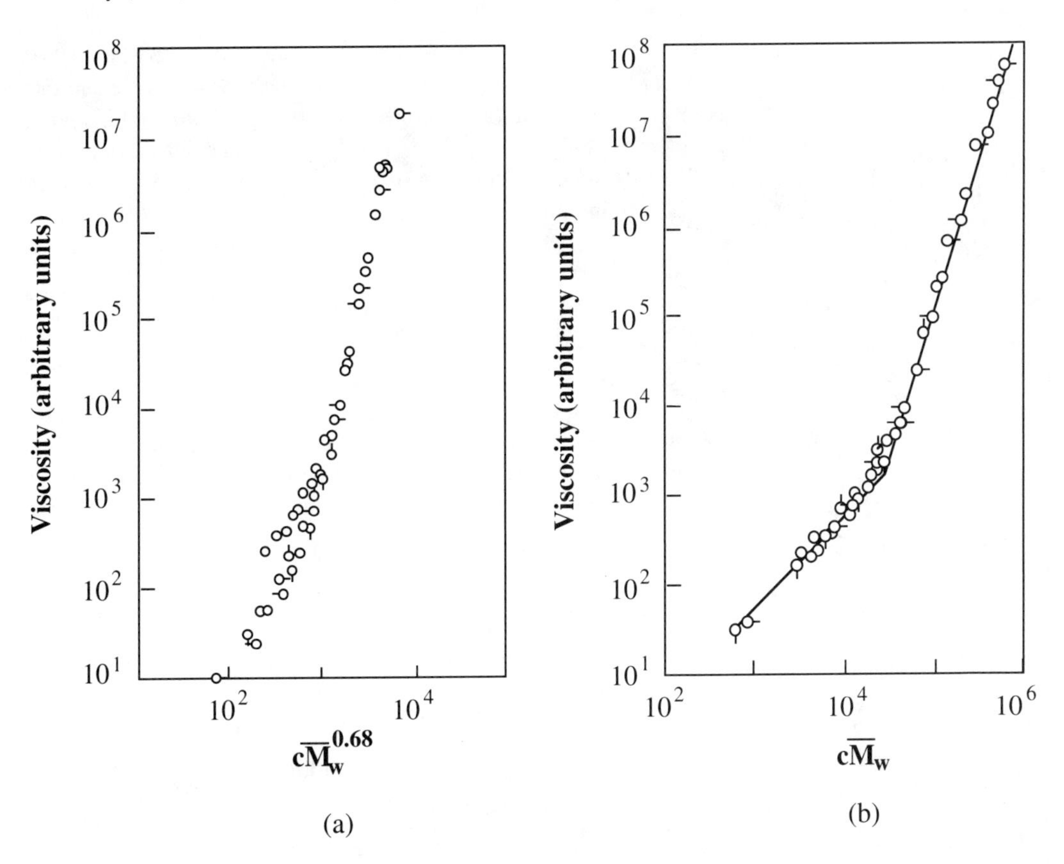

molecules. Two molecules may be thought of as approaching each other in a shear field. When they are sufficiently close, entanglements begin to occur at a finite rate. As the molecules pass, disentanglement occurs. An entanglement density for the bulk material may be defined to characterize the number of entanglements that exist at any instant, averaged over the material. For an entanglement to exist, two molecules must first be within a certain distance of each other. Second, the molecules must remain within this sphere for a finite time; otherwise, no entanglement will occur. The greater the shear rate, the more rapidly two molecules move relative to each other. Hence, the entanglement density is reduced by high shear rate, because fewer molecules will remain in the "entanglement sphere" for a sufficiently long time at high shear rate.

This model, which was among the very first proposed to explain some aspects of entangled polymers, is mainly aimed at understanding some of the important and dramatic, *nonlinear* rheological properties, particularly the shear rate dependence of viscosity (Graessley, 1974). The approach was difficult to extend to viscoelastic effects quantitatively and has been abandoned in favor of the reptation model.

11.5.2 Reptation Model

In 1971 DeGennes proposed a new model for molecular motion in concentrated polymer systems, one that now dominates all theoretical considerations (Lodge et al., 1990). His model is known as "reptation" because it describes macromolecular motion much like that of a snake moving in a contorted "tunnel" formed by the surrounding polymer molecules (Figure 11.5.7). The basic idea is

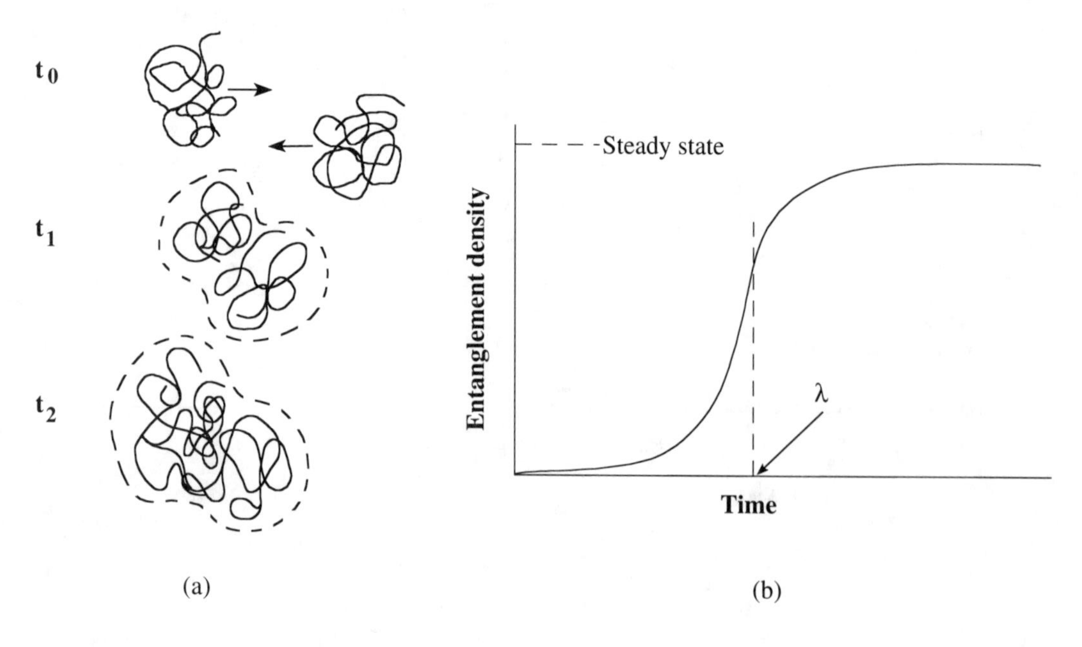

Figure 11.5.6.
(a) Schematic of an entanglement and (b) plot of entanglement density versus time.

that in an entangled polymer fluid a monomer unit can move only in one direction, either by Brownian motion or in response to an applied force: that is, along its axis (see Figure 11.1.1). If we describe the probability P as finding a particular segment at some position s at time t, in a coordinate system where s is measured along the chain axis (an arc length coordinate), then P satisfies this simple one-dimensional diffusion equation:

$$\frac{\partial P}{\partial t} = D_o \frac{\partial^2 P}{\partial s^2} \tag{11.5.1}$$

with the boundary conditions

$$P(s, 0) = \delta(0)$$

$$P(s, t) = 0 \text{ as } s \to \pm\infty$$

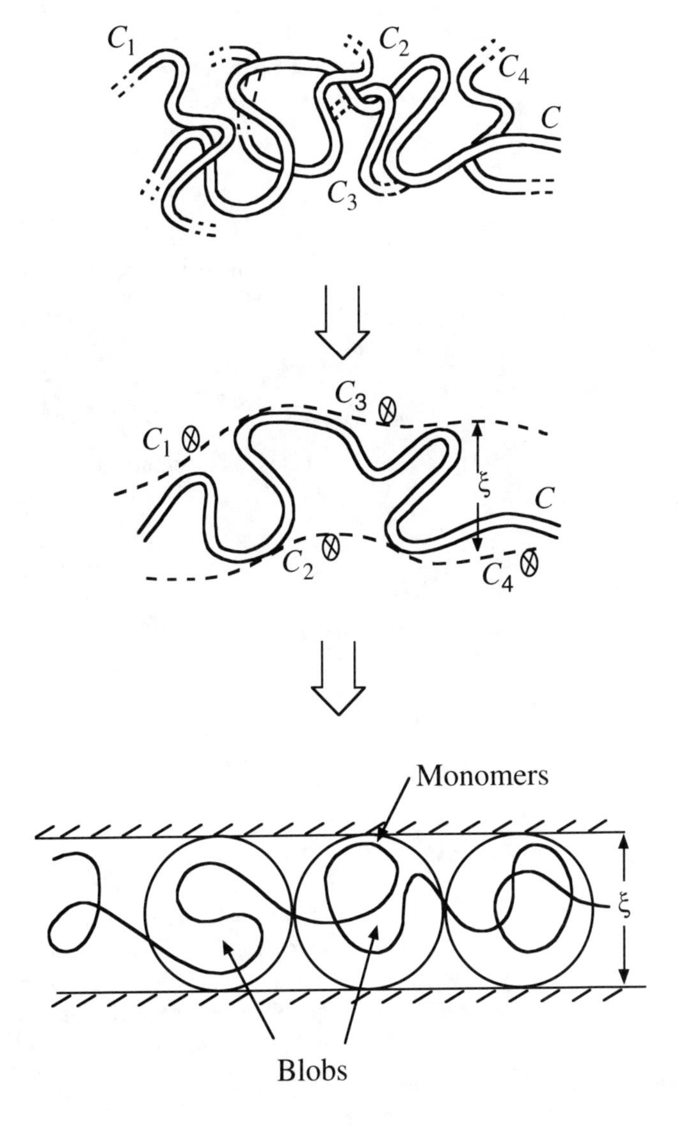

Figure 11.5.7.
The idea of reptation in polymer solutions (taken from articles by Klein, 1978, and de Gennes, 1979). A given polymer chain C entangled with other chains $C_1 - C_4$ (top) may be regarded as enclosed within a virtual pipe (bottom), defined by the locus of the constraints imposed on its motion by the other chains. ξ is the mean permitted displacement of a segment of C in a direction normal to the "pipe axis," and the points ($\otimes$) in the middle figure represent cross sections through $C_1 - C_4$ in a plane parallel to the paper. Since each of the chains $C_1 - C_4$ may itself be regarded as being in a similar pipe, the mean separation of the cross sections ($\otimes$) is also $\sim \xi$.

where δ is the delta function, meaning the first segment is at the origin, and D_o is the diffusion coefficient for motion along the chain. We expect $D_o \sim 1/M$; that is, the resistance to motion of a "rope" along its axis is proportional to the length of the rope. The solution to eq. 11.5.1 is

$$P(s,t) = \frac{1}{(4\pi D_o t)^{1/2}} e^{-s^2/4D_o t} \tag{11.5.2}$$

The mean-square displacement of s at time t is

$$< s^2 > = \int_{-\infty}^{\infty} s^2 P(s,t) ds = 2D_o t \tag{11.5.3}$$

We define a time λ_{rep} as the time required for a chain to completely renew its configuration, that is, the time for the chain to diffuse one chain length L along s. We recognize that this will also be the longest relaxation time of the polymer liquid. From eq. 11.5.3 we see

$$\lambda_{\text{rep}} = \frac{L^2}{2D_o} \tag{11.5.4}$$

From which the molecular weight dependence follows directly $(L \sim M)$

$$\lambda_{\text{rep}} \sim \frac{M^2}{M^{-1}} \sim M^3 \tag{11.5.5}$$

This is to be compared with the prediction of the Rouse theory for the relaxation times λ_{Rouse} (eq. 11.4.25)

$$\text{Rouse}: \ \lambda_{\text{Rouse}} \sim N^2$$

$$\text{deGennes}: \ \lambda_{\text{rep}} \sim N^3 \tag{11.5.6}$$

One can also estimate the self-diffusion coefficient

$$D_s = \frac{\bar{r^2}}{6\lambda_{\text{rep}}} \sim \frac{M}{M^3} \sim M^{-2} \tag{11.5.7}$$

This molecular weight dependence of D has been seen experimentally in melts by Klein (1978) and in concentrated solutions by Leger et al. (1981). Reviews of diffusion behavior are available (Tirrell, 1984; Kausch and Tirrell, 1989). One can also deduce the molecular weight dependence of the viscosity by a nonrigorous but plausible argument. Suppose the entire fluid behaves as a simple viscoelastic solid (Maxwell element); then its relaxation time would be

$$\lambda = \frac{\eta}{E} \tag{11.5.8}$$

We expect the modulus E to be independent of overall molecular weight—to depend only on the molecular weight between "entanglements"—that is, M_b, the weight of a blob. We reach the conclusion

$$\eta \sim E\lambda \sim M^o M^3 \sim M^3 \tag{11.5.9}$$

Of course this is not the 3.4 power observed experimentally (Figure 11.5.2), but it is the closest of any molecular theory. Many suggestions have been made to explain the exact molecular weight dependence (Lodge et al., 1990).

This derivation of the viscosity has been made much more rigorous (but still reaches the same conclusion) by Doi and Edwards (1978). They derive a constitutive relationship from this "reptation" molecular picture (Larson, 1988). Their result can be expressed as a complete constitutive equation in the form

$$\boldsymbol{\tau} = G_o M(t) \mathbf{Q}(\mathbf{E}) \tag{11.5.10}$$

where

$$\tau_{22} - \tau_{33} = n_o H[< x_2^2 > - < x_3^2 >] \tag{11.4.8}$$

Average values are calculated in the usual way by integrating over the distribution function:

$$G_o = 3cNkT \quad \text{(equilibrium or plateau modulus)}$$

$$M(t) = \sum_{p_{\text{odd}}} \frac{8}{p^2\pi^2} e^{-tp^2/\lambda_{\text{rep}}} \quad \text{(memory function)}$$

$$\mathbf{Q}(\mathbf{E}) = \frac{1}{< \rho >} < \frac{\rho\rho}{\rho} > \quad \text{(strain measure)}$$

and where $\boldsymbol{\rho} = \mathbf{E} \cdot \mathbf{u}$ and $\mathbf{E}$ is an affine deformation tensor and $\mathbf{u}$ is a unit vector along the chain axis. This constitutive equation resembles a BKZ or Wagner-type model (Chapter 4) in that the time and strain portions are separable. It models stress relaxation well [although $M(t)$ is rather too close to a single exponential]. It predicts a non-Newtonian viscosity that depends too strongly on shear rate ($\eta \sim \dot{\gamma}^{-3/2}$ at high $\dot{\gamma}$) and predicts an infinite extensional viscosity at finite extension rate. Marrucci and Hermans (1980) discuss some possible improvements.

11.5.3 Effects of Long Chain Branching

Branched macromolecules obviously cannot reptate the way linear macromolecules do, since the long branches anchor the chain.

For chain motion or relaxation to occur, the branched chain must pull out pieces of its arms as illustrated in Figure 11.5.8. This is much slower than ordinary reptation. DeGennes (1979) has shown that the longest relaxation time now depends exponentially (!) on molecular weight instead of varying like M^3. Doi and Kuzuu (1980) have derived a constitutive equation for branched polymers that parallels the Doi–Edwards model. The strain-dependent function **Q(E)** does not change. However, the relaxation function $M(t)$ is changed markedly. Compared with linear polymers, the maximum relaxation time is longer and the relaxation spectrum is more diffuse. See Figure 11.5.9. Exactly this has been observed experimentally (Graessley et al., 1981; Roovers and Graessley, 1981).

In terms of viscosity, branched polymers have lower zero shear viscosity than linear polymers of the same total molecular weight ($M > M_e$) when the branches are short ($M_b < M_e$). When $M_b > M_e$, the branched chain viscosities overtake those of the equally massive linear chains.

11.5.4 Effect of Molecular Weight Distribution

It is worthwhile to mention, at last, the effect of molecular weight distribution (MWD) on rheological properties. We will limit our discussion to the steady shear viscosity η and the recoverable shear compliance J_e^0.

Viscosity of polymer systems approaches a limiting constant value at low shear rates. The onset of shear rate dependence is usually quite sharp for monodisperse polymers. However, the broader

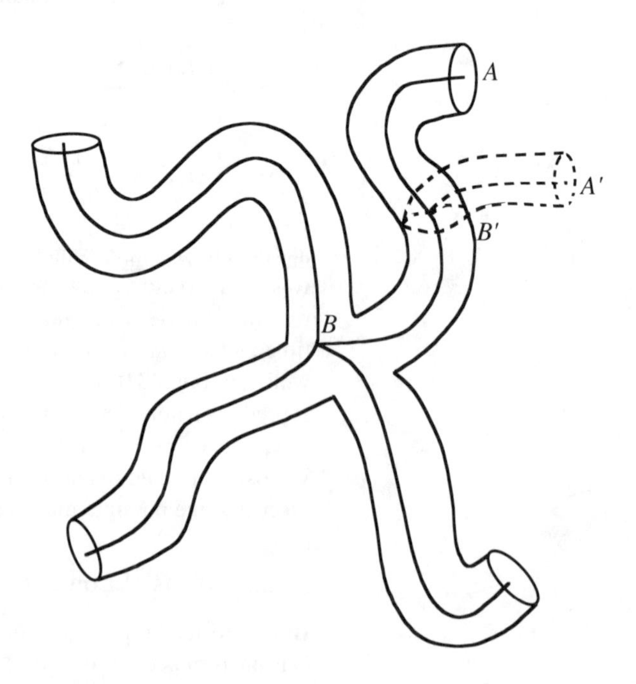

Figure 11.5.8.
Star polymer relaxation process. Individual arms cannot reptate; rather, they relax by retracting from the tubes created by entanglements with surrounding polymers.

Figure 11.5.9.
G' and G'' for branched (solid lines) and linear (dashed lines) polymers $\alpha = M_{arm}/M_e$. From Doi and Kuzuu (1980).

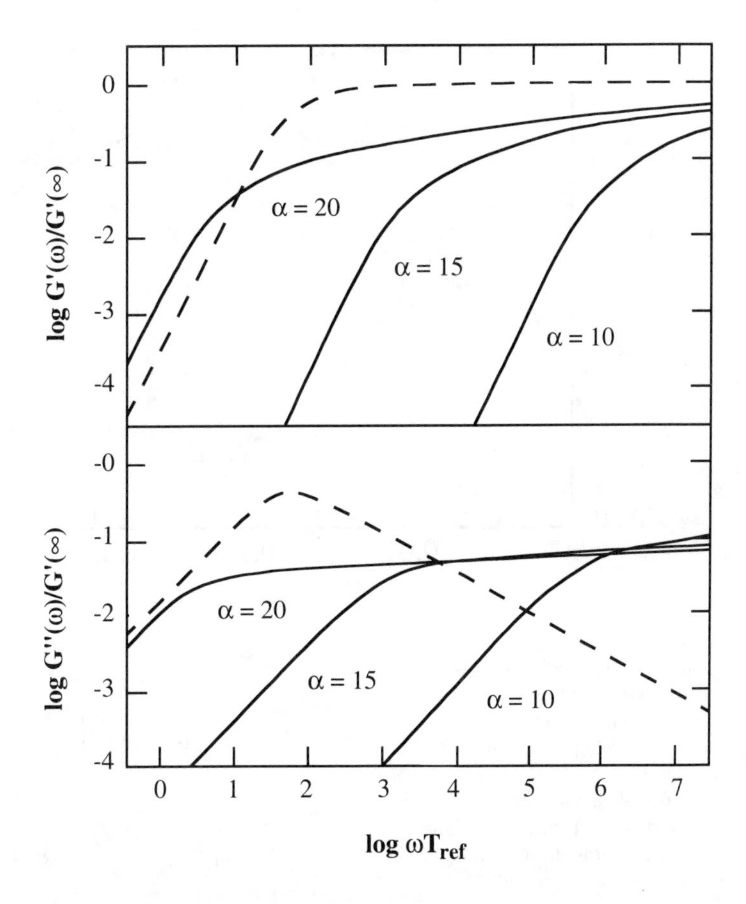

the molecular weight distribution, the more diffuse the onset of shear rate dependence. This may actually present problems in determining the zero shear viscosity, for example, in a broad MWD polyethylene. Figure 11.5.10 illustrates this behavior.

Equation 11.4.26 shows that J_e^0 is governed by the relative spacings or breadth of the relaxation time distribution (RTD). Certainly, as the MWD gets broader so will the RTD. Consequently, the compliance goes up. This means that for the same average molecular weight, a sample with broader MWD will be more compliant. According to the Rouse theory, $J_e^0\, c\, RT/M$ should be constant at 0.4 for monodisperse polymers. Figure 11.5.11 shows some experimental data that are better characterized by (Graessley, 1974)

$$J_e^0 = \frac{0.4M/cRT}{[1 + 0.08\,(cM/\rho M_c)^2\,]^{1/2}} \tag{11.5.11}$$

The Rouse theory also predicts that J_e^0 should increase with polydispersity as:

$$J_e^0 \sim \frac{\bar{M}_z \bar{M}_{z+1}}{\bar{M}_w^2} \tag{11.5.12}$$

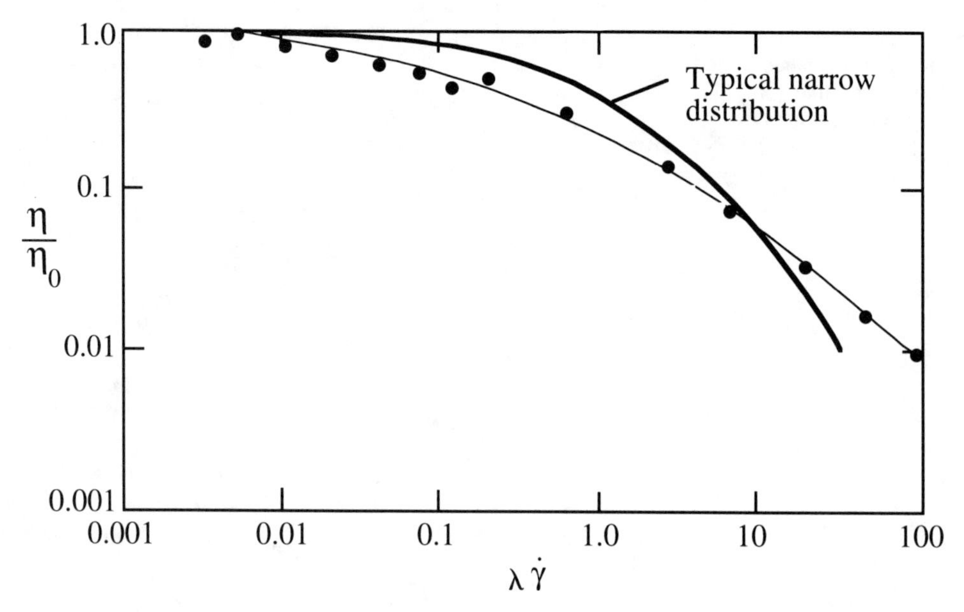

Figure 11.5.10.
Comparison of reduced viscosity versus reduced shear rate for a broad (•) and a narrow MWD polyethylene (Graessley, 1974). Notice how polydispersity broadens the transition from Newtonian to non-Newtonian behavior.

Figure 11.5.11.
Recoverable shear compliance for several polymers. Lines represent experimental data (Graessley, 1974). Asymptote of 0.4 is the Rouse number.

Experimentally, it is found that

$$J_e^0 \sim \left(\frac{\bar{M}_z}{\bar{M}_w}\right)^2 \tag{11.5.13}$$

This uncertainty may be due in part to the difficulties of obtaining "equilibrium" values for J_e^0 with the different techniques used to measure it.

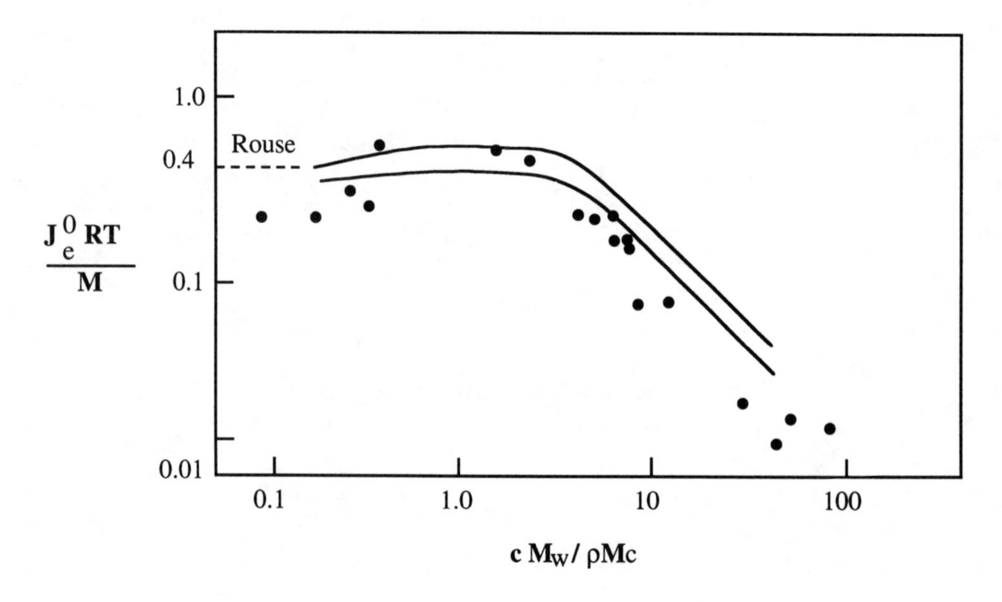

TABLE 11.6.1 / Parameters Characterizing Temperature Dependence of a_T for Various Polymer Systems*

Polymer	T_0 (K)	C_1^0	C_2^0 (deg)	T_g (K)
General				
Polyisobutylene	298	8.61	200.4	205
Polyvinyl acetate	349	8.86	101.6	305
Polyvinyl chloroacetate	346	8.86	101.6	296
Polystyrene	373	12.7	49.8	370
	373	13.7	50.0	373
Poly(α-methyl styrene)	445	13.7	49.3	445
	441	16.8	53.5	441
Polymethyl acrylate	324	8.86	101.6	276
	326	8.86	101.6	
Polyhexene-1	218	17.4	51.6	218
Polydimethylsiloxane	303	1.90	222	150
Polyacetaldehyde	243	14.5	24	243
Polypropylene oxide	198	16.2	24	198
Zinc phosphinate polymer	373	6.94	66.6	324
Styrene-*n*-hexyl methacrylate copolymer	373	7.11	192.6	277
Styrene-*n*-hexyl methacrylate copolymer	373	6.56	156.4	287
Rubbers				
Hevea rubber	248	8.86	101.6	200
	298	5.94	151.6	
Polybutadiene, *cis-trans*	298	3.64	186.5	172
Polybutadiene, high *cis*	298	3.44	196.6	161
Polybutadiene, *cis-trans*-vinyl	263	5.97	123.2	205
Polybutadiene, high vinyl	298	6.23	72.5	261
Styrene–butadiene copolymer	298	4.57	113.6	210
Butyl rubber	298	9.03	201.6	205
Ethylene–propylene copolymer	298	5.52	96.7	242
Ethylene–propylene copolymer	298	4.35	122.7	216
Polyurethane	283	8.86	101.6	238
Polyurethane	231	16.7	68.0	221
Methacrylate Polymers				
Methyl (atactic)	381	34.0	80	381
Methyl (isotactic)	323	8.90	23.0	
Methyl (conventional)	388	32.2	80	388
	493	7.00	173	378
Ethyl	373	11.18	103.5	335
n-Butyl	373	9.70	169.6	300
n-Hexyl	373	9.80	234.4	268
n-Octyl	373	7.60	227.3	253
2-Ethyl hexyl	373	11.58	208.9	284
Diluted Systems				
Polystyrene in decalin 62%	291	8.86	101.6	
Polyvinyl acetate in tricresyl phosphate 50%	293	8.86	101.6	
Poly(*n*-butyl methacrylate) in diethyl phthalate 50%	273	9.98	153.1	206

*Adapted from Ferry, 1980

11.6 Temperature Dependence

Finally, it is worthwhile to discuss the temperature dependence of polymer melt flow properties, because much useful structural information can be obtained from it. Temperature dependence also was discussed in Section 2.6.4.

As with all liquids, the viscosities of synthetic polymer melts decrease with increasing temperature. However the form of the temperature dependence is rather complex. At elevated temperatures, well above any transition temperatures, the viscosity follows a simple exponential relation (the Andrade–Eyring equation):

$$\eta = \eta_\infty \exp\left[\frac{E_\eta}{RT}\right] \tag{11.6.1}$$

where E_η is an activation energy for viscous flow. Here viscosity is governed by the barriers to relative slip of molecular planes. For melts of glassy polymers in the region of T_g to $T_g + 100$, an equation of the Williams–Landel—Ferry (WLF) form is usually appropriate:

$$\log \frac{\eta T_r \rho_r}{\eta_r T \rho} = \frac{-C_1(T - T_r)}{C_2 + T - T_r} = \log a_T \tag{11.6.2}$$

where subscript r refers to conditions at some arbitrary reference condition, often taken as T_g. C_1 and C_2 are constants. Values for many polymers are tabulated in Table 11.6.1. This equation is a direct result of the dependence of viscosity on free volume for

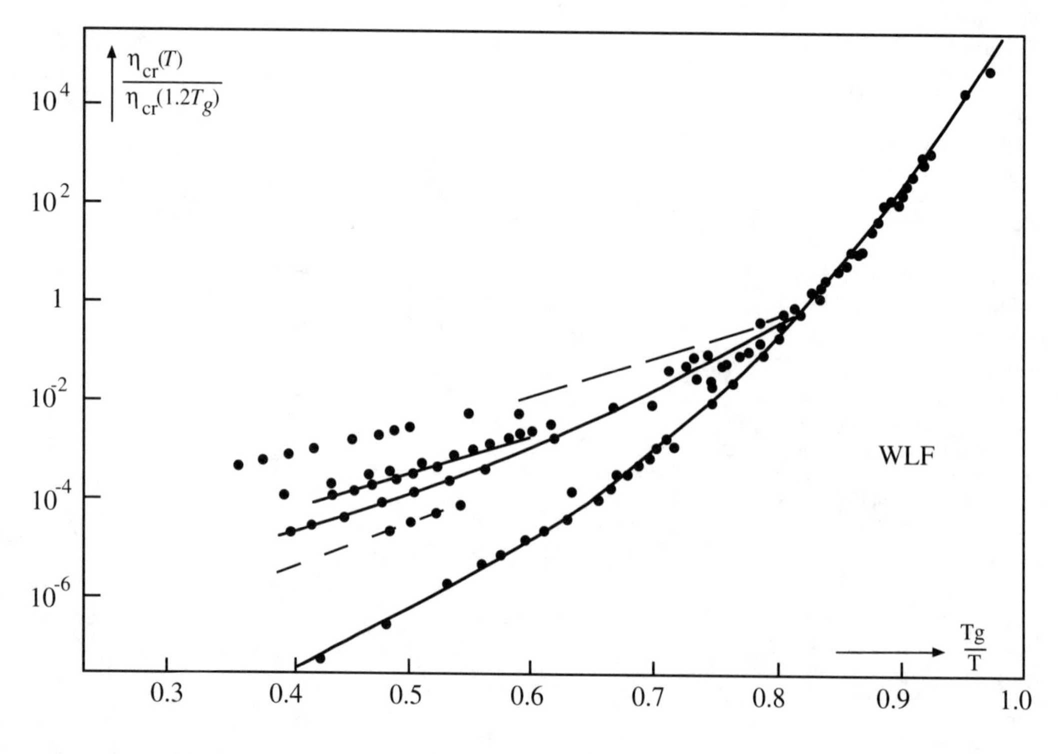

Figure 11.6.1.
η_{cr} is zero shear viscosity at break point in Figure 11.4.2. Note the transition to exponential T dependence far from $T_g[T_g/T \cong 0.85]$. Adopted from Van Krevelen (1990).

Figure 11.6.2.
Storage compliance of poly (*n*-octyl methacrylate) in the transition zone between glasslike and rubberlike consistency plotted logarithmically against frequency at 24 temperatures as indicated. The shifting of one data point from 80 to 100°C, the reference temperature, using eq. 11.6.2 is illustrated. Adopted from Ferry (1980).

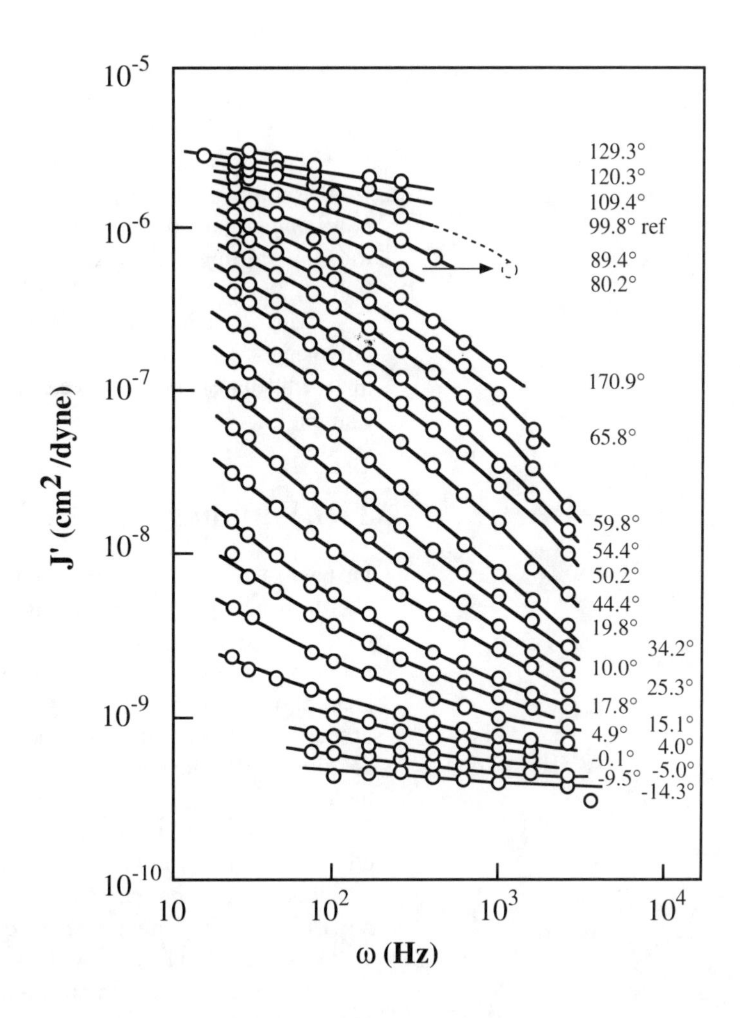

Figure 11.6.3.
Composite curve obtained by plotting the data of Figure 11.6.2 with reduced variables, representing the behavior over an extended frequency scale. Adopted from Ferry (1980).

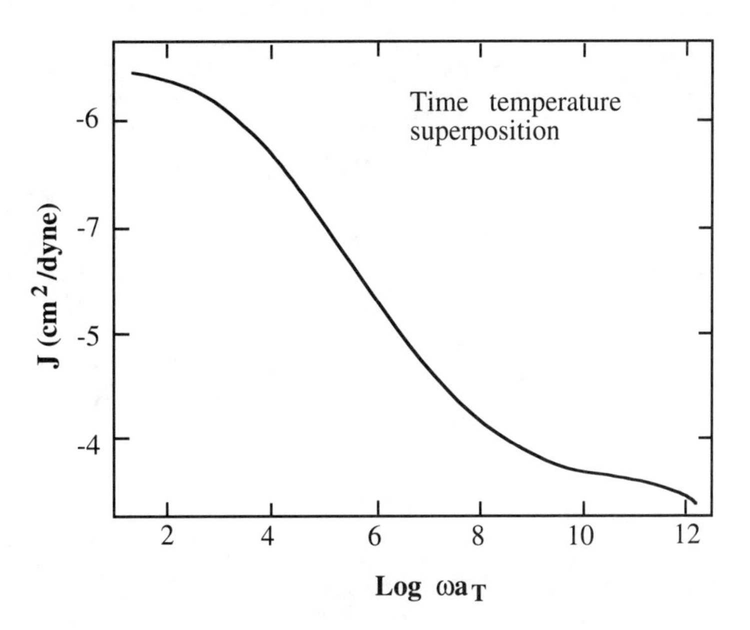

glass-forming materials. The transition between the two types of behavior is shown in Figure 11.6.1 (Van Krevelen and Hoftyser, 1976).

This WLF temperature dependence forms the basis for horizontal shifting of linear viscoelastic properties to the so-called *time–temperature superposition*. Figures 11.6.2 and 11.6.3 illustrate this shifting. Other examples of time–temperature superposition are given in Figures 2.4.1, 4.2.3, 4.4.2, and 4.4.4.

The activation energy for viscous flow E_η for many linear polymers is about 6–7 kcal/mol. Introduction of long chain branching seems to just about double this value (Small, 1975).

11.7 Summary

We have seen in this chapter that considerable insight can be gained into the behavior of polymers by treating them through a succession of simple models. Einstein sphere models capture global hydrodynamic properties such as intrinsic viscosity. To mimic internal degrees of freedom, and ultimately some aspects of linear viscoelastic properties, the bead–spring models are powerful. In concentrated solutions and melts it appears useful to consider entangled macromolecules moving as snakes (reptation) rather than as spheres.

With these generally accepted, but not necessarily accurate, conceptual models in hand, major efforts are going into molecular modeling of more complex real behavior. This is the state of the art. Some important areas of current work include nonlinear viscoelasticity, branched polymers, blends of different molecular weights, and chemical composition. Deep problems remain, such as the definitive explanation of the 3.4 power law for the molecular weight dependence of melt viscosity and proper description of concentrated solution rheology.

References

Berry, G. C.; Fox, T. G., *Adv. Polym. Sci.* 1968, *5*, 261.

Bird, R. B.; Curtiss, C. F.; Hassager, O.; Armstrong, R. C., *Dynamics of Polymeric Liquids:* Vol. 2, *Kinetic Theory*, 1st ed. 1976; 2nd ed.; Wiley: New York, 1987.

Bird, R. B.; Warner, H. R.; Evans, D. C., *Adv. Polym. Sci.* 1971, *8*, 1.

Birshtein, T. M.; Ptitsyn, O. B., *Conformations of Macromolecules*; Wiley-Interscience: New York, 1966.

Brandrup, J.; Immergut, E. H., Eds., *Polymer Handbook*, 2nd ed.; Wiley: New York, 1989.

Daoud, M.; Cotton, J. P.; Farnoux, B.; Jannink, G.; Sarma, G.; Benoit, H.; Duplessix, R.; Picot, C.; de Gennes, P. G., *Macromolecules* 1975, *8*, 804.

de Gennes, P. G., *Scaling Concepts in Polymer Physics*; Cornell University Press: Ithaca, NY, 1979.

des Cloizeaux, J.; Jannink, G., *Polymers in Solution: Their Modelling and Structure*; Clarendon Press: Oxford, 1990.

Doi, M.; Edwards, S. F., *J. Chem. Soc., Faraday Trans II* 1978, *74*.

Doi, M.; Edwards, S. F., *The Theory of Polymer Dynamics*; Clarendon Press: Oxford, 1986.

Doi, M.; Kuzuu, N., *J. Polym. Sci. Lett.* 1980, *18*, 775.

Ferry, J. D., *Viscoelastic Properties of Polymers*; 3rd ed.; Wiley: New York, 1980.

Flory, P. J., *Principles of Polymer Chemistry*; Cornell University Press: Ithaca, NY, 1953.

Flory, P. J., *Statistical Mechanics of Chain Molecules*; Wiley-Interscience: New York, 1969.

Flory, P. J.; Fox, T. G., *J. Am. Chem. Soc.* 1951, *73*, 1904.

Fujino, K.; Senshu, K.; Kawai, H., *J. Colloid Sci.* 1961, *16*, 262.

Graessley, W. W., *Adv. Polym. Sci.* 1974, *16*, 1.

Graessley, W. W., *Adv. Polym. Sci.* 1982, *47*, 67.

Graessley, W. W., *Polymer* 1980, *21*, 258.

Graessley, W. W.; et al., *Macromolecules* 1981, *14*, 1668.

Happel, J.; Brenner, H., *Low Reynolds Number Hydrodynamics*; Prentice-Hall: Englewood Cliffs, NJ, 1965.

Huggins, M. L., *J. Am. Chem. Soc.* 1942, *64*, 2716.

Kausch, H. H.; Tirrell, M., *Annu. Rev. Mater. Sci.* 1989, *19*, 341.

Kimmich, R.; Schnur, G.; Kopf, M., *Prog. Nuclear Magr. Resonance Spectrosc.* 1988, *20*, 385.

Kirkwood, J. G., *Macromolecules*; Gordon and Breach: New York, 1967.

Klein, J., *Nature* 1978, *271*, 143.

Larson, R. G., *Constitutive Equations for Polymer Melts and Solutions*; Butterworths: Boston, 1988.

Leger, L.; Hervet, H.; Rondelez, F., *Macromolecules* 1981, *14*, 1732.

Lodge, A. S., *Trans. Faraday Soc.* 1956, *52*, 120.

Lodge, T. P.; Rotstein, N.; Prager, S., *Adv. Chem. Phys.* 1990, *3*.

Marrucci, G.; Hermans, J. J., *Macromolecules* 1980, *13*, 380.

Noda I.; Yamada Y.; Nagasawa M., *J. Phys. Chem.* 1968, *72*, 2890–2898.

Onogi, S.; Masuda, T.; Kitagawa, K., *Macromolecules* 1970, *3*, 109.

Roovers, J. E. L.; Graessley, W. W., *Macromolecules* 1981,*14*, 766.

Rouse, P. E., *J. Chem. Phys.* 1953, *21*, 1272.

Small, P. A., *Adv. Polym. Sci.* 1975, *18*, 1.

Tirrell, M., *Rubber Chem. Technol.* 1984, *57*, 523.

Van Krevelen, D. W., *Properties of Polymers*, 2nd ed.; Elsevier: Amsterdam, 1976.

Van Krevelen, D. W., *Properties of Polymers*, 3rd ed.; Elsevier: Amsterdam, 1990.

Van Krevelen, D. W.; Hoftyser, P. J., *Angew. Makromol. Chem.* 1976, *52*, 101.

Volkenshtein, M. V., *Configurational Statistics of Polymer Chains*; Wiley-Interscience: New York, 1963.

Williams, M. C., *AIChE J.* 1975, *21*, 1.

Wiltzius, P.; Haller, H. R.; Cannell, D. S.; Schaeffer, D. W., *Phys. Rev. Lett.* 1983, *51*, 1183.

Yamakawa, H., *Modern Theory of Polymer Solutions*; Harper & Row: New York, 1971.

Zimm, B. H., *J. Chem. Phys.* 1956, *24*, 269.

APPENDIX SOLUTIONS TO EXERCISES

Chapter 1

1.10.1 Tensor Algebra

(a)

$$T_{ij}v_j = \begin{bmatrix} 3 & 2 & -1 \\ 2 & 2 & 1 \\ -1 & 1 & 0 \end{bmatrix} \cdot \begin{bmatrix} 5 \\ 3 \\ 7 \end{bmatrix} = (14, 23, -2) \quad \text{(a vector)}$$

(b)

$$v_i T_{ij} = (5, 3, 7) \cdot \begin{bmatrix} 3 & 2 & -1 \\ 2 & 2 & 2 \\ -1 & 1 & 0 \end{bmatrix} = (14, 23, -2) \text{ since } \mathbf{T} = \mathbf{T}^T \text{ then } \mathbf{v} \cdot \mathbf{T} = \mathbf{T} \cdot \mathbf{v}$$

(c)

$$T_{ij}T_{ij} = \begin{bmatrix} 3 & 2 & -1 \\ 2 & 2 & 1 \\ -1 & 1 & 0 \end{bmatrix} : \begin{bmatrix} 3 & 2 & -1 \\ 2 & 2 & 1 \\ -1 & 1 & 0 \end{bmatrix} = \begin{matrix} 9 & +4 & +1 \\ +4 & +4 & +1 \\ +1 & +1 & +0 \end{matrix} = 25 \text{ (a scalar)} = \text{tr}\mathbf{T}^2$$

(d)

$$v_i T_{ij} v_j = \text{(using the vector result from (a))} = (5, 3, 7) \cdot \begin{bmatrix} 14 \\ 23 \\ -2 \end{bmatrix} = 125 \text{ (a scalar)}$$

(e)

$$v_i v_j = \begin{bmatrix} 25 & 15 & 35 \\ 15 & 9 & 21 \\ 35 & 21 & 49 \end{bmatrix} \text{ or, showing the unit } \mathbf{vv} \text{ dyads} = \begin{bmatrix} \hat{\mathbf{x}}_1\hat{\mathbf{x}}_1 25 & \hat{\mathbf{x}}_1\hat{\mathbf{x}}_2 15 & \hat{\mathbf{x}}_1\hat{\mathbf{x}}_3 35 \\ \hat{\mathbf{x}}_2\hat{\mathbf{x}}_1 15 & \hat{\mathbf{x}}_2\hat{\mathbf{x}}_2 9 & \hat{\mathbf{x}}_2\hat{\mathbf{x}}_3 21 \\ \hat{\mathbf{x}}_2\hat{\mathbf{x}}_1 35 & \hat{\mathbf{x}}_1\hat{\mathbf{x}}_2 21 & \hat{\mathbf{x}}_3\hat{\mathbf{x}}_3 49 \end{bmatrix}$$

(f)

$$T_{ij}x_1 = \begin{bmatrix} 3 & 2 & -1 \\ 2 & 2 & 1 \\ -1 & 1 & 0 \end{bmatrix} \cdot \begin{bmatrix} 1 \\ 0 \\ 0 \end{bmatrix} = (3, 2, -1) \quad \text{(a vector)}$$

(g)

$$T_{ij}I_{jk} = \begin{bmatrix} 3 & 2 & -1 \\ 2 & 2 & 1 \\ -1 & 1 & 0 \end{bmatrix} \cdot \begin{bmatrix} 1 & 0 & 0 \\ 0 & 1 & 0 \\ 0 & 0 & 1 \end{bmatrix} = \begin{bmatrix} 3 & 2 & -1 \\ 2 & 2 & 1 \\ -1 & 1 & 0 \end{bmatrix}$$

($\mathbf{T} \cdot \mathbf{I} = \mathbf{T}$ is a definition of **I**, see eq. 2.2.33)

1.10.2 Invariants

Recall the definition of the invariants (eqs. 1.3.6–1.3.8)

$$I_T = \text{tr}\mathbf{T} = \text{ sum of the diagonal components of } T = 3 + 2 + 0 = 5$$

$$II_T = \frac{1}{2}\left(I_T^2 - \text{tr}\,\mathbf{T}^2\right)$$

$$T^2 = \mathbf{T} \cdot \mathbf{T} = T_{ij}T_{jk} = \begin{bmatrix} 14 & 9 & -1 \\ 9 & 9 & 0 \\ -1 & 0 & 2 \end{bmatrix} \qquad \text{tr}\,\mathbf{T}^2 = 14 + 9 + 2 = 25$$

$$II_T = \frac{1}{2}(25 - 25) = 0$$

$$III_T = \det \mathbf{T} = 0 - 2 - 2 - 3 - 0 - 2 = -9$$

1.10.3 Determination of the Stress Tensor

(a) This exercise is very similar to Example 1.2.2. $1\,\text{N}/1\,\text{mm}^2$ is 1 MPa.

Normal to Test Surface	*Stress Vector* (MPa)
$\hat{\mathbf{x}}_1$	$\mathbf{t}_1 = \hat{\mathbf{x}}_1$
$\hat{\mathbf{x}}_2$	$\mathbf{t}_2 = -2\hat{\mathbf{x}}_3$
$\hat{\mathbf{x}}_3$	$\mathbf{t}_3 = -2\hat{\mathbf{x}}_2$

$$\mathbf{T} = \hat{\mathbf{x}}_1\mathbf{t}_1 + \hat{\mathbf{x}}_2\mathbf{t}_2 + \hat{\mathbf{x}}_3\mathbf{t}_3 = \hat{\mathbf{x}}_1\hat{\mathbf{x}}_1 - 2\hat{\mathbf{x}}_2\hat{\mathbf{x}}_3 - 2\hat{\mathbf{x}}_3\hat{\mathbf{x}}_2$$

Thus the state of stress or stress tensor at the test point is

$$T_{ij} = \begin{bmatrix} 1 & 0 & 0 \\ 0 & 0 & -2 \\ 0 & -2 & 0 \end{bmatrix}$$

(b) What is the net force on the $1\,\text{mm}^2$ surface whose normal is $\hat{\mathbf{n}} = \hat{\mathbf{x}}_1 + \hat{\mathbf{x}}_2$?

$$\mathbf{t}_n = \hat{\mathbf{n}} \cdot \mathbf{T} = \left(\frac{1}{\sqrt{2}}, \frac{1}{\sqrt{2}}, 0\right) \begin{bmatrix} 1 & 0 & 0 \\ 0 & 0 & -2 \\ 0 & -2 & 0 \end{bmatrix} = \frac{1}{\sqrt{2}}(1, 0, -2)$$

$$\mathbf{f}_n = a_n\mathbf{t}_n = \frac{1}{\sqrt{2}}(1\hat{\mathbf{x}}_1 - 2\hat{\mathbf{x}}_3) \quad \text{in newtons}$$

(c) The normal component of $\mathbf{f}_n$ normal to $\hat{\mathbf{n}}$ is

$$F_{nn} = \hat{\mathbf{n}} \cdot \mathbf{f}_n = \left(\frac{1}{\sqrt{2}}, \frac{1}{\sqrt{2}}, 0\right) \begin{bmatrix} 1/\sqrt{2} \\ 0 \\ -\sqrt{2} \end{bmatrix} = \frac{1}{2} N$$

(d) Invariants of $\mathbf{T}$ are

$$I_T = 1\,\text{MPa} \qquad II_T = \frac{1}{2}(1{-}9) = -4(\text{MPa})^2 \qquad III_T = -4(\text{MPa})^3$$

1.10.4 C as Length Change

Use the definition of the deformation gradient tensor, eq. 1.4.3, to substitute for $d\mathbf{x}$

$$\alpha^2 = \frac{(\mathbf{F} \cdot d\mathbf{x}') \cdot (\mathbf{F} \cdot d\mathbf{x}')}{d\mathbf{x} \cdot d\mathbf{x})}$$

Using the transpose, eq. 1.2.27, to change order of operations, we obtain

$$\alpha^2 = \frac{(d\mathbf{x}' \cdot \mathbf{F}^T) \cdot (\mathbf{F} \cdot d\mathbf{x}')}{d\mathbf{x}' \cdot d\mathbf{x}')} = \frac{(d\mathbf{x}' \cdot (\mathbf{F}^T \cdot \mathbf{F}) \cdot d\mathbf{x}'}{|d\mathbf{x}'|^2} = \hat{\mathbf{n}}' \cdot \mathbf{C} \cdot \hat{\mathbf{n}}'$$

since $\hat{\mathbf{n}}' = d\mathbf{x}'/|d\mathbf{x}'|$.

1.10.5 Inverse Deformation Tensors

(a) From eq. 1.4.22

$$\frac{1}{\alpha^2} = \frac{d\mathbf{x}' \cdot d\mathbf{x}'}{d\mathbf{x} \cdot d\mathbf{x}}$$

Substituting $d\mathbf{x}' = \mathbf{F}^{-1} \cdot d\mathbf{x}$, eq. 1.4.29

$$\frac{1}{\alpha^2} = \frac{(\mathbf{F}^{-1}\dot{d}\mathbf{x}) \cdot (\mathbf{F}^{-1} \cdot d\mathbf{x})}{d\mathbf{x} \cdot d\mathbf{x}} = \frac{(d\mathbf{x} \cdot (\mathbf{F}^{-1})^T) \cdot (\mathbf{F}^{-1} \cdot d\mathbf{x})}{|d\mathbf{x}|^2} = \hat{\mathbf{n}} {\cdot} \mathbf{B}^{-1} {\cdot} \hat{\mathbf{n}}$$

since $\hat{\mathbf{n}} = d\mathbf{x}/|d\mathbf{x}|$ and $\mathbf{B}^{-1} = (\mathbf{F}^{-1})^T \cdot (\mathbf{F}^{-1})^T$ by eq. 1.4.30.
(b) From eq. 1.4.14.

$$\frac{1}{\mu^2} = \frac{d\mathbf{a} \cdot d\mathbf{a}}{d\mathbf{a}' \cdot d\mathbf{a}'}$$

From eq. 1.4.17. for an incompressible material

$$d\mathbf{a}' = d\mathbf{a} \cdot \mathbf{F}$$

The inverse reverses this operation (recall eq. 1.4.29)

$$d\mathbf{a} = d\mathbf{a}' \cdot \mathbf{F}^{-1}$$

Substituting

$$\frac{1}{\mu^2} = \frac{(d\mathbf{a}' \cdot \mathbf{F}^{-1}) \cdot (d\mathbf{a}' \cdot \mathbf{F}^{-1})}{d\mathbf{a}' \cdot d\mathbf{a}'} = \frac{(d\mathbf{a}' \cdot \mathbf{F}^{-1}) \cdot ((\mathbf{F}^{-1})^T \cdot d\mathbf{a}')}{|d\mathbf{a}|^2} = \hat{\mathbf{n}}' \cdot \mathbf{C}^{-1} \cdot \hat{\mathbf{n}}'$$

1.10.6 Planar Extension of a Mooney–Rivlin Rubber

(a) (b) The boundary deformations will be the same as in Example 1.8.2. Thus, by eq. 1.8.8 **B** will be

$$B_{ij} = \begin{bmatrix} \alpha^2 & 0 & 0 \\ 0 & 1 & 0 \\ 0 & 0 & \alpha^{-2} \end{bmatrix} \quad \text{and} \quad B_{ij}^{-1} = \begin{bmatrix} \alpha^{-2} & 0 & 0 \\ 0 & 1 & 0 \\ 0 & 0 & \alpha^2 \end{bmatrix}$$

(c) With these tensors we can readily calculate the stresses for a Mooney–Rivlin rubber. Rewriting eq. 1.6.3 in terms of $g_1 = 2C_1$ and $g_2 = -2C_2$, we have

$$\begin{aligned} \mathbf{T} &= -p\mathbf{I} + 2C_1\mathbf{B} - 2C_2\mathbf{B}^{-1} \\ T_{11} &= -p + 2C_1\alpha^2 - 2C_2\alpha^{-2} \\ T_{22} &= -p + 2C_1 - 2C_2 \\ T_{33} &= -p + 2C_1\alpha^{-2} - 2C_2\alpha^2 = 0 \qquad \text{free surface} \\ p &= 2C_1\alpha^{-2} - 2C_2\alpha^2 \\ T_{11} &= (2C_1 + 2C_2)(\alpha^2 - \alpha^{-2}) \end{aligned}$$

This result has exactly the same functional dependence as the neo-Hookean model. Thus measurements of T_{11} in planar extension could not differentiate between the two. However

$$T_{22} = 2C_1(1 - \alpha^{-2}) + 2C_2(\alpha^2 - 1)$$

which has a dependence on α that differs from the neo-Hookean.

1.10.7 Eccentric Rotating Disks

Note that in the literature this geometry is called the Maxwell orthogonal rheometer or eccentric rotating disks, ERD (Macosko and Davis, 1974; Bird, et al., 1987, also see Chapter 5). Usually, the coordinates $\hat{\mathbf{x}}_2 = y$ and $\hat{\mathbf{x}}_3 = z$ are used.

(a)

$$F_{ij} = \frac{\partial x_i}{\partial x_j'} = \begin{bmatrix} c & -s & \gamma \\ s & c & g \\ 0 & 0 & 1 \end{bmatrix} \text{ where } c = \cos\Omega t \text{ and } s = \sin\Omega t$$

$$B_{ij} = F_{ik}F_{jk} = \begin{bmatrix} c^2 + s^2 & cs - sc & 0 \\ sc - cs & c^2 + s^2 + \gamma^2 & \gamma \\ 0 & \gamma & 1 \end{bmatrix} = \begin{bmatrix} 1 & 0 & 0 \\ 0 & 1 + \gamma^2 & \gamma \\ 0 & \gamma & 1 \end{bmatrix}$$

Note that there are shear and normal components of the strain. Also

note that this is the same deformation as simple shear of eq. 1.4.24 with slightly different notation.
(b) Using eq. 1.5.2., we can readily evaluate the stresses

$$T_{11} = T_{33} = -p + G$$
$$T_{22} = -p + G(1 + \gamma^2)$$
$$T_{23} = T_{32} = G\gamma$$

The stress components acting on the disks will be $\mathbf{T} \cdot \hat{\mathbf{x}}_3 = \mathbf{t}_3$

$$\mathbf{t}_3 = \begin{bmatrix} T_{33} & 0 & 0 \\ 0 & T_{22} & T_{23} \\ 0 & T_{23} & T_{33} \end{bmatrix} \cdot \begin{bmatrix} 0 \\ 0 \\ 1 \end{bmatrix} = (0, T_{23}, T_{33})$$

The force components can be calculated by integrating these stresses over the area of the disk.

$$\mathbf{f} = \int \mathbf{t}_3 dA$$
$$f_{x_1} = 0$$
$$f_{x_2} = \pi R^2 G\gamma$$
$$f_{x_3} = 0$$

Macosko and Davis discuss using the boundary conditions to evaluate f_{x_3}.

1.10.8 Sheet Inflation

(a) From a right triangle formed with the bubble radius, R, as the hypotenuse and the initial sheet radius, R_0, as the base, we obtain $R^2 = R_0^2 + (R - h)^2$ and thus $R = (R_0^2 + h^2/2h)$.
(b) *Deformation in Membrane*

$$\alpha_1 = \frac{\Delta x}{\Delta x_o} \quad \alpha_2 = \frac{\Delta y}{\Delta y_o} \quad \alpha_3 = \frac{\delta}{\delta_o}$$

$\alpha_1 = \alpha_2$ near the pole because the bubble is symmetric

$\alpha_1\alpha_2\alpha_3 = 1$ for an incompressible solid

Thus $\alpha_3 = 1/\alpha_1^2$ or $\delta/\delta_o = (\Delta x_o/\Delta x)^2$. We can determine the thickness of the bubble by measuring the stretch near the pole.
(c) *Stresses in the Membrane.* Applying the neo-Hookean model

$$T_{ij} = \begin{bmatrix} -p + G(\alpha_1^2 - 1) & 0 & 0 \\ 0 & -p + G(\alpha_1^2 - 1) & 0 \\ 0 & 0 & -p + G(\alpha_3^2 - 1) \end{bmatrix}$$

$T_{33} = 0$ for a thin membrane; therefore $p = G(\alpha_3^2 - 1)$. Thus $T_{11} = T_{22} = G(\alpha_1^2 - 1/\alpha_1^4)$.

T_{11} and T_{22} can be treated as surface tensions where Γ is the stress in the membrane times unit thickness $\Gamma = T_{11}\delta$. Using the membrane balance equation, eq. 1.8.5.

$$p_{\text{in}} - p_{\text{out}} = p = \frac{\Gamma_1}{R_1} + \frac{\Gamma_2}{R_2} = \frac{2\Gamma}{R} = \frac{2T_{11}\delta}{R}$$

since $R_1 = R_2 = R$ for a sphere. Substituting gives

$$p = \frac{2G(\alpha_2^2 - 1/\alpha_2^4)\delta}{R}$$

1.10.9 Film Tenter

(a) Equate the volumetric flow rate at the entrance and exit.

$$v_{\text{in}} A_{\text{in}} = v_{\text{out}} A_{\text{out}}$$
$$(1\,\text{m/s})(0.5\text{m})(150 \times 10^{-6}\text{m}) = (3\,\text{m/s})(1\,\text{m})h$$
$$h = 25\,\mu\text{m}$$

(b) Find the stress on the last pair of clamps.
The extensions are fixed by the tenter

$$\alpha_2 = \frac{\Delta x_2}{\Delta x_2'} = 1/0.5 = 2$$
$$\alpha_3 = \frac{\Delta x_3}{\Delta x_3'} = \frac{25 \times 10^{-6}}{150 \times 10^{-6}} = \frac{1}{6}$$

Because the material is incompressible, the volume will be constant:

$$\Delta x_1' \Delta x_2' \Delta x_3' = \Delta x_1 \Delta x_2 \Delta x_3$$
$$\alpha_1 \alpha_2 \alpha_3 = 1$$
$$\alpha_1 = \frac{1}{\alpha_2 \alpha_3} = \frac{1}{2\,(1/6)} = 3$$

Recall that

$$x_i = \alpha_i x_i' \quad \text{and} \quad F_{ij} = \frac{\partial x_i}{\partial x_j'}$$

Thus

$$F_{ij} = \begin{bmatrix} \alpha_1 & 0 & 0 \\ 0 & \alpha_2 & 0 \\ 0 & 0 & \alpha_3 \end{bmatrix}$$

To evaluate $\mathbf{B}$, we use $\mathbf{B} = \mathbf{F} \cdot \mathbf{F}^T$ because $\mathbf{F} = \mathbf{F}^T$; then $\mathbf{B} = \mathbf{F}^2$ and

$$B_{ij} = \begin{bmatrix} \alpha_1^2 & 0 & 0 \\ 0 & \alpha_2^2 & 0 \\ 0 & 0 & \alpha_3^2 \end{bmatrix}$$

For the neo-Hookean solid

$$\mathbf{T} = -p\mathbf{I} + G\mathbf{B}$$

Substituting for **B** gives

$$T_{11} = -p + G\alpha_1^2$$
$$T_{22} = -p + G\alpha_2^2$$
$$T_{33} = -p + G\alpha_3^2$$

but $T_{33} = 0$, no external forces acting in $\hat{\mathbf{x}}_3$ direction, perpendicular to the film. Therefore

$$p = G\alpha_3^2$$
$$T_{22} = -G\alpha_3^2 + G\alpha_2^2 = G(\alpha_2^2 - \alpha_3^2)$$
$$= 5 \times 10^5 (4 - \frac{1}{36}) = 1.99 \times 10^6 \text{ Pa}$$

Thus the force exerted per unit area in the last pair of clamps is

$$1.99 \times 10^6 \text{ Pa}$$

(c) Assume that the torque needed to turn the roller is due only to the force required to stretch the film in the $\hat{\mathbf{x}}_1$ direction. The force is the stress component $\mathbf{t}_1$ times the film cross-sectional area a_1.

$$\text{torque} = \mathbf{R} \times a_1 \mathbf{t}_1 = \left(\frac{0.3}{2}\text{m}\right)\hat{\mathbf{x}}_3 \times ([1\text{ m}][25 \times 10^{-6}\text{ m}])T_{11}\hat{\mathbf{x}}_1$$

$$\text{torque} = (0.15\text{ m})(25 \times 10^{-6}\text{ m}^2)(G(\alpha_1^2 - \alpha_3^2))\hat{\mathbf{x}}_3 \times \hat{\mathbf{x}}_2$$

$$\text{torque} = (0.15)(25 \times 10^{-6})(5 \times 10^5 (9 - \frac{1}{36}))\{-\hat{\mathbf{x}}_1\}$$

$$\mathbf{M} = -16.25\hat{\mathbf{x}}_1 \text{N} \cdot \text{m}$$

Chapter 2

2.8.1 B and D for Steady Extension

An extensional flow is steady if the instantaneous rate of change of length per unit length is constant.

$$\frac{1}{\ell}\frac{dl}{dt} = \dot{\epsilon} = \text{constant}$$

or

$$\frac{1}{\alpha}\frac{d\alpha}{dt} = \dot{\epsilon} \quad \text{where } \alpha = l/l_0 = \text{extension ratio}$$

Integrating with the intitial condition $l = l_0$ at $t = t_0$, we obtain

$$\int_{l_0}^{\ell} \frac{d\ell}{\ell} = \int_{t_0}^{t} \dot{\epsilon}\, dt \qquad \alpha = \frac{l}{l_0} = e^{\dot{\epsilon}(t-t_0)}$$

For a general extension

$$B_{ij} = \begin{bmatrix} \alpha_1^2 & 0 & 0 \\ 0 & \alpha_2^2 & 0 \\ 0 & 0 & \alpha_3^2 \end{bmatrix}$$

Therefore, for a general steady extensional flow

$$B_{ij} = \begin{bmatrix} e^{2\dot{\epsilon}_1(t-t_0)} & 0 & 0 \\ 0 & e^{2\dot{\epsilon}_2(t-t_0)} & 0 \\ 0 & 0 & e^{2\dot{\epsilon}_3(t-t_0)} \end{bmatrix} \tag{2.8.1}$$

The rate of deformation tensor is just the first time derivative of $\mathbf{B}$ evaluated at $t_0 = t$

$$2D_{ij} = \frac{\partial B_{ij}}{\partial t}\Big|_{t_0=t} = \begin{bmatrix} 2\dot{\epsilon}_1 e^{2\dot{\epsilon}_1(t-t_0)} & 0 & 0 \\ 0 & 2\dot{\epsilon}_2 e^{2\dot{\epsilon}_2(t-t_0)} & 0 \\ 0 & 0 & 2\dot{\epsilon}_3 e^{2\dot{\epsilon}_3(t-t_0)} \end{bmatrix}_{t_0-t} = \begin{bmatrix} 2\dot{\epsilon}_1 & 0 & 0 \\ 0 & 2\dot{\epsilon}_2 & 0 \\ 0 & 0 & 2\dot{\epsilon}_3 \end{bmatrix}$$

Recall the definitions of the invariants from eqs. 1.3.6–1.3.8.

$$I_{2D} = \text{tr}\mathbf{D}(= 0 \text{ for incompressible})$$
$$II_{2D} = \frac{1}{2}\left((\text{tr}\, 2\mathbf{D})^2 - \text{tr}\,(2\mathbf{D})^2\right)$$
$$III_{2D} = \det 2\mathbf{D}$$

Now apply these results to each of the special cases.

(a) *Steady Uniaxial Extension.* For the special case of uniaxial extension, symmetry gives

$$\alpha_2 = \alpha_3$$

while for an incompressible material (conservation of volume) it gives (recall eq. 1.4.6)

$$\alpha_1\alpha_2\alpha_3 = 1 \qquad \alpha_1 = \frac{1}{\alpha_2^2}$$

Thus

$$B_{ij} = \begin{bmatrix} \alpha_1^2 & 0 & 0 \\ 0 & 1/\alpha_1 & 0 \\ 0 & 0 & 1/\alpha_1 \end{bmatrix}$$

and for steady uniaxial

$$B_{ij} = \begin{bmatrix} e^{2\dot{\epsilon}_1(t-t_0)} & 0 & 0 \\ 0 & e^{-\dot{\epsilon}_1(t-t_0)} & 0 \\ 0 & 0 & e^{-\dot{\epsilon}_1(t-t_0)} \end{bmatrix} \tag{2.8.2}$$

Now we can solve for the invariants

$$I_B = \text{tr}\,\mathbf{B} = \alpha_1^2 + \frac{2}{\alpha_1}$$

or

$$= e^{+2\dot{\epsilon}_1(t-t_0)} + 2e^{-\dot{\epsilon}_1(t-t_0)} \tag{2.8.3}$$

For II_B we need

$$(I_B)^2 = \alpha_1^4 + 4\alpha_1 + 4/\alpha_1^2$$

and

$$\text{tr}\,(\mathbf{B}^2) = \alpha_1^4 + \frac{1}{\alpha_1^2} + \frac{1}{\alpha_1^2}$$

$$II_B = \frac{1}{2}[(I_B)^2 - \text{tr}\,\mathbf{B}^2] = 2\alpha_1 + \frac{1}{\alpha_1^2}$$

or

$$= 2e^{\dot{\epsilon}(t-t_0)} + e^{-2\dot{\epsilon}(t-t_0)}$$

and

$$II_B = 1 \qquad \text{(for all incompressible materials)}$$

For the rate of deformation we can take the time derivatives of B_{ij} or reason directly. Again by symmetry $\dot{\epsilon}_2 = \dot{\epsilon}_3$ and for an incompressible material $I_{2D} = \text{tr}\,2\mathbf{D} = 0$. Thus

$$\dot{\epsilon}_1 + \dot{\epsilon}_2 + \dot{\epsilon}_3 = 0$$

which gives

$$\dot{\epsilon}_1 = -2\dot{\epsilon}_2$$

Thus

$$2D_{ij} = \begin{bmatrix} 2\dot{\epsilon}_1 & 0 & 0 \\ 0 & -\dot{\epsilon}_1 & 0 \\ 0 & 0 & -\dot{\epsilon}_1 \end{bmatrix} \tag{2.8.4}$$

and the invariants are

$$\begin{aligned} I_{2D} &= \text{tr}\, 2\mathbf{D} = 0 \\ II_{2D} &= 0 - \frac{1}{2}\text{tr}\,(2\mathbf{D})^2 = -3\dot{\epsilon}_1^2 \\ III_{2D} &= 2\dot{\epsilon}_1^3 \end{aligned} \tag{2.8.5}$$

(b) *Steady Equal Biaxial Extension.* This is the reverse of uniaxial extension $\alpha_b = \alpha_1^2$ and $\alpha_2 = 1/\alpha_b$.
Thus

$$B_{ij} = \begin{bmatrix} 1/\alpha_1^2 & 0 & 0 \\ 0 & \alpha_1 & 0 \\ 0 & 0 & \alpha_1 \end{bmatrix}$$

and for steady equal biaxial

$$B_{ij} = \begin{bmatrix} e^{-2\dot{\epsilon}_1(t-t_0)} & 0 & 0 \\ 0 & e^{+\dot{\epsilon}_1(t-t_0)} & 0 \\ 0 & 0 & e^{+\dot{\epsilon}_1(t-t_0)} \end{bmatrix} \tag{2.8.6}$$

The first invariant is

$$I_B = \frac{1}{\alpha_1^2} + 2\alpha_1$$

or

$$I_B = e^{-2\dot{\epsilon}_1(t-t_0)} + 2e^{\dot{\epsilon}_1(t-t_0)}$$

and the second becomes

$$II_B = \alpha_1^2 + 2\alpha_1 \tag{2.8.7}$$

or

$$II_B = e^{2\dot{\epsilon}_1(t-t_0)} + 2e^{-\dot{\epsilon}_1(t-t_0)}$$

D_{ij} can be evaluated from the derivatives of B_{ij}

$$2D_{ij} = \begin{bmatrix} -2\dot{\epsilon}_1 & 0 & 0 \\ 0 & \dot{\epsilon}_1 & 0 \\ 0 & 0 & \dot{\epsilon}_1 \end{bmatrix} \qquad 4D_{ij}D_{jk} = \begin{bmatrix} 4\dot{\epsilon}_1^2 & 0 & 0 \\ 0 & \dot{\epsilon}_1^2 & 0 \\ 0 & 0 & \dot{\epsilon}_1^2 \end{bmatrix} \tag{2.8.8}$$

$$II_{2D} = 0 - \frac{1}{2}\,\text{tr}\,\mathbf{D}^2 = -3\dot{\epsilon}_1^2 \tag{2.8.9}$$
$$III_{2D} = 2\dot{\epsilon}_1^3$$

We note that although equal biaxial extension is just the reverse of uniaxial, the invariants of **B** are different. Therefore we would expect material functions measured in each deformation to be different in general. Another common approach to equibiaxial extension is to let $\alpha_b = \alpha_1^{-2}$ and $\dot{\epsilon}_b = 2\dot{\epsilon}_1$, basing length change on the sides rather than the thickness of the samples.

(c) *Steady Planar Extension.* In this case, as we saw in Example 1.8.1, $\alpha_3 = 1$. Then from conservation of volume $\alpha_1 = 1/\alpha_2$, and thus

$$B_{ij} = \begin{bmatrix} \alpha_1^2 & 0 & 0 \\ 0 & 1/\alpha_1^2 & 0 \\ 0 & 0 & 1 \end{bmatrix}$$

and for steady planar extension

$$B_{ij} = \begin{bmatrix} e^{2\dot{\epsilon}_p(t-t_o)} & 0 & 0 \\ 0 & e^{-2\dot{\epsilon}_p(t-t_o)} & 0 \\ 0 & 0 & 1 \end{bmatrix} \tag{2.8.10}$$

$$I_B = e^{2\dot{\epsilon}_p(t-t_0)} - e^{-2\dot{\epsilon}_p(t-t_0)} + 1$$
$$\tag{2.8.11}$$
$$II_B = e^{-2\dot{\epsilon}_p(t-t_0)} + e^{2\dot{\epsilon}(t-t_0)} + 1$$

and

$$2D = \begin{bmatrix} 2\dot{\epsilon}_p & 0 & 0 \\ 0 & 2\dot{\epsilon}_p & 0 \\ 0 & 0 & 0 \end{bmatrix}$$

$$I_{2D} = 0, \quad II_{2D} = -4\dot{\epsilon}_p^2, \quad \text{and} \quad II_{2D} = 0$$

2.8.2 Stresses in Steady Extension

(a) *Power Law Fluid.* Apply eq. 2.4.12 to the kinematics found in Exercise 2.8.1. The results are:

Uniaxial extension

$$T_{11} - T_{22} = 3m| - 3\dot{\epsilon}^2|^{(n-1)/2}\dot{\epsilon} = 3^{(n+1)/2} m\dot{\epsilon}^n$$
$$T_{22} - T_{33} = 0$$

Biaxial

$$T_{11} - T_{22} = 0$$
$$T_{22} - T_{33} = 3^{(n+1)/2} m\dot{\epsilon}^n$$

Planar

$$T_{11} - T_{22} = m| - 4\dot{\epsilon}^2|^{(n-1)/2}\, 4\dot{\epsilon} = 2^{n+1} m\dot{\epsilon}^n$$
$$T_{22} - T_{33} = m| - 4\dot{\epsilon}^2|^{(n-1)/2}\, (-2\dot{\epsilon}) = -2^n m\dot{\epsilon}^n$$

(b) *Bingham Plastic.* We can use the constitutive equation to rewrite the yield stress criteria in terms of $\mathbf{B}$. Since $\boldsymbol{\tau} = G\mathbf{B}$

$$II_\tau = \frac{1}{2}[(\text{tr}\, G\mathbf{B})^2 - \text{tr}\, G^2\mathbf{B}^2] = G^2 II_B$$

Bingham Plastic results are summarized in the Table 2.8.1.

2.8.3 Pipe Flow of a Power Law Fluid

You need to increase the pipe diameter. Recall eq. 2.4.21

$$Q = \frac{\pi R^2}{(1/n) + 3}\left[\frac{\Delta p R}{2mL}\right]^{1/n}$$

Let $Q_1 = Q_2$; $\Delta p_1 = \Delta p_2$, $2L_1 = L_2$; $m, n =$ constants and solve for R_2/R_1.

$$\left(\frac{R_2}{R_1}\right)^{1/n+3} = 2^{1/n}$$

From eq. 2.4.22 the ratio of shear rates in the two pipes will be

$$\frac{\dot{\gamma}_{w_2}}{\dot{\gamma}_{w_1}} = \left(\frac{R_1}{R_2}\right)^n$$

TABLE 2.8.1 / Bingham Plastic Results

	Hookean for $II_\tau < \tau_y^2$	*$\boldsymbol{\tau}_y$ Criteria II_τ*	*Newtonian for $II_\tau > \tau_y^2$*
Uniaxial	$T_{11} - T_{22} = G(\alpha_1^2 - 1/\alpha_1)$	$\geq G^2(2\alpha_1 + 1/\alpha_1^2)$ $< 3\eta_0^2\dot{\epsilon}^2$	$T_{11} - T_{22} = 3\eta_0\dot{\epsilon} + \sqrt{3}\tau_y$
Equal biaxial	$T_{11} - T_{22} = G(1/\alpha_1^2 - \alpha_1)$	$\geq G^2(\alpha_1^2 + 2/\alpha_1)$ $< 3\eta_0^2\dot{\epsilon}^2$	$T_{11} - T_{22} = 3\eta_0\dot{\epsilon} + \sqrt{3}\tau_y$
Planar	$T_{11} - T_{22} = G(\alpha_1^2 - 1/\alpha_1^2)$ $T_{22} - T_{33} = G(1/\alpha_1^2 - 1)$	$\geq G^2(2\alpha_1^2 + 1/\alpha_1^2 + 2)$ $< 4\eta_0^2\dot{\epsilon}_p^2$	$T_{11} - T_{22} = 4\eta_0\dot{\epsilon}_p + 2\tau_y$ $T_{22} - T_{33} = -2\eta_0\dot{\epsilon}_p - \tau_y$
Simple shear	$T_{12} = G\gamma$ $T_{11} - T_{22} = G\gamma$	$\geq G^2(\gamma^2 + 3)$ $< \eta_0^2\dot{\gamma}^2$	$T_{12} = \eta_0\dot{\gamma} + \tau_y$ $T_{11} - T_{22} = 0$

2.8.4 Yield Stress in Tension

Using the results of Example 2.8.2 we obtain

$$\tau_y = \frac{10}{\sqrt{3}}\ \text{Pa}.$$

Chapter 3

3.4.1 Relaxation Spectrum

Substitute for $G(s)$ into $\eta'(\omega)$

$$\eta'(\omega) = \int_0^\infty \left[\int_0^\infty \frac{H(\lambda)}{\lambda} e^{-s/\lambda} d\lambda \right] \cos \omega s \, ds$$

Rearranging gives

$$\eta'(\omega) = \int_0^\infty \frac{H(\lambda)}{\lambda} \left[\int_0^\infty e^{-s/\lambda} \cos \omega s \, ds \right] d\lambda$$

Using standard integral tables, we find

$$\int_0^\infty \cos\ \text{at}\ e^{-bt}\, dt = \frac{b}{b^2 + a^2} \qquad \text{for } b > 0$$

Thus

$$\eta'(\omega) = \int_0^\infty \frac{H(\lambda)\, d\lambda}{1 + \lambda^2 \omega^2}$$

3.4.2 Two-Constant Maxwell Model

The two-constant integral linear viscoelastic model is

$$\tau = \int_{-\infty}^{t} \{G_1 e^{-(t-t')/\lambda_1} + G_2 e^{-(t-t')/\lambda_2}\} \dot{\gamma}(t') dt'$$

Differentiating gives

$$\frac{\partial \tau}{\partial t} = -\int_{-\infty}^{t} \left\{ \frac{G_1}{\lambda_1} e^{-(t-t')/\lambda} + \frac{G_2}{\lambda_2} e^{-(t-t')/\lambda_2} \right\} \dot{\gamma}(t')dt' + (G_1 + G_2)\dot{\gamma}(t)$$

Multiplying by $(\lambda_1 + \lambda_2)$ yields

$$(\lambda_1 + \lambda_2)\frac{\partial \tau}{\partial t} = -\int_{-\infty}^{t} \{G_1 e^{-(t-t')/\lambda_1} + G_2 e^{-(t-t'/\lambda_2}\}\dot{\gamma}(t')\, dt'$$

$$- \int_{-\infty}^{t} \left\{ \frac{\lambda_2}{\lambda_1} G_1 e^{-(t-t')/\lambda_1} + \frac{\lambda_1}{\lambda_2} G_2 e^{-(t-t')/\lambda_2} \right\} \dot{\gamma}(t')dt'$$

$$+ (\lambda_1 + \lambda_2)(G_1 + G_2)\dot{\gamma}(t)$$

Differentiating again,

$$\frac{\partial^2 \tau}{\partial t^2} = \int_{-\infty}^{t} \left\{ \frac{G_1}{\lambda_1^2} e^{-(t-t')/\lambda_1} + \frac{G_2}{\lambda_2^2} e^{-(t-t')/\lambda_2} \right\} \dot{\gamma}(t')dt' - \left\{ \frac{G_1}{\lambda_1} + \frac{G_2}{\lambda_2} \right\} \dot{\gamma}(t) + (G_1 + G_2)\frac{\partial \dot{\gamma}(t)}{\partial t}$$

or

$$\lambda_1 \lambda_2 \frac{\partial^2 \tau}{\partial t^2} = \int_{-\infty}^{t} \left\{ \frac{\lambda_2}{\lambda_1} G_1 e^{-(t-t')/\lambda_1} + \frac{\lambda_1}{\lambda_2} G_2 e^{-(t-t')/\lambda_2} \right\} \dot{\gamma}(t')dt'$$

$$- (G_1\lambda_2 + \lambda_1 G_2)\dot{\gamma}(t) + \lambda_1\lambda_2(G_1 + G_2)\frac{\partial \dot{\gamma}(t)}{\partial t}$$

Combining the three equations, we find that

$$\tau + (\lambda_1 + \lambda_2)\frac{\partial \tau}{\partial t} + \lambda_1\lambda_2\frac{\partial^2 \tau}{\partial t^2} = (\lambda_1 G_1 + \lambda_2 G_2)\dot{\gamma}(t) + \lambda_1\lambda_2(G_1 + G_2)\frac{\partial \dot{\gamma}(t)}{\partial t}$$

or since $\eta_1 = G_1\lambda_1$, $\eta_2 = G_2\lambda_2$, and

$$\tau + (\lambda_1 + \lambda_2)\frac{\partial \tau}{\partial t} + \lambda_1\lambda_2\frac{\partial_2 \tau}{\partial t^2} = (\eta_1 + \eta_2)\dot{\gamma}(t) + (\lambda_2\eta_1 + \lambda_1\eta_2)\frac{\partial \dot{\gamma}(t)}{\partial t}$$

3.4.3 Derivation of G' and G''

For sinusoidal oscillations the shear rate is $\dot{\gamma} = \dot{\gamma}_0 \cos \omega t$, eq. 3.3.27. Substituting

$$\tau = \int_{-\infty}^{t} G(t - t')\dot{\gamma}_0 \cos \omega t'\, dt'$$

$$= \dot{\gamma}_0 \int_0^\infty G(s) \cos \omega(t - s) ds$$

Using the trigonometric relation for $\cos(x - y)$, we can write

$$\tau = \dot{\gamma}_0 \int_0^\infty G(s) \cos \omega s \, ds \ \cos \omega t + \dot{\gamma}_0 \int_0^\infty G(s) \sin \omega s \, ds \ \sin \omega t$$

From eq. 3.3.17 we see that

$$\tau_0' = \dot{\gamma}_0 \int_0^\infty G(s) \ \sin \omega s \, ds \text{ and } \tau_0'' = \dot{\gamma}_0 \int_0^\infty G(s) \cos \ \omega s \, ds$$

Thus, from eqs. 3.3.28 and 3.3.29

$$\eta' = \int_0^\infty G(s) \cos \ \omega s \, ds \text{ or } G'' = \omega \int_0^\infty G(s) \cos \ \omega s \, ds$$

$$\eta'' = \int_0^\infty G(s) \sin \ \omega s \, ds \text{ or } G' = \omega \int_0^\infty G(s) \sin \ \omega s \, ds$$

We can obtain these quantitites in terms of the discrete exponential relaxation times by substituting in for $G(s)$ with eq. 3.2.8 or 3.2.10 and solving the definite integrals of the exponentials (check any standard integral table). For example, with eq. 3.2.8

$$G' = \omega \int_0^\infty G_0 e^{-s/\lambda} \ \sin \ \omega s \, ds = G_0 \frac{\omega^2 \lambda^2}{1 + \omega^2 \lambda^2} \text{and } \eta'' = \eta \frac{\omega \lambda^2}{1 + \omega^2 \lambda^2}$$

Using eq. 3.2.10 gives

$$G' = \sum_{k=1}^{N} G_k \frac{\omega^2 \lambda_k^2}{1 + \omega^2 \lambda_k^2} \qquad 3.3.31$$

$$G'' = \sum_{k=1}^{N} G_k \frac{\omega \lambda_k}{1 + \omega^2 \lambda_k^2} \qquad 3.3.32$$

3.4.4 Energy Dissipation

Recall that the rate of energy lost by viscous dissipation per unit volume is

$$\text{rate of energy dissipation } = \boldsymbol{\tau} : \mathbf{D}$$

and that the energy dissipated over a length of time t is

$$\text{energy dissipated} = \phi = \int_0^t \boldsymbol{\tau} : \mathbf{D}\, \text{dt}$$

For small amplitude sinusoidal oscillations, this expression becomes

$$\phi = \int_0^{2\pi/\omega} \tau \dot{\gamma}\, dt$$

According to eq. 3.3.15, $\gamma = \gamma_0 \sin \omega t$, so $\dot{\gamma} = \omega \gamma_0 \cos \omega t$.
Then from eq. 3.3.17, $\tau = \tau_0' \sin \omega t + \tau_0'' \cos \omega t$.
Then

$$\phi = \int_0^{2\pi\omega} (\tau_0' \sin \omega t + \tau_0'' \cos \omega t)\gamma_0 \cos \omega t\, dt$$

$$= \omega\gamma_0 \int_0^{2\pi/\omega} (\tau_o' \sin \omega t \cos \omega t + \tau_o'' \cos^2 \omega t) dt$$

$$= \omega\gamma_0 \left[\tau_0' \left(\frac{1}{\omega} \sin^2 \omega t \right) + \tau_0'' \left(\frac{t}{2} + \frac{1}{4\omega} \sin 2\omega t \right) \right] \Big|_0^{2\pi/\omega}$$

$$= \omega \gamma_0 \tau_0'' \frac{\pi}{\omega}$$

Since

$$\tau_0'' = G'' \gamma_0$$

Then

$$\phi = \pi\, G'' \gamma_0^2$$

3.4.5 Zero Shear Viscosity and Compliance from G', G''

Recall that

$$\eta_0 = \int_0^\infty G(s)\, ds \quad \text{and} \quad J_e^0 = \frac{\int_0^\infty sG(s) ds}{[\int_0^\infty G(s) ds]^2}$$

Also from Exercise 3.4.3

$$G'' = \omega \int_0^\infty G(s) \cos \omega s \, ds$$

$$G' = \omega \int_0^\infty G(s) \sin \omega s \, ds$$

Expand sin ωs and cos ωs in a Taylor series around $\omega s = 0$

$$G'' = \omega \int_0^\infty G(s)[1 - \frac{(\omega s)^2}{2} + \cdots] \, ds$$

$$G' = \omega \int_0^\infty G(s)[\omega s - \frac{(\omega s)^3}{6} + \cdots] \, ds$$

Then in the limit as $\omega \rightarrow 0$

$$\lim_{\omega \rightarrow 0} G'' = \omega \int_0^\infty G(s) ds \text{ and } \lim_{\omega \rightarrow 0} G' = \omega^2 \int_0^\infty sG(s) ds$$

Thus

$$\eta_0 = \lim_{\omega \rightarrow 0} \frac{G''}{\omega} \quad \text{and} \quad J_e^0 = \lim_{\omega \rightarrow 0} \frac{G'/\omega^2}{(G''/\omega)^2} = \lim_{\omega \rightarrow 0} \frac{G'}{G''^2}$$

Chapter 4

4.6.1 Relaxation After a Step Strain for the Lodge Equation

The shear stress is given by eq. 4.3.19, and $\gamma(t, t')$ is given for a step shear in eq. 4.3.20. From these two equations we find

$$\tau_{12} = \int_{-\infty}^{0} \frac{\eta_0}{\lambda^2} e^{-(t-t')/\lambda} \gamma_0 dt'$$

$$= \frac{\eta_0}{\lambda^2} \gamma_0 \int_{-\infty}^{0} e^{-(t-t')/\lambda} dt' = \frac{\eta_0}{\lambda^2} \gamma_0 e^{-t/\lambda} \lambda = \frac{\eta_0}{\lambda} \gamma_0 e^{-t/\lambda}$$

The portion of the integral from zero to t is zero because $\gamma(t, t') = 0$ when $t' > 0$.

To obtain the first normal stress difference, $N_1 = \tau_{11} - \tau_{22}$, from eq. 4.3.18, we must obtain the components $B_{11}(t, t')$ and $B_{22}(t, t')$ for the strain tensor **B**. We find from eq. 1.4.24 that

$$B_{11}(t, t') = \gamma^2(t, t') + 1; \quad B_{22}(t, t') = 1$$

and therefore

$$B_{11}(t, t') - B_{22}(t, t') = \gamma^2(t, t')$$

As before, $\gamma(t, t')$ is given by eq. 4.3.20. Carrying out the same manipulations as we did for the shear stress, therefore, yields

$$N_1 = \tau_{11} - \tau_{22} = \int_{-\infty}^{0} \frac{\eta_0}{\lambda^2} e^{-(t-t')/\lambda} \left[B_{11}(t, t') - B_{22}(t, t') \right] dt' = \frac{\eta_0}{\lambda} \gamma_0^2 e^{-t/\lambda}$$

The ratio N_1/τ_{12} is then

$$\frac{N_1}{\tau_{12}} = \gamma_0$$

which is the Lodge–Meissner relationship given by eq. 4.2.8.

4.6.2 Stress Growth After Start-up of Steady Shearing for the Lodge Equation

For steady shearing that began at time zero, the history of the strain tensor is given by eq. 4.3.21. Therefore, according to eq. 4.3.19 the shear stress is

$$\tau_{12} = \int_{-\infty}^{0} \frac{\eta_0}{\lambda^2} e^{-(t-t')/\lambda} \dot{\gamma} t \, dt' + \int_{0}^{t} \frac{\eta_0}{\lambda^2} e^{-(t-t')/\lambda} \dot{\gamma}(t - t') dt'$$

$$= \frac{\eta_0}{\lambda^2} \dot{\gamma} t e^{-t/\lambda} \int_{-\infty}^{0} e^{t'/\lambda} dt' + \frac{\eta_0}{\lambda^2} \dot{\gamma} e^{-t/\lambda} \int_{0}^{t} e^{t'/\lambda}(t - t') dt'$$

$$= \frac{\eta_0}{\lambda} \dot{\gamma} t e^{-t/\lambda} + \frac{\eta_0}{\lambda^2} \dot{\gamma} e^{-t/\lambda} t \int_{0}^{t} e^{t'/\lambda} dt' - \frac{\eta_0}{\lambda^2} \dot{\gamma} e^{-t/\lambda} \int_{0}^{t} e^{t'/\lambda} t' dt'$$

$$= \frac{\eta_0}{\lambda} \dot{\gamma} t e^{-t/\lambda} + \frac{\eta_0}{\lambda^2} \dot{\gamma} e^{-t/\lambda} t \lambda \left(e^{t/\lambda} - 1 \right) - \frac{\eta_0}{\lambda^2} \dot{\gamma} e^{-t/\lambda} \left\{ \lambda t e^{t/\lambda} - \lambda \int_{0}^{t} e^{t'/\lambda} dt' \right\}$$

$$= \frac{\eta_0}{\lambda} \dot{\gamma} t e^{-t/\lambda} + \frac{\eta_0}{\lambda^2} \dot{\gamma} e^{-t/\lambda} t \lambda \left(e^{t/\lambda} - 1 \right) - \frac{\eta_0}{\lambda^2} \dot{\gamma} e^{-t/\lambda} \left\{ \lambda t e^{t/\lambda} - \lambda^2 \left(e^{t/\lambda} - 1 \right) \right\}$$

Most of the terms above cancel out, leaving

$$\eta^+ = \frac{\tau_{12}}{\dot{\gamma}} = \eta_0\left(1 - e^{-t/\lambda}\right)$$

This is the same result that we obtained with the UCM model; see eq. 4.3.14.

INDEX